ARTIFICIAL INTELLIGENCE IN DESIGN '92

ARTIFICIAL INTELLIGENCE IN DESIGN '92

edited by

J. S. GERO

University of Sydney, Australia

Associate Editor:

FAY SUDWEEKS

Department of Architectural and Design Science,
University of Sydney, Australia

SPRINGER SCIENCE+BUSINESS MEDIA, B.V.

Library of Congress Cataloging-in-Publication Data

Artificial intelligence in design '92 / edited by J.S. Gero.
p. cm.
"Papers ... from the Second International Conference on Artificial Intelligence in Design held in June 1992 in Pittsburgh"--Pref.
Includes index.
ISBN 978-94-010-5238-2 ISBN 978-94-011-2787-5 (eBook)
DOI 10.1007/978-94-011-2787-5
1. Engineering design--Data processing--Congresses. 2. Computer-aided design--Congresses. 3. Artificial intelligence--Congresses.
I. Gero, John S.
TA174.A793 1992
620'.0042'C28563--dc20 92-15405

ISBN 978-94-010-5238-2

Printed on acid-free paper

CONTENTS

PREFACE

Design has now become an important research topic in engineering and architecture. Design is one of the keystones to economic competitiveness and the fundamental precursor to manufacturing. However, our understanding of design as a process and our ability to model it are still very limited. The development of computational models founded on the artificial intelligence paradigm has provided an impetus for current design research.

This form of design research has only been carried out in the last decade so in the temporal sense it is still immature. Notwithstanding its immaturity noticeable advances have been made both in extending our understanding of design and in developing tools based on that understanding. Whilst many researchers in the field of artificial intelligence in design utilise ideas about how humans design as one source of concepts there is no attempt to model human designers. Rather the results of the research presented in this volume demonstrate approaches to increasing our understanding of design as a process. The goal in most of this research is to make the computer more useful in design since it is clear when looking at designs produced by unaided humans that they often fail to perform satisfactorily. The expectation is that computer-aided human designers will produce better designs. The research methods employed are closely linked to the scientific method but that does not imply that the activity of designing is scientific.

The papers in this volume are from the Second International Conference on Artificial Intelligence in Design held in June 1992 in Pittsburgh. They represent the state-of-the-art and the cutting edge of research and development in this field. They are of particular interest to researchers, developers and users of computer systems in design. This volume demonstrates both the breadth and depth of artificial intelligence in design and points the way forward for our understanding of design as a process and for the development of computer-based tools to aid designers.

The forty-five papers are grouped under the following headings:

Configuration design
Constraint processes in design
Cooperative design
Design object modeling
Shape design
Integrated design environments
Design rationale
Case-based reasoning in design
Design analysis
Architectures for design knowledge
Learning in design
Conceptual design
Design processes

All papers were extensively reviewed by three referees drawn from a large international panel. Thanks go to them for the quality of these papers depends on their efforts. They are listed below. Fay Sudweeks, working superhuman hours, gave shape to this volume in her inimitable fashion from the variegated material submitted by the authors. As always, special thanks are due to her.

John S. Gero
University of Sydney
March 1991

International Panel of Referees

1

CONFIGURATION DESIGN

A SYSTEM FOR SUPPORTING DESIGN CONFIGURATION

K. J. MACCALLUM, B. YU
The CAD Centre
University of Strathclyde
Glasgow G1 1XJ UK

A. FREDERIKSEN
NEI Control Systems
227 Kingsway, Team Valley
Gateshead NE11 0QJ UK

and

D. MCGREGOR
Department of Computer Science
University of Strathclyde
Glasgow UK

Abstract. In the development of new products, configuration is the task of determining a complete set of components and their relationship such that the resulting product structure satisfies all requirements and constraints. This paper presents an approach to configuration design based on interactive decision support. A prototype system based on nonmonotonic propagation through a network is presented and illustrated using an example from the field of large power generation systems.

1. Introduction

In the development of new products, configuration is the task of determining a complete set of components and their relationships such that the resulting product structure satisfies all requirements and constraints. For many industries where products are complex, but innovative solutions are required, configuration can present a significant problem. Reuse and adaptation of previous products is a major feature of complex design. Examples are engine variations in a number of generations of car designs, or "stretched" aircraft designs. This type of product development always takes place under pressure of time and budget. Development strategies which consider financial, commercial, manufacturing and technical issues simultaneously are essential.

The cost of errors in configuration is high. Omission of necessary components results in poor costing and planning for product introduction; it also delays the design time by introducing changes at a late stage of product introduction. These difficulties inhibit the opportunities for exploring alternative configurations. There is, therefore, a need to be able to carry out configuration rapidly, to explore alternatives, and to use all the knowledge and expertise which exists within an organisation about product configuration.

There have been many attempts over the years to produce computer systems which support the generation of configurations. Typically these involve the use of checklists, decision trees, and other algorithmic methods. In most cases such methods are unsuitable. Search, iteration, redesign, and use of company expertise

J. S. Gero (ed.), Artificial Intelligence in Design '92, 3–19.

are all necessary aspects for effective configuration. Classical rule-based expert systems have been more successful (see, for example the well documented history of Digital's XCON). However, even here, knowledge maintenance has been shown to be a problem, and there has been a move towards better structuring of knowledge.

In the field of Computer Aided Design we have seen the emergence of products which successfully combine a rich base of knowledge and expertise representation and reasoning mechanisms with more conventional CAD geometric modelling for simultaneous engineering. These systems have already demonstrated individual applications of configuration generation. However, approaches to configuration have relied on building specific rule bases for configuration generation and have overlooked the importance of having a general module which generates and subsequently manages a configuration throughout the product development process. This implies supporting partial configuration decisions, the ability to explore alternatives, and the ability to cope with the implications of design change all within the structure of a general knowledge-based reasoning system.

2. Aims

The aim of this paper is to present an approach to configuration design and management, based on a concept of interactive decision support. The work is concentrated on the configuration problems at the early stages of design, although the overall approach has relevance to other aspects of configuration management.

The idea in this paper was developed and tested in the industrial context of large power generation systems. In this field products exhibit all the characteristics which make configuration management a critical problem, i.e. large number of sub-systems and components, multi-functional system, high degree of complexity, requirement for rapid and accurate design response, and innovative design with re-use of previous design solutions.

3. About the Problem

3.1. The Role of Configuration in Design

The process of design can be considered as a progression from an understanding of design requirements, into conceptual design, through embodiment design, and on to detail design (Pahl, 1984). For many practical design processes, past designs and design expertise are used extensively to assist in the formation of concepts and the embodiment. However the novelty of the requirements together with the needs for innovation and design competitiveness mean that past solutions undergo modifications which are sometimes fundamental. For the classes of designed products being considered here, configuration decisions take place during both the concept and embodiment stages with previous designs or part-designs providing a stimulus to developing a configuration structure. This structure then forms the skeleton on which the design solution will develop.

Because of the nature of design at this stage, and because a configuration provides only one solution out of many, configuration should not be viewed as merely a "once through" process of generation. Instead it should allow iteration, adaptation and reuse. The existing common view of configuration as a process of selection and assembly should be modified to cover removal and replacement as well as selection of components, partial disassembly as well as assembly of systems, and selection, adaptation and reuse of complete partial structures. This will enable configuration to take an integral part in a design system and hence be incorporated as part of the overall design process.

The design of the turbine generator plant for power stations, is a typical example of a configuration design problem. These machines are expensive and complex and, in a very tight market, the designs offered to match the customers' specifications, and to potential customers must be competitive in forms of both price and performance. Even when the details of the plant have been quite tightly defined by the customer, there is normally considerable scope for variation of the parameters.

The designer must optimise the design carefully, evaluate alternative configurations where possible, and carefully trade off a number of possibly configuring factors, such as efficiency, cost, complexity, and reuse of existing components to reduce design and tooling costs. There is a great deal to be gained from getting the design right. However, the time available for design is often limited by tight tendering and delivery schedules.

Ideally, in this class of design problem, the tasks should be performed concurrently, but traditionally they have been performed sequentially. First the basic plant cycle is laid out, then a rough design prepared using assumed values for the characteristics fed back to the cycle design, which is then designed in detail.

Some components which have long delivery times, such as large forgings, have to be ordered very early in the manufacturing process, often before the design has been finalised, which may impose unnecessary constraints on the design. A faster design process would enable more of the detailed design to be performed much earlier on in the design and manufacturing cycle, possibly at the tender stage, which would reduce or eliminate these artificial constraints.

3.2. Existing Configuration Systems

A number of computer based configuration systems have been developed for computers and for other products including robots, hydraulic pumps, chemical plant, electrical switchgear and electrical substation remote control equipment. Many of these are conventional expert systems used by salesmen to configure customers' orders which incorporate little in the way of novel techniques and do not approach configuration from a generic point of view. However, systems which have been identified as being of special interest include XCON, MAPLE, COSSACK, PIPPA.

- **XCON**

  XCON was developed by DEC to configure their VAX and PDP series computers and has a number of claims to fame: it was the first successful large commercial expert system, has had a wealth of information published about it, and is currently the largest knowledge based system in the world.

XCON is a rule based system, initially written in OPS5, which checks the completeness of orders, groups the components into the cabinets in the correct sequence and prepares connections further (Kraft, 1984). It uses a technique known as match, based on Winston's "bag packing" algorithm, in which each step in the execution is determined locally from the previous step. This presupposes that it is always possible to determine each step locally and that actions do not affect previous ones. It is efficient because it performs the task in a single pass without search or backtracking, but is only applicable to problems that are procedural in nature. If several rules are applicable at any given time the most specific rule is attempted first. This is known as specificity ordering.

When introduced in 1980 XCON consisted of about 700 rules and had a database of a few hundred parts, with typically eight attributes per part (Bachang, 1984; Polit, 1985). However, by 1987 it had 6,200 rules and a database of 20,000 parts, each with anything up to 125 attributes. Maintenance was becoming increasingly difficult even when reimpletemented in XCON-in-RIME. Consequently

maintainability when developing large, commercial applications is a critical consideration (Soloway, 1987).

MAPLE

MAPLE is a hardware configuration system developed at Reading University for dedicated microprocessor applications (Bowen, 1983; Bowen, 1985).

MAPLE determines the user's specifications by interview, designs the system, then produces a report which includes an estimate of development cost and time, production costs, level of difficulty and likelihood of success. It maintains a database of previous applications which can be used to guide new designs, in effect a simple form of learning.

MAPLE is based on the technique of functional reasoning (Freeman, 1971). It uses a depth-first, bottom-up search with backtracking, and performs some optimisation by using heuristic pruning to reduce the search space so that an exhaustive search is not required. Functional reasoning is very close to both constraint formalisation techniques and to the methods used by planning systems.

The main limitation of this approach is that it works by matching the functions required by the configuration to the functions supplied by the components. It therefore relies on both the overall specification and the functional parameters of the components being defined at the outset. This is possible for computer configuration design, but would break down in many configuration design application, where the components and overall requirements are not fully defined at the outset, and may change during the process.

- **COSSACK**
  COSSACK is a microcomputer configuration system developed by Xerox, with the objective of gaining a general understanding of the configuration problem (Frayman, 1987). Configuration is seen as a special case of design, involving a fixed set of components and a predefined architecture for fitting them together.
  Configuration knowledge is grouped into four categories: knowledge about components, knowledge about relations between components; constraints on properties of, or relations between components and user preferences.
  COSSACK uses constraint reasoning with a partial choice strategy and optimises the individual components of the system. It does not optimise the whole system because the evaluation criteria for an optimum overall solution are often very subjective, but it may generate several valid alternative configurations, leaving the final choice to the user.
  The representation of inter-component constraints is regarded as an important aspect of configuration. In COSSACK, constraints are attached to the components to which they relate. Whilst this strategy is satisfactory for a stand-alone configuration system for microcomputers, it does not support nonmonotonic reasoning, so would not be applicable to configuration design. Also, with the constraint representation used, it is possible to specify that one component requires another to be present, but not to specify explicitly that two given components are incompatible.
- **PIPPA**
  PIPPA (Professional Intelligent Project Planning Assistant) is developed by Rediffusion Simulation and Brighton Polytechnic to automate the generation of tender component breakdowns and process plans for flight simulators, and thus enables estimates to be prepared quickly and accurately (Lim et al., 1987; Lim and Wrigglesworth, 1987; Marshall, 1987).
  PIPPA is built in the rule and frame based shell "RBFS", which allows the user to define the types and characteristics of links between frames. It uses a backward chaining strategy, and determines the firing order by the certainties attached to the rules and by their physical order in the knowledge base. The knowledge base is displayed using the graphical editor, "VEGAN", and a strong emphasis is placed on the use of graphics tools for model generation, system management and use. The developers claim that this allows much more complex models to be managed than would otherwise be possible.

3.3. Requirements for a Configuration System

Design modellers should be able to support the design of configurations of components, as well as being able to design the components themselves, if they are to be applicable to configuration design problems. They should allow models of complex systems to be configured and complete systems, or components of them, to be manipulated as configurations. Design models should not have to be configured individually for each application, but the system should allow components of the configuration to be created, stored and retrieved. Throughout components should be maintained in a consistent state, so that configurations can be manipulated easily without sacrificing the integrity of the model. A further benefit of being able to create and maintain consistent configuration information during the design process is that it can also be used to assist in the definition and management of the design and manufacturing process.

This points to the need to support the management of configuration for design system models. In this context, configuration management can be defined as the capability to model configurations by assembling, dismantling and reconfiguring components according to a set of constraints and requirements, but without necessarily being tied to a fixed problem solving strategy. It includes the ability to store, recall and reuse partial or complete configurations. In short, configuration management is nonmonotonic modelling of configurations. Without it, advanced design modellers will offer little advantage over existing algorithmic design methods where configuration is involved, because any benefits gained will quickly be swallowed up by the need to configure and reconfigure the models individually, with no built-in consistency checks.

A number of methods and approaches used by existing configuration system are identified as candidates for a configuration management system architecture.

These are:

- A graphical or frame based decision representation in the form of a network or multiple hierarchies;
- A mechanism for representing constraints between arbitrary points in the network;
- A separate, explicit control strategy incorporating nonmonotonic reasoning;
- Graphical development and user interfaces.

4. The Configuration Management System

An architecture and knowledge representation for a configuration management system, based on the approach of the previous section, is proposed here. trialimplementation is also described and evaluated, using a turbine generator configuration problem.

4.1. THE SYSTEM ARCHITECTURE

Fig.1 shows the architecture for the system. It comprises the decision network processor, the configuration knowledge bases with the knowledge base editor and loader, the output processor and the user interface.

The core of the system is the decision network which represents the decisions available to the user and the interactions between them. It allows the user to make

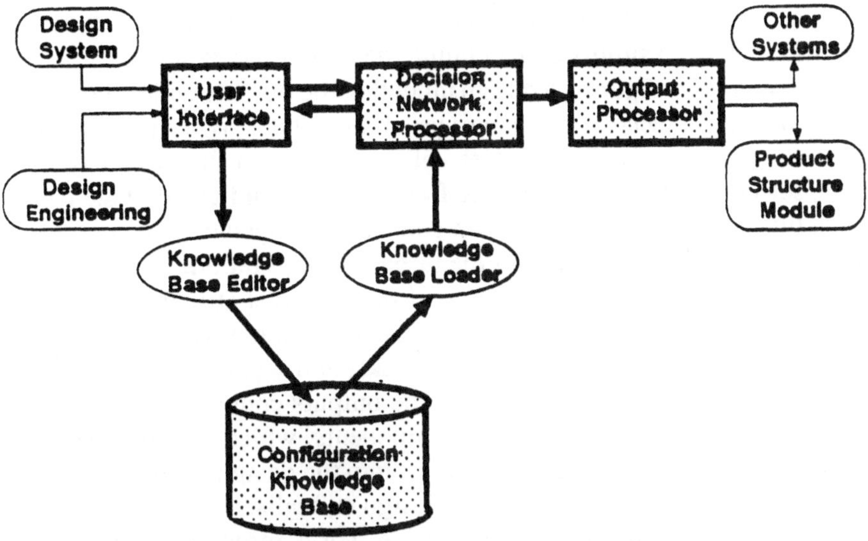

Fig. 1. Architecture of the configuration management system

or undo decisions in any order, propagates the effects of any decisions throughout the network, selecting any choices implied by a decision and removing any options invalidated by it and reverses the process when a decision is undone. it also prevents the user making any decisions that conflicts with ones already made, and can explain the status of any option.

The decision graph definitions for each product or decision process handled by the system are stored in knowledge bases. This separates the domain knowledge from the domain independent control strategy and implementation details. Applications can be changed by creating and loading the appropriate knowledge bases, which need only contain domain knowledge. This makes the system generic, and applicable to a range of applications, rather than restricted to one. The components of the system that relate to the interface and knowledge base must also, by implication, be generic.

Knowledge bases are created in a high level definition language using the knowledge base editor, and loaded into the decision network processor by the knowledge base loader. the loader converts the definitions contained on the knowledge base files into the internal decision network representation, which can then be manipulated by the decision network processor. This means that knowledge base development and editing should by considerably quicker and easier than would be the case if they were coded in the implementation language, and also ensures more consistent knowledge base definitions and formats.

The implications of the decisions and selections represented in the model can be communicated to the user, or to another system through the output processor. Since the outputs will depend on the application, by separating them from the decision process, the same knowledge base and configuration information can be used in several different ways simply by redefining the output procedures.

If the system is configuring a design modeller, the outputs will be procedured to create and link the objects on the design system's database. When parts of the model are defined, the corresponding objects will be created in the design system. For example, selecting a turbine with three feedheaters would create three feedheater modules, link their feedwater and drain connections together, and couple them to the turbine in the correct manner for the turbine configuration.

Alternatively, the outputs could be part of a product structure, which can be passed to a project management system as the configuration is defined. This enables project management networks to be set up quickly and accurately, so that the project can be controlled from an earlier stage. The outputs could also initiate planning and cost estimating procedures for the components, which would enable the cost and planning data to be created and fed back into the design so that it can be taken into account in the evaluation before the design is finalised.

The system communicates to the outside world through the user interface. Ideally this should have full WIMP functions and display a graphical representation of the knowledge base, allowing the user to select and activate points of the network and view the state of the model directly. The system can also be driven from a design system. In this case the user interface would be replaced by an interface to the design system, from which it would take inputs. It could return inferences about the product configuration through the user or output interfaces.

4.2. Knowledge Based Organisation

The knowledge is represented as a decision network, which is similar in concept to an AND/OR network. It is defined by the developer in the definition language, which specifies the required network primitives and the links between them. The network is created in the system's internal representation by parsing and compiling a file containing the required definitions.

The main network primitives are shown in Fig.2. Attributes, groups and values are all types of node, and constraints are a type of link. Attributes are OR nodes with alternative values, and are linked to groups or values by AND relationships. Groups can be used to cluster attributes together to allow the knowledge base to be divided into manageable sections. Constraints represent additional exclusive OR relations across the network. The other primitive is the output module. These are procedures that can be attached to paths through the network, and are executed when the respective path is activated.

The user creates models from the knowledge base by traversing the network through the interface, selecting and activating or deactivating portions of it by choosing attribute values. The decision network processor automatically propagates the effects of each user action, and executes the implications, or consequences, as defined by the output procedures, in response.

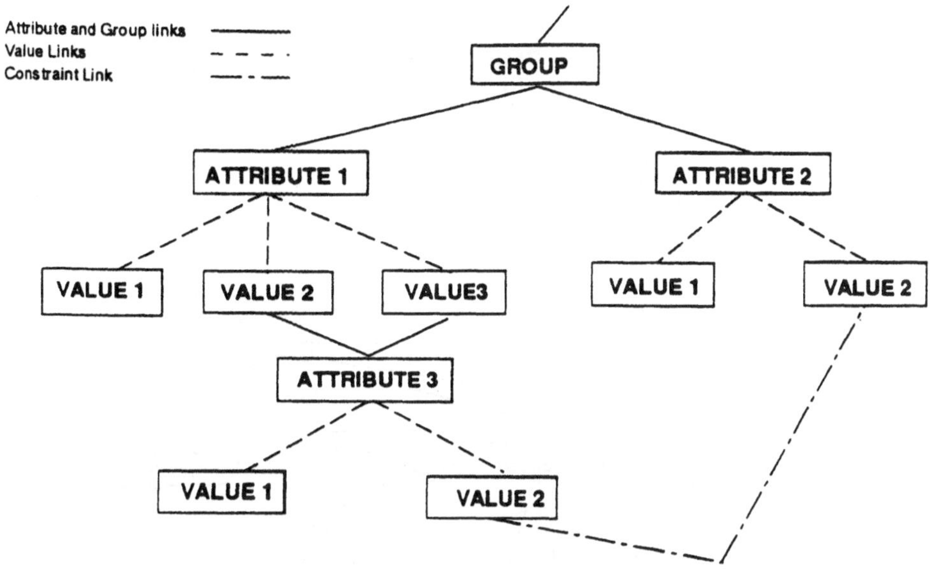

Fig. 2. Decision network primitives

An attribute represents a decision that can be made by the user. Each has a number of alternative values, only one of which may be true at any time. An attribute is therefore equivalent to an "exclusive or" node in a decision network. Attributes may be attached to one or more nodes, which may be either groups or values of other attributes. They are identified by names, and have text strings attached to them, which are displayed as a prompt to the user when they are selected. The user can select values of the attributes and set them true or false.

A group represents a set of decisions, or attributes, that must all be active when that group is active. This is equivalent to an "and" node in a decision network. A group may be attached to another group or to a value of an attribute, in which case it is only active when its parent value is active. It is identified by a name, and has a text string attached to it, which is displayed as a prompt to the user when it is selected.

A constraint represents mutually exclusive pairs of values, which may belong to any attributes in the network. When one of the values becomes true, the other must be excluded, and if that value is subsequently undefined, the constrained value again becomes available for selection.

4.3. Operation of Knowledge Bases

The decision network processor manipulates the knowledge base in response to the user's actions. As attributes are selected and their values defined, it models the effects of these decisions, adopting a forward chaining strategy to prevent any inconsistent selections being made, thus removing the need for backtracking.

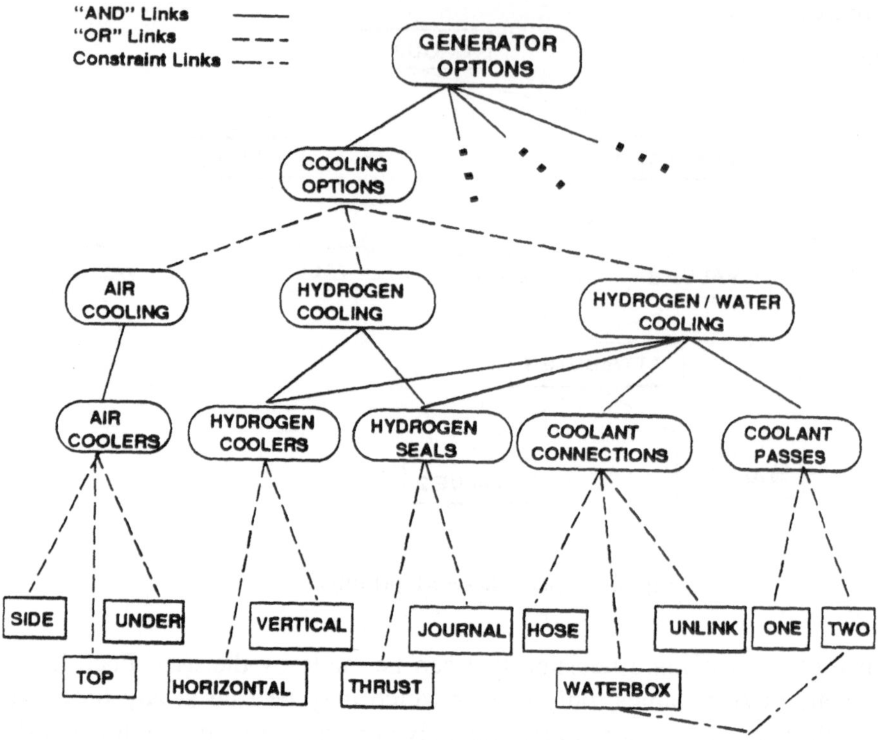

Fig. 3. Generator cooling options

When an attribute has one of its values set to true, the other values of that attribute are automatically excluded by being set to false. Any values that are linked to the selected value by a constraint are also excluded. If it is linked to two other values by a three-way constraint and another of the values has been set to true, then the third will be excluded. For example, in the generator cooling example (Fig.3), if the value "waterbox" is selected for "coolant connections", then "hose" and "unilink" are excluded. The value "two" for "coolant passes" is also excluded by a constraint which represents the fact that waterbox connections are incompatible with two coolant passes.

When an attribute has had a value selected, if it has only one parent and that parent has not yet been selected, or if it has several parents, only one of which has not been excluded, then that parent will also be set to true. Any other subnodes belonging to that parent will then have to have a value selected before that part of the configuration can be completed. In the above generator cooling example, as "coolant connections" has only one parent, hydrogen/water cooling, this is selected, and values have to be selected for each of the other three subnodes to complete this part of the configuration.

If an attribute's status is set to true, for example when its parent is set to true, and only one of its values has not been excluded either by the user or by a constraint, that value will be automatically set to true by default. For example, in the generator cooling model, when “coolant connections” is set to “waterbox”, “coolant passes” is set to true and, because one of the two values for coolant passes is already excluded by a constraint, the other value, “one”, is selected automatically.

If a value that was set to true is set to undefined these processes are reversed, but all of the other decisions made by the user, and their implications, will be left in place.

If a value is selected for an attribute, which is then set to true, but the path from that attribute to the origin is ambiguous, a routine is invoked to determine the effects of that selection. For example, in the generator cooling model, if a value is selected for the attribute "hydrogen seals", and the type of cooling is not known, the path to the origin is ambiguous. In this case, the system must recognise that any of the available paths could be true and some nodes will be true and have to be completed whichever path is chosen, but some others will be excluded, together with their subnodes, whatever choices are made subsequently. In the generator example, the machine could be either hydrogen or water cooled, but not air cooled, and a value must be defined for hydrogen seals whichever type of cooling is subsequently chosen. The attributes “coolant connections” and “coolant passes” should be left undefined, and the attribute “air coolers” excluded. No values for “air coolers” can be selected because the system prevents the user selecting an invalid combination of design features.

To explain the status of any defined value, the system finds and displays any values defined by the user that are affecting its status. These will be the values that the user may have to change in order to change its status. If the value was defined by the user the system will indicate this. Otherwise if it is true, it could either be because all the alternative values are false and the attribute to which it belongs is true, or because one of its subnodes is true. A value could be false because an alternative value is true, because it is constrained, because all of the values of one of its subnodes are false, or because the attribute to which it belongs is false. The system attempts to explain each of these in turn, terminating when it finds a value that was defined by the user.

The user is thus able to explore the configuration space freely, but is prevented from generating inconsistent configurations. The user can ask the system to explain any of the values that have been selected by the system as a result of any choices he has made.

Output can be attached to nodes in the network, and are activated as the nodes are selected . This could be performed as a batch process once the decision process has been completed, or continuously, the outputs being activated as their respective nodes are selected and removed as they are unselected.

4.4. Implementation of the System

The full implementation of the architecture described above, which would meet all the requirements of a configuration management system, depends on nonmonotonic reasoning through a decision and constraints network. The main purpose of the implementation was to test the critical issue on which the whole architecture and knowledge representation depends: nonmonotonic propagation through a network.

The system was implemented in POP-11 (Barrett, 1985) and ODDS (Bennett, 1986). The software was chosen because the proposed architecture lent itself well to implementation in an object-oriented development environment, using demons to implement the forward chaining propagation strategy. It also has flexible data structures, incremental compilation and a readable syntax, which assist rapid prototyping.

A distinction is made between values set by the user and values set by the system in response to actions taken by the user. Values set by the user can not be undone by the system, and the user cannot change any values set by the system. This distinction makes it easier to prevent the model becoming inconsistent, or the system looping, and simplifies constraint relaxation. The explanation facility also uses the values set by the user to terminate the explanation chains.

5. Application of the system

As an example, the configuration of a design programme for heat balance calculations for a steam turbine plant cycle is presented. The heat balance example is based on the turbine cycle performance calculation, and is intended to configure a design modeller to perform these calculations on a variety of cycle configurations. There are six basic steam cycles: reheat, non-reheat, double-reheat, combined cycle, nuclear and pass-out. As the bulk of tender design work is for reheat cycles, which are also the most complex to configure, initial development was restricted to reheat cycles, and a number of simplifications have been made to the model. However, it covers the main variations on the reheat cycle, and is adequate to demonstrate the technique, including the link to a design modeller.

The turbine has three cylinders: HP, IP and LP, and electric boiler feed pumps are used. The designer is able to vary:

- the number high and low pressure feedheaters
- the arrangement of the low pressure heater drains
- the locations of some feedheater extractions on the turbine
- the destinations of the turbine gland leak-offs
- the method of extracting air from condenser
- the presence of boiler attemperator sprays.

The decision graph is shown in Figure 4.

On entering the system the user is presented with four choices: load a knowledge base, use the knowledge base currently loaded, create outputs, or quit the system.

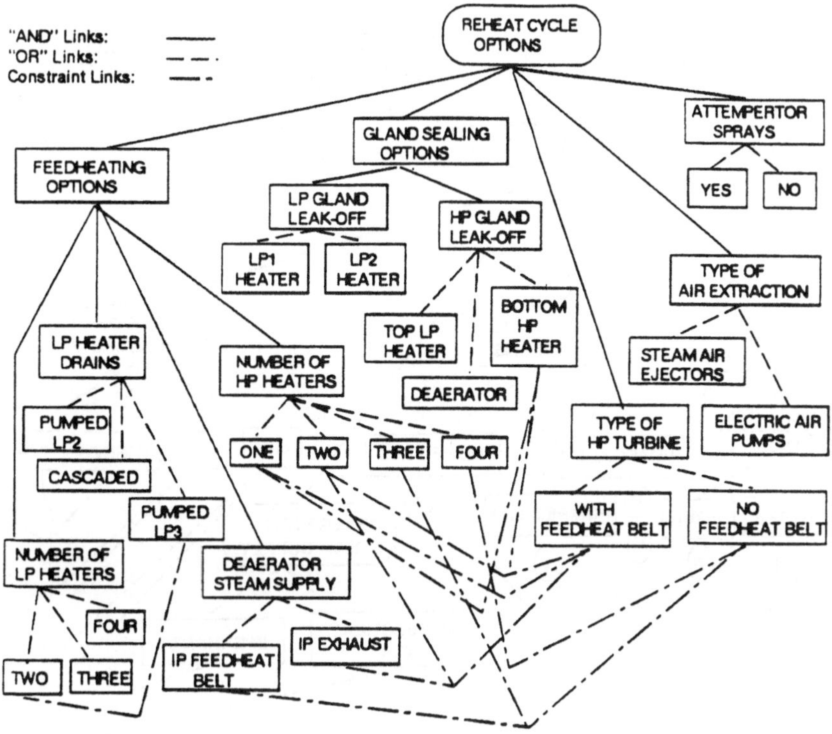

Fig. 4. Network for heat balance knowledge base

The heat balance model, is loaded, and the system displays POP messages as the output procedure definition files are loaded.

The user is now able to select the Use option and a list of available options is presented. There are twelve nodes (groups and attributes, not values) on the top list (see Fig.4).

To show how the user can step through the knowledge base selecting values, the origin, "REHEAT CYCLE OPTIONS" (indicated by capitalisation), is selected. The system responds by displaying a list of its subnodes.

This first item on the menu: Feed heating options, is chosen, and a list of its subnodes is presented. The first of these: Number of LP heaters, is selected and displayed. There are three possible values for the number of LP heaters, none of which have yet been defined. When value "two" is selected from the menu, and set to true, the system responds by setting the other values to false, and redisplays the menu, indicating the status of the values, and showing by means of the "U" that the value "two" was defined by the user (see Fig. 5).

On quitting from this menu, the system returns to the previous menu, and indicates that the attribute "Number of LP heaters" has been completely defined by means of the "C" (see Fig. 5).

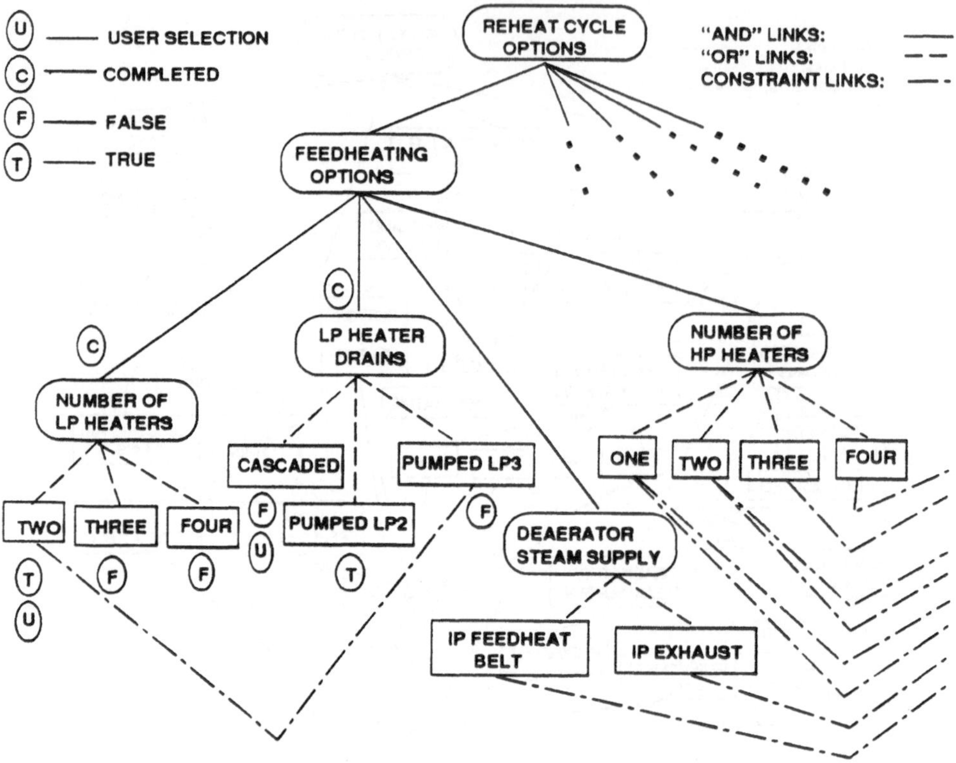

Fig. 5. Decision network with constraint

To show how constraints and the mechanism for selection of items by default operate, another attribute, "Type of LP drains" is now selected.

The third item on the menu, "Forward pumped on LP3", has already been set to false by the system. One of the undefined values, "Cascaded", is set now to false (as may be required if this option is specifically excluded for a technical reason) leaving only one possible value for this attribute. The system sets this to true.

To ascertain why the system has set this value to true, an explanation of its status can be requested. The system indicates that it has been selected because one of the other values has been set to false by the user, and the other is excluded by a constraint. (It would not be possible to fit drain pumps on the third heater when there are only two heaters!)

On leaving this menu, two of the attributes are now fully defined.

The user can continue this process of stepping through the menus until all the necessary attributes to define a configuration have been completed.

On returning to the top level menu, the outputs can be created by choosing the "Output" option. On this model these are commands to create and link the modules to the database of a design system.

6. Discussion

The system described in this paper shows the feasibility of developing a domain independent engine which manages a configuration of a product design based on a decision network. Important ideas in the system were the non-monotonic reasoning processes allowing configurations to be modified throughout the design process, and the potential links to an overall design environment.

The control strategy, which relies partly on preventing the user changing values set by the system, is a simple and efficient way of ensuring that the model is always in a consistent state. Unlike the ATMS and some machine-manipulated representations, it is not prone to serious scale or performence problems, because it does not rely on explicit calculation of all possibilities. Also, as it does not perform any lengthy searches involving backtracking, a reasonable performance can be expected. However, when there are a number of constraints in the model, attempting to change a system-set value can be a complicated process. The user must first determine what needs to be changed before a value set by the system can be altered, which means that an explanation of the value must be requested. The attributes and values that are constraining it must then be found and changed before returning to the original value and changing it. The lack of a graphical interface does not help in this respect. This problem could be alleviated if constraints can be weighted, or defined as "hard" or "soft", representing those that are technically impossible, so cannot be violated, and those that are simply non-preferred options which can be overridden.

As well as an efficient mechanism for nonmonotonic reasoning through decision graphs, the prototype incorporates two enhancements taken from the pilot systems that pointed to the final architecture, that are not commonly found in other configuration systems. These are decoupling the decision process from the product structure, first used in the flowchart method, and the use of separate, externally defined knowledge bases.

By decoupling the consequences of the decisions, such as object and product structure definitions, from the decision process, this representation is considerably more compact than would otherwise be the case. This makes the decision process easier to represent and manipulate, and enables intelligent help and explanation facilities to be incorporated because the line of reasoning can be deduced by tracking paths through the decision network. Another advantage of decoupling the decisions from their consequences is that the same decision process can be adapted for a number of applications simply by changing the output definitions.

Separate knowledge base files allow the knowledge base to be built in a pre-defined format or language, which can then be translated into object and link definitions, rather than forcing the user to code these directly. These are used in some form in almost all expert system shells, but have not yet found their way into many decision graph based configuration systems.

These two enhancements greatly increase the flexibility of the system, enabling it to be adapted quickly to handle a wide range of configuration and other applications, and to be linked to a design system on a generic basis. They also increase the speed and ease of knowledge base development. Apart from these features and the nonmonotonic reasoning capability, the architecture is based on techniques that are widely accepted as the most suitable for this class of problem, and are used in a number of systems.

The prototype has been tested on two knowledge bases of widely differing characteristics, one having a complex decision graph, and the other having complex output generation procedures. Although it cannot cope with all problems because the multiple path propagation mechanism only works over one level, and because the constraints representation is limited, there is no fundamental reason why these features cannot be fully implemented. This would allow the system to handle configuration and other problems of arbitrary complexity, including non-configuration applications, such as the motor selection and pension scheme choice examples for which flowcharts were successfully developed.

The example in this paper is one of a number of application tests which have been carried out to test the validity and feasibility of the ideas. In all cases, practising designers endorsed the approach and confirmed the potential value in time saving, but more particularly, error reduction.

The approach can be seen as an individual design support tool, or as part of a larger design support environment. Already there is experience of using configuration as a component within a blackboard design architecture. The links to product structure on the output also provide the opportunity to integrate with systems for detail design, design change management, and project planning.

In its present form there are many aspects of the idea which require further test, development and refinement. However, initial results and feedback confirm out belief that the approach can offer important support in early design decision making.

Acknowledgements

The work presented here was carried out as part of a postgraduate research programme sponsored by NEI Parsons Ltd.

References

Bachant, J. and McDermott, J.: 1984, R1 Revisited, *AI Magazine*, **5**(3), 21-32.

Barrett, R., Ramsay, A. and Sloman, A.: 1985, *POP-11: A Practical Language for Artificial Intelligence*, Ellis Horwood Ltd., ISBN 0-85312-940-1.

Bennett, M.E.: 1986, *ODDS User Manual*, Computer Science Group, Cambridge Consultants Ltd.

Bowen, J.A. and Smith, M. F.: 1983, Expert systems for analysis and design of microprocessor applications, *Journal of Microcomputer Applications*, **6**(2), 156-161.

Bowen, J.A.: 1985, Automated configuration of hardware for dedicated microprocessor application systems, *in* J. S. Gero (ed.) *Knowledge Engineering in Computer-Aided Design*, North-Holland, Amsterdam, pp. 351-365..

Frayman, F. and Mittal, S.: 1987, COSSACK: A constraints-based expert system for configuration tasks, *Knowledge-Based Expert Systems in Engineering: Planning and Design*, Computational Mechanics Publishers, Southampton, pp. 144-166.

Freeman, P. and Newell, A.: 1971, A model for functional reasoning in design, *Proceedings IJCAI-71*, pp. 621–633.

Pahl, G. and Beitz, W.: 1984, *Engineering Design*, The Design Council, London.

Kraft, A., Winston, P. H. and Prendergast, K. A.: 1984, XCON: An expert configuration system at Digital Equipment Corporation, *The AI Business*, MIT Press, Cambridge, Mass., ch 3, pp.41-49.

Lim, B. S., Marshall, G., Kellett, J. M., Kuczora, P. W., Boardman, J. T., Murray, P. M., Wrigglesworth, D. and Whitehouse, M.: 1987, A foundation for a knowledge based computer integrated manufacturing system, *Proceedings 4th European Conference on Automated Manufacturing*, Birmingham, 458-474.

Lim, B. S. and Wrigglesworth, D.: 1987, A prototype intelligent manufacturing assistant in the manufacture of flight simulators, *Proceedings IEE Colloquium Expert Planning Systems: A New Application for Control Theory*, pp.4/1-4/10.

Marshall, G., Lim, B. S., Kellett, J. M., Boardman, J. T.: 1987, PIPPA: An expert project planning system in manufacturing engineering, *Proceedings KBS 87*, London, pp. 199-205.

Polit, S.: 1985, R1 and beyond, *AI Magazine*, 5(4), 76-78.

Soloway, E., Bachant, J. and Jensen, K.: 1987, Assessing the maintainability of XCON-in-RIME: coping with the problems of a very large rule base, *Proceedings AAAI-87*, pp. 824-829.

DOMAIN-INDEPENDENT DESIGN SYSTEM

Environment for rapid development of configuration-design systems

J. T. RUNKEL, W. P. BIRMINGHAM, T. P. DARR, B. R. MAXIM
and I. D. TOMMELEIN
Department of Electrical Engineering and Computer Science
University of Michigan
Ann Arbor MI 48109
USA

Abstract. This paper describes the Domain-Independent Design System (DIDS). DIDS provides a set of tools capable of rapidly constructing configuration-design systems from a library of reusable software elements, called mechanisms. The power of DIDS comes from its model of configuration design that enables reusable mechanism to be identified. DIDS contains four components. The first component, the Problem-Solving-Method (PSM) Editor builds PSMs by combining mechanisms. The Code Generator, DIDS's second component, generates a problem solver from the PSM description created in the editor. The third component, the Knowledge-Acquisition Tool Generator builds a knowledge acquisition (KA) tool that interviews the domain expert to gather the knowledge required by the DIDS-generated problem solver. The final component, the Debugging Tool, monitors the execution of the problem solver to uncover errors made during KA, and to improve the performance of the design tool. This paper presents a scenario demonstrating how DIDS will be used to build configuration systems.

1. Introduction

Knowledge-based design systems (KBDS), such as R1 (also known as XCON) (McDermott, 1980), M1 (Birmingham *et al.*, 1989) and VT (Marcus, 1987), have demonstrated the benefits of automating design tasks where essentially the same artifact is repeatedly designed with only minor variations, *i.e.*, the specifications vary slightly from design to design. These systems have significantly reduced both the cost of and time required to produce a design. For example, R1, has reduced the time required to produce a design, the number of errors in a design, and given a large number of organizations access to the design expertise encoded with in it (Barker and O'Conner, 1989).

Although the benefits of the design systems are great, the cost and the level of expertise required to build them impedes their widespread application. A development team, consisting of both domain experts and experienced knowledge engineers, requires several man-years to build a system. Designers require the assistance of knowledge engineers because they are not versed in AI concepts and techniques. The long development time results from two factors. First, the designers must communicate all relevant domain concepts to the knowledge engineers. This communication is slow, and requires numerous iterations.

J. S. Gero (ed.), Artificial Intelligence in Design '92, 21–40.

Second, design systems tend to be large programs, which are difficult to construct (Boehm, 1987).

The Domain-Independent Design System (DIDS), by automating the process of building KBDSs, will reduce the time and expertise required to build them. DIDS provides an environment facilitating the rapid development of systems that perform a restricted form of design called configuration (Mittal and Frayman, 1989). DIDS, which currently is a partially complete prototype, will provide tools that configure design systems from a library of design-process software elements. Each element represents a fundamental design process that is shared across numerous domains. Systems can be easily created and revised since users will manipulate high-level elements, instead of programming language constructs. DIDS will enable users with limited knowledge of artificial intelligence (AI) to build systems, because most those techniques will be encoded in the elements and, therefore, hidden from the user.

The success of DIDS depends on the development of an element library that has a manageable size, but which provides enough elements to cover a significant percentage of configuration tasks. We believe that such a library can be built because of the insights gained from the DIDS model of configuration design. The model enables elements that are highly reusable and easily combined to be identified. It describes the types of knowledge used during configuration and defines mechanisms, the elements in the library, as operators on these knowledge types. This definition enables reusable mechanisms to be distinguished from non-reusable ones, and defines the appropriate grain size of each mechanism. In addition, viewing mechanisms as operators on knowledge types provides the insight that enables DIDS to automatically generate a knowledge-acquisition tool from a set of mechanisms.

The design literature contains some evidence suggesting that a library of reusable elements can be created. Researchers, such as Chandrasekaran (Chandrasekaran, 1986), Tong (Tong, 1987), and Balkany *et al.* (Balkany *et al.*, 1991) have identified fundamental elements of the design process that are shared across domains. These researchers, and others, have proposed models of design that place structure on the design process. The authors of these models, by proposing them, implicitly assume commonalty between design processes in different domains. Other researchers have demonstrated that design tools intended for one domain can be ported to new domains (Birmingham and Tommelein, 1991), (Langrana *et al.*, 1986), (Johnson *et al.*, 1988), (Maher, 1987), (Brown and Chandrasekaran, 1987).

The remainder of the paper is organized as follows. Section 2 defines configuration-design tasks. Section 3 describes the DIDS model of configuration design, and supports the reusability claims of the model. Section 4 describes the architecture of the DIDS system, and the functionality of each of its subsystems. Section 5 presents a scenario of how we envision DIDS will be used to construct design systems. Section 6 compares DIDS to other systems that aim to automate the construction of design tools. Finally, Section 7 summarizes the paper.

2. Configuration Design

The DIDS model and system have been formulated for configuration-design problems, although clearly many of these ideas can be applied to design in general. Configuration is sufficiently complex to be interesting, but restricted enough to enable a detailed model of design to be developed. The configuration task can be characterized by the following (adapted from Mittal and Frayman, 1989):

> Given a fixed library of components, some specifications on functionality, performance, and cost, construct an artifact using these components that satisfies the specifications. The artifact must obey either rules of interconnection of components, or rules on topology, or both.
>
> Optionally, a set of preference or optimization criteria can be given. The artifact must conform to these criteria.

Configuration design can be decomposed into four very broad classes of tasks: part selection, structure selection, backtracking, and arrangement. Most existing configuration systems only perform two or three of these tasks. The part-selection task chooses parts from the library of elements that perform the desired functions while satisfying the design constraints. The structure-selection task determines the connections between the parts. This task identifies the input-output relationships between parts without considering topological, or geometric considerations. The backtrack task modifies previous decisions in response to constraint violations. The arrangement task determines the topological or geometric relationships between components.

3. The DIDS model of Configuration Design

The success of the DIDS research depends on the development of a library containing reusable mechanisms (Klinker, 1990) that cover a substantial percentage of all configuration tasks. Our approach has been to study configuration-design tasks and the KBDSs that perform them (Balkany et al., 1990) at both the knowledge-level (Newell, 1981) and the knowledge-use-level (Steels, 1990). Like Chandrasekaran, Steels, Gero (1990), Klinker, Balkany et al., we believe configuration design can be decomposed into a set of tasks that are shared across domains. In our view, any set of configuration-design tasks can be decomposed, with regard to function, into a set of subtasks that are atomic. A subtask is atomic if the intersection between the subtask and each of the other subtasks is either empty or equal to the subtask. In DIDS, a mechanism implements an atomic task. A construct called a problem-solving method (PSM) (McDermott, 1988) represents a sequence of mechanisms proven to be useful for solving a specific design problem.

Unfortunately, there are numerous atomic tasks, and the reusability of each corresponding mechanism is difficult to measure. We believe that reusable mechanisms can be identified by viewing them as operators on knowledge. An emphasis of DIDS research has been to identify the types of knowledge used by configuration-design systems. A mechanism takes one or more types of knowledge as inputs, and produces some result that can be described in terms of these knowledge types. Since a mechanism described in this manner can be applied to any configuration task where the requisite knowledge types are available, these mechanisms are highly reusable. In addition, this approach ensures that mechanisms have standardized inputs and outputs that can be readily combined, defines the types of tasks that mechanisms should perform, and establishes the grain size of mechanisms.

The DIDS model of configuration design, which encapsulates these reusability ideas, consists of a set of constructs, which include mechanisms and PSMs, that can be used to represent design problem solvers and the domain knowledge they operate over. The model contains two layers that decompose configuration-design knowledge into two classes. The Knowledge Layer (KL), identifies the types of declarative knowledge that can be used during configuration. This knowledge, which, for the most part, is search-control knowledge, consists of domain concepts and design expertise in a non-operational form. The knowledge in the KL is of no value without a set of processes to interpret and to determine when and how to use it. The Mechanism Layer (ML) contains the mechanisms and PSMs that use the knowledge.

The basic elements of the KL are the thirteen knowledge types that recur in configuration-design systems (Figure 1). All systems studied so far use a subset of these thirteen types. At this point we believe that the KL is complete. For example, *required-component knowledge*, one of the thirteen knowledge types, expresses a relationship between a set of parts. In particular, knowledge of this type lists the set of additional parts that must be added to the design to support the operation of some part. These additional parts do not increase the functionality of the artifact, but simply ensure that the original part will perform properly. Systems, such as M1 and Cossack (Frayman and Mittal, 1987), use this type of knowledge. Figures 8, 9, 10, 11 and 12 depict some of these knowledge types.

1. Library of Parts
2. Type Decomposition
3. Functional Decomposition
4. Task Decomposition
5. Required Component
6. Constraint
7. Preference
8. Error Recovery
9. Task Ordering
10. Connection
11. Arrangement
12. Dependency
13. Formula

Fig. 1. The thirteen knowledge types

The KL of a particular design system contains a set of constructs that are defined in terms of the thirteen knowledge types. Each construct consists of one or more knowledge types, which are combined to represent a class of domain concepts. Instances of the constructs represent specific concepts, and links between them express their interrelationships. For example, M1 contains a construct called templates. Templates, which consist of constraints, required-component knowledge, and a set of connections, represent a legal way of connecting a set of components. An instance of a template might describe the connections among three parts: A, B, and C.

The ML contains both PSMs and mechanisms. PSMs, which consist of a sequence of steps contained within a loop, describe the high-level design strategy used by a system (Figure 3 shows the PSM for M1). They model the repetitive sequence of steps typically performed by design problem solvers. A PSM determines the order in which mechanisms are executed, since a sequence of one or more mechanisms implements each step of a PSM.

The DIDS model establishes a criterion for identifying reusable mechanisms. The question of a methodology for discovering the set of mechanisms that covers a significant percentage configuration tasks still remains. Existing configuration-design systems, however, can be studied to determine the set of mechanisms useful for solving problems. Analysis of a large number of configuration systems should result in a library of mechanisms capable of performing most configuration-design tasks. We believe that 80% to 90% of the mechanisms necessary for configuration can be found this way. This is because our rate of adding mechanisms to the library is shrinking rapidly to less than one per new system studied.

4. DIDS ARCHITECTURE

4.1 ARCHITECTURE OVERVIEW

The DIDS model is the basis for the design of the DIDS system. Therefore, the bottom layers of DIDS implement the KL and the ML. All components of DIDS that interact with the user and that perform analysis operate by inspecting, modifying, and reporting the results of inferences made in these two layers. The KL implementation consists of a knowledge-representation scheme to support simple inferences over the knowledge types. The ML contains constructs that represent mechanisms, problem-solving methods, and the data and control relationships among them.

DIDS consists of four major components: a Problem-Solving-Method Editor, a Knowledge-Acquisition-Tool Generator (KATG), a Debugging Tool (DT) and a Code Generator (CG) (Figure 2). The PSM Editor allows a DIDS user to construct a design system by selecting mechanisms and pre-defined PSMs from libraries. The CG takes the PSM description and generates the C++ code that implements the PSM. The KATG uses the PSM description to create a tool that gathers the knowledge from a domain expert, necessary for correct operation of the PSM. The DT, which monitors the execution of mechanisms during problem solving, allows the user to find errors. These errors include missing mechanisms, mechanisms inappropriate for the current problem-solving task, and incorrect information entered during knowledge acquisition (KA).

Design systems will be constructed in DIDS using an iterative process. Users will select mechanisms and PSMs without necessarily understanding their complete behavior, and the effect these constructs will have on the design system. The first design system generated by DIDS will be regarded as a prototype. In order to evaluate this prototype, the user will use the DIDS-generated KA tool to perform preliminary knowledge acquisition. The initial session will encode enough knowledge so that the performance of the DIDS-generated design system can be evaluated. The DT monitors the execution by displaying each mechanism's inputs and outputs, and by explaining its operation. Using information from the DT, the user can fine-tune the design system by using the PSM Editor. Such modifications may include removing mechanisms that do not perform any useful action during the design. This process of editing, code generation, knowledge acquisition and testing will continue until DIDS generates a design system that efficiently performs the user's design task.

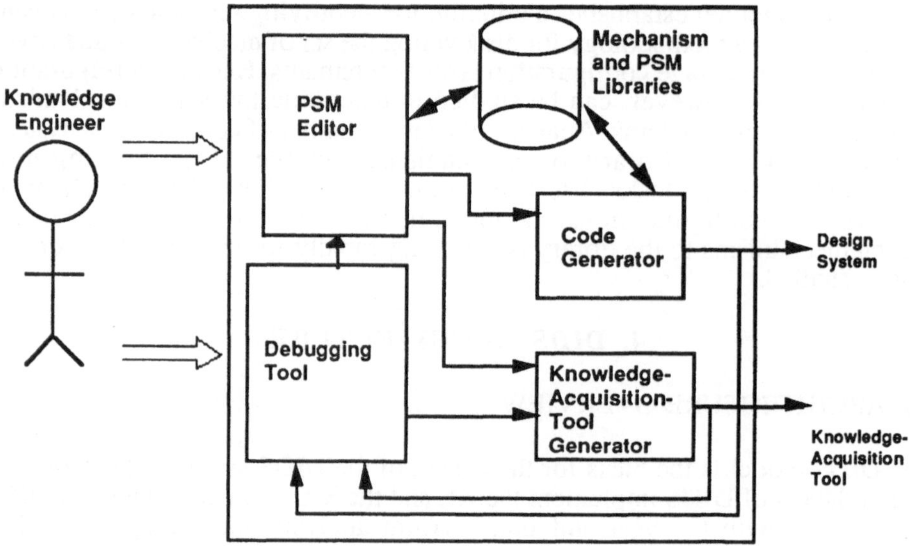

Fig. 2 DIDS architecture

The next sections describe the DIDS components in more detail.

4.2 PROBLEM-SOLVING-METHOD EDITOR

The PSM Editor, a windowed editing environment, provides a set of tools that allows new design systems to be constructed by combining mechanisms identified by the DIDS model. To facilitate system construction, DIDS contains predefined PSMs. A user will select one of these PSMs, and then modify it by adding or deleting mechanisms to improve the design system's performance on the particular problem. The editor consists of a group of menus that provide editing functions, buttons that allow the user to browse the mechanism and PSM libraries, and a graphics window where PSMs are built (Figure 3). Modifications to the PSM in the graphics window simply involve cutting and pasting mechanisms from the library.

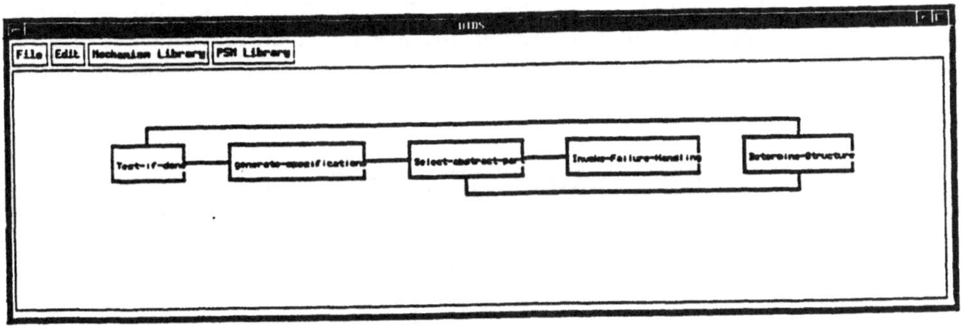

Fig. 3. PSM Editor

DIDS provides several facilities for describing data and control flow between mechanisms. Connecting two mechanisms by a line specifies sequential control flow, and a diamond icon is used to specify while, repeat-until, case, and if-then-else control relationships (Figure 3). Users specify data-flow information by selecting two mechanisms, and then choosing the connect button. This brings up a new window listing the parameters of the two mechanisms (Figure 4). The CG uses this information to match parameters in the PSM. Once the user completes the construction of a PSM, a C++ code instance of that PSM is created. The CG combines the C++ code implementation of each mechanism in the mechanism library according to the users' data and control flow specifications.

4.3 KNOWLEDGE-ACQUISITION-TOOL GENERATOR

Any reasonably complex design problem will require a large knowledge base. Acquiring the requisite knowledge by hand is not only difficult, but it is contrary to the notion of speeding the development of design systems. Thus, DIDS has a KA-tool generator.

Each mechanism has an associated knowledge-acquisition mechanism (KAM). These KAMs know how to acquire the knowledge needed by its associated mechanism. The KAMs contain a model of the knowledge types used by the mechanism, and a set of strategies for acquiring the knowledge. A model of a KAM for the mechanism *select-parts* is given in Figure 5. This mechanism applies constraints to various candidates to determine the non-dominated set of parts. The KL facilitates the development of KAMs because of its regularity, *i.e.* the types of knowledge used are known and the representation scheme is fixed.

Each KAM can be pieced together to form the basis of a KA tool. This, however, is not sufficient. To be an effective gatherer, the tool must be able to analyze the input knowledge. For example, it must be able to determine if sufficient knowledge has been entered to allow the PSM to operate properly. Furthermore, the tool must develop an overall strategy for acquiring knowledge. In other words, the PSM does not provide the sequencing information necessary for a meaningful interaction with a domain expert. For example, it makes more sense to acquire knowledge about parts before acquiring constraints to select among parts. Armed with this knowledge, the KA tool can, for instance, ensure that the domain expert has provided sufficient constraints to differentiate between similar parts.

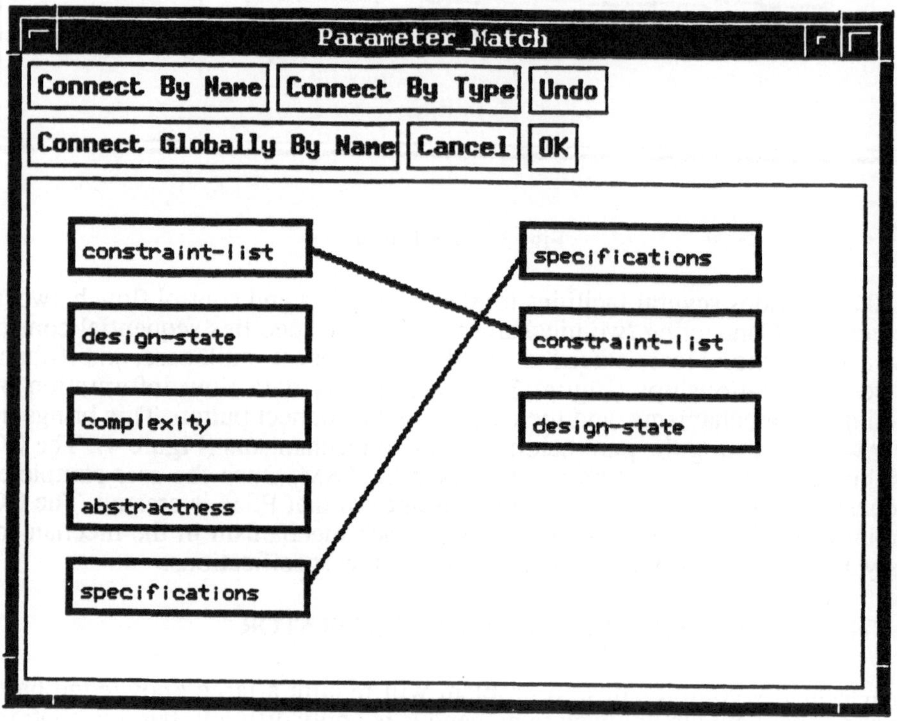

Fig. 4. Specifying data flow between two mechanisms

Mechanism: *select-parts*

Knowledge used:

Constraints (C) : $\{< > \leq \geq = \neq\}$

Specifications

Attributes

Form of knowledge in mechanism

specification CONSTRAINT candidate's attribute

Strategies for acquisition:

Ask user to supply all knowledge

Fig. 5. KAM model for select-parts mechanism.

Thus, the KATG contains not only KAMs, but also heuristics for building KA tools. Some of these heuristics analyze the knowledge used by the mechanisms in the PSM and suggest an ordering for the KA session. For example, if part information is used and calculations are performed based on the parts' attributes, then acquire the parts first, and then acquire the calculation information. Another heuristic is *reduce the tool-expert dialog*: Favor generating information automatically (using deduction from already given facts, for instance) to asking the domain expert. Finally, the KATG contains knowledge about analyzing the structure of the acquired knowledge. As explained in Section 5, the tool can analyze a graph and determine whether backtracking is necessary.

4.4 DEBUGGING TOOL

The DT will monitor the performance of a PSM as it attempts to solve a particular design task. The environment shows the execution of the DIDS-generated design system at the mechanism level (Figure 6). The DT displays the four classes of information that will be of interest to the user: the current point in the PSM, the mechanism being executed, the knowledge being operated on by the mechanism, and the evolving design. For each mechanism, the DT environment will also show the knowledge being operated on by the mechanism, and how that knowledge is being used to produce the mechanism's results. The user can examine the PSM, the knowledge base and the evolving design. If an error is detected, then the user can invoke the other DIDS tools (PSM Editor, KA Tool) to make modifications.

The DT will provide most of the features found in debuggers for programming languages. For example, breakpoints can be set on mechanisms and elements of the knowledge base. Alternatively, the design state can be monitored so that the user is notified when mechanisms inspect or modify it.

DT divides the screen into four windows. As the PSM is executed, the PSM window highlights the current mechanism, and information about its execution is displayed in the other three windows. The Knowledge-Base Window (KBW), displays a graphical representation of the knowledge base created by the DIDS generated KA tool. The KBW highlights those portions of the knowledge base currently being examined by the mechanism highlighted in the PSM Window. The Design Window displays a graphical representation of the artifact being designed. It highlights the modifications made to the design by the execution of the current mechanism. The Mechanism Window, describes decisions made by the mechanism highlighted in the PSM Window. It displays information that explains how the mechanism produced its output given its inputs.

5. Scenario—Creating a DIDS-Generated System to Design a Bicycle

This section demonstrates how DIDS might be used to create a design system that configures mountain, touring, and (street) racing bicycles. The current version of the DIDS prototype can execute this scenario, except for some of the knowledge acquisition features. The design task involves configuring a bicycle that meets the user's specifications while minimizing the cost of the bicycle. The part catalog includes frames, rims, forks, derailleurs, chains, and so forth. In this scenario, the DIDS user is both a knowledge engineer, and a bicycle expert. For more complicated problems, both a domain expert and a knowledge engineer will collaborate while using DIDS. In addition, this scenario assumes that the DIDS user, who will be referred to here as the designer, has limited knowledge of the PSM and mechanism libraries, and has a working knowledge of the DIDS interface. Using DIDS, a domain expert performs the following steps:

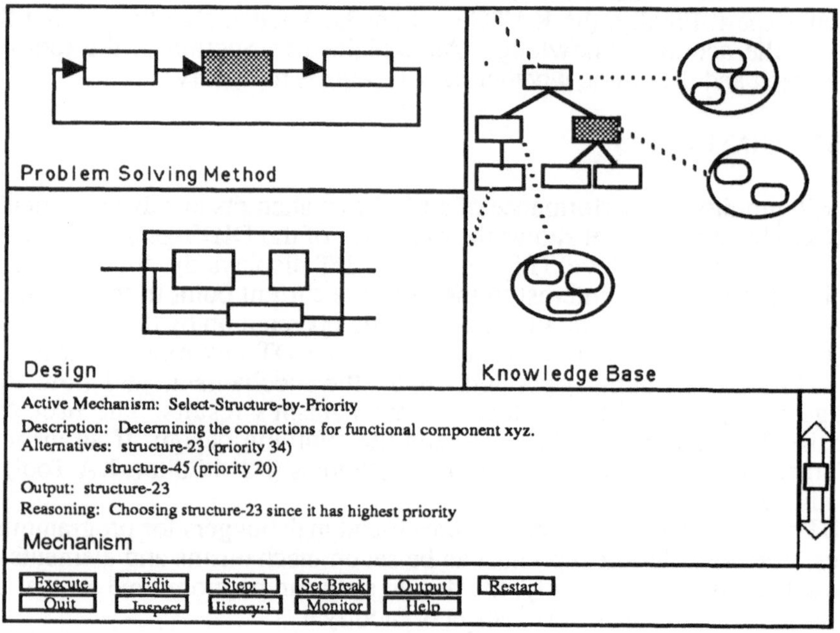

Fig. 6. DIDS Debugging Tool

1. Select a PSM: a library of PSMs is presented, and the domain expert chooses one which is most appropriate for the task to be performed.
2. Construct a prototype system: based on the chosen PSM, a KA tool is produced. This tool acquires from the expert enough knowledge to construct a prototype, working system. This prototype is analyzed, and the PSM is modified as needed.
3. Full-scale KA: once the prototype has been tuned, the expert uses the KA tool to describe completely the design domain .

5.1 SELECT A PSM

To save time in constructing a system, the designer selects a PSM from the library, instead of developing one from mechanisms. The designer browses the PSMs that were archived by previous DIDS users, and examines a high-level characterization of the types of tasks solved by each PSM. Figure 7 shows a characterization of four of the PSMs in the library. The figure distinguishes PSMs by the subtasks they perform.

PSM	Part Selection	Structure Selection	Backtracking	Arrangement
M	X	X	X	
R	X		X	X
S				X
V	X		X	

Fig. 7. PSM comparisons

The designer selects PSM M for several reasons, which are based on personal experience in the domain. First, the task requires part selection. Second, backtracking is necessary because incorrect part selections are often made. Third, since the bicycles will eventually have to be assembled, the design system should reason about the connections among components (structure). Fourth, the structure implicitly arranges the components. The designer's knowledge may or may not be accurate; in this case, it is not. Figure 3 shows PSM M as it appears in the PSM Editor.

5.2 CONSTRUCT PROTOTYPE SYSTEM

After the PSM has been selected, the designer has DIDS generate both the problem solver and KA tool for PSM M. To test the prototype, the designer performs some KA using the generated tool. PSM M makes assumptions concerning the structure of the domain knowledge that it uses. DIDS generates a KA tool that acquires knowledge in the form required by the PSM.

PSM M assumes that the KL contains three constructs: parts, abstract parts, and templates. Each construct, which consists of one or more types of knowledge, represents a different type of domain concept. The part construct, which contains part knowledge, represents the components in the part catalog. The abstract part construct, which contains type decomposition, formula, and constraint knowledge, successively organizes the parts into classes that share common properties, forming a type hierarchy (Figure 8). In this hierarchy, each abstract part contains the set of properties shared among its children. Finally, templates, which contain connection and constraint knowledge, describe alternate ways of connecting parts (Figure 9). Associated with each template is a precondition that determines when the template should be selected.

The abstract parts in PSM M's KL can be distinguished by their set of attributes. Each abstract part contains both specification and characteristic attributes. Characteristic attributes describe properties of the part that are intrinsic to its behavior. Typically, these attributes are given in a part catalog. These include properties such as cost, size, speed, performance, and so forth. Specification attributes, which are associated with formulas, represent properties that the part must satisfy for a particular design. PSM M uses these attributes, whose values are calculated from the requirements and from other attributes, to select parts from the library. Constraints express the relationship between the specification attributes of an abstract part and the characteristics of its child parts in the type hierarchy.

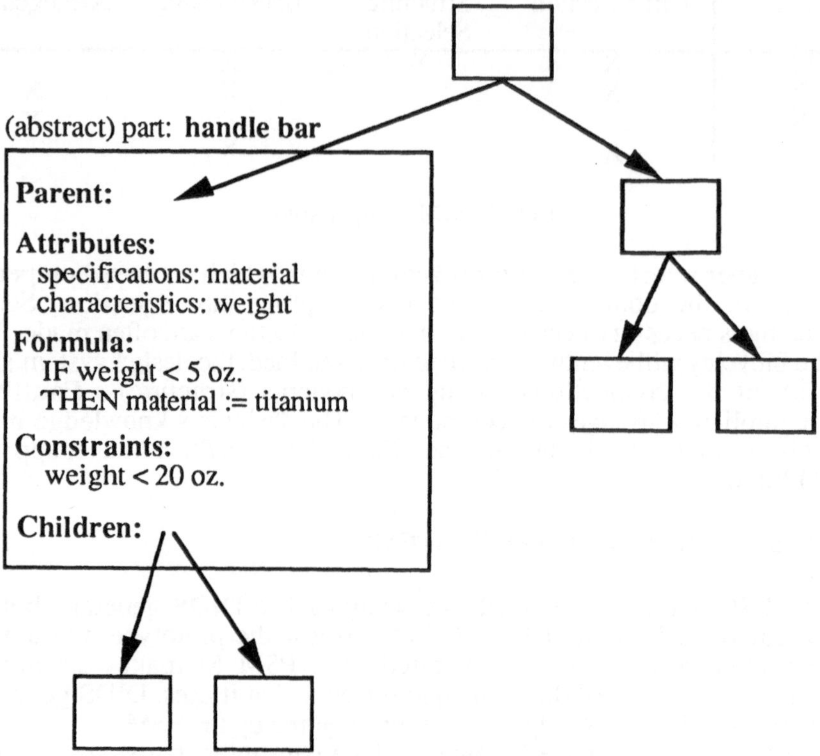

Fig. 8. Abstract part hierarchy.

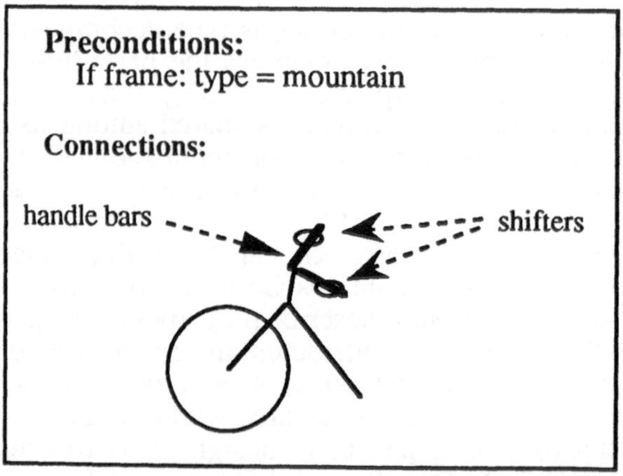

Fig. 9. Template showing connections between mountain bike shifters and handle bars.

PSM M operates by comparing the specification attributes of an abstract part to the characteristic attributes of each of its children using constraints. The child part that satisfies all the constraints is selected. If no such part exists, then PSM M must backtrack. Backtracking involves determining the set of specification attributes that need to be relaxed to allow selection to proceed. Once these attributes have been identified they are traced back to the requirements from which they were calculated. The PSM then allows the designer to modify one or more of these requirements. If selection is successful, then PSM M determines how to connect the selected part with the set of parts already in the design. PSM M makes the connections by selecting a template using the preconditions associated with each template. Before the next part can be selected, PSM M calculates the values of any specification attributes on abstract parts whose values are dependent upon the value of the part just added to the design.

The KATG uses the description of PSM M's KL and ML to generate a customized KA tool. This KA tool will build a KL consisting of parts, abstract parts, and templates. The KATG determines the best order to ask the designer for information by analyzing the structure of the KL. For PSM M, the analysis is straightforward because templates are defined in terms of abstract parts and parts. Therefore, a template cannot be specified without first specifying the abstract parts. The KA tool begins by asking the designer to describe the parts and abstract parts for the bicycle domain.

The designer begins the preliminary KA phase by entering parts listed in the bicycle-component manufacturers' catalogs. Once these have been entered, the designer creates abstract parts by organizing the parts by their properties. Figure 10 shows a few of the hierarchies created by the domain expert. Portions of these hierarchies are not terminated; this is to show that the hierarchies ground-out to a large number of parts.

The KA tool prompts the designer for the attributes of each part and abstract part. The values of characteristic attributes will be supplied by the designer, and the values of specification attributes will be calculated during execution. For specifications, the designer provides formulas that calculate their values, and constraints that restrict the set of acceptable values (Figure 11). The relationships between the attributes expressed by the formulas form a directed graph. Figure 12 shows a portion of the graph created by the domain expert. The ovals, which represent attributes, contain the abstract part name, followed by a colon, and then the specification name.

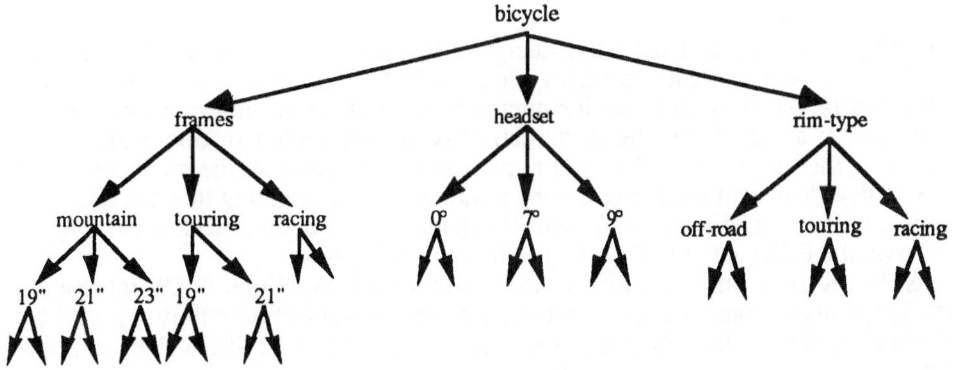

Fig. 10. Bicycle domain type hierarchy.[1]

[1] All parts shown are abstract.

After the parts, abstract parts, formulas, and constraints have been entered, the KA tool analyzes the formula graph. It determines that this graph is a tree, and therefore has a width of 1. Therefore, the design system will not need to perform backtracking to find the solution, if the KA tool ensures that the graph is arc-consistent before problem solving begins (Dechter and Pearl, 1988). After being advised of this by the KA tool, the designer uses the PSM editor to remove the backtracking step from PSM M. DIDS ensures that all future versions of the KA tool contain a component that makes the graph arc-consistent after each KA session.

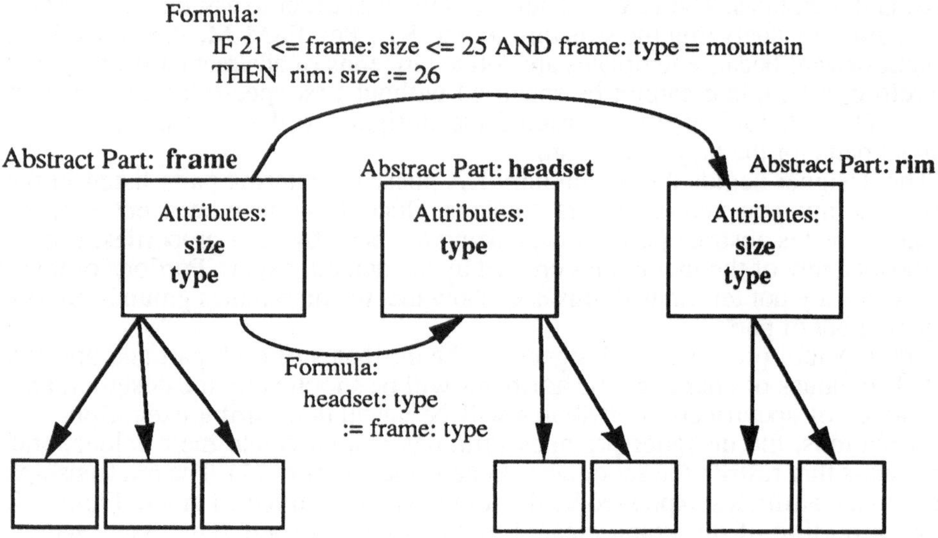

Fig. 11. Formulas between abstract parts

Finally, the KA tool asks the designer to specify the templates. The tool explains that each template specifies a legal way of connecting a set of abstract parts or parts. At this point, the designer realizes that, in this domain, alternatives do not exist; a set of bicycle parts can only be connected in one useful way. Therefore, the bicycle task does not require templates, and all the mechanisms in PSM M that deal with templates can be removed. After realizing this, the designer brings up the PSM editor and removes the determine-structure mechanism. Figure 13 shows the PSM at the end of the preliminary KA phase.

PSM modifications do not result in information loss. Since DIDS represents knowledge using standard forms, which are independent of how they are used by individual systems, the knowledge base remains stable. If a PSM is changed, new knowledge may need to be acquired, and previously acquired knowledge may become irrelevant, but the contents of the knowledge base will not be lost.

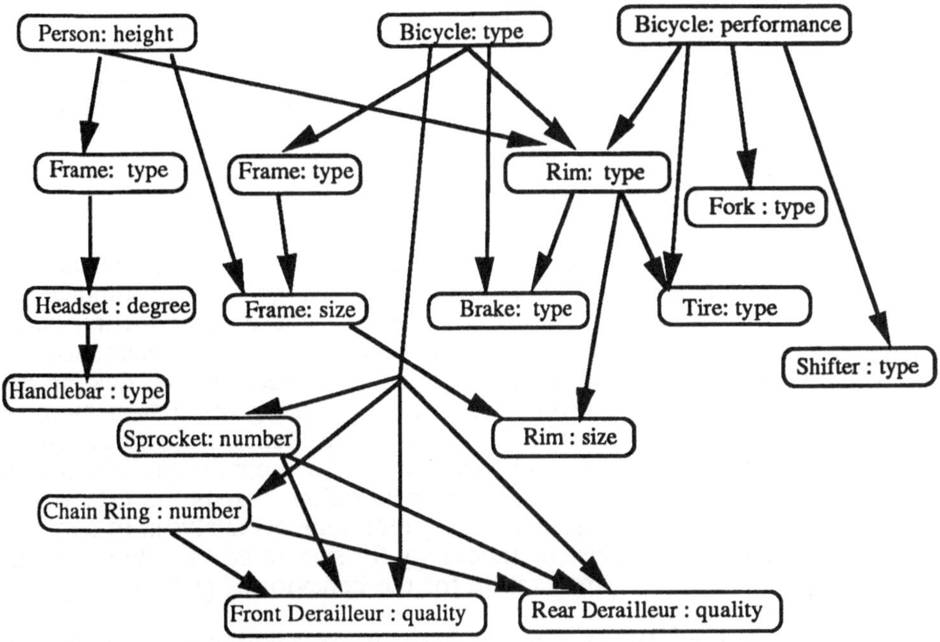

Fig. 12. Directed graph formed by the formulas

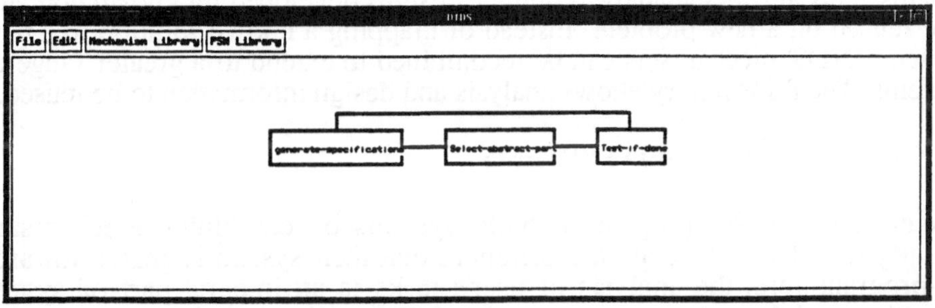

Fig. 13. PSM at the end of preliminary knowledge acquisition

The bicycle design task presented in this scenario is admittedly simple when compared to the task performed by existing KBDSs. This example, however, has demonstrated the power of the DIDS system and suggests how DIDS will be used to develop more complex KBDSs. Specifically, it has demonstrated that DIDS will allow designers to rapidly construct KBSDs from a library of mechanisms using an iterative methodology. It also highlights some of the support that DIDS will provide during this process.

6. Related Work

This section compares DIDS to two classes of systems. The first class of systems automate programming in general, and the second class of systems facilitate the construction of knowledge-based systems.

6.1 AUTOMATIC PROGRAMMING

The Programmer's Apprentice (PA) (Rich and Waters, 1988), an assistant to a software engineer, facilitates program development using reusable components in all phases of the software-development process. These components, called clichés, represent commonly used combinations of programming elements. The PA, which contains clichés that represent familiar specification, design, and implementation constructs, develops software by using inspection methods. During inspection, the PA helps the user to recognize clichés in the specifications and to choose among the lower-level clichés that implement the specification. In contrast, DIDS completely automates code generation, assists the user during design, and provides little assistance during requirements analysis. In addition, DIDS uses a propose-and-revise methodology for system development whereas the PA, which has a rich representation for the behavior of clichés, supports a refinement methodology.

Draco (Neighbors, 1984) automates software development by reusing software components. Draco not only supports the reuse of code, but also the reuse of analysis and design information. Draco libraries contain domains for which the typical problem statements and implementation alternatives are known. Users develop new systems by describing requirements in terms of known domains. This allows the analysis and the designs developed for known domains to be reused on a new problem. Instead of mapping a known solution to a new problem, DIDS' mechanisms can be recombined to extend to a greater range of problems. The PSM library allows analysis and design information to be reused.

6.2 KNOWLEDGE-BASED SYSTEMS

Klinker *et al.* (1990) propose to build systems by combining mechanisms similarly to DIDS. The approach differs in that their system is geared towards non-programmers, the analysis of user's tasks is an integral part of system generation, and the task type is not restricted. This makes it difficult to determine *a priori* the types of knowledge and the set of mechanisms required to construct systems. Instead, they have developed a *shared* vocabulary of task activities that can be used to describe tasks in domain-independent terms. The system analyzes the user's task, and helps to describe it in terms of the shared vocabulary. Each activity in the shared vocabulary is associated with a set of mechanisms that can be used to implement it. The shared vocabulary helps to make mechanisms usable, *i.e.*, understandable by users, and reusable.

Neches *et al.* (Neches *et al.*, 1991) propose to build new knowledge-based systems through knowledge sharing. This involves building tools that enable the knowledge base of one system to be used by another, and that facilitate the communication between knowledge-based systems. A standard knowledge-representation system, a standard knowledge-base query language, a standard concept ontology, and a standard language for expressing knowledge form the heart of this approach. DIDS shares the most with the standard concept ontology, since the mechanism and PSM libraries can be viewed as ontologies of problem-solving components. The philosophy of DIDS differs significantly from the other parts of Neches *et al.* approach. DIDS supports the reuse of fundamental software components by identifying mechanisms shared across domains instead of reusing existing knowledge bases. We believe the task specific bias of most knowledge bases makes their reuse difficult.

PROTEGE II (Puerta *et al.*, 1991), a system similar to DIDS, generalizes the capabilities of PROTEGE (Musen, 1989) and combines them with a mechanism-based model. The restriction to skeletal-plan refinement PSMs in PROTEGE II, and a weaker model of expertise distinguish it from DIDS. PROTEGE II will generate KA tools for any skeletal-plan refinement PSM specified as a set of mechanisms. PROTEGE II will contain a library of tasks and a library of mechanisms. A mechanism's description will include a description of the data used by the mechanism and links to tasks. These links will be used to suggest mechanisms capable of performing a user's task. Once the mechanisms have been selected, PROTEGE II generates a KA tool by looking at the data required by each mechanism. PROTEGE II's model of mechanisms and the data used by mechanism is weak when compared to the DIDS model. The set of mechanisms, the types of data operated upon by mechanisms, and a procedure for identifying mechanisms has not been established.

The CONGEN system (Sriram and Cheong, 1991) supports the development of KBDS by providing a graphical user interface for capturing the elements of a problem solver as well as its knowledge base. CONGEN differs from DIDS on several accounts. First, CONGEN users manipulate programming-language constructs. In contrast, DIDS users combine mechanisms, which perform tasks at a significantly higher level than program statements. Second, CONGEN makes no distinction between the process of acquiring the problem solver and acquiring the domain knowledge used by the problem solver.

DIDS can also be compared to design programming languages. Our work is aimed at understanding how design systems operate, namely the way in which they solve their particular problems. What makes these systems different is the design knowledge they use, and the domains in which they operate. Thus, our work is significantly different from those developing languages for constructing design systems, such as DSPL (Brown and Chandrasekaran, 1989), Edesyn (Maher, 1988), and DESCRIBE (Mittal and Araya, 1987). These languages provide programming constructs to easily capture design knowledge for specific tasks. All three, however, work at a different abstraction level than the mechanisms described in this paper. Furthermore, they provide simpler, albeit potentially more general, operators (mechanisms) than are assumed by our models. In fact, the mechanisms described here could be implemented in any of these languages, as they could in more traditional programming languages.

The analysis presented here is similar in intent to Tong's framework (1987) function level, which specifies the "problem-solving process". This is equivalent to our PSMs and mechanisms. Tong's paper presents a specific approach to design, goal-directed planning, while we are interested in uncovering similarities and differences between problem-solving systems.

SALT (Marcus, 1987), the knowledge-acquisition tool for the VT system, embodies many of the ideas that will be present in DIDS generated KA tools. SALT plays an active role during knowledge acquisition by detecting inconsistencies and missing knowledge. We believe that the strong model of mechanisms and knowledge forming the foundation of DIDS will enable the products of the KATG o have these features. SALT, however, was built to support a particular PSM. Thus, it is limited in applicability. DIDS removes these limitations and incorporates VT's mechanisms and PSM in its libraries.

Chandrasekaran proposes a model of design based solely on the concept of generic tasks (Chandrasekaran 1986, 1990). Generic tasks decompose design tasks hierarchically. Each task is defined by its position in the hierarchy, the method used to perform it, and the knowledge, both declarative and control, required to perform the task. Generic tasks have three principal weaknesses, which are shared with DIDS, but to a lesser degree. DIDS' model of configuration design, which is grounded in a careful study of configuration systems, helps to reduce these problems. First, the generic-task model does not establish which tasks are generic. Second, the design system's task may not decompose neatly into a disjoint set of high-level generic tasks even though there may exist a set of lower-level tasks that could implement the system. For example, it is possible that a design system might first perform half of the classification generic task, then do a critiquing task, and then finish the classification. Finally, the separate implementation of each generic task requires an environment to integrate and to allow communication between tasks. Also, a system implemented using generic tasks may contain multiple copies of the same piece of knowledge, since the knowledge used by two generic tasks may overlap. The environment must ensure that the knowledge contained in different generic tasks is consistent.

7. Summary

The DIDS system will dramatically reduce the time required to build configuration-design systems by providing a set of tools that can be used to construct systems from a library of mechanisms. The DIDS model of configuration design forms the basis of this approach. The model identifies reusable mechanisms by focusing on the types of knowledge used during configuration, and the operations performed on them. The ability to distinguish reusable mechanisms from non-reusable ones ensures that the library size will be manageable, and that the mechanisms in the library can be applied to a wide variety of problems.

Preliminary versions of the PSM Editor and CG have been implemented and can successfully execute the bicycle scenario presented here. Work has just begun on the KATG, and the current version of the debugging environment is just a specification document. Users can browse the mechanism library and build a PSM using the graphical editing features of the PSM Editor. The code for the resulting PSM can then be generated and executed. The small size of the current mechanism and PSM libraries limits the set of tasks that can be solved by the current version of DIDS. These libraries will grow as our analysis of existing configuration-design systems using the DIDS model continues.

The four components of the DIDS system—PSM Editor, KATG, CG, and DT—will be used in an iterative process. First, a prototype PSM will be created using the PSM Editor. Since DIDS users may not be familiar with the behavior of each mechanism or PSM, incorrect choices will be made. Therefore, the DIDS-generated KA tool and the DT will analyze the PSM, KL, and specific domain knowledge to help uncover any errors. When an error is found, the PSM Editor will be used to modify the PSM. The problem solver and KA tool will be regenerated. Since both the new KA tool and new problem solver will be consistent with the previously acquired knowledge, KA and the PSM tuning will continue.

Acknowledgment

Alan Balkany and the other members of the University of Michigan's Integrated Systems Design Group have made significant contributions to the DIDS research. Many of the ideas expressed in this paper are a result of their efforts. This work was funded, in part, by the National Science Foundation Grant MIPS-905781 and by Digital Equipment Corporation.

References

Balkany, A., Birmingham, W. P. and Tommelein, I. D.: 1991, A Knowledge-level analysis of several design tools, *in* J. S. Gero (ed.), *Artificial Intelligence in Design '91*, Butterworth-Heinemann, Oxford.

Barker, V. and O'Conner, D.: 1989, Expert systems for configuration at Digital: XCON and beyond, *Communications of the ACM* March.

Birmingham, W. P., Gupta, A. and Siewiorek, D.: 1989, The MICON system for computer design, *IEEE Micro*, October.

Birmingham, W. P. and Tommelein, I. D.: 1991, Towards a domain-independent synthesis system, in M. Green, (ed.), *Knowledge Aided Design*, Academic Press, London.

Boehm, B. W.: 1987, Improving software productivity, *Computer*, September.

Brown, D. and Chandrasekaran, B.: 1987, *Design Problem Solving—Knowledge Structures and Control Strategies*, Morgan Kaufman, San Mateo, CA.

Chandrasekaran, B.: 1986, Generic tasks in knowledge-based reasoning: high-level building blocks for expert system design, *AI Magazine*, Fall.

Chandrasekaran, B.: 1990, Design problem solving: a task analysis, *AI Magazine*, Winter.

Dechter, R. and Pearl, J.: 1988, Network-based heuristics for constraint satisfaction problems, *Artificial Intelligence*.

Frayman, F. and Mittal, S.: 1987, Cossack: A constraint-based expert system for configuration tasks, *Proceedings of the 2nd International Conference on Application of AI to Engineering*, August.

Gero, J. S.: 1990, Design prototypes: A knowledge representation schema for design, *AI Magazine*, Winter.

Johnson, M. V. Jr and Hayes-Roth, B.: 1988, Learning to solve problems by analogy, *Report No. KSL-88-01*, Department of Computer Science, Knowledge Systems Laboratory, Stanford University.

Klinker, G., Bhola, C., Dallemagne, G., Marques, D. and McDermott, J.: 1990, Usable and reusable programming constructs, *Proceedings of the 5th Knowledge Acquisition Workshop*, AAAI.

Langrana, N., Mitchell, T. and Ramachandran, N.: 1986, Progress towards a knowledge-based aid for mechanical design, *Symposium on Integrated and Intelligent Manufacturing*, The American Society of Manufacturing Engineers.

Maher, M.: 1987, Engineering design synthesis: a domain independent representation, *AI EDAM*, March.

Marcus, S, Stout, J. and McDermott, J.: 1987, VT: An expert elevator designer that uses knowledge-based backtracking, *AI Magazine*, Winter.

McDermott, J.: 1988, Preliminary steps toward a taxonomy of problem-solving methods, *in* S. Marcus (ed.), *Automating Knowledge Acquisition for Expert Systems*, Kluwer, Boston, MA.

McDermott, J.: 1980, R1: A rule-based configurer of computer systems, *No. CMU-CS-80-119*, Department of Computer Science, Carnegie Mellon University.

Mittal, S. and Araya, A.: August, 1987, A knowledge-base framework for design, *Proceedings of the 9th IJCAI*, August.

Mittal, S. and Frayman, F.: 1989, Towards a generic model of configuration tasks, *Proceedings of the 11th IJCAI*, August.

Musen, M.: 1989, *Automated Generation of Model-Based Knowledge-Acquisition Tools*, Pitman, London.

Neches, R., Fikes, R., Finin, T., Gruber, T., Patil, R., Senator, T. and Swartout, W.: 1991, Enabling technology for knowledge sharing, *AI Magazine*, Fall.

Neighbors, J.: 1984, The Draco aproach to constructing software from reusable components, *IEEE Transactions on Software Engineering*, September.

Newell, A.: 1981, The knowledge level, *AI Magazine*, Summer.

Puerta, A., Edgar, J., Tu, S. and Musen, M.: 1991, A Multiple-Method Knowledge-Acquisition Shell for the Automatic Generation of Knowledge-Acquisition Tools, *Personal Communication.*

Rich, C. and Waters, R.: 1988, Programmers apprentice: a research overview, *IEEE Computer*, November.

Sriram, D. and Cheong, K.: 1991, Engineering design cycle: a case study and implications for CAE, *in* M. Green (ed.), *Knowledge Aided Design*, Academic Press, London.

Steels, L.: 1990, Components of expertise, *AI Magazine*, Summer.

Tommelein, I. E., Levitt, R. E., Hayes-Roth, B. and Confrey, T.: 1991, SightPlan experiments: alternate strategies for site layout design, ASCE, *Journal of Computing in Civil Engineering*, January.

Tong, C.: 1987, Towards an engineering science of knowledge-based design, *Artificial Intelligence in Engineering*, 2(3).

FUNCTIONAL REASONING IN CONFIGURATIONAL DESIGN

A. CHAWLA and R. SANGAL
Department of Computer Science and Engineering
Indian Institute of Technology
Kanpur 208 016
India

Abstract. In this paper we describe a system being developed by us to solve the problem of configurational design, a kind of design that generates descriptions of artifacts containing a number of sub-parts. It maps the functional requirements to a structural description to be used in the detailed routine design of the artifact. The approach used by us relies upon the functional description of the requirements and on our part representation technique. Our representation strengthens the hierarchical representation used in routine design systems by introducing explicit information on connectivity between various parts. This paper describes how the system uses abductive reasoning to guide its search for artifacts that satisfy some of the requirements and how it can assemble these artifacts in a new configuration. It also describes how this description of freshly assembled artifacts generated by the system can be used by a routine design problem solver for the detailed design of the individual sub-parts. The abductive reasoning used by us differs from abductive hypothesis assembly since the causalities and assembly techniques for hypothesis and those for parts differ. The work reported in this paper forms a part of an attempt to tackle large real life design problems in an intelligent way. We have described how this work forms the basis of relating multiple levels of deep and shallow knowledge and generating novel failure recovery mechanisms. We further envisage that this work builds a foundation for using structural and functional analogy to solve the problem of configurational design.

1. Introduction

We have earlier defined (Chawla & Sangal 1991*b*, Chawla & Sangal 1991*a*) configurational design as a design class which forms a front end to routine design and is distinct from the configuration task as defined by (Mittal & Frayman 1988, Mittal & Falkenheiner n.d.). Design activity has been classified into three categories, viz class I, class II and class III (Brown 1985, Brown & Chandrasekaran 1985, Sriram & et al 1989). We treat configurational design as a typical class II design case and believe that in most real life large design problems, it forms an essential part of the design process. Briefly, a configuration is a description of an artifact such that it forms the basis of the next stage of the design process, viz, the detailed routine design of the artifacts involved. This description contains information on how the individual parts of the artifact are to be designed, how these parts are connected to one another and how the overall requirements of the design are broken down to the design specifications of the individual sub-parts of the artifact. The aspect of design that generates a configuration from a given set of requirements and a library of generalized part descriptions is what we have defined as configurational design (Chawla & Sangal 1991*b*).

We have earlier talked of various kinds of knowledge required to solve the problem of design (Chawla & Sangal 1991*b*). We have given a detailed analysis of the kinds of knowledge that are not being represented in other design systems and have proposed a formalism that takes care of these problems.

J. S. Gero (ed.), Artificial Intelligence in Design '92, 41–60.

The essential modification that we have made is the explicit representation of connectivity details. In addition to storing information on how parts are connected, additional constraints imposed by joining the parts are also explicitly stored. This information is stored in the arcs connecting the respective models (nodes) of the concerned parts. In (Chawla & Sangal 1991*b*) we have discussed the advantages such a representation gives and in (Chawla & Sangal 1991*a*) we have proposed an abduction based problem solving strategy that can solve the problem of configurational design.

The process of configurational design is guided by the functional requirements of the problem. A typical design specification shall consist of a set of functionalities to be met by the designed artifact. Relating these requirements with the artifacts is the task of the configuration designer. The obvious choice for solving the problem of configurational design is a function based approach. The abductive assembly process we use is also based on functional reasoning. In the past functional modeling has been used in a variety of ways (Goel & Chandrasekaran 1989, Joscowicz 1987, Sticklen, Goel, Chandrasekaran & Bond 1989, Ulrich & Seering 1988, Ulrich & Seering 1989). Unlike other domains, where the task is to use the functional model of the parts in the reasoning process, in configurational design the task is to use the functional description of the requirements and to generate the functional as well as a design model for the resulting artifact. This has been referred to as function to structure mapping in (Gero 1990). In this paper we describe in detail how functional reasoning has been used by us to solve this problem.

2. System description

A number of approaches have so far been tried in order to solve the problem of design. These include a plan based approach (Brown & Chandrasekaran 1985, Brown & Chandrasekaran 1986, Brown 1989, Brown 1985, Marcus 1987), a generic task approach (Chandrasekaran 1988, Chandrasekaran 1989), and a logic based approach (Ohsuga 1989*b*, Ohsuga 1989*a*, G Han & Yamauchi 1987). These essentially solve the problem of routine design. We choose a plan based hierarchical outlook of the routine design process. In this approach, the routine design of an assembled artifact is broken down to the design of its constituent artifacts in a pre-defined hierarchical manner. Design is done on the basis of plans and tasks which are stored with the model of the part. Since we view configurational design as a front end to the routine design stage, the output of a configurational design system is a model of the desired artifact so that it can be fed to a routine design problem solver for the detailed design of the artifacts involved. In other words the output of a configurational design system is a set of design plans complete with all dependencies among the various decision parameters. In (Chawla & Sangal 1991*b*) we have presented a description of the kind of knowledge needed to solve the problem of configurational design and have presented the advantages that we believe our formalism will offer. We now describe our system and some of its achievements in detail. We first describe the representation we have used to model the parts and their functionality.

2.1. Representation of the Parts

Configurational design is the task of mapping the required functions to the structure of the resulting artifact. As we have mentioned earlier it requires knowledge regarding part connectivity as well as a description of the functionality of the artifact. Further since configurational design forms a front end to routine design, information regarding how the routine design of the artifact shall be carried out also needs to be stored. In accordance with these general principles we represent the model of the part as follows.

2.1.1. Structural details:

The structural description of the part consists of the sub-parts of the part and the parts to which this part can be connected. The latter also includes details on how this part is connected to the other parts and the additional constraints imposed by such connections. This information is stored in arcs joining the nodes representing the parts. The model (the internal representation) of a part stores an arc for each part with which it can be connected in an actual configuration. This description captures the connectivity capabilities of the part and is used by the configurational design system in its problem solving process. When this part is used in a particular configuration, the system generates an instance of the part containing specific information regarding which parts it is connected to. In other words in forming a configuration the system shall generate an instance of the part and retain only those arcs that are relevant in the particular context. These arcs contain a set of constraints imposed by connecting the parts in question. These constraints form an essential component of the problem solving process and the routine design stage that follows. Before we go into the details of the problem solving process, we describe the representation in greater detail.

Consider Figure 1. The arcs field in the description of an artifact contains the arcs through which this artifact can be connected to others. In Figure 1a, the part *shaft* can be connected to a *gear* or to a *motor*. Therefore the model of the *shaft* contains arcs to the generalized model of the *gear* as well as to that of the *motor*. The detailed descriptions of these arcs, the constraints imposed by them, and a description of how the parts are connected has been omitted for clarity. However, when a *shaft* is connected to a *gear* an instance is created of both the parts and only the arc between these two parts is retained. Thus the models of the parts are *joined* to one another using an instance of the arc.

Figure 1b gives the detailed descriptions of the node corresponding to the model of the *shaft*, a description of the arc between a *motor* and a *shaft*, and the description of one of the properties of the *shaft*. The *sub-parts* field of the description of the parts contains the list of the sub-parts of the artifact, and the parent field contains the description of the assembly of which the artifact is a part of (both of which are nil in this case). The fields *props*, *conds* and *goals* and the description of the property representation are described further down in the paper.

Figure 1b also shows the representation of an arc between a *shaft* and a

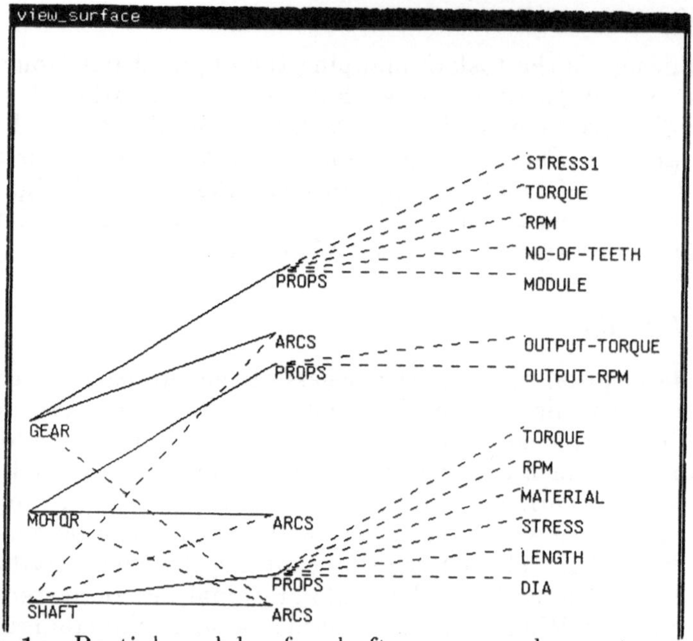

Figure 1a: Partial models of a shaft, a gear and a motor.

motor. The parts used in connecting these parts are stored under the *uses* field. Thus the *shaft* and the *motor* are connected using a *coupling*. The affect clause of the arc represents the additional constraints imposed when these parts are connected. The representations of these constraints is described in section 2.1.2. How these constraints affect the design process has been described later on.

```
(print-node 'shaft)
****description of the representation of a shaft*****
name : SHAFT
parent : NIL
sub-parts : NIL
arcs (to) : (GEAR MOTOR)
props : (DIA LENGTH STRESS MATERIAL RPM TORQUE)
conds : NIL
goals : NIL
>(print-arc (get-node 'shaft) (get-node 'motor))
****description of the representation of an arc******
between MOTOR and SHAFT
uses (COUPLING)
props : NIL
conds : NIL
parent : NIL
affect : ((= ?(OUTPUT-TORQUE MOTOR) ?(TORQUE SHAFT))
          (= ?(OUTPUT-RPM MOTOR) ?(RPM SHAFT)))
>(print-prop (car (node-props (get-node 'shaft))))
```

```
****description of the representation of a property**
name : DIA
conds :((< ?(DIA) 200))
generic-types :(LEN)
parent (type NODE) : SHAFT
definition : ((FUNC (?(LENGTH) ?(MATERIAL) ?(RPM) ?(TORQUE))))
uses : NIL
modifies  : NIL
```

Figure 1b: Sample description of a node, arc and a property (continued from previous page).

2.1.2. Routine design details:

Since the final output of the configurational design system is a model of an artifact to be used by a routine design system, the generic model of the artifact has to contain routine design information about the artifact. This information includes

- *Properties and their inter-dependencies :* These include the dependencies between various properties of the artifact and its sub-artifacts that are imposed by various constraints. When a new configuration is built the dependencies amongst the various properties change as fresh constraints are imposed by the connectivity details. The affect of these fresh constraints on dependencies shall be discussed later on.
- *Design plans and heuristics on selection of plans and tasks :* The routine design of an artifact consists of hierarchically decomposing the design decisions and then assigning values to the various design parameters. This knowledge can be stored in the form of design plans and tasks (as in DSPL (Brown 1985, Brown & Chandrasekaran 1986, Brown 1989, Brown & Chandrasekaran 1985) and SALT (Marcus 1987, Marcus & McDermott 1989)) and needs to be stored with the artifact models so that it can be modified depending upon the connectivities chosen in the configurations being generated. However since we do not have a routine design problem solver at this stage, we do not specifically represent this knowledge. At this stage we use information on constraints and demonstrate how it can be modified to generate new plans and tasks while building new configurations.
- *The pre-conditions on selection of the part :* This includes the set of conditions that need to be satisfied for selecting an artifact in a configuration. Even though this kind of knowledge affects the selection of the artifact in a particular configuration it has been placed under the routine design knowledge because it shall also affect dependencies in forming the routine design plans of the artifact.

To give an example of the routine design knowledge of an artifact consider Figure 1. The information regarding the various properties of the artifacts, their definitions, the constraints they are expected to fulfill etc all comes under this category. The *props* field under any entity stores a list of the properties that describe the entity. Thus, in the figure the *shaft* is described by the properties *diameter, length, stress, material* etc. The *conds* field stores the pre-conditions on an entity that have to be satisfied for that entity to be a part of the final design. In these constraints, the variables are represented as *?list* where *list* describes the location of the property with respect to an entity. *?(output-torque motor)* thus represents the property *output-torque* of the part *motor*. Similarly, *?(length)* in the description of the property *dia* refers to the property *length* of the part *shaft* (which is the the parent of the property *dia*). Thus the *conds* field in the property description in Figure 1b means that the property *dia* of the *shaft* (parent of the property *dia*) should be less than *200*.

Similarly the affect clause in the description of an arc in Figure 1b means that the *rpm* and *torque* of the *shaft* and those of the *motor* are the same. These are the additional constraints imposed by connecting the two parts.

The definition clause in the property description stores the function that is invoked and the list of variables to be used as the actual parameters in calling the function. Thus in Figure 1b, the property *dia* can be assigned a value by calling the function *func* with variables *length, material, rpm and torque* which are properties of the part *shaft*.

The dependencies among the various properties imposed by these constraints have not been shown in the figure. These along with the design plans to be used for the routine design of these artifacts shall also come under the category of routine design knowledge.

2.1.3. Functional details:

In addition to the structural information the configurational design process is also guided by the functionality of the parts and the functional requirements of the problem at hand. The model of the part, therefore has to store the functionality of the part. This functionality is a list of functions the part is expected to fulfill when used in practice. Thus when the system is looking for an artifact that carries out a particular function it gets a means of controlling the search for new artifacts. Also while building new configurations, the system needs to find out the requirements that have not yet been satisfied by the parts selected. The functional description of the part shall also be needed to decide how well a part matches the given requirements and in solving the problem of design diagnosis (Murakami & Nakajima 1989).

The functional description of the part is stored under the field *goals* as a list of functions the part can carry out. For instance in Figure 2, the part *crane-1* satisfies the functions *pos-mechnsm and smpl-mve-up*. These are stored as a list of functions. The representation of these functions is discussed in section 2.2.

```
***************description of a crane-1**********************
>(print-node 'crane-1)
name : CRANE-1
parent : NIL
sub-parts : NIL
```

```
arcs (to) : NIL
props : NIL
conds : ((AND (<= ?(OBJ-WT GOAL SMPL-MVE-UP) 100000)
       (>= ?(OBJ-WT GOAL SMPL-MVE-UP) 100)))
goals : (SMPL-MVE-UP POS-MECHNSM)
(NOTE: The 'conds clause above means that the property
object-weight of the goal 'smpl-mve-up of this part should
be between 100 and 100000.)
```

Figure 2: Model of a *crane-1* (continued from previous page).

2.2. Representation of the functions

We now describe how we represent the functionality of the various artifacts. The functional specifications are used primarily for two purposes. Firstly it has to be used to guide the search process in selecting artifacts that satisfy a given set of requirements. Secondly when no artifact is found that satisfies the given requirements, this description has to be used to decompose the functional requirements to finer details so as to be able to find artifacts that satisfy these requirements and then join (assemble) them. In addition this description has to be used to check as to how well an artifact matches a set of requirements, and to identify aspects of the functionality that are either not satisfied by the part or are superfluous. In accordance with these requirements the representation that we have chosen for the functional model of the parts consists of:

- *Function / task description :* This is the description of the function and includes details regarding the various parameters that completely specify a function and the various sets of sub-functions (sfsets) (Chawla & Sangal 1991*a*) that the function can be decomposed into. For instance consider Figure 3. The function *lift-obj* can be specified completely by specifying the *size, weight* and *shape* etc (stored under the field *props*) of the object to be *lifted.* At the same time, *lift-obj* can be carried out by breaking it up into *positioning the mechanism, holding the object* and *moving the mechanism up.* These are stored in the field *sfsets* of the function description. The sfsets are being shown as a list of functions Thus the function *hold* can be decomposed into *holding-it-magnetically, holding-in-a-hook, tieing-by-rope-and-holding-in-a-hook* or *holding-in-a-gripper*, each of of which forms an sfset of the function *hold.* The system can thus decompose this function into any of these function sets when need be.
- *Satisfiability details :* These shall include the parts that can satisfy a particular function. For instance in Figure 3, the parts that can satisfy the function of *positioning the mechanism* are the different types of *cranes* and a *robot-arm.* Depending upon the requirements and the pre-conditions on each of these parts, the system can select any of these. If this list is null or if the pre-conditions on none of the artifacts mentioned are met, then the system needs to break up the function into one of the sfsets.

- *Constraint details :* These include the set of constraints that have to be met to carry out the function. Similarly constraints shall also be there with the sfsets of the functions and are used to decide whether a particular function is being satisfied in a particular configuration or not. For instance in Figure 3, the function *gripper-hold* can be satisfied only if a particular set of conditions on the *object-shape, size, weight* and *features* are satisfied by the design requirements.

```
***************************
name : LIFT-OBJ
parts : NIL
props : (OBJ-WT OBJ-SIZE OBJ-SHAPE OBJ-FEATURES)
sfsets : ((POS-MECHNSM&HOLD SMPL-MVE-UP))
***************************
name : POS-MECHNSM&HOLD
parts : NIL
props : NIL
sfsets : ((POS-MECHNSM HOLD))
***************************
name : POS-MECHNSM
parts : (WALL-CRANE CRANE-1 CRANE-2 CRANE-3 ROBOT-ARM)
props : NIL
sfsets : NIL
***************************
name : HOLD
props : NIL
sfsets : ((MAG-HOLD) (HOOK-HOLD) (ROPE-TIE HOOK-HOLD)
  (GRIPPER-HOLD))
***************************
name : SMPL-MVE-UP
parts : (CRANE-1 CRANE-2 CRANE-3 WALL-CRANE ROBOT-ARM)
props : NIL
sfsets : NIL
***************************
name : GRIPPER-HOLD
parts : (GRIPPER)
props : NIL
sfsets : NIL
conds : ((<= ?(OBJ-SIZE) 15)
         (OR (MEMBER 'CYL ?(OBJ-SHAPE))
             (MEMBER 'SPHERE ?(OBJ-SHAPE)))
     (<= ?(OBJ-WT) 100))
(NOTE: The 'conds clause here means that the object size
(property of the function 'gripper-hold) should be less
than 15, the object shape should be either cylindrical or
a spherical; and the object-weight should be less than 100)
```

Figure 3: Functional description of the function *lift-obj*.

3. Building configurations

Having described the representations that we have chosen for modeling parts and their functionality, we now describe how we use these features to solve the problem of configurational design. In (Chawla & Sangal 1991*a*) we have proposed a three step problem solving strategy consisting of a part selector, extra function remover and an abductive assembler. We shall now describe this problem solving strategy as implemented by us and demonstrate from where the system draws its power.

Given a set of requirements, the system first searches for a viable artifact that satisfies the requirements. This search is guided by the satisfiability details stored with the functional descriptions. In other words, on seeing a particular requirement the system shall look up its functional description to see if there are any known artifacts that satisfy the requirement. Of these artifacts, an artifact is called viable if all its pre conditions for selection are met by the given requirements.

```
Sample run of the system for the requirements specifying
'design a part that can lift an object of weight 100.
**********************************************************
>created 212 as the functional representation of the
requirements.
trying to find artifacts that satisfy 212
  trying to find artifacts that satisfy the sfset
         (211 (LIFT-OBJ))
      trying to find artifacts that satisfy the sfset
                (192 (POS-MECHNSM&HOLD SMPL-MVE-UP))
        trying to find artifacts that satisfy
                POS-MECHNSM&HOLD
          trying to find artifacts that satisfy the sfset
                (193 (POS-MECHNSM HOLD))
            trying to find artifacts that satisfy
                POS-MECHNSM
            found (WALL-CRANE CRANE-1 CRANE-2
                CRANE-3) to satisfy POS-MECHNSM
            trying to find artifacts that satisfy HOLD
              trying to find artifacts that satisfy the
                        sfset (197 (HOOK-HOLD))
                trying to find artifacts that satisfy
                        HOOK-HOLD
                found (HOOK) to satisfy HOOK-HOLD
              found (HOOK) to satisfy (197 (HOOK-HOLD))
              trying to find artifacts that satisfy the
                        sfset (199 (ROPE-TIE HOOK-HOLD))
                trying to find artifacts that satisfy
                        ROPE-TIE
                found (HOOK) to satisfy ROPE-TIE
                trying to find artifacts that satisfy
```

```
                HOOK-HOLD
            found (HOOK) to satisfy HOOK-HOLD
          found (HOOK) to satisfy
                (199 (ROPE-TIE HOOK-HOLD))
          trying to find artifacts that satisfy the
                sfset (201 (GRIPPER-HOLD))
            trying to find artifacts that satisfy
                GRIPPER-HOLD
            found (GRIPPER) to satisfy
                GRIPPER-HOLD
          found (GRIPPER) to satisfy
                (201 (GRIPPER-HOLD))
        found (HOOK GRIPPER) to satisfy
                HOLD
        joining HOOK to WALL-CRANE
        created 241
        joining HOOK to CRANE-1
        failed
        joining HOOK to CRANE-2
        failed
        joining HOOK to CRANE-3
        failed
        joining GRIPPER to WALL-CRANE
        failed
        joining GRIPPER to CRANE-1
        failed
        joining GRIPPER to CRANE-2
        failed
        joining GRIPPER to CRANE-3
        failed
      found (241) to satisfy (193 (POS-MECHNSM HOLD))
    found (241) to satisfy POS-MECHNSM&HOLD
    trying to find artifacts that satisfy SMPL-MVE-UP
    found (CRANE-1 CRANE-2 CRANE-3
            WALL-CRANE) to satisfy SMPL-MVE-UP
  found (241) to satisfy
            (192 (POS-MECHNSM&HOLD SMPL-MVE-UP))
 found (241) to satisfy LIFT-OBJ
found (241) to satisfy (211 (LIFT-OBJ))
found (241) to satisfy 212
```

Figure 4a: Sample run for the requirements *design an artifact that can lift an object of weight 100.* (continued from previous page)

For instance consider Figure 4. The requirements in this case say that *design an artifact that can lift an object of weight 100.* The system on seeing the requirements starts its search for artifacts that can satisfy the requirements by looking at the satisfiability details in the description of the function *lift-obj.* Had it found such a part, the search would have ended. However,

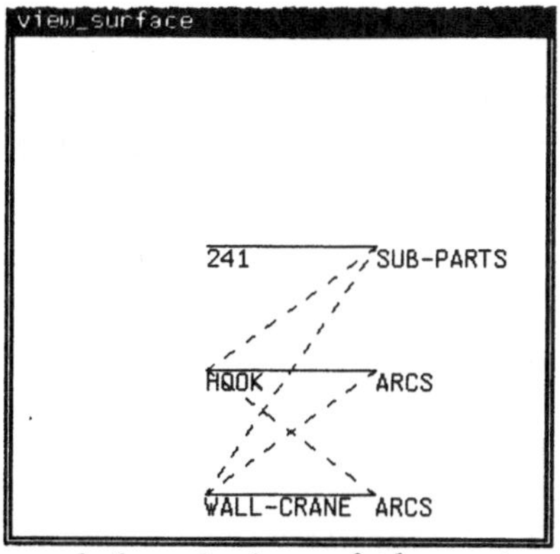

Figure 4b: A structural description of the output produced for the run shown in figure 4a.

since it does not find such a part it looks at all the viable sfsets of the function. For these sfsets it now tries to find artifacts that satisfy the individual functions in the sfset. The artifacts found to satisfy the functions in an sfset are then combined meanwhile ensuring that combining them does not introduce constraints that violate the requirements. It thus tries to satisfy the functions *pos-mechnsm&hold* and *smpl-mve-up*; and similarly breaks up the former into the functions *pos-mechnsm* and *hold*. Continuing similarly it finds artifacts that satisfy these functions and joins them by selecting appropriate arcs. Figure 4a illustrates how the various artifacts are selected and how the final artifact shown in Figures 4b and 4c obtained.

3.1. Assembling parts

The process of assembling the artifacts proceeds by first generating an instance of the two parts to be assembled, finding an arc that can be used to join the two artifacts and storing an instance of the arc with the model of the combined artifacts. For instance, in Figure 4 discussed above, the system tries to satisfy the requirements of *pos-mechnsm&hold* by breaking this function into the sfset *(position-mechanism hold-object)*. It then finds that the function *position-mechanism* can be satisfied by the part *crane-1*, short listed on the basis of the pre-conditions stored with it. Similarly it satisfies the function *hold* by selecting the part *gripper*. The system now tries to join the two parts and generate an assembled artifact. It does so by taking an instance of *crane-1* and an instance of *gripper* and storing with the resultant model an instance of the arc that joins the two parts.

```
*************************************************
name : WALL-CRANE
parent : 241
sub-parts : NIL
arcs (to) : (HOOK)
props : NIL
```

```
conds : NIL
part-types : (WALL-CRANE)
goals : (SMPL-MOVE-UP SMPL-MVE-UP POS-MECHNSM)
**************************************************
name : HOOK
parent : 241
sub-parts : NIL
arcs (to) : (WALL-CRANE)
props : NIL
conds : NIL
part-types : (HOOK)
goals : (HOLD HOOK-HOLD ROPE-TIE)
**************************************************
name : 241
parent : NIL
sub-parts : (WALL-CRANE HOOK)
arcs (to) : NIL
props : NIL
conds : NIL
part-types : NIL
goals : (212 LIFT-OBJ POS-MECHNSM&HOLD SMPL-MVE-UP
 POS-MECHNSM HOLD HOOK-HOLD ROPE-TIE)
```

Figure 4c: The output obtained from Figure 4a (continued from previous page).

However, the question that needs to be answered is that how does this process of assembling the part models help in generating a model of the resultant artifact. The eventual goal of the configurational design process is to generate a model of the artifact that is usable by the routine design stage that follows. In order to demonstrate how this has been achieved, we now describe the assembly of two simple parts, and show by simple constraint analysis how joining the parts affects the dependencies and hence the routine design plans of the assembled artifact.

3.1.1. Affect of joining the parts :

Consider two parts, a *shaft* and a *motor*, as shown in the Figure 1. The design of the shaft involves assigning values to the various decision parameters of the model of the shaft. These are *dia, length, torque, stresses, rpm etc.* These properties have certain constraints amongst themselves and values can be assigned to them only if they are consistent with these constraints. We therefore carry out a constraint analysis of the concerned constraints and decide the order in which values ought to be assigned to these properties. The method that we use for this purpose is discussed in (Serrano & Gossard n.d.). Briefly, it consists of representing the constraints as a constraint graph, getting a bipartite matching in a graph between constraints and variables and then converting the constraint graph into a directed graph. A topological sort on the resulting directed graph then gives a sequence of assignments

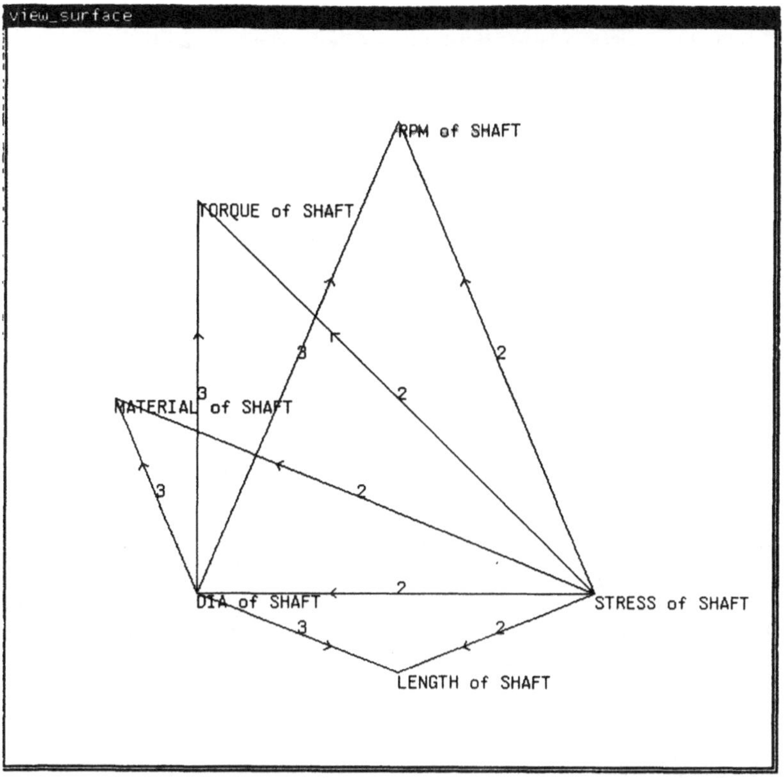

Figure 5a: Dependencies imposed by the constraints stored in the model of the *shaft*.

amongst the various design parameters. Strong components found during the topological sort give systems of simultaneous equations to be solved. This technique can also be used to find redundant and contradictory constraints. The details of this technique are given in (Serrano & Gossard n.d.). In our system we have used this system only to decide whether a system of constraints is solvable and to get the sequence in which the various variables have to be assigned the values. This information can then be used to generate design plans for the composite artifact by using the models for the constituent parts.

We use this technique to analyze the dependencies first on the model of the *shaft* before it has been joined to the *motor* and then after the two parts have been joined. The dependencies obtained in the two cases are shown in the figures 5a and 5b. In this figure an arrow labeled *no* from a property *A* to a property *B* indicates that *A* depends on *B* through the constraint number *no*. In the first case where we have analyzed the constraints in the model of the *shaft* before it is joined to the model of the *motor*, the system can decide the value of the property *dia* on the basis of the properties *material, torque, rpm* and *length*, and then the value of *stress* on the basis of these properties. Thus the system needs to know the values of the properties *material, torque,*

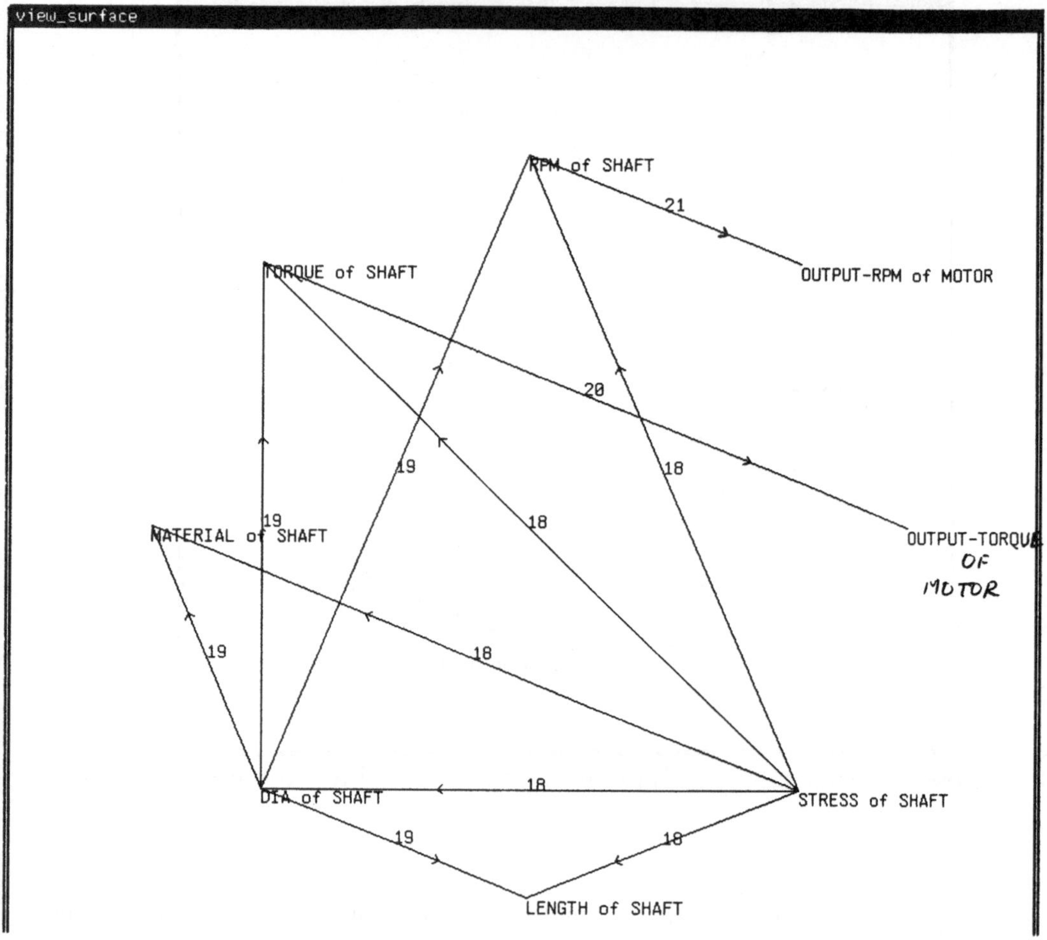

Figure 5b: Dependencies imposed when the *shaft* is joined to the *motor*.

rpm and *length* in order to solve the system of constraints.

In the second case, however, the model of the *motor* specifies some of these properties through the constraints in the arc joining the two parts. These properties are the *output-rpm* and the *output-torque* of the *motor*. As a result, the properties *rpm* and *torque* of the *shaft* can also now be assigned values through constraint numbers *21* and *20* respectively. (These constraints are the ones stored in the affect clause of the arc joining the parts.) The system of constraints can now be solved if the properties *material* and *length* of the *shaft* are known. Thus the model of the *shaft* is able to inherit properties from the model of the *motor*. If in a similar fashion, the model of the *gear* (Figure 1) had also been joined to the *shaft*, the *rpm* and the *torque* of the *gear* would also have taken values from the model of the *shaft*. Thus these properties would have been transferred from the model of the *motor* to those of the *shaft* and then to those of the *gear*. In (Chawla & Sangal 1991*b*) we have called this feature as the inheritance of properties across arcs.

The technique used by us demonstrates that joining the parts the way we do strengthens the system and gives enough knowledge to help generate dependencies of the composite artifact. These can then be used to generate the design plans. Some other approaches have also been tried in this regard (for instance, (Araya & Mittal 1987, Brown & Sloan 1987)) but they deal with generating design plans for simple artifacts where the constraints are predefined. Clearly, more work needs to be done in this area.

3.2. Disassembling parts

While joining two parts is an essential part of the configurational design process so is the process of removing (or disconnecting) two already joined parts. This is because connecting and disconnecting parts are the two basic processes by which an existing configuration can be modified. Since joining two artifacts consists of taking instances of the artifact and storing with them an instance of the arc that joins them, disconnecting the two parts simply consists of deleting the connectivity information. However, it is crucial to the configurational design process to correctly decide as to when two parts are to be disconnected. Parts already in a configuration need to be removed from the configuration when the role they are playing in the configuration becomes redundant. This can happen when another part selected in the configuration already does what this part is supposed to do or when the requirements in the given case do not need the functionality of this part. In (Chawla & Sangal 1991*a*) we have talked of the extra function remover as a part of the abductive design process which identifies redundant parts and removes them. We now describe this technique and discuss how it has contributed to our configurational design system.

```
The design requirements specify that :
((move (on road) (no-of-speeds (> 0)))
 (braking (type 'mech) (braking-dist 25))))
```

Figure 6a: The design requirements in this case demand that *design an artifact that can move on the roads, may have any number of different speeds, and has mechanical braking with a braking distance of (at most) 25.*

```
name : CAR
parent : NIL
subparts:(STRG-MCHNSM ENGINE MGBOX BODY DIFF W-ASLY BRAKE-ASLY)
arcs (to) : NIL
props : NIL
conds : NIL
goals : (MOVE TAKE-TURNS BRAKING)
```

Figure 6b: The model of a *car* as already known to the system.

Consider Figure 6. The requirements of the design are given in Figure 6a. Figures 6b and 6c give the functional and structural description of the

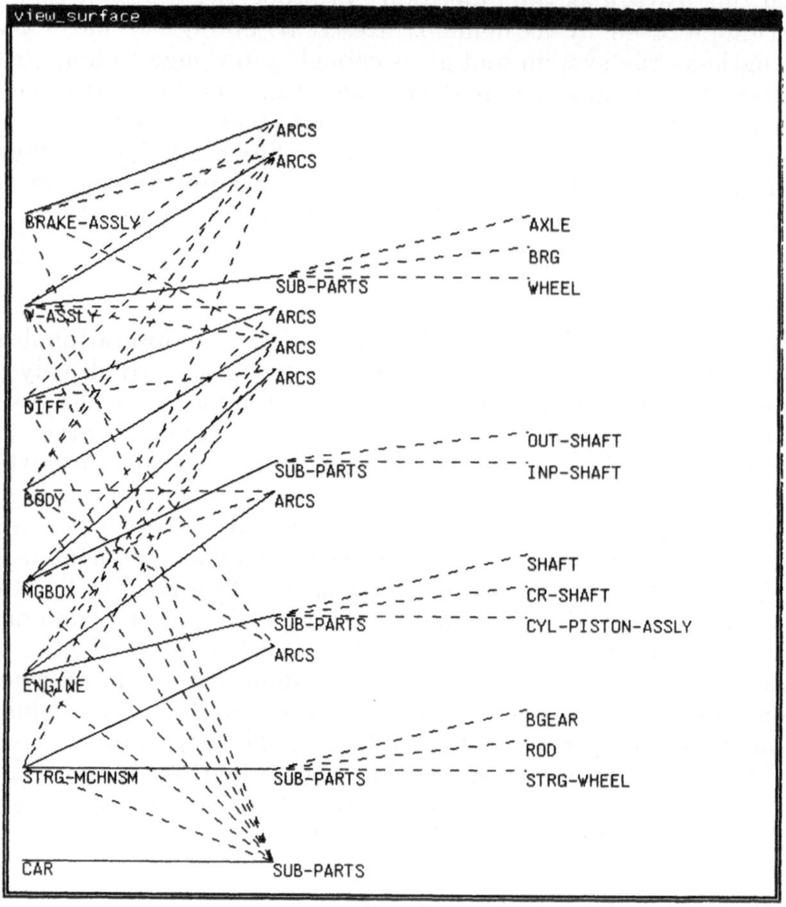

Figure 6c: The structural description of the original model of a *car* as shown in Figure 6b.

part *car*. Thus among other things, the *car* consists of a *body*, a *steering-mechanism*, an *engine*, a *multiple-speed-gear-box*, a *differential* and *wheel-assemblies* connected in a specified way. This configuration of the car carries out a number of functions, some of which are not required as per the requirements of the current case. Specifically, the *car* has the capability to *take turns* and has a *multiple-speed-gear-box* while the requirements of this case only demand that the final artifact be able to move on a straight path and have a *single-speed-gear-box*. In such a case the system first selects the *car* as an artifact that can fulfill the given requirements. While this artifact can suffice as the final output for the given set of requirements, it carries out some functions that are actually not required. The system therefore proceeds to find out the parts that can possibly be replaced by simpler parts so as to get an artifact that does not satisfy any extra functions.

In order to achieve this the system first finds the extra-functions being fulfilled by the part. In this particular example the extra-functions consists of *take-turns*. The next step is to find out the parts that carry out these extra functions. For this purpose it considers all the sub-parts of the model of the *car* and finds out that the part *steering-mechanism* carries out the function of *steering* and the part *differential* carries out the function of *generate-speed-difference-in-wheels*, both the functions being parts of an sfset of the function *take-turns*. On breaking the functions into their sfsets, it also realizes that the requirements ask for *no-of-speeds* $\geq$ *1* while the part *mgbox* has *no-of-speeds* = *5*. Even though this is not a design failure, the system can now look for parts that satisfy the requirements more closely.

At this stage the system gets two targets. Firstly to remove the parts that are responsible for the function *take-turns* and secondly to find a part instead of the *mgbox* so that the requirements are more closely met. As mentioned above the parts responsible for *take-turns* would be the *steering-mechanism* and the *differential*. However, the *differential* also carries out the function *(shift-rotation (by 90degrees))*, which is a part of the sfset of the function *move*. Therefore the system now tries to find a part that satisfies only the latter function and replace the *differential* with that part. Thus it initiates a fresh design task with a set of requirements so as to satisfy only the *shift-rotation* function. Let us say, that it succeeds in finding a *bevel-gear-box*. It therefore replaces the *differential* by the *bevel-gear-box* in the model of this *car*. Similarly for the *mgbox*, it might be able to replace it by a *single-speed-gear-box* (say). Once this is done, the system once again checks whether the new configuration satisfies the requirements. This is because, removing some parts may affect the functionality of the other parts. At the same time the system should also check whether in the new configuration the constraints in the artifacts are solvable (this particular check has not been implemented as yet, but can easily be incorporated). The system thus produces the output as shown in Figures 6d and 6e.

```
name : CAR
parent : NIL
sub-parts : (S-GBOX BEVEL-GBOX ENGINE BODY W-ASSLY BRAKE-ASSLY)
arcs (to) : NIL
props : NIL
conds : NIL
goals : (MOVE BRAKING)
```

Figure 6d: The model of a *car* as output for the requirements in Figure 6a.

This example demonstrates how in a given configuration, the system can identify parts that carry out functions that are not required, or parts that carry out functions that are more general than the requirements. Thus this example also demonstrates the differences between the abductive assembly of parts and abductive hypothesis assembly (Josephson, Chandrasekaran,

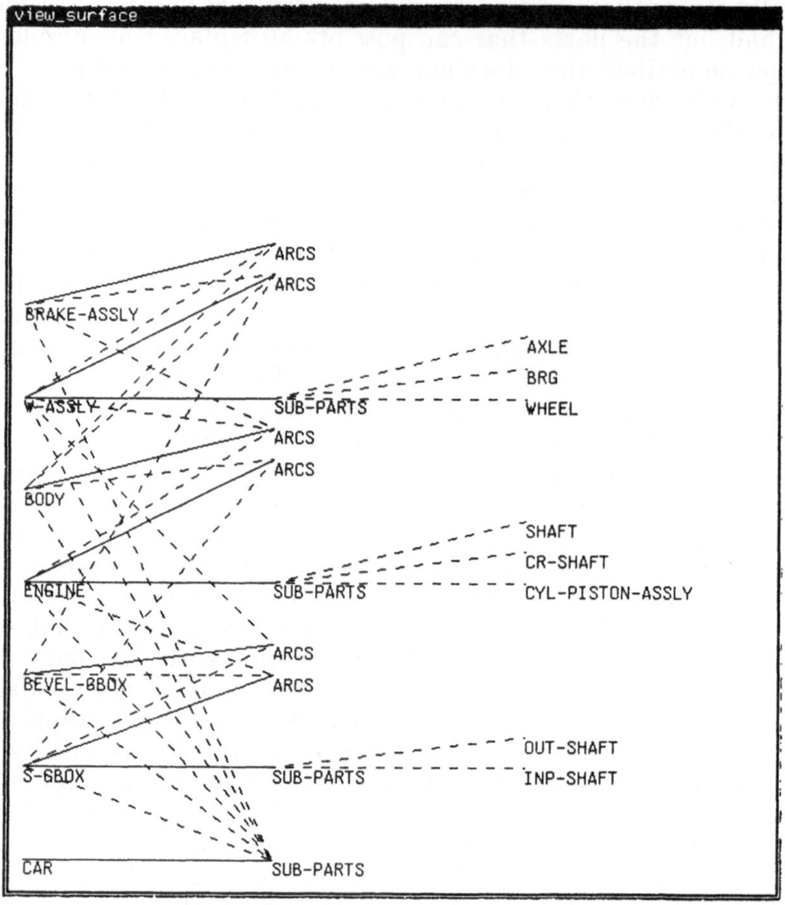

Figure 6e: The structural description of a *car* as output for the requirements in Figure 6a.

Smith & Tanner 1987). In abductive hypothesis assembly, hypotheses can be combined after looking at their causal implications and at the unexplained symptoms. Also hypothesis once combined into a composite can be separated easily if a hypothesis is found that takes care of the respective symptoms. Thus the abductive part assembly differs from hypothesis assembly on two grounds. Firstly, the satisfiability of the functions has to be investigated by looking at the sfsets of the given functions while in the case of symptoms there is no such decomposition into smaller units. Secondly, removing a part may affect the satisfiability of the functions of the other parts as also the validity of the configuration and the satisfiability of the requirements by the configuration. While composing hypothesis, on the other hand, removing an hypothesis from a composite affects neither the validity of the composite nor the symptoms being accounted for.

4. Future work and conclusions

To conclude we mention that we have demonstrated the use of functional dependencies in the process of configurational design. We have shown how functional reasoning guides the process of configurational design and how our notion of connectivity information helps us in generating models of assembled artifacts from the models of the constituent artifacts. The process of configurational design includes many other aspects, some of which we have discussed in (Chawla & Sangal 1991*b*). In any actual design situation configurational design and the routine design stages interact continuously among themselves and the designer has to repeatedly switch between these stages. Also, the design process is characterized by multiple levels of deep and shallow knowledge and for the purpose of doing intelligent failure recovery, the configurational design system needs to switch between these levels automatically. (Chawla & Sangal 1991*b*) discusses how these can be achieved. At this moment we are incorporating these techniques in our system so as to demonstrate the advantages gained by them.

The system developed by us only demonstrates the usefulness and viability of the techniques proposed. A complete system also requires other modules like a routine design problem solver and a constraint management system. These modules shall be incorporated in our system at a later date.

One thing that needs to be pointed out is that the problem solving strategy that we have proposed is by no means complete. There may be situations in which this strategy shall not be fruitful. On such occasions other strategies can be very helpful in the design process. For instance, analogy based on the requirements and the functional model of the artifact can be used to generate new configurations. We shall investigate these aspects in more detail at a later date, but believe that the techniques proposed by us are well suited for these strategies as well.

References

Araya, A. A. & Mittal, S. (1987), Compiling design plans from descriptions of artifacts and problem solving heuristics, *in* 'Proceedings of International Joint Conference on AI (IJCAI)'.

Brown, D. C. & Chandrasekaran, B. (1985), Expert systems for a class of mechanical design activity, *in* J. S. Gero, ed., 'Knowledge Engineering in Computer Aided Design', Elsiever Science Publishers B.V. (North Holland), pp. 259–290.

Brown, D. C. & Chandrasekaran, B. (1986), 'Knowledge and control for a mechanical design expert system', *IEEE Computer magazine* pp. 92–100. Special issue on expert systems for engineering problems, Se June Hong (ed).

Brown, D. C. & Sloan, W. N. (1987), Compilation of design knowledge for routine design expert systems : An initial view, *in* 'Proceedings of the ASME Conference on computers in engineering'.

Brown, D. C. (1985), Capturing mechanical design knowledge, Technical Report AIRG-DCB85-CIE, Computer Science Department, Worcester Polytechnic Institute.

Brown, D. C. (1989), Making design routine, *in* H. Yoshikawa & D. Gossard, eds, 'Intelligent CAD I', Elsiever Science Publishers B.V. (North Holland), pp. 187–200.

Chandrasekaran, B. (1988), 'Generic tasks as building blocks for knowledge based systems : the diagnosis and routine design examples', *Knowledge Engineering Review* pp. 183–210.

Chandrasekaran, B. (1989), A frame-work for design problem solving, Technical Report 89-BC-FRAMEDES, Department of Computer and Information Science, Ohio State University.

Chawla, A. & Sangal, R. (1991*a*), Abduction in design, *in* '4th UNB AI Symposium, University of New Brunswick, Canada'.

Chawla, A. & Sangal, R. (1991*b*), Capturing design knowledge, *in* 'Knowledge Acquisition Workshop, Banff, Canada'.

G Han, S. O. & Yamauchi, H. (1987), The application of knowledge base technology to CAD, *in* J. S. Gero, ed., 'Expert Systems in Computer Aided Design', Elsiever Science Publishers B.V. (North Holland), pp. 25–51.

Gero, J. S., ed. (1987), *Expert Systems in Computer Aided Design*, Elsiever Science Publishers B.V. (North Holland).

Gero, J. S. (1990), 'Design prototypes: A knowledge representation schema for design', *AI Magazine* pp. 26–36.

Goel, A. & Chandrasekaran, B. (1989), Functional representation of design and redesign problem solving, Technical Report 89-AG-REDESIGN, Department of Information and Computer Science, Ohio State University.

Joscowicz, L. (1987), Shape and function in mechanical devices, *in* 'Proceedings of the 6th National Conference on Artificial Intelligence', pp. 611–615.

Josephson, J. R., Chandrasekaran, B., Smith, J. W. & Tanner, M. C. (1987), 'A mechanism for forming composite explanatory hypotheses', *IEEE Transactions on Systems, Man and Cybernetics* **SMC-17**(3), 445–454.

Marcus, S. & McDermott, J. (1989), 'SALT : A knowledge acquisition language for propose-and-revise systems', *Artificial Intelligence* **39**, 1–39.

Marcus, S. (1987), 'Taking backtracking with a grain of SALT', *International Journal of Man Machine Studies* **26**, 383–398.

Mittal, S. & Falkenheiner, B. (n.d.), 'Dynamic constraint satisfaction problem', pre print.

Mittal, S. & Frayman, F. (1988), Towards a generic model of configuration tasks, Technical report, XEROX-PARC.

Murakami, T. & Nakajima, N. (1989), Design diagnosis using feature description, *in* H. Yoshikawa & D. Gossard, eds, 'Intelligent CAD I', Elsiever Science Publishers B.V. (North Holland), pp. 169–185.

Ohsuga, S. (1989*a*), A consideration to intelligent CAD systems, *in* H. Yoshikawa & D. Gossard, eds, 'Intelligent CAD I', Elsiever Science Publishers B.V. (North Holland), pp. 89–101.

Ohsuga, S. (1989*b*), 'Towards intelligent CAD systems', *Computer Aided Design* **21**(5), 315–337.

Serrano, D. & Gossard, D. (n.d.), Tools and techniques for conceptual design, *in* C. Tong & D. Sriram, eds, 'Artificial Intelligence approaches to Engineering Design', Addison-Wesley.

Sriram, D. & et al, G. S. (1989), 'Knowledge based system applications in engineering design : Research at MIT', *AI Magazine* pp. 79–96.

Sticklen, J., Goel, A., Chandrasekaran, B. & Bond, W. E. (1989), Functional reasoning for design and diagnosis, Technical Report 89-JS-FUNREAS, Department of Computer and Information Science, Ohio State University.

Tong, C. & Sriram, D., eds (n.d.), *Artificial Intelligence Approaches to Engineering Design*, Addison-Wesley.

Ulrich, C. T. & Seering, W. P. (1988), Function sharing in mechanical design, *in* 'Proceedings of The 7th National Conference on Artificial Intelligence', pp. 342–346.

Ulrich, C. T. & Seering, W. P. (1989), Achieving multiple goals in conceptual design, *in* H. Y. H & D. G. D, eds, 'Intelligent CAD I', pp. 213–222.

Yoshikawa, H. & Gossard, D., eds (1989), *Intelligent CAD I*, Elsiever Science Publishers B.V. (North Holland).

AUTOMATED CONFIGURATION DESIGN OF HYDRAULIC SYSTEMS

C-L. LEE, G. IYENGAR and S. KOTA
Design Laboratory
Department of Mechanical Engineering and Applied Mechanics
University of Michigan
Ann Arbor MI 48109
USA

Abstract. Configuration design is a type of design activity in which a set of pre-defined components can be combined in certain ways to design a system (Mittal and Frayman 1989). This paper focuses on the configuration design of power transmission systems in general and hydraulics systems in particular. We present a general framework for configuration design in which the design specifications are separated into basic functions, performance goals, and constraints. The design space is divided into a functional space and a physical space. Each are further organized into hierarchies of functional modules and generic physical devices respectively. Functional modules are representations of behaviors of physical devices and are *domain-independent*. Starting with design specifications, a skeletal design consisting of a network of essential functions is formed. Functions are mapped to physical devices that satisfy performance goals and constraints. Based on the framework presented in this paper, a knowledge-based design tool, called HYSYN (HYdraulics SYNthesizer), was developed. A design example of systematic configuration of hydraulic systems is also presented. Issues in automated configuration design issues, such as function-sharing, granularity of the building blocks, and combinatorial explosion, are also discussed.

1. Introduction

In configuration design a set of catalog components are combined in certain ways to design a system. Catalog components are the known, fully specified physical devices that one would, typically, choose from a manufacturer's catalog. Central to this definition are two major tasks: (1) selection of appropriate components and (2) determination of their interconnections. This definition of configuration design can be readily applied to configuration of computer systems or circuit design whether electronic, pneumatic, or hydraulic. In configuration design, each component generally has a well-defined function. although a particular function can be accomplished in many different ways by the synthesis of different components. In configuration design, geometric features of individual components are not central to the reasoning process.

J. S. Gero (ed.), Artificial Intelligence in Design '92, 61–82.

2. Related Research

According to Freeman and Newell (1971), functions are attached to objects and the functions are indivisible. Bowen (1986) noted that to perform systematic configuration design in a top-down manner, one would need *decomposition laws*, which would indicate how to decompose higher level functions into low-level functions. Bowen developed a program called MAPLE which automatically configures microcomputer hardware from board level components in a bottom-up fashion. Our methodology subscribes to the qualitative and quantitative approach to top-down design based on functions. Further, we address the issue of decomposition laws that are pointed out by Bowen (1986). In our methodology, we have explicitly addressed the abstraction of functionality of components into functionality of sub-assemblies More specifically, in our framework, (a) functions are arranged hierarchically and (b) links between abstract functions and physical components exist at *all* levels.

In accordance with Freeman and Newell's notion of "indivisible functions", other researchers have identified various elemental functions. For instance, listed 105 elemental functions. Collins' (1976) list includes abstract functions such as sensing, fastening, torque limiting, signal transmitting etc. Each of these functions may be accomplished in a number of different ways. French (1971), Rodenacker (1976), Roth (1982), Koller (1976), Pahl and Beitz (1988) and Hundal (1991) among others have made attempts to *classify* functions of various physical artifacts. For instance, Koller (1976) defines 12 basic physical functions and their inversions. Hundal categorizes functions as "store", "connect", "branch", "channel" etc. Crossley (1980) described the importance of reasoning with functions in conceptual design of systems. He presented a graphical technique for construction of function diagrams. According to Crossley, individual function boxes within a function diagram describe "what is to be accomplished" without any reference to "how" a particular function is implemented.

An expert system, called R1, was developed by McDermott (1982) to configure a VAX computer system to the customer specifications. It is a rule-based system that implicitly employs heuristics about various components in the VAX computer systems configuration and the relationships among these components. R1 constructs computer systems and checks for missing or incompatible components. Mittal and Frayman (1989) proposed a generic model for configuration tasks. Given one or more feasible functional architectures (functional networks), a systems configuration problem is viewed as a "generative" task and is solved by first implementing the primary functions with "key" components and then embodying the supportive components for the key components. An analog circuit design expert system, BLADES (Bell Laboratory Analog Design Expert System) was developed by El-Turky and Nordin (1986).

BLADES is based on circuit partitioning and the use of subcircuits as standard building blocks. Its knowledge base consists of a large collection of subcircuits frequently used in analog circuit design. Harjani et al. (1987) developed a prototype expert system for analog circuit synthesis. In their system, the alternatives for hierarchical circuit topologies to be selected, called design style selection, are built-in in advance. The design style selection process is applied to each sub-block of the topology until the process reaches the bottom of the hierarchy.

3. Configuration Design Methodology: A General Description.

Configuration design is a type of design activity in which a set of pre-defined components can be combined in certain ways to design a system (Mittal and Frayman, 1989). This definition of configuration design can be readily applied to design of electronic circuits, pneumatic systems, hydraulic systems, electro-mechanical power transmission systems, and a multitude of other domains. We present a methodology for configuration design as it applies to automatic configuration of hydraulic systems.

Given:

1. Functional requirements, such as inputs/outputs and desired sequence of output events.
2. Performance goals, such as cycle time, efficiency, reliability, and duty rating.
3. Constraints, such as cost, geometry, environment, and compatibility.

Task: To synthesize a system that performs the desired functions, meets the performance standards, and satisfies constraints.

3.1 ORGANIZATION OF KNOWLEDGE

In our framework, the given design specifications are first dissected into functional requirements, performance goals, and constraints as shown in Figure 1. A function describes the intended behaviors of the artifact to be designed. Constraints describe the circumstances under which the artifact is expected to operate, i.e., the limits of design variables, cost limits, space restrictions, and so forth. A design based purely on function might produce the desired behavior, but it might be too expensive to manufacture and maintain, or too bulky to install, or it might fall apart after a few hours of operation. Therefore, a functionally synthesized design is *not* a complete design solution in itself. Conversely, if the designed artifact cannot be operative (cannot function), it is immaterial whether it is cost-effective or easy to assemble or compact.

The fundamental hypothesis of our framework is that design begins with reasoning about functions before fulfilling constraints. We view this ordering of functions and constraints as essential, since we attempt to develop a systematic procedure for the conceptual design process. Whenever an order or discipline is imposed on an open-ended process such as conceptualization, the need inevitably arises to focus on certain aspects of a problem first and set aside the others until later stages. This does not mean that constraints are not important or even less important than functions. It only means that a functional design should first be created before constraints come into play.

In view of our hypothesis, we organize the design knowledge into three tiers. The top tier contains hierarchically organized, domain-independent functions. The middle tier contains generic physical devices, each representing a set of catalog components that are classified according to their working principles. The bottom tier contains descriptions of catalog components (points within a given design subspace). A gross outline of the knowledge-organization is shown in Figure 2. Starting with design specifications, basic functions are first identified, and each function is then mapped to one or more feasible generic devices that meet performance criteria and constraints. Specific catalog components are then identified for each generic device by taking into account constraints, tolerances, and weighting factors.

Functions. One of the key aspects of our framework is the hierarchical organization of *domain-independent* functions (Figure 2). Numerous functional modules are arranged in hierarchies (similar to a road map) and connected together by "AND" or "exclusive OR" junctions. An "AND" junction is a decomposition point that breaks down a function into its subfunctions. An "exclusive-OR" junction connects (two or more) functions that are *different* at their level of abstraction into a *single* higher level function. While a choice between the alternatives is only *relative* at an "OR" junction, the choice is *absolute* at an "exclusive-OR" junction. The choice between power switching and power multiplexing is absolute since the former implies exercise of control on how the power is to be distributed to various branches, and the latter implies no control whatsoever. In Figure 2, the *number* of suffixes denote the level of decomposition and the suffixes themselves indicate the path taken along the hierarchy. For example, function F_i is decomposed into F_{i1}, and F_{i2}. F_{i12} is the second node in the decomposition of F_{i1}.

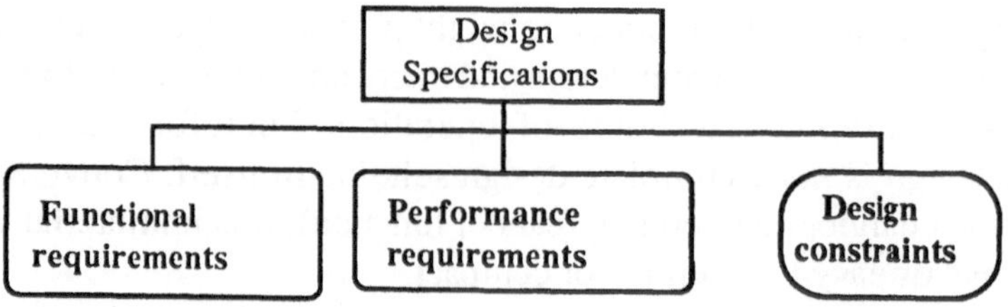

Fig. 1. Separation of functions, performance goals, and constraints in design specifications

The choice of the particular level at which mapping occurs depends on the prescribed constraints and performance criteria. It will be shown later in this paper that there is a direct correlation between the stringency of the design constraints and the level at which the mapping could take place. The more stringent the set of design constraints, the deeper the function-decomposition tree must be traversed.

Functions are abstracted in such as way that some of the higher level functions map to sub-systems while the lower-level functions map to individual components. Additionally, higher-level functions that map to subsystems or a self-contained multi-functional component under certain performance requirements and constraints might fail to do so under a different set of prescribed constraints. Mapping is therefore *context-dependent.* Links exist from functions at various levels to different generic devices and between functions and subfunctions. In actual implementation, the former links take priority over the latter links. Mappings of functions to generic devices exist, at least in theory, at all levels. This means, a higher level function can theoretically be mapped to a generic device. Such mapping could, however, be futile if the performance goals and constraints specified in a given task are not satisfied. The higher-level function should then be decomposed into lower-level subfunctions until a feasible map is established. The process of decomposition is interleaved with mapping and systems-level constraint evaluation.

The choice of the particular level at which mapping occurs depends on the prescribed constraints and performance criteria. It will be shown later in this paper that there is a direct correlation between the stringency of the design constraints and the level at which the mapping could take place. The more stringent the set of design constraints, the deeper the function-decomposition tree must be traversed before feasible generic devices are identified for implementation.

Generic Devices: This level also has a hierarchical structure within it. The working principles of existing devices serve as bridges that connect functional descriptions to physical domain. A working principle is a vehicle that delivers an intended function through a particular set of structural characteristics. Generic devices are classified according to their working principles. These are arranged hierarchically and are connected together by "OR" junctions only. An "OR" junction is a selection point that connects alternate physical implementations to their generic physical devices. For example, a function "Channeling Module" would have generic devices below it classified according to whether their working principles are hydraulic, electric or mechanical. Hydraulic devices would have components such as a multi-way spool valve under it and mechanical devices may have a planetary gear train under it.

Each generic device represents a *class* of catalog components. For instance, a linear actuation function can be accomplished by a solenoid or a hydraulic cylinder. A solenoid would be considered as a generic physical device that embodies electromagnetic principle. Similarly, a hydraulic cylinder would be a considered as a generic physical device that embodies hydrostatic principle. Each generic device represents a *range* of performance characteristics, cost, and environmental characteristics that it can deliver. Thus each generic device represents a subspace within a multi-dimensional design space. For instance, in terms of performance, a hydraulic cylinder could easily deliver up to several hundred pounds of output force, whereas a solenoid is limited to a few pounds of output force. Specific values of performance, cost, size, and so forth depend on the specific device under consideration (a point within the subspace) within a given class.

The notion of a generic device representing a subspace is carried into our computational model for identifying a *feasible range* for each design parameter (parametric design). The parametric design scheme takes as fundamental the notion of a discretized design space. In the hydraulics domain, certain characteristics are identified that allow all specifications and generic devices to be described in terms of their values. Now, each characteristic has a *domain* to which its values are restricted. Two types of domain are defined: interval-equipped domains and point domains. In the case of interval-equipped domains, a value in that domain is normally accompanied by an interval of values within which the value falls. Characteristics like "system pressure" have interval domains, since the pressure must be within a certain interval. Usually, such characteristics are defined by a specific desired value within the interval. In the case of point domains, a value is not accompanied by an interval. Characteristics like "type of fluid" have point domains. These characteristics are thus dimensions of a design space, which is discretized into a finite number of partitions, each represented by a generic device thought of as occupying its center. The n_C characteristics define an n_C- dimensional *design space*, which may be thought of as the cartesian product of the domain of the characteristics. The dimensionality of the design space is first reduced by matching certain critical characteristics and thus eliminating them. A search is then conducted in this reduced domain to find a generic device(s) that matches the given specifications most closely. The design paradigm assumes that a designer is presented with a specification that allows flexibility in the sense that various values may be renegotiated within fixed limits. Our design scheme screens various generic devices and performs initial selection. Given a specification (set of performance requirements and constraints), it finds one or more generic devices that are promising initial designs for meeting the specification.

A computation network implemented within our knowledge-based design tool, measures how well a generic device matches the given design specifications and returns the goodness-of-match value. The network propagates and consolidates individual goodness-of-match values that are computed for each of the characteristics in order to establish feasibility of a generic device to provide a desired function while meeting the performance requirements and constraints.

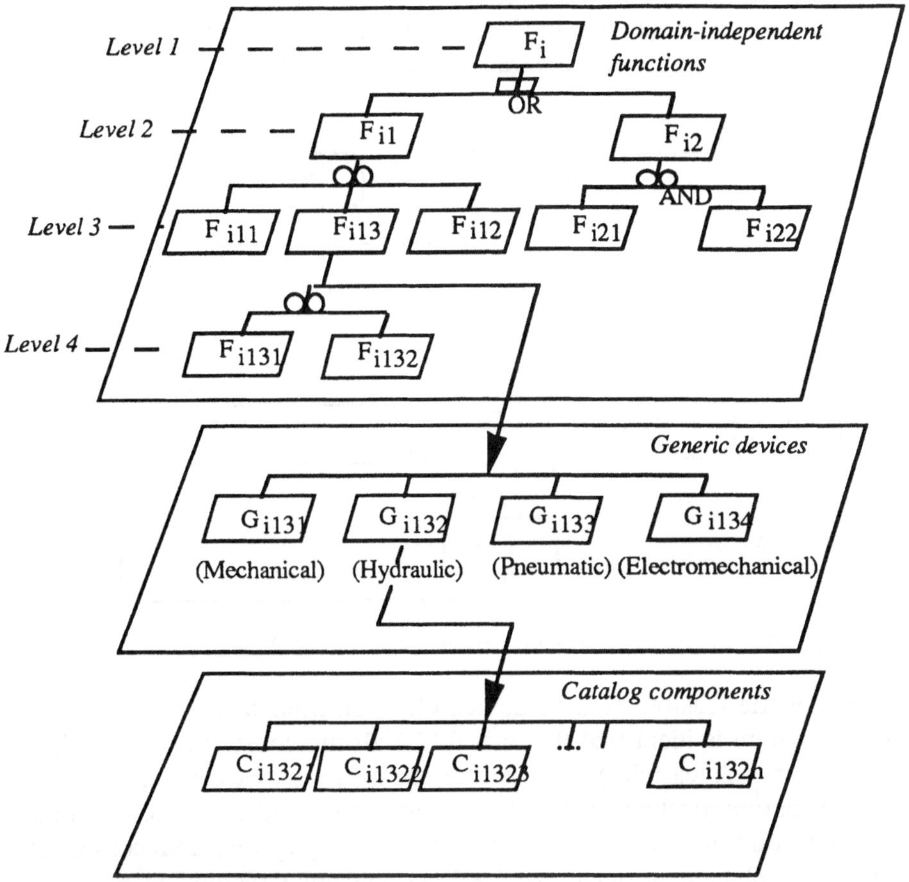

Fig 2: Organization of design knowledge

3.2 THE DESIGN PROCESS

The design process of our configuration design method can be summarized in the following three steps.

Step 1. Separate the functional requirements from the design constraints. Establish a basic network of domain-independent functional modules (Fi) based on functional requirements (Figure 2(a)). The functional space is meant to serve as a domain-independent unified representation of all working principles and existing devices. It, therefore, follows that a functional network, formed by choosing appropriate modules from the functional space, serves as a common semantic model for the physical design to be developed.

Step 2. For each functional module (Fi), identify a feasible generic physical device (Gij) that meets performance goals and satisfies constraints., or decompose it into simpler subfunctions (Fik) using domain-independent function decomposition hierarchies. A complete network of necessary functions, called a functional configuration (shown in Figure 2(b)), is thus developed by replacing some or all (depending on the constraints) of the original functions by simpler subfunctions.

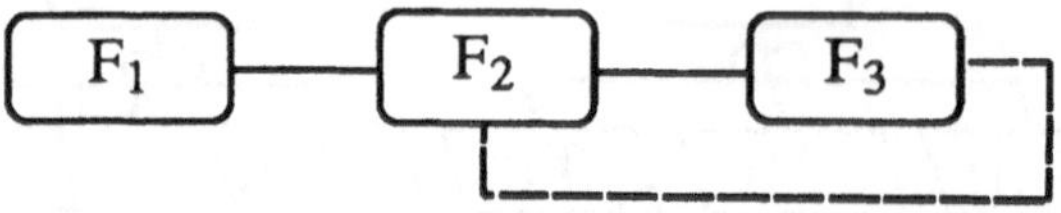

Fig 2(a) A preliminary functional network

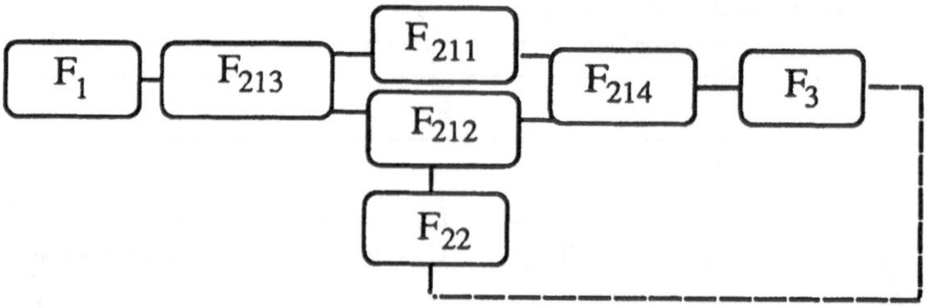

Fig 2(b) A network of basic functions as an initial skeletal design

Step 3. For the feasible generic device Gij identified in step 2, a specific catalog component is identified that can fulfill all the constraints and designer preferences. For each feasible generic device, the most suitable catalog component (Ci24) that satisfies absolute constraints is then determined using a computation network. The computation network computes, propagates, and consolidates goodness-of-match indices for each of the functional and structural characteristics and selects the "best" component. Feasibility of a generic device is established by matching individual constraints against the domain of generic devices. As each generic device is added to the design network, certain new constraints may come into play, which are then stored in the network and are used during component selection.

3.3 ALTERNATE CONFIGURATIONS

For a given task, functional requirements are first met by identifying the basic functions that are needed. These functions and their interconnections are represented as a functional network. For a given task, basic functions that were identified in step 1 remain unchanged. In principle, *alternate* physical designs can be generated in three distinct ways: (a) by enumerating alternate topologies of the same set of functions or (b) by choosing alternate working principles (generic devices) or (c) by selecting alternate components within a particular class of generic devices. However, generally the performance criteria and the design constraints impose severe restrictions on the number of feasible designs that fulfill the specifications. In our method, alternates at each stage are evaluated against performance criteria and constraints. Typically, when more than one alternate design path exist at any given stage, the path that seems "most promising" is pursued. The design process backtracks to an alternate path when it either fails to identify a feasible component or simply reaches a dead-end.

4. Application to the Hydraulics Domain

This methodology has been successfully implemented for automatic configuration of hydraulic systems. In the following sections, we will describe the methodology with design examples from the hydraulics domain.

4.1 THE HYDRAULICS DOMAIN

Hydraulics is one of the oldest branches of mechanical engineering and numerous text books at various levels of detail are available. Readers who are not familiar with hydraulics are encouraged to read the introductory chapter from (Henke 1983) or any other text on hydraulics.

Hydraulic power systems are used extensively in modern aircraft, automobiles, heavy industrial machinery, and many kinds of machine tools. Although the components of hydraulic systems are generally more modular than those of pure mechanical artifacts, they are less modular than elements of electronic digital circuits. The component choices in hydraulics domain are, in general, wider than in digital systems. We chose the hydraulics domain as a middle ground between modular digital systems and three-dimensional, non-modular mechanical systems.

Typically, hydraulic systems have an electric motor-driven or an engine-driven hydraulic pump (as input) that draws the fluid from a reservoir and delivers it to output devices such as linear (hydraulic cylinders) or rotary (hydraulic motors) actuators. The system may have multiple inputs and multiple outputs. The fluid is transported via pipes from one component to another. The distribution of power to the various output devices is controlled by various valves such as directional control valves, pressure control valves, flow control valves, and so forth. Valves may be controlled, among other ways, electrically by a signal that triggers a solenoid located inside the valve, mechanically by a cam, or manually. Pump-size determines the flow-rate (gallons per minute), and the external load (for example the output cylinder working against an inertial load) influences the system-pressure (for example, 2000 psi). Protective devices such as filters, relief valves (which bypass the fluid if the system pressure reaches the maximum limit due to sudden overload), and emergency push buttons ensure safe and reliable operation. Figure 5 shows a very simple hydraulic circuit consisting of a motor, a pump, a three-position-four-way directional control valve, and an output double-acting cylinder. As the fluid flows to one-end of the cylinders, the piston rod extends. As the piston rod reaches the end of the stroke, a limit switch transmits a signal via feed-back loop to actuate the solenoid of the valve causing the valve to shift. The fluid is then directed into the cylinder, out again, and back to the reservoir or tank.

There are hundreds of different types of hydraulic components listed in commercial product catalogs. They are generally categorized as pumps, motors, rotary actuators, cylinders, accumulators, filters, directional control valves, pressure control valves, flow control valves, solenoids, torque motors, pilot valves, manifolds, fixed or variable orifices, limit switches, pressure sensors, and so forth. In addition to individual components, a standard set of subcircuits, such as regenerative circuits, counterbalance circuits, and synchronizing circuits, are also described in fluid-power handbooks. In our knowledge-based system subcircuits are not stored in memory but rather are generated from first principles (using the generalized operators described in detail later) in order to accommodate a much wider variety of design constraints.

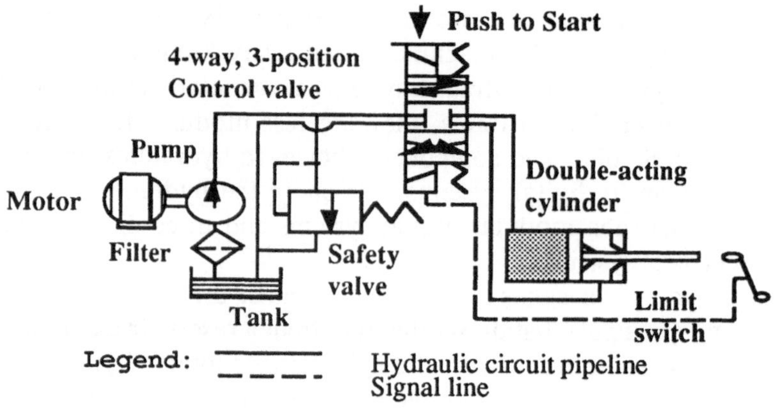

Fig. 5. A simple hydraulic circuit

Functional modules in the hydraulics domain. The functional modules in the hydraulics domain are essentially abstractions of existing physical devices. The seven basic domain-independent functional modules represented at the highest level of abstraction are:

1. Energy Conversion module: represents different means of converting energy from one form to another and includes devices such as: hydraulic pumps, hydraulic motors, rotary actuators, linear actuators, electric motors etc.
2. Motion actuation module: represents various output actuators such as linear and rotary actuators.

3. Power distribution module distributes the power from the input(s) to the motion actuation modules. Some of the major subfunctions within this module include: power routing, power channeling, and power multiplexing.
4. Power regulation module regulates the power level (pressure, torque, and force) and power rate (flow rate and speed). This function represents diverse physical devices such as orifices, intensifiers, relief valves, hydraulic fuses, resistors, fluid couplers, torque convertors etc.
5. Sensing module represents a variety of activity sensing devices such as limit switches, pressure sensors, etc.
6. Power storage module represents various passive power supply devices such as hydraulic accumulators, reservoirs, flywheels, capacitors etc.
7. Contamination control module represents the function of preventing or containing the contaminants from entering the system and it represents various filters, and trapping devices.

Each of these top-level functions is decomposed into a number of subfunctions to form function-decomposition hierarchies. For instance, the function decomposition tree for power distribution function is shown in Figure 6.

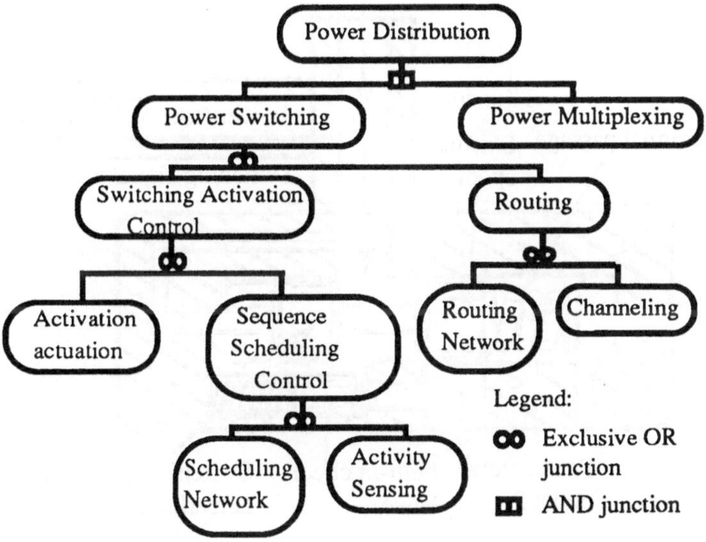

Fig. 6. Functional decomposition of the power distribution module

4.2 MAPPING APPLIED TO THE HYDRAULICS DOMAIN

A unique feature of the architecture of our knowledge representation scheme is that the functions at different levels in the function hierarchy can be mapped to different physical devices (or structures). In the context of a given design task, a particular function might have to be decomposed into lower-level subfunctions when a higher level function fails to map directly to any existing physical device. It is the nature of the constraints and performance requirements that dictate the success or failure of establishing a feasible map. Once a feasible generic device is identified, there is no need to decompose the function. The process of mapping from different levels in the function hierarchy will be illustrated by using the power switching subfunction in a hydraulic system.

A power switching function is needed if the input power has to be routed to *n* different output actuators in a prescribed sequence depending on the occurrence of a certain physical conditions. Figure 6 shows a hierarchical representation of the power switching function. The switching function can be decomposed into lower-level functions such as "switching activation module" and "routing module". Each of these subfunctions can be further decomposed into lower-level functions and so forth. If *n* equals two, and if fluid pressure is required to trigger the "switching action," then the power switching function can directly be mapped to a hydraulic sequence valve shown in Figure 7, in which case there is no need to decompose the function any further.

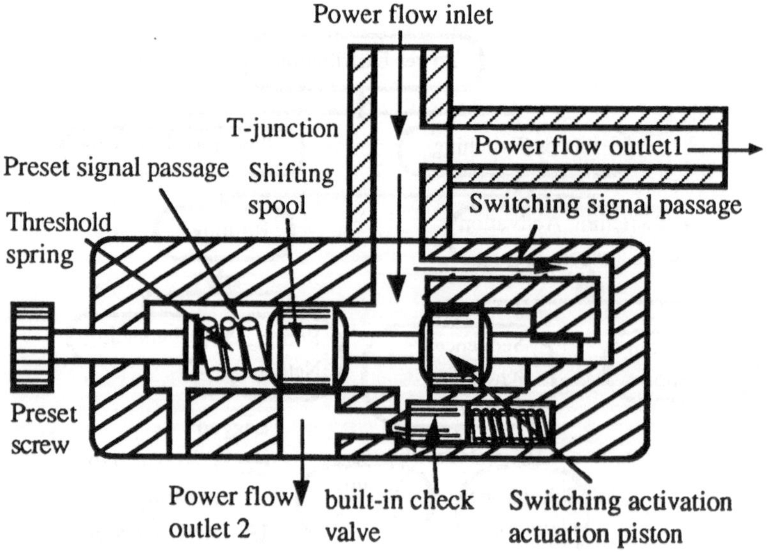

Fig 7: A two-way sequence value as a physical implementation of power switching function

A sequence valve blocks the hydraulic power flow through it when the pressure is below a preset threshold value and will open the flow route when the pressure reaches or exceeds the preset value. Careful analysis of a sequence valve allows us to relate various subfunctions to its different physical features. A two-way sequence valve can be considered a multifunctional component in the sense that it contains all the functions that are shown below level 2 in the power switching hierarchy (Figure 6). However, if the desired number of output paths *n* is greater than 2, then the power switching function cannot be readily mapped to any existing physical device. In this event, the power switching function would have to be decomposed into lower levels until each subfunction can be identified with a separate physical device.

For instance, if four different power switching routes are desired, then the power switching assembly should be synthesized systematically by incorporating channeling modules, activity sensing modules, activity actuation modules, and electrical signal transmission paths in to a routing network (generated by a generalized operator) as shown in Figure 8.

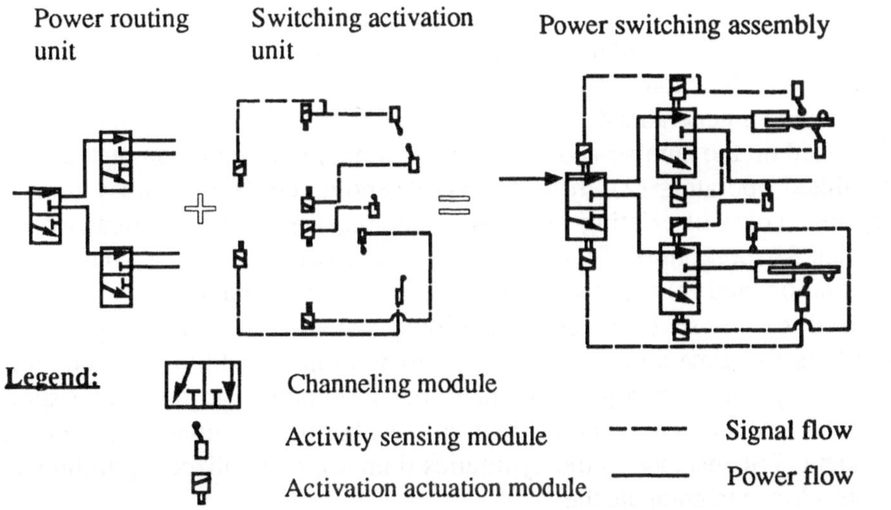

Fig. 8. A four-way power switching function must be decomposed since no readily available devices exist. The resulting design is an assembly of various components corresponding to respective subfunctions.

5. Generalized Operators

The methodology described in the previous sections in conjunction with several generalized algorithms has been implemented in a knowledge-based system called HYSYN (HYdraulic SYNthesizer) for configuration of hydraulic circuits. These algorithms are called generalized operators or GOs. The GOs are synthesis algorithms and are used like macros. The knowledge represented by these Go's is the knowledge required for inference. GOs have been developed for the following tasks:

1. to construct the preliminary functional network for any given task
2. to configure a power routing network
3. to construct scheduling networks to connect sensors to switching actuators
4. to generate synchronizing circuits
5. to create regenerative circuits.
6. to generate *alternate functional configurations.*

The operators provide systematic procedures to (a) relocate functions from upstream to downstream or vice-versa,(b) add/delete power regulation and power conversion modules within a system to streamline the power flow through various components. These are called adapters and are discussed in section 7.

Generalized operators at the function level help construct functional networks. Each operator provides an operational guideline on ways to connect different functional modules according to a given overall function. The operators are *problem-independent,* and they are derived at the time the generalized knowledge is compiled as function-decomposition trees. One of the generalized operators in our system is a generalized routing network algorithm that determines the flow paths based on the number and the nature of input / output motion segments. The generalized operators (2) thru (5) listed above configure subsystems. By analyzing scores of existing hydraulic systems, we have identified two basic strategies for combining functions. These are: (a) in series or common-flow configuration, and (b) in parallel or common-effort configuration.

Based on these basic configurations, we have developed generalized algorithms for generating a variety of subsystems. In addition to generalized operators, quantitative algorithms have been developed to determine specific parameter values. These algorithms are problem-independent and domain-dependent. For instance in the hydraulics domain, generalized algorithms have been developed to compute the

1. size of the actuator that is needed,
2. capacity and type of the accumulator, if necessary,
3. capacity of the pump,
4. capacity of the reservoir, and so forth.

6. Design Example

(The purpose of this design example is to illustrate a systematic method for configuration design and therefore, the details of parameter design are not explained due to space considerations.)

The user enters the design specification via interactive menus of HYSYN. The design specifications consist of descriptions of inputs and outputs, including the sequence of output motions, operator interface, operating environment, reliability, and safety factors.

6.1 DESIGN SPECIFICATIONS (FIGURE 9)

User enters the design specifications through a series of menu prompts supplied by HYSYN. The following is a brief summary of the user defined specifications. For instance, the user specifies the number, nature and sequence of desired outputs motions. Although numerical values are not given below for the sake of simplicity, they must be specified for loads, speeds, cycle time, duty rating, geometric restrictions, desired system efficiency etc. Based on load cycle, HYSYN not only establishes the number and type of hydraulics power sources or pumps that are required but also determines if an accumulator is necessary.

Power source: Electrical 120 volts, 60 Hz.
Input: Operator pushes a button to start the cycle.
Output: Two linear reciprocating and coordinated output motions per following sequence.
Cycle time: N seconds
Output motions and their sequence
A graphic illustration of the two motion sites S1 and S2 along with the time sequence of output motions is shown in figure 8.

1. Inertia load L_1 lbs should be moved through a distance x_1 at speed (power rate) R_1 in the direction a_1 at the site S_1.(motion segment O_1 in figure 8)
2. The load L_1 must be held stationary while
   (i) another inertial load L_2 is advanced through distance x_2 at speed (power rate) R_2 in the direction a_2 at site S_2.(motion segment O_3 in figure 8)
   (ii) the inertial load L_3 is advanced at speed (power rate) R_3 through a distance x_3 in the direction a_2 at the site S_2 (motion segment O_4 in figure 8).
   (iii) the load L_2 rapidly through a distance x_2+x_3 in the negative direction a_2 at the site S_2 (motion segment O_5 in figure 8)
3. Retract the inertial load L_1 through a distance x_1 in the negative a_1 direction at the site S_1.(motion segment O_2 in figure 8)

6.2 SELECTION AND DECOMPOSITION OF FUNCTIONAL MODULES

The first step in our configuration design process is to analyze the functional requirements and identify a set of basic functional modules. Templates are picked on the basis of the type of design; whether synchronizing, sequential or regenerative. The basic functional modules are the root nodes of the function decomposition hierarchies that are stored in our knowledge-base. In the design example, the basic functional modules are the power supply module, the power distribution module, and the motion actuation modules as shown in Figure 10. Since the desired output motions are specified at two different sites, it is inferred that two separate motion actuation modules are needed. Among these three basic functional modules, only the motion actuation modules are readily implementable and hence require no further function decomposition. The other two other modules are decomposed into their subfunctions to seek proper physical implementations. The function decomposition process is guided by the precompiled function decomposition trees (similar to road maps). The power distribution hierarchy shown in Figure 6 is just one such tree and is used to expand the power distribution node of Figure 10.

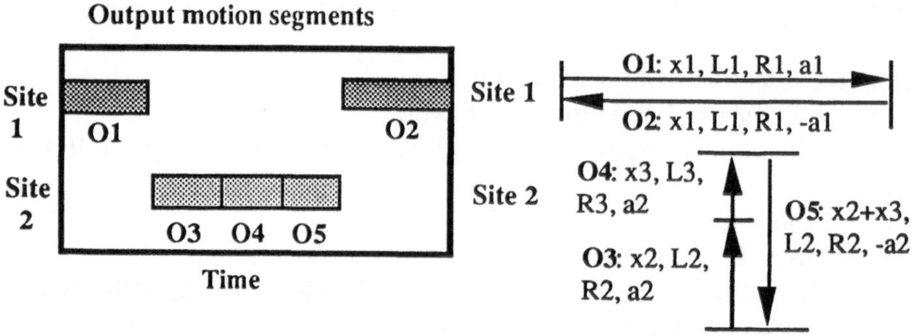

Fig. 9. Design specifications for a hydraulic systems design task

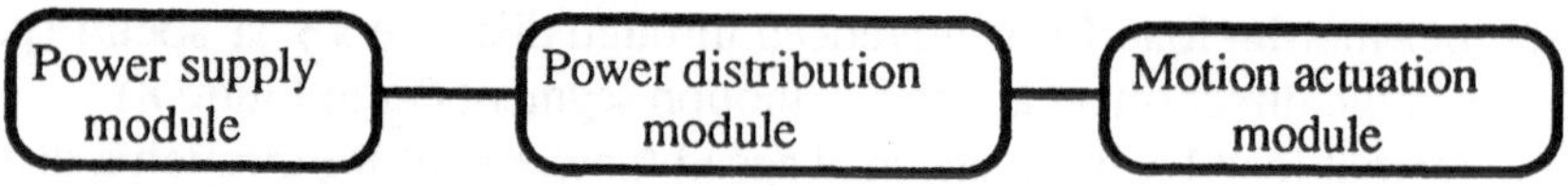

Fig. 10. A preliminary functional network of basic functions

Since the desired output motions are required to be sequenced, the power switching branch of the power distribution tree is selected and is further explored. The power switching node is not readily implementable and is decomposed into its subfunctions: a routing module and an activation module. The routing module is decomposed into a network of channeling modules by a generalized operator. The channeling modules are implemented by channeling devices (or multi-way valve cavities) as generic physical devices. The activation control module is decomposed into a scheduling control module and a set of activation modules. In this particular design example, a network of activity sensing modules is algorithmically constructed to partially fulfill (scheduling) the function of the activation control module. Individual activity sensing modules are implemented as limit switches or other alternatives, such as hydraulic pressure sensor. The final result is the complete functional network shown in Figure 11. All the nodes in the functional network have been evaluated against performance criteria and each node is identified with a feasible generic device.

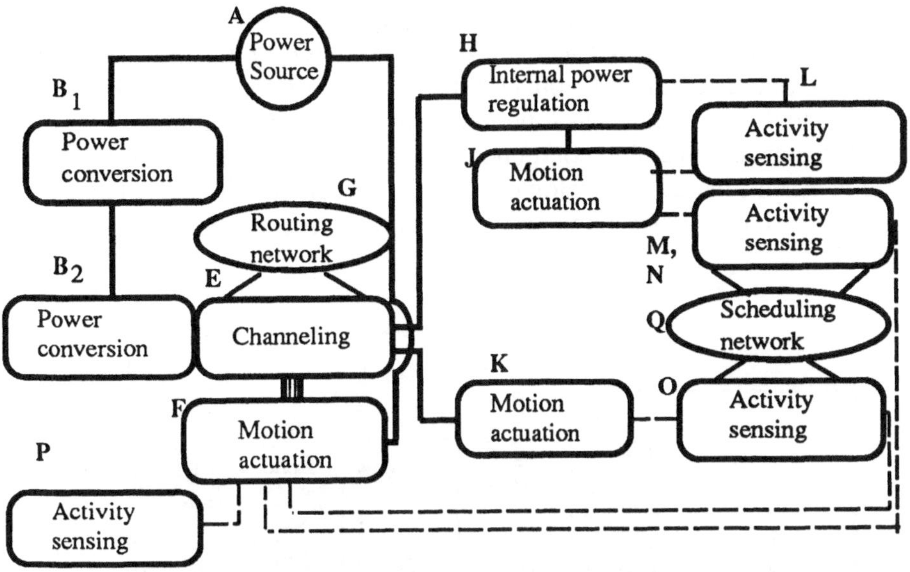

Fig. 11. A complete network of essential functions

6.3 SELECTION OF CATALOG COMPONENTS

Based on the prescribed load-cycle, quantitative inferences are made to (1) estimate the system pressure and flow rate and (2) estimate the size of the pump and auxiliary power supply units. These systems-level parameters (flow-rate, pressure, etc) help constrain the number of choices in the down-stream design process.

The component characteristics are grouped into two factors (1) *systems-level* performance requirements, consisting of medium (fluid-type), power flow rate, power level or operating pressure, and (2) component-specific requirements, consisting of cost, reliability index, duty rating, response frequency, pressure drop, port size, mounting type, size, weight, and so forth. Systems performance requirements are derived from quantitative inferences made during the configuration design stage, and these requirements must be satisfied by all the components in the system. The component-specific requirements vary for different components. For instance, the component-specific requirements for linear actuators include: single/double acting, stroke length, mounting type, port size, rod-end type, etc. The design specifications are matched against the characteristics of the stored design models (classes of catalog components). Each design characteristic is associated (1) with a value and a tolerance-band for specifications and (2) a value and an "achievable region" for stored design models. The interval matching process involves computation of a value between zero and one indicating the goodness-of-match (one being a perfect match).

Individual goodness-of-match values are propagated to the solution nodes and an overall "best" component is selected. Weights are used to indicate the designer's preferences. See Esterline and Kota (1990) for details on the computation network for component selection. Each functional module in the network is eventually mapped to a specific catalog component. Figure 12 shows the final design of a hydraulic system including catalog components. Each component is labeled (A - U) and these labels correspond to labels shown in the functional network of Figure 11.

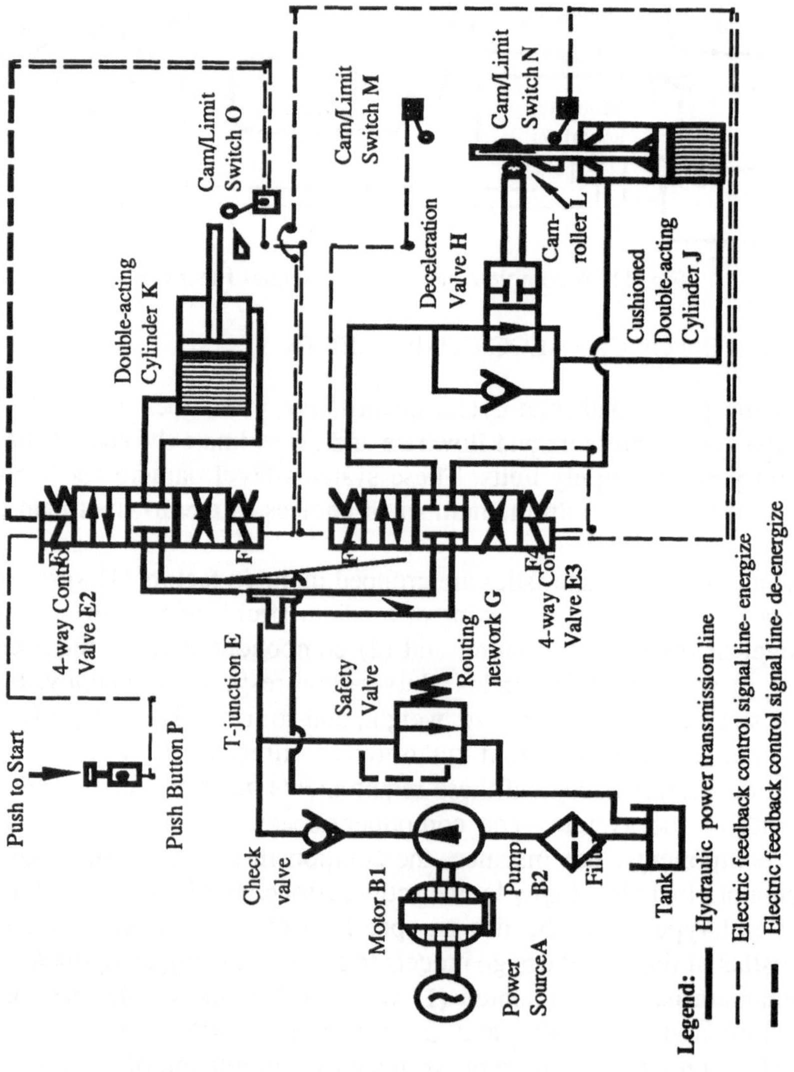

Fig. 12. Hydraulic system configuration derived from the functional diagram shown in Fig. 11, alternate configurations derived from the same functional diagram not shown

7. Issues in Automated Configuration Design

In the paradigm of top-down configuration of hydraulic systems, we view the design of a physical system as a process of selection and synthesis of existing building blocks. The functional modules and the generic devices, described in this paper, carry notions similar to the terms "functional building blocks" and "physical building blocks" respectively. Functional building blocks are simply abstract representations of existing physical devices. Some of the important questions that arise in any top-down design scheme that synthesizes systems from building blocks are:

1. What is the proper granularity of building blocks?
2. How is the problem of combinatorial explosion addressed?
3. How can the notion of function-sharing be incorporated into the design framework?
4. How does the design scheme handle failures?

We have made an attempt to address these issues in the following paragraphs in the context of our hydraulic systems configuration. The solution schemes presented can hopefully be applied to other systems design paradigms as well.

7.1 GRANULARITY AND COMBINATORIAL EXPLOSION

The term granularity in this context refers to the size of the building blocks. If granularity is too fine, too many building blocks will be needed to synthesize a physical system; hence combinatorial explosion will soon occur if all possible combinations of building blocks need to be explored. If granularity is too coarse, two problems might arise; (i) the novelty in synthesis will be hindered, and (ii) in the case of more demanding design tasks, none of the building block combinations will yield a solution. However, we believe that combinatorial explosion can be controlled by dealing with a few important attributes at a time. In our configuration framework, desired functions are considered first before constraints and performance goals are taken into account. In doing so, the granularity of the functional modules is kept *coarsest* in any given instance in that the process of function-decomposition is halted as soon as a feasible generic device is identified. In actual implementation, the links that connect a function to generic devices are given higher priority over the ones that connect a function to its subfunctions. This way the building blocks are no finer than they have to be.

As mentioned earlier, functions at all levels in the decision trees have links with one or more generic devices. The granularity decreases as one traverses down the function decomposition trees Subfunctions that belong to lower levels of the decomposition hierarchy can generally be mapped to physical device implementations in many more domains (hydraulic, electrical, mechanical etc) than the functions at the higher levels. Whenever a difficult design problem with stringent performance requirements is encountered, higher level functions fail to map to feasible generic devices. This requires decomposition to deeper levels resulting in a physical system that may be novel but involves more than usual number of components. The increase in the number of components in the system is due to the reduced size of the building blocks.

7.2 FAILURE-HANDLING IN 'HYSYN'

In our method, failures are handled by (a) backtracking and (b) use of "adapters". The top-down design process progresses from abstract functions to catalog components using taxonomies such as function decomposition trees, and device classification trees. During a design session, if the design constraints and performance goals are highly restrictive, the reasoning process might fail to map a function that belongs to the bottom-most level of the decision tree This results in failure and needs some "re-thinking" or undoing previously made design decisions. In HYSYN, whenever a design decision needs to be undone, the reasoning process backtracks along the same decision path up to the nearest selection point ("OR" junction) and pursues an alternate path, if one exists. Sometimes, depending on the given problem,the backtracking process would have to retrace all the way up to functional level, and might exhaust all alternatives or reach a dead-end. If this happens, HYSYN uses "adapters" (described below) in an attempt to salvage the design. These adapters are simply power regulators and energy conversion modules. HYSYN configures the flow, pressure or physical characteristics of adapters based on a particular mismatch between two existing devices.

To summarize our configuration design framework,

1. Functions, performance criteria, and constraints prescribed in given design specifications are handled sequentially in a top-down fashion.
2. *Problem-independent* decision trees are precompiled and stored in the knowledge base. Functional trees are *domain-independent* decision trees. Generic physical devices are classified according to the working principles. Each generic device represents a class of catalog components.
3. Creation of a functional network as a skeletal design is the first stage in our top-down design process. Performance goals govern the process of mapping individual functions to generic devices. Function decomposition and mapping are interleaved. In the final stage, specific catalog components within a generic class are selected based on design constraints.
4. Alternate physical designs can be generated by (a) enumerating alternate functional topologies or (b) selecting alternate working principles, or (c) choosing alternate components.

An important feature of our methodology is that mapping abstract functions to physical devices is permitted at *all levels* in a function hierarchy. A particular level at which such a mapping can successfully take place depends on the nature of performance requirements and constraints. The extent of originality of resulting design is directly related to the depth to which function decomposition hierarchies are traversed before a function can be successfully mapped to a physical device. Issues in configuration design such as function sharing, combinatorial explosion, granularity of building blocks, and backtracking are discussed (section 7). Use of adapters to handle inconsistencies in the physical system (when backtracking fails) is also discussed.

8. Summary and Conclusions

We have identified and abstracted design building blocks for power transmission problems in general and hydraulic systems in particular. Based on the configuration framework presented in this paper, a prototype knowledge-based system (HYSYN) has been developed. HYSYN automatically configures hydraulic circuits by reasoning with the functions and selects catalog components based on performance goals, constraints, and priorities. Our knowledge-based system currently has about ten functional modules, eight generic devices, and detailed description of about 130 (different types and sizes) catalog components. Additionally, we have developed and implemented numerous algorithms for determining (a) network topologies for routing, scheduling, and feed-back signal transmission functions and (b) sizes of pumps, cylinders, accumulators etc. HYSYN's knowledge-base is being expanded to include about twenty functional modules, eighteen generic devices representing about five-hundred catalog components in order to handle most industrial hydraulics applications. HYSYN was written in the Common Lisp Object System on an Apollo Workstation, and it has been tested for configuration of motion synchronizing systems, multiple-input, multiple-output systems, hydraulic press designs, and so forth. Interested readers are encouraged to contact the authors for details.

Another point to consider is the originality of the design obtained from HYSYN. When a higher-level function is mapped to a feasible generic device, the resulting design is less likely to be original. If a higher-level function can be successfully mapped to a physical device the resulting design is more likely to have multi-functional components. Consequently, it can be stated that at least first-cut original designs are less likely to be multi-functional.

Although, the hydraulics domain is clearly more modular than general mechanical systems, configuration of hydraulic systems is by no means trivial. The configuration design framework presented in this paper can be extended to other domains as well, such as electromechanical power transmission and braking systems design. Discretization of design space in terms functional and physical design building blocks is the driving force behind our computational model. The key to generalization of the configuration methodology presented here to other less-modular domains depends on one's ability to describe design building blocks, both functional and physical, at different levels of abstraction.

Acknowledgements

The authors wish to thank the National Science Foundation for providing the financial support for this research under grant # DDM 890180.

References

Brown, D. C. and Chandrasekaran, B.: 1985, Expert systems for a class of mechanical design activity, *in* Gero, J. S. (ed.), *Knowledge Engineering in Computer-Aided Design*, North-Holland, Amsterdam, pp. 259-282

El-Turky, F. M. and Nordin, R. A.: 1986, BLADES: An expert system for analog circuit design, *IEEE Symposium on Circuits and Systems*, pp. 552-555

Esterline, A. and Kota, S.: 1990, A general paradigm for routine design—theory and implementation (submitted to *AI EDAM*).

Freeman, P and Newell, A.: 1971, A model for functional reasoning in design, *Proc. of the 2nd International Joint Conf. Artificial Intelligence*, pp. 621-633

Harjani, R, Rutenbar, R. A. and Carley, L. R.: 1987, A prototype framework for knowledge-based analog circuit synthesis, *24th ACM/IEEE Design Automation Conference*, pp. 32-49.

Henke R. W.: 1983, *Fluid Power Systems and Circuits*, Penton Publishing, Inc.

Hundal, M. S. and Byrne, J. F.: 1990, Computer-aided generation of function block diagrams in a methodical design procedure, *in* Rinderle, J. R. (ed.), *Design Theory and Methodology*, Chicago, Illinois, pp.251-257.

Kinoglu, F., Riley, D. and Donath, M.: 1987, Knowledge-based system model of the design process, *Proc. of the 1987 ASME International Computers in Engineering Conf. and Exhibition*, pp. 181-191

Kota, S., Erdman, A. G., Riley, D., Esterline, A. and Slagle, J. R.: 1988, A network-based expert system for intelligent design of mechanisms, *AI EDAM* **2**(1).

Kota, S. and Lee, C-L.: 1989, A computational model for the conceptual design of hydraulic circuits, *International Computers in Engineering Conference.*

McDermott, J.: 1982, R1: A rule-based configurer of computer systems, *Artificial Intelligence*, **19**, 39-88.

Mittal, S. and Frayman, F.: 1989, Towards a generic model of configuration tasks, *Proc. of the International Joint Conf. Artificial Intelligence*, Detroit, August, pp. 1395-1401

Pahl, G. and Beitz, W.: 1984, *Engineering Design*, Design Council, Springer-Verlag.

2

CONSTRAINT PROCESSES IN DESIGN

SUPPORTING MULTIPLE PERSPECTIVES

A constraint-based approach to concurrent engineering

J. BOWEN and D. BAHLER
Department of Computer Science
North Carolina State University
Raleigh NC 27695-8206 USA

Abstract. In on-line systems intended to support Concurrent Engineering it is necessary to support the multiple perspectives that are taken by members of the various engineering disciplines involved. We report on Galileo2, a constraint programming language that enables constraint networks to be divided into different (possibly overlapping) regions, called fields of view. Engineers from different disciplines, who have different perspectives on the product development process, are assigned different fields of view when they interact with a program written in the language. Different fields of view may be presented through different types of interface, as dictated by the type of user who is expected for each field of view. Users of different fields of view may be given different levels of permission to edit the constraint network which they all share. The run-time system for the language facilitates interaction between users of different perspectives in the event of incompatible decisions by these users.

1. Introduction

Concurrent Engineering is an approach to design which takes into account not just the functionality of a product but also its manufacturability, testability, maintainability, and so on. One way to do Concurrent Engineering is to use design teams, comprising professionals interested in each phase of the product life-cycle. Throughout the process of generating the design, the designer receives comments from the other members of the team (test engineers, manufacturing engineers, etc.) on his evolving design. However, although some companies are starting to use them, design teams present many logistic, scheduling and other management difficulties.

Consequently, we are developing a programming language to enable the easy construction of on-line design advisors which would help ensure that designed products perform as well as possible in all phases of their life cycle. There are four key requirements for such a technology. It should:

- facilitate the representation of such entities as the artifact being designed, the components from which the artifact is configured and the environment in which the product will be manufactured, tested and deployed;
- facilitate explicit representation of the mutually constraining influences that these entities exert on each other;
- facilitate the construction of application programs which can take account of the differing interests of various members of a product development team, by presenting each member with an interface appropriate to his needs; and
- support interaction between these users when their decisions conflict.

J. S. Gero (ed.), Artificial Intelligence in Design '92, 85–96.

Research to date (Bowen and Bahler, 1991a, 1991b, 1992; Bowen, Bahler and Dholakia, 1992) has indicated that frame-based constraint networks are a suitable basis on which to build a language for Concurrent Engineering applications. A *constraint* is a declarative statement which specifies some requirement that must be satisfied by the values assumed by a set of related parameters. A constraint *network* (Mackworth, 1987) is a set of such constraints which are interconnected by virtue of sharing parameters. A *frame-based* constraint network is a constraint network in which parameters need not be scalars; specifically, in these networks, parameters can also be frames (Minsky, 1975) that is, data structures which are organized in an inheritance hierarchy.

In a frame-based constraint network, frames can be used to represent: the artifact being designed; the components from which the artifact is configured or the materials from which it is made; and the life-cycle environment in which the artifact will be manufactured, tested and deployed. Constraints can be used to express in an explicit way the mutual restrictions exerted on each other by artifact functionality, component/material properties, and life-cycle processes.

A major attraction of constraint networks for Concurrent Engineering is that constraints support multi-directional inference: information can flow in any direction through a network. Thus, for example, the impact of a design decision on the options available to a test engineer can be determined by propagating the design decision and its consequences throughout the network. Equally, if testing decisions are made early on, the impact of these decisions can be reflected in restrictions placed on the designer's freedom. By supporting multi-directional inference, one constraint network can support both of these forms of interaction with equal ease.

Several researchers have investigated or proposed the use of constraints in design-related tasks, in different application areas, including architecture, electrical and electronic circuits and systems, experiment planning, job shop scheduling, process planning, mechanical design, road design, and space planning. We provided an extensive application bibliography in (Bowen, O'Grady and Smith, 1990), and there is an excellent review in (Serrano and Gossard, 1988).

Several constraint programming languages have been developed; we provided a review in (Bowen, O'Grady and Smith, 1990). However, most of these languages, for example (Sutherland, 1963; Borning, 1981; Gosling, 1983), were developed for specific application areas, such as graphics and simulation, which are quite different from Concurrent Engineering; as a result, these systems are not appropriate for our area of interest. Even systems developed for design applications, such as CONSTRAINTS (Sussman and Steele, 1980) which was used for analysis and synthesis of electronics, or the Concept Modeler (Serrano and Gossard, 1988) which was used for Mechanical Engineering applications, do not offer the required facilities. Nor do languages which attempt to be general purpose, for example CLP($\Re$), CHIP or Prolog III (Heintze, Michaylov and Stuckey, 1986; Graf, Van Hentenryck, Pradelles and Zimmer, 1989; Colmerauer, 1987).

This situation motivated us to develop a generic constraint-based programming language for Concurrent Engineering applications. In this paper, we report on certain aspects of Galileo2, the latest version of our language. In Galileo2, frames are used to represent artifacts, components, materials and life-cycle environments, while constraints are used to explicitly represent the mutual restrictions that these entities exert on each other. In this paper, however, we focus on how Galileo2 supports the multiple perspectives that are held by the different members of a product development team.

There are three aspects of the language which support multiple perspectives. First, Galileo2 enables constraint networks to be divided into different (but possibly overlapping) regions, or fields of view, each of which contains only the parameters that are of interest to the holder of one type of perspective. Second, the language allows these different fields of view on the same network to be presented through different styles of interface, including spreadsheets and feature-based CAD. Thus, for example, if one parameter belongs in two different fields of view, it may be shown as a graphical icon in one field of view and be presented as a spreadsheet cell in the other. Third, in the event of conflicting decisions by users from different perspectives, the run-time system for Galileo2 encourages interaction between these users in order to overcome the conflict.

In this paper, we introduce Galileo2 by presenting extracts from an example scenario which shows some interactions between a Galileo2 application program for printed wiring board (PWB) design and various members of a Concurrent Engineering team.

In Section 2, we introduce some aspects of the PWB application that have been chosen to form the basis of the example scenario. In Section 3, we briefly discuss the architecture of application programs in Galileo2 and illustrate our remarks with examples from the PWB application program, thereby setting the scene for the example scenario. In Section 4, we present extracts from the example scenario. In Section 5, we make some concluding remarks.

2. Aspects of the Example Application

To illustrate the language and the way in which it is intended to support the multiple perspectives held by different members of a Concurrent Engineering design team, we will consider some extracts from a series of interactions with a Galileo2 program that assists in the Concurrent Engineering of printed wiring boards.

In small companies, this program, which is called KLAUS, could function as a "virtual product design team," by monitoring the decisions made by a designer, as they are being made, and interjecting the kinds of comments and suggestions that would have been made by the members of a design team if the company had been able to establish such teams. In larger companies, where product design teams have been established, a distributed version of KLAUS could function as a clearing house for decisions made by members of a product design team who are unable to meet very often, perhaps because they have conflicting schedules or because they are located in different plants.

KLAUS considers many aspects of the functionality, manufacturability, reliability, testability and cost of the board being designed. However, we restrict the example interactions that we present below to some extracts that consider issues of testability, as follows. It is often appropriate to introduce into a circuit components that are not needed for product functionality, but which are instead provided for testing purposes. For example, if a board contains a crystal that oscillates at a rate faster than that which can be accommodated by the available bed-of-nails tester, then an ancillary divider circuit for the crystal should be designed into the board. This means that, from the test engineer's perspective, the salient features of the life-cycle environment include attributes of the available test equipment, most notably in this scenario the maximum frequency testable by that equipment.

3. Programs in Galileo2

A program in Galileo2 comprises a set of declarations and definitions, which can appear in any order. A Galileo2 program is interactive, user input being in the form of additional declarations. Users can input their declarations in any order and a user can withdraw, at any time, any declaration that he has made in the past. The program specifies an initial constraint network, while user inputs extend this network.

In a program, the declaration statements include declarations of the parameters that exist in the initial network, the constraints that exist between these parameters, and which types of parameter, if any, users may add to the network at run-time. (There is no restriction on the type of constraint that may be added by users at run-time.) The definition statements in a program include definitions for the application-specific domains, functions and predicates that are used in parameter or constraint declarations, as well as boundaries for the various subnetworks (called "fields of view") that may be seen by different members of a design team. By default, a field of view is presented to its users through an interface based on a single-column spread-sheet or "scrollsheet." However, if any field of view is to be presented through a feature-based CAD interface, the definition statements in a program must include definitions for the icons that will be used to represent the parameters in that field of view.

In KLAUS, the parameters in the network represent salient attributes of the board, the circuit on the board, the components of this circuit and the environment in which the board will be manufactured and tested. The constraints in the network represent the mutual restrictions that the board, the circuit, the components and the environment exert on each other. The region of the network that is seen by a KLAUS user (i.e., his field of view) depends on whether he is a test engineer, circuit designer or manufacturing engineer; a field of view is defined for each such class of user. The test and manufacturing engineers see their fields of view through a scrollsheet interface while the circuit designer is given a CAD interface. Each user can modify only that part of the network that is within his field of view. Since parameters in one field of view are connected by constraints with parameters in the other fields of view, decisions made by users flow across the boundaries between fields of view.

3.1. Building a Network

In Galileo2, a parameter declaration names the parameter, specifies the corresponding domain of legal values and provides an optional long synonym. For example, the following declaration from KLAUS

```
'the facility where the board will be tested'
(the_test_facility) : test_facility.
```

introduces a parameter called **the_test_facility**, assigns to it the long synonym "**the facility where the board will be tested**" (which is how the parameter will be named in all interactions with users), and specifies that the parameter takes its values from the domain **test_facility**. Long synonyms, which are arbitrary strings delimited by apostrophes, enable the Galileo2 run-time system to construct meaningul natural language utterances whenever it is outputting textual information to users.

If the set of domains that are predefined in Galileo2 are not adequate for a program, application-specific domains can be defined. Consider, for example the domain **test_facility** from which the parameter **the_test_facility** declared above takes its values. This domain could be defined as follows:

```
domain 'description of the available test facility'(test_facility)
=::=
('the name of the equipment at'(name) : tester_name,
 'the maximum clock frequency testable at'(maxfreq) : frequency,
 'the maximum number of test points testable at'(maxtpoints)
```

Any well-formed sentence in first-order logic is a well-formed constraint in Galileo2, including atomic, compound and quantified sentences. These sentences can use either predefined or application-specific predicate and function symbols. See (Bowen and Bahler, 1992) for details about how Galileo2 supports the definition of application-specific predicates and functions.

In writing a Galileo2 program, constraints are written using the short names for the parameters, predicates and functions involved but, if these symbols have been given long synonyms, the long names are used in explanations to the user. For example, given the long synonyms declared in the program for **the_test_facility** (see above) and other parameters, as well as the long synonyms given in the definitions for the **test_facility** (see above) and other domains, the constraint

```
the_test_facility.maxtpoints >= the_circuit.totaltstpts.
```

would be paraphrased by the Galileo2 run-time as follows:

```
the maximum number of test points testable at
the facility where the board will be tested >=
the total number of test points in the
circuit being designed.
```

3.2. Partitioning the Network

A field of view definition specifies the subset of the network parameters that may be seen within that field of view. Several fields of view are defined in KLAUS, one for each type of user. The following statement, for example, specifies that in the field of view called **testability** (which is the field of view that will be seen by test engineers) there is only one parameter, namely the one called **the_test_facility** whose definition we saw earlier:

```
field testability =::= {the_test_facility}.
```

On the other hand, the definition

```
field configuration =::= {X : component(X)}.
```

specifies that in the **configuration** field of view (which is the one seen by the circuit designer) all, and only, the parameters of domain **component** will be seen.

The number of parameters present in a network can increase at run-time. Users can declare new parameters and, as we we will see in the example scenario, new parameters can also be created because the necessity of their existence is inferred by the run-time system.

We can specify which users are allowed to declare new parameters and can specify what types of parameter they are allowed to declare. For example, in the KLAUS application, the statement

```
permission({X:component(X)}, configuration).
```

specifies that the user of the **configuration** field of view, i.e., the circuit designer, can add new parameters of domain **component**, to represent components that he decides to add to the circuit.

There is no restriction on the types of constraint that may be input by a user at run-time, except that each constraint must reference at least one parameter within his own field of view. That is, a user can specify a constraint between a parameter in his own field of view and parameters which may exist in other fields of view. But he must mind his own business, at least to the extent that he may not specify a constraint which references only parameters that "belong" to other users.

Galileo2 support three kinds of interface: feature-based CAD interfaces, scrollsheet interfaces and topological interfaces. Different fields of view in a network can be presented to users through different types of interface. KLAUS provides a feature-based CAD interface to the `configuration` field of view which is seen by circuit designers. It provides a scrollsheet interface to the other fields of view, which are seen by test and manufacturing engineers. Scrollsheet interfaces are provided automatically by the run-time system and require no programming. The icon-related definitions and declarations necessary to provide a feature-based CAD interface include: icon paintings; mappings between the icons and the network parameters they represent; and, if icons must change appearance when parameters change state, specification for how icon appearances and parameter states are related. Although our example scenario will include interactions with a feature-based CAD interface, the necessary icon-related definitions and declarations are beyond the scope of this paper and will not be considered.

4. An Example Scenario

A life-cycle advisor written in Galileo2 can be used for multiple design projects. To use the advisor on a particular project, we activate the advisor and then do a database read to load the network state with data specific to the project. To move on to another project, we save the current network state to the database, clear the network state, and load data on the next project from the database.

In the following scenario, we assume that a project leader has set up a database entry for a new project. We take up the story when the test engineer, who in this case is the first team member to make some decisions about this new project, starts to interact with KLAUS about the project. (We emphasize, however, that this scenario is only one of many that are possible: users from different perspectives can interact with KLAUS in any order desired.) Fig. 1 shows the interface presented by KLAUS to the test engineer after he has selected the test equipment to be used for the project. The largest window in this screen is a scrollsheet, in which each cell occupies one or more lines.

[ ]Help [ ]File [ ]New [ ]Utilities [ ]Search [ ]Up [ ]Down [ ]Focus [ ]Toggle	
[ ]the name of the equipment at the facility where the board will be tested	erdsys
[ ]the maximum clock frequency testable at the facility where the board will be tested	9.8
[ ]the maximum number of test points testable at the facility where the board will be tested	200
>>>	
KLAUS - a PWB Design Advisor (Testability)	

Fig. 1

Fig. 2 shows the perspective of the circuit designer. Notice that this perspective provides a feature-based CAD interface rather than the scrollsheet interface provided to the test engineer. The drawing in this screen represents the circuit after the designer has introduced a CPU, and KLAUS has used its knowledge to infer the need for an associated pullup resistor and oscillating crystal. KLAUS introduced these components into the circuit and the designer has decided to accept these suggestions as part of his design.

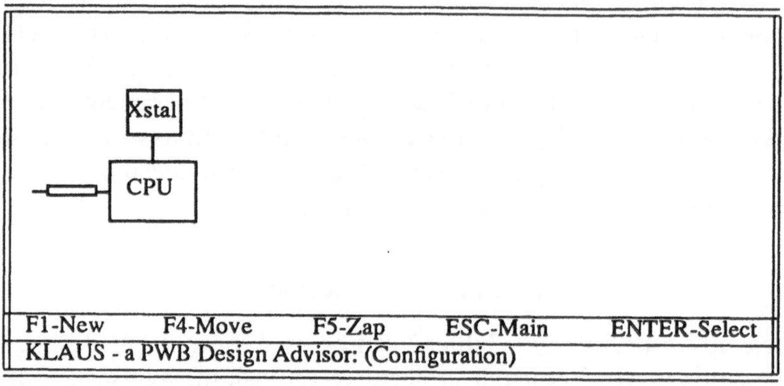

Fig. 2

Suppose that the designer, having accepted the crystal, specifies that it should oscillate at 25 MHz. Now, however, the following universally quantified constraint comes into play:

```
all X : osc_crystal(X) and the_test_facility.maxfreq < X.freq
implies
exists(X.'an ancillary divider circuit for'
       (anc_divider) : divider).
```

This universally quantified constraint specifies that every crystal oscillator must satisfy the following: if the maximum clock frequency testable at the facility where the board will be tested is less than the the oscillation frequency of the crystal oscillator, then an ancillary divider circuit for the crystal oscillator must exist. Although Galileo2 supports both universal and existential quantification, the symbol **exists** in this constraint is not an existential quantifier. It is a free logic existence specifier (Lambert and Van Fraassen, 1972), as explained in (Bowen and Bahler, 1991a), which discusses the theoretical basis for the dynamic introduction of new parameters into constraint networks at run-time.

Because of this constraint and because 25 MHz exceeds the maximum testable frequency of 9.8 MHz specified earlier by the test engineer (Fig. 1), the system introduces an ancillary divider circuit for this oscillator. The result can be seen in Fig. 3, where KLAUS is suggesting that the new component on the screen should be added to the circuit.

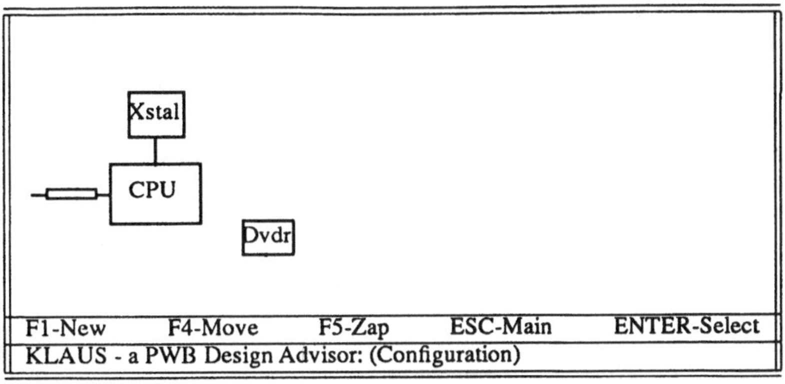

Fig. 3

The designer is surprised by the introduction of this component, so he asks KLAUS to justify it. In KLAUS's explanation (Fig. 4), the constraint given above, which introduced the component, is paraphrased in natural language; note the use in this paraphrase of the natural language synonym "**`an ancillary divider circuit for`**" for the parameter "**`anc_divider`**," which was defined in the constraint (see above) that inferred the necessity of this parameter's existence.

Despite this justification, the designer decides to reject this ancillary component. Now, however, the constraint which wanted to introduce the ancillary divider is violated, leading to the message shown in Fig. 5.

Justification

an ancillary divider circuit for the oscillator for the cpu exists
because of the following constraint
(38) every crystal oscillator must satisfy the following:
if the maximum clock frequency testable at
the facility where the board will be tested <
the oscillation frequency of the crystal oscillator
then an ancillary divider circuit for the crystal oscillator
must exist and must be a frequency divider;
and because of the following parameter value(s):
the maximum clock frequency testable at the facility where
the board will be tested = 9.8;
the oscillation frequency of the oscillator for the cpu = 25.

the maximum clock frequency testable at the facility where the board will
be tested was established according to the perspective taken by
test engineers.

the oscillation frequency of the oscillator for the cpu = 25
because you said so.

Fig. 4

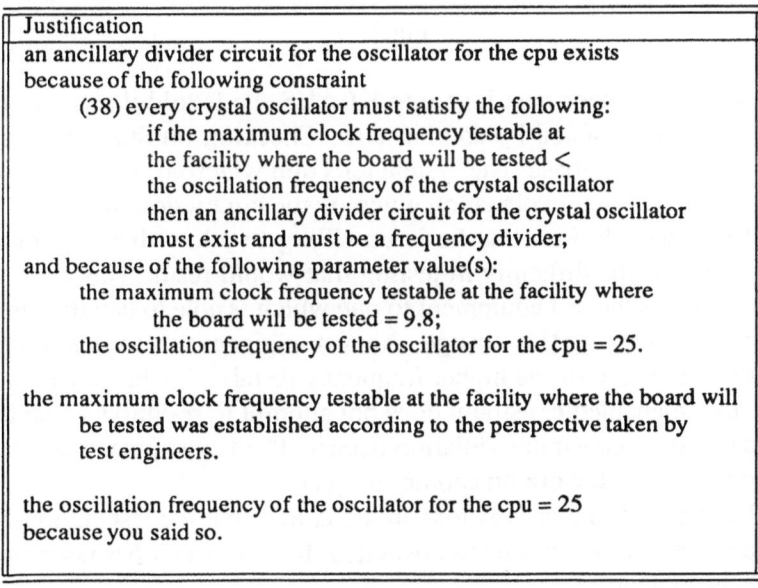

Choosing among the suggestions offered in this message, the designer decides to disable constraint (38). However, this constraint also refers to a parameter in the testability field of view, so the decision to disable the constraint must be accepted by the test engineer. Whenever a user disables a constraint other than one he previously asserted himself, he is required to enter a free-text explanation of his action, which is saved for possible use in a design audit. These free-text explanations are also relayed by the system as a kind of electronic mail between users of different perspectives.

```
VIOLATION:
an ancillary divider circuit for the oscillator for the cpu
should exist, but is prohibited.

SUGGESTIONS:
(1) Retract the constraint
      (104) an ancillary divider circuit for the oscillator
               for the cpu should not exist.
(2) Disable the constraint
      (38) every crystal oscillator must satisfy the following:
              if the maximum clock frequency testable at
              the facility where the board will be tested <
              the oscillation frequency of the crystal oscillator
              then an ancillary divider circuit for the crystal oscillator
              must exist and must be a frequency divider.
(3) Change the oscillation frequency of the oscillator for the cpu.
(4) Request that, in the perspective taken by test engineers,
      a change be made to
               the maximum clock frequency testable at the facility where
               the board will be tested.
```

Fig. 5

When the test engineer next logs into KLAUS, he is told that a constraint of interest to him was disabled by another user. Checking on this, he calls up the design state from the database. The system tells him which constraint was disabled and produces the free-text explanation given by the circuit designer.

The test engineer decides that he is unwilling to allow this constraint to be disabled because of the difficulty in testing that would result. However, to compromise, he changes the test equipment to one which is able to handle a frequency of 25 MHz. After making this change, the test engineer reactivates the disabled constraint (38). Because of the higher frequency testable by the new piece of test equipment, the re-enabled constraint does not attempt to re-introduce an ancillary divider circuit, so no constraint violation occurs. The test engineer saves the new design state and starts to work on another project.

When the circuit designer next logs in, he is told that the test engineer has re-enabled the constraint but, since the unwanted divider circuit has not reappeared, the designer is content.

Suppose, however, that no such easy compromise was possible. For example, the designer might have selected an oscillation frequency that exceeded even the upper limit of the fastest available tester. In this case, the test engineer and designer will successively disable and reenable their shared constraint (38), offering free-text explanations to each other until one or the other gives way or appeals to the project leader. In this case, the test engineer could give give way by deciding tc build a special divider test fixture; the circuit designer could give way by using a lower oscillation frequency.

5. Concluding Discussion

In Concurrent Engineering advice systems, it is necessary to support the multiple perspectives on life-cycle design that are taken by members of the various engineering disciplines involved and to facilitate interaction between these parties when they make conflicting decisions.

In this paper, we have shown how Galileo2 enables constraint networks to be divided into different fields of view to reflect the different perspectives of various members of the Concurrent Engineering design team. We have shown that these different fields of view can be presented to users through different styles of interface, including feature-based CAD interfaces and scrollsheets. We have also shown that the run-time system for the language facilitates interaction between the various members of a product development team when they make conflicting decisions.

From the wide variety of experimental applications on which we have applied Galileo2, the language seems to offer a suitable environment for building Concurrent Engineering systems. However, it still remains to be validated in large-scale industrial applications. This process has now started, with the development of a Design For Solderability advisor which has been commissioned by a local manufacturer of printed wiring boards. Several theoretical questions also remain open. In particular, we are investigating the use of negotiation techniques from the field of distributed AI (Werkman, 1991).

References

Borning, A.: 1981, The programming language aspects of ThingLab, a constraint-oriented simulation laboratory, *ACM Transactions on Programming Languages and Systems*, **3**(4), 353-387.

Bowen, J, O'Grady, P. and Smith, L.: 1990, A constraint programming language for life-cycle engineering, *International Journal of Artificial Intelligence in Engineering*, **5**(4), 206-220.

Bowen, J. and Bahler, D.: 1991a, Conditional variable existence in generalized constraint networks, *Proceedings of AAAI-91, National Conference of the American Association for Artificial Intelligence*.

Bowen, J. and Bahler, D.: 1991b, Supporting cooperation between multiple perspectives in a constraint-based approach to concurrent engineering, *Journal of Design and Manufacturing*, **1**, 89-105.

Bowen, J. and Bahler, D.: 1992, Frames, Quantification, perspectives and negotiation in constraint networks for life-cycle engineering, *International Journal of Artificial Intelligence in Engineering* (in press).

Bowen, J., Bahler, D. and Dholakia, A.: 1992, An AI constraint network-based approach to bed-of-nails DFT for digital circuit design, *Computers and Electrical Engineering* (in press).

Colmerauer, A.: 1987, *An Introduction to Prolog III*, Draft, Groupe Intelligence Artificielle, Universite Aix-Marseille II, November.

Gosling, J.: 1983, Algebraic constraints, *Technical Report CS-83-132*, Computer Science Department, Carnegie-Mellon University, Pittsburgh.

Graf, T., Van Hentenryck, P., Pradelles, C. and Zimmer, L.: 1989, Simulation of hybrid circuits in constraint logic programming, *Proceedings of the 11th International Joint Conference on Artificial Intelligence*, 72-77.

Heintze, N., Michaylov, S., and Stuckey, P.: 1986, CLP($\Re$) and some electrical engineering problems, *Proceedings of the 4th International Conference on Logic Programming*.

Lambert, K. and Van Fraassen, B.: 1972, *Derivation and Counterexample: An Introduction to Philosophical Logic*, Dickenson Publishing Company, Enrico, CA.

Mackworth, A.: 1987, Constraint satisfaction, *in* S. Shapiro (ed.), *Encyclopedia of Artificial Intelligence*, Wiley, New York, pp. 205-211.

Minsky, M.: 1975, A framework for representing knowledge, *in* P. Winston (ed), *The Psychology of Computer Vision*, McGraw-Hill, New York, pp. 211–277.

Serrano, D. and Gossard, D.: 1988, Constraint management in MCAE, *in* J. Gero (ed), *Artificial Intelligence in Engineering: Design*, Computational Mechanics Publications, Southampton, pp. 217-240.

Werkman, K.: 1991, Evaluating alternative connection designs through multiagent negotiation, *in* D. Sriram (ed.) *Computer Aided Cooperative Product Development*, Springer-Verlag, New York.

Sussman, G. and Steele, G.: 1980, Constraints—a language for expressing almost-hierarchical descriptions, *Artificial Intelligence*, **14**, 1-39.

Sutherland, I.: 1963, SKETCHPAD: a man-machine graphical communication system, *IFIPS Proceedings of the Spring Joint Computer Conference*, Spartan, Baltimore.

CONSTRAINT SATISFACTION AS A PLANNING PROCESS

B. FROMONT and D. SRIRAM
Intelligent Engineering Systems Laboratory
Department of Civil Engineering, I-253
Massachusetts Institute of Technology
Cambridge MA 02139
USA

Abstract. Constraint satisfaction problems (CSPs) are closely related to a large class of engineering problems, especially planning and design problems. Many attempts have been made to address CSPs and several techniques have been developed for that purpose. These techniques widely differ from each other and usually they impose a specific representation of the problem: a unified formalism is still lacking.
In this paper, we discuss how the planning paradigm can be used to solve a CSP. A planner is used as a top-level control process, guiding the search for a solution and producing an appropriate *solution plan* when the problem is solvable. The CSP is described by a *goal* stating which constraints should be satisfied. The planner produces a non-linear *plan* at an abstract level where the different steps needed to achieve the goal are partially ordered. At the bottom level, numerical and symbolic methods are chosen in the order defined by the plan and can be executed to produce a solution to the CSP. This is very efficient in the case where one wants to vary a parameter over a certain range and to study its influence on other values for a given CSP.

1. Introduction

Much of the knowledge in a large class of engineering problems can be represented as a set of mathematical relations between symbols or real-valued quantities. When trying to maintain these relationships and to perform inferences on them, one faces a constraint satisfaction problem. Constraints provide an attractive way to represent many problems in AI; they provide a *declarative* formalism where one specifies *what* things should be without specifying *how* to achieve this goal. However, this is a non-trivial task and there have been many attempts in addressing the problem.

In many cases, constraint solvers are part of bigger systems. They serve purposes such as "smart" graphic interfaces (e.g., Sutherland's Sketchpad (1963), Borning's ThingLab (1979), Gosling's Magritte (1983)) or vision systems (the most famous is Waltz's filtering algorithm (Waltz 1975) which was applied to scene analysis). Typical planning systems — developed by researchers in artificial intelligence (AI) — use "constraint posting" as a means to insure the correct execution of a plan (Wilkins, 1988; Chautard and Honnorat 1991), or to prune the search space (Wilkins, 1988; Chapman, 1987). In most of these cases special purpose constraint posting algorithms have been developed to address specific problem types. Although several researchers have developed constraint satisfaction algorithms to address arc-consistency, path-consistency, and then K-consistency in constraint networks (Mackworth, 1977; Mohr and Henderson,

J. S. Gero (ed.), Artificial Intelligence in Design '92, 97–117.

1986; Han and Lee, 1988), these algorithms were developed solely from a computer science perspective and do not reflect the way engineers tackle problems. The propagation algorithm, for example, is fast and well suited to some kinds of problems but its nature is pure number crunching. In many cases it is desirable to have an insight on how the machine has tackled the problem, especially when it fails!

It has long been recognized that the constraint paradigm is essential to the design process (Sriram and Maher, 1986), and the need for a higher-level constraints solving process has been acknowledged (Tsang, 1990). Essentially in design automation, constraint satisfaction forms the core of the reasoning (together with refinement). Even if specialized algorithms are efficient we feel that relying on a more abstract mechanism would be more appropriate for engineering problem solving.

In this paper, we present a new approach for solving CSP problems. We use a goal-directed solver to capture the constraint posting process, i.e., a planner to guide the solution strategy, selecting appropriate methods to be used and, more important, keeping track of its reasoning. We view the CSP solution as a path from an initial world where variables have given values to a final world where the constraints are satisfied. The planner paradigm is especially useful since it reasons about actions and is thus able to report its strategy to an end user. Also, locations of failure are more easily found and the richness of the paradigm facilitates the integration of many CSP algorithms. Different forms of constraints can be handled, provided that the corresponding operators used to solve them are defined. Moreover, when a solution plan has been found it can be reused with different numerical values. Our implementation compiles these plans into procedures, thus leading to an extremely efficient solution scheme.

Organization of the paper. In Section 2, we review some constraint satisfaction techniques and definitions. In Section 3, we show how the planning paradigm provides a more general and abstract way of solving CSPs. A description of our framework — called COPLAN — will be provided in Section 4. An example of COPLAN in action is provided in Section 5.

2. Background

2.1 BASIC DEFINITIONS

An informal definition of a constraint has been given by Gosling (1983) as: "A *constraint* is a relation stating what *should* be true about one or more *objects*." This general definition has been redefined by many researchers in the field, each transposing the definition to his own problem. Most of the time, the constraint expresses a relationship between *variables*. These variables may represent a physical parameter, the attribute of an object, etc. Each variable has a domain which represents the set of different variable bindings.

Definition 1: An *assignment* is a pair (X,d) (also denoted X = d), where X is a variable whose domain is D, and $d \in D$.

Our definition of a constraint is the following:

Definition 2: Given a set of variables $V=\{X_i\}$ and their associated domains D_i, an n-ary constraint $C_{i_1...i_n}$ on V is a sub-set of the cartesian product $\Pi_{i \in \{i_1...i_n\}} D_i$.

A constraint, thus, specifies the values of a variable which are compatible with the others. The subset need not be specified: for instance it could be given implicitly by an equation. We do not presuppose the nature of the domain — discrete or continuous.

Discrete domains arise in many problems such as graph coloring, logical puzzles, scene and edge labeling. The search space is finite, therefore this type of problem is *decidable*. Nonetheless, solving a network of constraints with finite domains is NP-complete. In practice, one looks for efficient algorithms, which may utilize some form of heuristic pruning techniques. *Logic programming* has several features that makes it a suitable tool for solving CSP with discrete domains (Colmerauer, 1990; Cohen, 1990). When one adds *lookahead* or *forward-checking* facilities, it can also be quite efficient (see (Van Hentenryck, 1989) for a good overview).

Continuous domains can be viewed as "closer" to real life, where physical parameters vary continuously over a real domain. In this case, the search space is infinite. In general, the problem can be viewed as a set of simultaneous algebraic equations to solve. Propagation, relaxation, graph algorithms are more suitable for this type of problem, although logical constraint programming languages — such as CLP(R) (Heintze, Michaylov and Stuckey, 1989)— handle real arithmetic. Engineering problems deal with both continuous and discrete domains; the continuous domains are not well addressed in the artificial intelligence (AI) literature. However, several researchers have tackled the discrete CSPs (e.g, Mittal (1987;1990)).

Given a set of variables and a set of constraints, constraint satisfaction consists in finding a set of *assignments* for each variable *consistent* with the set of constraints. If we represent the CSP as a graph where nodes represent variables connected by constraints, the solution of the CSP is also called a *labeling* of the graph consistent with the set of constraints.

There are two techniques for representing a CSP as a graph:

1. The graph is such that its nodes represent variables and its edges are the constraints between these variables (we will discuss this representation later).
2. A node can be either a variable or an elementary operator. Constraint networks of this type look like electrical circuits. This representation is convenient for illustrating many constraint satisfaction techniques.

Consider the following constraint: $A*X + A*Y=Z$. According to the latter formalism this constraint can be represented as shown in Figure 1.

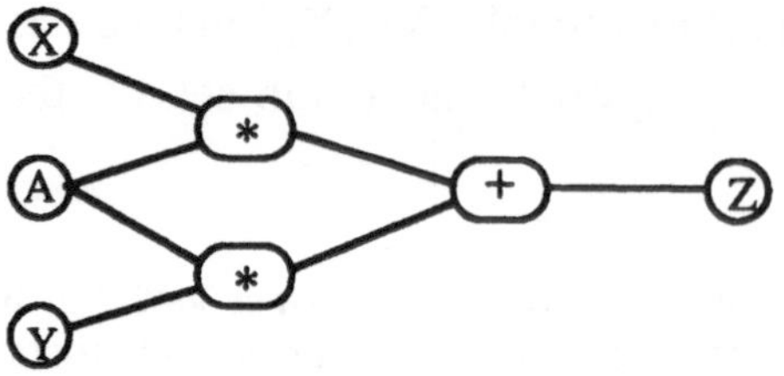

Fig. 1. Network Representing A*X+A*Y=Z

Each operator node can compute one of its missing parameters. Suppose that A and X are known. The multiplier can deduce the value of its output, which is A*X. Moreover, if X and the output O are known, the multiplier will infer that A=O/X. There is no *directionality* in the behavior of those operators. The early work of Sussman and Steele (1980) makes use of this formalism to study *propagation*..

2.2 PROPAGATION

One of the fastest and simplest algorithms to solve a constraint network is called *propagation*. It involves firing any operator which has enough information about its entries. The process is repeated until no node can fire.

Consider the example in Figure 1, and assume that A=2, X=3, Y=1. The output of the first multiplier is set to A*X=6. The adder cannot fire since two of its entries are unknown. However, the second multiplier fires to give A*Y=2. Finally, the adder produces the output Z=8.

The computation flow can also go in the other direction. For example, assume that A=2, X=3, Z=10. The output of the first multiplier is set to A*X=6. Then the adder is able to set one of its entries to Z-6=4. And the value of Y is updated by the second multiplier: Y=4/2=2.

The propagation algorithm reduces the global CSP to a local constraint problem on an operator node. Although the propagation algorithm is fast and simple, it does not provide a general mechanism to address nodes that have multiple outgoing links. For example, assume that X, Y and Z are known. The propagation algorithm cannot infer a value for A.

Despite this major drawback, propagation is widely used to solve CSPs. It provides a fast algorithm, which is easy to implement. Moreover, propagation is not restricted to propagation of numerical values. For example, Davis (1987) has shown how propagation could be used to reason about intervals. Steele's PhD thesis (1980) provides many techniques for implementing an explanation mechanism in a constraints solver based on propagation. Nevertheless, the difficulties encountered show that propagation is still a low level reasoning paradigm.

2.3 GRAPH TRANSFORMATION

Graph transformation allows the system to rewrite parts of a complex network into simpler sub-graphs. This aims at modifying the *topology* of the graph. Figure 2 shows a graph representing the constraint shown in Figure 1 after factorization.

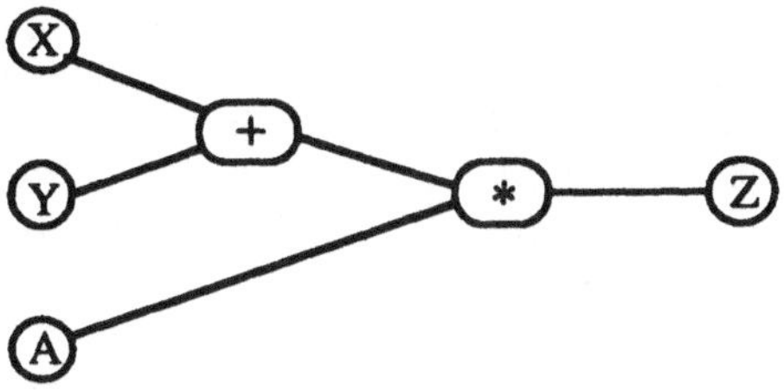

Fig. 2. Modified Network Representing A*(X+Y)=Z

The effect of this transformation has been to break the loop in the network. Now, *propagation* is a sufficient technique to determine the value of A, given X, Y, and Z. This technique is also called *term rewriting*. A set of *rewriting rules* is specified which describes the authorized transformations. For example:

$X*1 \Rightarrow X$

$X+X \Rightarrow 2*X$

$X*Y+X*Z \Rightarrow X*(Y+Z)$

A graph transformation allows algebraic manipulations. It is more powerful than *propagation* for it has a more global view of the CSP. However, long cycles such as those produced by simultaneous equations cannot be solved by rewriting. The programming language Bertrand (Leler, 1988) is based upon augmented term rewriting. The major drawback of this approach is that it does not handle many types of constraints, except equalities.

2.4 A HIGHER-LEVEL REPRESENTATION

In this representation, a node represents a variable of the problem. The edges of the graph are the constraints between variables. To study the constraint network, it is necessary to determine the interdependencies between parameters. A distinction is made between *known* variables and *unknown* variables. The known and unknown
variables are linked through constraint relationships.

As mentioned in the previous section, propagation starts as soon as a variable becomes known. When a *deadlock* occurs, it is hard to find its cause. Serrano (1987) introduced the idea that propagation should be activated only when the nature of the network is clearly identified.

According to him, one may want to determine:

1. whether the network is consistent or not (is there any loop or conflict ?)
2. if the network is consistent, a path for propagation to determine all the unknown values.

These two goals are achieved by the following steps:

1. Remove all the redundant edges in the network. This filtering on the edges is achieved by *bipartite matching*. Basically it results in binding each unknown variable to a unique constraint. This constraint will be used to derive a value for the variable. At the end of this step, the graph is transformed into a *directed* graph with a smaller number of edges.

2. Identify the loops in the network, and the strong components. At this stage it is possible to determine if the network is consistent or not. The strong components link several variables together. Their value can be derived only by solving a system of equations (using either Gauss or Newton-Raphson method, or symbolic processing such as Macsyma or Mathematica). At the end of this step, the directed graph is transformed into a *tree*.
3. Find one or many paths for propagation. This is achieved by *reverse topological sort* on the tree.

Figure 3 illustrates the process:

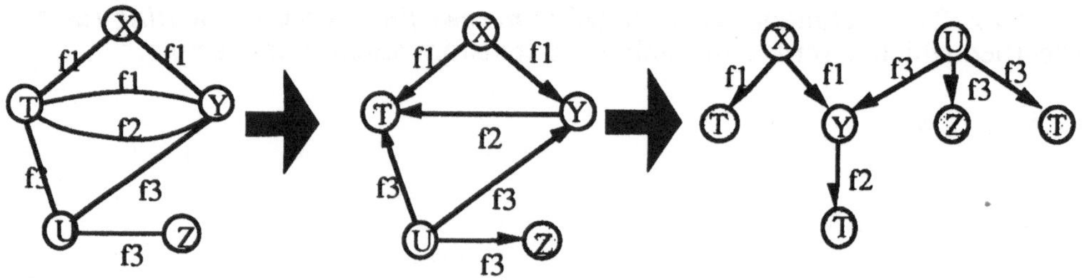

Fig. 3. Graph Processing

The variables T and Z are supposed to be known. The bipartite matching establishes the following relations: "X will be solved using f1", "U will be solved using f3", "Y will be solved using f2". The next step gives the ordering by which these variables will be instantiated. We see that Y must be solved first, then U or X.

Note that the method does not presuppose a particular technique of solution for each of the variables. Sometimes the bipartite matching gives several solutions. It may happen that a particular one is easier to solve than the other. Backtracking may be necessary.

This method suffers from several problems.
- constraints cannot be added in an incremental manner;
- inequalities are not handled properly; and
- finding all the solutions for a particular CSP is a cumbersome process.

However, the hidden idea behind this work is that constraint satisfaction may be viewed as a planning process. By controlling the propagation flow, one can determine the nature of the problem and an adequate solution plan (if it exists).

This kind of representation has also been studied extensively by Dechter and Pearl (1989). In their work they express the same idea: transforming the constraint network into a directed tree to achieve a backtrack-free propagation. The method (*tree-clustering*) involves triangulating the constraint graph, identifying the maximal cliques of the triangulated graph, solving the constraints associated with each clique and organizing the solutions in a tree structure.

3. CSP as a Planning Process

3.1 MOTIVATION

Most constraint solving algorithms do not capture the essence of the problem. Propagation, for example, fires constraints blindly until it stops (see previous section). The fact is that these algorithms lack *abstraction*. The use of abstraction is an effective approach to reducing search in problem solving. It usually suggests ignoring details and concentrating on the most important aspects of the problem. More precisely, at the higher level of the abstraction hierarchy, a constraint satisfaction scheme should not depend on the nature of the constraints. For example, a design generation program should be able to handle different types of constraints (Sriram and Maher, 1986), e.g., synthesis, parametric, interaction, and causal constraints. Such a program would have the following advantages: 1) it would provide a unifying framework to describe constraints and reason about them, 2) it would search a smaller space and it could choose the most efficient algorithms available to solve for each subset of constraints.

Moreover, a nice thing about abstraction is that it facilitates reusability. For example, when you solve the problem: 2x=1, 2x+y=3, you know that x will be determined by the first equation and that you will be able to deduce the value of y. If the numerical values change you still follow the same procedure. Another key idea follows: when solving similar problems, abstract reasoning can be done once for all; only number crunching needs to be repeated.

3.2 THE CONSTRAINT WORLD

In the "blocks world", which is a benchmark for testing classical AI planners, a state of the world consists of a set of objects and facts or axioms about these objects. The problem is to construct a plan, consisting of an appropriate sequence of actions, which transform the initial state into the goal state. In a similar manner, we introduce a space where the objects manipulated are constraints and variables. In this space, we use special purpose *operators* which transform states of the world. A state of the world is described in clausal form by a list of assertions about constraints and variables, such as *(known x)*, *(satisfied constraint1)*, etc. The planner tries to find a path from the initial world to a state where all constraints are satisfied (see Figure 4).

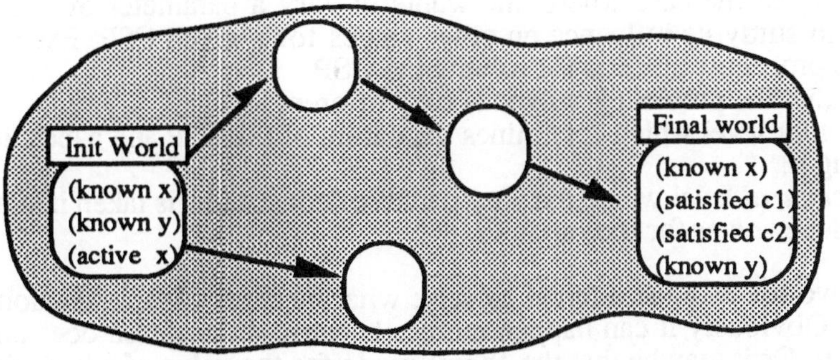

Fig. 4. Planning in the Constraints-world

Operators have a precondition-list, an effects-list, a delete-list, and a *procedure* slot that describes the actual routine which modifies the world. Procedure slots describe how to expand the plan and, in this sense, are similar to *plot* slots of SIPE's operators (Wilkins, 1988). Only the *procedure* slot of an operator is related to the kind of constraint which must be solved. Thus, we can define operators for solving linear equations, checking inequalities, solving algebraic equations, and so on. This makes the planning paradigm a good candidate for smoothly integrating different CSP techniques.
The overall strategy for solving a CSP is shown in Figure 5.

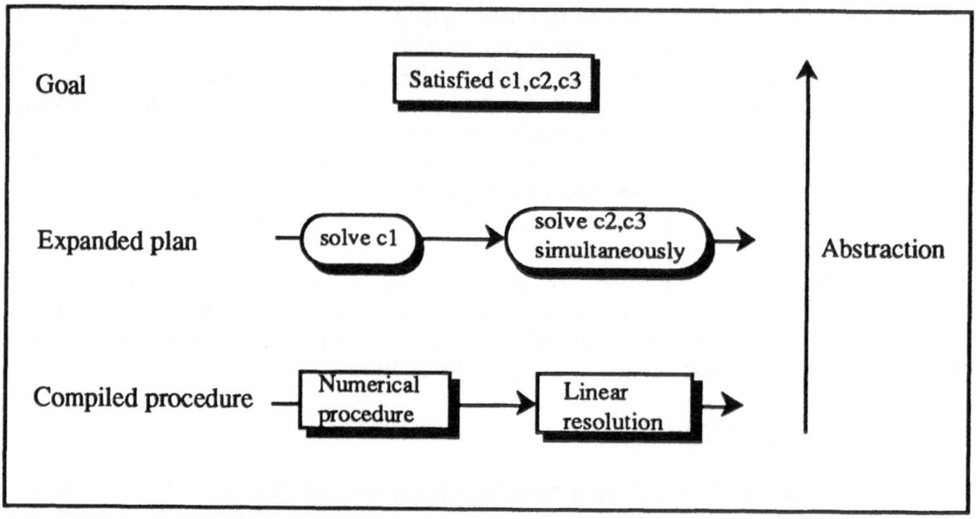

Fig. 5. Solution Strategy for Solving a CSP

The CSP is described by a goal. Usually the goal states which constraints should be satisfied but is more generally a list of assertions that should be true in the final world. The planner produces a non-linear plan at an abstract level where the different steps needed to achieve the goal are partially ordered. At the bottom level, numerical and symbolic methods are chosen in the order defined by the *plan*.. The execution of a plan consists in executing the above procedure. This is very efficient in the case where one wants to vary a parameter over a certain range and to study its influence on other values for a given CSP. Execution of the generic procedure suffices for solving the CSP.

We see that the problem is solved at two different levels:

- a *planning* level which determines the nature of the CSP and a technique for solving it.
- a *procedural* level where the very nature of constraints is taken into account to deduce *values* for the variables.

These two levels must interact closely: what happens if the execution of the plan fails? Obviously it can happen merely because an equation does not have any solution. Or it may be that the choice made for the value of a variable turns out to be incorrect. In this case, the execution is suspended and the planner tries to find *another plan* to achieve the goal. If this is possible, the execution is resumed, otherwise a solution is not possible.

At any stage, one can add a new constraint or ask to achieve a new goal. The planner begins its search from the state of the world just reached. This formalism supports easily *incremental* addition of constraints: this can be of considerable help in many problem solving situations, such as design.

3.3 NONLINEAR PLANNING

Planning is a very difficult problem in AI. Usually, AI planners such as STRIPS (Fikes and Nilsson, 1981) and NOAH (Sacerdoti, 1977) assume that the world is in a certain *state*. Changes which occur in a state must be explicitly listed among the effects of the operator applied to the world (STRIPS assumption). The main advantage of NOAH was to allow non-linearities, e.g., concurrent execution of actions not interacting with each other. This leads to a smaller search space and a gain in efficiency. Unfortunately, NOAH did not use backtracking, and many problems could not be solved. Perhaps the most important novelty in planning was the introduction of the *modal truth criterion* by Chapman (1987). It resulted in a planner that proved *complete* and gave some meaning to planning.

In what follows, we will use the terminology discussed in (McAllester, 1990) to describe a planning algorithm. A plan is composed of *steps*, each one being associated with an operator. A step establishes some properties (effects of the operator) in the current world. A step s is the *causal source* of a proposition P if it is the latest step which establishes P.

Definition 3: A causal link is a triplet (s, P, w) where: P is a proposition, w is a step which has P as prerequisite, and s is a step which is the causal source of P.

For example, s could be *(solve c1)*, P could be *(known x)*, and w could be *(known c2)*, where x is evaluated using c1 and is used in the constraint c2. Assume there exists a step t which adds or deletes P; in (McAllester, 1990), t is called a *threat* to the causal link defined above. The threat vanishes if one of the conditions $t<s$ or $t>w$ holds (where $<$ is a precedence order on steps).

Definition 4: A complete nonlinear plan consists of a set of steps S, a set of causal links L, and a set of ordering conditions C such that:

- every step name appearing in C and L is in S
- for every prerequisite P of every step w, there exists a causal link (s,P,w).
- for every link (s,P,w) and for every threat to this link t, the plan contains an ordering condition $t<s$ or $t>w$.

To produce a *complete* non-linear plan corresponding to a given CSP, we begin by a seed plan which contains two steps: 1) an initial step which asserts the initial propositions of the CSP; 2) a final step which has the goal of the CSP as a prerequisite.

Our algorithm works as follows:

1. If the current plan is order inconsistent, find a loop in the plan. Try to merge the operators in the loop. If impossible FAIL.
2. If the plan is complete, return the plan

3. If there is a causal link (s, P, w) in the plan and a threat to this link t such that no safety condition t<s or t>w is in the plan then nondeterministically add t<s or t>w to the plan and go to first step.
4. There must exist a step w whose precondition P is not satisfied. Nondeterministically do one of the following:
   a) Find an existing step s which adds P. Add the link (s, P, w) to the plan and go to Step 1.
   b) Find an operator o which adds P. Create a new step s associated with o, add the link (s, P, w) to the plan and go to Step 1.

Basically, all the ordering of steps takes place in Step 3 of the algorithm. In Step 4 choices of new operators are made. What happens in Step 1 is more "tricky". In traditional planners any inconsistency in the plan is considered as a failure. This comes from the assumption that *only one* operator can be associated with one step. In other words, a planner cannot do two things at the same time.

In the "constraint world", it is common to solve equations simultaneously and thus loops in the plan often occur. Consider the trivial case of two equations both depending on the variables x and y (see Figure 6). If the planner tries to solve these equations separately an inconsistency will occur because two operators depend on each other.

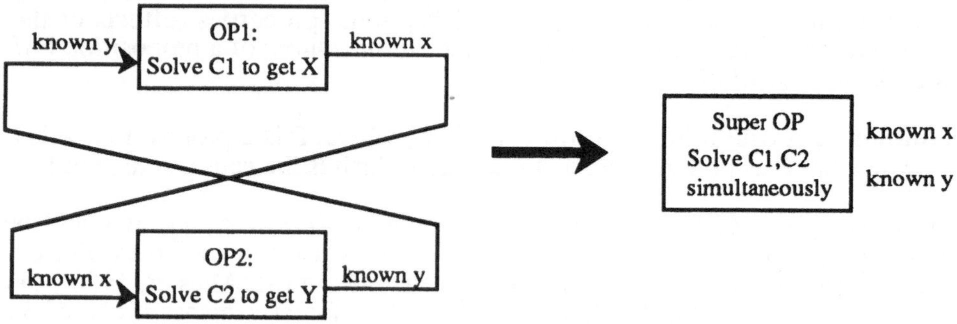

Fig. 6. Merging of Two Operators

The solution adopted is to merge dynamically these two operators when it is possible. When the merging does not succeed, the current plan is rejected and the planner backtracks to the previous choice point. Note that in Serrano's approach (Serrano, 1987) simultaneous equations are formed by "collapsing" strong components in a directed graph. This is implicit in our algorithm.

We have described how the planning paradigm could provide an abstract and efficient way of solving CSPs. The next section gives an overview of our prototype — called COPLAN — which is based on the above ideas.

4. COPLAN: COnstraint PLANer

4.1 SCHEMATIC OVERVIEW

COPLAN is a prototype implemented to show that ideas presented in Section 3 can be made to work. On the standpoint of AI technology, COPLAN is hardly novel. But the uniqueness of this program, and source of its power, is that it uses the classic planning paradigm to exploit specific knowledge about constraints. Planning is done at an abstract level where only symbolic values are taken into account. Then, when a plan is completed, this plan is executed and numerical procedures are called. If the plan fails during execution COPLAN backtracks to find alternate plans. This kind of high-level reasoning allows to clearly identify reasons of failure and to give explanations on the method used to solve a given problem. Furthermore, the plan obtained can be compiled into a generic procedure so that solutions of similar problems are quickly obtained: it only involves executing the plan.

Figure 7 illustrates the architecture of COPLAN.

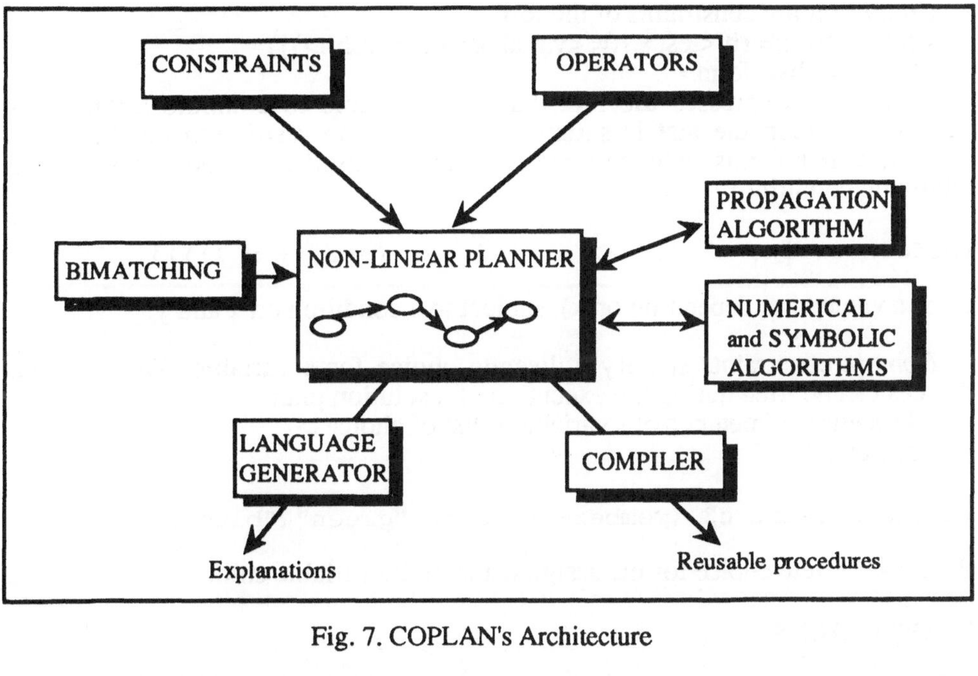

Fig. 7. COPLAN's Architecture

Note that the planner can call the standard propagation algorithm when the problem is "simple" and is not worth planning. The following sections review the different components of COPLAN.

4.2 TYPES OF CONSTRAINTS

The types of constraints that are to be supported in COPLAN are: *equality*, *inequality*, *condition*, and *domain*. The first two types restrict the expressions to be *polynomial*. The reason is simple: it is much easier to deal with polynomials in symbolic computations. Polynomials are translated into an internal form which is exploited by COPLAN to generate operators. Basically, all the methods to compute a variable given the others are generated. We use a generalization of Gosling's method for linear equations (Gosling, 1983).

The general syntax for the definition of constraints in COPLAN is given hereafter in BNF:

<constraint>::= (defconstraint <identifier> <constraint-form>)
<constraint-form>::=<equality> | <inequality> | <condition> | <domain>

Let us elaborate on the different constraints:

<equality::= (= <polynomial><polynomial>)
<inequality>::= (<op> <polynomial><polynomial>)
<op>::= > | < | >= | <=
For example, the constraint $yx^2+3 = 0$ is represented in COPLAN as:

```
(defconstraint c1 (= 0 (+ (* y (expt x 2)) 3)))
```

Conditions are constraints of the form:
<condition>::= (if <test> {(= <variable> <lisp-form>)})
<test>::= <lisp-form>

Conditions expresses the fact that it is possible to compute a value for <variable> when the test is satisfied. Note that no restriction on the forms applies. Other forms than polynomials may be used here. For example the following condition:

```
(defconstraint c2 (if (> x (+ 2 y)) (= z (+ x 1))))
```

gives a value to z (depending on x), subject to a condition on x and y.

Domain constraints specify a discrete choice for a variable. The different choices can be tried during the execution of a solution plan.
<domain>::= (member-of <variable><list of choices>)
For example,

```
(defconstraint c3 (member-of color "green" "blue"))
```

allows a discrete choice for the assignments of the variable color.

4.3 OPERATORS

Operators serve as a basis for COPLAN's reasoning. Essentially, operators provide COPLAN with elementary pieces of knowledge about constraint solving. The task of the planner is to choose the most relevant ones, for a given problem, and to order them.

As outlined in the previous section, most of the operators are generated during the definition of a constraint. For example, if we define the constraint y=x+3, COPLAN generates three operators:

1. One operator whose precondition is *(known x)* and whose effects are *(known y)(satisfied c)*. The subgoal will be to find a numerical operator which actually solves the corresponding y-polynomial: *(solve c for y)*.

2. One operator whose precondition is *(known y)* and whose effects are *(known x)(satisfied c}*. The subgoal will be to find a numerical operator which actually solves the corresponding x-polynomial: *(solve c for x)*.
3. One *check* operator whose preconditions are *(known x)(known y)* and whose effect is *(satisfied c)*. This operator will simply evaluate the constraint with given values of x and y.

The structure of an operator is given below.

OPERATOR:	
pname:	the name of the operator (generated by the system)
cnames:	name(s) of the constraint(s) solved by the operator
precond:	list of preconditions in clausal form
f-precond:	LISP form which computes the preconditions
effects:	list of effects in clausal form
delete:	list of negative effects in clausal form
procedure:	LISP procedure which achieves the effects of the operator
subgoal:	clausal form

Most of the slots are described by a clausal form. A clausal form is simply a list of atoms and/or variables which describes a fact. Any operator which conforms to this syntax is allowed. In the current implementation, we use fairly simple predicates such as *(known x) (solve c for x)*, etc. But an extension of COPLAN is currently under way to take into account interval values for variables (Alefeld and Hetzberger, 1983). For this purpose a new set of operators is defined with a richer set of predicates to express facts.

Some other operator's attributes need some explanation:

- *f-precond* is used by generic operators to compute dynamically their preconditions. Consider the operator ALGEBRAIC-SOLVE which finds the roots of any polynomial of degree lesser than 4. The effect of such an operator is *(solve $c for $x)*. This clause can be matched during the planning process against the subgoal of an operator. At this stage, the variables $c and $x will be *bound*. Only then can the operator generate its precondition which is *(< (degree c) 4)*.
- *procedure* is the actual function to computes values for variables.
- *subgoal* is a clause which describes another goal that the planner must achieve. In COPLAN, subgoals are usually satisfied by generic operators like ALGEBRAIC-SOLVE described above.

An example of an operator generated by COPLAN is:

OPERATOR:	
pname:	solve-x-c1-987
cnames:	(c1)
precond:	((known y))
f-precond:	nil
effects:	((known x)(satisfied c1))
delete:	nil
procedure:	nil
subgoal:	(solve c1 for x)

There are also some generic operators which solve a particular numerical or symbolic problem. In our initial prototype (implemented in CommonLISP), we only have operators to solve: simple algebraic equations; simultaneous equations; and an equation by Newton-Raphson method. We are in the process of adding other operators as an extension of our prototype.

4.4 NONLINEAR PLANNER

The problem of COPLAN's performance, in terms of reliability and breadth of scope, is twofold:

- the size of the operators' set must be sufficient to handle many types of constraints; and
- the planner should be *complete*. In other words, if we define a set O of operators — such as the one mentioned above — and if we are given a CSP such that there exists a sequence of operators in O which solves the CSP, then the planner should find it.

At the heart of the planner is the algorithm outlined in Section 3.3. A proof of completeness is being investigated (based on the work of McAllester (1990)) but like in many planning problems is usually difficult to obtain (see (Chapman, 1987)). However, testing of COPLAN on various examples revealed no failure of our planning algorithm.

The rest of this section illustrates how operators are chosen during the planning process. Consider the circuit represented in Figure 8.

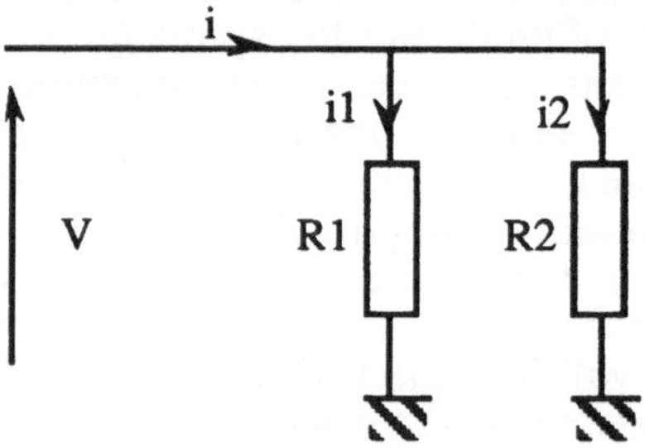

Fig. 8. Simple Circuit

We have the following constraints:

C_1: $i = i_1 + i_2$
C_2: $V = i_1 R_1$
C_3: $V = i_2 R_2$

Our goal is to satisfy these constraints. COPLAN starts from an *initial world* where V, i, R_2 have a value. The initial world contains the assertions (known V), (known I), (known R2). The planner begins with a plan containing a step *start* — whose effects are the above clauses — and a step *finish* — whose prerequisites are the goals *(satisfied C1)(satisfied C2)(satisfied C3)*. Since the goal *(satisfied C1)* is not already true in the initial world, COPLAN will look for an operator which has this clause in its *effects* slot. Among the possible operators, COPLAN can choose the one which solves for i_1 using C_1. COPLAN will create a new step s_1 associated with this operator (Step 3.a of the algorithm). The first prerequisite of this operator is *(known i)* which is *established* by the step *start*. COPLAN adds the link *(start, (known i),* s_1*)* — Step 4.a of the algorithm — and the safety condition (*start* < s_1) to the plan. The second prerequisite of s_1 is *(known i2)*. Since there is no existing step establishing this clause, the planner looks for another operator whose effects contain the above clause — Step 4.b of the algorithm.

The process continues until a complete plan is found. In this case we can get the following sequence: *(solve for i_2 using C_3)(solve for i_1 using C_1)(solve for R_1 using C_2)*. At this stage, the planner is called recursively to solve for the subgoals of the various operators. Here, the same operator which solves algebraic equations will be used. Then, when no more subgoals remain to be achieved, the planner looks for the procedure slot of each operator and executes them in sequence.

In another version of the circuit problem, it may be that only R_1, R_2, and i are known. In this case, some operators may depend each other and the problem will be solved by merging together the operators. The resulting operator achieves the simultaneous solution of the system of constraints. The procedure associated with this operator solves the system of equations. Note that this kind of problem cannot be solved using the standard propagation algorithm.

4.5 PRUNING THE SEARCH SPACE

If we use only the searching abilities of the planner to solve a CSP, we have a system that works but is generally not very efficient. Pruning the search space can be obtained by bipartite matching. This technique is used by Serrano in his work (Serrano, 1987). A bipartite matching is a surjective mapping from the set of unknown variables to the set of constraints. The *maximum* bipartite matching gives an idea of which constraint to use to compute a particular unknown variable. Suppose we are given the three following equations:

$$y+zx = 0$$
$$2z+a = 0$$
$$xz-5 = 0$$

Assume that x, y, z are unknown. A maximum matching is shown in Figure 9:

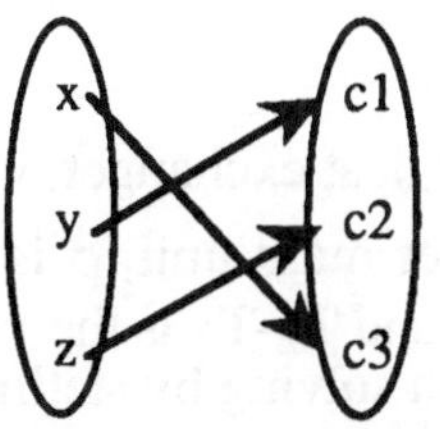

Fig. 9. Maximum Bipartite Matching

This matching is used to give a priority for some operators to be used by the planner. In this case, the operators selected would be displayed by Figure 10.

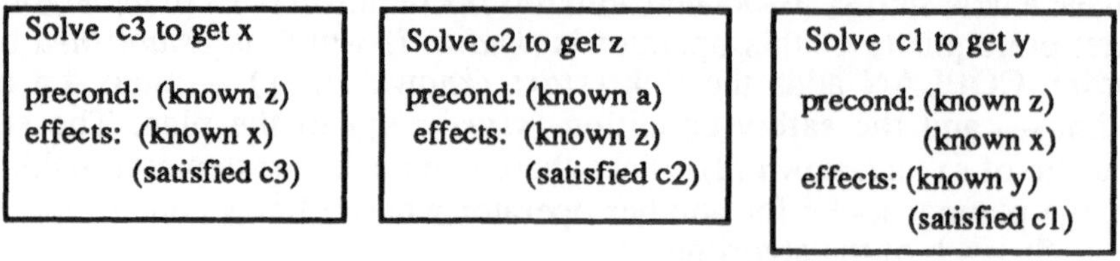
Solve c3 to get x
precond: (known z)
effects: (known x)
(satisfied c3)
Solve c2 to get z
precond: (known a)
effects: (known z)
(satisfied c2)
Solve c1 to get y
precond: (known z)
(known x)
effects: (known y)
(satisfied c1)

Fig. 10. Operators Selected in Priority

We cannot rely only on this technique to solve all CSP problems. In most of CSPs, the number of unknown variables and the number of constraints are not equal. The constraint solver must have a mechanism to take care of the remaining constraints or variables. Furthermore, there are often many maximum matchings for a given problem. The choice of the "good one" can be tricky! (see for example (Serrano, 1987)). In COPLAN, the result of the bipartite matching is only used to give "a good start" to the planning process. In the prototype version of COPLAN we have used the Hopcroft and Karp's algorithm to find the maximum matching (Hopcroft and Karp, 1973).

5. Example

We illustrate COPLAN with an example, which involves the design of a cooling system for a powerful engine. This engine radiates a power P, which we want to extract with a tube of circulating water (see Figure 11).

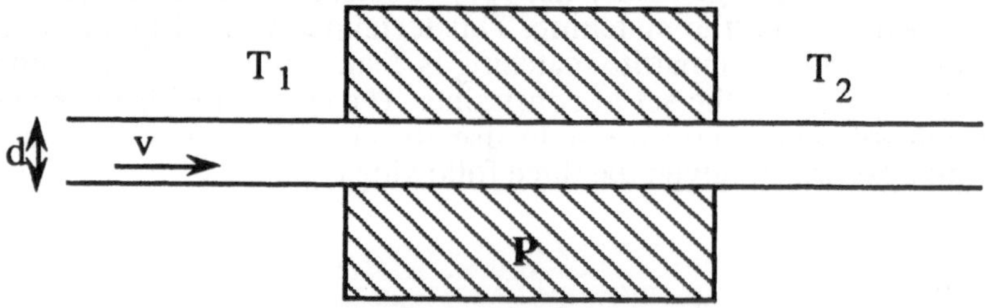

Fig. 11. Cooling System

The equation describing the heat exchange is given by:

$$P\eta/v = C_w \rho S(T_2 - T_1)$$

where: η is the efficiency of the heat exchanger, v is the velocity of the fluid, C_w is the heat capacity of water per mass unit, ρ is the density of water, S is the cross sectional area of the tube, and T_2-T_1 is the variation of temperature.

We begin the design problem solving by stating the following constraint:

```
>(defconstraint heat-exch (= (* P eta)(* v Cw ro S (- T2
T1))))
DONE.
```

However, we do not want the water to reach its boiling point. Hence, we type in the constraint *water-temp*, as described below:

```
>(defconstraint water-temp (< T2 373))
DONE.
```

where the temperature is expressed in Kelvins.

Suppose we can use two types of tubes which have different properties. We express this choice using a DOMAIN constraint as follows:

```
>(defconstraint tube-types (member-of tub-type "type1"
"type2"))
DONE.
```

In particular "type1" tubes have a 2 centimeters diameter, whereas "type2" tubes have a 5 centimeters diameter. These constraints are expressed as follows.

```
>(defconstraint d1 (if (eq tub-type "type1")(= d 0.02)))
DONE.
>(defconstraint d2 (if (eq tub-type "type2")(= d 0.05)))
DONE
```

Finally, we express the relationship between the diameter and the cross sectional area S, and then define the CSP by the `defcsp` command:

```
>(defconstraint diameter (= S (* 3.14 (expt (/ d 2) 2))))
DONE.
>(defcsp mypb (satisfied diameter d1 d2 heat-exch water-
temp tube-types)(= P 40000)(= Cw 4180)(= v 0.2)(= T1 300)(=
ro 1000)(= eta 0.5))
CSP MYPB DEFINED.
```

In the definition of the CSP, we have stated the constraints to satisfy and the known variables of the problem.

To solve this simple case, COPLAN first computes a maximum matching between unknown parameters and the set of constraints. There are several maximum matchings possible. For example: *(solve d_1 to get d)(solve heat-exch to get T_2)(solve diameter to get S)(solve tube-types to get tube-type)*. COPLAN will try to find a plan using the above operators. Since there are more constraints to satisfy than unknown variables, COPLAN must then use *check* operators to satisfy all the constraints.

The trace is shown below:

```
>(csp-solve 'mypb)
THIS PROBLEM CAN BE SOLVED AS FOLLOWING:
FIRST, SOLVE FOR TUBE-TYPE USING EQUALITY TUBE-TYPES
THEN, SOLVE FOR D USING CONDITION D1
      SOLVE FOR S USING EQUALITY DIAMETER
      SOLVE FOR T2 USING EQUALITY HEAT-EXCH
EVENTUALLY, CHECK THE INEQUALITY WATER-TEMP
EXECUTING...
```

Note that at this stage all the operators have been chosen.

```
GETTING VALUE FOR TUBE-TYPE "TYPE1" USING OP <CHOICE1>
GETTING VALUE FOR D 0.02 USING OP <COND1>
GETTING VALUE FOR S 3.14E-4 USING OP <ALGEBRAIC-SOLVE1>
GETTING VALUE FOR T2 376 USING OP <ALGEBRAIC-SOLVE2>
SOLUTION FAILURE: WATER-TEMP IS VIOLATED
PLANNING...

GETTING VALUE FOR TUBE-TYPE "TYPE2" USING OP <CHOICE2>
SOLUTION FAILURE: CANNOT GET D FROM D1
PLANNING...
```

Eventually, COPLAN reaches a correct solution:

```
FIRST, SOLVE FOR TUBE-TYPE USING EQUALITY TUBE-TYPES
THEN, SOLVE FOR D USING CONDITION D2
      SOLVE FOR S USING EQUALITY DIAMETER
      SOLVE FOR T2 USING EQUALITY HEAT-EXCH
EVENTUALLY, CHECK THE INEQUALITY WATER-TEMP
EXECUTING...
GETTING VALUE FOR TUBE-TYPE "TYPE2" USING OP <CHOICE2>
GETTING VALUE FOR D 0.05 USING OP <COND2>
GETTING VALUE FOR S 1.96E-3 USING OP <ALGEBRAIC-SOLVE1>
GETTING VALUE FOR T2 312.2 <ALGEBRAIC-SOLVE2>
CHECKING WATER-TEMP USING OP <CHECK1>
DONE.
```

At the end of a successful solution COPLAN generates a procedure to solve the CSP. This procedure is a concatenation of all the numerical routines used in the solution. This facilitates us to study a problem with several parameters. For example, we can compute the output temperature if η=0.4 instead of η= 0.5, as follows:

```
>(csp-solveagain 'mypb (= P 40000)(= Cw 4180)(= v 0.2)(= T1
300)(= ro 1000)(= eta 0.4))
((TUBE-TYPE "TYPE2")(D 0.5)(S 1.96E-3) (T2 309.7))
```

This mechanism is very powerful since most of the time consuming operations are performed only once.

6. Contributions and Future Work

The research challenge of engineering problem solving, and especially design automation, largely relies on constraint satisfaction. Current constraint satisfaction algorithms are usually not suited to handle multiple kinds of constraints and to provide explanations of their results. The research reported here articulated the idea that the planning paradigm can overcome these difficulties. In particular, our approach provides:

- a higher-level formalism to represent CSPs;
- a mechanism to deal with multiple constraint types and solution methods;
- incremental addition of constraints.

Such a formalism is promising since the problem solving mechanism is less dependent on the types of constraints. Furthermore, understanding of the reasoning is made easier.

We applied these principles to implement a prototype called COPLAN and demonstrated that the planning paradigm could be made to work in a significant way. In particular, a key feature of this program is that it can use previously computed plans to solve similar problems. A next step would be to investigate whether COPLAN could be provided with a learning mechanism which would exploit a base of abstract plans.

An important issue concerns the efficiency of our approach. A planner working on a large CSP could be somewhat slower than traditional solvers (provided they could solve the problem). On the other hand, in COPLAN, time is spent mostly on high-level tasks. Execution of a plan itself is very fast, and if one wants to vary a parameter over a certain range for a given CSP (which often happens in practise), COPLAN will provide an answer quickly. Furthermore, the explanations generated can be of tremendous help for users facing a large CSP.

Currently, we are planning to undertake the following extensions to COPLAN:

- **Implement COPLAN in C++.** COPLAN will be reimplemented in C++ and will form a part of our computer supported collaborative engineering framework.
- **Develop Complete Operator Semantics.** A complete list of operators and solution procedures that deal with a variety of symbolic and numeric relationships will be developed. For example, numerical procedures for dealing with inequalities will be encoded, as well as means to deal with interval analysis.
- **Develop a Graphical User Interface** A graphical user interface will be developed in Motif/X Windows for entering constraints, browsing through constraint networks, for performing what-if analysis, and for explaining the results of computation.

References

Alefeld, G. and Herzberger, J.: 1983, *Introduction to Interval Computation*, Academic Press.

Borning, A.:1979, THINGLAB - A Constraint Oriented Simulation Laboratory, PhD Thesis, Department of Computer Science, Stanford University.

Chapman, D.: 1987, Planning For Conjunctive Goals, *Artificial Intelligence*, Vol. 32.

Chautard, J. C. and Honnorat, C.: 1991, PEX: A Reactive Procedure Based Decision Maker, *Proc. IEEE CAIA*, Vol 1.

Cohen, J.: 1990, Constraint Logic Programming Languages, *Communication of the ACM*, Vol 33, no 7.

Colmerauer, A.: 1990, An Introduction to PROLOG 3, *Communication of the ACM*, Vol 33, no 7.

Davis, E.: 1987, Constraint Propagation with Interval Labels, *Artificial Intelligence*, Vol. 32.

Dechter, R. and Pearl, J.: 1989, Tree Clustering for Constraint Networks, *Artificial Intelligence*, Vol. 38.

Fikes, R. and Nilsson, N.: 1981, Learning and Executing Generalized Robot Plans, *in* Nilsson and Webber (Eds), *Readings in Artificial Intelligence*, Morgan Kaufmann Publishers, Inc.

Freeman-Benson, B. N., Maloney, J., and Borning, A.: 1990, An incremental Constraint Solver, *Communications of the ACM*, Vol 33, no 1.

Gosling, J.: 1983, Algebraic Constraints, Phd Thesis, Department of Computer Science, Carnegie-Mellon University.

Han, C. C. and Lee, C. H.: 1988, Comments on Mohr and Henderson's Path Consistency Algorithm, *Artificial Intelligence*, Vol. 36.

Heintze, N., Michaylov, S., and Stuckey, P.: 1989, CLP(R) and some Electrical Engineering Problems, Technical report CMU-CS-89-139, Dept of Computer Science, Carnegie-Mellon University.

Hopcroft, J. E. and Karp, R.: 1973, An $n^{5/2}$ Algorithm for Maximum Matchings in Bipartite Graphs, *SIAM J. Computing*, Vol 2, no 4.

Hua, K., Faltings, B., Haroud, D., Kimberley, G., and Smith, I.: 1990, Dynamic Constraint Satisfaction in a Bridge design System, in *Proc. Intl. Workshop on Expert Systems in Engineering*, Vienna.

Leler, W.: 1988, *Constraint Programming Languages. Their Specification and Generation*, Addison-Wesley.

Mackworth, A. K.: 1977, Consistency in Networks of Relations, *Artificial Intelligence*, Vol. 8.

McAllester, D.: 1990, *Course Notes for Artificial Intelligence (6.824)*, Department of Electrical Engineering and Computer Science, M.I.T.

Mittal, S. and Frayman, F.: 1987, Making Partial Choices in Constraint Reasoning Problems, *in Proc. of AAAI*, pp. 631-638, Morgan Kaufmann Publishers.

Mittal, S. and Falkenhainer, B.: 1990, Dynamic Constraint Satisfaction Problems, *inProc. of AAAI*, pp. 25-32, Morgan Kaufmann Publishers.

Mohr, R. and Henderson, T. C.: 1986, Arc and Path Consistency Revisited, *Artificial Intelligence*, Vol. 28.

Montanari, U. and Rossi, F.: 1991, Constraint Relaxation May be Perfect, *Artificial Intelligence*, Vol. 48.

Navinchandra, D. and Rinderle, J. R.: 1989, Interval Approaches for Concurrent Evaluation of Design Constraints, in *Proc. Symposium on Concurrent Product and Process Design*, San Francisco.

Sacerdoti, E.: 1977, *A Structure for Plans and Behavior*, Elsevier, North-Holland.

Serrano, D.: 1987, Constraint Management in Conceptual Design, Phd Thesis, Department of Mechanical Engineering, M.I.T.

Sriram, D. and Maher, M.: 1986, The Representation and Use of Constraints in Structural Design, *in* Sriram and Adey Eds., *Applications of Artificial Intelligence in Engineering Problems*, Springer-Verlag.

Sriram, D., Cheong, K., and Kumar, L.: 1992, Engineering Design Cycle: A Case Study and Implications for CAE, *Knowledge Aided Design*, Academic Press.

Steele, G. L.: 1980, The Implementation and Definition of a Computer Programming Language Based on Constraints, M.I.T - AI TR 595, Department of Electrical Engineering and Computer Science, M.I.T.

Sussman, G. J. and Steele, G. L.: 1980, CONSTRAINTS: A Language for Expressing Almost-hierarchical Descriptions, *Artificial Intelligence*, Vol. 14.

Sutherland, I.: 1963, SKETCHPAD: A Man-Machine Graphical Communication System, in *IFIPS Proceedings of the Spring Joint Conf.*

Tsang, J. P.: 1990, Constraint Propagation Issues in Automated Design, *in* Gottlob and Nejdl (Eds), *Expert Systems in Engineering*, Springer-Verlag.

Van Hentenryck, P.: 1989, *Constraint Satisfaction in Logic Programming*, MIT Press.

Ward, A. C.: 1989, A Theory of Quantitative Inference Applied to a Mechanical Design Compiler, Phd Thesis, Department of Mechanical Engineering, M.I.T.

Waltz, D. L.: 1975, Understanding Line Drawings of Scenes with Shadows, in *The psychology of computer vision*, McGraw-Hill, NY.

Wilkins, D. E.: 1988, *Practical Planning*, Morgan-Kaufmann, Inc.

A LARGE SCALE, MULTIPLE CONSTRAINT NETWORK SYSTEM FOR DESIGN FOR TESTABILITY FOR PRINTED WIRING BOARDS

C. KIM, P. O'GRADY and R. E. YOUNG
Group for Intelligent Systems in Design and Manufacturing
Department of Industrial Engineering
North Carolina State University
Raleigh NC 27695-7906 USA

Abstract. Design for Testability involves including a consideration of testing in the design of a Printed Wiring Board. This is important in that testing can account for one-third of the manufacturing cost of a Printed Wiring Board (PWB). Testing can also lead to increased manufacturing lead times. Both the cost and lead times associated with testing Printed Wiring Boards can be reduced by using Design for Testability. In this paper the constraining influences from the testing operation are placed in a constraint net form. The architecture of a comprehensive Design for Testability system, called TEST, is presented. This consists of multiple constraint networks that address different perspectives of Design for Testability. TEST has been implemented using the SPARK constraint net system. The implementation and operation of TEST are described. The overall result is a powerful system that gives advice to the designer on improvements that can be made to the design of the PWB so as to improve it's testability. TEST has a number of advantages: it has a relatively large number of constraints, it is flexible enough to be expanded in a relatively straightforward fashion, it naturally aligns with the overall design process used by the designer without restricting the designer to follow a fixed path, advice is given on improvements that can be made to the design, TEST is bi-directional, and other models, including cost models and manufacturing constraint models, can be readily added.

1. Introduction

Testing involves determining the type and location of faults in Printed Wiring Boards (PWBs). It has been reported that the costs of testing can account for one-third of the total manufacturing cost (Arabian, 1989). Testing therefore contributes a major portion of the manufacturing cost of a PWB. This proportion of cost is likely to increase with the increasing complexity of PWBs. In particular the use of Surface Mount Technology (SMT) has resulted in PWBs that have a higher density of components, have fewer natural test points, and which may have components on both sides of the PWB (Kakani, 1987; Stillwell, 1989; Ginsberg, 1989). Under these circumstances testing will necessarily become a more demanding exercise (O'Grady, Kim and Young, 1991).

The purpose of testing is to detect flaws in the PWBs as early as possible in the manufacturing process since early detection will reduce the cost of

J. S. Gero (ed.), Artificial Intelligence in Design '92, 119–137.

rectifying the fault. Typical faults can be associated with interconnections (including shorts, opens, and open solder joints), components (including mis-oriented or missing components), the overall functionality of the PWB, or with the thermal/mechanical properties of the PWB (O'Grady, Kim and Young, 1991). Of these, it has been stated that the open solder joint is the primary fault source in PWBs (Hansen, 1989).

A loaded PWB can be tested by three main methods: a functional test, an in-circuit test (ICT), or by using Boundary Scan Architecture (BSA) (O'Grady, Kim and Young, 1991). The functional test is carried out by connecting test equipment to the edge connectors of the PWB and checking if the board carries out its intended functions. While this can give an indication of a pass or fail, functional testing does not identify the source of a fault. ICT by contrast does indicate the source of a fault. In ICT, the PWB is tested by using a "bed of nails" which is placed against the PWB. Electrical signals, passing between the PWB and the "bed of nails", test individual components. An ICT helps to ensure that there are no shorts or opens on the PWB and that the components function as intended. However, the increasing density of PWBs has resulted in a major problem in the accessibility of test points. This has prompted efforts to ensure that test considerations are incorporated directly into the components. One result of this has been the development of Boundary Scan Architecture (BSA) (adopted as IEEE Standard P1149.1). In the BSA approach a cell is placed within the component at each pin. This cell can be switched to allow the testing of the connections or the internal logic. Advantages of BSA over ICT include the avoidance of the need for test point access, and that the elimination of backdriving. There are, however, a number of hurdles that remain before BSA can be regarded as a viable test approach. These hurdles include the cost of BSA (due to the increased circuit size), and the low implementation rate of BSA. These hurdles mean that ICT will remain as the main PWB loaded board test method (O'Grady, Kim and Young, 1991).

We therefore have a situation where the increasing complexity and density of PWBs has led to difficulties in ensuring that a PWB can be economically tested by ICT and where the BSA approach, which can mitigate some of these difficulties, is unlikely to be fully implemented in the foreseeable future.

This paper proposes an approach to this problem by assessing the design during the design process and making recommendations on improvements that can be made to the design to improve it's testability by ICT. A PWB designed with testability firmly in mind will have fewer access difficulties and will consequently be easier to test using ICT. The format of this paper is as follows. First, in section 2, the overall architecture of the proposed design advising system, TEST, is explained with its role in the PWB design environment. Section 3 gives an overview of the constraint network approach which is the underlying methodology of TEST. Sections 4 and 5 give details of the architecture and implementation of TEST. Finally, an example operation of TEST is presented.

2. Test and the PWB Design Process

The design of a PWB can be considered to be divided into a number of categories [Fig. 1, adapted from Ohr (1990)]. The process is begun with a *schematic diagram*, which is a schematic capture of the circuit information. This schematic diagram would usually be obtained by a detailed consideration of the electrical and electronic characteristics desired in the PWB. The schematic representation is then analyzed (*engineering analysis*) to determine the initial electronic correctness of the design. The PWB designer then converts the symbols in the schematic diagram to real physical components (*packaging* or *component selection*). Since the schematic diagram specifies only the functionality of a component, several alternatives may exist for selecting an appropriate component and the PWB designer has to make a selection among those available. The selected components are then positioned on the board (*placement*) and the interconnections are routed (*routing*). The designer can then step back and check the physical attributes of the design (*overall analysis*).

TEST links into this design process by providing a detailed *analysis* of the design from the perspective of testability [Fig. 1]. For example, the designer may select a component during the *component selection* phase that would require special test equipment not currently available in the manufacturing facility. As another example, the design may not have access to suitable test points for ICT. TEST analyses the design and then provides feedback to the designer on improvements that can be made so as to improve the design from the perspective of testability. This analysis is carried out at each of the stages of *component selection, placement,* and *routing*. An *overall analysis* is also carried out to give a final check on the design. Advice is given to the designer on improvements that can be made to the design, and the designer can then revise the design using a CAD system. TEST is designed to take data from the CAD system. The overall design process with TEST becomes one of iterating through designs until a suitable design is produced. The final result is the production of *artwork* and *other documents* associated with a good design. The artwork is a scaled configuration of the PWB from which the master pattern is photographically produced (Lindsey, 1979). Supporting documents include part lists, drilling instructions, assembly plans, and test preparation data.

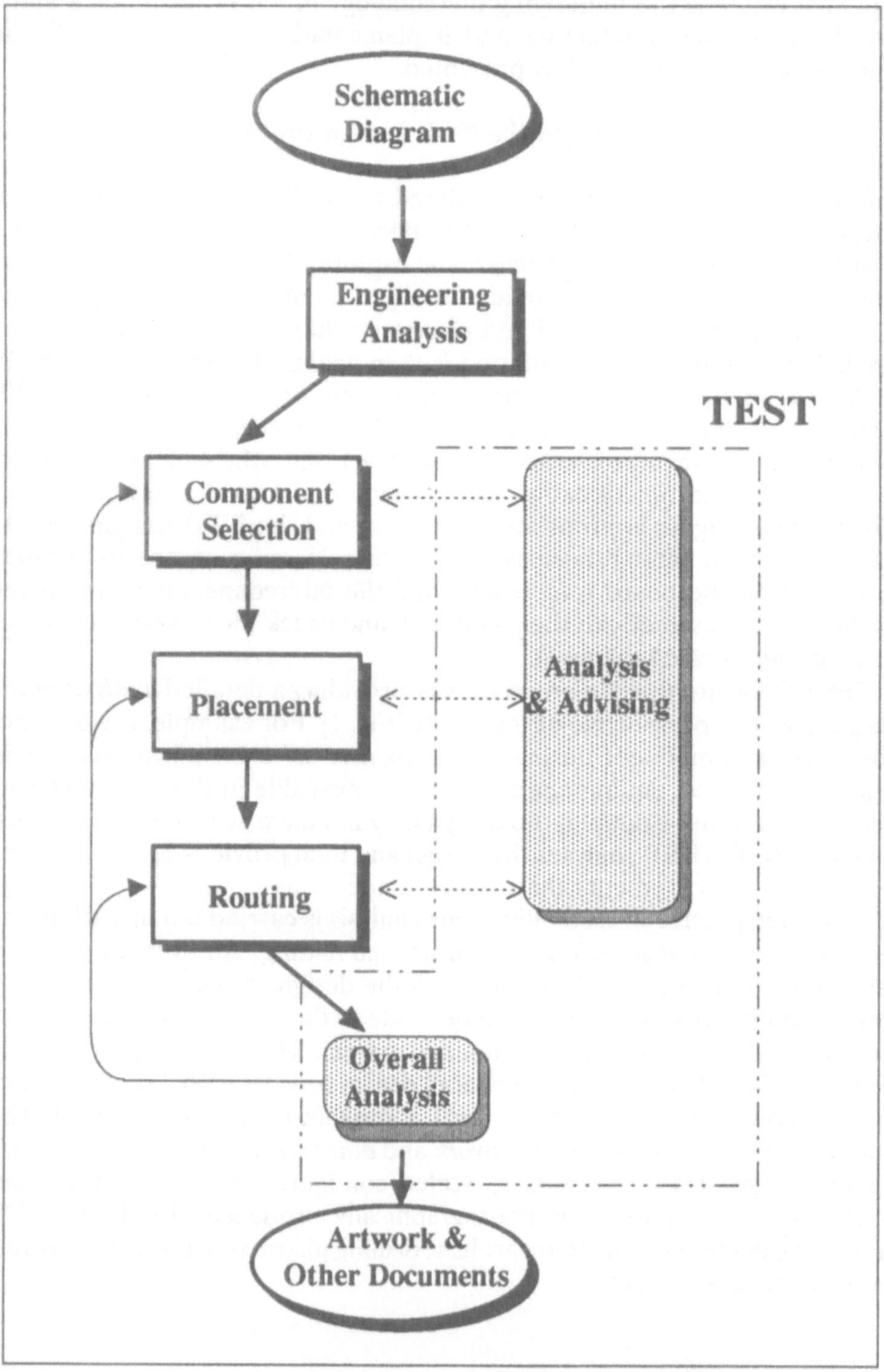

Fig. 1. Overall Design Process using TEST

3. Constraint Networks

Most design monitoring and advice systems developed to date use rule based expert systems. While there seems to be general agreement that such techniques can be useful, a cautionary note is sounded by Mayer and Lu (1988) who note some limitations to using the techniques. The first limitation is that the knowledge engineers may distort the knowledge. The second is that there is a wide variety of information in design analysis, including equations/procedures from engineering science, rules from accepted practice, intuition and judgment from personal experience, and information proprietary to the company and this wide variety of information may be difficult to represent (Shah and Wilson, 1988). The third is that the user typically plays a rather passive role in a knowledge based system.

These limitations are the result of a mismatch between the structural properties inherent in an expert system and the way designers attack problems. It is widely accepted that design is a constraint-based activity involving the recognition, formulation and satisfaction of constraints (Serrano and Gossard, 1987; Gross et al., 1987; Wu, 1988). This view of design has been applied to a wide variety of design domains including electrical circuits (Tong, 1987), mechanical design (Ishii et al., 1988) and VLSI design (Wu, 1988). Constraint based modeling is consistent with this viewpoint and has recently led to the concept of *constraint networks*, a constraint-based modeling approach which specifically aims to match the way designers attack problems. This problem approach views a set of constraints, linked through shared variables, as a conceptual network. Constraint networks represent relationships among variables using a logical form and is focused on finding a feasible solution.

Constraint networks trace their history from constraint satisfaction in which search techniques were developed to find optimal solutions to equation sets linked through shared variables (see Mackworth, 1987, for a discussion of constraint satisfaction). Examples include Sketchpad (Sutherland, 1963), ThingLab (Borning, 1979), IDEAL (Van Wyk, 1981), CONSTRAINTS (Sussman and Steele, 1980), and TK!Solver (Konopasek and Jayaraman, 1984), and CLP(R) (Jaffar and Lassez, 1987). However, the NP-complete nature of constraint satisfaction means that fully automated constraint satisfaction is impossible for most practical applications. A constraint-based system which does not require heavy computation is needed.

The recognition of the need for a constraint-based modeling system specifically tailored to design has led to the development of SPARK (Young, Greef, and O'Grady, 1991a, 1991b), which is a successor to CADEMA (O'Grady et al., 1988), and Galileo/LEO/Genesis (O'Grady et al., 1990; Bowen, O'Grady, and Smith, 1990). SPARK is a constraint network programming language, with frame-based inheritance. This allows users to model design advice systems as constraint networks, where the constraints are interconnected through shared variables. The objective of a SPARK session is to find a set of variable values that does not violate any of the constraints. Values are propagated bi-directionally through the network. The values are determined automatically by the system or are input by the user. Consequently, with a SPARK constraint network, the user is utilized *a priori* as part of the inference mechanism. From the user's point of view, SPARK provides a powerful environment to construct application-specific networks. It does not require structural programming, and the syntax is relatively straightforward. As with the constraint satisfaction approach, the existence of a solution is not guaranteed. It depends on the consistency of the constraints. If the constraints are reasonably determined, there will be high probability of a solution. In the current version of SPARK, run-time relaxation of a constraint is not allowed. However, one or a set of constraints can be easily deleted or re-inserted in the editor. The constraint network approach is a recent development; however it has been applied to a wide range of design and concurrent engineering problems across a wide range of industries (Kim, O'Grady, and Young, 1991; Liau and Young, 1991; O'Grady and Oh, 1991; O'Grady, Young and Greef, 1991; Oh, O'Grady and Young, 1991).

4. Architecture of TEST

The overall architecture of TEST consists of eight inter-linked constraint networks: Test Equipment Capacity, Test Vector Generation, Test Point Position, Component Selection, Component Placement, Distribution of Test Points, Size of Test Points, and Spacing Between Test Points. Each of the constraint networks contains constraints associated with the particular area of consideration and parameters [shown as ovals in Fig. 2] that link the constraints. A full description of the constraint networks is beyond the scope of this paper but description of the role of each of the constraint networks is as follows:

The <u>Test Equipment Capacity</u> Constraint Network contains the constraints associated with the limitations of the test equipment. Typical parameters include the maximum board width, maximum board length, production rate, and the multiplexing ratio (mux_ratio).

The <u>Test Vector Generation</u> Constraint Network contains the constraints associated with the generation of the test vector. The constraints include those imposed by the circuit logic, the backdrive limit, and the length of the test vector.

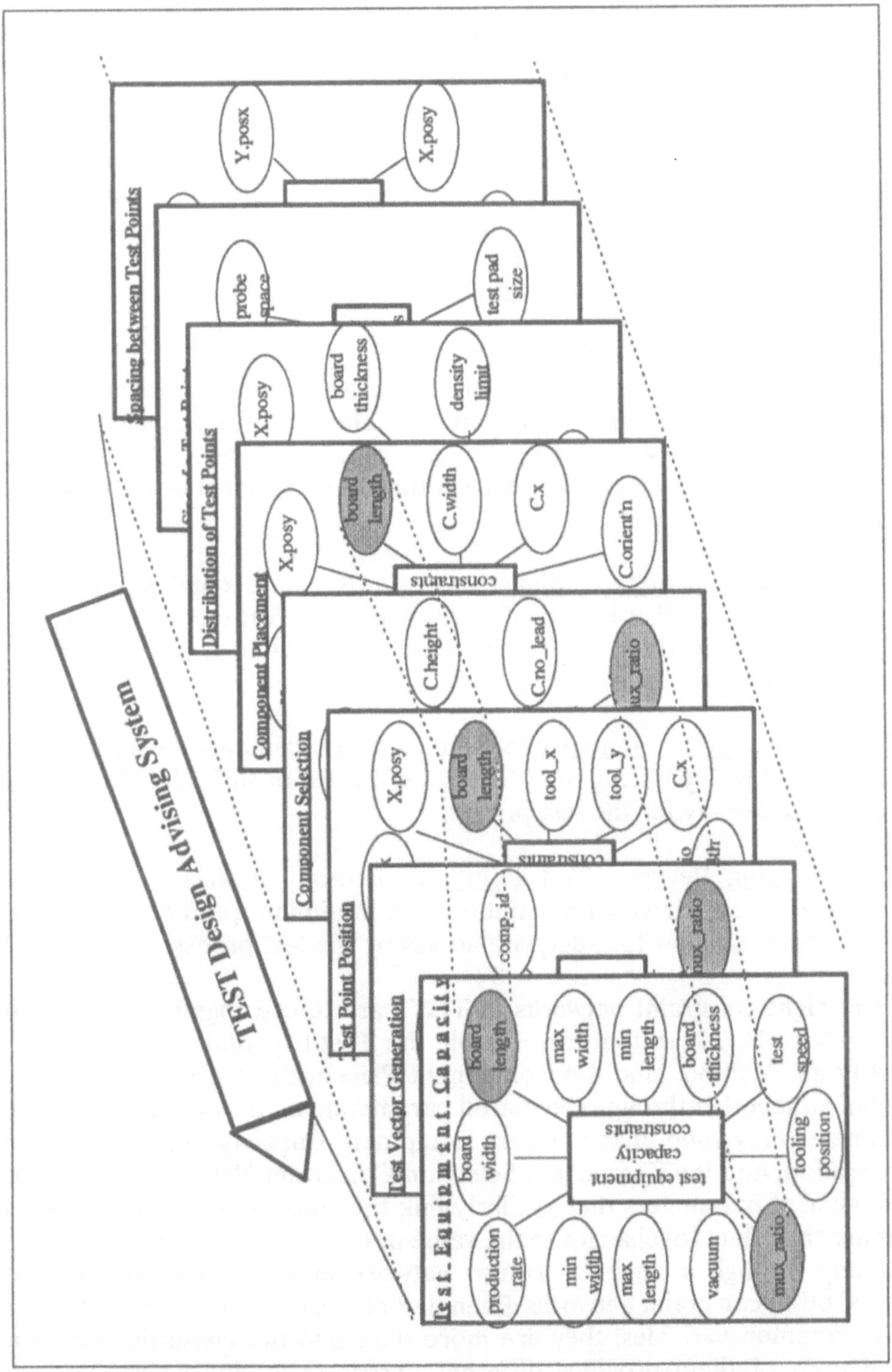

Fig. 2. Example Constraint Net Links in TEST

The Test Point Position Constraint Network contains the constraints that are associated with the position of the test point. These constraints include those imposed by the requirements for a minimum distance between the test point and the edge of the board, or for a maximum distance between a component pin and the test point.

The Component Selection Constraint Network contains the constraints associated with the desire to limit the number of varieties of components on the PWB. Increased variety can raise costs, increase lead times, increase the complexity of testing, and cause manufacturing and testing difficulties.

The Component Placement Constraint Network contains the constraints that limit the position of components. The constraints include those associated with maintaining a minimum distance between the components, and those that ensure that the components are correctly oriented.

The Distribution of Test Points Constraint Network contains constraints surrounding the distribution of the test points on the PWB. If the test points are grouped together, the stresses imposed during ICT can distort, and possibly damage, the PWB.

The Size of Test Points Constraint Network contains the constraints associated with ensuring that the test points are of sufficient size to give adequate leeway for the test probes.

The Spacing Between Test Points Constraint Network contains the constraints associated with maintaining a minimum spacing between the test points to allow for adequate access by the test probes.

The eight constraint networks in TEST are linked together by sharing parameters [Fig. 2] so that for example the Test Equipment Capacity, the Test Point Position, and the Component Placement Constraint Networks are linked together through the shared parameter *board length*. Similarly the parameter *mux_ratio* links the Test Equipment Capacity, the Test Vector Generation, and the Component Selection Constraint Networks. There are a number of parameters that similarly link the constraint networks. In this manner the effects of changes to the value of one parameter may propagate not only through a single constraint network in TEST but also through several other constraint networks. Even though the eight constraint networks share common variables, they are more than one flat constraint network since each of them provides different perspectives, and when they are immigrated into a multi-user platform, they can provide multitasking environment.

5. Implementation

TEST has been implemented in SPARK (Young, Greef, and O'Grady, 1991a, 1991b). Design parameters of components and test points are represented as structures, making use of the inheritance property of SPARK. Other design parameters are included as SPARK variables. The boards which TEST deals with at present are double-sided, single layered, and populated with both surface-mounted and plated-through hole devices. However, the TEST system can be easily extended to cover other types of PWBs.

5.1 Data Structures

Components and test points are represented within TEST as *structures*, which are frame-like data structures containing such slots as an identification number, a component name, location, dimensions, and mounting side [Fig. 3]. The dimensions and location are represented as separate substructures, with these substructures being automatically linked to the main structure by an inheritance mechanism. Multiple test points can be assigned to one component, with the usual arrangement of the number of test points of a component being the same as the number of output pins of the component. The structure of a test point is linked to the component structure by the component number slot [Fig. 3].

This can be directly represented in SPARK as a structure, so that, for example, the component and testpad structures are represented as:

```
STRUCTURES

component[
    'component number'(compno):number,          /* component ID no.*/
    'device type'  (devtyp):device_type,         /* SMT or PTH device */
    'device name'  (devname):device_name,        /* device name */
    /* inherited */ position :comp_pos,
    /* inherited */ dimension:comp_dim,
    'rotation'     (rotn) :rotation,             /* mounting rotation */
    'mounting side' (side):board_side ].         /* mounting side */

testpad[
    'test pad number' (padno),                   /* test pad ID no.*/
    'position X'   (posx),                       /*  x coordinate */
    'position Y'   (posy):number,                /* y coordinate */
    'test pad type' (tptype):pad_property,
```

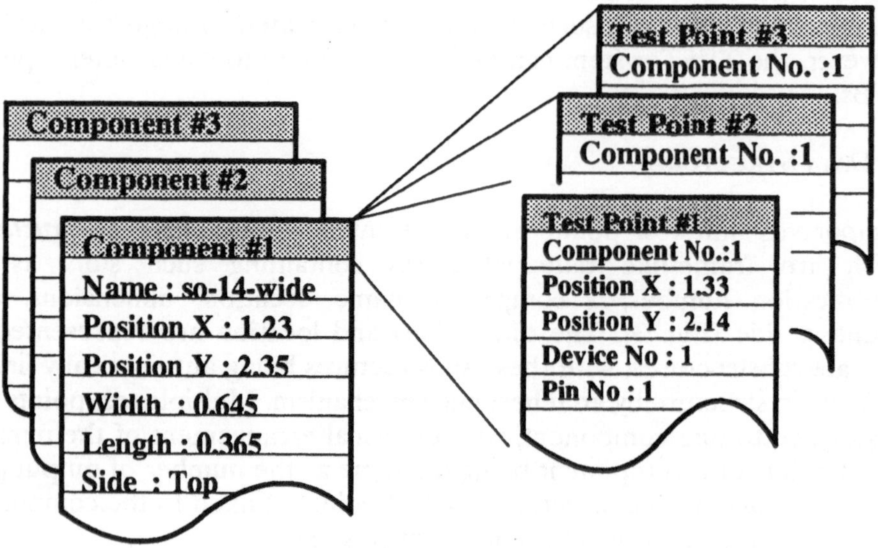

Fig. 3. Data Structure of TEST

```
'test pad shape'(tpshp):pad_shape,
'test pad size' (tpsize):number,
'probing side'  (tpside):board_side,          /* board side of test pad */
'device number' (devno),                      /* device ID no.of test pad */
'pin number'    (pinno):number ].             /* pin ID no.of test pad */
```

5.2 KNOWLEDGE BASE

The knowledge base of TEST contains the reference data for the electronic components. This is kept in each constraint network in TEST to allow for ready access by the constraint network. The reference data is comprehensive and includes, for example, the information on the allowable domains for a component as well as the relationship between the component and its attributes. For example the relationship between the component type, name, width, length, height, number of pins, and pin offset is expressed in the Component Selection Constraint Network in TEST as a *relation* as follows:

RELATIONS

```
compatible2(device_type,device_name,device_width,device_length,device_height,pin_
number,pin_offset) =
{      (plated_thru,dip_30,1.250,0.270,0.125,24,0.036),
       (plated_thru,dip_60,1.050,0.540,0.105,20,0.054),
       (surface_mounted,sot_23_high,0.115,0.051,0.034,3, 0),
         |
         |
       (surface_mounted,pcc_28,0.450,0.450,0.173,28,0),
       (surface_mounted,pcc_44,0.650,0.650,0.173,44,0),
       (surface_mounted,pcc_68,0.950,0.950,0.173,68,0),
       (surface_mounted,pcc_84,1.150,1.150,0.173,84,0),
       (surface_mounted,pcc_124,1.650,1.650,0.190,128,0) }.
```

5.3 CONSTRAINTS

The prior section on the architecture of TEST gives an overview of the constraint networks that form TEST. A wide variety of constraints are included in TEST to represent the mutually constraining influences that are integral to the design. These constraints are taken from discussions with industry personnel, from standards, and from Kakani (1987), Bullock (1987), and Barnes (1983). Some of these constraints that relate to the Size of Test Point Constraint Network and to the Spacing Between Constraint Network are shown in Table 1.

TABLE 1
Example Constraints from the Size of Test Point Constraint Network and from the Spacing Between Constraint Network

CONSTRAINT
The edge length of a square test pad should be between 0.035 and 0.060 inch.
The diameter of round test pads should be between 0.040 and 0.065 inch.
Each test pad should be apart from the board edge by more than 0.125 inch.
Each component should be apart from the board edge by more than 0.125 inch.
The spacing between any two test pads should be more than 0.1 inch if they are on the same side of a board.
Each test pad should be apart from any SMT component by at least 0.2 inch if the component is more than 0.2 inch tall.
Each test pad should be apart from any SMT component by at least 0.15 inch if the component is less than 0.2 inch tall.
The test pad should be within 0.3 inch of the component border if they are on the same side.
Each test pad should be within 0.1 inch of the component border if they are on opposite sides.
The test pad side should be on the opposite side to the component side for plated through hole components.
The holes of a plated through hole component are to be used as test pads.

These can be shown graphically in constraint network form as shown in Fig. 4 (this is actually a portion of the Test Point Position constraint network).

An example of a constraint, relating to the spacing between two test pads, written into TEST is as follows:

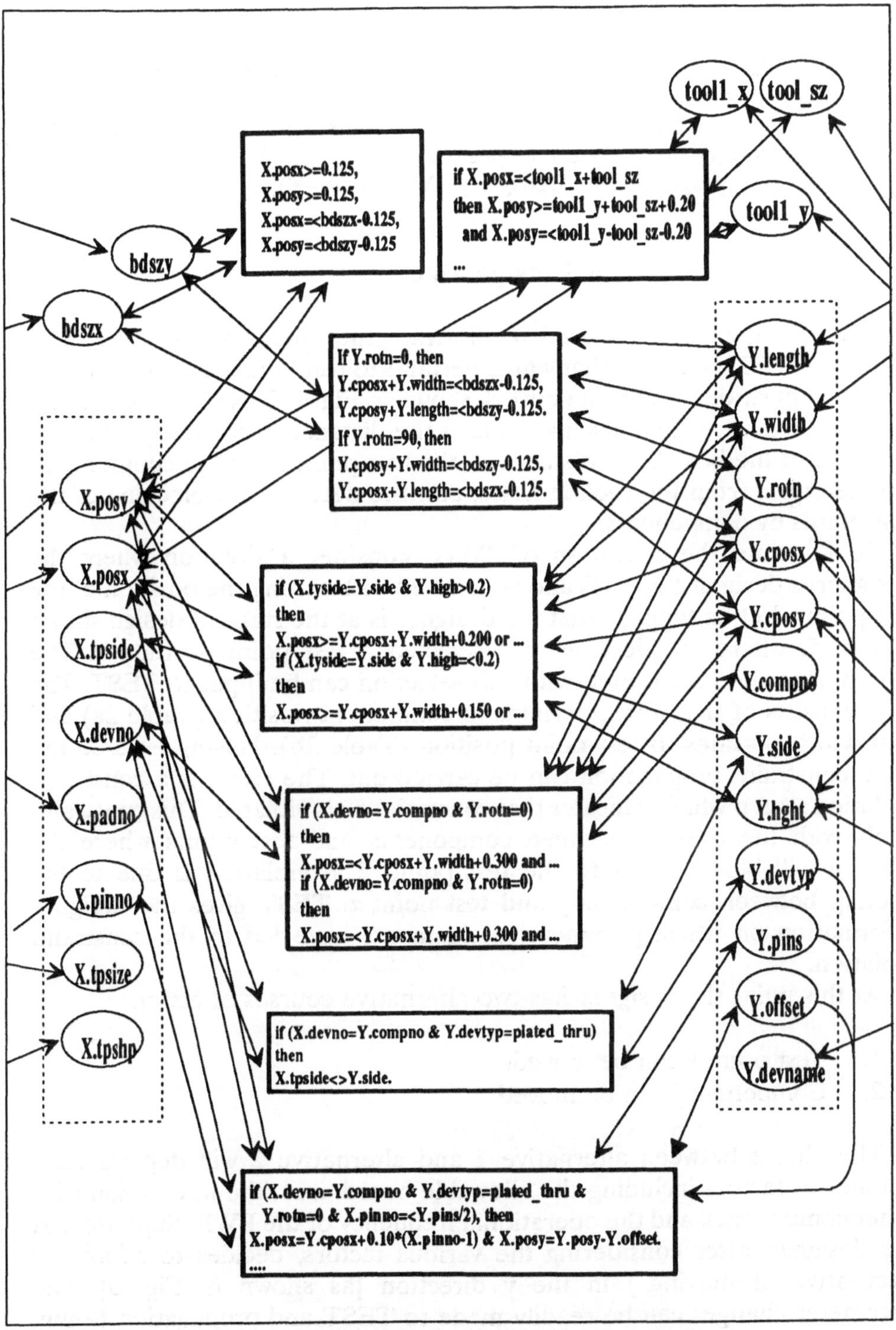

Fig. 4. Test Point Position Constraint Network

```
All X,Y (testpad(X),testpad(Y)) [
/* spacing between two test pads X and Y */
41          [0,5]     ((X.tpside=top & Y.tpside=top) or (X.tpside=bottom &
        Y.tpside=bottom)) = = >
            ((((abs(X.posx-Y.posx)=<0.100) or ((abs(X.posy-Y.posy)=<0.100))) or
            (sqrt((X.posx-Y.posx)*(X.posx- Y.posx)
            +(X.posy-Y.posy)*(X.posy-Y.posY)) >= 0.100))).
]
```

6. Example Operation

The following example is used to illustrate the operation of TEST. The portion of TEST shown is that which pertains to test point position, with the test point position constraint network shown in Fig. 4. This network includes constraints that relate to such aspects as the distance of the test point from the edge of the board, the distance of the test point from the components, the clearance from the tooling holes, and the detection of blocking of the test points by components.

To illustrate the operation of TEST, consider a situation where the designer is designing a PWB and is involved with setting the position of the test points. Let us suppose that the designer is at the stage of design shown in Fig. 5, where the designer has selected a test point position *x*. The parameter values associated with this selection can be input to TEST. The data consists of that which pertains to component position (Table 2a) and that which pertains to test point position (Table 2b). Having entered the data, constraint propagation can be carried out. The result is a constraint violation screen which indicates that constraint 52 is violated. This constraint deals with the overlap between components and test points where the overlap will block access to the test point. The violation is due to the overlap between component *j* and test point *x*. TEST gives the designer information on which parameter assignments have led to the constraint violation.

At this point the designer has two alternative courses of action:

1. Test point *x* can be moved.
2. Component *j* can be moved.

The choice between alternative 1 and alternative 2 will depend on a number of factors including the allowable board area, the interaction with other components, and the operational frequency of the PWB. Suppose that the designer, after considering the various factors, decides to adopt the alternative of moving *j* in the y direction [as shown in Fig. 5]. The parameter changes can be readily made to TEST and propagation begun. With the move of component *j*, the constraint violation is removed and no

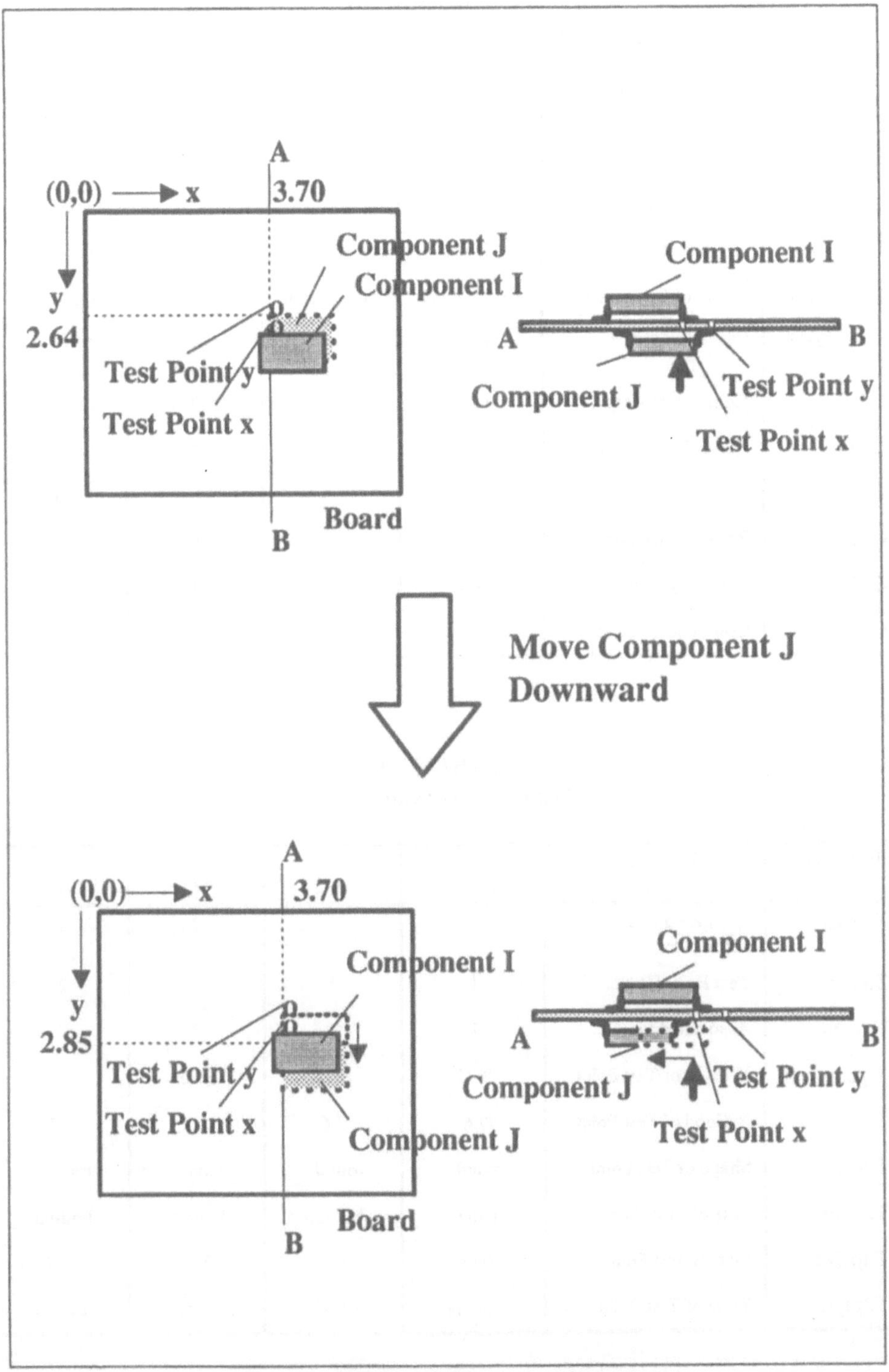

Fig. 5. TEST Operation Example

TABLE 2 (a)
Data on Component Position

Parameter	Description	comp1 (No)	comp1 (Yes)	comp2 (No)	comp2 (Yes)
C.compno	Component ID No.	1	1	2	2
C.cposx	X Cood of Left Upper	3.5	3.5	3.7	3.7
C.cposy	Y Cood of Left Upper	2.75	2.75	***2.64***	***2.85***
C.devname	Name of Component	so_14_wide	so_14_wide	so_16_wide	so_16_wide
C.devtype	Package Type	surface_ mount	surface_ mount	surface_ mount	surface_ mount
C.hght	Component Height	0.098	0.098	0.098	0.098
C.length	Component Length	0.295	0.295	0.295	0.295
C.pins	No.of Pins	14	14	16	16
C.rotn	Rotation (Degree)	0	0	0	0
C.side	Mounting Side	top	top	bottom	bottom
C.width	Component Width	0.354	0.354	0.404	0.404

TABLE 2 (b)
Data on Test Point Position

Parameter	Description	tp1 (No)	tp1 (Yes)	tp2 (No)	tp2 (Yes)
T.devno	Component ID it belongs to	1	1	1	1
T.padno	Test Point ID No.	1	1	2	2
T.pinno	Associated Pin No.	1	1	2	2
T.posx	X Cood of Test Point	3.75	3.75	3.75	3.75
T.posy	Y Cood of Test Point	2.6	2.6	2.7	2.7
T.tpshp	Shape of Test Point	round_	round_	round_	round_
T.tpside	Side of Test Point	bottom	bottom	bottom	bottom
T.tpsize	Size of Test Point	0.04	0.04	0.04	0.04
T.tptype	Type of Test Point	art_via	art_via	art_via	art_via

violation screen appears.

This example has shown how the designer may select an undesirable test point position and how TEST can detect this, thereby guiding the designer towards designing a PWB that can be readily tested. One possible improvement would be that the system recommends possible values. However, the recommended set of values may lead to another inconsistency at another part of the constraint network, although it may satisfy the violated constraints. Thus, it would be more helpful to show the designer which constraints are violated, rather than recommending locally satisfying solution.

7. Summary and Conclusion

This paper has presented a new approach to Design for Testability by placing the constraining influences from the testing operation into a constraint net form. The architecture of a comprehensive Design for Testability system, called TEST, has been presented. TEST has been implemented using the SPARK constraint net system. The implementation and operation of TEST have been described. The overall result is a powerful system that gives advice to the designer on improvements that can be made to the design of the PWB so as to improve its testability. Specific advantages of TEST are:

- it is flexible enough to be expanded in a relatively straightforward fashion,
- it naturally aligns with the overall design process used by the designer without restricting the designer to follow a fixed path,
- advice is given on improvements that can be made to the design,
- TEST is bi-directional so that, for example, a designer can start at a desirable feature of a design and determine what parameters need to be changed to obtain that feature or, alternatively, the designer can start at the parameters and determine the resulting feature,
- other models, including cost models and manufacturing constraint models, can be readily added.

TEST is currently being re-implemented in SATURN (Young, Greef, Murali, and O'Grady, 1991). SATURN is a constraint modeling environment that has several major advantages over other constraint network systems including the inherent ability to represent hierarchies.

Arabian, J.: 1989, *Computer Integrated Electronics Manufacturing and Testing*, Dekker.

Barnes, R.N.: 1983, Fixturing for surface mount devices, *IEEE International Test Conference.*

Borning, A.: 1979, ThingLab—A constraint-oriented simulation laboratory, *Stanford Technical Report STAN-CS-79-746.*

Bowen, J., O'Grady, P. and Smith, L.: 1990, A constraint programming language for life-cycle engineering, *International Journal for Artificial Intelligence in Engineering*, **5**(4).

Bullock, M.: 1987, Designing SMT boards for in-circuit testability, *IEEE International Test Conference.*

Ginsberg, G.: 1989, *Surface Mount and Related Technology*, Dekker.

Gross, M., Ervin, S., Anderson, J., Fleisher, A.: 1987, *Designing with Constraints. Computability of Design*, John Wiley, USA.

Hansen, P.: 1989, Testing conventional logic and memory clusters using boundary scan devices as virtual ATE channels, *IEEE International Test Conference Proceedings.*

Jaffar, J. and Lassez, J.: 1987, Constraint logic programming, *Proceedings of the Fourteenth Annual ACM Symposium on Principles of Programming Languages*, pp. 111-119.

Kakani,T.: 1987, Testing of surface mount technology boards, *IEEE International Test Conference.*

Kim, J., O'Grady, P. and Young, R. E.: 1991, Feature taxonomies for rotational parts: a review and proposed taxonomies, *International Journal of Computer Integrated Manufacturing* (to be published)

Konopasek, M. and Jayaraman, S.: 1984, *The TK!Solver Book*, Osborne/McGraw-Hill, Berkeley, CA.

Liau, J. S. and Young, R.: 1991, Applying processing knowledge to printed circuit board design, *Technical Report*, Department of Industrial Engineering, North Carolina State University.

Lindsey, D.: 1979, *The Design and Drafting of Printed Circuits*, Bishop Graphics,Inc.

Mayer, A. K. and Lu, S. C-Y.: 1988, An AI-Based approach for the integration of multiple sources of knowledge to aid engineering design, *Journal of Mechanisms, Transmissions, and Automation in Design* **110**(3), 316-323.

Mackworth, A.: 1987, Constraint satisfaction, *in* S.Shapiro (ed), *The Encyclopedia of Artificial Intelligence*, Wiley, NY, pp. 205-211.

O'Grady, P., Kim, C. and Young, R. E.: 1991, Issues in the testability of printed wiring boards, *Technical Report*, Department of Industrial Engineering, North Carolina State University.

O'Grady, P., Young, R. E., Greef, A. and Smith, L.: 1991a, An advice system for concurrent engineering, *International Journal of Computer Integrated Manufacturing*, **4**(2), 63-70.

O'Grady, P. and Young, R. E.: 1991, Issues in concurrent engineering systems, *Journal of Design and Manufacturing* (to be published).

O'Grady, P., Bowen, J. and Smith, L.: 1990, Advice systems for life-cycle engineering, *Technical Report*, Department of Industrial Engineering.

O'Grady, P. J., Ramers, D. and Bowen, J.: 1988, Artificial intelligence constraint nets applied to design for economic manufacture and assembly, *Computer Integrated Manufacturing Systems* **1**(4).

Oh, J., O'Grady, P. and Young, R. E.: 1991, An artificial intelligence constraint network approach to design for assembly, *Technical Report*, Department of Industrial Engineering, North Carolina State University.

Ohr, S.A.: 1990, *CAE: A Survey of Standard Trends and Tools*, Wiley.

Serrano, D. and Gossard, D.: 1987, Constraint management in conceptual design, *in* D. Sriram and R. Adey (eds), *Knowledge-based Expert Systems in Engineering: Planning and Design*, Computational Mechanics Publications, pp. 211-224.

Shah, J. J. and Wilson, P. R.: 1988, Analysis of knowledge abstraction, representation and interaction requirements for computer aided engineering, *in* V. A. Tipnis and E. M. Patton (ed.), *Computers in Engineering*, ASME, pp. 17-24.

Stillwell, H.R.: 1989, *Electronic Product Design for Automated Manufacturing*, Dekker.

Sussman, G. and Steele, G.: 1980, CONSTRAINTS—A language for expressing almost-hierarchical descriptions, *Artificial Intelligence*, **14**, 1- 39.

Sutherland, I.: 1963, SKETCHPAD: A man-machine graphical communication system, *IFIP Proceedings of the Spring Joint Computer Conference.*

Tong, C.: 1987, Toward an engineering science of knowledge-based design, *Artificial Intelligence in Engineering*, **2**(3), 133-166.

Van Wyk, C.: 1981, IDEAL Users Manual, Bell Labs, *Computer Science Technical Report 103*.

Wu, P.: 1988, Design for testability, *Proceedings, National Conference of the American Association for Artificial Intelligence*, pp. 358-363.

Young, R. E., Greef, A. and O'Grady, P.: 1991a, Spark: An artificial intelligence constraint network system for concurrent engineering, *in* J. S. Gero (ed.), *Artificial Intelligence in Design '91*, Butterworth-Heinemann, Oxford.

Young, R. E., Greef, A. and O'Grady, P.: 1991b, An artificial intelligence constraint network system for concurrent engineering, *International Journal of Production Research* (to be published).

3

COOPERATIVE DESIGN

MODELLING HETEROGENEOUS ENGINEERING KNOWLEDGE AS TRANSACTIONS BETWEEN DELEGATING OBJECTS

J. ZUCKER and A. DEMAID
Knowledge Based Systems in Engineering Research Group
Faculty of Technology
The Open University
Milton Keynes MK7 6AA UK

Abstract. This paper details features of a knowledge representation scheme useful to modelling a design world of discourse. A model is produced by factoring the world of engineering design into separate information structures — contexts or "perspectives" — which may be complex or simple, short- or long-lived.

Our software technique maintains the organizational independence of these design perspectives by keeping them separately represented. The problem is that separate perspectives overlap in their contents, and may be mutually consistent or inconsistent to varying degrees. This is known as the multiple perspectives problem.

Solving this problem is made difficult by the message-interpretation strategy of the more commonly available object-oriented languages — i.e. inheritance systems oriented to instances of classes. To populate a conceptual model of design, our own knowledge programming language, Splinter, adopts inheritance to convey a process of specialization (continual refinement) between objects which are *not* constructed as members of classes. In addition to inheritance, Splinter uses a communication process which is derived from the scheme of object-to-object message delegation, as devised by the Actor languages and Actor model of computation. Delegation is used in our representational scheme to model the process by which communication takes place between separate perspectives.

The paper describes features for evolutionary design of perspectives, for complex modelling and for combining multiple perspectives. The paper's contention is that evolution- and communication-based software architectures are more adequate to the dynamic world of design representations than the template-based object-oriented languages currently available.

1. Introduction

This paper describes features of a knowledge representation scheme adequate to the complexity and heterogeneity of engineering-design information. Complexity is largely a question of the compound nature of engineering information, as evidenced by a movement to enhance the semantic features of engineering databases (Hartzband, 1985). Heterogeneity arises in practical knowledge engineering from the different specialist analyses which may be applied to a single domain. In the design of engineering products it is exemplified by the frequent need in the literature to specify contexts within which to compare like with like and so investigate detail differences. (See, e.g.: French, 1971.)

J. S. Gero (ed.), Artificial Intelligence in Design '92, 141–159.

This search for context applies to conceptual design as well as to variational design. For example, it is possible to show all the kinematic routes to a rotational internal combustion engine on one graph. The mapping of kinematic routes to component groupings within rotary engines provide one way of classifying "family resemblances" with performance implications. In general, combinatorial approaches are at their most powerful when the context is mathematically clean, where they can be shown to support conceptual design decision making at a high level. Variational design provides the context of a previously designed component or component-class within which a designer works with confidence.

We expect a computerized world of design information to be populated with appropriate orderings and groupings: for example, manufacturing perspectives, materials perspectives, functional perspectives, and so on. Our modelling of materials information proceeds by factoring the knowledge into materials families according to strict physical domain principles (Demaid and Zucker, 1988). Inevitably the groupings we describe are not the product of one principle, conveniently embodied in a programming language, rather they are ad hoc associations of varying longevity, complexity, and power. An association might be represented in a deep, rich hierarchy, such as the chemical groupings of different materials which are expected to support a number of different product applications, or they might be formed on the fly to explore a particular route to an artefact's manufacture and be garbage collected when that purpose has been fulfilled.

We postulate therefore a world of design information consisting of many, different sized groupings. This can be modelled by an object-oriented technique which possesses the representational adequacy to create a model containing separate hierarchical representations for distinct empirical concerns in a product engineering domain. Such distinct information structures can then be made to interrelate, dynamically, without mixing of perspectives. Our core thesis is that such different perspectives must be maintained as separate partitions of the one knowledge base and allowed to relate through the policy of message delegation.

2. Engineering design and object-oriented modelling

Engineering designers are known to be prolific users of engineering information, processed through "elaboration and exploration as opposed to any specific, well-formulated end" (Newton, 1989). In the realm of computing theory, techniques of object-oriented programming and object-oriented information modelling hold promise as a vehicle for more adequately addressing the information needs of an engineering designer.

Objects, as modular devices in a computational system which provide appropriate services in response to messages which they are able to interpret, form the basis of a programming architecture (Rentsch, 1982) and, increasingly under commercial development, a database architecture (Dittrich and Dayal, 1986). In general, there are three main features which are often held to be essential to object-oriented programming:

- The ability to construct objects as a set of operations and a memory.
- The classification of objects, i.e. each object as an instance of a class.
- An inheritance mechanism defining superclass-subclass relationships.

These features form the "classification paradigm" proposed by Wegner (1987) .

Although inheritance systems **oriented to instances of classes** inform the model adopted by the better known object-oriented software mechanisms, the next section explains that this is ***not* the scheme adopted** in the language, Splinter, implemented by Zucker (1989).

Object-oriented software systems are advantageous for modelling engineering design activities because of their support for complex relationships and evolutionary processes. The strongest example of complex relationships is the provision of nested (descriptions of) properties of objects so that properties of objects are, themselves, recursively composed of subordinate objects. This problem is at the forefront of representing the design process in CAD and CAE, since in design it is the general rule that components are composed of subcomponents (Brown, 1991).

The design process is characterized by continually evolving specifications which need to be managed efficiently. The ability to change behaviours and property-structures raises some thorny questions for most object-oriented mechanisms, which require that classes of object be re-defined for the objects' property-structures to be mutated. However, the interactive modification of classes and automatic updating of their instances (involving migration of data values), is a capability which several object-oriented mechanisms promote. In the Smalltalk-80 programming language, for example, interactive modification of class definitions is a feature of the language's support environment (Goldberg and Robson, 1988). In the field of object-oriented databases, a corresponding system of "dynamic schema evolution" has been championed by Banerjee et al. (1987). Splinter has strong complex-object modelling and evolutionary capabilities, in support of design reasoning (Demaid and Zucker, 1992a).

3. A representation scheme for complex domain knowledge

Wegner's classification paradigm rests on the identification of object-oriented programming with special systems featuring default property inheritance and based on instances of classes which connote real-world objects in the knowledge domain. Classes act as **templates**. They control the formats of instances and govern the sharing of property information according to a class ordering.

Several mechanisms for object-oriented programming, however, do *not* adopt the approach that generalized knowledge should be conveyed by classes. The

Actor model of Hewitt et al. (1973) is the earliest example: some later mechanisms are due to Lieberman (1986), Borning (1986), Lalonde et al. (1986) and Ungar and Smith (1987). These adopt a **class-free knowledge representation scheme** in which objects are constructed through refining property-structures of existing objects, rather than through "instantiating" a definition of a class of such objects. (For a survey treatment of these concepts see: Nierstrasz, 1988).

The particular class-free representation scheme implemented in our own programming language Splinter is a scheme for **inheritance between propertied objects**. The scheme allows the programmer to describe inheritance as taking place directly between object-representations.

Specifically, Splinter's class-free *default* property inheritance corresponds to the principle that any given object has a property structure, no matter how general or special, and an inheritance link (Touretzky, 1986) may be formulated as a specialization relation on two such propertied objects. The properties of one object, in this relation, are derived from the properties of the preceding object, according to a description of the more specialized object as inheriting the property-structure of the original object, but with possible property modifications. (Modifications are interpreted by the language as subject to consistency constraints, as discussed in the final paragraphs of our section 5.) The object which was previously the more specialized object may now, in turn, be used to generate an even more specialized object. A Splinter knowledge base is a hierarchical inheritance structure grown through successively describing specialization relations upon objects (Zucker and Demaid, 1992a).

Two essential features of object-oriented *modelling* are the same in either classification-based or class-free representional styles:

- Objects may convey abstract or concrete domain concepts (following standard practice in object-oriented modelling).
- Any object embodies operations and a memory which it may use to respond to messages which it receives (by theoretical definition).

The essential feature of the class-free style, in distinction to the classification-based style, is that there are *no* templates to be described *separately* from each object. The fundamental differences between classification-based and class-free theories of object orientation were at one stage disputed in the literature but have since been clarified, by consensus: see Stein et al. (1988). Our paper reports some particular experience in object-oriented knowledge description and modelling in the class-free tradition.

Models composed of instances generated from classes differ from class-free models in their strategies for accommodating **evolutionary change**. In the class-free representational style, changes are made directly to an object and those changes are *automatically* respected by existing objects which inherit its properties. In a system of instances of classes, however, changes to a class definition have to be enforced over instances of the affected class and of its subclasses.

The use of Splinter to program evolutionary knowledge bases is illustrated in Zucker and Demaid (1992b). Splinter's grammatical structure consists of the Scheme language embedded alongside a "postal system" for message-control. It also extends Scheme by adding "objects" and "descriptions" as first-class datatypes (Zucker, 1989). Issuing messages to nominated objects is syntactically straightforward: the nominated object is to the left of a colon symbol, **:**, and the message is any legal expression to its right. Such a legal expression is free to mention the names of the receiver's properties and perform calculations.

It is paramount that we may express **complex objects**, i.e. objects having interdependent attributes, in a knowledge description language for engineering design. The problem is that design entities often consist of several assembled component entities or parts, each of which can consist of several sub-assemblies.

4. A complex design perspective

For illustrations' sake, we consider the problem of representing the basic designs of polymer-bodied electric kettles. We shall see that the evolution of designs of this engineering artefact belie the apparent simplicity of a complex modelling mechanism.

We begin with a representation of a plastic kettle. From this representation we elaborate a more specific variant of 'complex bodied kettle', which is complex in that its 'handle' property becomes (referenced as) a property of its body.

```
➡ (plastic_kettle :
    (a container                          ;;container might be an object in a hierarchy of
                                          ;;product designs
      with
       handle (a structural_artefact with material)
                                          ;;structural_artefact might be an object with
                                          ;;mechanical properties such as stiffness and strength
       heating_element  (an electrical_device)))
                                          ;;electrical_device will be visualized in section 6

➡ (complex_bodied_kettle :
    (a plastic_kettle                     ;;container has already been described as having a
      with                                ;;property named body, so the property description
       body                               ;;is now being refined
        (a moulding
              with
               handle == (handle for complex_bodied_kettle))))
                                          ;;this uses the language's co-reference operator ==
                                          ;;with the effect that complex_bodied_kettle's
                                          ;;handle property is also known as the handle
                                          ;;property of its body
```

As an immediately useful refinement to this model, we require that the handle material and the body material are the same thing. This can be programmed:

```
➡ (body for complex_bodied_kettle :
    (material for handle) == (material for body))
```

It will be apparent that the above descriptions `(a container with ...)` and `(a plastic_kettle with ...)` achieve two purposes as part of the object-oriented language mechanism. First, they convey the particular properties special to each concept, and the properties of those property-objects, and so on. Second, the series of descriptions accumulates into a hierarchy of objects, where one object augments and specializes the properties of the object it is derived from. Such an arrangement implements the representational scheme of single ("one-parent") default property inheritance between objects.

The engineering designer, who is the potential builder of a knowledge base concerning kettles, knows that plastic kettles are composed from obvious sub-components: bodies, handles, lids, electrical switch covers, and leads. In our representation scheme, each sub-component of `plastic_kettle` is an independent entity which can, in principle, be designed according to its own individual specifications of strength, stiffness, colourability, electrical resistance, and so on.

We allow that, at any stage in the engineer's elaboration of domain concepts, properties may be described as expressing literal values, either behavioural (algorithmic) or direct mappings onto a single state (e.g. a number). If a property has not yet been described as value-expressing, it is still an object: i.e. it is roughly equivalent to an "entity pointer" in frame languages (or to an "unbound slot" if it does not, itself, have properties). Complex objects arise where the private data space memorized by a given object may be further subdivided into further name-property associations.

As complex designs develop further, however, so the representation of simply divisible name-property associations turns into a considerably more difficult modelling problem for realistic engineering-design domains.

For example, there have been two different stages in the commercial development of the plastic jug kettles. We have already discussed the first design, in which the handle and the body were produced from one, complex injection moulding. As separate entities the handle and the body are differentiated by the need for the body to withstand boiling water, a severe environment for a plastic. Both items have similar structural requirements, described in terms of stiffness and strength. Aesthetic requirements are also somewhat similar, with the much larger body moulding dominating.

Once a complex object is created which contains the body and the handle then the specifications become interrelated. Essentially the specifications become the highest common denominator, so the handle must be manufactured from a polymer capable of resisting exposure to boiling water. We have illustrated such a design as represented by `complex_bodied_kettle` in which both the body and its handle are ascribed the same material.

The second design in the commercial evolution of the jug kettle combined the handle with the switch cover, leaving the body as a separate entity. The structural and electrical properties which are combined do not give rise to radically different specifications for the material requirements as polymers are inherently good insulators and the structural integrity of the handle is achieved through geometry rather than absolute material properties. We analysed have two conceptually different design approaches to plastic-kettle production:

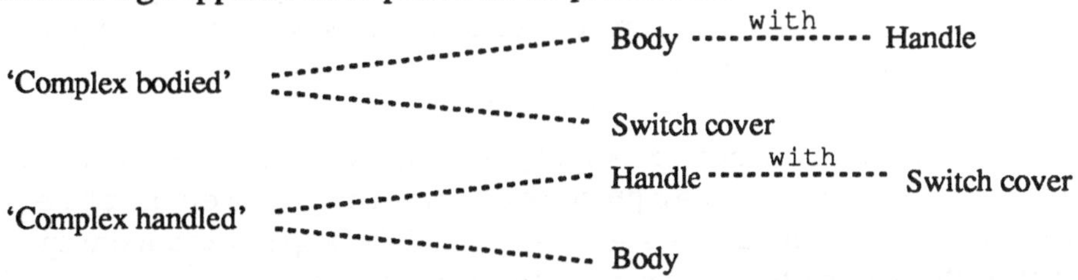

The two historic designs are dichotomous representations in the sense that the two concepts convey different property structures rather than variant refinements. We therefore choose to depict the second design concept, `complex_handled_kettle`, as a fresh variant of `plastic_kettle`.

```
➡ (complex_handled_kettle :
    (a plastic_kettle
       with                    ;;plastic_kettle has already been described with
        handle                 ;;handle, so the property is now being refined
           (an Object
                with switch_cover)))
```

The outstanding problem, in this basic knowledge-description exercise, is that we have still not accounted for switch covers of the earlier design of kettles. Conceptually there are two, subtly different possibilities, which might be programmed:

```
➡ (complex_bodied_kettle :
    (a plastic_kettle
        with switch_cover))

➡ (plastic_kettle :            ;;plastic_kettle is re-described here in a self-
    (a plastic_kettle          ;;referential manner: recursive processing of
        with switch_cover))    ;;descriptions is convenient for continued re-
                               ;;elaboration of fresh property information about an
                               ;;existing object
```

In accordance with the semantics of inheritance between propertied objects, the two programming expressions both imply that `complex_bodied_kettle` acquires the property of `switch_cover`. In the first case this is a special property with which `complex_bodied_kettle` is described in its own right: in the second case it is inherited — i.e. automatically transmitted — from `plastic_kettle`.

These two expressions reflect different domain semantics. In the first case, the first design, `complex_bodied_kettle`, has a `switch_cover` property which is quite independent of the property of the same name as developed in the second design, complex_handled_kettle.

In the second case, where switch_cover is inherited, complex_bodied_kettle is donated the new property by plastic_kettle. It therefore acquires the description of its switch_cover property-object as exemplified by all variants of plastic_kettle in general. The second design, complex_handled_kettle, may now be refined by making its existing switch_cover property a property of its handle. This is effected by typing:

```
➡ (complex_handled_kettle :
    (switch_cover for handle) == switch_cover)
```

In this case, the switch_cover property of complex_handled_kettle's handle and the switch_cover property of complex_bodied_kettle are related. complex_handled_kettle's handle's property is a more special variant of the switch_cover property of plastic_kettle and of its property-recipients. This will have important repercussions as the descriptions of this part of the product design hierarchy are further elaborated.

The different ways of expressing the attribution relationships between our representations of plastic kettle design correspond to different basic modelling intentions. The ability to express, programmatically, different product designs through continually re-describing complex property-structures leads us to some fundamental problems of change management.

5. Evolutionary development of complex objects

Evolutionary development is conceptually straightforward in a class-free knowledge representation scheme. When a donor object, such as plastic_kettle in our inheritance hierarchy, is re-described the property details are automatically transmitted to its recipients. Thus

```
➡ (plastic_kettle :
    (a plastic_kettle with switch_cover (a moulding)))
                                  ;; again employing a "self-referential" description,
                                  ;; where by "self" we mean the description's receiver
```

has the effect of incrementally adjusting the attributes of plastic_kettle. Automatically the effects are seen by its recipients when they come to look-up a property of the name switch_cover in response to a message.

By way of contrast, Gero (1991) uses the idea of "prototype instantiation" to indicate the advantages, in a system of instances and inheriting classes, of storing highly re-usable class templates. These templates, or schemas, are prolifically instantiated: in effect they provide a repository of prototypical knowledge of an accomplished design characterizing a variety of specific exemplars.

On this occasion, our plastic_kettle representation has served a similar purpose, i.e. as a generator from which successive variant designs are repeatedly derived. However, there are major, principled differences of representation strategy:

1. The prototypical information conveyed by the `plastic_kettle` "template" may be extracted very simply. Either `plastic_kettle` or its design variants are equally available to respond to messages, e.g. about `switch_cover`.

```
➡ (plastic_kettle : switch_cover)
  (A  MOULDING)          ;;program execution response
➡ (complex_bodied_kettle : switch_cover)
  (A  MOULDING)
```

2. In turn, the variants of the `plastic_kettle` prototype may themselves serve as generative patterns during the evolutionary refinement of product designs. Thus there may be variants produced from descriptions of the form

```
      (a complex_bodied_kettle with ...)
or    (a complex_handled_kettle with ...).
```

   In principle, our knowledge representation scheme appeals to the idea that no matter how specialized a particular design concept may be, further refinements of it may always be produced. As a result, a knowledge base is simply an instantaneous snap-shot of the design process. Evolution continues indefinitely: the "terminal objects" of today's knowledge base will be the templates of tomorrow's.

The computational contrast between class-free and class/instance styles of representation is very strong in this respect. In Splinter, all objects simultaneously convey generative templates and concrete representations of prototypical knowledge. As computing devices they serve as generators when they occur as the subject of descriptions, and they are concrete in their response to messages.

Our evolutionary structure, based on default property-inheritance relations on complex objects, introduces the problem of control over the consistency of property descriptions. One approach to consistency, in the object-oriented field, is that taken by the language Eiffel (Meyer, 1988). Eiffel ensures that all component attributes of S, a subtype of type T, can only be refined if those attributes are described as being subtypes of their original types, as defined by their types in T. A broadly analogous policy is adopted in Splinter, except that such checking of specialization relationships may have to be done retrospectively. For example, if `plastic_kettle` were to be re-described as having a specialized specification of its `body` — currently inherited from `container` and donated to `complex_bodied_kettle` — then that new specification would have to be reconciled with the existing property's specification in container ("up" the hierarchy) and with its effects upon the modified property in `complex_bodied_kettle` ("down" the hierarchy).

This activity is undertaken by a variant of truth maintenance system, running behind the scenes during each object's execution of messages which contain descriptions. The system manages configurational change to the knowledge base's inheritance structure, affecting objects and their properties, and gives advisory diagnostics (Zucker and Demaid, 1989).

6. A problem in organizing heterogeneous knowledge

This paper's illustrations have considered the representational adequacy of a knowledge-programming language mechanism for evolution of complex objects. In order to model the dynamic activity of design, we shall now address a wider aspect of the problem of complex modelling in relation to heterogeneous knowledge.

Figure 1 depicts part of a Splinter knowledge base. The important property of this single-inheritance tree, from the representational point of view, is that it contains two subtrees which convey very distinct perspectives on the engineering-design domain.

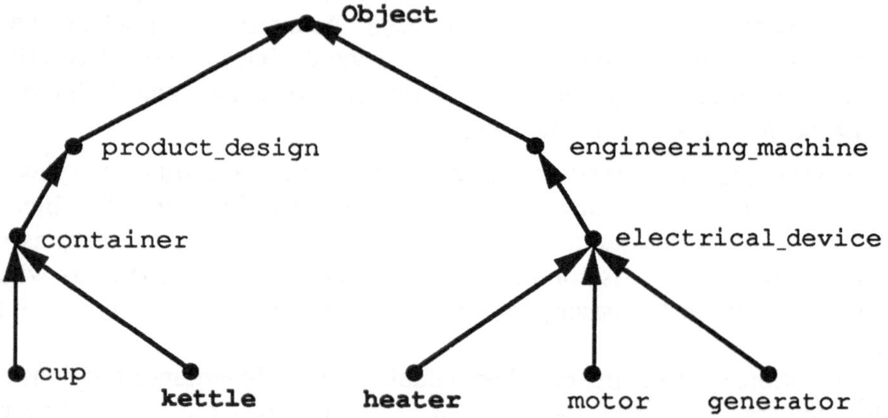

Fig. 1. Skeleton of an inheritance hierarchy
with distinct 'product design' and 'engineering machine' subtrees

One subtree embodies information about the designs of products, in particular domestic utensils, the other embodies information about engineering machines. Since knowledge of product designs and of engineering machinery is the result of different specialist skills, they reflect different empirical concerns. The purpose of the hierarchical system of organization is to maintain a separation of these concerns. For convenience, we depict the subtrees as being rooted at the top of the complete inheritance structure, i.e. at `Object`. However, their exact roots are unimportant as the root may change through evolution (e.g. `product_design` or `engineering_machine` may be demoted in the hierarchy) and the two branches maintain independence because of the different brands of messages that they are able to answer. Our complete knowledge base comprises many additional, substantial fragments of independent knowledge description; e.g. a subhierarchy which organizes polymer materials according to a classification of chemical families.

Empirically however, perspectives are *not* necessarily independent. In our knowledge base they may need to combine in many ways. For example in Figure 1, the objects, 'kettle' and 'heater' ought to be contingently related in the sense that a kettle, although described principally in terms of its component structures, is also an electrically heated device. This interaction is generally known as the **multiple perspectives problem**.

Clearly, there needs to be some manner of **liaison** between `kettle`, under the product-design perspective, and `heater`, under the engineering-machine perspective. Such liaison needs to accomplish property sharing, in the sense that relevant properties which `heater` might possess, such as electrical- and fire-safety calculations, need to be accessible to `kettle` for the purposes of answering questions. One solution to the problem is therefore to say that `kettle` inherits from `container` and also inherits from `heater`, i.e. that our concept of kettle receives an overlapping of the properties of `container` and `heater` through inheritance, as in Figure 2.

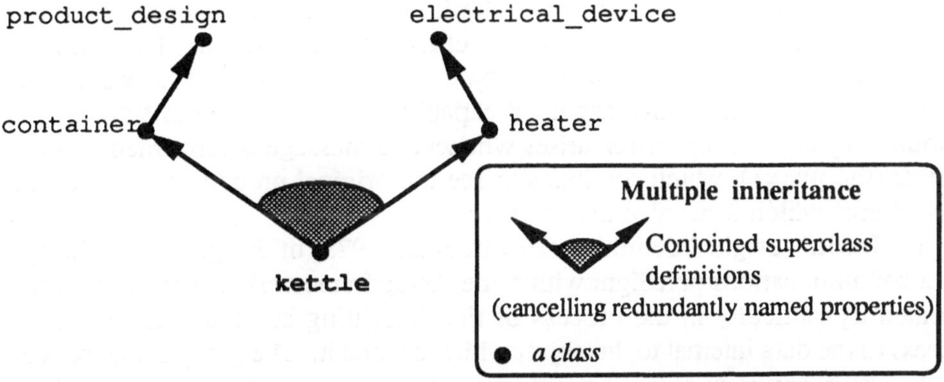

Fig.2. A multiple inheritance solution to combining perspectives on the kettle concept, as often implemented in class/instance systems

Prima facie, multiple (or overlapping) inheritance, often adopted in class/instance programming systems, solves the problem of sharing property information across "perspectives". It represents instances of the `kettle` class in terms of the conjoined definitions of the `container` class and the `heater` class. However, this implies that `kettle` has absorbed all the property information of `container` in particular, and inherited from `product_design` in general, all the property information of `electrical_device` in particular, and of `engineering_machine` in general. The multiple inheritance mechanism, therefore, at best achieves precisely an overlap: in effect, it represents kettles by a homogeneous attribute space in which properties about engineering machinery are as significant as properties about domestic utensils. The different perspectives which provide the **context** for understanding what each sort of property — design-related or machine-related — has not been distinguished in the multiple-inheritance representation.

This simple example heralds an extraordinary potential for complex expansion since useful design perspectives are numerous. It is easy to add the perspective of a manufacturing method to those of product designs and engineering machines and so, using multiple inheritance, continually absorb merger upon merger of inheritance pathways. The size of each object affected by the multiplicity of superclasses would lead to great structural complexity.

We address the problem of the liaison between the two perspectives affecting `kettle` and `heater` by an approach which (i) retains a single-inheritance hierarchy, (ii) translates the problem into a question of allowing perspectives to communicate with one another.

7. Principles of message delegation

Starting from the idea of property-inheriting objects, extra representational capacity is achieved using object-to-object **message delegation**. Delegation is a regime for control over object-to-object communication. It is based upon allowing one object to have extended capabilities by acting as an extension of another object. This extension arises whenever a message is forwarded to a new object (the **proxy**), which continues to see the original properties of the object (the **client**) which forwarded the message.

In Hewitt's original definition (Hewitt et al., 1973) of delegation, established as a communications paradigm within the Actor framework, the proxy object is created dynamically in the process of the delegating communication and has access to the data internal to the object which created it. The proxy is independent of the client, however, as it is not influenced by concurrent changes to the client.

Object-to-object delegation is practically helpful in modelling engineering-design decision-making. It permits the submission of a query which needs to be answered through accessing some of the properties embodied by objects *other* than the original receiver of the message. As a knowledge encoding methodology, this use of delegation differs from inheritance manifestly because the latter provides an organization of objects through anticipated connections whereas the former is a run-time technique to program dynamically established relationships.

Despite this, both delegation and inheritance provide methodologies for programming the **sharing of properties** between objects. The computational difference between delegation and inheritance lies in the **localization of processing**. In delegation, responsibility for processing a message passes from a client to its proxy. We may illustrate this principle, for the case of our own language mechanism, as follows.

A message, in Splinter, is any Lisp-structured expression, and is permitted to reference objects and the receiving object's properties. Since objects — including property-objects — are first-class values, according to the language specification, we are allowed to refer to objects by name or any other variable identifier visible in the scope of the expression. The receiving object's private data space counts as part of the scope of visibility in this language design. Further, the attribute space of a property-object is nested within the attribute space of the object of which it is a property. (By an object's attribute space we mean its property-structure paired with the corresponding data values, where these are known.) In addition to the examples already seen, i.e. processing descriptions and returning single property associations, we can formulate message-processing expressions such as:

```
➡ (plastic_kettle :                   ;; arithmetic addition using the global Scheme
    (+ 1 1))                          ;; procedure-value named +
  2
➡ (complex_bodied_kettle :
    (equal?
        (donor (self))                ;; look up the object's "parent"
        plastic_kettle))
  #T                                  ;; Scheme literal constant meaning true
```

In place of such message transactions with a single receiving object, message delegation extends the transaction to allow it to be forwarded from one object to another. A trivial extension might be

```
➡ (plastic_kettle :                   ;; reference to the client object
    (complex_bodied_kettle :          ;; reference to its proxy object
        (+ 1 1))
  2
```

The communication pattern is revealed more clearly if we type

```
➡ (plastic_kettle :
    (complex_bodied_kettle :
        (name (self))                 ;; obtain name of the object given by (self)
        (name (client))))             ;; give name of object which is currently the client
  (COMPLEX_BODIED_KETTLE              ;; self's (proxy's) name
        PLASTIC_KETTLE)               ;; client's name
```

In the process of forwarding the message, the proxy object is created, based on `complex_bodied_kettle` — but specially initialized so that the private data space of the original `complex_bodied_kettle` is nested within the private data space of `plastic_kettle`. The message's previous receiver, here `plastic_kettle`, is appended to the transaction, so as to be returned by the procedure `(client)` while the proxy object is yielded by `(self)`.

In Splinter, delegation is therefore an extension to the language's message-processing algorithm which involves the creation of a proxy — here the one "based on" `heater` — i.e. an object which

- dies after the message has been serviced;
- is "special" in the sense that it possesses the attributes of an existing object (`heater`) initialized upon message-reception so as to see the client's (`kettle`'s) non-inherited properties appended onto it;
- has the ability to access the previous receiver (`kettle`) explicitly as its client.

Discourse about objects of knowledge is supported in the language by allowing objects to interrelate, one to another, through communicating by forwarding messages. Each time that this happens, the re-routed message is serviced by the newly contacted object as an agent, specially prepared for the task by the original object.

The semantics of inter-object communication, illustrated above, offer a rich, practical scheme of delegation, i.e. passing of responsibility for evaluation from client to proxy during message processing, which is intended to satisfy the needs of a design representation language and complement an evolutionary inheritance policy for complex modelling.

8. Delegation and inheritance in alliance for modelling heterogeneous knowledge

Splinter's delegation feature models a process of addressing extra attribute spaces incidental to the attribute spaces of objects determined through inheritance relations. In representing design information, our particular use of delegation has been to model **perspective combination**, as in the case of our problem with the electrical characteristics of `kettle`. Our solution is that the newly contacted proxy object, specially created from `heater`, has not been described as related to the client object through inheritance, but nevertheless contains information relevant to the characteristics associated with a query which we may ask about its client. So we use inheritance and delegation techniques, within the same software architecture, to accomplish different representational purposes.

The delegation link of Figure 3 works by forwarding a message from `kettle` to a proxy, specially created from the `heater` object. Notice that `kettle` and `heater` remain in the hierarchies with which they were described. The delegation link is a run-time communications link, in the sense that it is effective for the duration of a particular message.

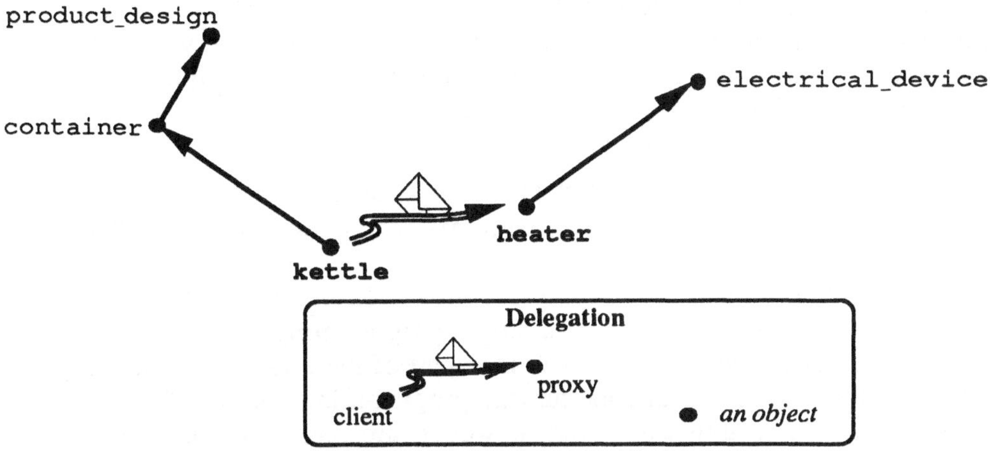

Fig. 3. A delegation solution to solution to combining perspectives on the kettle concept

A simple example of using delegation might be the single message-processing statement:

```
➡ (kettle :
     (heater :
       (<? mains-voltage
         (breakdown-voltage-of
             ((client) : (material for body))))))
```

Here `kettle` is the object returned by `(client)` in the message serviced by `heater`, because `heater`, suitably initialized, is the object to which `kettle` transfers control (`heater` appearing after the colon as the receiver of the message), i.e. `heater` is acting as the proxy of `kettle` as client.

The purpose of the message received by `heater`, in this illustration, is to test for voltage safety, say. If it is found to be an appropriate piece of code, then the knowledge-base builder's next step would generally be to turn this complete message-processing activity into a behavioural property memorized as a property of `kettle`. Suppose it is described as being a behavioural property of `kettle` named `voltage-safety?`.

Code to describe the new behavioural property, or operation, of `kettle` could be:

```
➡ (kettle :
     (a kettle
       with
       voltage-safety?            ;; an operation is a (value-expressing) "data object"
         (an Operation            ;; which expresses something "procedural", i.e. a
                                  ;; Lisp lambda expression
            =  (lambda ()
                                  ;; this procedure is the delegatory behaviour, tested above
                 (heater :
                   (<? mains-voltage
                      (breakdown-voltage-of
                         ((client) : (material for body))))))
```

The behaviour which we had originally developed "on the fly" is now remembered as a property of `kettle`. Once `voltage-safety?` has been described as above, the knowledge base supports, say, a simple query of `kettle`, or a descendant of `kettle` (e.g. `plastic_kettle`),

```
➡ (kettle : (voltage-safety?))
```

with the effect that the delegatory link is used automatically as part and parcel of the behavioural characteristics of `kettle` (or `plastic_kettle` which inherits this behaviour).

Our mechanism therefore supports both (i) expressions giving rise to delegation within the execution of a single programming statement; (ii) the memorization of the delegated behaviour as part of the established knowledge base. The former models an opportunistic programming policy. The latter is more formal and established. Both, however, are *more dynamic* forms of interobject relationship than the inheritance structure into which knowledge is factored in the first instance. Inheritance, although highly reconfigurable in the Splinter language, establishes a fully anticipated connectivity between objects. A delegatory behaviour, whether expressed opportunistically or as a memorized property, depends for its effects exclusively on the current status of the proxy object as determined at run-time. The proxy might, at any moment, be further modified with side-effects as to how the properties which are referenced in the delegated message get looked up: the proxy might even re-route the message to yet another object. Delegation is therefore, deliberately, a less predictable programming technique than inheritance. Splinter's combination of inheritance and delegation is directed at supporting the discipline of a stepwise elaboration of domain concepts along with freedom of experimentation.

As a final example, `plastic_kettle`'s own `heating_element` (as described in the code at the beginning of section 4) may be given its own remembered `voltage-safety?` calculation. E.g. the knowledge-base builder might submit

```
➡ (heating_element for plastic_kettle :
    (an electrical_device          ;; elaborating from code at beginning of section 4
        with voltage-safety?       ;; delegatory operation appropriated from kettle's
         (an Operation             ;; description — alternatively re-coded or modified
         == (voltage-safety? for kettle)))
```

Now as a result, in the expression

```
➡ (plastic_kettle : (heating_element: (voltage-safety?)))
```

there are two delegatory processes occurring. One is the memorized delegation used by the `voltage-safety?` operation. The other is `plastic_kettle`'s act of delegation to its own `heating_element` property.

The latter example illustrates an interaction between our approaches to supporting multiple perspectives and to complex modelling. The interaction arises because Splinter's message interpretation strategy provides the use of a property "sub-object" (`heating_element`) of an object (`plastic_kettle`) as a receiver, once initialized as a proxy, of a message forwarded through the object of which it is a property. In turn, the property "sub-object" has recourse to its own properties — here, a delegatory behaviour which it has memorized as `voltage-safety?`.

In principle, such interactions typify the use of Splinter for exploratory programming. They are underpinned by the representation scheme of inheriting and delegating objects: although inheritance and delegation are different computational principles for property-sharing, once combined into a single, consistent scheme they afford support to short-, medium- and long-lived modelling relationships. This is summarized in Table 1.

TABLE 1.
Short-, medium- and long-lived information sharing
in a representation scheme of inheriting and delegating objects

long-lived interobject relationships	**property inheritance** between objects
medium-lived interobject relationships	**message delegation** between objects, memorized as the behaviour of a particular object or objects
short-lived interobject relationships	**message delegation** between objects, during the execution of a single message

The construction of a single computational formalism, combining inheritance and delegation, is further discussed in Zucker and Demaid (1992a).

9. Conclusions

The paper reports a scheme for knowledge representation and its realization in the in-house programming language Splinter. The purpose of the language is to describe and manipulate engineering concepts so as to create knowledge bases to help model the activity of product design.

Adequate support to modelling design leads us to provide an evolutionary strategy for elaborating domain concepts. Our knowledge description scheme is based on a system of inheritance relationships between propertied objects, directly. We have illustrated how this facilitates evolutionary domain descriptions at the level of fundamental language mechanism, rather than through coercion of an underlying representation by a program-support environment. The programming technique we employ is oriented to objects which are concrete representations and, simultaneously, serve to generate more specialized objects. We have contrasted this class-free representational scheme for object-oriented programming with class/instance schemes and demonstrated the relative flexibility of the former in creating high-level descriptions of the design domain.

Separate design perspectives may be elaborated, in our scheme, as separate inheritance structures, which evolve and are supported by complex-object modelling so that properties are themselves propertied "sub-objects" nested within the property-structure of the object through which they are accessible.

A contingent problem with design modelling is that there are *many* such possible perspectives. While they may, at first, be described independently, in reality they overlap. This problem, associated with heterogeneous information, is known as the multiple perspectives problem.

The language mechanism of Splinter allows the represented perspectives, while separately evolving, to communicate with one another. The liaison between perspectives is accomplished by the technique of message delegation, which models a more fluid form of exploratory relationship than that achieved using inheritance connections, alone, in the structure of a knowledge base.

Delegation and inheritance are unified in Splinter into a higher order formalism for property-sharing objects. The formalism underscores a modelling policy for relationships between objects in which long-term relationships are conveyed by inheritance and more short-term relationships by delegation, as summarized in Table 1.

Better to satisfy the requirements of design modelling we have devised a programming language for high-level domain-specific knowledge description. The language supports a strategy of continual re-description and property mutation, as a logical model of evolutionary reconfiguration of perspectives in a developing knowledge base. It also provides for a liaison between these separate information-structures as part of the language's algorithm for message interpretation.

References

Banerjee, J., Kim, W., Kim, H. and Korth, H.: 1987, Semantics and implementation of schema evolution in object-oriented databases, *Proc. ACM SIGMOD Annual Conf. Management of Data*, 5(1), 311-322

Borning, A.H.: 1986, Classes versus prototypes in object-oriented languages, *Proc. IFIP Fall Joint Computer Conf.*, 36-40

Brown A.: 1991, *Object-Oriented Databases: Applications in Software Engineering*, McGraw Hill

Demaid, A. and Zucker, J.: 1988, A conceptual model for materials selection, *Metals and Materials*, 4(5), 291-297

Demaid, A. and Zucker, J.: 1992a, Evolutionary inheritance and delegation as mechanisms in knowledge programming for engineering product design, *in* G. Rzevski and R.A. Adey (eds), *Applications of Artificial Intelligence in Engineering VI, AIENG/91*, Computational Mechanics Publications 269-285

Demaid, A. and Zucker, J.: 1992b, Prototype-oriented representation of engineering design knowledge, *Artificial Intelligence in Engineering*, 17 (In press.)

Dittrich, K.R.. and Dayal, U.: 1986, Object-oriented database systems: The notions and the issues, *in* K.R. Dittrich. and U. Dayal (eds), *International Workshop on Object-Oriented Database Systems*, IEEE, pp. 3-6

French, M.J.: 1971, *Engineering Design: The Conceptual Stage*, Heinemann

Gero, J.S.: 1991, Design prototypes: A knowledge representation scheme for design, *AI Magazine*, 11(4), 26-36

Goldberg, A. and Robson, D.: 1988, *Smalltalk-80: The Interactive Programming Environment*, Addison-Wesley

Hartzband, D.: 1985, Enhancing knowledge representation in engineering databases, *IEEE Computer*, Fall issue, pp. 39-48

Hewitt, C., Bishop, P. and Steiger, R.: 1973, A universal, modular Actor formalism for Artificial Intelligence, *Proc. 3rd Intl. Joint Conf. Artificial Intelligence*, 235-245

Lalonde, W.R., Thomas, D.A. and Pugh, J.R.: 1986, An exemplar based Smalltalk, *Proc. OOPSLA'86 ACM Conf. Object-Oriented Programming Systems Languages and Applications*, SIGPLAN Notices 21(11), 322-330

Lieberman, H.: 1986, Using prototypical objects to implement shared behaviour in object oriented systems, *Proc. OOPSLA'86 ACM Conf. Object-Oriented Programming Systems Languages and Applications*, SIGPLAN Notices 21(11), 214-223

Meyer, B.: 1988, *Object-Oriented Software Construction*. Prentice-Hall.

Newton, S.: 1989, The irrelevant machine, *Design Studies*, 10(2), 118-123

Nierstrasz, O.: 1988, A survey of object-oriented concepts, *in* W. Kim and F. Lochovsky (eds), *Object-Oriented Concepts, Databases and Applications*, Addison-Wesley, pp. 3-21

Rentsch, T.: 1982, Object oriented programming, *ACM SIGPLAN*, 17(9), 51-57

Stein, L.A., Lieberman, H. and Ungar, D.: 1988, The Treaty of Orlando: A shared view of sharing, *in* W. Kim and F. Lochovsky (eds), *Object-Oriented Concepts, Databases and Applications*, Addison-Wesley, pp. 31-48

Touretzky, D.S.: 1986, *The Mathematics of Inheritance Systems*, Pitman / Morgan Kaufmann Research Notes in Artificial Intelligence series

Ungar, D. and Smith, R: 1987, Self: The power of simplicity, *Proc. OOPSLA'87 ACM Conf. Object-Oriented Programming Systems Languages and Applications*, SIGPLAN Notices 22(12), 227-242

Wegner, P.: 1987, The Object-Oriented Classification Paradigm, *in* B. Shriver and P. Wegner (eds), *Research Directions in Object-Oriented Programming*, MIT Press, pp. 479-560

Zucker, J.: 1989, *Engineering design computed by prototypes and descriptions*, PhD Thesis, Jenny Lee Library, Open University, Milton Keynes, UK

Zucker, J. and Demaid, A.: 1989, A software machine designed for selection, *Knowledge-Based Systems*, 2(3), 178-184

Zucker, J. and Demaid, A.: 1992a, Prototype-oriented knowledge representation for design reasoning, *Artificial Intelligence Review* (Under revision.)

Zucker, J. and Demaid, A.: 1992b, The rôle of evolutionary inheritance in developing knowledge bases for conceptual design of engineering products, *European Jnl. Engineering Education*, 17(2) (April issue in press.)

MULTIPLE AGENT COOPOERATIVE DESIGN EVALUATION USING NEGOTIATION

K. J. WERKMAN
IBM Corporation/FSC
Mail Drop 0210
Owego Laboratories, Route 17C
Owego NY 13827
USA

Abstract. This paper presents a form of multiagent cooperative problem solving where computational agents, paralleling human design, manufacturing and assembly participants, evaluate a design and comment on it from their unique expert perspectives. The evaluation process utilizes a novel form of *incremental negotiation* called ***knowledge-based negotiation*** where agent conflicts are resolved through the use of shared knowledge representations called ***shareable perspectives***. Two agents' shareable perspectives can be linked together to form an ***interagent issue relation*** which are grouped into a relational network and maintained by a third-party ***arbitrator agent***. The arbitrator uses the relations to develop alternatives during conflict resolution between agents performing a design review. Through the use of arbitrator suggested shareable perspectives, a form of ***issue unlinking*** occurs and this allows the agents to consider additional ***viable alternatives*** which they might have dismissed earlier. This behavior is similar to ***integrative bargaining*** found in human negotiation where conflicting parties are enlightened about secondary benefits of a particular proposal. The resulting research is demonstrated as an implementation within a knowledge-based tool called the ***Designer Fabricator Interpreter (DFI) System***.

1. Introduction

An important goal in cooperative design evaluation is to develop a conflict resolution methodology that enables multiple expert to resolve problems in a cooperative fashion. The benefit of obtaining a solution produced by input from differing viewpoints of several domain expert is that the resulting solution should be better than a solution produced by a single expert. This is especially true in cases where multiple experts' viewpoints can be combined to produce a more robust critique of a design's impact on "down stream" participants in the manufacture and field assembly process of a product's life cycle. Unfortunately, given the fact that experts tend to be busy people, critiquing expertise is not always available when it might most positively influence the design. The greatest impact of design evaluation, as stated by researchers promoting

[1]Author's current address. The following work was performed while the author was employed by the NSF Sponsored ATLSS Engineering Research Center at Lehigh University.

J. S. Gero (ed.), Artificial Intelligence in Design '92, 161–180.

concurrent or *simultaneous engineering* (Sprague, *et. al.*, 1991), would be at preliminary design step. Indeed, several researchers in the field of Distributed Artificial Intelligence (DAI) have suggested that cooperative multiagent design systems that allow for multiple perspectives and perform conflict resolution through negotiation are extremely beneficial as a critiquing tool for designers (Bond, 1990; Sycara, 1991). Therefore, motivation exists for the development of a design evaluation tool that uses a knowledge-based approach and integrates multiple experts' perspectives to generate robust critique from each experts viewpoint. This paper describes such a system.

During design evaluation, each design expert is considered an intelligent "agent" which maintains its own perspective of the design process. During this dynamic interaction process, agents analyze problems proposed by other agents and respond to them by generating solutions from their own perspectives. This paper looks at both the negotiation mechanisms for knowledge-based reasoning among agents as well as a description of an implementation of these mechanisms in the Designer Fabricator Interpreter (DFI) system. The DFI system, based in the domain of structural design of building connections, allows a structural design engineer to review alternative steel connections proposals which have been cooperatively evaluated from other agent viewpoints. This allows the designer to select the connection design which is the most feasible and economic to produce before committing to any specific connection design. This paper focuses on the problem-solving aspects where agents evaluate the designer's initial connection, propose alternative configurations based on issues relevant from their viewpoints, and negotiate the outcome of their proposals among the other agents in the system.

1.1. DFI OVERVIEW

One objective of the DFI research project is to develop a computer tool that allows structural design engineers to bring construction knowledge to bear during the preliminary design stage of beam-to-column connections in buildings. Connections were chosen as a focus because they have been noted as "hot spots" for problems in structures. The most memorable example of a structural failure which directly resulted from improperly designed connections was the Hyatt Regency skywalk collapse. Though this is an extreme example, the need for systems which can point out potential downstream (i.e., construction) problems to engineers during preliminary design has the additional benefits of allowing engineers to design connections that are feasible and more economical to construct. The primary objectives of systems like DFI include assisting construction projects by managing the complexity of the agent interaction to allow them to be completed closer to the proposed completion date and within the projected budget.

This paper is structured in the following fashion. In the following section, an overview of the current construction industry design practice is presented along with its deficiencies and how the DFI system attempts to address these problems. An illustrative example is given in Section 3. A discussion of the knowledge-based negotiation model and knowledge-based negotiation systems (KBNS) are presented in Section 4. Section 5 describes the various agent proposal strategies used in DFI's knowledge-based negotiation scheme including key issues, agent issue, and a description of a negotiation cycle. A discussion about the role of arbitration within DFI is presented in Section 6 including resolution techniques when agent proposals fail (mediation and arbitration strategies). Finally, a summary and future work directions are outlined in Section 7.

2. Present Practice

Construction projects frequently include several construction agents, at geographically distributed locations. A major problem is that vital engineering information is seldom communicated in a timely fashion among agents during the construction process (Simpson & Cochran, 1987). This usually results in delays and cost overruns for the project. Thus, the present practice of the industry suggests that there is a need for computer tools which act as catalysts to promote more coordination of activities. Through the development of integrated design and construction systems, construction agents can interactively utilize data and knowledge from other agents' perspectives throughout the construction process as questions and problems arise. By integrating multiple agents' perspectives with a method of distributed problem-solving using a negotiation scheme to resolve agent conflicts, a cooperative design evaluation environment can be provided.

2.1. The DFI Approach

Computer-based methods for computer integrated construction (CIC) are intended to alleviate basic information flow problems and promote coordinated problem-solving among agents at all stages of the construction process. Unlike many civil engineering design knowledge-based systems which attempt to optimize structural design (Maher, 1985) based on one aspect, e.g., minimum steel weight, the Designer Fabricator Interpreter, similar to work done by others (Sriram, *et. al.*, 1989), is an attempt at developing a framework for distributed cooperative problem-solving between construction agents. The DFI system reflects the distributed nature of the construction industry by providing a multiagent architecture, on a single computational platform, which models design, fabrication and erection processes. Even though the problem being addressed is inherently distributed, all agent in the DFI system reside on a single computer platform and communicate through a shared memory with messages. Given the modularity of the agents and the messaging scheme, the system could be distributed across a variety of platforms. Instead of addressing the many issues of multiple platform distribution, this report discusses the architecture required for multiagent cooperative design evaluation through means of negotiation. The reader is directed to (Werkman, 1991), an earlier report which gives the basic overview of the DFI system.

3. An Illustrative Example

This section provides a brief example of how the agents in the DFI system evaluate a design after the user (a structural design engineer) selects a particular connection. Each agent utilizes unique knowledge about connections including a standardized qualitative rating scheme for its related connection issues. After selecting a connection, the user is asked for a single, most important *key issue* to be maintained by all agents during their evaluation of the given connection. In this example, the user specifies an *endplate* connection with a *key issue* of *strength*. The key issue, as described in Section 5.1, consists of a single non-negotiable agent issue which every agent must consider before proposing an alternative. During the evaluation, other agents will attempt to suggest alternatives that are of the same connection type as the key issue connection proposal. Initially, the arbitrator takes control from the user and ***commands***[2] the design agent to ***accept*** the user's *endplate* connection proposal using *strength* as the positive supporting (key) issue because the user is the designer in the first cycle of the negotiation. The design agent then ***informs*** all agents of the key issue and ***requests*** that the proposed connection be evaluated. The designer's request is shown graphically in the *designer's window* in Figure 1.

Before each agent's evaluation, the arbitrator reviews all proposed connections to determine which agent is most detrimentally affected and hence should go next. In this case, the erector agent is most severely affected by the designer's *endplate* proposal. The erector determines that the designer's proposal is unacceptable because the *endplate* connection is too difficult in terms of *erection ease*. Therefore, the erector ***refuses*** the designer's connection and looks to the fabricator in hopes that it might have proposed an acceptable connection. At this stage, the fabricator has not yet proposed anything, so the

---

[2]Command in the DFI agent communication language appear in ***bold italics***.

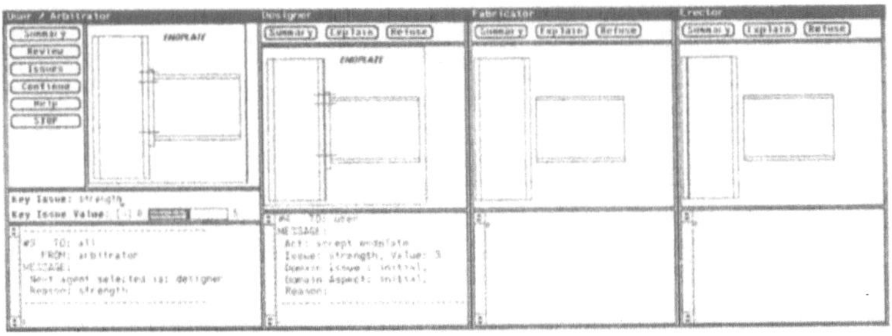

Fig. 1. Designer's Initial Configuration

erector selects a connection from the set of possible connections it knows about which satisfies it's *erection ease* issue as well as satisfying the user specified *key issue.* Thus, the erector ***request***s the *plates-tee* as seen in the *erector's window* in Figure 2.

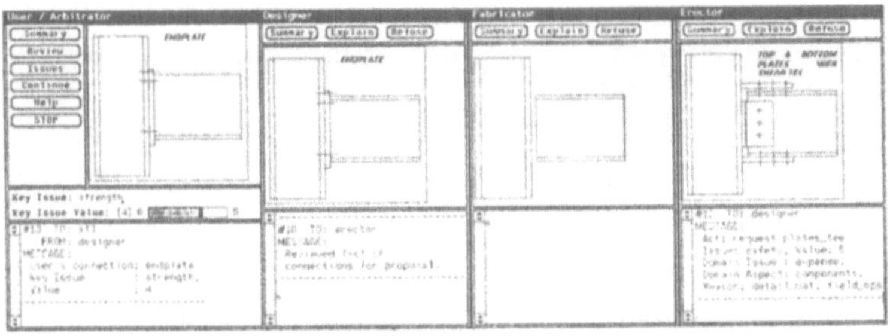

Fig. 2. Erector Replies With Plates-Tee

It is important to note that the erector has directed it's proposed connection back to the designer for review since the designer was the last agent to make a proposal. In this case, the designer ***accept***s the erector's proposed connection because it exceeds the designer's *key issue* of strength as well as meets the designer's criterion for the *endplate* connection. Note that the erector's proposal has increased the value of the *key issue* to the new value associated with the erector's proposed connection. By increasing the value of the *key issue*, the search space of possible connection alternatives is reduced, thus causing the agents to converge more quickly on a set of agreeable proposals. The designer's acceptance is seen graphically in the *designer's window* in Figure 3.

Next, the arbitrator assumes control and notices that two agents have proposed the same connection. Usually, this would cause the arbitrator to ***inform*** all agents of a halting condition. This is not so in this case because an ''unfair'' evaluation condition has occurred -- the fabricator has not yet had a chance to evaluate any connection designs. Thus, the arbitrator gives control to the fabricator who reviews the last proposal (the designer's acceptance in this case) and immediately notices that the issue of *material cost* causes a problem for the

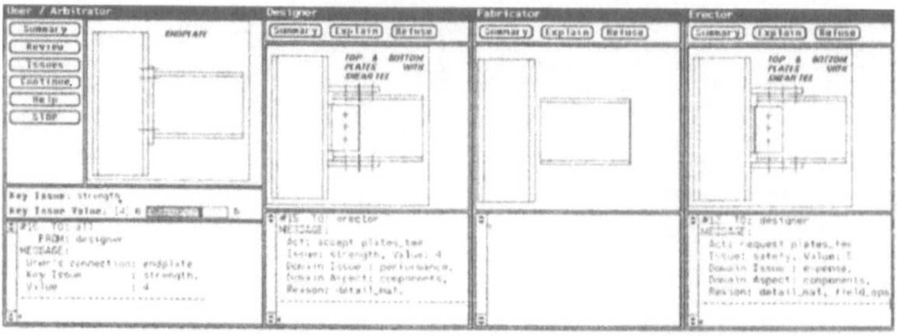

Fig. 3. Designer Agrees With Erector

fabricator agent. Since both the designer's and erector's connection are the same, the fabricator performs only one review. Here, the next best connection that maintains the *key issue* as well as improves the fabricator's *material cost* issue is the *direct flange weld with shear plate* as seen in the *fabricator's window* in Figure 4.

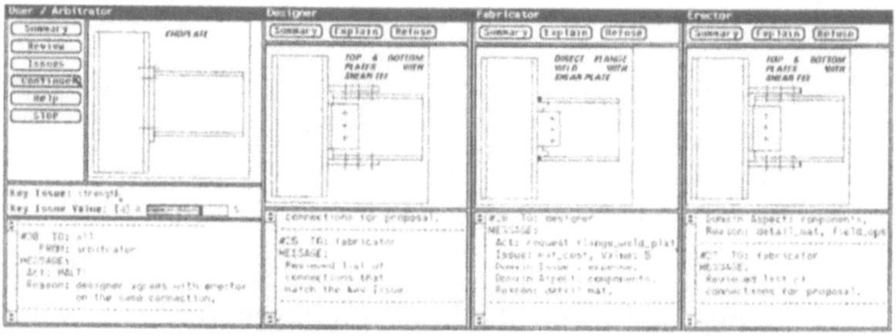

Fig. 4. Fabricator Proposes Flange Weld

Again, the arbitrator reviews the evaluation process and notices that two agents have agreed on a connection and that each agent has had a chance at suggesting an alternative. The arbitrator ***inform***s all agents of the halting condition and control is returned to the user. At this point the user can ask any agent to ***explain*** its proposed connection or ***continue*** with the evaluation. If the user ***continue***s, the arbitrator reviews the situation and notices that no particular agent is in "peril", and allows the agents to determine their own control sequence. Whichever agent received the last message is given a chance to respond to that message. In this case, the fabricator proposed a connection to the designer. The design agent, upon reviewing this connection, notices that the fabricator's connection satisfies all issues of the designer and thus ***accept***s the fabricator's proposal as seen in the *designer's window* in Figure 5. This causes another halting condition upon where the arbitrator returns control to the user[3].

[3]Control is returned to the user after each proposal cycle.

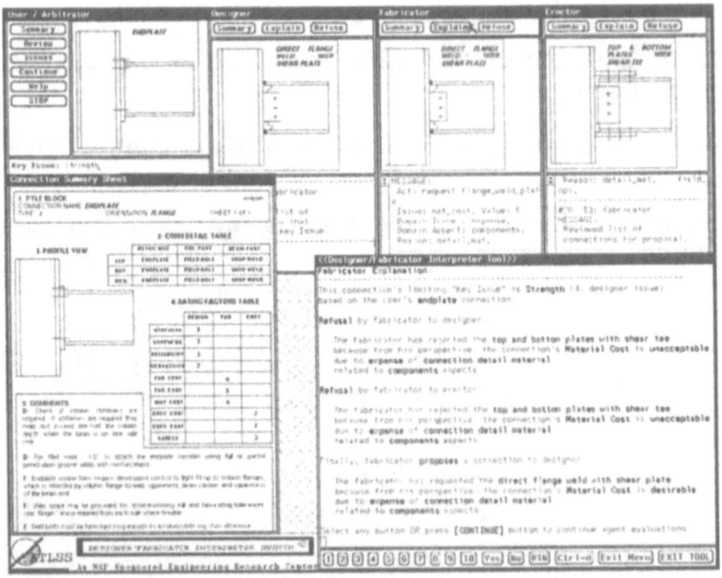

Fig. 5. Designer Agrees With Fabricator After Continuing

In Figure 5, the user has the option of selecting buttons from the *User/Arbitrator window*. This allows the user to obtain a ***summary*** of the initial connection, ***review*** the agent dialog of the entire negotiation process from start to finish listing each agent's proposal and related issues. Also, the user can change the overall key ***issue*** which focuses the negotiation, ***continue*** the agent evaluation, ask for ***help***, or ***stop*** the agent evaluation and exit. Moreover, the user can select buttons from the *Agent windows* and obtain a ***summary*** description for each agent's proposed connection or ask the agent to ***explain*** its last proposal action. The explanation appears in the system's *Input/Output window* beneath the *Agent windows*, as seen in Figure 5, and includes the key issue, the connections under review, the agent's response to the reviewed connections, i.e., the acceptance or rejection of other agent's connection, the reasons for the action, and the agent's proposed connection with its justifications. At times, it is also useful for the user to be able to ***refuse*** and remove a connection from the evaluation process if the user knows that a particular connection is not desirable.

3.1. GENERAL CONTROL STRATEGIES

This section briefly describes the arbitrator agent's control functions in the DFI system in addition to mediation and arbitration which will be described below. A flow chart of the arbitrator control strategies is depicted in Figure 6. The arbitrator takes an active control during the agent evaluation process when:

- its help has been requested by an agent,

- it determines that a halting condition exists,
- it notices a similar proposal/counterproposal sequence (repetitive or flip-flop conditions)
- it notices that the user has changed the key issue in mid-negotiation, or
- it determines which agent is the ***worst issue agent*** and passes control to that agent.

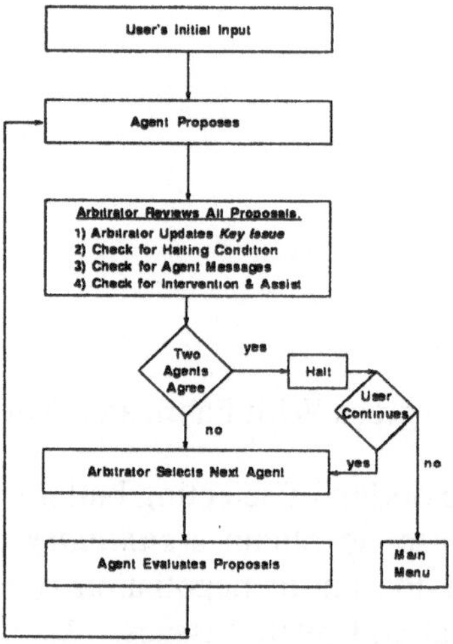

Fig. 6. DFI Basic Control Loop

4. The Knowledge-Based Negotiation Model

A novel form of conflict resolution is used in the DFI system known as knowledge-based negotiation. Such knowledge-based negotiation systems (KBNS) require that agents be able to communicate their issues about the problem, reason from their own and other agent's perspectives, and finally generate acceptable counterproposals. For agents to be able to contemplate the effects of their proposals on others, they need to share a common background of the problem domain. This is done through the sharing of agent perspectives with other agents. Though sharing common domain knowledge and agent perspectives, agents can review the dialog of the negotiation and use this knowledge to make better informed counterproposals at future steps in the design evaluation process. The history of the agent dialog also provides a basis for explanations for the user. The negotiation dialog describes the agents' behaviors at each point in time during the negotiation process. The user can use the resulting dialog-based explanations to make a better decision about which final solution to select.

4.1. AGENT COMMUNICATIONS

The DFI system includes both an interagent communications language and a centralized communications medium called a *blackboard* (Nii, 1986). The blackboard scheme in DFI allows agents to send and receive messages among agents during the negotiation process. Conceptually, the DFI blackboard maintains five partitions, three of which are used to maintain agent messages related to ***requested proposals***, ***accepted proposals*** and ***rejected proposals***. In addition, an ***ask and reply*** partition exists where agents post requests for review and receive feedback on their potential proposals. Finally, a ***shared knowledge*** partition exists on the blackboard which contains labels of domain objects,, all known proposals, and a network of agent issues called the DFI Relational Network.

The interagent communications language in the DFI system is composed of a set of message primitives is based on *speech act theory* (Allen & Perrault, 1980) taken from natural language research. The basic goal of speech acts is to model a listener's responses to speaker's questions in an intelligent fashion. The listener tries to infer the intentions of the speaker's speech plan and offer assistance if they can by noting *obstacles* in the speaker's plan. A similar plan-based approach of response generation is used in the DFI system which makes for relatively short, explicit messages, thus reducing extraneous message overhead within the system. In addition, by selecting the right message primitives combined with the right historical dialog, agents can communicate abstract levels of intentions and therefore reason about the beliefs of other agents (Tenney & Sandell, 1981).

4.2. KBNS KNOWLEDGE REQUIREMENTS

This section details the types of knowledge required by a knowledge-based negotiation system (KBNS). In a KBNS, three types of knowledge are maintained. The first type of knowledge is ***shared knowledge*** which is accessible to all agents including the arbitrator agent. The shared knowledge includes *domain object knowledge* as well as knowledge about the *history* of the negotiation dialog. Object knowledge includes such things as the names of objects known to all agents and is predefined and static in the system. In DFI, this includes a list of connections labels (connection proposals), a list of connection component pieces (proposal attributes) and a list of agent issue labels (aspects of proposal attributes). The negotiation history includes all agent proposals, rejections, and counterproposals made by the agents and is dynamically generated.

The second type of knowledge maintained by the DFI system is ***unique agent knowledge***. In DFI, this includes specialized knowledge of construction processes maintained by each agent which is also statistically defined in the system.

The third form of knowledge in the DFI system includes ***knowledge maintained by the arbitrator***. The arbitrator agent maintains both *local* unique agent knowledge related to mediation and arbitration strategies as well as *shared* agent knowledge including object knowledge, agent issue labels and the negotiation dialog. In addition to this shared knowledge, the arbitrator also maintains each agent's ***shared perspective*** (related agent issue) for all domain objects. The arbitrator maintains this knowledge in a relational network of ***shareable agent perspectives*** called the DFI Relational Network and can be seen in Figure 7.

4.2.1. *DFI Relational Network*

The ***DFI Relational Network*** consists of a series of interagent issue relations between agents and domain objects in the system. Each interagent issue relation is a mapping between two *agent perspectives* and a single *domain object*. Further, each agent perspective consists of a single agent issue related to a given domain object and a context. Through the proper selection of interagent issue relationships, the arbitrator can suggest alternative agent perspectives to designer, fabricator and erector agents in terms of functional, component, and fastener domain aspects.

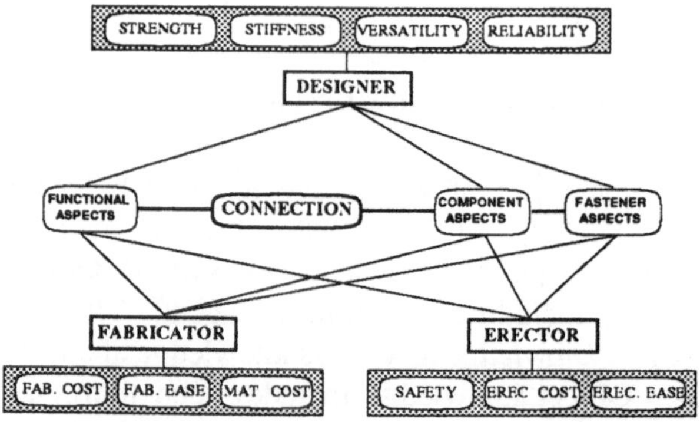

Fig. 7. DFI Relational Network

It should be noted that the *DFI Relational Network* only provides the arbitrator with an apriori abstract description of issues and how they relate to one another within the connection domain. Even though, this provides the arbitrator with enough information to detect interagent issue conflicts and assist in the negotiation when necessary. The arbitrator does not contain any knowledge about any agent's unique operations knowledge. Therefore, in order for the arbitrator to append a proposed solution with additional arguments, the arbitrator has to query each agent directly as to the reason and explanations behind the issue relationship under consideration.

Figure 8 depicts a the generic structure of the DFI relational network. The network contains three ***agent nodes*** each with its own local ***issue nodes*** (grey block of issues), linked by *agent issue links*. The two remaining types of nodes in the figure are the ***domain aspects*** and the ***domain object*** nodes (boxes with rounded corners). Each ***agent node*** is linked to a ***domain object*** by means of a commonly accepted *domain issue link* (such as expense) which also passes through a ***domain aspect*** node. This linking of nodes comprises a ***shareable agent perspective*** in the relational network. When two unique agent perspectives are linked to the same domain object, the results is an ***interagent issue relation***. This is seen in Figure 8 as a heavy black line.

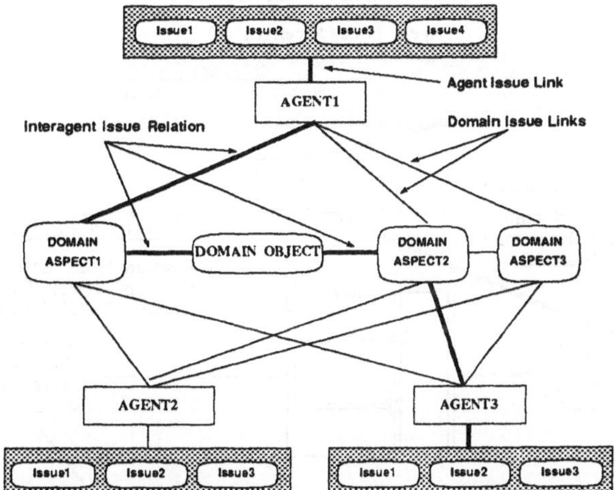

Fig. 8. Example Relational Network

4.2.2. *Interagent Issue Relations*

An example of an *interagent issue relation* which links two agent's unique perspectives to a common domain object is shown in Figure 9. In the figure, a design agent with an issue of *strength* and an erector agent with an issue of *erection cost* are linked to the *endplate* connection via their own unique agent perspectives. The designer's perspective on the endplate connection is seen through the *functional domain aspect* based on the global domain issue of *performance* (domain aspects are described in Section 4.3.2). The erector agent views the endplate connection through the *fastener domain aspect* with a domain issue of *expense.* From the designer's viewpoint, the endplate connection is favorable based on the supporting issue of *strength* (noted as a "+" in the figure) given the domain issue of *performance.* From the erector's perspective, the endplate connection is undesirable based on the detracting issue of *erection cost* (noted as a "-" in the figure) given the domain issue of *expense.* Associated with each agent's issue is a *reason* describing why that agent considers the issue as supporting or detracting from the agent's goal. The supportive reason for the *strength* issue of the endplate connection from the designer's perspective is because of the *detail material* used in the connection (the endplate detail material is strong). From the erector's viewpoint, the endplate connection is costly (*erection cost*) because of the *field operations* required to assemble the connection in the field (alignment problems for bolting the endplate).

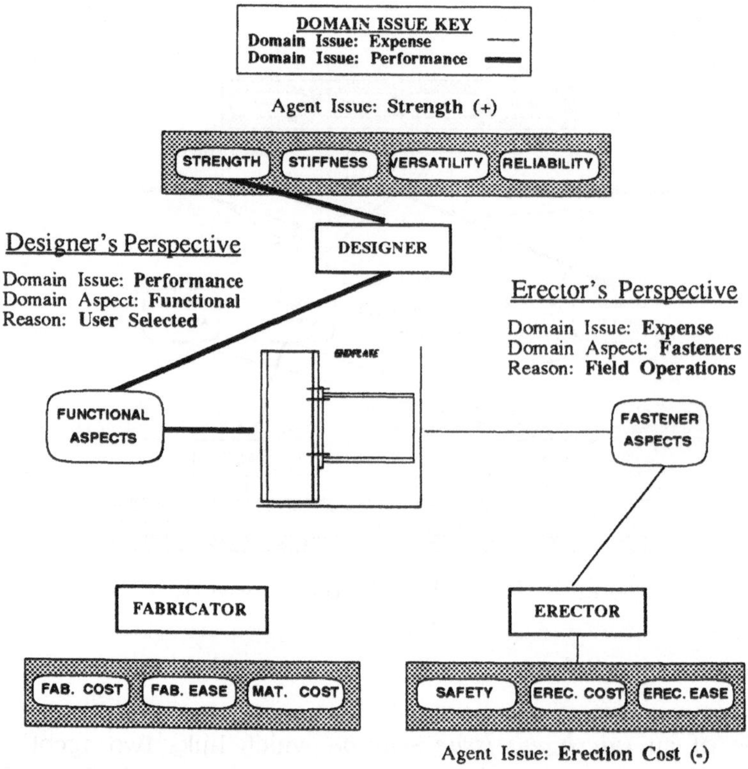

Fig. 9. Example Interagent Issue Relation

4.3. SHAREABLE AGENT PERSPECTIVES

An agent perspective includes a combination of agent issues and other domain factors that are related to an object being evaluated given the context of the evaluation. An example of an agent perspective for a design agent would include one of the designer's agent issues of *strength*, *stiffness*, *versatility* or *reliability* related to one of the connections known by the design agent. An agent perspective is different from the concept of *object perspective* as described by Bobrow in the KRL language (Bobrow & Winograd, 1977). Object perspectives usually include the selection of specific object attributes that are inherited from an object taxonomy depending on the focus of the perspective. Instead, agent perspectives are a separate knowledge structure orthogonal to the object taxonomy, much like the concept of *domain perspective* presented in work by McCoy (McCoy, 1990). Domain perspectives include *salience values* to indicate which perspectives are highlighted and which are suppressed. Because all agent issues are made available to a third party arbitrator in the form of agent perspectives and interagent issue relations, agent perspectives can be considered ''shareable.'' Once an agent obtains an interagent issue relation between itself and another agent, it can use this newly acquired knowledge to develop a counterproposal which might be more acceptable than if the agent had not had this knowledge available.

4.3.1. *An Agent Perspective*

Each agent's ***perspective*** of a given domain object is composed of the agent's *local issue*, a *domain issue link*, and a *domain aspect*, as seen previously in Figure 8. These three elements make up a context in which the agent views a domain object (that agent's perspective). The first element, the agent's *local issue*, consists of ordinal values which allow the agent to rank the issue's level of desirability. The second element, the *domain issue link*, links the agent's *local issue* to a shared *domain aspect* and thus provides a way of relating (grouping) common domain concepts (domain issues) among the agents. Domain issues could be used to effect the outcome of the negotiation by causing each agent to focus on a shared issue during its proposal cycle[4]. Domain issues can also be used to develop general explanations that are understood by all agents. The third component of a ''shareable'' agent perspective is the *domain aspect* and is described below.

---

[4]To some extent, a more limited approach is used initially in the DFI system where one agent's *local issue* (referred to as the *key issue*) is used to focus all other agents' proposals. This is described below in Section 5.

4.3.2. *Domain Aspects*

Domain aspects are similar in concept to *domain perspectives* as described in work by McCoy on correcting dialog misconceptions by use of agent perspectives (McCoy, 1990). Both DFI's domain aspects and McCoy's domain perspectives are similar in their four major characteristics. Both concepts include *salience values* to indicate which perspectives are to be highlighted and which are suppressed over the entire domain of objects. Second, both are separate data structures that are orthogonal to the system's object generalization hierarchy. Third, any number of domain objects can be viewed from the same domain aspect or perspective. Some domain aspects might not be applicable (make sense) to every domain object, but this is acceptable considering that in real life the same thing happens when one tries to apply a particular viewpoint to all possible objects in the world. The fourth major characteristic of McCoy's domain perspectives states that only one domain perspective is active at any one time during an evaluation. In the DFI system, one domain aspect is active for any given agent at one time. During an evaluation, two agents might be viewing the same domain object (connection) from their own agent perspective and hence unique domain aspect. Therefore, an object can be considered from several viewpoints at one time in DFI. Also, since domain aspects are ranked by salience values, they can be thought of as "filters" through which agents view domain objects. A domain aspect with a high salience value will highlight a set of agent perspectives (makes them more important). These perspectives can then be used by an agent to determine the benefit of another agent's proposal based on the highlighted aspect. This is similar to McCoy's summation of salience values used to determine the importance of domain objects in a given context.

5. Agent Proposal Strategies

In knowledge-based negotiation systems, an agent might formulate a counterproposal in one of three ways. First, the agent may generate a counterproposal in response to an earlier agent's proposal ignoring any side effects on other agents *(help single agent approach)*. This is a somewhat *competitive* approach in that the agent's counterproposal may hinder another agent's potential proposal during the next negotiation cycle. A second possibility is that the counterproposing agent might act more *cooperatively* by considering the effect its counterproposal has on the majority of agents and *ignore* the agent that requested the earlier proposal *(help other agents approach)*. The third possibility is that the counterproposing agent (also know as the reviewing agent) could just ignore all others and propose what it feels would best improve its own position *(help myself approach)*.

In the DFI knowledge-based negotiation approach, a combination of all three of the above extreme positions are used to allow the agents the maximum degree of flexibility during the counterproposal process. By using the *help single agent approach* combined with the *help myself approach*, the agent can produce a more *competitive evaluation mode*. A *cooperative evaluation mode* is obtained when the *help other agents approach* is used in conjunction with the *help myself approach*. In either case, the *help myself* approach forces the reviewing agent to always improve its position. This constrains the search space on possible alternatives and hence focuses the negotiation proposals. Depending on the level of coordination (agent evaluation mode), the reviewing agent will be constrained by one (*help single agent approach*) or several (*help other agent approach*) agent issues and behave more competitive or cooperative, respectively and is discussed in detail below.

5.1. KEY ISSUE DIRECTED PROPOSALS

In DFI, the focus of the agent proposal and negotiation process is controlled by a ***key issue***. The key issue consists of a single (local) agent issue which every agent must consider before proposing an alternative connection. Since the key issue is non-negotiable and must be maintained throughout the negotiation process, the *key issue agent's* viewpoint tends to become dominant. Because the key issue agent is given the power to maintain a non-negotiable issue position, it attains a higher bargaining power over the other agents. This elevated bargaining power helps constrain and focus the proposals made by the other agents. In the DFI system, the key issue is selected by the user and reflects the user's interest in a particular agent issue for a given connection evaluation.

5.2. AGENT ISSUE DIRECTED PROPOSALS

The general ***agent issue directed*** proposal scheme seen in Figure 10 has been implemented for distributed problem-solving among semi-autonomous agents in the DFI system. Each agent (called the *reviewing agent* when reviewing a connection) reviews all of the other agents' proposals and then proposes the best alternative possible (by either accepting an existing proposal or generating an alternative counterproposal). Agents review other agents' proposals based on the first (reviewing) agent's unique set of issues and how they are affected by the characteristics of the proposal. In DFI, characteristics of an agent's proposal include such things as component elements and fastener methods (bolting or welding).

The process outlined in Figure 10 starts with the ***reviewing agent*** evaluating the ***proposing agent's*** connection to determine which issue is most problematic

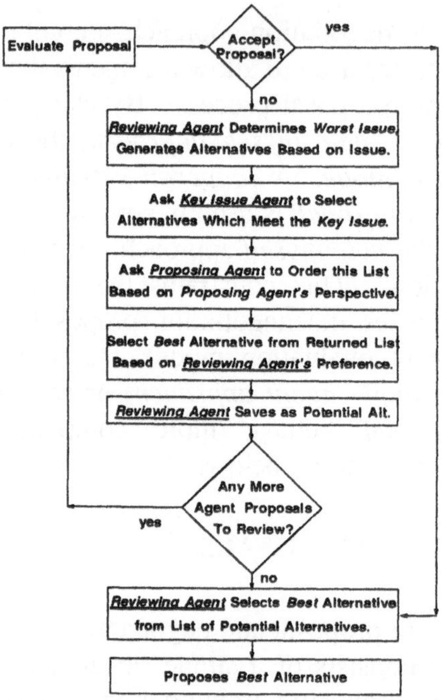

Fig. 10. Agent Evaluation and Proposal Process

based on the connection's characteristics. If the ***reviewing agent*** finds that the current connection proposal is acceptable, then the ***reviewing agent*** accepts the ***proposing agent's*** connection and proposes this as its own solution. If on the other hand the ***reviewing agent*** notices that one of its particular issues is effected, the ***reviewing agent*** will search for a counterproposal which improves its effected issue while maintaining or improving the value of the user's initial *key issue*. The ***reviewing agent*** then generates alternatives that enhance the worst issue and submits this list for review to the ***key issue agent***. The ***key issue agent*** selects only the connections that meet or exceed the user specified *key issue* and returns this new list back to the ***reviewing agent***.

In an attempt to provide a coordinated behavior among the agents (cooperative solution), the ***reviewing agent*** sends the list of alternatives which meets the key issue to the ***proposing agent*** for its review. This gives the ***proposing agent*** a chance to rank the list of alternatives based on its preferences. The ***reviewing agent*** then takes this ordered list and selects its best possible connection counterproposal in response to the ***proposing agent's*** initial proposal. The ***reviewing agent*** saves this specific counterproposal and performs a similar evaluation for all other agent proposals. Upon evaluating all other agent proposals, the ***reviewing agent*** selects its best alternative from its *best alternatives list* and proposes that alternative to the initial ***proposing agent*** whose proposal is to be replaced by the ***reviewing agent's*** proposal.

6. Arbitration In DFI

At some point during the agent proposal evaluation and negotiation process, a proposal from a ***proposing agent*** might exceed the acceptable limits of the issues of the group. This may prevent a ***reviewing agent*** from being able to generate an alternative. One agent will have to concede or relax an issue for the negotiation to proceed. It is possible that neither agent may be able to concede any issue because it would be too costly for each agent. When agents can not generate counterproposals, a "deadlock" situation occurs which has the effect of halting the negotiation process without producing an agreeable solution. The methods used to address the "deadlock" are important to system *efficiency* (the amount of time and resources required to resolve the deadlock) and of *quality of the solution* (the resulting usefulness of the solution produced). In such cases, a third-party arbitrator agent is called upon to mediate a solution.

The arbitrator addresses this problem through using its local mediation and arbitration knowledge as well as the globally shared knowledge of the *DFI Relational Network* and the negotiation dialog. The arbitrator's help is enlisted when a ***reviewing agent*** is unable to generate a proposal that meets its minimum limit of acceptability (usually one of its issue salience values). When this happens, the ***reviewing agent*** sends a **help** message to the arbitrator. The arbitrator responds to this help request by initially attempting to *mediate* among the agents. If the agents still can not agree, the arbitrator performs *arbitration* and force one or several agents to behave in a specific fashion in order to produce a solution.

6.1. MEDIATION STRATEGIES

The arbitrator employs two mediation techniques to assist the agents in reaching any ***viable alternatives*** that they may have overlooked or ignored during the process of negotiation. The two mediation techniques include:

1. Issue relaxation techniques used mainly for relaxation of the *key issue*, and

2. Intelligent proposal generation using the *DFI Relational Network* and the negotiation dialog.

In the first method, the arbitrator **commands** an agent to relax the salience value of a particular issue by one (1) salience value. Even though this approach might seem like an arbitration strategy because of its binding command directive, the motivation behind the arbitrator's action is for indirect mediation purposes. The goal is to force the dominate ***key issue agent*** to "relax" its key issue so that the ***reviewing agent*** can resubmit its list in hope that one of its alternatives will now meet the relaxed key issue. The ***key issue agent*** allows the key issue to be "relaxed" back to a limit salience value which was the salience value of the user's initial key issue selection.

In the second method, the arbitrator takes a more computationally difficult approach and attempts salvage the current state of the proposal process by suggesting an alternative that is somewhat acceptable to both conflicting agents. The arbitrator accomplishes this by using both *interagent issue relations* and the negotiation dialog. The algorithm consists of three main steps:

1. Search for all *interagent issue relations* between the two conflicting agents. There will be several pairs which can be grouped by *domain aspect, domain issue*, or *agent issue.*
2. Separate the interagent issue relations into *shareable agent perspectives* and propose the relative agent perspective to each agent for their review. The arbitrator notes all pairs of agent perspectives that received *supportive votes* (both agents considered the given connection acceptable).
3. Suggest a possible connection alternative to *each* agent based on the maximum number of *supportive votes* each alternative connection receives from each agent.

6.2. ARBITRATION STRATEGIES

When both arbitrator mediation routines fail, the arbitrator enters into *arbitration mode.* The two arbitration strategies used by the arbitrator include:

1. Setting a time limit for an agreement to be reached (implicit), and
2. Suggesting alternatives taken from previous negotiation cycles found in the negotiation dialog.

In the first case, the arbitrator maintains a count of the maximum number of counterproposals (called negotiation cycles). If the agents' proposals do not converge after six iterations (default in DFI), the arbitrator stops the evaluation and returns control to the user.

In the second case, the arbitrator takes a somewhat more aggressive position and attempts to provide a solution to the agent requesting help by analyzing each agent's *proposal preference list* throughout past negotiation cycles from the negotiation dialog. This approach, similar to the *Heuristic Trial and Error* approach, is a last resort attempt at selecting possible alternatives that both agents might agree upon.

7. Summary and Future Work

In summary, this paper presented an incremental form of negotiation between multiple agents called ***knowledge-based negotiation*** which utilized a shared knowledge representation called ***shareable agent perspectives***. This allowed the agents to perform negotiation in a manner similar to cooperating (or competing) experts who share a common background of domain knowledge[5]. A grouping of two of more shareable agent perspectives results in an ***interagent issue relation***. A relational network of these relations is maintained by a third party ***arbitrator*** agent which utilizes them during its mediation phase of conflict resolution. The arbitrator then uses the network of relations to develop alternative proposal suggestions. If this fails, the arbitrator then arbitrates by setting time limits and forcing the negotiation to acceptance. The resulting research is demonstrated as an implementation within a knowledge-based tool called the ***Designer Fabricator Interpreter (DFI) System*** which structural design engineers may use to evaluate preliminary designs from the additional perspectives to discover and avoid potential ''down stream'' problems, thus reducing costs.

Current and future work includes the reimplementation of the negotiation scheme in the DFI system into a more generic tool for use in other application domains. Currently, work is being done to modify the DFI system to be used in a concurrent engineering setting where the system will act as a coordinator of cooperative problem solving between human and computer agents.

Acknowledgments

Special thanks to Mr. Marcello Barone who developed the construction domain relations used in the *DFI Relational Network* and Ms. Stephanie Wagaman for her assistance in discussing aspects of agent negotiation. Also, thanks to Dr. Donald Hillman and Dr. John Wilson for their guidance and dedication to this project.

References

Allen, James F. and Perrault, C. Raymond: 1980, Analyzing Intention in Utterances, *Artificial Intelligence*, **15**(3), pp. 441-458.

Bobrow, D.G. and Winograd, T.: 1977, An Overview of KRL: A knowledge representation language, *Cognitive Science*, **1**, pp. 3-46.

---

[5]A comprehensive overview of shareable agent perspectives can be found in (Werkman, 1990).

Bond, Alan H. 1990, A Computational Model for Organization of Cooperating Intellitent Agents. *Conference on Office Information Systems.* Cambridge, Massachusetts: ACM, ACM.

Maher, Mary Lou: 1985, HI-RISE and Beyond, *Computer-Aided Design*, **17**(9), pp. 420-427.

McCoy, Kathleen F.: 1990, Generating Context-Sensative Responses to Object-Related Misconceptions, *Artificial Intelligence*, **41**, pp. 157-195.

Nii, H. Penny: 1986, Blackboard Systems, Part 2, *AI Magazine*, 7(3), pp. 82-106.

Simpson, G.W. and Cochran, J.K.: 1987, An Analytic Approach to Prioritizing Construction Projects, *Civil Engineering Systems*, **4**(4), pp. 185-190.

Sprague, R.A., Singh, K.J., and Wood, R.T.: 1991, Concurrent Engineering in Product Development, *IEEE Design & Test of Computers*, **8**, pp. 6-13.

Sriram, D., Logcher, R.D., and Cherneff, J. 1989, *DICE: An Object Oriented Programming Environment for Cooperative Engineering Design* (Tech. Rep.)). Massachusetts Institute of Technology. Intelligent Engineering Systems Laboratory.

Sycara, Katia P. 1991, Cooperative negotiation in concurrent engineering design, In Sriram, D. (Ed.), *Computer Aided Cooperative Product Development, MIT/JSME Workshop, Lecture Note Series in Computer Science.* NY, NY: Springer-Verlag.

Tenney, Robert R. and Sandell, Nils R., Jr.: 1981, Strategies for Distributed Decisionmaking, *IEEE Transactions on Systems, Man and Cybernetics*, **SMC-11**(8), pp. 527-538.

Werkman, Keith James 1990, *Multiagent Cooperative Problem Solving Through Negotation and Perspective Sharing.* Doctoral dissertation, Lehigh University.

Werkman, K.J. 1991, Evaluating Alternative Connection Designs Through Multiagent Negotiation, In Sriram, D. (Ed.), *Computer Aided Cooperative Product Development, MIT/JSME Workshop, Lecture Note Series in Computer Science.* NY, NY: Springer-Verlag.

BUILDING MODELING BASED ON CONCEPTS OF AUTONOMY

J. GAUCHEL, S. VAN WYK, R. R. BHAT
Center for Building Performance and Diagnostics
Department of Architecture
Carnegie Mellon University
Pittsburgh PA 15213-3890 USA

and

L. HOVESTADT
Institut für Industrielle Bauproduktion
Universität Karlsruhe, Karlsruhe, Germany

Abstract. A knowledge-based approach to building modeling is described. It is formulated on concepts of autonomy, and based on the idea that asynchronous design decisions in distributed organizations and among concurrent decision processes cannot be accomplished by a centralized software approach. Also described are initial efforts to implement the approach in a prototype integrated, distributed building design environment for modeling and performance evaluation applications. Included are overviews of (1) previous experience with large-scale building modeling efforts; (2) a specification for user-interfaces as tools to explore design spaces; and (3) a data communication module that interconnects distributed and concurrent applications for the exchange of data and recognition of conflicts between different decisions.

1. Building Modeling

Building modeling involves many different participants, each having a unique mental model of the process and varying in education, skill, expertise, and role. These models are partially represented by their individual communication tools e.g. drawings and specifications, which try to balance the information needs and requirements of the professions with a need to provide an analogy between the models and the real-world objects they represent. CAD systems failed to offer common data platforms to different professions for integrated work as their data structures are very specialized. Recent efforts, such as STEP (Wilson 1988), are trying to provide more obvious analogies with real world systems which make up a building, by proposing a common data standard for interchange of information among CAD systems and other software. But, the development of building models, which provide analogues to real world objects, is as much an art as a science (Ziegler 1978) and cannot be based on standards.

New approaches to building modeling are based on advances in artificial intelligence and object-oriented programming that allow the declaration of domain and design knowledge to support designers with a means to test design decisions, to create new aspects of designs, and to enhance existing aspects. Some of the projects such as ARMILLA, IBDE, ICADS (Haller 1985; Fenves 1990; Pohl 1992: Bjoerk 1989). investigate the possibility that building models

J. S. Gero (ed.), Artificial Intelligence in Design '92, 181–197.

can become capable assistants to building designers, and can also serve as intelligent kernels of future CAD systems to integrate a wide variety of computer-aided tools used in building design.

Any implementation of building models as software tools must take into consideration the fact that building modeling has characteristics that form a unique set of constraints. These are discussed briefly below:

- *Building modeling is multi-disciplinary.* It involves synthesis of design knowledge from many different sources and institutions. The participants in the process vary in education, skill, expertise, and role. Their activity level varies with building type, the stage of design, parts and aspects of the building, the degree of automation, and the nature of the problem-solving task at hand.
- *Building modeling is distributed, concurrent and asynchronous.* Participants are often located at different sites, are not often aware of the contributions of other, distributed design decisions that may affect their efforts, and often do not participate in team decision-making that makes other participants aware of their decisions and design contributions.
- *Building modeling is a problem-solving enterprise.* At the beginning of the modeling process only a few precise descriptions of important objectives, goals, and constraints are usually specified. Throughout the process, many of them will change or stay vague. Thus, the process is non-sequential and non-monotonic, as decisions made at later stages in the design process may affect and require re-evaluation of decisions taken earlier.
- *Building modeling is large-scale information management.* The different participants in building modeling problem-solving are assisted by tremendous amounts of data and information of a diverse nature. A developing building design is itself a very large information management problem where participants require a common design description and must test their design decisions against those of other participants.
- *Building modeling is a life-cycle activity.* Building modeling is not limited to phases that precede construction of the artifact. Buildings are operated, re-designed, remodeled, maintained, monitored, managed, and ultimately, dismantled and perhaps recycled. Since these activities are considered to be highly innovative modeling tasks, all life-cycle activities may be covered by a broader understanding of design.

While these features describe the real-world problems of building modeling, much work in the area of computer-supported building modeling seems to be based on the premise that these features can be accommodated by tools based on general, formalized descriptions of design problems, or on standard empirical descriptions, or by powerful, knowledge-based database management approaches, etc. In contrast, the effort presented in this paper accepts these features as initial constraints of all building modeling activities, and seeks to specify and develop software tools that can support such activities in a computer environment. It is therefore focused on problems of distributed, concurrent and asynchronous decision making.

2. Concepts of Autonomy

The concept of autonomous objects is one of the paradigms that can be applied to model the distributed, concurrent and asynchronous nature of the design process. The concept is most developed in the literature of the disciplines of concurrent programming languages and database management systems (DBMS). In concurrent languages, the actor model (Agha 1986) is an approach for modeling shared objects with changing local states and dynamic reconfigurability and inherent concurrence. Other paradigms in concurrent programming languages include process-models (Hoare 1978), functional programming (Backus 1978), and logic programming (Clark 1978). In database management systems, the paradigm of data object autonomy in a distributed persistent database, has been suggested (Kemper 1990), which moves away from traditional DBMS and even object-oriented DBMS approaches.

The following general characteristics of autonomy are drawn from the literature of both concurrent languages and distributed database management systems.

- *Object structure and behavior.* Autonomous objects are non-divisible entities that represent models of the real world object, and thus have encapsulated knowledge of and control over many aspects of their internal structure and behavior. Autonomous objects may hide their interior structure and behavior from the external world, but may also choose to present "views" to the external world and to change them according to life-cycle phases.
- *Object control.* In autonomous objects, the concept of hierarchical control is replaced by a built-in "communication interest" with other objects.
- *Object communication.* Autonomous objects communicate with each other via sensory mechanisms that accommodate "message passing" in order to resolve decisions involving inconsistency and conflict in their world. These mechanisms are dynamic and can change according to life-cycle phases. Because of the need to communicate, the object cannot be totally autonomous; there must be some common understanding about the communication media and the structure or content of the communication. Autonomous objects function asynchronously unless there is a specific request to act synchronously, e.g., to "conference" with other objects in order to share internal information or resolve inconsistent belief systems.
- *Object life cycle and transformation.* Autonomous objects are dynamic and can change their internal structure and their internal and external behavior according to life-cycle phases. In addition, they can change their physical location or migrate according to need or as a result of transformation.
- *Object integrity and consistency.* Traditional models of logic that do not tolerate inconsistency, seek to resolve inconsistent beliefs through reasoning

and conflict resolution mechanisms. In models with autonomous objects, different viewpoints and beliefs can exist without destroying the integrity of the overall model.

3. A Case of a Knowledge-based Building Model

The number of design objects in a building model can become very large. Since building modeling by its very nature involves modeling changes of state in design objects, the behavioral changes of design objects over the life cycle of the design process impose severe complexity and performance limitations for traditional CAD systems. Experience with such large scale building modeling environments suggests that existing approaches do not work in real practice. The following section describes one aspect of building modeling, namely, spatial layout of mechanical systems, and documents experiences of a research effort to model it in a computer on a large scale.

3.1. MECHANICAL SYSTEMS LAYOUT

The spatial layout of mechanical subsystems in buildings is a highly complex problem, even if undertaken only for a single building plenum. For each mechanical subsystem, the task can be viewed as a problem of routing services from a central delivery point to several supply points. The constraints of the routing problem are the cross-sections of the service (pipes or ducts), technical concerns (required slope for wastewater drainage, etc)., and the geometry of the building design. The problem, however, is a dynamic one, since the possible routing alternatives are dependent upon the routing of other subsystems, and the resulting cross-section of the service is dependent on the length of run and number of service points. Goals of the system design can be expressed as constraints too, e.g., minimize the length of the service, minimize the number of changes in service direction, and make the changes in direction as "smooth" as possible.

Even in one mechanical subsystem, these constraints can lead to severe spatial allocation problems. But a highly complex building contains many such mechanical subsystems, and so the relevant task is to find an integrated solution. Constraints for the integrated design of such subsystems are, in principle, the same as described above, though additional goals or constraints are imposed, e.g., satisfy an overall layout, loosely pack all services, leave large contiguous spaces for future remodeling, etc.

3.2. THE ARMILLA PROJECT

In 1987, Fritz Haller formed the ARMILLA Project at the Institut für Industrielle Bauproduktion, Universität Karlsruhe, Germany, to implement a computer-aided, knowledge-based design environment for the spatial layout of mechanical systems (Haller 1974; Haller 1985; Haller 1988; Mathis 1988; Raetz 1989; Hovestadt 1989; Gauchel 1990). The effort is based on sets of design strategies, which allows the spatial layout of mechanical systems in coordination with the physical structure of modular building components. The strategies were tested in the design of several buildings of Fritz Haller, most notably, the education center of the National Railway Company of Switzerland in Murthen. In 1990, the ARMILLA Project was joined by researchers at the Center for Building Performance and Diagnostics at Carnegie Mellon University (Hartkopf 1991; Van Wyk 1991). Recently, the focus of the joint effort has become the development of computer-aided design environments to support life-cycle activities of the building. The on-going research effort resulted in four prototype design tools.

Since the experience gained from the ARMILLA/A2 prototype which was completed in 1989 (Hovestadt 1990 ; Gauchel 1991) is an important source of inspiration and direction for the efforts underway, the significant features and problems of this prototype are discussed in the following sections.

3.3. THE ARMILLA/A2 PROTOTYPE

The A2 prototype is a design environment based on layout strategies, which are defined for each mechanical subsystem--e.g. pipe, duct, wiring or cable network. Central to the strategies is a system of three-dimensional orthogonal grids and a system of production rules, which checks layout alternatives against the others to resolve the spatial routing between the mechanical subsystem's supply and delivery points. The result is an integrated and systematic layout, which is restricted only by the configuration of the proposed building geometry and the ingenuity of the proposed layout strategies. The system architecture of A2 has the following features:

- The "design world" is represented as a semantic net, whose nodes are object-oriented descriptions of *building objects*: spaces, physical parts, assemblies; *modeling objects*: design tasks, intentions, production rules, constraints; and *graphic interface objects*: graphic primitives, display primitives, and interactive tools, etc.
- The design process is represented as a set of tasks that can be expressed at any level of abstraction from conceptual layout sketches to parts lists, and accessed non-sequentially or in stages.
- The implementation is based on a system kernel that supports the interactive definition and manipulation of *building objects*. The system kernel can be made more "intelligent", with domain-specific, user-supplied modules of knowledge-based generative and diagnostic tools.
- The user interface supports three different levels of user interaction with design tasks: (1) in the traditional CAD sense, the user interface provides graphic interface and modeling objects to define and manipulate building

objects *interactively*; (2) in the expert diagnostic design sense, the interface acts as a design assistant by providing "background" consistency and plausibility checks on actions initiated by the user; and (3) in the expert generative design sense, the interface acts as design assistant by providing rule- and knowledge-based automated design actions that react to initial changes by the user. The latter two levels provide the user with knowledgeable design assistants, whose actions can be overridden at any time.

- The graphic interface is based on the concept of design tasks. Each design task is configured to provide (1) a specific view or representation of involved design objects; (2) a set of graphic and interactive tools to define and change the state of design objects; and (3) a set of user-activated generative and diagnostic tools corresponding to the design modes discussed above.
- The system kernel is supported with utilities to import and export the three-dimensional data so that AutoCAD could be used to define the geometry of building objects and other advanced CAD systems could be used to visualize, render, and animate the building objects defined in ARMILLA/A2.

3.4. LIMITATIONS OF ARMILLA/A2

Though ARMILLA/A2 was specifically designed to handle the spatial coordination of mechanical systems, it was always viewed by its designers as a "mini", generic, and dynamic building model, which could be enhanced and applied to other real-world building design problems. That view was illusory, as was proved by an experiment to design all mechanical subsystems that were required in a single plenum of a single floor of a highly complex laboratory building. The resulting model, composed of several thousand instances of more than a hundred building objects, led to some major problems and these are discussed further here.

- *Complexity*. The design world, as described earlier, can be understood as a set of interacting objects (Fig. 1). An obvious implementation of this understanding is a semantic net: the objects can be described as nodes and the edges as connections between the objects, in terms of pre-defined bindings, which serve as channels for message passing. In a densely connected net the number of connections would tend towards the square of the number of nodes, so that the introduction of new nodes leads to increased density of connections. It became increasingly difficult to model, implement, predict and effectively test the behavior of A2, and there was a sharp degradation in performance.

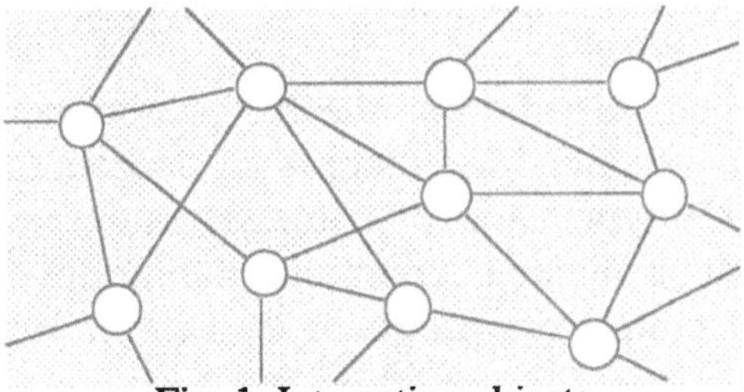

Fig. 1. Interacting objects

- *Multi-User Systems.* Though the A2 prototype demonstrates important features as to how different experts can work together, it was a single user system. For A2 to function as a multi-user system, different users would have to work concurrently on the entire net, or on separate partial nets defined for special purposes. For both alternatives, the designers of A2 found no promising solutions. Working concurrently on the entire net led to unpredictable results due to the inherent connectivity. Defining partial nets led to problems in handling interactions between connected nodes in different partial nets as well as the problem of overlapping partial nets (Fig. 2).

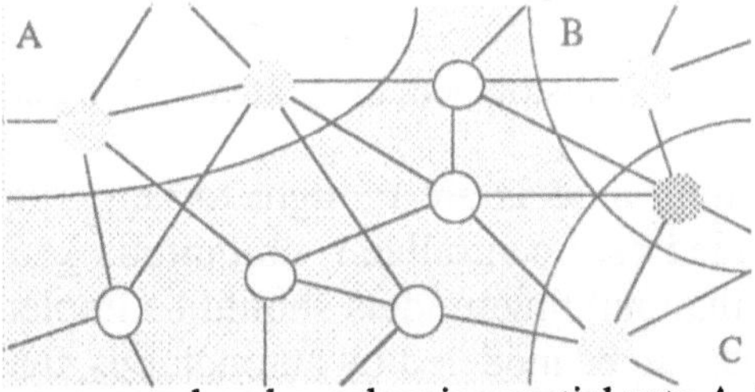

Fig. 2. Interconnected and overlapping partial nets A, B and C .

- *Unpredictability.* Another problem was that of making modifications to a tightly connected net. Adding or deleting nodes or edges led to erratic and unanticipated behavior in other parts of the system, leading to unpredictable design states of the system. Maintaining the integrity of a semantic net and the inheritance properties of the various objects in such a dense net became extremely difficult.

Evaluating these limitations, some members of the ARMILLA group became convinced that most of the problems of the A2 prototype arose from the fundamental decision to choose a semantic net as implementation principle. They decided to search for a different approach, which would allow one to easily describe and modify separate partial models while retaining useful features to design semantic nets, e.g., class structures and inheritance mechanisms.

4. The ARMILLA/A4 Prototype

Based on the experience gained from the A2 prototype, and influenced by discussions of concepts of autonomy, there is an on-going effort to implement the ARMILLA/A4 prototype as a distributed, integrated building design environment, which will be able to support all building life cycle activities. The principles of its architecture were proposed by Ludger Hovestadt (Hovestadt 1991).

A4 includes two types of components, (1) applications that synthesize, analyze, evaluate, visualize, etc., and (2) a data communication module. Among the applications are user-interfaces, building models and third party evaluation applications (Fig. 3).

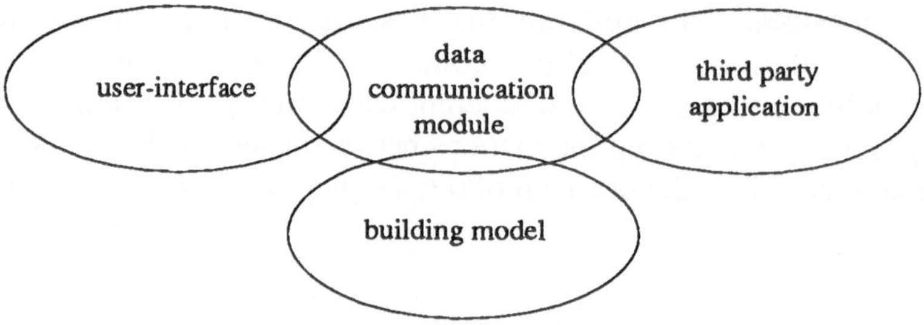

Fig. 3. ARMILLA/A4 - a distributed, integrated building design environment.

Traditionally, building models are thought to be partial models of a global model, and implemented as a centralized, ubiquitous knowledge source. Our experiences suggest that building models should be implemented as distributed knowledge sources that can be modeled by autonomous, interacting components.

Thus, in A4, building models are sets of "building modules", which are selected to serve special purposes. Modules can support users with tools to develop descriptions of aspects of real-world buildings, or to create data for third party applications such as a building performance simulation tool. As a result, building models are always paired with at least one other application. Though these applications are separate tools, the building models can overlap simply by sharing building modules.

5. Building Modules

Building modules are pre-defined, knowledge-based systems, that support users with tools to develop building descriptions and to represent their design decisions. The architecture and interaction of building modules are based on the paradigm of autonomy.

5.1. THE APPROACH TO BUILDING MODULES

The approach to building modules answers the following questions: How can highly-connected building descriptions be split into separate modules, which can interact with one another to re-achieve the quality of a highly connective description? And, in search for an appropriate answer: Can there be a system architecture, which is common to all modules, and a unified interaction principle between modules? These questions suggest a close relation between the approach to building modules and the approach to a data communication module.

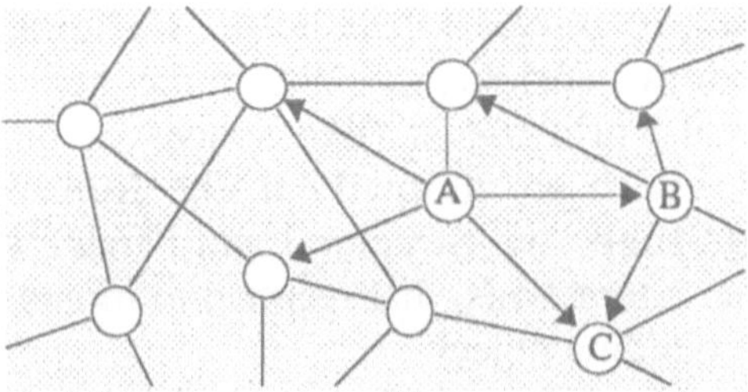

Fig. 4. Two building objects as sources of interactions with other building objects

Imagine two building objects A and B that may represent spaces or building components and are sources of interactions with some other building objects, for example with object C (Fig. 4).

The basic approach to building modules is to split highly connected building descriptions into simple interaction patterns, as shown in Fig. 4, and to encapsulate them into separate modules, as shown in Fig. 5: While module A encapsulates the interaction pattern from object A to objects B, C and others, module B encapsulates the pattern from object B to objects C and others. Since the principle of splitting highly connected descriptions into interaction patterns is not dependent on the nature of building objects and their connections, it allows all building modules to be structured in the same way

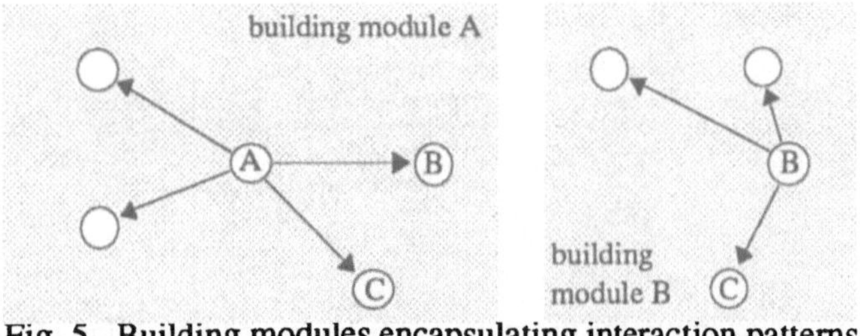

Fig. 5. Building modules encapsulating interaction patterns.

Encapsulating the interaction patterns A and B into separate modules is only possible by introducing redundant representations of the shared building objects B and C. That leads to the basic principle for interactions among modules: modules may interact if they share building objects. Since this general principle is also not dependent on the nature of building objects and their connections, a single unified interaction or communication mechanism for all modules can be developed.

5.2. BUILDING OBJECTS

The architecture of building modules is based on building objects that are the nodes of the encapsulated interaction pattern. Building objects are sets of structural and behavioral attributes. While the structural attributes are descriptions of the real-world object (like spaces or building components), behavioral attributes, which are normally in the form of functions, can be characterized as design supports. Structural and behavioral attributes may be linked to create internal interactions, that respond to inputs, i.e. changes that are initiated outside of the building object

In principle, all initial changes are aimed at a single structural attribute of a building object, which may be linked to a behavioral attribute. In this case, the change of state of this attribute will trigger the function, which will change the state of another structural attribute, and so on, until the changed structural attribute has no link to a behavioral attribute. Thus, building objects can be thought of as systems, reacting on inputs, seeking stable, internal states or data levels, and having no output. Building objects can be implemented as "frames and slots" and the internal linkage between structural and behavioral attributes in the form of "daemons" (Fig. 6).

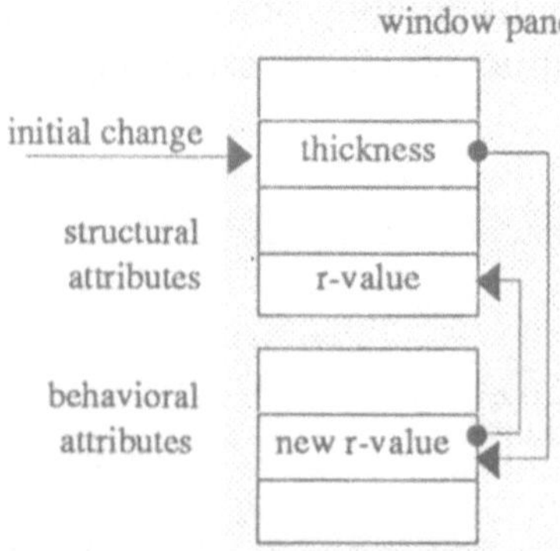

Fig. 6. An example for the state and behavior of a building object

5.3. CENTRAL AND SURROUNDING OBJECTS

Inside a building module, building objects have different roles. There is one central object and one or more surrounding objects. Only the central object can get inputs from outside the module. By its changes, it can interact with the surrounding objects, allowing them to respond with changes of their own. Thus, a building object is a central object of its own building module and can be a surrounding object in any other building modules.

5.4. INTERACTIONS BETWEEN BUILDING OBJECTS

The interactions of the central object with its surrounding objects are modeled as production rules. These rules are "watching" the changes of structural attributes of the central object, and are automatically triggered by special changes that results in changes of structural attributes in surrounding objects. This may include creation, deletion or modification of the surrounding objects (Fig. 7).

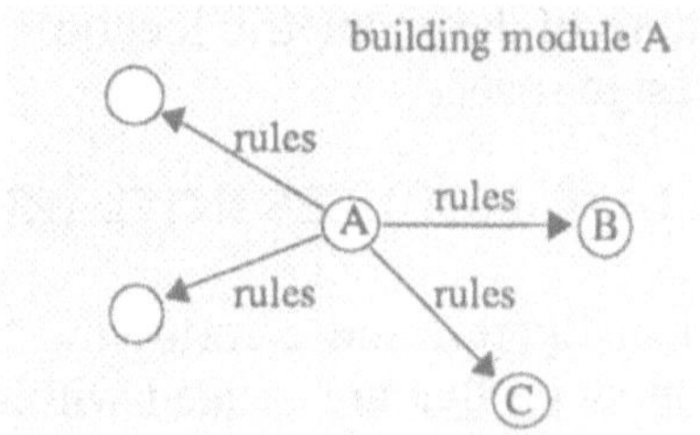

Fig. 7. The architecture of a building module, based on interactions of a central object with its surrounding objects.

5.5. CONCEPTS, INSTANCES, INHERITANCE

The design of a building module is based on concepts of the included building objects, and therefore a building module has the character of a concept. Thus, it is possible to create instances of building objects, as well as instances of building modules. The instances of a building module inherit all characteristics of the concept.

6. Building Models

Building models are "shells" that allow users (1) to select sets of building modules for special purposes; (2) to declare "areas of interest"; and (3) to load instances of building modules into these areas. While building modules can be viewed as domain applications, these customized building models can be viewed as design applications.

6.1. DESIGN SPACES AND AREAS OF INTEREST

The ARMILLA/A4 prototype is based on a concept of a multi-dimensional design space that may cover all building life-cycle activities. For example, since all spaces and components of buildings are located in a geometric space, which can be described by three dimensions, the geometric attributes form one dimension of the design space; since building modeling is dependent on time, time is another dimension. Another dimension is that of "expert views", which represent interests of experts in special aspects of building descriptions, according to their expertise and role in the building process. Developing this approach further, at least thirteen dimensions of a design space can be declared as relevant for building modeling (Hovestadt 1991a, 1991b, 1992).

Implementations of this concept must have two aspects: (1) All building objects must have structural attributes, that describe their location in the multi-dimensional design space. (2) Areas of interest have to be described as contiguous sections of such a design space. Probably, the implementation of different dimensions has to follow different strategies. Our current experience is gained from implementations of the geometric location, the time, and the expert-view dimensions of the design space.

6.2. LOADING BUILDING MODULES INTO AREAS OF INTEREST

A declaration of an area of interest and a following "load" command causes instances of the selected modules that are located within the declared area to be loaded to form an actual design application. This selection of the instances is automatically done by the loading mechanism.

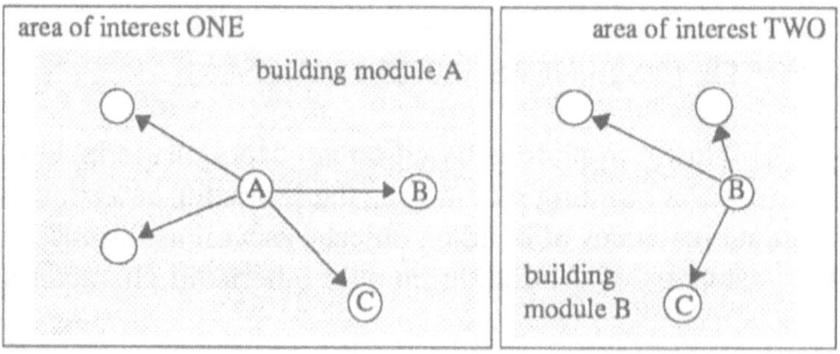

Fig. 8. Building Modules A and B loaded into different areas of interest

There are two alternatives when loading different building modules like modules A and B: their instances can be loaded into different or into the same area of interest. Loading them into different areas of interest would have the result in redundant representations of shared building objects (Fig. 8). This is the case even if the areas overlap in the design space or even if they are identical. Loading them into the same area of interest would result in breaking up the modules and in a union of instances brought in by each of the modules. This is automatically done by the loading mechanism.(Fig. 9).

Working within an area of interest makes it impossible to manipulate instances that are outside this area. Calling for new information from within a current area of interest has the result that only those instances are loaded, which may have been changed or created concurrently in other areas of interest, but which are within the current area.

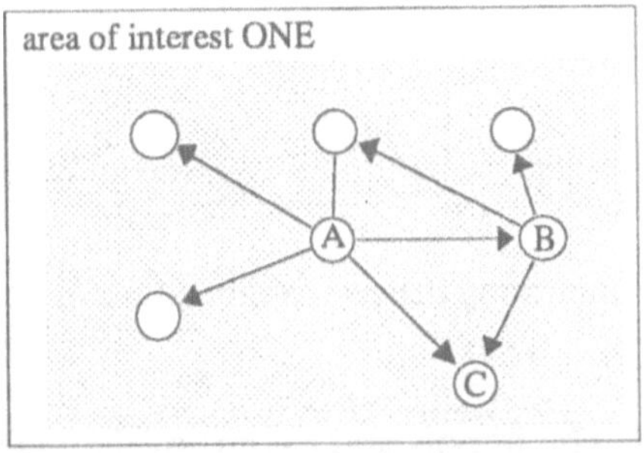

Fig. 9. Building modules A and B loaded into one area of interest

Users are represented to the system as unique areas of interest. By changing their personal area of interest during the course of design, the users can navigate through the system.

7. Principles of the Data Communication Module

In A4, many applications (user-interfaces, building models, third party applications) may be concurrently active, may be on one machine, but mostly on different machines and on different sites. To achieve an integrated design, the applications have to communicate, which is the task of the data communication module. This task includes (1) data transfer between applications, (2) data import-export tools and (3) mechanisms for conflict recognition for the applications (Fig. 10).

Data Transfer. We propose to implement the data transfer between applications by drawing on principles of electronic mail systems across networks using the client/server model. A similar implementation for autonomous object oriented systems such as KNOS (Casais, 1988) has been described. Since transfering data across networks is decentralized, there is no means to recognize and resolve data integrity conflicts. Since the data transfer mechanisms have no design knowledge, they cannot recognize and resolve conflicts due to distributed design decisions.

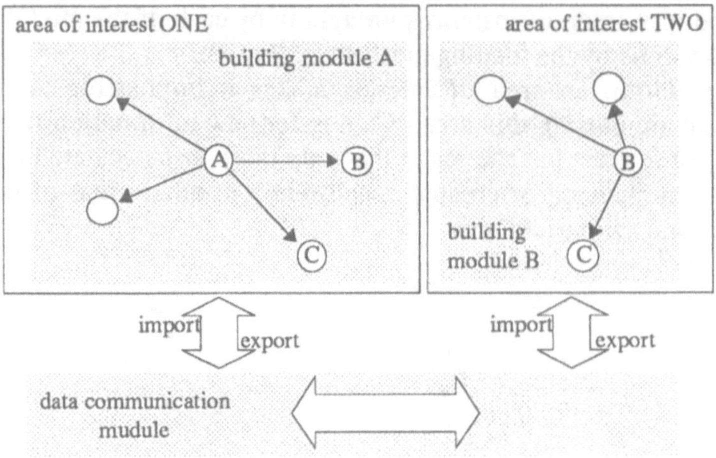

Fig. 10. Communication between applications or areas of interest

Data Import-Export. The import and export of data is individually controlled by applications. Since there are situations, in which communication is not helpful, and others, in which any internal change should be exported and any external change should be imported immediately, all applications are free to decide when to import and export.

Conflict Recognition and Resolution. Conflict recognition and resolution can only happen inside applications. Conflict recognition is supported by import tools by sending appropriate signals to the users. Thus, the import of a completely new instance into the application, or the updating of an existing instance results in the detection of conflicting internal states that can be signaled by highlighting the relevant objects. It is the user's task to find out, if these signals indicate design conflicts.

Since all design knowledge is encapsulated in an area of interest or application, there cannot be an automated conflict resolution, when conflicts are imported from outside. But since signalling of possible conflicts may also include the "name and address" of the other application that may be involved in a conflict, "conferences" between the involved applications can be arranged to solve the problem.

8. Principles of User-Interfaces

User-interfaces should enable users to explore design-spaces. Exploration means that interfaces should allow users to create objects and their interaction dynamically, in a way that the design of data structures and automated procedures and using these pre-definitions should become closely related tasks or even one task, Software that is based on this principle will enable users to move into sections of design spaces,that are not covered by strict pre-definitions

(Müller 89). It is obvious that this possibility is very helpful to deal with some of the features of building modeling described earlier.

Currently, we are using an user-interface that allows us to develop hierarchical organized, highly overlapping,but completely distributed "patterns" of areas of interest, that describe positions and relations of spaces and building components on the xyz-dimension of a design space. It can manipulate three types of areas of interest: (1) zones, as vague descriptions of locations of spaces and building components; (2) virtual hulls or envelopes, as spatial reservations for one space or one component; and (3) contours as precise descriptions of spaces and components. Since all building objects can be identified by their contours, they all can be viewed as areas of interest.

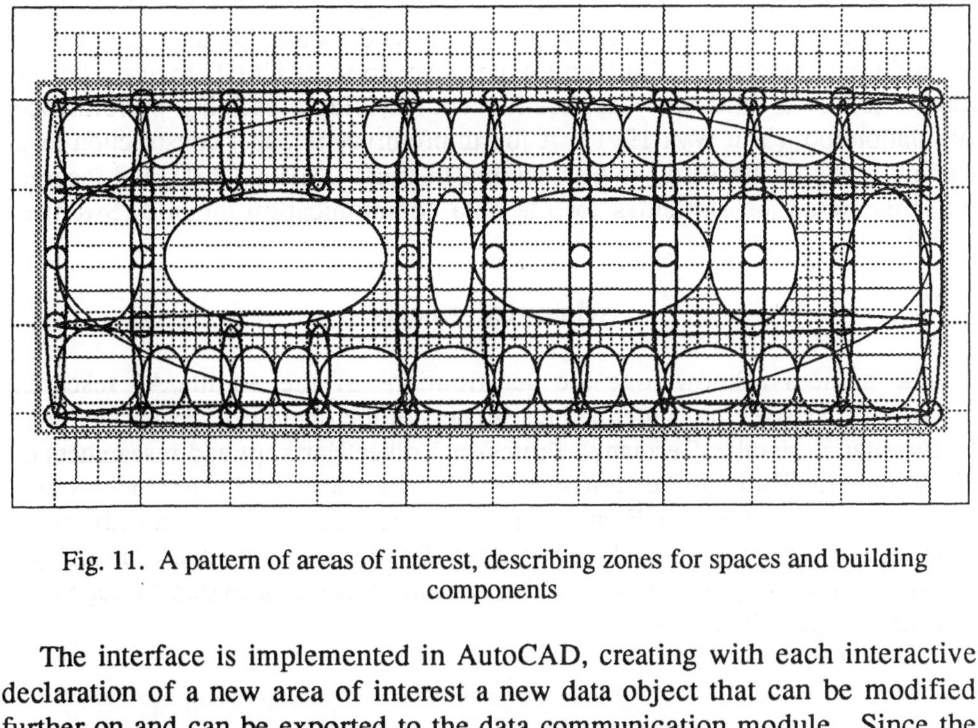

Fig. 11. A pattern of areas of interest, describing zones for spaces and building components

The interface is implemented in AutoCAD, creating with each interactive declaration of a new area of interest a new data object that can be modified further on and can be exported to the data communication module. Since the used software has some limitations, it is not possible to extend the prototype to have editors which can develop the data objects to more powerful descriptions of real-world objects.

9. Conclusions

Levels of autonomy. The basic approach of A4 can be described as "encapsulating scenarios of objects - in terms of object-oriented programming - into sections of design spaces, to be handled as autonomous objects"

Beside this, there are "lower" levels of autonomy in A4, especially when looking at the concept of modular building models: (1). Building Models are sets of single building modules, which are small environments that can be described without any reference to other building modules. (2) The "primitives" of building modules are building objects, which can be designed as systems of their own. Instances of building objects can receive "inputs" from outside that can trigger internal activities, but have no pre-defined linkages to external activities and (3) these activities are modeled by production rules, which are software tools, without any pre-defined linkages to their surroundings.

Prototypes. Currently, the development of A4 is based on four prototypes. One focuses on user-interfaces, and is done by Ludger Hovestadt at the University of Karlsruhe, Germany. Another prototype is done by the other authors at Carnegie Mellon University (CMU) in Pittsburgh, USA, and focuses on the development of building models. A third prototype, also done at CMU, focuses on the integration of a third party application with DOE2, a building performance simulation tool (Van Wyk 1991). A fourth prototype is under construction by a student group of the Software Engineering Institute (SEI) at CMU, and focuses on implementation strategies for the data communication module (Swonger 1991).

Acknowledgements

The authors acknowledge the contributions of the armilla/A4 research project: Professor Fritz Haller and researchers of the Institut für Bauproduktion, Universität Karlsruhe, Germany; Professor Volker Hartkopf and researchers of the Center for Building Performance and Diagnostics; graduate students and researchers of the Software Engineering Institute at Carnegie Mellon University; funding and support by the RETEX project (BMFT, Germany and Company of ROM); and funding and support of The Advanced Building Systems Integration Consortium (ABSIC).

References

Agha, G.A.: 1986, *ACTORS: A Model of Concurrent Computation in Distributed Systems*, MIT Press, Cambridge, MA.

Backus, J.: 1978, Can programming be liberated from the Von Neumann style? A functional style and its algebra of programs, *CACM*, **21**(8), 613-641.

Bjoerk, B. C.: 1989, A proposed structure of a building product model., *Computer Aided Design*, **21**(2), 71-78.

Clark, K.L., McCabe, F.G and Gregory, S.: 1982, IC-Prolog, *Logic Programming*, Academic Press, London.

Fenves, S. J., Flemming U., Hendrickson, C., Maher, M. L. and Schmitt, G.: 1990, Integrated software environment for building design and construction, *Computer Aided Design*, **22**(1).

Gauchel, J. (ed.): 1990, *KI-Forschung im Baubereich*, Ernst and Sohn, Germany.

Gauchel, J: 1991, ARMILLA—A knowledge-based design tool for an integrated layout of complex mechanical systems in buildings, Carnegie Mellon University, Pittsburgh (unpublished).

Haller, F.: 1974, *MIDI—ein offenes system für mehrgeschossige bauten mit integrierter medieninstallation*, USM Bausysteme Haller, Munsingen, Switzerland.

Haller, F. et al.: 1985, *ARMILLA—ein installationsmodell*, Universität Karlsruhe, Germany.

Haller, F.: 1988, *Fritz Haller—Bauen und Forschen*, Solothurn, Switzerland

Hartkopf, V. and Gauchel, J.: 1991, A computer-aided design approach for total building performance. *Proceedings, Building Systems Automation-Integration Conference*, Madison, WI

Hoare, C.A.R.: 1978, Communication sequential process, *CACM*, **21**(8), 666-677.

Hovestadt, L.: 1989, *Ein 'intelligentes CAD-System' für die Planung und Verwaltung von Leitungsnetzen in Hochinstallierten Gebäuden*, VDI-Berichte 775, Germany.

Hovestadt, L.: 1990, *The Armilla Film*, MacroMind Animation for Macintosh, Universität Karlsruhe, Germany.

Hovestadt, L.: 1991a, A4—a model for an extensive use of computers in architecture, *Proceedings of the 3rd International Symposium on Systems Research, Informatics and Cybernetics*, Baden-Baden, Germany.

Hovestadt, L.: 1991b, BO—A Generative Model for An International, Intergated and Distributed Computer Environment for The Building Industry, *Proposal*, ESPRIT III , R&D TASK IV.2.3.

Hovestadt, L.: 1992, *A4—Digitales Bauen*, draft document, PhD Dissertation, Universität Karlsruhe, Germany (unpublished).

Kemper, A., Lockemann, P.C., Moerkotte, G., Walter, D.D. and Lang, S.M.: 1990, Autonomy over Ubiquity: Coping with the Complexity of a Distributed World, *Proceedings, International Conference on Entity-Relationship Approach*, Lausanne, Switzerland.

Mathis, C.: 1988, *Installationsplanung mit einem Wissensbasierten System*, VDI-Verlag, Germany.

Müller, J. and Fanz, L.: 1989, Anforderungen des Konstrukteurs an die Entwicklung von Wissenssystemen für das Konstruieren, in *VDI-Berichte* 775, Germany.

Pohl, J., et al: 1992, *A Computer-Based Design Environment—Implemented and Planned Extensions of The ICADS Model*, California Polytechnic and State University, CA.

Raetz, P.: 1989, *XNET: Ein intelligentes CAD-System für die Planung von lokalen Netzwerken in Gebäuden*, VDI-Verlag, Germany.

Swonger, R., Asada, T., Bounds, N. and Duerig P.: 1991, *The APED Project*, Software Engineering Institute, Carnegie Mellon University (unpublished).

Van Wyk, S., Bhat, R., Gauchel, J. and Hartkopf, V.: 1991, A knowledge-based approach to building design and performance evaluation, *Proceedings, ACADIA '91 Reality and Design*.

Wilson, P. R.and Kennicott, P. R. (eds): 1988, ISO STEP baseline requirements document (IPIM), ISO/TC184/SC4/WG1 *Document N284*. First Working Draft of STEP 1.0.27.10.1988, 607p.

Ziegler, B.: 1978, *Theory of Modeling and Simulation*, John Wiley, New York.

4

DESIGN OBJECT MODELING

REPRESENTING DESIGN OBJECTS IN SORAC

A data model with semantic objects, relationships, and constraints

B. MACKELLAR*
Department of Computer and Information Science
New Jersey Institute of Technology
Newark NJ 07102 USA

and

J. PECKHAM**
Department of Computer Science and Statistics
University of Rhode Island
Kingston RI 02881 USA

Abstract. In this paper, we apply our experience in developing ArchObjects, an intelligent architectural design system, towards the development of a general data modeling language appropriate for design representations. An adequate design representation language is crucial to the ability to reason about designs. Current object-oriented and semantic data models are not well-suited to complex design domains, however. The built-in relationship types of the traditional semantic data models do not adequately support the needs of design applications nor do they provide adequate behavioral support. Although object-oriented models are popular, and provide encapsulation of behavior, these do not support built-in relationships or reuse of relationship semantics. Neither model includes constraints, which are a fundamental concept in ArchObjects and other intelligent design systems. To solve these problems, we are building SORAC, an object-oriented model with semantic relationships and constraints. This model has several major features. Relationships are represented as objects, with a relationship class hierarchy. In addition, relationship classes are derived from a core set of semantic relationships by selecting certain semantic options. Complex constraints are also represented as objects; these are constraints over several relationships and can represent domain knowledge such as fire safety codes. First, we discuss the requirements imposed on this data model from the design domain. Next, we present the core semantic relationships of SORAC, optional semantics, and an overview of the data modeling language. Finally, some examples of using relationship semantics to simplify triggering and evaluation of domain constraints are presented.

1. Introduction

The ability to adequately represent a design is crucial to the ability to reason about it. One of the most important tools in the task of representing designs is the data model (or knowledge representation language). This is the language in which templates for designs and knowledge about designs must be expressed. We call this template a *schema* in this paper. There has been quite a flurry of interest in building languages for representing designs among both database researchers and AI (Artificial Intelligence) researchers. The database research has been concentrated in the area of building object-oriented (OO) models that can

* SBR Grant #421120, New Jersey Inst. of Technology
** Proposal Development Grant #537-061, University of Rhode Island

J. S. Gero (ed.), Artificial Intelligence in Design '92, 201–219.

incorporate the behavioral needs of design applications, and exploring efficient ways to implement persistent design objects. Within the AI community, research has been concentrated in the area of developing richer representation languages, typically without persistence, that support intelligent computational methods. The two views, however, have been converging as the need for representations that are both rich enough to support advanced design applications and capable of supporting persistence is recognized.

In general, database-style representations tend to be more formally defined, and to provide greater support for the user. There is a large body of literature describing schema design tools, methods for propagating updates, version control systems, and query languages. On the other hand, most data models are still not rich enough for adequate support of design objects. In particular, most data models are weak in their support of relationships between objects, which are essential in design representations. The relational model imposes a flat structure on every piece of data, making explicit representation of relationships impossible. The semantic models of the early 80's (Hammer and McLeod 1981, Mylopoulos, Bernstein and Wong 1980) provided strong capabilities in the area of relationship modeling. Collections of built in relationships, typically generalization and aggregation, were supported. Unfortunately these models do not provide a rich enough set of constructs to support the complex structures needed in design systems. In particular, part relationships are not well understood or supported. Also, most semantic data models do not support encapsulation of behavior or complex constraints, both necessary for the representation of design domain knowledge.

Some of these deficiencies were addressed with the object-oriented (OO) models by associating methods with object classes, thus supporting encapsulation. These models also support specialization and inheritance. This paradigm presents clear advantages for representation of design objects because:

- Design objects are large and complex. OO concepts offer a way to structure this complexity and decrease the number of errors.
- The concepts of inheritance and specialization are natural and useful in this domain.
- The use of classes promotes reusability of component designs. A class library of design components can be developed.
- Because of specialization, constraints can be stated at the highest possible level. For instance, in an architectural design database, a constraint on door objects may be stated at that level rather than repeated for specialized classes such as SlidingGlassDoor or HingedDoor.

Despite this appropriateness, the object-oriented model does not provide support for relationships other than IS-A: again, part relationships are not well supported. Therefore, a obvious improvement to the object-oriented model is to augment it with built-in semantic relationships, producing an "object-oriented semantic data model".

Several researchers have addressed this idea of representing relationships in object-oriented models. (Nguyen and Rieu 1991) have addressed the issue of semantic complexity in design objects in their design model SHOOD. Kim (Kim, Chou and Banerjee 1987, Kim 1989) has investigated the semantics of composite objects extensively in the context of ORION. An object-oriented data model called ADAM(Diaz 1990) is close to SORAC. Both ADAM and SORAC are based on earlier work by (Rumbaugh 1987), who proposed an OO language in which relationship types form classes. ADAM has greatly influenced the present model. It does not, however, provide as strong a notion of built-in relationship types.

Many of these difficulties with relationship representation have also been observed in the course of building ArchObjects (MacKellar and Ozel 1991), (MacKellar 1992). ArchObjects is an intelligent design assistant that functions in the domain of architecture. It includes a graphical user interface, a knowledge representation, a set of complex constraints, and a persistent backend DBMS. ArchObjects provides design verification services by evaluating the design against a set of design codes, particularly legally required rules such as fire code. Semantic integrity constraints are used to model the design codes and design requirements. Relationships between objects function as a central organizing concept within the knowledge representation.

Although ArchObjects was based on a set of relationships, its semantics were embedded in the various methods defined on the object classes, making interactions between relationships difficult to predict and manage. In addition, these relationships were specific to ArchObjects. If this system were to be significantly extended, or another application built, a more general set of relationships and semantics will be needed in order to provide a cleaner separation between the application and the modeling language. In such an environment, the designer of an intelligent design system (we will call this person the "schema designer"), could fine-tune the semantics of each of the standard relationships by choosing from a "menu of semantics". Correctly maintaining those semantic choices would then be the responsibility of the data model rather than individual method programmers. To this end, we are developing a a data model called SORAC (Semantic Objects, Relationships, and Constraints), in which the second version of ArchObjects will be implemented.

In this paper we describe SORAC and show its suitability for knowledge-based design representations. This model supports the following types of objects: objects, relationships, and constraints. The central purpose of this model is to elevate relationships to the same level of importance as objects and to provide the same level of encapsulation and reusability normally accorded to objects. Rather than merely provide built-in semantic relationships, SORAC provides built-in core relationship types with associated sets of update rules. An update rule is an operational means of supporting relationship semantics. The schema designer may pick and choose among the update rules in order to construct a new *derived relationship type* with appropriate behavior. The process of deriving such a relationship type is accomplished by automatically generating a derived relationship type description and associating the update rules which maintain the chosen semantics with the description as methods. The derived relationship types are arranged in an inheritance hierarchy in order to support reuse of update rules.

2. Characteristics of Design Data Models

Design representation is a great challenge for data models because of the structural, semantic, and behavioral complexity of designs. Much of the structural complexity occurs at the instance level. In traditional data representation problems, most attention is focused on complexity at the schema or type level. For example, in constructing a representation within the object-oriented model, the structure of instances is determined by the structure of their type. In the design domain, however, the structure of an object is determined dynamically through its relationships with other objects. Thus, it is impossible to state at the type level that all doors will contain windows, even though some instances might. In general, all that can be predicted about an object type such as a door is that it will contain a set, possibly empty, of components.

The problem is that the designer may wish to associate specialized behavior with the part relationship between the door and the window *in the event* that a door does have a window. For example, we may wish the system to automatically delete a window that is contained by a door if the door is deleted, but to retain the door's locking hardware for reuse. This distinction is impossible if all that is known at the type level is that the door contains a set of components. On the other hand, if we explicitly associate the lock and window parts with the door type, we run into the problem of sparse relationships(Rumbaugh 1987). At the instance level space will be reserved for each of these potential part relationships, whether or not a given instance contains such a part.

Another characteristic of design representations is that more relationship types than <is_a> and <has_part> are often needed. Sets and geometric relationships may be needed; ArchObjects contains both these types of relationships. ArchObjects uses a role relationship to associate instances with various functional roles, because functional roles occur independently of component types(MacKellar and Ozel 1991). Therefore, an adequate design data model must not only support relationship semantics, but provide a way to design new relationship types as well. Incorporating relationship types into an IS-A hierarchy as suggested by (Rumbaugh 1987), permits reuse of relationship descriptions.

A final characteristic is the need for *complex constraints*. In design domains, much of the knowledge that must be represented is in the form of constraints across groups of related objects, and the subject of the constraint may be the relationships themselves. For example, a constraint may state that a particular size of elevator must be contained in a shaft of another particular size. This can be alternatively viewed as constraining the elevator instance, the shaft instance, or the ability to insert a "contains" relationship between them. Therefore, an adequate design representation needs to support complex constraints as an entity independent of object classes. Such constraints are a major feature of ArchObjects, where they are used to verify buildings compliance with fire code. In addition, since complex constraints are often defined on relationships, interactions between the constraint and relationship semantics may occur. These must be taken into account as well. Since there may be many complex constraints, a way to selectively evaluate only those impacted by a change to the database is needed. The approach used by ArchObjects is that of triggers, which tie a database update action to appropriate rules to be evaluated.

3. Relationships are Central

In our view, a relationship is a different kind of entity from an object, since it involves two or more objects. In the design domain in particular, relationships assume a central role. For example, in an architectural plan, the individual components are not usually as interesting as the relationships among them. We may wish to determine if two components are next to each other, or if one component is part of another. In ArchObjects, the complex constraints representing fire code use defined relationships between objects as their main building blocks. Complex constraints in ArchObjects consist of quantified IF-THEN rules, which may contain only relationships and methods. A complex constraint can be viewed as an existence constraint defined between several relationships. For example, this complex constraint states that a locked door in the means of egress must have a warning sign. This rule,

```
forall D,L
 (IF inst(D,door) AND
     role(D,MOEC) AND
     has-part(D,L) AND
     inst(L,lock) AND
     has-attr(L,keyreq,true)
  THEN
    (exists S (inst(S, sign) AND
               has-part(D,S) AND
               attr(S,msg,"Warning..."))
```

is evaluated by finding relationship instances.

A simpler type of constraint is defined over a single relationship and specifies the relative states of the related objects. We refer to these as *relationship constraints* in order to distinguish them from the domain-specific complex constraints. Cardinality constraints and existence constraints express the relative legal

states of related objects in the database. For example, suppose that in ArchObjects we have a type `has-part(room, enclosureset)`. The semantics of the `<has_part>` relationship may specify that an instance of `room` cannot be created without being associated with a set of related enclosure instances. This is an existence constraint. Therefore, we associate with every constraint an *update rule*, which specifies the exact operational means by which a relationship constraint will be maintained in the database
(Peckham, Maryanski, Beshers, Chapman and Demurjian 1989). For example, if a ROOM instance is created, the associated update rule dictates a set of actions that either prompt the user to create an instantiation of the set of enclosures, or automatically By providing relationship types as a modeling construct in SORAC, update rules to maintain constraints may associated with each derived relationship type. Furthermore, incorporating derived relationship types into an IS-A hierarchy permits reuse of update rules.

4. The SORAC Model

Therefore, relationship types and the update rules that maintain their semantics are crucial to a design representation language. The usual technique of embedding relationship semantics within hand coded methods on object types means that it is difficult to ensure that the correctness of the relationship is upheld. In addition, repeatedly ensuring the same semantics across all methods is a waste of the developer's time. For example, in ArchObjects, the `<has_part>` relationship is central to the representation of design objects, appears across all object types, and must always be maintained correctly. Therefore, a data model that is adequate for the task of specifying ArchObjects must support these features :

1. built in relationship types
2. built-in constraints and automatically maintained update rules on the relationship types.
3. the ability to define new relationship types.
4. the ability to define complex constraints involving the built-in relationships.

These goals are met by SORAC which supports semantic relationships as a central construct. Our modeling of relationships as classes is based on Rumbaugh's idea of distinct association classes(Rumbaugh 1987). A primary difference, however, is that our model is also based on a core set of semantic relationship types and incorporates complex constraints between relationship types. In SORAC, we provide :

– A set of core semantic relationships which provide a conceptual framework for the schema designer.
– A coherent set of update rules associated with each core relationship type, ensuring that the chosen behavior of each kind of derived relationship type conforms to the concept represented by that relationship.

– Each core relationship type can serve as a generator of derived relationship types. Method templates for optional update rules can be associated with these relationship generators. When a specific relationship type is generated from a core type, the method templates for the chosen update rules are instantiated and associated with the new relationship class.

First, we describe the set of core relationship types. Then, a language which incorporates these core types as metaclasses is presented.

4.1. Core Semantic Relationship Types

Core relationship types are "generic" and are used as templates from which the actual relationship types used in the schema are derived. The core relationship types are defined minimally. There are two relationship constraints imposed on all instances of relationship types. Instances of relationship cannot be established between objects without the presence of the objects. This is called the *general relationship insertion rule*. Also, we assume that if an object participating in a relationship instance is deleted from the database, the relationship instance also automatically disappears. This is called the *general relationship deletion rule.* Additional relationship constraints can then be imposed by the schema designer on the derived relationship type. Each core relationship type has an associated set of possible update rules, which make sense in the context of the supported relationship. In this paper, we will present only three of the core relationship types in SORAC.

The core IS-PART-OF relationship type is crucial for the support of design databases. The core IS-PART-OF relationship type contains only structure and cardinality constraints. Further optional semantics are then provided which can be selected by the schema designer when deriving a relationship type from the generic IS-PART-OF. The optional semantics are fully system-supported. The <has_part> relationship type in ArchObjects is derived from this core relationship type. The semantics required to support this derived relationship type in ArchObjects are

1. If an object `p` pf type `T1` exists, then it must participate in `has-part(q,p)` with an object `q` of type `T2`. To support this constraint, these update rules are needed :
   - MANDATORY DELETION (RDM) :
     ```
     If delete(has-part(ob1,ob2))
     then delete(inst(ob2,T))
     ```
   - CONDITIONAL DELETION (DC):
     ```
     If delete(has-part(ob1,ob2))
     then delete(inst(ob2,T))
     ```
     for all `ob2` that are not participating in some other `has-part(o,ob2)`.
2. If an object `p` of type `T1` exists, it must participate in `has-part(p,q)` with another object `q` of type `T2`.

To support this constraint :

- MANDATORY DELETION (RODM) :
  `If delete(has-part(ob1,ob2))`
  `then delete(inst(ob1,T))`
- CONDITIONAL DELETION (DOC) :
  `If delete(has-part(ob1,ob2))`
  `then delete(inst(ob1,T))`
  as long as there is no other relationship `has-part(ob1,o)`.
- CONDITIONAL INSERTION (IC)
  If `insert(inst(o1,T1))` then for all `T2` such that `has-part(T1,T2)` is in the type definition of `T1`, either
  `insert(inst(o2,T2))`
  `insert(has-part(o1,o2))`
  or `insert(has-part(o1,o2))` for some already existing `inst(o2,T2))` identified by the user.

These constraints and update rules are in fact only a subset of those available in SORAC. The construction of the <`has_part`> relationship in ArchObjects using this core relationship type is presented in Section 4.1.1.

The COLLECTION core relationship type allows objects consisting of collections or sets of objects of a given type to be described. The semantics of collection objects may vary depending upon the needs of the particular objects being modeled. For example, there are some modeling circumstances in which we might not wish to delete the collection object if it becomes empty. An example of this is a builder's definition of a deck in which the parts represent actual resources used to build standard parts of the deck. We might temporarily take the pieces of lumber away from the deck for a more immediate project. In this case, we wish the defined deck object to remain, even though the details of its resources are not yet available. The optional relationship constraints are :

1. If `p` is an object of type `T1`, it must participate in at least one membership relationship with some object `m` of type `T2`, where `m` is a member.
   To support this :
   - MANDATORY COLLECTION DELETION (MCD): This means that upon deletion of all of the members of the collection object, the collection object is deleted.
   - CONDITIONAL COLLECTION DELETION (CCD): This means that the user may decide whether to to permit the the empty collection remain or to delete the collection. The system notifies the user of the situation and permits a run time choice of the appropriate action.
2. If `m` is an object of type `T`, then it must be a member of a collection object `q`, of type `T1`.

To support this :

- CONDITIONAL MEMBER DELETION (CMD): Upon deletion of the collection object, all member objects which are not part of any other existing is-part-of or collection relationship are also deleted.

The core derivation relationship type is used whenever an attribute of an object type `O.A` is derived from other attributes from other objects. We refer to `O.A` as a *derived attribute* and to the objects on which it depends as *domain objects*. It is essential that we view this interdependency among objects as a relationship, since the manipulation of domain objects affects objects attributes derived from them. The following semantics can be selected for the derivation relationship semantics.

1. UPDATE SEMANTICS:
   a) VIRTUAL (VU): The attribute value is never stored. It is computed upon access by the user. Appropriate messages are sent to the user in the absence of data needed for computation of the derived value.
   b) IMMEDIATE (VI): The attribute value is updated whenever one of the values from which it is derived is updated. This is the default.
2. EXISTENCE SEMANTICS:
   a) (DL) Upon deletion of one of the domain objects, the derived attribute is set to *null*.
   b) (DD) The user is denied permission to delete a domain object. The derived object or the derivation relationship link between the domain object and the derived object must first be removed

Since derivation dependencies may occur in chains, we need also to consider the semantics of derivation chains. This is an example of such a derivation chain:

$F \leftarrow O \leftarrow P$

We need to consider what happens if intermediate objects in the chain are deleted. Therefore, in the deletion semantics, we need to also include these possibilities :

1. If deletion of domain objects is permitted and O,P, and F are objects and O depends on P and F depends on O, then, if O is deleted, break the derivation chain and set F to *null*. (This can be accomplished by choosing the DL rule above.)
2. Alternatively, reconnect the derivation chain by specifying that F depends on P.

4.1.1. Example of Derived Relationships in ArchObjects

In the ArchObjects system, the `<has_part>` relationship is used to model complex architectural objects. An object type containing the `<has_part>` relationship type can be modeled using a combination of the IS-PART-OF and COLLECTION core relationship types. For example,

```
has-part(room, wallset)
has-part(room,floor)
has-part(room,ceiling)
has-part(room.openingset)
has-part(room, elementset)
has-attr(room,maxheight,M)
has-geometry(room,solidrep)
derive(room.maxheight,[solidrep])
```

is an ArchObject definition which defines an object class called `room`. `Elementset` is a set of elements which represent possible architectural components which may be added to a `room` instance. The abstract representation of this structure in terms of the core relationships is be as follows:

ELEMENTSET is-collection-of (MCD, CCD) ELEMENT
WALLSET is-collection-of (CMD, CCD) WALL
OPENINGSET is-collection-of (CMD, CCD) OPENING

ROOM <has_part>: is-part-of^{-1} (∞, C) WALLSET
ROOM <has_part>: is-part-of^{-1} (∞, DC, IC) FLOOR
ROOM <has_part>: is-part-of^{-1} (∞, DC, IC) CEILING
ROOM <has_part>: is-part-of^{-1} (∞, DC, IC) OPENINGSET
ROOM <has_part>: is-part-of^{-1} (∞, DC) ELEMENTSET
ROOM.MAXHEIGHT: derive (VI) SOLIDREP

Here we have created derived <has_part> relationship types between ROOM, FLOOR, CEILING, OPENINGSET, and OBJECTSET. These are derived from the core relationship IS-PART-OF. The symbol, ∞, indicates that there is no limit on the number of <has_part> relationship instances in which a part, such as a FLOOR instance, may participate. The inverse notation for the is-part-of relationship is used to indicate the direction of the is-part-of relationship. This is important since the insertion and deletion semantics on the owning and part types are not necessarily symmetric. The specification that the type of the <has_part> relationship is is-collection-of(CMD, CCD) indicates the constraints and update rules of the relationship. The `has-attr` and `has-geometry` relationship types are derived from the core HAS-ATTRIBUTE relationship type, which is not discussed here due to space limitations.

4.2. A Language to Implement SORAC

In this section, we outline the framework for an object-oriented language that includes semantic relationships as central concepts. This language is similar to ADAM(Diaz 1990), particularly in the idea of relationships as object types in the

IS-A hierarchy. This is largely because of the common influence of Rumbaugh's model(Rumbaugh 1987). It extends these languages, however, by including the group of core semantic relationships described earlier as metaclasses. The metaclasses serve as generators of derived relationship types by including methods for the construction of derived relationship types. This provides a way to integrate one of the chief contributions of semantic modeling : the idea that a few fundamental abstract relationships could be used to build a wide variety of models. The result is a language that is a union of semantic and OO modeling concepts.

4.2.1. Objects

This model incorporates a fairly standard notion of an object class, which is defined in terms of structure and behavior. An object consists of an identifier, a set of instance variables, and a set of methods. The instance variables are used to record the values of attributes or simple properties of an object, or to refer to a relationship that the object participates in. An example of an attribute would be properties such as height or color. Relationship references are maintained with objects mainly for convenience. An example of such a class definition is

```
TYPE door
  thickness : inches
  height: feet
  made-of : material
  has_color : color
  fire_resistance : F(made-of,thickness)
  geometry : has_geometry(door,solidmodel)
  contained_in : has_part(door_assembly,door)
  components : has_part(door,bldg_components)

  derive(door.height, solidmodel)
  method(door,display_icon)
```

Here, `thickness`, `height`, `made-of`, and `hascolor` are instance variables containing simple variables.

4.2.2. Relationships

The extensional definition of a binary relationship class is that it consists of a set of pairs of object instances. Each pair is an instance of the relationship. The intensional definition includes the object classes that participate in the relationship class, the semantics, and the methods defined on the relationship, including query methods. The basic query types are :

– add a relationship instance

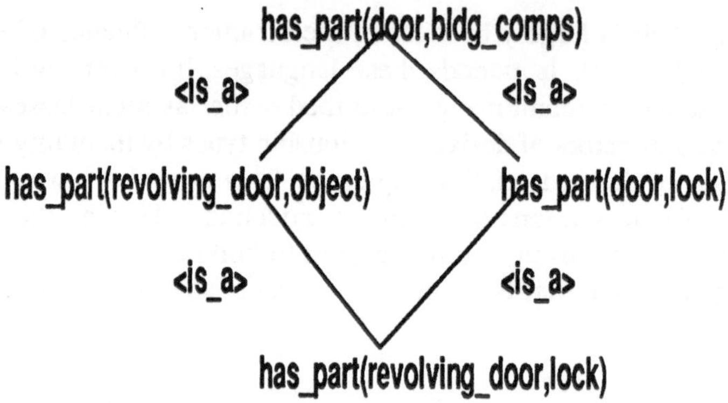

Fig. 1. Relationship Class Hierarchy

- delete a relationship instance
- select a set of relationship instances

Relationship classes can be organized into an IS-A hierarchy based on the participating object classes. For example, Figure 1 shows such a hierarchy. Sparse relationships are more easily modeled using the relationship class hierarchy. For example, the `has_part(door,bldg_components)` relationship type establishes that a `door` instance contains a possibly empty collection of components. A particular door may contain a `lock` object : `has_part(door13,lock22)`. This relationship is an instance of both `has_part(door,lock)` and `has_part(door,bldg_component)`. This means that any specialized semantics associated with the type `has_part(door,lock)` can be applied to this instance, but at the object type level, the possible components of a door do not have to be enumerated. Instead, the possible components are all instances of `has_part(door,bldg_components)`.

Each relationship class is an instantiation of a relationship metaclass. The metaclasses correspond to the core relationship types described in Section 4.1. Each relationship metaclass includes constructor methods that provide support to the schema designer in selecting semantics for constructed relationship classes. For example, the metaclass corresponding to the core IS-PART-OF relationship type provides methods for the creation of part relationship classes, including support for the selection of semantics and analysis of the correctness of these choices. In order to create a particular relationship class, the schema designer calls a create method associated with the IS-PART-OF metaclass :

```
create_reltype(is-part-of,room,floor,
               {-1,INFINITE,DC,IC})
```

Here, the designer has chosen conditional insertion and conditional deletion. The metaclass selects method templates that enforce the chosen semantics, instantiates them, and places them in the constructed relationship class definition. At this point, the metaclass may also check the interactions of the semantics with already existing relationship classes.

The update rules of a relationship class are implemented by special methods which are triggered by database updates. In the preceding example, this relationship class description was created :

```
TYPE has_part(room,floor)
   {instance variables}
   {set of enforcement methods}
```

with the following associated triggers and enforcement methods :

- `delete(O:floor)` : Deletion of a `floor` instance triggers a *relationship existence* method, which deletes the <has_part> relationship instance in which the `floor` object participated. Since in our model, deletion of a participating object always implies deletion of the relationship instance, this method is inherited by all relationship classes.
- `delete(O:room)` : The deletion of a room object triggers the relationship existence method. The action also triggers the DC-preservation method, which checks if the `floor` object associated with the deleted `room` object through the <has_part> relationship participates in any other part or collection relationships. If not, the `room` object is also deleted.
- `insert(O:room)` : Insertion of a `room` object triggers the IC-preservation method, which creates <has_part> relationship instances for all pairs consisting of the `room` object and other objects specified by the designer as components of the room.

These triggers and methods are generated and instantiated when the relationship class is created, as described earlier.

4.3. Complex Constraints

Relationship classes contain reference variables, which refer to constraint objects that include the relationship class. The constraint objects implement complex constraints. We will not go into details of the constraint model here, except to state that constraint classes have this form :

```
CONSTRAINT C1 : forall R,F
 IF has_part(R:room,F:floor) AND
   has_attr(R:room,loadbearing,yes) AND
   has_attr(R:room,construction, light_wood_frame) AND
   has_attr(R:room,fire_rating, 1 hour)
 THEN
   exist FR (fire_rating(F) = FR AND
             FR  1 hour)
```

Any instantiation of this constraint type must evaluate to true; this is checked by associated constraint methods. Triggers are used to invoke evaluation of a particular complex constraint. Triggers are based on update actions: insert a relationship instance, delete a relationship instance, modify an attribute value. Relationship classes contain references to the constraint classes in which they are included. Such references permit all constraints involving a given relationship class to be easily located. For example,

```
has_part(room1:room,floor1:floor)
```

refers to constraint `C1` above. A graphical view of the interactions between relationship metaclasses, relationship classes, constraint classes, and object classes may be seen in Figure 2.

4.4. Relationship Semantics and Complex Constraints

Triggers are generated during the *analysis* stage, which occurs when the constraint is first specified. Constraints are triggered by updated actions and evaluated using variable bindings from the update action. Complex constraint evaluation interact with the semantic properties of relationships at several points :

1. At the constraint analysis stage, when triggers are generated. In the absence of a notion of explicit relationships, it is difficult to derive triggers correctly. For example, derived attributes appearing in a constraint should generate triggers for the domain objects of the derived attribute. Also, the specialization hierarchy is needed at this stage - triggers based on subtypes of any types appearing in a constraint must be generated.
2. At the constraint evaluation stage. Here, relationships function as predicates returning a true or false value. Explicit relationships support this view. In addition, all transitivity properties of the specialization, functional, and part hierarchies must be supported, as well as connectivity and adjacency.

Explicit representation of relationship semantics allows for analysis of relationship interactions at schema design time. The update rules associated with specific relationship types implement simple constraints themselves. Unlike the complex domain constraints, however, update rules must always be satisfied. By taking the update rules into consideration, the number of triggers produced during complex constraint analysis can be reduced. A trigger has the general form `trigger(action(X:T1 R Y:T2))` where `X` and `Y` are variables, `T1` and `T2` are their respective types, and `R` is the relationship between them. In a rule of the form

```
if C1 and C2 and C3
then D1 and D2
```

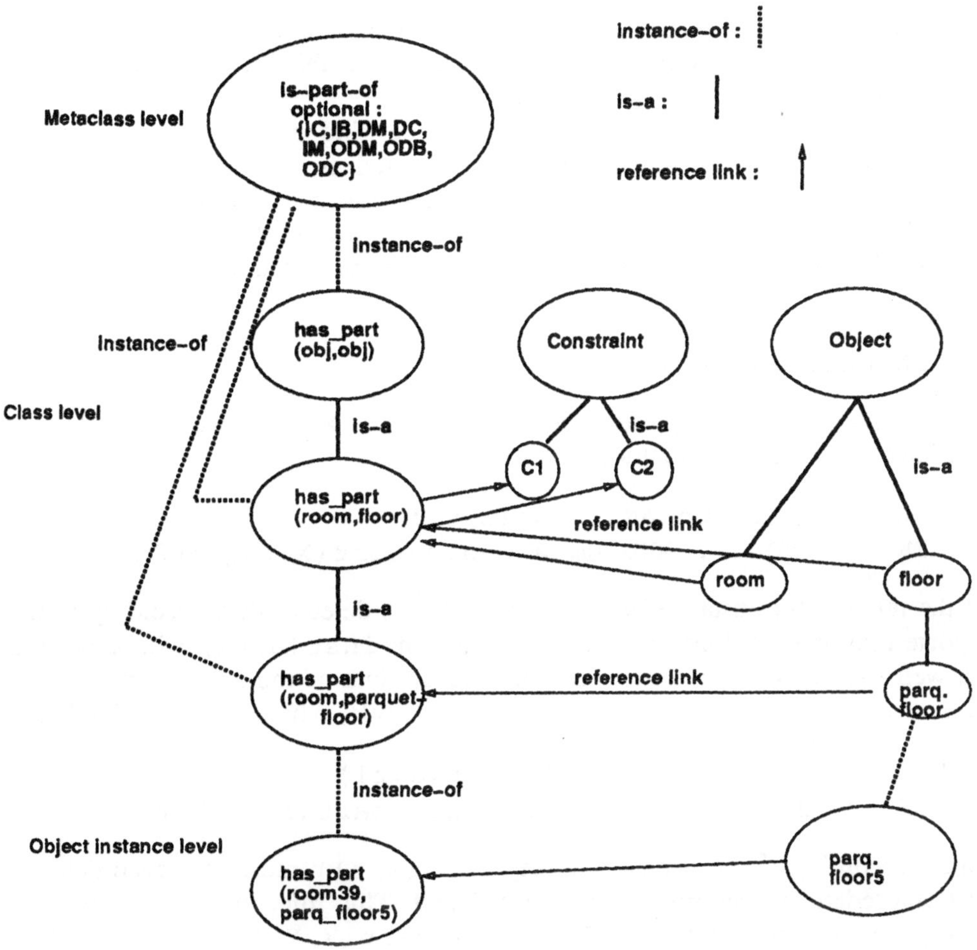

Fig. 2. Relationship Classes and Metaclasses

triggers to evaluate this rule are generated for insertions of C1, C2, and C3, and for deletions of D1 and D2. This method is based on (Nicolas 1982) and is detailed in (MacKellar and Ozel 1991). It assumes that all constraints are satisfied before the database update occurs, so that only the new inserted fact needs to be checked. Because of IS-A transitivity, the potential number of triggers that can be generated is very large; therefore, any reduction in their numbers is desirable. In addition, elimination of triggers in cases where the rule will never apply can significantly reduce time spent needlessly evaluating rules.

Update rules fall broadly into mandatory insertions, conditional insertions, mandatory deletions, and conditional deletions. In addition, there are update rules governing the behavior of derived values. Each of these types of update rules interact with specific types of complex constraints, and govern the cases where a particular update may or may not cause a rule to fail. In this paper, we will only consider one type of optional update rule.

A mandatory deletion rule states that if one kind of object or relationship is deleted, then the related object must be deleted as well. An example is the RODM option that may be associated with `has_part` relationships. This states that if a part relationship is deleted, the associated owner object must be deleted as well. For example, the RODM update rule can be stated as

```
IF   delete(has_part(X:t1,Y:t2))
THEN delete(inst(X,t1))
```

where `inst(X,t1)` represents an object instance `X` of type `t1`. Generally, if we have a complex constraint of the form

```
IF C1 AND ... Cn AND inst(X,t1)
THEN D1 AND ... AND Dm AND has-part(X:t1,Y:t2)
```

if the relationship instance `has-part(a,b)` is deleted, then according to the update rule, the relationship instance of the form `inst(a,t1)` will be deleted as well. At the same time, any complex constraints of the above form will be triggered. Evaluation will be on the instantiated rule, which will have the form

```
IF C1 AND ... AND Cn AND inst(a,t1)
THEN D1 AND ... AND Dm AND  has-part(a:t1,b:t2)
```

Since according to the update rule, any `inst(a,t1)` must have been deleted, the antecedent of this rule cannot be satisfied. This rule, in other words, will always remain satisfied for a `delete(has-part(X,Y))` action. Therefore, a trigger for this type of delete should not be maintained, since there is no reason to recheck the rule in this case.

For example, let us assume that the relationship type

```
has_part(staircase,risers)
```

has been specified with the option RODM, which means that if the `has_part` relationship between a particular staircase and a particular set of risers is deleted, the staircase will be deleted as well. Now, assume this complex constraint is specified:

```
FORALL S
IF inst(S:staircase) AND
   has_role(S:staircase,means_of_egress_component)
THEN  EXISTS R
   has_part(S:staircase,R:risers) AND R.height  7 inches
```

which states that if a given staircase is part of the means of egress, it must have a set of risers whose height is no more than 7 inches. For this rule, the following triggers are generated:

```
insert(has_role(S:staircase,MOEC))
modify(has_attr(R:risers,height))
```

No trigger is generated for

```
delete(has_part(S:staircase,R:risers))
```

because of the RODM update rule. In addition, no trigger is generated for `insert(inst(S:staircase))`, because according to the general relationship insertion rule, `has_role(S:staircase,means_of_egress_component)` cannot be inserted until after `inst(S:staircase)` is inserted.

4.4.1. Constraints and derived values

Derived attributes present a particular challenge for triggering complex constraints, because a change in a value not mentioned in the rule must trigger the rule. In a design domain, however, many values will be derived from others. Such values may be updated when their domain attribute is modified, or they may not be updated until needed. The second case is specified by the VIRTUAL option and is the case that cause complications for the trigger generating analysis stage. If a given attribute depends on other attribute values, this is represented as

```
has_attr(X, A, V) = F(C,D,E)
```

where `X` is an object, `A` is an attribute name, `V` is the attribute value, `F` is the function which specifies how A must be derived from attributes `C`, `D`, and `E`. If a complex constraint C2 exists that contains

```
C2 :  IF has_attr(X:t,A,V) AND...
```

and `A` is defined as {VU,DL}, then the set T_{C2} of triggers for C2 contains

```
{modify(C), modify(D), modify(E), delete(C),
  delete(D), delete(E)}'.
```

The delete triggers are needed because according to the DL constraint, if a domain object is deleted, the derived attribute is set to *null*. The situation becomes more complex if these domain attributes are themselves derived. Consider the derivation chains in Figure 4.4.1. Here, `A` depends on `B` and `C`. In turn, `B` depends on `D`, and `C` depends on `E`. In this case, the bottommost VIRTUAL link in a chain determines for which attributes triggers are generated. For example, if

Fig. 3. A derivation tree

the `A = F(B)` link is VIRTUAL but the `B = F(D)` link is IMMEDIATE, then T_{C2} will contain `modify(B)`, but not one for `modify(D)`. This is because a change to D will immediately cause a change to B, and thus cause a `modify(B)` action to occur. This holds true for delete triggers as well. If on the other hand, updates are IMMEDIATE on the `A=F(C)` link, but VIRTUAL along the `C=F(E)` link, then T_{C2} will contain `modify(E)` and `delete(E)`.

It can be seen from these examples that the potential for many kinds of interactions exist between optional semantics and complex constraints. These interactions can profoundly affect the correctness of constraint evaluation. A planned next step is to integrate analysis tools into the schema design phase in order to check the consistency of the update rules and complex constraints, and to govern the generation of triggers according to the update rules.

4.5. Summary

Our experiences with ArchObjects have led us to believe that the typical object-oriented paradigm of embedding relationships into individual methods is not powerful enough for complex design representations. It is very difficult to ensure correctness and consistency of updates with this type of ad hoc implementation. Therefore, we have developed a data model, SORAC, that is based on the idea of built-in semantic relationships as defined in earlier semantic models. This model is being developed at The University of Rhode Island and The New Jersey Institute of Technology, using C++ and Ontos(Ont 1990). A second version of ArchObjects will be implemented in this modeling language and will serve as a test implementation.

Much of the meaning of the design model is conveyed through an understanding of the relationship semantics. Therefore, the relationship semantics are made explicit, rather than buried in method code. This frees the schema designer and method writers from worrying about maintaining the semantics of such relationships. A design model may require modeling concepts other than composition; for example, ArchObjects is based on eight relationship types. Therefore, in the SORAC model, we have provided a semantic basis for a *set* of core relationship types : is-a, part, attribute, collection, and derivation, from which more specific relationship types may be derived. Each core relationship type has an associated set of semantic choices, a kind of "menu". In each case, the set of choices is designed to be appropriate to the concept expressed by the relationship. Because the constraints and update rules supporting the relationship semantics are encapsulated as part of the relationship class, it is easy to to comprehend the behavior of the various relationship types. Therefore, this approach provides semantic support for the intricate relationships between objects that are typical of architectural modeling. Other complex domains, such as geographic information systems or molecular design systems, could benefit as well from the SORAC model because of the integration of semantic modeling with object-oriented modeling.

In addition, complex constraints that involve relationships between different objects are supported. These constraints are defined on the relationship types and are triggered by insert or delete actions on relationship instances. By explicitly associating relationship behavior with each relationship type, it is possible to analyze any interactions with the complex constraints. In the future, a schema design tool that will automatically perform this analysis is planned.

Acknowledgements

Many thanks to Mike Doherty, Zhenghong Dong, and Falguni Vora of the SORAC group at URI, who provided many helpful comments during the writing of this paper.

References

Diaz, O.: 1990, Semantic-rich user-defined relationships as a main constructor in object oriented databases, *Proceedings of IFIP TC2 Conference on Database Semantics : object-Oriented Databases*.

Hammer, M. and McLeod, D.: 1981, Database description with SDM : A semantic data model, *ACM Transactions on Database Systems* **6**(3).

Kim, W., Chou, H. and Banerjee, J.: 1987, Operations and implementations of complex objects, *Proceedings of the Third International Conference on Data Engineering*, pp. 626–633.

Kim, W.: 1989, Composite objects revisited, *Fifth International Conference on Data Engineering*.

MacKellar, B. and Ozel, F.: 1991, ArchObjects : Design codes as constraints in an object-oriented KBMS, *First International Conference on Artificial Intelligence in Design*.

MacKellar, B.: 1992, A constraint-based model of design object versions, *3rd International Conference on Data and Knowledge Systems for Manufacturing and Engineering*. to appear.

Mylopoulos, J., Bernstein, P. and Wong, H.: 1980, A language facility for designing interactive database-intensive applications, *ACM Transations on Database Systems* **5**(2), 185–207.

Nguyen, G. and Rieu, D.: 1991, Representing design objects, *Proceedings of Artificial Intelligence in Design*.

Nicolas, J.: 1982, Logic for improving integrity checking in relational databases, *Acta Informatica* **18**(3), 227–253.

Ont: 1990, *ONTOS Object Database Documentation, Release 1.5*.

Peckham, J., Maryanski, F., Beshers, G., Chapman, H. and Demurjian, S.: 1989, Constraint based analysis of database update propagations, *Proceedings of th Tenth International Confernece on Information Systems*, pp. 9–18.

Rumbaugh, J.: 1987, Relations as semantic constructs in an object oriented language, *ACM OOPSLA Proceedings*, pp. 466–481.

SHOOD: A DESIGN OBJECT MODEL

G. T. NGUYEN and D. RIEU
*IRIMAG**
Laboratoire de Génie Informatique
IMAG-Campus
BP 53 X
38041 Grenoble Cedex
France

Abstract. This paper is an overview of a three years effort aimed at providing a flexible and powerful model for design applications. The result is a data model called SHOOD based on object-oriented concepts and frame-based knowledge representation. SHOOD implements sophisticated features, such as:

- object persistence,
- multi-methods along a specific specialization hierarchy, which is independent of the class hierarchy,
- sophisticated semantic relationships e.g, dependency relationships between objects, which are totally independent of the composition relationship,
- multiple object representations, allowing the users to manipulate the objects from several points of views simultaneously,
- the systematic use of a powerful meta-object kernel, which is used to implement a reflexive architecture.

This paper is an informal overview of SHOOD. It focuses on the last two issues.

1. Introduction

SHOOD is an object-oriented data model designed to support highly dynamic applications. Examples of such applications are found in design, genetics and weather forecast, where the challenge is to merge partial amounts of information into a consistent and flexible computerized model.

Yet, SHOOD is not just another exotic data model. It includes features which are seldom found simultaneously in existing data models. Some of its salient characteristics are :

- the support for evolving data, both in their definitions and values,
- the support for flexible user defined semantic relationships between objects e.g, composition relationships, specific dependency relationships (existential, exclusive, etc.),
- the support for multiple object representations, which can be concurrently defined by different users on the same objects e.g hydraulic and electric points of view for an engine,
- ad-hoc object persistence, although SHOOD does not provide at the present time all the functionalities of a complete database management system e.g, concurrency control and restart facilities,

J. S. Gero (ed.), Artificial Intelligence in Design '92, 221–240.

- multiple inferencing capabilities, allowing for several alternatives to be defined for attribute computations e.g, depending on the arguments availability.

From a historical perspective, SHOOD was designed to support the requirements of mechanical and VLSI CAD/CAM applications (Rieu, 1992; Nguyen, 1991). It elaborates on knowledge representation techniques found in AI and data models found in the database area (Minsky, 1975). It also imports recent advances in object-oriented databases and programming languages (Kim, 1990; Gabriel, 1991). It includes such notions as meta-classes, methods, inferences, and dependencies. They are merged in a powerful and flexible data model which is currently tested on full size applications in mechanical engineering design. SHOOD is implemented in Le_Lisp™ on Sun SPARCstation 2™. A user-friendly interface is built on top of the system to provide an easy access to its functionalities. It is based on a menu-driven windowing system implemented in Aida™, a graphic interface development tool running under X-Window.

The paper is organized as follows. Section 2 is an in depth analysis of the design requirements for SHOOD. Section 3 describes the meta-object kernel. It includes the meta-classes which are used to implement the concepts in the model. Section 4 is an overview of the multiple representations supported for the user objects, with indications on the classification mechanism available. Section 5 is a conclusion.

2. Rationale for the design of SHOOD

2.1 EXTENSIBILITY

Among the issues raised by design applications, the evolution of objects values and structure—grossly speaking database evolution - is probably the most challenging. It questions one of the most securing aspect of data storage and manipulation, i.e object stability. Because design applications are constructive, they require the definitions of the objects to evolve over time. This departs dramatically from business applications which manipulate large amounts of information with relatively few data definitions. Design in contrast manipulate smaller amounts of data because the goal is to define artifacts rather than query the database.

The need for flexible data models is therefore fundamental. The commonly accepted relational model of data is for this matter all but flexible. User relations flatten object attributes which are disseminated in multiple pieces among relations without explicit semantic relationships.

Recent advances in object-oriented data modeling have opened new perspectives for the definition and manipulation of complex composite objects. Besides their ability to encapsulate and reuse existing definitions - which are of first importance in design - they benefit from research in the field of object-oriented database systems (Silberschatz, 1990; Unland, 1990).

Severe shortcomings limit however their use in design applications. First there is a crucial lack of design methodologies for object-oriented systems' applications. More fundamentally, they suffer from inherent limitations in their ability to model user defined semantic relationships (Escamilla, 1990; Giacometti, 1990). Specific dependency relationships need to be defined between parts of a composite object e.g, there is an existential dependency between the airframe and an aircraft built around it, but not between the engines and the aircraft that uses them, as well as attribute propagation : the color of some airframe parts of a red aircraft must be red.

Further, fundamental concepts in object-oriented paradigms such as encapsulation, message passing, instantiation and method selection are not always well-suited for design environments. For example, encapsulation must be broken because it is usually necessary for the designers to work on the object definitions. Instantiation also requires an instance to belong to exactly one class, whereas designers require the instances to belong simultaneously to several points of views. Methods are usually attached to one class definition, which commands method selection whatever the designer intent may be.

To overcome such limitations without throwing away the object-oriented approach, we decided to define and implement an extensible object-oriented data model. This means that SHOOD provides a limited number of concepts with which the designers may customize their application model. To achieve this goal, a powerful and flexible meta-object kernel is provided.

The model is reflexive and meta-circular, i.e it is self descriptive and is implemented using its own concepts (Cointe, 1987). Thus, classes are instances of generic meta-classes, class attributes are also instances of generic meta-classes describing attributes. This approach is used systematically, allowing specific classes and attributes to be defined by a single instantiation mechanism, provided the appropriate meta-classes are added to the meta-object kernel by the designers. For example, methods, inferences and constraints are also defined by instances of meta-classes. Key attributes and optional attributes are similarly defined by instances of appropriate meta-classes. This is detailed in Section 3.

2.2 OBJECT EVOLUTION

In our approach, the extensibility of the object model is the support for object evolution. We believe that a powerful, flexible and maintainable system can only be achieved this way, rather than adding specific functionalities to a limited or outdated data model, in an everlasting process.

Reuse and redesign being standard in design applications, the model—when required—still benefits from the outstanding advantages of the object-oriented approach e.g, encapsulation, sharing, inheritance, etc. It also overcomes their most stringent assumptions. For example, object instances may belong simultaneously to several classes, providing a straightforward implementation of multiple points of view (Nguyen, 1990). Objects instances may also be partially incomplete and inconsistent (Nguyen, 1992). Hence, an instance may not necessarily be an exact representative of a specific class. These last two features depart from fundamental assumptions in existing object-oriented models. They are indeed necessary in design applications because the definition of an adequate object structure, even prior to any valuation of its attributes, is what engineering design is all about.

The support for incomplete and inconsistent objects is therefore the basis for object evolution (Nguyen, 1989). It is complemented by a classification mechanism and a migration facility. They both implement the assistance given to the designers when asking : "To what object definition(s) does this partial design correspond best ?" and "Attach this partial design to the most suited existing definition(s)". The classification mechanism searches in the class graph the most resembling classes for the given instances, depending on specific conditions e.g, attribute values. The similarity is based on the degrees of completeness and consistency of the instances with respect to the class definitions (Nguyen, 1991). The migration facility allows the designers to override the classification mechanism and force instances to belong to less specific classes than those resulting from the system's classification process. This emphasizes user's authority over the system's automatic assistance. Overall, modifications of the object instances can be monitored by the designers with the help of the system, which may track automatically the evolution of the objects being modified.

Another facility provided by SHOOD is a powerful inferencing capability, allowing object to be computed and updated using other objects as arguments to appropriate methods. Multiple inferences may be defined for an attribute, and may be triggered independently depending on the arguments availability. This will be the basis for a powerful method combination mechanism (Keene, 1989).

2.3 SEMANTIC RELATIONSHIPS

A critical issue in design application is the support for user-defined relationships between objects. Whereas many object-oriented systems now provide some form of composite objects (Keller, 1991; Kim, 1990), an open question is the implementation of versatile semantic relationships. Semantic networks and knowledge representation languages like SRL (Fox, 1986) provide powerful mechanisms for this aspect. It is clear that the usual inheritance relationship is not sufficient to model complex artifacts. Therefore a number of proposals propose to enhance their capability to support dependency links and attribute propagation, sometimes called "selective inheritance" (Carré, 1990). SHOOD implements a systematic approach to the concept of relationship by allowing user-defined relationships to be defined through the meta-object level. A semantic relationship - say, an existential dependency between a composite object and one of its components—is modeled by a class which is an instance of a generic "relationship" meta-class. The particular semantics of the relationship is attached to the relationship class, i.e in the (user-defined) methods made available to it. Hence a potentially versatile relationship implementation framework. An in depth analysis of the relationships in SHOOD is given in (Escamilla, 1990). It will not be detailed any further in this paper.

A particular form of semantic relationship given for free with SHOOD is the disjoint class relationship. It states that if two classes A and B are declared disjoint, then no instance of A or any sub-class of A, may be simultaneously an instance of B or any sub-class of B. This helps clarifying the class graph for applications involving a large number of classes. It also simplifies solving some inheritance conflicts : attributes with similar names but different semantics defined in disjoint classes cannot cause conflicts. Last, it speeds up the classification of instances in the class graph (section 4.4).

2.4 SUPPORT FOR EVOLVING APPLICATIONS

A challenging issue is that SHOOD is designed to support not only evolving application requirements, sometimes due to coarse modeling or rapid prototyping, but also evolving knowledge on the application domain : this is a more general, more challenging goal. Example can be found in genome sequencing. In such applications, there is a need to integrate partial pieces of information whose interactions are not known in advance and whose structure may evolve over time. This is the case when refined knowledge is acquired after a previous application model has been implemented. Thus, particular forms of knowledge acquisition must be supported : the discovery of new relationships between objects, new object structures and attributes, or new inferences between object values. This calls for powerful and flexible modeling concepts. The approach adopted by SHOOD is to provide a sophisticated meta-object kernel which is used to generate the appropriate application concepts when required. This is described in section 3. This approach is used throughout the literature to implement reflexive architectures (Cointe, 1987).

3. The meta-object kernel

The meta-object kernel is the core of the system from both a conceptual and implementation point of view. As a conceptual framework, it is used as a foundation for extensibility of the model. As an implementation tool, it forms the bootstrap of the model SHOOD itself. It serves the purpose of generating the whole model from a reduced set of meta-classes. This approach bears similarities with ObjVlisp and CLOS (Cointe, 1987; Keene, 1989).

3.1 META-CLASSES

All the classes defined in SHOOD are instances of meta-classes. The latter define the structure and behavior of the classes. For example the class Aircraft is defined as an instance of the meta-class Meta as depicted in figure 1. The meta-class Meta defines the minimal attributes and behavior of classes. It includes the attributes "instance_of", "class_name", "super", "sub_classes" "instances" and "attributes".

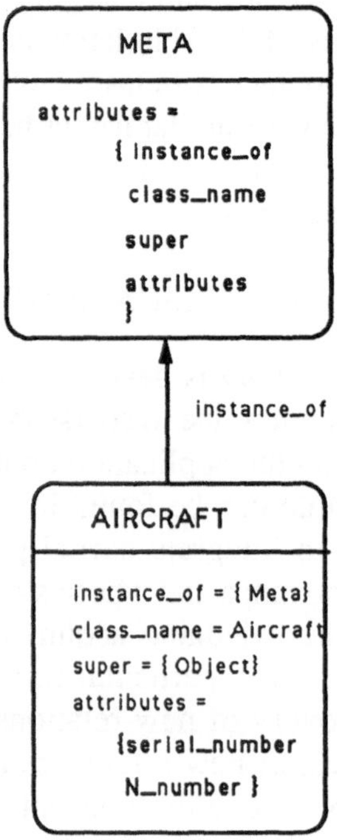

Fig. 1. Classes are instances of Meta

Note that classes and instances bear an attribute called "instance_of" the domain of which is a set of classes. The value of the attribute "instance_of" is the set of classes to which the class or instance considered belongs. This means that the model SHOOD allows instances to belong to several classes simultaneously. This is called here multiple instantiation. This departs from usual object-based models where instances are allowed only one single class membership (Goldberg, 1983). As will be explained in section 4, this allows a straightforward implementation of multiple object representations.

An attribute "instances" is also defined for classes. It is the inverse of the attribute "instance_of". It gives the set of instances actually attached to the class or meta-class considered. This attribute is automatically maintained by the system. It is not depicted in figure 1.

An attribute "class_name" defined in Meta allows for classes to be referenced by a unique name.

The attribute "super" defines the set of super-classes of a class. The model therefore supports multiple inheritance. A specific class named Object is the default super-class for all objects. The super-classes are not ordered. The domain of the attribute "super" is therefore a set of classes. Sets are enclosed by "{" and "}" in the figures. Names conflicts for attributes are addressed in section 3.3.1.

The attribute "sub_classes" is the inverse of "super". Sub-classes implement the inheritance relationship. The semantic of inheritance in SHOOD is structural and set inclusion. A class inherits the attributes of all its superclasses. It can define new attributes. It can also refine inherited attributes. However, due to the set inclusion semantics, it cannot override inherited attribute definitions. The set of instances of a class belongs to the intersection of the sets of instances of all its super-classes.

Meta-classes can generate classes. Therefore, all meta-classes must inherit the structure and behavior of Meta, i.e they must be direct or indirect sub-classes of Meta (Figure 2).

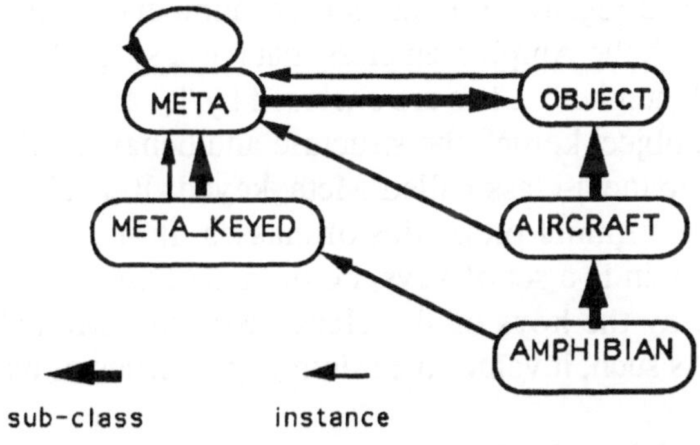

Fig. 2. Examples of inheritance and instantiation graphs

Classes generate instances. They must therefore instantiate the minimal properties of classes defined in Meta e.g, "class_name", "super". They must also define their specific attributes by instantiation of "attributes" in Meta. All classes must therefore be instances of Meta or, by set inclusion, of one of its sub-classes.

Meta-classes are classes generating particular instances, i.e the application classes. They must therefore inherit the minimal attributes and behavior of classes e.g, "attributes" defined in Meta. All meta-classes must therefore be sub-classes of at least one meta-class. Being classes, they must also be instances of at least one meta-class. An example is given in figure 3 for the meta-class Meta_keyed.

The class Object defines the minimal properties of instances e.g, their object identifier (oid). All instances, including classes which are instances of the meta-classes, are therefore direct or indirect instances of the class Object. A consequence is also that Meta is a sub-class of Object. Further, Object being a class, it is an instance of Meta. Similarly, Meta is an instance of itself. The single exception to this approach is that Meta is a sub-class of Object, which is not a meta-class.

A consequence of inheritance in SHOOD is that classes inherit the structure and behavior of all their super-classes. They must therefore be instances of all the meta-classes of all their super-classes. For example, the meta-classes of the class Amphibian must be the same as, or sub-classes of, the meta-classes of Amphibian's super-classes, i.e Aircraft (Figure 2).

3.2 OBJECT IDENTIFICATION

In object-oriented systems, objects are associated with a unique system-wide identifier usually known as "oid". For user convenience, SHOOD supports also an explicit notion of "key", which is user-defined. It is a list of attributes in a class defined as an access key to object instances. It is implemented as a bijective function between key values and oids. Several keys may be defined for a class e.g, serial number and registration number (N_number) for an aircraft. Figure 3 gives an example with the Amphibian class that has a single key formed with one attribute, i.e "serial_number". Lists are enclosed by "(" and ")" in the figures.

Using the meta-object kernel, the structure and behavior of keyed classes are defined in a specific meta-class called Meta_keyed. It a sub-class of Meta. As such, it inherits the minimal properties of classes. It also defines a new "Key" attribute which domain is a set of keys, i.e of lists of key-attributes. This "Key" attribute is valued by the keys of the class instances. Meta_keyed is also an instance of Meta. As such, it values the minimal properties of classes.

3.3 META-CLASS INSTANTIATION

3.3.1 *Attributes.* Class attributes are modeled in SHOOD by a specific meta-class called Meta_attribute. This is the reason why the section "3.3 Meta-class instantiation" includes this section (3.3.1 Attributes). An attribute is defined in SHOOD by a name and a set of attribute descriptors. They include a type descriptor as well as inference and constraint descriptors. This section is concerned only with type descriptors. The inference and constraint descriptors are described in sections 3.4.

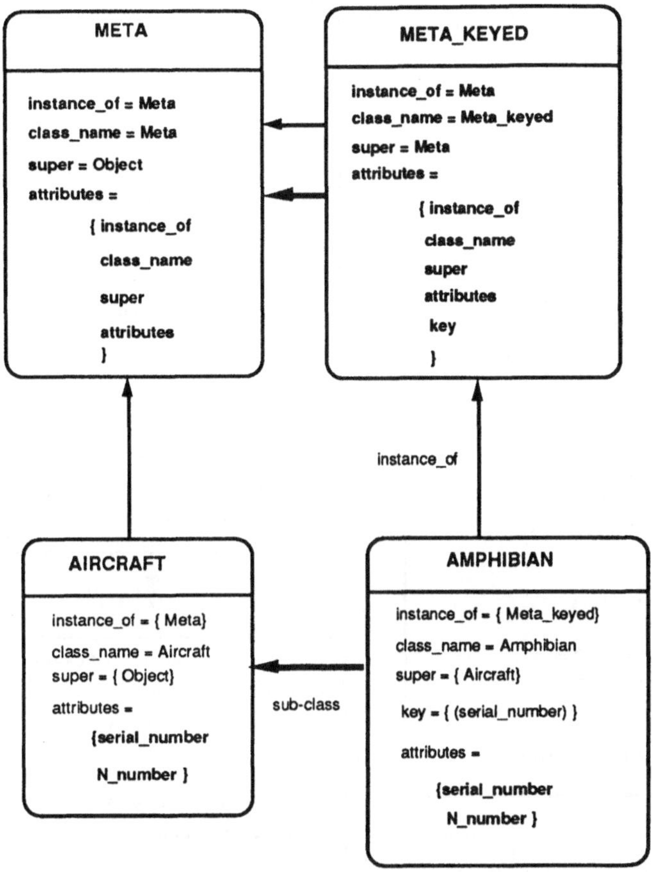

Fig. 3. Example meta-classes and keyed class.

Name conflicts arising from multiple inheritance are solved by using complete attribute names. A complete attribute name is a triplet <class_name, attribute_name, origin_class>, where "class_name" is the name of the class using the attribute "attribute_name", as defined by the class "origin_class". The class where an attribute is defined first is called the origin class. The attribute is inherited by all the sub-classes. Two attributes with the same name but with different origin classes are considered different. All attributes, whatever their origin class, are inherited. A partial example of attribute definition in the meta-object kernel is given in figure 4.

Type descriptors for attributes include a mandatory part and an optional part. The mandatory part is a domain definition. It can be one of : "a", "set_of" and "list_of" key-word, followed by a class name. For example the "powerplant" attribute in the class Aircraft can be defined as a "set_of Engine", where the Engine class is defined elsewhere. The optional part is a set of domain restrictions. They allow specific enumerations of values, intervals and excluded values to be specified.

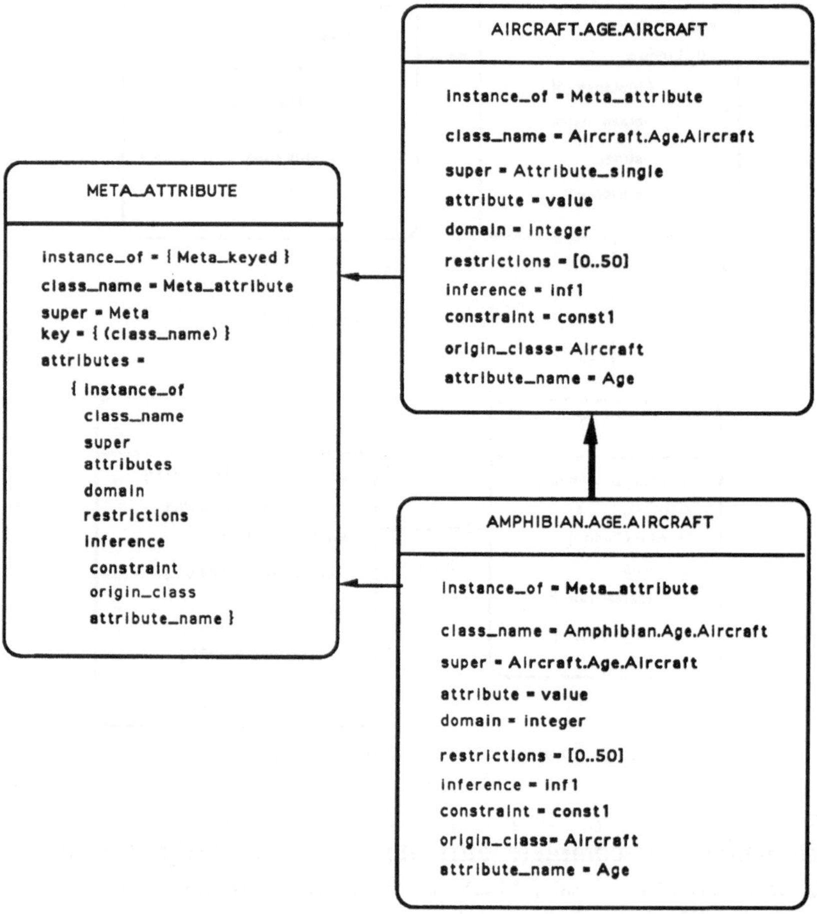

Fig. 4. Attribute definition

Attributes are modeled by classes defining their name, origin class, domain, restrictions, constraints and inferences. They are instances of the Meta_attribute class. Giving a value to the attribute A of an instance X in class C is therefore amenable to the creation of an instance Y in the class B modeling the attribute A. The instance X refers to its attribute's value for A by the oid of Y in the class B.

If a sub-class C' of C defines a refined attribute for A, say A', then a sub-class B' of B is automatically generated. Refining an attribute is therefore a specialization process for the class defining the attribute.

A particular form of attribute declaration is the "deferred attribute" definition. It bears some similarities with the notion of deferred feature in Eiffel (Meyer, 1988). It restricts the multiple declarations of an attribute in the sub-graph which starts at the class where the deferred declaration takes place. In all the classes of the sub-graph, all the occurrences of an attribute with the same name refers to the deferred declaration. There is therefore no other possible semantics for an attribute with the same name in that sub-graph. This helps solving some inheritance conflicts due to syntactic and semantic ambiguities - in French for sure.

3.3.2 *Methods*. Methods are defined outside classes. In contrast with the original object-oriented approach e.g, Smalltalk, but like recent systems like CLOS, SHOOD relies on a specific graph for methods, which is independent of the class inheritance graph. The organization of the method graph is based on all the arguments types. Methods are modeled by classes. A class defining a method is an instance of a particular meta-class called Meta_method. The attributes of the class are the "code" that implements the procedure and its arguments. Method overloading is implemented by defining sub-classes in the method graph. Method selection first generates an instance of the method class, then classifies it in the method graph. At present, the first selected method is returned. Extensions are planned to take into account multiple answers to the method selection process.

As a consequence, method selection is based on all the arguments, not only on one selector, as implemented in most object-oriented environments.

3.4 INFERENCES AND CONSTRAINTS

Methods are used by application programs and user interfaces to manipulate the objects. They are also used by the inferences and constraints defined in the attribute descriptors.

Inferences may be defined to compute attribute values. Several inferences may be defined for one attribute. They are ordered by default by their order of appearance in the descriptor. They are executed in their order of appearance in the list until a executable one succeeds. An executable inference is one for which all arguments are instantiated. If all the executable inferences abort, an error is returned. A particular inference called "user" is provided which can end the inference list. If all inferences abort and the "user" inference is defined, the attribute value is asked to the user.

Constraints may be defined on attributes to restrict their values. Several constraints may be defined for an attribute. They call methods. Constraints are considered so far as mandatory, i.e they cannot be violated. Extensions are planned to authorize optional constraints that instances may violate under user control. This is interesting for design applications because objects may often violate temporarily constraints upon user manipulations, although they are dedicated to the current class. They could be treated as exceptions, because they do not strictly adhere to the class definitions. However, this would contradict the set inclusion semantics of inheritance in SHOOD. They are therefore explicitly considered as incomplete and/or inconsistent objects. The model support these objects by maintaining consistency and completeness degrees for every instance in the database. Monitoring them allows to keep track of object evolution and can be smoothly integrated in the versioning mechanism.

4. Multiple object representations

A crucial issue in design applications is the management of multiple object representations. Cooperative work by which different designers work on complementary fields to achieve the global design project is the rule. For example, aircraft design involve aerodynamicists to design the airframe, powerplant specialists to design the engines, electronics specialists to design the monitoring and "fly-by-wire" systems, hydraulics specialists, etc. Each one works on particular representations of the aircraft. The various representations coexist simultaneously and contribute to the design process.

Existing object-oriented systems usually lack the powerful support for multiple object representations (Sciore, 1989; Richardson, 1991). A fundamental requirement of software engineering environments has emphasized the need for several concurrent representations of software modules : source code, compiled code, executable modules, versions, etc. This has given rise to specific implementations and proposals (Ahmed, 1991; Banerjee, 1987). They usually call for explicit concepts e.g, adding new notions to a data model. For example, "representation relationships" have been proposed in the literature (Carré, 1990), as well as "roles" (Pernici, 1990) and specific algorithms to maintain parallel class hierarchies (Marino, 1990).

Another approach is to use only concepts that already exist in the data model. This implies an implicit definition of object representations. It elaborates on existing notions e.g specialization, aggregates and multiple instantiation. There is no need for specific operators to define and manipulate the representations. This greatly simplifies the concepts and mechanisms involved in the implementation of the data model. This approach is used for SHOOD (Nguyen, 1992; Rieu, 1992). It uses the notion of multiple instantiation, by which an instance is allowed to belong simultaneously to several classes. They are not necessarily related by a specialization relationship. An instance can therefore belong simultaneously to the Aircraft class and to the Museum_collection class (Figure 5). This approach has the fundamental property of preserving object identity throughout representations. It is complemented by a classification mechanism called MIC—an acronym for multiple instantiation and classification.

In SHOOD, representations are objects themselves. They are modeled by classes. This provides a consistent framework in which the reflexive architecture of the model is fulfilled.

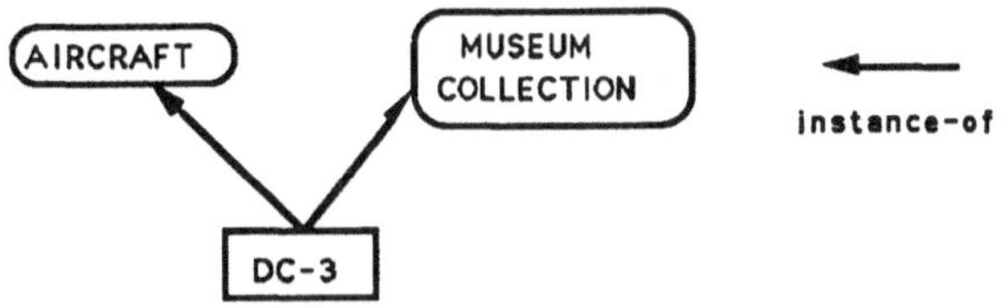

Fig. 5. Multiple instantiation

The use of multiple instantiation for multiple object representations is addressed in the next section (section 4.1). Section 4.2 discusses the use of existing concepts. It focuses on specialization and aggregates. The respective merits of aggregates, specialization and multiple instantiation are then discussed (section 4.3). Section 4.4 presents a classification tool.

4.1 MULTIPLE INSTANTIATION

Most object-oriented programming languages do not actually implement implicit instantiation (Goldberg, 1983) : instances of a class are not instances of its super-classes. This is usually dependent on the semantics of the inheritance relationship implemented : the most common form is the structural and behavioral inheritance, by which sub-classes inherit the structure and behavior of their super-class(es). Other form of inheritance exist, among which is set inclusion and specialization. Using specialization, sub-classes may also refine the constraints imposed on attributes. With set inclusion, the sub-classes represent sets of instances which are sub-sets of their super-class(es). This last form prohibits exceptions. It also prohibits attribute redefinition, except by constraint refinements. We assume that implicit instantiation is given for free using the set inclusion semantics of inheritance.

We are interested here in the explicit form of instantiation i.e, instances can be simultaneously attached to classes which are not related by a specialization relationship. Multiple instantiation supports in our approach different semantics attached to the same objects. It therefore enhances the semantics of the classes because the various representations of a particular object may bear information which have - at least at first sight - little in common. For example the aircraft structural design includes the descriptions of the various fuselage sections and wings. This defines a particular aircraft representation. The computerized flight management system for the aircraft includes in turn a set of redundant computers which have nothing in common with the previous representation, except that, when flying, the computers output signals that will drive the aircraft controls through actuators. Note that these actuators are usually hydraulic or electric systems which are not described in either of the representations. There is therefore no explicit requirement that both representations be related.

4.2 SPECIALIZATION AND AGGREGATES

An crude approach to the implemention of multiple object representations is to define a common sub-class that will hold the corresponding instances. The super-classes of this inter-representation class are the classes that define the various representations. If implicit instantiation is available, this also preserves object identity. Another advantage is that this approach allows an easy location of the semantic relationships existing between the different representations, i.e in the common inter-representation class.This simplifies somewhat the class graph. In the example, an inter-representation class called Inter_Rep_Aircraft is defined as a common sub-class for the classes Structure and Flight_Management. It holds the constraints and relationships connecting both representations (Figure 6).

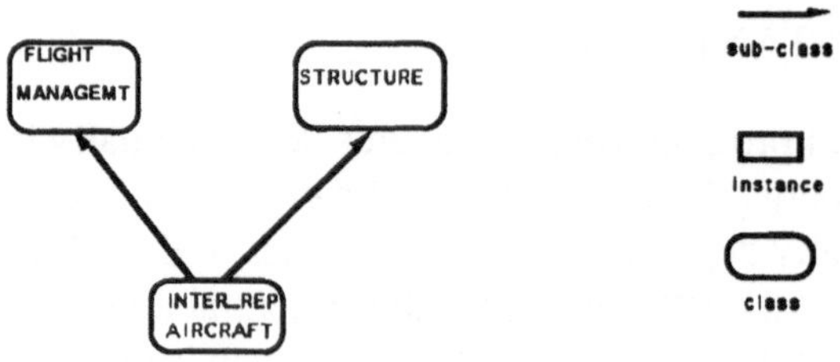

Fig. 6. Multiple representations using specialization

It is our opinion however that this is a misuse of the specialization relationship. Indeed, an aircraft is not a sub-class of its structural or electronic representations. These should be better defined as components of the aircraft. The latter is a large composite object which includes, among others, a structural part and lots of avionics sub-parts. If available, this can be implemented by a composition relationship (Figure 7). Aggregates of components however do not preserve object identity. The aggregate classes define inter-representation classes e.g, the class Aircraft in Figure 7. Each of their attributes models a specific representation e.g, the attributes f and s. The domain of these attributes are the classes that define each representation e.g, Flight_management and Structure. An object instance X that belongs to the aggregate class is therefore an aggregate of references to the particular instances Y, Z, ... of the representations. These references are object identifiers. Aggregates cannot therefore implement object identity.

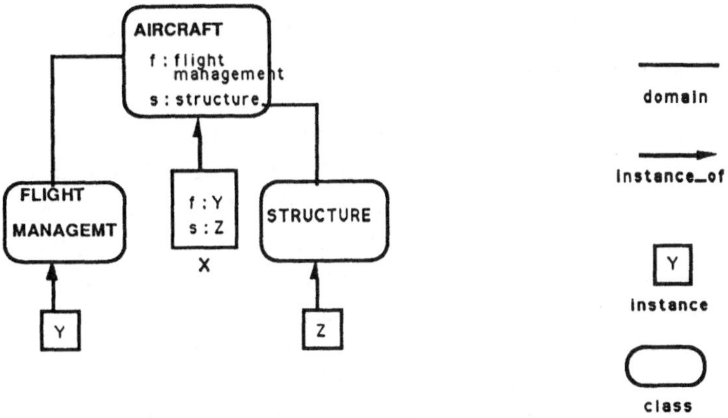

Fig. 7. Multiple representations using aggregates

Nonetheless, aggregates bear specific advantages. First, the various representations may be defined independently of each other. This advantage is also obvious with multiple instantiation. The class graph can therefore be made simpler and is easier to maintain. This departs from the specialization approach, where sub-classes can be defined to refine the super-classes (the standard usage of specialization) but also to group representations. In this case, the sub-class is an inter-representation class (Figure 8). This is quite confusing. It is our opinion that this approach should be avoided. In contrast, aggregates can be used while keeping the original semantics of specialization, i.e refining super-classes. This has an interesting impact on the effectiveness of classification mechanisms. The simpler the specialization graph, the better the classification efficiency, because it is not disturbed by the inter-representation classes. This is discussed in Section 4.4.

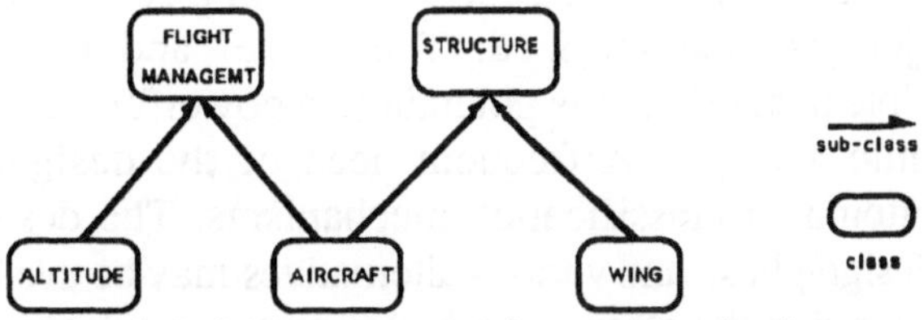

Fig. 8. Mixing representations with sub-classes.

4.3 AGGREGATES AND MULTIPLE INSTANTIATION

As mentioned above, multiple instantiation is effective when the representations have no explicit semantic relationships. This seems counter-intuitive, but the example in section 4.1 gives such an example. It is also interesting if the users focus on individual instances and not on whole classes. This is usually the case in engineering design environments.

If either of these requirements is not met, aggregates can be combined in SHOOD with multiple instantiation. The existence of inapplicable representations is then modeled by the type constructor "OR". If type A and type B are defined, the type "A OR B" denotes the union of the sets of instances of type A and of type B. An aircraft is the union of its structural and electronic representations : "electronic OR structural". But a commercial aircraft e.g, "a1", has both structural and electronic representations. It is therefore defined by the type "electronic AND structural". This means that both representations are mandatory for this kind of aircraft. This approach has the advantage of preserving the specialization and inheritance semantics in SHOOD because the set of aircraft with type "A AND B" is a subset of the set of aircraft of type "A OR B".

It is possible to define inter-representation relationships and constraints in the aggregates e.g, the total weight of the aircraft is the sum of the weight each of the parts in the various representations and it cannot exceed X tons.

Combining aggregates and multiple instantiation has the following advantages :

- object identity is preserved,
- multiple representations can be worked out in parallel,
- sequencing the design is expressible among representations,
- consistency within representations is standard,
- consistency among representations is expressible and precisely located, i.e in the aggregates.

4.4 CLASSIFICATION

The rationale for design applications is to define precisely artifacts, starting with general user requirements and refining the object until a complete and consistent model of the object is characterized. The model and the values evolve continuously. Flexible instantiation is potentially a powerful tool in order to take these evolutions into account. A frequent need of the designers is also the requirement for automated classification mechanisms. The design object may reach a particular design phase and various alternatives may be available to resume the process. Characterizing the classes to which a particular object resembles most can therefore be a valuable tool to meet some design constraints if experience already exists for similar objects. Hence the increasing role of classification mechanisms. They yield basically two results :

- refinement in the knowledge pertaining to the objects,
- modification of existing knowledge.

This is described in the next section (Section 4.4.1).

4.4.1 *Object migration.* Refinement in the knowledge corresponding to the objects is achieved by attaching them to new specialized classes. For example, incomplete objects may see the addition of attribute values which enable the designers to attach them to sub-classes which are more detailed : if "my_aircraft" is an instance of the classes Recreational and Home-built, and if it is now fitted with floats, it can now be attached to the class Amphibian (Figure 9).

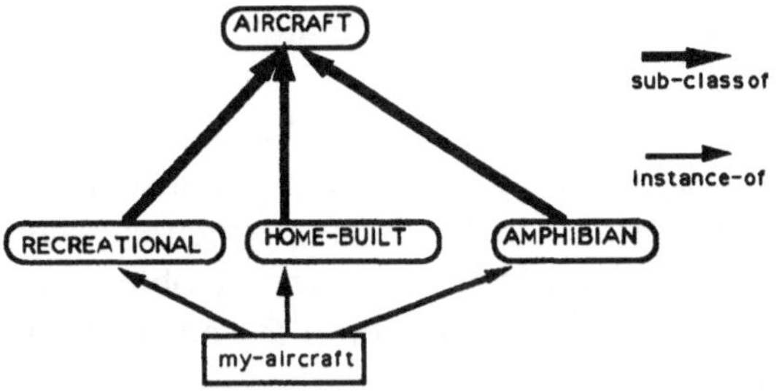

Fig. 9. Classification of modified instances

Should the attribute values change in the objects, the classes to which they belong may change. For example, my DC-3 used to belong to the Commercial aircraft class only. Being too old now, it has to migrate to the class Collection_item.

A similar requirement exists when the class graph is modified (Casais, 1990). This issue is often referred to as schema evolution (Nguyen, 1989). Should the designers require the addition or deletion of attributes in the classes or relationships between classes, the existing instances may be required to migrate because they do not conform to the new class definitions anymore. For example, if the concept of business aircraft evolves from aircraft with at most 12 passengers to aircraft carrying 8 passengers at most, all the business aircraft with 9 to 12 passengers must migrate from the class Business to the class Commercial (Figure 10).

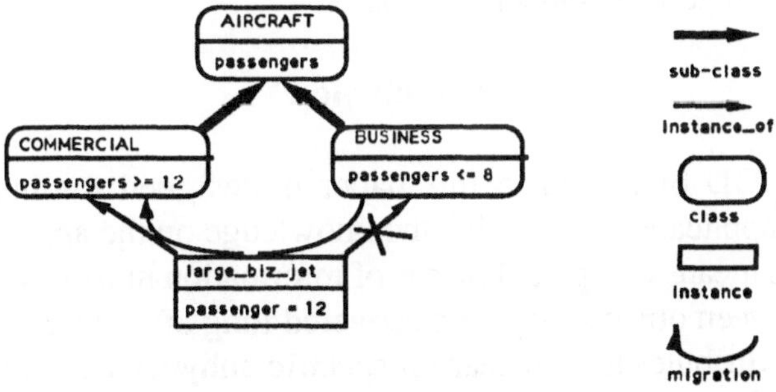

Fig. 10. Migrating instances due to class modifications

4.4.2 *Creating objects and representations*. On the one hand, classification helps to take into account object evolution. On the other hand, multiple instantiation is a means to take into account multiple object representations. Basically, attaching objects to classes is the result of a classification process. This is true even when creating the objects : it is first questionable whether a given instance, when just created, conforms to any class definition. A classification process must tell whether it can be an instance of the specified class(es). If it can, the instantiation mechanism implements the "instance_of" relationship between the instance and the corresponding class(es).

Generalizing the classification approach in the context of multiple representations, we consider the creation of an object instance as:

- a classification of the instance in the class(es) specified when it is created,
- the existence of a particular representation for the object, i.e its creation class.

Taking into account a new representation for an object instance requires a subsequent classification:

- the new representation must be a potential class,
- if cleared, another "instance_of" relationship can be implemented between the object and the new representation. Note that this does not require the various representations to be related by any specialization relationship.

Classification mechanisms allows therefore the following statements to be fully supported:

- let my DC-3 be an instance of the classes "Aircraft" and "Collection" simultaneously,
- attach my DC-3 to the most specific sub-class of Aircraft,
- should my DC-3 be painted in red, make it an instance of the "Red_aircraft" class, which is a sub-class of Aircraft. My DC-3 will subsequently be an instance of the classes "Red_aircraft" and "Collection" simultaneously.

Classification mechanisms and multiple instantiation are therefore unifying concepts that are used here to implement such different operations as :

- object creation and modification,
- representation creation and modifications.

5. Conclusion

The model SHOOD presented in this paper is designed for highly dynamic applications, i.e applications for which the knowledge on the application domain may evolve. This means that partial pieces of information must be integrated, new relationships between objects may be discovered long after their first encounter, new unexpected attributes may appear for specific unhypothesized characteristics of the objects and so on. Examples of such applications are engineering design and genetics. It is often the case that redesign of existing artifacts, usually practised as a refinement process, is not sufficient. This departs dramatically from previous assumptions in the literature.

To support this dynamicity, SHOOD provides features that are seldom found simultaneously in a single and unifying framework. It elaborates on object-oriented paradigms and knowledge representation languages. Among the features supported are:

- complex composite objects,
- user-defined semantic relationships,
- incompletely defined or partially inconsistent objects,
- multiple object representations,
- multiple inferencing,
- a powerful although flexible method mechanism,
- a classification mechanism,
- object persistence.

The model is implemented with a meta-class kernel which defines all the basic concepts and provides the necessary features to implement the application level. It is also the basis for the extensibility required by the applications. New concepts and new relationships can be implemented by definition of ad-hoc meta-classes holding the appropriate functionalities, i.e attributes and methods. They can later be instantiated to provide the user with the desired objects.

This proposal is based on experience gained in mechanical CAD/CAM for GSIP (Groupement Scientifique Interdisciplinaire de Productique). It is a joint research effort in Computer Integrated Manufacturing from the Universities of Grenoble with industrial partners.

Acknowledgements

The authors wish to thank J.M BONNEFOND, who implemented the user interface, F. BOUNAAS, who is in charge of object evolution, J. ESCAMILLA who implemented the meta-object kernel, and M. TOLLENAERE from Institut de Mécanique de Grenoble for many invaluable discussions. This work is supported by INRIA and IMAG for the project SHERPA and by GSIP.

References

Ahmed, R., Navathe, S.: 1991, Version management of composite objects in CAD databases, *Proceedings of ACM SIGMOD '91 Conference*, Denver, USA.

Banerjee, J., Kim, W., Kim, K. J. and Korth, H.: 1987, Semantics and implementations of schema evolution in object-oriented databases, *Proceedings of ACM SIGMOD '87 Conference*, San Francisco, USA.

Carré, B. and Geib, J. M.: 1990, The point of view notion for multiple inheritance, *Proceedings ECOOP/OOPSLA '90*, Ottawa, Canada.

Casais, E.: 1990, Managing class evolution in object-oriented systems, Centre, Universitaire Informatique. Université de Genève, Switzerland.

Cointe, P.: 1987, Metaclasses are first class: the ObjVlisp Model, *Proceedings OOPSLA'87*.

Escamilla, J. and Jean, P.: 1990, Relationships in an object knowledge representation Model, *Proc. 2nd Int'l Conference Tools for Artificial Intelligence*, Washington, USA.

Fox M., Wright J. M. and Adam D.: 1986, Experiences with SRL: an analysis of a frame-based knowledge representation. *Proc. 1st Intl. Conf. Expert Database Systems*, Charleston, USA.

Gabriel, R. P et al.: 1991, CLOS: integrating object-oriented and functional programming, *Comm. ACM*, **34**(9).

Giacometti, F. and Chang, T. C.: 1990, Object-oriented design for modelling parts, assemblies and tolerances, *Proceedings of the 2nd International Conference on TOOLS '90*, Paris, France.

Goldberg, A. and Robson, D.: 1983, *Smalltalk 80: The Language and its Implementation*, Addison-Wesley, Reading.

Keene, S. E.: 1989, *Object-Oriented Programming in Common Lisp. A Programmer's Guide to CLOS*, Addison-Wesley, Reading.

Keller, T. et al.: 1991, Efficient assembly of complex objects, *Proc. ACM SIGMOD '91 Conference*, Denver, USA.

Kim, W.: 1990. Object-oriented databases: definition and research directions, *IEEE Trans. on Knowledge and Data Engineering*,. **2**(3).

Marino, O. et al.: 1990, Multiple perspectives and classification mechanism in object oriented representation, *Proceedings of the ECAI Conference*, Stockholm, Sweden.

Meyer, B.: 1988, *Object-Oriented Software Construction*, Prentice-Hall, Englewood Cliffs, NJ.

Minsky, M.: 1975, A framework for representing knowledge, *in* P. Winston (ed.), *The Psychology of Computer Vision*, McGraw-Hill, New York, pp. 211-277.

Nguyen, G.T. and Rieu, D.: 1989, Schema evolution in object-oriented database systems, *Data and Knowledge Engineering*, North-Holland, **4**(1).

Nguyen, G.T and Rieu, D.: 1990. Heuristic control on dynamic database objects, *in* Meersman (ed.), *Information Processing '90*, North-Holland, Amsterdam.

Nguyen, G. T and Rieu, D.: 1991, Representing design objects, *in* J. S. Gero (ed.), *Artificial Intelligence in Design '91*, Butterworth Heinemann, Oxford.

Nguyen, G.T. and Rieu D.: 1992. Multiple object representations, *Proceedings of the 20th ACM Computer Science Conference*, Kansas City, USA.

Pernici, B.: 1990, Objects with roles, *Proceedings of the International Conference on Office Information Systems*, Boston, USA.

Richardson, J. and Schwarz, P.: 1991, Aspects: extending objects to support multiple, independent roles, *Proc ACM SIGMOD Conference*, Denver, USA.

Rieu, D. and Nguyen, G. T.: 1986, Semantics of CAD Objects for Generalized Databases, *Proc. 23rd Design Automation Conference*, Las Vegas, USA.

Rieu, D. and Nguyen, G. T.: 1992, Object views for engineering databases, *Proceedings of the 3rd International Conference on Data and Knowledge Systems for Manufacturing and Engineering*, Lyon, France.

Sciore, E.: 1989, Object specialization, *ACM Trans. on Office Information Systems*, **7**(2).

Silberschatz, A., Stonebraker, M. and Ullman, J. D.: 1990, Database systems: achievements and opportunities, *Laguna Beach Report. TR 90-22*, Department of Computer Science, University of Texas, Austin, USA.

Unland, R. and Schlageter, G.: 1990, Object-oriented database systems: concepts and perspectives, *Lecture Notes in Computer Science*, Springer-Verlag.

BASIC STRUCTURE OF A BUILDING MODEL FOR REPRESENTING AND USING KNOWLEDGE OF BUILDINGS IN CAAD SYSTEMS

H. SHIMODAIRA
FM Service Department
Nihon MECCS Co Ltd
22-10, Nishishinbashi 1-Chome, Minato-ku
Tokyo 105 Japan

Abstract. This paper presents a basic structure of a building model for representing and using knowledge of buildings concerning the geometry in the field of architectural design based on the object-oriented concept. We construct a building model using knowledge of location relationships between building components and we are concerned with the type of buildings which have the concepts of reference lines and stories with horizontal floors.

The salient features and effectiveness of the proposed model are as follows. Classes of objects are classified by generalization abstraction. A conceptual building model is constructed by the aggregation abstraction using classes of building components and relationships between them. The location of each component is defined by a disposition object which includes general rules used to specify relative location relationships between building components in the practice of architectural design. The connection relationships between building components are automatically derived from attributes of instances for defining the location of components. A room is represented by objects for surfaces different from building components. Knowledge of the way of representing a building component in an application model can be included in methods of the component. Because of the effect of such methods: (i) the data structure of the proposed model is simple and precise and the processing of room-related data is easy; (ii) the modification of the model is automatically carried out according to the general rules of location relationships between building components; (iii) the representations in an application model can be automatically generated and changed; and (iv) the proposed model has high extensibility and stability.

1. Introduction

Usually, the development of a design from initial sketches to a completed final solution is performed in a process of iterative step-wise refinement, by using external media for recording and representing the results of the intermediate stages and the final solution of design. CAD systems have provided designers with means of changing the design and trying new ideas and alterations. However most current CAD systems do not have sufficient facilities for supporting such processes (Peraza, 1990). The major issues are as follows (Eastman, 1988).

1. Design refinement needs to be facilitated.
2. Drawing production needs to be automated, reducing time and costs.
3. Use of analysis, visual evaluation and other types of application need to be facilitated.

One of the key ideas for solving these problems is to equip the CAD system with intelligent facilities by including knowledge of buildings in the model in a CAD system. For this purpose, CAAD systems based on a building model are promising, because we can automatically perform the generation and modification of the model and the application models such as drawings by including knowldge of buildings in the building models.

J. S. Gero (ed.), Artificial Intelligence in Design '92, 241–263.

Building models are conceptual structures specifying what kind of information is used to describe buildings and how such information is structured. Building models include data describing the geometry, function, and situation of uses concerning the buildings throughout their life cycles. Building model definition data comprise the geometry, relationships, and attributes necessary to completely define a component part or an assembly of parts for purposes of design, analysis, construction, operation, and maintenance.

Various research and development has been carried out to develop building models. Internationally, the work for developing the Standard for Exchange of Product Model Data (STEP) is presently carried out in the US PDES project and the International Standards Organization. In the activities for developing STEP, various attempts have been made to define building product models (Gielingh, 1988; Willem, 1988). In these researches, the concepts and methodologies for defining building product models have been proposed, but up to now concrete methods for describing geometrical information have not been proposed.

Law (1986) presented a generalized abstraction hierarchy suitable for modelling building design data. The model represents the building design information from three perspectives: topology, structural engineering, and architectural design. The geometry of buildings is represented by the topological representation, which is analogous to the boundary representation of solid objects, and structural members are associated with this topological entity. The description of geometry in this model is somewhat graphic-oriented. Watanabe (1991) presented an architectural knowledge representation model. In his model, architectural elements are arranged in a generalization abstraction hierarchy structure and a building is defined by the relationship of other objects. The structure of his model is considerably more complicated and not easy to underatand, because the principle for representing relationships between architectural elements is not precise. Yamada (1991) presented a model which represents an architectural element as a combination of a geometrical model which represents 2-D and 3-D geometries and a semantics model which represents the role of the element. The system enables the users to assign the semantics to an element and to separate the semantics from the element. The model seems to be versatile, but the advantage of such a structure is not given.

This paper presents a basic structure of a building model for representing and using knowledge of buildings concerning the geometry in the field of architectural design. Because we construct a building model by using knowledge of location relationships between building components, we are concerned with the type of buildings which have the concepts of reference lines and stories with horizontal floors. Arbitrarily-shaped architectures such as domes are not dealt with in this paper, because we believe at present that it is difficult to construct a comprehensive building model including such architectures. The issues and the basic ideas employed in this paper are as follows.

1. We need effective concepts and disciplines to make the model as simple and precise as possible. We use the object-oriented concept (Dittrich, 1986) to

construct a building model, because it enables us to directly represent building components and relationships between them. Major concepts used in the model presented here are data abstraction, class and instance, message and method, and encapsulation. In the abstraction process, we use data abstraction techniques which are formulated as a combination of generalization and abstraction to represent a conceptual building model by using the classes of building components and the relationships between them.

2. We need effective methods for representing geometries of buildings and their related knowledge. In the model presented here, the geometries of a building and the parts of it are represented as assemblies of building components which have attributes for defining the shapes and sizes. The geometry of a building component can be defined by a small number of data (for example, the location and the thickness for a wall). The location of a building component is determined by referring to the location of other components and specifying relative distances from them. By this method, the general rules used to specify buildings in the Japanese practice can be included in the model to facilitate the operations in generating and modifying the model. The data for application models such as drawings and solid models can be generated from the assemblies of building components.

3. A room is usually a space enclosed by wall, floor, and ceiling surfaces, and walls and floors usually exist between adjoining two rooms. The major issue in representing a room is therefore how to simply and precisely describe the relationship between a space and the surrounding walls. In most of the models (Watanabe,1991), because a room is represented by pointers to building components such as walls, the structure of the models is complicated. The model presented here uses a wall surface as well as a wall as an object. It makes the model simple and precise by describing a room by its surrounding wall surfaces. The attributes of the walls such as finishes are defined for the wall surfaces.

4. We need effective methods for automatically generating and modifying a building model and the application models derived from it. In the model presented here, connection relationships between building components are automatically derived from the attributes defining the locations of the comonents. Furthermore, the knowledge of the way of representing a building component in an application model can be included in the methods of its class. By these methods, the generation and modification of the building model and the application model such as a drawing is automatically carried out.

2. Intelligent Facilities in CAAD Systems

2.1. NEED FOR INTELLIGENT FACILITIES IN CAAD SYSTEMS

Usually, the development of design from initial sketches to a completed final solution is performed in a process of step-wise refinement, by using external media for recording and representing the results of the intermediate stages and the final solution of design. CAD systems have provided designers with means of changing the design and trying new ideas and alterations. Peraza (1990) investigated the relationship of design changes and their respective graphical changes on drawings in a church project as the case. He used a two-dimensional drawing CAD system to record all the stages in the design development and analysed design changes through graphical representations on the drawings. Based on these recorded drawings, he reviewed the procedures of graphical changes performed by the CAD systems and evaluated the facilities of CAD system. The results showed that some single change in an architectural element and the introduction of a new architectural element inevitably causes changes of sizes and/or positions in the related elements and much work was repeatedly performed to update the drawings. Thus, he concluded that the future CAAD system that supports design studies requires such facilities: first to keep record of rules of positional relationships and connectivities between architectural elements, and second to adjust the size or even the shape of related elements automatically, following the recorded rules.

2.2. INTELLIGENT FACILITIES IN CAAD SYSTEMS USING BUILDING MODELS

To utilize a building model in architectural design, the building model needs to be transformed into various application models such as 2-D drawings and solid models using appropriate transformation prcesses. The processing related to building models in such CAAD systems consists of two stages: formation of a building model by data input and generation of application models from the building model. Taking the research of Peraza mentioned above into consideration, intelligent facilities required for CAAD systems using building models are as follows.

1. In the stage of formation of a building model, facilities for efficiently performing generation and modification of the model are needed. In generating the model, facilities for efficiently performing the definition of shapes, sizes, and locations of building components are needed. In modifying the model, facilities by which the influence of changes of shape, size, and location of building components is as automatically as possible exercised on the related building components are needed.
2. In the stage of generation of application models, facilities by which application models are automatically generated from and corrected according to the change of the building model are needed. For example, concerning drawings, drawings corrected in correspondence with changes of the building model need to be automatically generated.

The considerable part of such intelligent facilities is realized by incorporating knowledge of buildings into a building model. Knowledge of buildings is represented by the attributes of respective building components and the relationships between them. In this paper, we present a basic structure of a building model for incorporating the major relationships between them and examples of methods for realizing intelligent facilities.

3. Building Model Definition Data

3.1. BUILDING COMPONENTS AND THE RELATIONSHIPS BETWEEN THEM

A building consists of such components as spaces, walls, columns, beams, floors, windows, doors, partition walls, etc, and of building equipment. For brevity, in this paper, we refer to walls, columns, beams, floors, etc. as main elements and windows, doors, partition walls, etc. as secondary elements, according to their roles and the order of assembling in the construction process. In this case, walls in main elements means walls for load-bearing. To represent these building components, such data items as name, use, shape, location, material, strength, load, cost, related provision of law, etc. are needed.

These building components have interrelations which derive from their geometry and functions. These relationships contain architectural knowledge concerning buildings. We classify these interrelations into the six relationships as follows.

1. The generalization relationship specifies that a set of similar components is generally thought of as a single named component. For example, a set of columns, beams, and walls can be regarded as structural members.
2. The location relationship specifies that the location of a particular component is determined by referring to another component.
3. The constitution relationship specifies that a particular assembly or aggregate of components constitutes another component. For example, a room is perceived as the space enclosed within, and thus constituted by, surfaces of walls, floors, and ceilings.
4. The connection relationship specifies that a particular component is connected to or adjoins other components. For example, a beam is connected at the both ends to columns. A room adjoins a corridor or another room through intermediate walls.
5. The part-of relationship specifies that a particular component is a part of another component. For example, a sink is part of a kitchen.
6. The function relationship specifies that particular components have a close functional relationship to each other which depends on their functions, and thus the existence and location of one component depends on the existence and location of another component. Of these six relationships, this relationship represents the highest level of architectural knowledge about buildings. That a beam is supported by the columns at the ends and the load on the beam is transmitted to the columns is an example of this relationship.

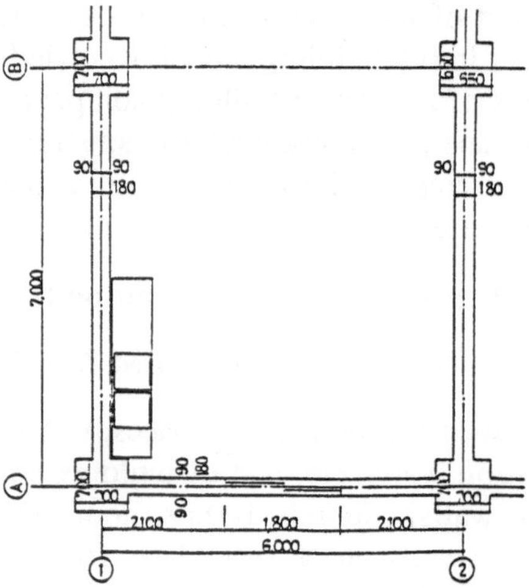

Fig. 1 Example of representation of building in plan (dimension: mm)

3.2. Rules of Location Relationships Between Building Components

The geometry of a building consists of the assembly of shapes of its components and the location of its respective component is specified by a location relationship between building components. Peraza (1990) pointed out, in his research, that rules of location relationships between building components need to be incoporated in future CAAD systems. Rules of location relationships are classified into two categories: general rules commonly used by most designers and particular rules set up by a specific designer.

General rules have been formed in the practice of architectural design in each country and so each country may have different rules. Most of general rules are found in drawings, because drawings are represented according to general rules. Fig. 1 shows an example of a plan, which is simplified for brevity. Major general rules of location relationships between building components in the Japanese practice are outlined in Table 1. The numbers in Table 1 indicate the approximate order in which building components are specified. This order corresponds to the order of assemblying building components in the construction process. The characteristic of these rules is that the location of a building component is determined sequentially by referring to the locations of other components and by specifying relative distances from them. In this paper, we utilize this characteristic to construct a building model.

It would be a significant advance, if computer-based design tools would enable us to record the design relationships we intend, along with the specific decisions that accomplish these intentions. Such facilities will ease and reduce the operations for iterative design exploration especially in the phase of conceptual design. Gross (1989) presented such a modelling environment in a two-dimensional drafting

TABLE 1
Location relationships between building components

Building components		Order	Datums for determining the location
Reference line		1	Main reference line
Main elemant	Column	2	Reference line
	Beam	3	Reference line. Surface of column. wall. and floor
	Wall	3	Reference line. Surface of column. beam. and floor
	Floor	4	Reference line. Surface of beam. wall. and column
Secondary element	Window	5	Surface of wall and column. Reference line
	Door	5	Surface of wall. floor. and column. Reference line
	Partition wall	5	Reference line. Surface of wall. floor. and column
Ceiling		6	Surface of floor. wall. and column
Finish		7	Surface of floor. wall. and column
Building equipment		5 or 8	Reference line. Surface of wall. floor. and column

system. Such intelligent facilities is also required in CAAD systems using building models. However, we have to perform more research to realize such facilities. In this paper, we do not deal with rules set up by designers.

4. Basic Ideas for Constructing a Building Model

Because we construct a building model by using knowledge of location relationships between building components, we are concerned with the type of buildings which have the concepts of reference lines and stories with horizontal floors. But reference lines in the two directions in a plan need not necessarily to intersect orthogonally.

We use the object-oriented concept to construct a building model. Major concepts used in the model presented here are data abstraction, class and instance, message and method, and encapsulation. In the data abstraction process, we use data abstraction techniques formulated by Smith (1977) as a combination of generalization and aggregation. Generalization is an abstraction in which a class of individual objects of similar types is regarded as a single named object. For example, a set of rectangular columns, circular columns, etc. is regarded as a column (see Table 3). In this abstraction hierarchy, we can efficiently manage data by using common attributes which descendant objects inherit from a generic object. Aggregation is an abstraction in which the relationship between objects is regarded as a higher-level object. This allows many details of the relationship to be ignored. For example, a story can be viewed as the aggregate of spaces, building components, and building equipment on the story. Once a story object has been defined, the details of its components can be ignored, but they are accessible when they are needed.

Objects defined by the generalization and aggregation, as well as individual

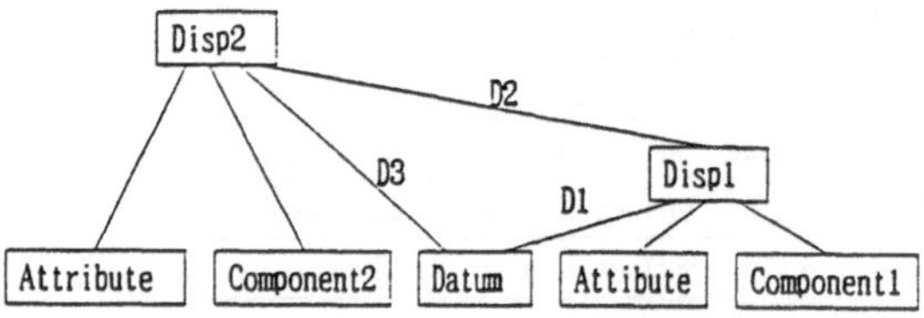

Fig. 2 Function of a disposition object

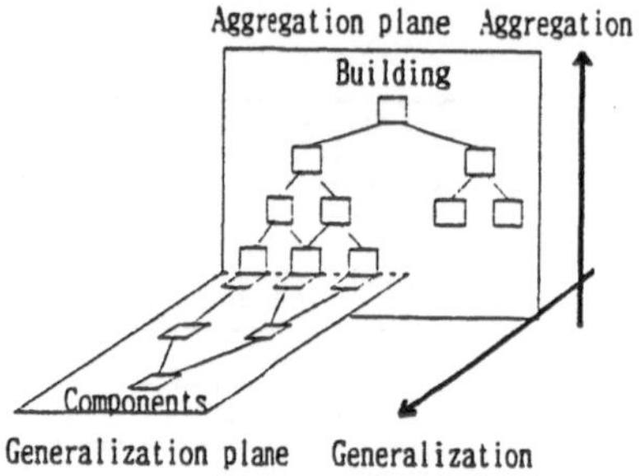

Fig. 3 Generalization and aggregation

objects, are used to represent the real world. Smith proposed a structuring discipline in which generalization and aggregation abstractions are carried out in two orthogonal planes. This method has been applied to a mechanical CAD system (Sponer, 1986). Such a structuring discipline has the following benefits:

1. the resulting model is easy to understand;
2. a structured and manageable model can be obtained;
3. a more systematic approach to database design can be developed;
4. stability of models can be provided under several kinds of evolutionary change.

To construct a building model, we first select classes of objects from building components and classify them by using the generalization abstraction. Class hierarchies can be developed to represent the generalization and classification processes. The data items for representing the building components described in Section 3.1 are included as the attributes of these objects. The generalization abstraction lets us efficiently classify a vast volume of building components.

Next, we select classes of objects to represent relationships between building components by the aggregation abstraction. In the following way, we represent the general rules described in Section 3.2. We use a disposition object to represent a relative location relationship (see Fig. 2). This relationship defines the location of an instance of a class of a building component by relative distances from datums and other components (D1, D2, D3). In Fig. 2, the Disp1 indicates that the location of a Component1 is determined by referring to the locations of datums (which are defined before they are referred to). The Disp2 indicates that the location of a Component2 is determined by referring to the locations of datums and the Component1 (defined before they are referred to).

Finally, we construct a conceptual building model as an aggregate of classes of building components and of location and constitution relationships in the aggregation plane (see Fig. 3). The generalization plane, in which building components are represented as generalization hierarchies, is set perpendicular to the aggregation plane so that each component of the lowest level in the generalization plane corresponds to a component in the aggregation plane.

By giving the data on a real building to the conceptual building model, a specific

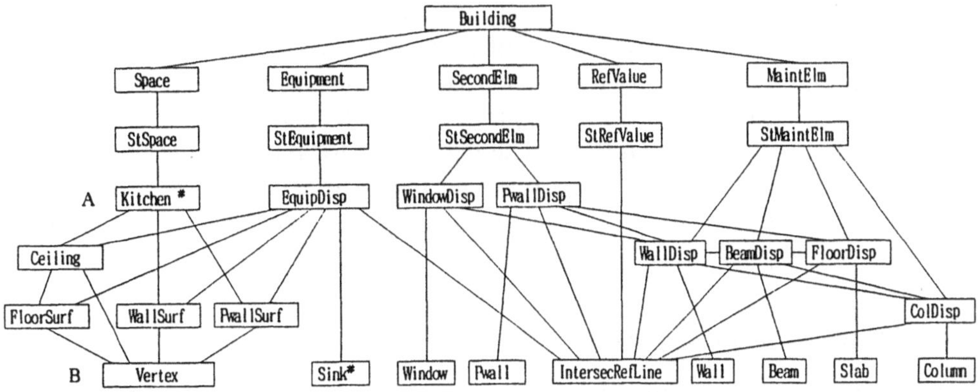

Fig. 4 Outline of the overall structure of a conceptual building product model in the aggregation plane

model of the real building is obtained as an instance of the conceptual model. At this stage, the real building model is represented by the combination of hierarchies and networks in which the instances of the classes of the building components are linked to each other by the instances of the classes of the location and constitution relationships.

The generalization, location, constitution, and part-of relationships are incorporated into this model. Connection relationships between building components are automatically derived from the attributes for defining the location of a component. We believe that most of function relationships can be derived from instances of object attributes and these five relationships by reasoning based on knowledge of buildings. How to set up function relationships is not dealt with in this paper and is a subject for future study.

5. Basic Structure of a Building Model

5.1. Overall Structure of a Conceptual Building Model

Fig. 4 outlines the overall structure of a conceptual building model in the aggregation plane. The generalization plane is omitted, but its explanation is given below. The objective of this paper is to present a basic structure of a building model, so unnecessary details are omitted. The # denotes that the object is indicated as a representative of similar objects defined in the generalization plane.

We decompose a building into reference values (RefValue), main elements (MainElm), spaces (Space), secondary elements (SecondElm), and building equipment (Equipment). The RefValue, etc. denotes the abbreviation of reference values, etc. Reference values, which refer to intersections of reference lines, let us construct a building model by using location relationships. In addition, because we are concerned with the type of buildings with the concept of stories, we decompose RefValue, MainElm, Space, SecondElm, and Equipment into their parts, StRefValue,

StMainElm, StSpace, StSecondElm, and StEquipment, on each story. We can deal with floors which exist in small parts of a plan in the same way, but devices to distinguish them from ordinary floors are needed. Hence, we describe below how to represent these objects on a story as aggregates of individual building components and the relationships between them. The data items for representing the building components described in Section 3.1 can be included as attributes of individual building components. Within the objective of this paper, the details of attributes of each building component are described below.

5.2. Representation of Reference Values

In a building with rectangular geometry, in which reference lines are straight and intersect orthogonally to each other, the locations of the reference lines on each story are determined by the distances between them. In this model, reference lines in the two directions in a plan are divided into segments in each span and the segments are represented by the intersections of reference lines at each end, so as to describe arbitrary planar shapes. Intersections of reference lines are denoted by IntersecRefLine. Thus, an object StRefValue is defined as an aggregate of IntersecRefLine. The location of intersections of reference lines is specified in a rectangular coordinate system whose origin is at an appropriate point (in this paper, the intersection of reference lines 1 and A in Fig. 1). To determine the vertical coordinates of building components, the attributes of StRefValue include a story reference line calculated from story heights. A story reference line is usually unique to each floor.

Table 2 shows examples of the attributes for a class of intersections of reference lines. The instance variables indicate variables used for the instances of the class. The attributes are classified into two parts: attributes for defining the class and attibutes for representing connection and reference relationships. The IntersecRefLine-Id Riid is an unique identification number given to a instance of IntersecRefLines. The IntersecRefLineRelat indicates the connection relationships with other IntersecRefLines. The Riid' indicates a Riid which is referred to represent the connection relationships. "( )" indicates an array in which the combinations of values in ( ), which are indefinite in number, are stored. "$( )^2$" indicates that the number is two.

5.3. Representation of Main Elements

Individual main elements are classified and represented as a hierarchy in the generalization plane. Fig. 5 shows a simplified example of a hierachy of main elements in the generalization plane. The generalization plane is set perpendicular to the aggregation plane so that each element of the lowest level in the generalization plane corresponds to a element on the level B of the aggregation plane in Fig. 4.

The location of such main elements as columns, beams, and walls is determined

TABLE 2
Attributes for the class of intersections of reference lines

Abbreviation	Instance variable
''definition''	
IntersecRefLine-Id	Riid
RefLineName	$(Rlnm)^2$
Coordinates	x y
''connection relationship''	
IntersecRefLineRelat	(Riid')
''reference relationship''	
ColDispRelat	Cdid'
WallDispRelat	(Wdid')
BeamDispRelat	(Bdid')
FloorDispRelat	(Fdid')
WindowDispRelat	(Mdid')
VertexRelat	(Rid' Vid')
WallDefnPointRelat	(Wdid' Wdpno')

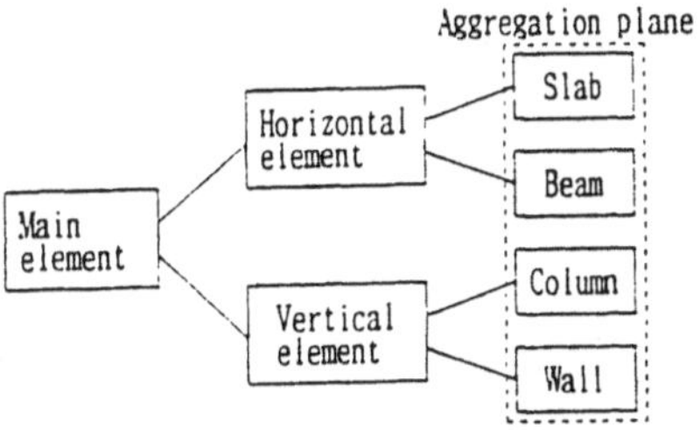

Fig. 5 Example of hierarchical representation of main elements in the generalization plane

by referring to intersections of reference lines. Hence, we define the relationship between an object column (Column) with attributes of shape and size and an intersection of reference lines as an object column disposition (ColDisp). In the Column, the type object is used to treat columns of various sectional shapes. In Fig. 4, the ColDisp indicates that after a Column and IntersecRefLines are defined, the location of the Column is determined by referring to the IntersecRefLine. In the similar manner, we define objects beam disposition (BeamDisp), wall disposition (WallDisp), etc. For a floor, we define the relationship between an object slab (Slab) with an attribute of thickness and IntersecRefLines (which describe the shape of the floor) as an object floor disposition (FloorDisp). An object StMainElm is thus defined as an aggregate of ColDisp, BeamDisp, WallDisp, FloorDisp, etc.

Tables 3 and 4 show examples of the attributes for classes of columns and walls. Sectional dimensions of columns and walls need to be specified between the story reference lines. For brevity, this paper assumes that sectional dimensions of columns and walls are uniform between the story reference lines. In the WallDisp, the WallEccent Wec indicates an eccentricity of the wall. For example, in case of

TABLE 3
Attributes for the classes of Column and ColDisp

(a) Column

Abbreviation	Instance variable
''definition''	
Column-Id	Cid
Column Type	Ct
''reference relationship''	
ColDispRelat	(Cdid')

(a.1) Rectangular Type

Abbreviation	Instance variable
Column-Id	Cid
Column Width	Cb
Column Height	Cd

(a.2) Circular Type

Abbreviation	Instance variable
Column-Id	Cid
Column Diameter	Cd

(b) ColDisp

Abbreviation	Instance variable
''definition''	
ColDisp-Id	Cdid
Column-Id	Cid
IntersecRefLine-Id	Riid
''connection relationship''	
BeamDispRelat	(Bdid')
WallDispRelat	(wdid')
FloorDispRelat	(Fdid')
''reference relationship''	
VertexRelat	(Rid' Vid')

TABLE 4
Attributes for the classes of Wall and WallDisp

(a) Wall

Abbreviation	Instance variable
''definition''	
Wall-Id	Wid
Wall Thickness	Wt
''reference relationship''	
WallDispRelat	(Wdid')

(b) WallDisp

Abbreviation	Instance variable
''definition''	
WallDisp-Id	Wdid
Wall-Id	Wid
DefnPoint	(Riid)[2] or (Wdid Wdpno)[2]
WallEccent	Wec
WallDefnPoint	(Wdpno Point D)
''connection relationship''	
WallDispRelat	(wdid')
PwallDispRelat	(Pdid')
ColDispRelat	(Cdid')
BeamDispRelat	Bdid'
WallsurfRelat	(Rid' Wsid')
''reference relationship''	
VertexRelat	(Rid' Vid')

We = 0, the wall is located on a reference line, and in case of Wec = −1, the wall is located so that the surface on the left or lower side of the wall coincides with the surfaces on the left or lower sides of the columns at the both ends. The WallDefnPoint is a point on the wall and used as a point for defining another wall or partition wall which exists at a distance from reference lines. The D indicates a distance from a Point which represents a Riid or another Wdpno. The shape of a floor is defined by a polygon whose vertices correspond to intersections of reference lines.

5.4. Representation of Secondary Elements

The hierachy of secondary elements in the generalization plane is set on the level B of the aggregation plane shown in Fig. 4. Windows and doors are installed in the opening of walls. The location of windows and doors is determined by the distances from surfaces of walls and foors, and intersections of reference lines. Hence, we define these relationships as an object WindowDisp (see Fig. 4). Because vertical coordinates of windows and doors are determined by vertical distances from the story reference line, these objects include the vertical distances from the story reference line among their attributes. In the same way, an object PwallDisp is defined as shown in Fig. 4. Because the horizontal and vertical locations of these building components with the shapes and sizes are determined by these instances, we can calculate such geometrical data as the shape of sections and surfaces of the building parts on the story. We can thus generate drawings, surface models, etc., if necessary, by using appropriate application programs. Data on the whole building is constructed by assembling data on each story.

5.5. Representation of Rooms

A room is an internal space with certain functions and which is surrounded by the surfaces of walls, floors, and ceilings. In this sense, a wall has two meanings (Willem, 1988): one is a physical body which stands for a building component wall; and another is a surface which encloses an internal space except for the particular case where walls exist partially. To construct a building model, we must therefore consider the following:

1. When viewing the wall surfaces inside a room, they are perceived as the surfaces which enclose the room.
2. If a wall adjoins plural rooms, each part of the surface is perceived as a surface which encloses each room.
3. The same finishes and colors are frequently applied to the plural surfaces enclosing the room.
4. A wall surface is used as a reference surface to determine the location of finishes, building equipment, etc.

These considerations also apply to a floor. Furthermore, the upper surface of a floor is a surface which encloses a room on the upper story. The lower surface of a floor is a surface which encloses a room on the lower story if there is no ceiling and is a reference surface for installing ceilings in the room.

To make the structure of the model simple and precise, therefore, we define the surface of walls and floors as objects of a wall surface (WallSurf) and a floor surface (FloorSurf), as well as defining their building components. The shape of wall, floor, and ceiling surfaces is represented by a polygon whose vertices are defined as an object Vertex. Wall and floor surfaces are respectively defined as aggregates of Vertex and finish (Finish) objects. An object ceiling (Ceiling) is defined as an aggregate of Vertex, FloorSurf (which is referred to in case of ceiling's installation) and Finish. Thus, as shown in Figs. 4 and 6, we define a object room (Room) as an aggregate of WallSurf, PwallSurf, FloorSurf, and Ceiling. Various rooms are classified and represented as hierarchies in the generalization planes, which is set on the level A of the aggregation plane shown in Fig. 4.

Tables 5 and 6 show examples of the attributes for classes of the Room and Vertex. The RoomRelat in TABLE 5 indicates that a room adjoins another room Rid' through the WallSurf Wsid1' (PwallSurf Psid1') of this room and the Wsid2'

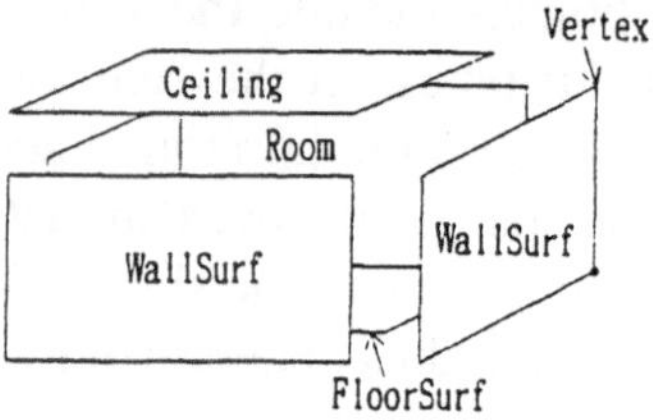

Fig. 6 Constitution of a room

TABLE 5
Attributes for the class of Room

Abbreviation	Instance variable
''definition''	
Room-Id	Rid
RoomName	Rnm
Surface	(SurfName Id)
''connection relationship''	
RoomRelat	(Wsid1' Rid' Wsid2')
	or (Psid1' Rid' Psid2')

TABLE 6
Attributes for the class of Vertex

Abbreviation	Instance variable
''definition''	
Room-Id	Rid
Vertex-Id	Vid
IntersecRefLine-Id	Riid
Coordinates	x y z
ElmDisp	(ElmName Id)
''reference relationship''	
WallSurfRelat	(Wsid')
PwallSurfRelat	(Psid')
FloorSurfRelat	(Fsid')
CeilingRelat	(Ceid')

TABLE 7
Attributes for the class of WallSurf

Abbreviation	Instance variable
''definition''	
Room-Id	Rid
WallSurf-Id	Wsid
Vertex-Id	(Vid)
Finish-Id	(Fid)
''connection relationship''	
WallDispRelat	Wdid'
''reference relationship''	
EquipDispRelat	(Edid')

TABLE 8
Attributes for the classes of Equipment and EquipDisp

(a) Equipment

Abbreviation	Instance variable
''definition''	
Equipment-Id	Eid
Specification	Espec
''reference relationship''	
EquipDispRelat	(Edid')

(b) EquipDisp

Abbreviation	Instance variable
''definition''	
EquipDisp-Id	Edid
Equipment-Id	Eid
Room-Id	Rid
Distance from walls	(Wsid D)
Distance from floor	Fsid D

(Psid2') of another room. The coordinates of Vertices represent relative coordinates in a rectangular coordinate system whose origin is a neighboring IntersecRefLine Riid. The coordinate z represents a vertical distance from a story reference line. The ElmDisp indicates the elements which intersect at the point (Vertex). Table 7 shows examples of the attributes for the class of the WallSurf.

5.6. Description of Building Equipment

The hierachy of building equipment in the generalization plane is set on the level B of the aggregation plane shown in Fig. 4. There are various kinds of building equipment, and their relationships with building components, which are referred to in case of their installation, are classified into several types. Pipings and ducts are installed by referring to reference lines, ceilings, and wall surfaces, etc. during or after the excecution of structural element work. Sinks, wall sockets, fluorescent lamps, etc. are installed by referring to surfaces of walls, floors, and ceilings, and they are frequently perceived as a part of a room. We need therefore to define these relationships as individual disposition objects according to the types of the relationships with building components. For brevity, in Fig. 4 we define an object of building equipment disposition (EquipDisp) as an aggregate of individual building equipment and IntersecRefLine, WallSurf, PwallSurf, FloorSurf, and Ceiling.

Table 8 shows examples of the attributes for a class of building equipment like a sink which is installed by referring to wall and floor surfaces.

6. Methods for Generating Objects and Changing their Attributes

6.1. AN EXMAPLE OF DESCRIPTIONS OF OBJECT CLASSES

In this Chapter, we present examples of methods for generating objects and changing their attributes. In the description, the notation like Smalltalk, shown in Fig. 7, is used. As an example of object classes, Fig. 8 outlines the class WallDisp. Instance variables (8(3)) represent attributes of instances of the class. 8(3) indicates a statement (3) in Fig. 8. The class variable WallDispDB (8(2)) is the database in which values of instance variables for instances of WallDisps are stored. The class methods are defined for generating instances of the class 8(4), changing a value of instance attributes (8(15)), and generating data required for application models (8(24)). As an example of generation of application models, we present methods for generating line segment data for drawing a plan. In the examples presented in this paper, only methods necessary for the explanation are shown.

6.2. METHODS FOR GENERATING OBJECTS

The outline of the method Generate for generating a new instance of the WallDisp is described in Fig. 8 (8(4)). A Wt and Riids of both ends are given as input data. A new number is given to a Wdid (8(6)). A message Generate with arguments Wt and Wdid is sent to the class Wall and a Wid is returned from the Wall (8(7)). A message GenerateSendData is sent to itself (8(8)). A message WallConnect is sent to itself to examine whether the wall is adjoining to other walls or not and a WallAnsw which contains (Wdid') of the adjoining walls or Nil is returned (8(9)). In the same way, messages WallConnects are sent to the ColDisp and the BeamDisp (8(10) and 8(11)). The (Cdid') and Bdid' indicate the instances of the ColDisp and BeamDisp which are adjoining to the wall. A message GenerateLines is sent to the class Plan (8(12)) to generate line segments for representing the generated wall and to change line segments representing related components in the plan. Finally, the values of the instance variables are stored in the WallDispDB (8-13).

The outline of the method Generate for generating a new instance in the Wall is described in Fig. 9 (9(1)). First, an instance of the Wall with the same Wt is searched in the WallDB (9(2)). If such an instance does not exist, then a new instance is generated and a new Wid is returned (9(3) and 9(4)). If such an instance exists, then the Wid of the instance is returned (9(4)).

When a new instance of a building component is generated, connection relationships between the adjoining components are examined and preparations for correcting the representation in the plan are made. As an example, the outline of a method for examining connection relationships between walls and columns in the class ColDisp is decribed in Fig. 10 (10(1)). First, an instance of the ColDisp with the same Riid is searched in the ColDispDB (10(3)). If such an instance exists, then the Wdid is stored in the WallDispRelat of the ColDispDB and a message GenerateSendData which instructs itself to generate graphic data on the column and to send them to the Plan is sent (10(4)).

When an instance of a building component is generated or the size or location of an instance is changed, the representations of the instances of the component and the adjoining components in the plan must be added or changed. For this purpose, the methods GenerateSendDatas in the WallDisp and the ColDisp are defined (8(24) and 10(5)). By these methods, each class generates the graphic data on the instance (for example, a rectangle for a column section) which are necessary to represent itself in the plan and sends them to the Plan. In generating these graphic data, knowledge of how to represent building components in plans is used. These graphic data are generated without taking the boundaries and overlaps between the adjoining components into consideration. Their coodinates are represented in an absolute rectangular coordinate system whose origin is at an appropriate point (for example, the intesection of reference lines 1 and A in Fig. 1) by referring to the coodinates of IntersecRefLines and using relative distances between components.

Fig. 11 outlines the class Plan which generates plans. The instance variable (Lsx1 Lsy1 Lsx2 Lsy2) is used to represent the coordinates of both ends of a line segment. (Wdid PointerToLine), etc is used to represent pointers to line segments which are produced from an instance Wdid of a WallDisp. In this object, graphic data sent from classes of various components are kept in the work area by the method ReceiveWallData, etc. (11(4) and 11(7)). When the Plan receives a message GenerateLines, then the method GenerateLines generates coordinates for defining line segments from graphic data receiving from instances of components, updates the line segment data, and stores them in the PlanDataDB (11(11) and 11(12)). These line segment data are produced by considering and eliminating the boundaries and overlaps between adjoining components. Their coordinates are represented in the absolute rectangular coordinate system which is used to represent the graphic data on the compoments.

```
Generate: Wt Riid Riid      Definition of method. Method-name: argument ......
Wall Generate: Wt Wdid      Definition of message. Receiver-object message-name: argument ......
| .... .... .... .... |     Temporary variables
.......... ← ..........     Substitute
( .... .... .... .... )     Array in which the combinations of values in ( ) are stored
self                        The object itself in which the statement exists
↑ ..................        Return value
[ ..... .......... ...... ] Set of processings
.......... = ..........     Equal
```

Fig. 7 Notation for representing the outline of object classes

```
① class name          WallDisp
② class variables     WallDispDB
③ instance variables Wdid Wid (Riid)
     (Wdpno Point D) (Wdid') (Cdid') Bdid'

   ''class methods''
   ''generation of a new WallDisp ''
④ Generate: Wt Riid Riid
⑤  | WallAnsw WallData |
⑥ Wdid  ← Newno.
⑦ Wid ← Wall Generate: Wt Wdid.
⑧ self GenerateSendData: Wdid.
⑨ WallAnsw← self WallConnect: Riid Riid.
⑩ (Cdid') ← ColDisp WallConnect: Riid Riid Wdid.
⑪ Bdid' ← BeamDisp WallConnect: Riid Riid Wdid.
⑫ Plan GenerateLines.
⑬ self Store: Wdid Wid WallAnsw (Cdid') Bdid'.
⑭ ↑Ok.

   ''change of wall thickness Wt''
⑮ ChangeWt: Wt Wdid
⑯ Wid (Wdid') (Cdid') Bdid' ←self Retreive: Wdid.
⑰ Wid ← Wall ChangeWt: Wid Wt Wdid.
⑱ self GenerateSendData: Wdid.
⑲ ColDisp GenerateSendData: (Cdid').
⑳ BeamDisp GenerateSendData: Bdid'.
㉑ Plan GenerateLines.
㉒ self Store: Wdid Wid.
㉓ ↑Ok.

   ''generate and send graphic data for plans''
㉔ GenerateSendData: (Wdid)
㉕  | WallData |
㉖ (WallData) ← [generate graphic data on walls
                   for plans].
㉗ Plan ReceiveWallData: (Wdid WallData).
㉘ ↑Ok.
```

Fig. 8 Outline of the class WallDisp

```
   ''generation of a new Wall''
① Generate: Wt Wdid
② Wid ←[search for an instance of Wall
        with the same Wt in WallDB ].
③ Wid isNil
     ifTrue: [Wid← Newno.
              self Store: Wid Wt Wdid]
④ ↑Wid.

   ''change of wall thickness Wt''
⑤ ChangeWt: Wid Wt Wdid.
⑥ (Wdid') ←self Retrieve: Wid.
⑦ (Wdid') =Wdid
     ifTrue:  [self DeleteWid: Wid]
     ifFalse: [self DeleteWdid: Wid Wdid].
⑧ Wid ←[search for an instance of Wall
        with the same Wt in WallDB ].
⑨ Wid isNil
     ifTrue: [Wid← Newno.
              self Store: Wid Wt Wdid]
⑩ ↑Wid
```

Fig. 9 Outline of methods Generate and ChangeWt in the class Wall

```
   ''connectivity of wall and column''
① WallConnect: Riid Riid Wdid
②  | ColumnData |
③ (Cdid)←[search for instances of ColDisp with
            the same Riid in ColDispDB].
④ (Cdid) notNil
     ifTrue: [self Store: (Cdid) Wdid'.
              self GenerateSendData: (Cdid).
              ↑(Cdid)]
     ifFalse: [↑Nil ].

   ''generate and send figure data for plans''
⑤ GenerateSendData: (Cdid)
   | ColumnData |
⑦ (ColumnData)←[generate graphic data on columns
                   for plans].
⑧ Plan ReceiveColumnData: (Cdid ColumnData).
⑨ ↑Ok.
```

Fig. 10 Outline of methods WallConnect and GenerateSendData in the class ColDisp

6.3. Methods for Changing Sizes of Building Components

When a size of a building component is changed, the attributes of the instance of the component, and the representations of the instances of the component and the adjoining components in the plan must be changed. As an example, the outline of the method ChangeWt for changing a Wt in the WallDisp is described in Fig. 8 (8(15)). The Wt and Wdid are given as input data. The values of the instance variables of the given instance Wdid are retrieved from the WallDispDB (8(16)). A message ChangeWt is sent to the Wall and a updated Wid is returned (8(17)). Messages GenerateSendData are sent to itself (8(18)) and the classes of the adjoining components according to the connection relationship data (Cdid'), etc. (8(19) and 8(20)). Finally, a message GenerateLines is sent to the Plan (8-21) and the new Wid is stored in the WallDispDB (8(22)). The outline of the method ChangeWt for changing a wall thickness in the Wall is described in Fig. 9 (9-5). If another WallDisp which refer to the given instance of the Wall do not exist, then the instance is deleted from the WallDB, else the WallDispRelat in the instance Wid is updated (9(6) and 9(7)). The processings in (8), (9), and (10) are the same as (2), (3) and (4).

6.4. Methods for Changing Locations of Building Components

When a location of a building component is changed, the attributes of the instance of the component, and the representations of the instance and the instances of locally related components in the plan must be changed. In this model, the location of a building component is specified by relative distances from other components except for IntersecRefLines. Because we need to change only these relative distances, the influence of the changes is confined to the local part, when we change the location of a building component.

As an example, the outline of the method ChangeCoodRiid for changing a location of an IntersecRefLine is described in Fig. 12 (12(1)). The Riid and the new coordinates x and y are given as input data. The new coordinates are stored in the class variable IntersecRefLineDB (12-2). The connection relationships with the related components are retrieved from the IntersecRefLineDB (12(3)). Messages GenerateSendDatas are sent to itself and the classes of the related components according to the connection relationship data cdid' etc (12(4), (5) and (6)). Because the new coordinates of IntersecRefLines are referred in the methods which generate the graphic data on these related components, the graphic data are updated according to the change of these coordinates. Finally, a message GenerateLines is sent to the Plan (12(7)).

6.5. Methods for Generating Room-Related Data

A room and the walls associated with it can be automatically identified by an appropriate algorithm (Franklin, 1986) which searches a closed loop formed by

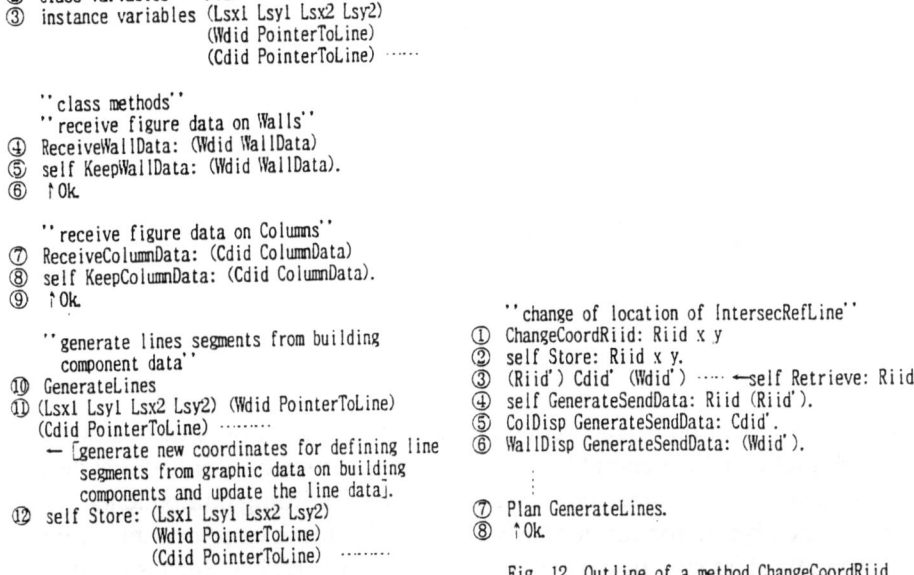

```
①  class name          Plan
②  class variables     PlanDataDB
③  instance variables  (Lsx1 Lsy1 Lsx2 Lsy2)
                       (Wdid PointerToLine)
                       (Cdid PointerToLine) ......

    "class methods"
    "receive figure data on Walls"
④  ReceiveWallData: (Wdid WallData)
⑤  self KeepWallData: (Wdid WallData).
⑥  ↑Ok.

    "receive figure data on Columns"
⑦  ReceiveColumnData: (Cdid ColumnData)
⑧  self KeepColumnData: (Cdid ColumnData).
⑨  ↑Ok.

    "generate lines segments from building
     component data"
⑩  GenerateLines
⑪  (Lsx1 Lsy1 Lsx2 Lsy2) (Wdid PointerToLine)
    (Cdid PointerToLine) .........
    ← [generate new coordinates for defining line
        segments from graphic data on building
        components and update the line data].
⑫  self Store: (Lsx1 Lsy1 Lsx2 Lsy2)
               (Wdid PointerToLine)
               (Cdid PointerToLine) .........
```

Fig. 11 Outline of the class Plan

```
    "change of location of IntersecRefLine"
①  ChangeCoordRiid: Riid x y
②  self Store: Riid x y.
③  (Riid') Cdid' (Wdid') ...... ←self Retrieve: Riid.
④  self GenerateSendData: Riid (Riid').
⑤  ColDisp GenerateSendData: Cdid'.
⑥  WallDisp GenerateSendData: (Wdid').
      :
      :
⑦  Plan GenerateLines.
⑧  ↑Ok.
```

Fig. 12 Outline of a method ChangeCoordRiid in the class IntersecRefLine

consecutive walls by using connection relationships in WallDisps. For example, in the plan of a building shown in Fig. 13, we search paths in a planar graph which is obtained by replacing walls with line segments. A path is a cyclic sequence of directed edges which marks the boundary of an area. In this algorithm, connection relationships between WallDisps are used to produce the graph. As a result, the insides of these paths are rooms and we can calculate the coodinates of Vertices for defining the WallSurfs, etc. of the room based on the data on the related instances of the WallDisps and the Walls, etc. The area of the room can also be calculated from these data. In addition, we can derive connection relationships between rooms by searching neighbor paths which have one or more line segmentes in common. In this example, the rooms A and B are adjoining to each other between points '2' and 'a' and the rooms A and C between 'a' and '5'. This algorithm holds for the case where rooms are completely enclosed by walls. When there are walls partially, we need to use another scheme for reasoning the functions of the spaces.

7. Effectiveness of the Proposed Model

1. *Simple and precise structure*. In this model, each object is unique and identifiable from all other objects. The unique identity of each object allows a direct correspondence between an object in the data representation and a building component. Classes of objects are classified by the generalization abstraction. A conceptual building product model is represented by the aggregation abstraction by using classes of building components and relationships between

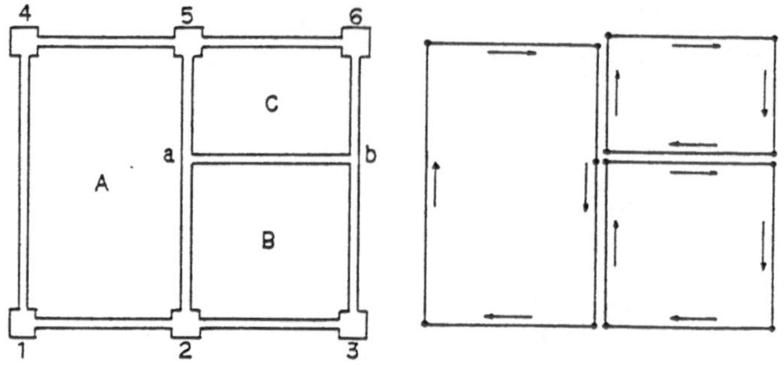

(a) An example of a plan b) Pathes formed by consecutive walls

Fig. 13 Rooms and the related walls

them. The location of each component is determined by a disposition object which includes the general rules used in the practice of architectural design to specify relative location relationships between building components. Because we define objects for surfaces as well as for building components to represent a room, the data structure of a room is simple and precise. Because of the effect of such methods, the structure of the proposed model is simple, precise, and easy to understand.

2. *Easy processing of room-related data.* A room and the walls associated with it can be automatically identified by an appropriate algorithm by using connection relationships between walls. Then we can calculate the coordinates of Vertices for defining WallSurfs, etc. and derive connection relationships between rooms. Because we can thus process data on rooms and surrounding wall surfaces etc. independently of data on building components, the processing of room-related data is easy. For example, we can easily define common finishes on the walls surrounding a room and generate finish shedules for that room.
3. *Ease of data modification.* In this model, the location of a building component is specified by relative distances from other components as shown in Fig. 2. Because of the effect of such a method, when we change only a relative distance between two components (for example, the distance D1 between a Datum and a Disp1), we need to change only the relative distance and thus the change is confined to the local part. The influence of the change of the location of the component (Disp1) is automatically exercised on the sizes and the locations of other components (Disp2), which refers to the Disp1, by the relative distances between them (D2). These changes are carried out according to the general rules used to specify building components in the practice of architectural design. We can thus reduce operation numbers for data input, when we change the location of a building component. For example, as described in the example of the method ChangeCoordRiid in the class IntersecRefLine in Section 6.4, we need to change only the coordinates of the instance, when we intend to change

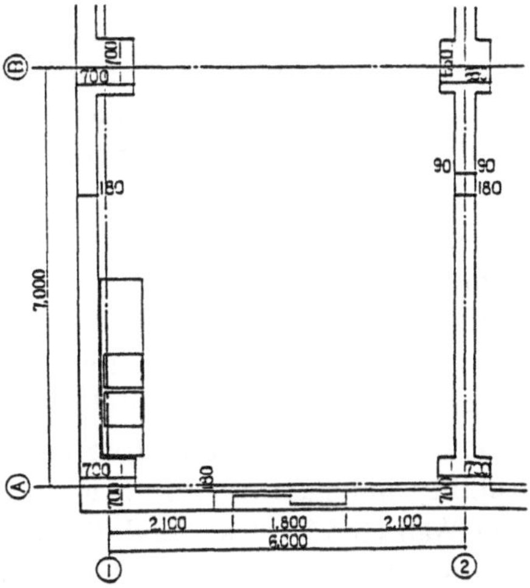

Fig. 14 Example of changes of locations of building components (see Fig. 1)

the location of the IntersecRefLine. According to the change of the location of the IntersecRefLine, the sizes (for example, wall length) and locations of other components such as WallDisps, WallSurfs, Equipdisps, ets. are automatically changed. Fig. 14 shows another example. In this case, the eccentricities Wes in the instances of the two walls located between 1-2 and A-B in Fig. 1 are changed from 0 to -1. According to the changes of the locations of the walls, the locations of the window and the sink are automatically changed keeping the same relative distances from the walls.

4. *Automatic generation and modification of application models.* In this model, because the location of a building component is specified by using a disposition object which represents its relative location relationship with other components, the connection relationships between building components are automatically derived from the definition attributes of the instances. Knowledge of the way of representing a building component in an application model can be included in methods of the class of the component. When an instance of a component is generated or the attributes of the size or location of a instance are changed, the corresponding representations in the application model can be automatically generated and changed by using these connection relationships and the knowledge of the way of representing building components in the application model, as described in the examples of the methods of Generate and ChangeWt in the WallDisp and of ChangeCoordRiid in the IntersecRefLine in Chapter 6.
5. *Extensibility and stability* In this model, the instance of each object is unique and identified by its identification number. Each object can refer to other objects by their identification numbers to define it. The information on the data for defining an object and the methods of operations for generating the instances

and changing their attribute are encapsulated into a single unit. The generation of an instance and the changes of attributes of an instance are carried out by receiving messages from and sending messages to the classes of the related components. Encapsulation hides the details of the data representation and the methods, making the result of the operation available. Because of such advantages of the object-oriented methods, we can add a new object class and change attributes or methods of an object without exercising much influence on other objects. In addition, because the model is represented by a hierachy structure based on the generalization and aggregation abstractions, an alteration in the details of an objects which is ignored at higher levels of the model will leave these higher levels unaffected. We therefore believe that the proposed model has high extensibility and stability. For example, when an attribute of the IntersecRefLine and Wall (see Tables 2 and 4) are altered, the WallDisp (see Table 4) and therefore the WallSurf (see Table 7) need not be altered as long as the method interfaces are not changed.

8. Conclusions

This paper has presented a basic structure of a building model for representing and using knowledge of buildings concerning the geometry in the field of architectural design based on the object-oriented concept. The salient features and effectivenesses of the proposed model are as follows.

1. The proposed model is constructed by the aggregation abstraction by using building components and the relationships between them and includes the general rules used to specify buildings in the Japanese practice. A room is represented by objects for surfaces different from building components. Because of the effect of such methods, the data structure of the proposed model is simple and precise and the processing of room-related data is easy.
2. In this model, the location of a building component is specified by relative distances from other components. Because of the effect of such a method, when we change a relative location between components, we need to change only the relative distances and then the influence of the change is automatically exercised on other components. We can thus reduce operation numbers for data input in generating and modifying the model.
3. The connection relationships between components are automatically derived from the definition data on the locations of the instances. The representations in an application model can be automatically generated and changed by using these connection relationships and the knowledge of the way of representing a component in the application model which is included in the methods of the class of the component.
4. Because the details of the data representations and the methods of a class is encapsulated, we can add a new object class and change the attributes or methods of an object without exercising much influence on other objects. Moreover,

because the model is represented by abstraction hierarchy, an alteration in the details of an object will leave the higher levels unaffected. The proposed model therefore has high extensibility and stability.

We are now planning to develop a prototype of a CAAD system based on the building model presented in this paper in C++ . The building model presented in this paper can be implemented by using the means of classes, objects, and message-passings provided in C++. The key issue is how to implement the class databases in which values of instance attributes are stored. The characteristic feature is that the objects use other objects for their descriptions. Consequently updating and retrieval of an object may include other objects which in turn may have other objects. Various storage schemes for implementing such complex data structures have been proposed (Valdriez, 1986). We are now performing the research on storage schemes appropriate for the building model.

References

Dittrich, K. R.: 1986, Object-oriented database systems: the notion and the issues, *Proceeding of International Workshop on Object-oriented Database Systems*, 2-4.

Eastman, C. M.: 1988, Conceptual modeling in architectural design, *Proc. CIB W74+W78 Seminar*, Lund, Sweden, 79-93.

Franklin, W. R.: 1986, Prolog and geometry projects, *IEEE CG & A*, 46-55.

Gielingh, W.: 1988, General AEC Reference Model (GARM), *Proc. CIB W74+W78 Seminar*, Lund, Sweden, 165-178.

Gross, M.: 1989, Relational modeling, *Proc. CAAD Futures'89 Conference Computer Aided Design Education.*

Law, K. H.: 1986, Data modelling for building design, *Proc. ASCE Fourth Conference on Computing in Civil Engineering*, 21-36.

Peraza, I. et al.: 1990, Tracing the changes in the design development by CAD, *Journal of Archit. Plann. Environ. Engng, AIJ*, **412**, 31-40.

Smith, J. M. et al: 1977, Database abstractions: aggregation, *Communications of ACM*, **20**(6), 405-413.

Smith, J. M. et. al: 1977, Database abstractions: aggregation and generalization, *ACM Transactions on Database Systems*, **2**(2), 105-133.

Spooner, D. L., et al.: 1986, Modeling mechanical CAD data with data abstraction and object-oriented technics, *Int. Conf. on Data Engineerin IEEE*, 416-424.

Valduriez, P. et al: 1986, Implementation techniques of complex objects, *Proc. Twelfth International Conference on Very Large Data Bases*, 101-110

Watanabe, S.: 1991, Architectural concept modeling in the framework for representing knowledge, *Fourth International Conference on Computing in Civil and Building Engineering*, Tokyo, Japan, 27

Willem, P.: 1988, A meta-topology for product modeling, *Proc. CIB W74+W78 Seminar*, Lund, Sweden, 213-221.

Yamada, S.: 1991, Architectural design support system DELTA, *Fourth International Conference on Computing in Civil and Building Engineering*, Tokyo, Japan, 35.

5

SHAPE DESIGN

TWO-DIMENSIONAL STRUCTURAL COMPONENT SHAPE IMPROVEMENT VIA CLASSIFIER SYSTEM

R. A. RICHARDS and S. D. SHEPPARD
Department of Mechanical Engineering
Stanford University
Stanford CA 94305
USA

Abstract. In this work the applicability of machine learning techniques in mechanical engineering design is investigated. Specifically, we explore how a classifier system can be used for two-dimensional component shape improvement. A classifier system learns rules, postulated as if/then statements, in order to improve its performance in an arbitrary environment. Classifier systems' versatility includes their adaptability and generality which overcome many of the obstacles facing today's conventional optimization techniques. After a brief review of mechanical engineering shape improvement methods, the classifier system is introduced with its underlying genetic algorithm. With this theoretical background we then tackle the problem of marrying the mechanical engineering and machine learning worlds by demonstrating how a classifier system can learn to improve the shape of a two-dimensional structural component. Furthermore, the learned classifier system performs the shape improvement without the benefit of sensitivity information which has been a virtual prerequisite to all prior shape improvement or optimization techniques.

1. Introduction

In the corpus of mechanical engineering, many methods have been developed for optimizing or improving designs of engineering systems. As pointed out by Haftka, Grandhi (Haftka & Grandhi, 1986) and others, despite the numerous advances and increased number of available tools, no one method has proven to be satisfactory across the broad spectrum of design problems faced by the modern engineer. Various methods such as dynamic programming guarantee the solution in their realm of applicability, but computational requirements render them useless for large problems. Other methods such as calculus based gradient search procedures converge nicely, but are limited to problems possessing properties such as continuity and unimodality. Obviously, there is a need for further development of efficient methods that are applicable over a broad spectrum of problems.

This investigation discusses one such promising technique known as a learning *classifier system.* In this investigation, we apply a relatively simple classifier system to a straightforward problem of improving the shape of a solid

J. S. Gero (ed.), Artificial Intelligence in Design '92, 267–286.

metal bar under tension loading. Although this particular problem has been solved by other techniques, it provides a useful point of departure for investigating the mechanics and power of the classifier system in engineering component design. Furthermore, the methodology presented here scales up to larger and more complex problems with no change required in the basic procedure. This cannot be said of sensitivity analysis or most of the other techniques which have solved this problem. As a mathematician with no knowledge of engineering would perform sub-par to an engineer in the application of mathematics to engineering problems, one needs engineering acumen to successfully apply classifier systems to engineering shape improvement or any other engineering problem. This point does not always seem obvious in the reading, but is abundantly clear when one attempts the application of classifier systems to the engineering realm. Some of the critical areas that require engineering background are discussed in the conclusion section.

Although classifier systems and their underlying genetic algorithm are not a daily part of the mechanical engineering vernacular, they have already proven their worth in real world applications. For example, General Electric has used a proprietary genetic algorithm and expert system package called *Engineous* to help design the engine for the upcoming Boeing 777, increasing engineer productivity by an order of magnitude (Antonoff, 1991; Ashley, 1991). Other applications include designing optimum welds (Deb, 1990), aircraft design and manufacturing (Bramlette et. al., 1990), and optimum paths in spacecraft rendezvous (Freeman et. al., 1990). An overview of the genetic algorithm and classifier system field is provided in Goldberg (Goldberg, 1989), while a more in depth theoretical foundation is provided by Holland (Holland, 1975).

First, some relevant shape optimization background will be presented. This is followed by an overview of the mechanics of classifier systems. Third, shape improvement using classifier systems is delineated and we show how a particular classifier system successfully improves the shape of a two-dimensional component. Finally, we present an overview of the current research and suggest possible directions for future applications of classifier systems in mechanical engineering.

2. Background—Shape Optimization

Shape optimization is generally associated with structural optimization where the objective is to minimize the mass with constraints imposed on stress, displacement, buckling and/or natural frequency. There are three distinct classes of shape optimization problems. In order of increasing computational complexity, these are (Sandgren, et al, 1991): cross-sectional, geometrical, and topological optimization.

- Cross-sectional optimization refers to the determination of specific geometric dimensions for a preselected design class such as the thickness of a shell or the diameter of a circular stress element. This class of problems has been under investigation for decades (Schmidt, 1960). Applications include truss structures.
- Geometric optimization introduces additional design variables which allow for boundary movement. Due to its increased difficulty relative to cross-sectional optimization, the geometrical changes are generally limited to a small region of the design such as around a fillet or hole. While geometric optimization has been applied with some success, there is much work to be done in the area of geometric modeling, analysis, and nonlinear programming methods before this class of shape optimization can become an integral part of the design process. Our current investigation is applicable to geometric optimization.
- Topological optimization involves topological as well as geometrical cross sectional modifications. Little work has been done in this area despite the importance of the concept. It will not be touched upon during this investigation.

As the complexity of optimization problems has increased, the *finite element* (FE) method (Zienkiewicz, 1980) has become an indispensable tool. However, the implementation of the finite element method presents its own limitations to many optimization techniques. For example, a particular FE representation may be suitable for the initial design, but as the optimization progresses and the design is modified, the original finite element representation degenerates in accuracy. The boundary may be modified by moving actual node locations or by moving *control points* on a spline curve or surface. The control point method allows for mesh and geometry independence because the control points define the curve or surface and when they are modified the curve or surface is modified. From the modified geometry a new mesh can be generated which is appropriate for the new shape. Unfortunately, the control point method is much more difficult to implement and automate than simply moving nodes. Thus, in the past, a completely automated progression of cycles through the optimization process was either heavily constrained by the finite element analysis or impossible. Another tool advancing optimization research is the boundary element method (Zhao, 1991). The boundary element method has been successfully used to solve shape optimization problems by removing some of the limitations of the finite element method. Just as many previous optimization techniques were hampered by the finite element state-of-the-art computer hardware or software technology available at the time, this research owes much of its success to the advancement of mechanical engineering computer-aided tools including finite element analysis. In the future, the boundary element method may prove more useful than the finite element method in the application of classifier systems.

3. Background—Classifier Systems and Genetic Algorithms

A *classifier system* (CS) is a machine learning system that learns syntactically simple string rules, called classifiers, to guide its performance in an arbitrary environment (Holland & Reitman, 1978; Goldberg, 1989). A classifier system has three major components:

- Rule and message system
- Apportionment of credit system
- Genetic Algorithm

A classifier system is similar to a control system (Dorf, 1983) in many respects. As a control system uses feedback to "control" or "adapt" its output for an environment, a classifier system uses feedback to "teach" or "adapt" its classifiers for an environment.

The classifier system has developed out of the merging of expert systems (Charniak & McDermott, 1985; Waterman 1986) and genetic algorithms (Holland, 1975; Goldberg, 1989). This synthesis has overcome the drawback to expert systems, namely, the long task of discovering and inputting rules. Using a genetic algorithm the CS learns the rules needed to perform in an environment; in this case the environment is two-dimensional structural component shape improvement.

3.1 RULE AND MESSAGE SYSTEM

Each classifier consists of a rule and an associated strength. The rule portion has the template

if <condition> then <action>

where,

<condition> is encoded as a string from the alphabet {0, 1, #}

<action> is encoded as a string from the alphabet {0, 1}.

The "#" symbol acts as a wild card or "don't care" in the condition.

The *strength* portion of the classifier gives a measure of the rule's merit in relation to the environment in which it is learning. That is, the higher a classifier's strength the better it has performed and the more likely it is to reproduce, this is covered in Section 3.3 and equation (1) below.

The *messages*, generated either from the environment or from the action of other classifiers, match the condition part of the classifier. For a condition to match a message, every part of the condition string must match every part of the message string. Therefore the message,

`011001`

would match all of the following conditions

```
0110#1
011001
##100#
######
```

as well as others.

To illustrate, the following table shows samples of strings which are valid forms for classifiers, (with the ":" symbol denoting the break between the condition and action, i.e. <condition>:<action>), in the first column, and their associated strength in the second column.

```
011:101                              23.2
011001##10#110:11                    17.3
10101000110011##100#:11100001        32.9
####:1                                7.1
100##00100##0011##:011001            29.0
```

The alphabet is restricted to allow for the power of genetic algorithms to be applied to the rule set as described below. The alphabet in no way restricts the representational capacity of the classifiers.

3.2 APPORTIONMENT OF CREDIT SYSTEM

The *apportionment of credit system* deals with the modifications in strength of classifiers as they learn. As the classifier system receives messages from the environment, certain classifiers will fire and modify the environment. The modifications will be deemed beneficial or detrimental to the environment. With this feedback the apportionment of credit system appropriately increases or decreases the strength of the classifiers that caused the modifications.

Empirical studies have shown that the exact mechanism for the apportionment of credit system is not critical to the learning ability of the CS. That is, the apportionment of credit system can have many forms and the CS will still learn in the environment. This is an example of one of the many CS variables which need to be set. The exact values to which the variables should be set is not critical but are guided by biological analogy and empirical results. Many times the variables are manipulated during the learning process to determine if such manipulations can enhance the learning process.

3.3 GENETIC ALGORITHM

A *genetic algorithm* (GA) is a search algorithm based on the mechanics of natural selection and natural genetics (Holland, 1975). Reproduction in GA theory, as in biology, is defined as the process of producing offspring (Melloni et al, 1979). However, mating may occur between any two classifiers, as there is no male-female distinction. The basic genetic algorithm operators involved in reproduction are:

- Selection
- Crossover
- Mutation

Selection is assortative*, with determination being probabilisticly proportional to the classifier's strength. During selection, then, high strength classifiers have a greater probability of producing offspring for the next generation than lower strength classifiers. There are many different ways to implement the selection operator, with most methods which bias selection towards high strength proving successful (Goldberg & Samtani, 1986). In this investigation, the probability that a classifier, x, will be selected for mating is given simply by the classifier's strength divided by the total strength of all the classifiers:

$$P(x) = \frac{Str(x)}{\sum_{k=1}^{n} Str(k)} \tag{1}$$

where,

$P(x)$	is the probability of selection for classifier x
$Str(x)$	is the strength of the classifier, x
n	is the total number of classifiers.

This gives every member of the population a finite probability of becoming a parent, but the stronger classifiers have a better chance of becoming parents.

After selection, the classifiers are copied into a mating pool and *crossover* occurs on the copies. First, the copies in the mating pool are paired at random. Second, each pair of copies undergoes crossing over as follows: an integer position k along the string is selected uniformly at random on the interval (1, L-1), where L is the length of the classifier. Two new classifiers are created by swapping all characters between positions L and k inclusively.

To visualize how this works, consider two strings A and B of length 7 mated at random from the mating pool created by previous reproduction:

$$A = a_1\, a_2\, a_3\, a_4\, a_5\, a_6\, a_7$$
$$B = b_1\, b_2\, b_3\, b_4\, b_5\, b_6\, b_7$$

Consider the random selection of k is four. The resulting crossover yields two new classifiers A' and B' following the partial exchange.

---

* assortative mating: mating which is not random, but involves individuals of specific characteristics

$A' = b_1\ b_2\ b_3\ b_4\ a_5\ a_6\ a_7$
$B' = a_1\ a_2\ a_3\ a_4\ b_5\ b_6\ b_7$

Although the mechanics of the selection and crossover operators are simple, the biased selection and the structured, though randomized, information exchange of crossover give genetic algorithms much of their power.

There is also the *mutation* operator, which performs a role in the reproduction process. Mutation is needed to guard against premature convergence, and to guarantee that any location in the classifier's search space may be reached. Mutation is the random alteration of a string position. In the classifier's tertiary code, a mutation could change a 0 to a 1 or #; a 1 to a 0 or #, or a # to a 0 or 1. By itself, mutation is a random walk through the classifier space. The frequency of mutation to obtain good results in empirical studies is on the order of 1 mutation per thousand position transfers. Mutation rates are analogously small in natural populations.

Despite the simplicity of the classifier system and the underlying genetic algorithm which it uses, it has been found to find near optimal solutions in a variety of environments (Goldberg, 1989).

4. Shape Optimization and Classifier Systems

For engineers to use a classifier system, in this or other types of engineering problems, they must adjust their thinking. Classifier systems are quite different from the conventional search methods encountered in engineering optimization in at least the following ways. They

- work with a coding of the parameter set, not the parameters themselves,
- search from a population of rules, not from a single rule,
- learn from experience,
- use probabilistic operators to guide their search. By contrast, most common engineering search schemes are deterministic in nature.

As a point of departure, one can think of gradient search as a classifier system with one rule encoding the gradient search technique. If one then created a set of rules, one being gradient search and the others being randomly generated, it would be possible to train the system to perform more efficiently or to perform where gradient search could not.

In this investigation, the classifier system is taught from a random set of rules in an environment in which gradient search could not perform because no derivative information is provided.

For structural design problems, the objective function is generally related to the mass, while constraints may include stress, buckling, displacement, or natural frequency limits. In this investigation the *Total Normalized Stress Error* (TNSE) compliments the mass as an evaluator. The TNSE is defined as:

$$TNSE = \frac{\sum_{k=1}^{n} \left| \sigma_k - \sigma_o \right| \Delta x_k}{\sum_{k=1}^{n} \Delta x_k} \tag{2}$$

where,

n	is the number of control points on a spline
σ_k	is the stress at control point k
σ_o	is the maximum allowable stress
Δx_k	is half the cord length between x_{k+1} and x_{k-1} along the spline.

The TNSE may be thought of as the entropy of stress, and the optimal design will be one of minimum entropy. Therefore, the change in mass and the change in TNSE per cycle are the dominant feedback components.

The optimization of a two dimensional solid component may be thought of as:

> a search through the space of determining the best boundary
> enclosing the material which best meets the design criterion
> while simultaneously satisfying all the design constraints.

With this in mind, the representation scheme implemented in this investigation modifies the boundary of the design in its quest for improvement.

To allow for modification of the boundary, the modifiable portions of the boundary are discretized into a series of points which will be allowed to move. The boundary is then represented by a series of cubic splines which passes through all the points. The cubic spline was selected for it has two continuous derivatives everywhere and possesses minimum mean curvature (Haftka & Grandhi, 1986). Finite-element analysis is used to determine the stress throughout the component. The stress results after being coded in binary form are fed into the classifier system which will then determine which points to move and by what magnitude. Figure 1 illustrates these points. The enclosed boundary in Figure 1 represents the boundary of a component. The boundary on the right side of the component is modifiable and therefore is discretized as represented by the circles. Moving to the column titled "Normalized Von Mises Stress" we see the stresses at the discretized points k-1, k, and k+1. The next column shows the same stress converted to their binary representations. The final column shows the IF part of a classifier that matches the binary representation. The classifier rule consists of a condition which has three parts, and an action which represents the magnitude of the modification of the middle node if this classifier fired. So the complete classifier is shown boxed, notice the left side of the classifier in the box is the concatenation of the three IF parts on the right column.

The points are moved via the geometry representation and not the finite element representation, as has been the case in many previous optimization investigations. This provides for mesh and geometry independence, so for each cycle a new mesh is generated allowing for a better mesh throughout the improvement process.

The CS has two major *modes* in its application to most problems, including this one. These are; the learning mode, and application mode. During the learning mode the system learns to operate in the structural component shape improvement domain. After learning, the CS is applied to a problem using its learned rules just as an expert system operates. The mechanics of these to modes are outlined below.

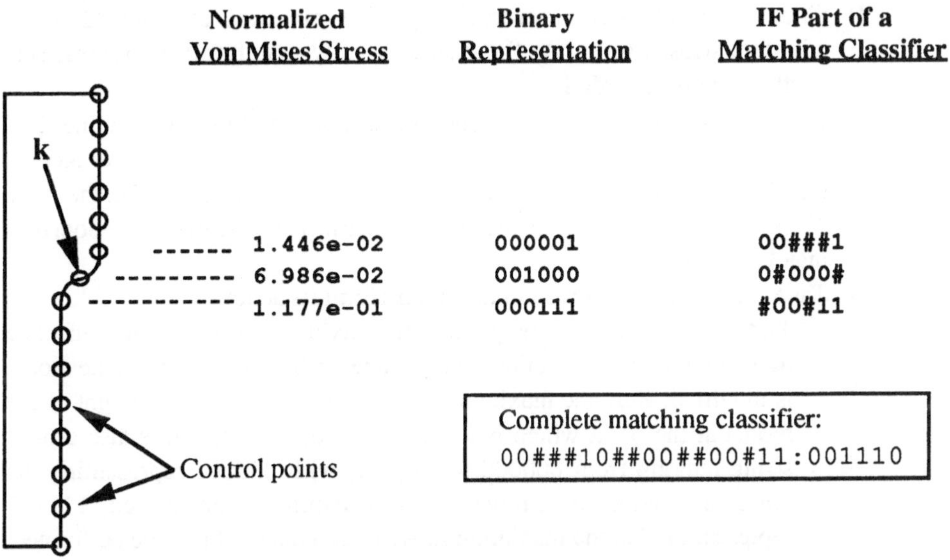

Fig. 1. Example of a Classifier Matching the Stress Condition at Control Point *k*

4.1 LEARNING MODE

The outline of a learning cycle of the classifier system is given below.

I. Analyze the structure and determine the stress at the control points for modification on the boundary of the component. Also determine the mass.

II. Compare this cycle's mass and TNSE to that of the previous cycle (except for the first in which case there is not previous cycle). Determine if the design has improved or deteriorated. With said information send feedback to the classifier which caused the most recent changes. The feedback will change the strength of the classifier that made the most recent change, either increasing it if the design is improved, or decreasing it if the design has degraded.

III. Modify the current design via the following sub-steps.

1) Convert stresses to their binary representation.

   In this investigation stresses are represented by a six bit string (e.g. 010111). The range of stresses covered is zero Von-Mises stress to two-times the optimal Von-Mises stress with stresses beyond two-times the optimal being mapped back to two-times the optimal. Therefore, zero Von-Mises stress is represented as, 000000; the optimal stress condition is, 100000; and stresses of two-times the optimal Von-Mises or more is represented as 111111.

2) Check each classifier determining if it matches any set of three adjacent stress points. As the classifier may match more than one set, each match is recorded.

3) Have all the classifiers that matched in (2) compete to see which one shall be permitted to execute its action. Factors that influence the winner include the classifier's strength and how well it matched the stress set (i.e. the more "#" symbols in a classifier the more general it is and the more poorly it matched the stress set).

4) Take the classifier that won in (3) and perform its action.

   The action is a percentage of the maximum move permitted. The maximum move is a user defined parameter, but can change as the model is modified. Here the maximum move permitted is 1/2 the height of the model at the point which is to be modified. The action is also a 6-bit string ranging from 000000 to 111111; with 000000 representing the maximum move to remove material should be performed, 111111 representing that the maximum move to add material is to be performed, and 100000 representing that no change should be performed.

IV. Record which classifier made the modifications.

V. Take the new design and return to step I.

An *epoch,* or block of learning cycles, is performed so that the present population of classifiers can be ranked. After an epoch has completed the classifiers are bred via a genetic algorithm to hopefully discover a better set of classifiers. After the GA is applied the new population starts another epoch of learning cycles. The entire process is repeated until the population performs to some standard.

4.2 APPLICATION MODE

The application mode is a subset of the learning mode. Since there is no learning the classifiers do not receive feedback on their performance, so steps (II) and (IV) above are eliminated. Furthermore, epoches do not exist for no GA is applied. The stop criterion is the same as in other improvement or optimization techniques; for example it may be based on, stopping after improvement per cycle is below a threshold for some number of cycles, stopping after a specified amount of computational resources are expended, or stopping when the optimum is reached if the optimality condition is determinable.

5. Application

Problem definition:

Goal:	Minimize mass
Constraints:	Maximum allowable stress = 36,000 psi
Modifications permitted:	Present boundaries not under traction or constrained
Loading:	Uniform tensile loading of 4,800 psi applied in the x-direction as shown in Figure 2.
Material:	UNS G10350 Cold-drawn Steel.
Initial shape:	As shown in Figure 2, with a height in y-direction of 1.25 inches on the constrained end and 0.75 inches on the end with the tension applied. The length in the x-direction is 6 inches. The thickness is 0.25 inches throughout.

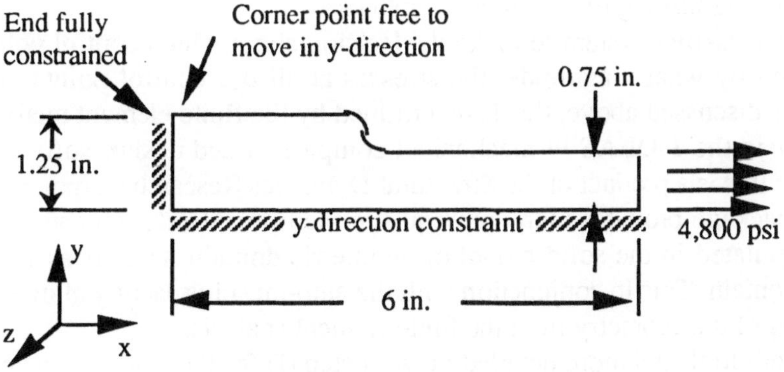

Fig. 2. Initial Configuration

The initial design is over designed and meets all the design constraints. As stated above, this particular problem has been solved by other techniques, and is rather trivial. Since there is a known solution, a benchmark is set for determining how well the classifier system has performed.

The solution for this problem is a rectangular bar with a height in the y-direction of 0.4 inches. The optimal shape is shown in Figure 3 to the same scale as in Figure 2.

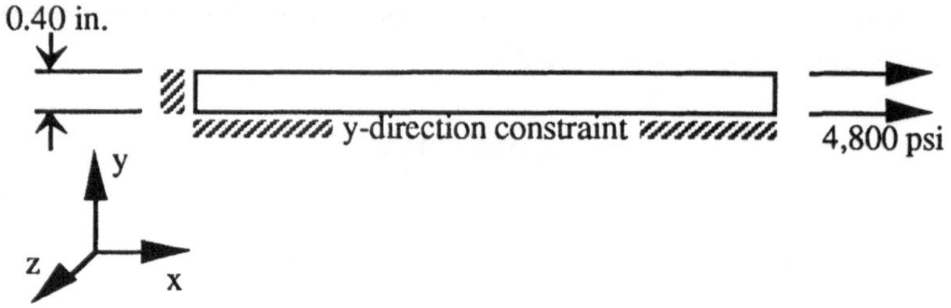

Fig. 3. Optimal Configuration

From the problem definition it is deduced that the sole boundary which may be modified is the "top" boundary as shown in Figure 2 for it has no applied traction and is under no constraints. To make the problem compatible to the input format that the classifier system expects, the boundaries which may be modified (in this case, one) must be discretized into a set of points defining the boundary that the classifier system may act on. For this problem the modifiable boundary was discretized into 15 evenly spaced control points as shown in Figure 1. Through the 15 points is fitted a set of cubic splines with the tangent at the end points being set to be perpendicular to the connecting edge, as is the case in the initial design. The boundary is modified each cycle by moving one or more of the control points and then re-establishing the set of cubic splines.

For the classifier system to make its decision about which control point(s) to modify and by what magnitude, the stresses at all the control points must be known. As discussed above, this is determined by the finite element method. For this analysis the I-DEAS^{tm} mechanical computer aided design software was employed; this is a product of the Structural Dynamics Research Corporation. The I-DEAS package provides many prerequisite facilities; the design can be stored and manipulated in the solid model or geometric domain as well as the finite element domain. This in conjunction with the auto-meshing facility allows for the uncoupling of the geometry from the finite element analysis.

Now let's look at a more detailed view of step (I) for the learning cycle of the classifier system which was first described in Section 4.0.

1) Represent the geometry in the solid model domain and determine the component's mass.
2) Map the geometry to the finite element domain.

3) Create a finite element mesh.
   This is performed by the automatic mesh generation facility available in I-DEAS. The automatically generated mesh goodness is checked with the element distortion method and the mesh is modified until the element distortion is low enough throughout.
4) Perform finite element analysis to determine stresses throughout the design.
   The I-DEAS internal linear elastic FEM solver is employed.
5) Map the stresses on the boundary nodes back to the control points.
   This operation is not provided by I-DEAS and is handled by an author written software module.
6) Check stresses throughout the design to determine if any are greater than the maximum allowed and record. This information is used in the determination of whether design at cycle x is better than design at cycle x-1.
7) Convert the stresses at the control points to their binary representation.
   At this point the system proceeds as outlined in steps II, III, and IV in section (4).

There are many variables that need to be set for a classifier system. As already stated, a random initial population is employed. This is the worst case scenario for the classifier system and thus fully tests its learning capacity. In any real world scenario a mixture of random and pre-determined rules would be used as the initial generation, for any engineer could quickly input a set of reasonable rules. The population size is 1,000 classifiers; though this may initially appear large, the total possible number of classifiers that could be generated from a 24 position string with a tertiary alphabet in the 18 positions and a binary alphabet in 6 positions make 1,000 appear too minuscule. Luckily genetic algorithm theory shows that 1,000 classifiers can search the space of all possible classifiers incredibly effectively (Holland, 1975). GA theory also shows that 1,000 is only one of a broad range of population sizes that can work effectively. Anywhere from a few hundred classifiers to many thousand classifiers could be used in the population. If the population becomes too large then the power of genetic algorithms is just being wasted.

Another major variable is the epoch or number of cycles that are to be performed before the application of mating (i.e. the genetic algorithm). In this study two epoches have been used, one is an epoch of 100 cycles and the other is an epoch of 250 cycles. These are relatively short epoches for a population of 1,000.

As the magnitude of the numbers above suggest, the learning phase is computationally expensive requiring many thousands of cycles before valuable learning should be expressed. This should be expected by such a general learning technique which is not given domain specific information and is also handicapped by starting with a random set of classifiers. When compared to the learning necessary for a human to be able to solve such a problem one could easily be drawn to conclude that this is an exercise in futility. The results show otherwise.

Furthermore, after learning has completed the use of the learned classifier system on a problem should be no more computationally expensive than other optimization schemes (if any other are available).

The complete suite of classifier system variables used for this problem are shown in Table 1 with the variables that have changed in Population II shown in bold. As discussed above the variables only need to fall into a range of values. The ranges are biologically motivated and/or have been determined through theoretical and empirical study. The interested reader may which to peruse the Goldberg, 1989 reference for complete discussion on the meaning and usage of all the variables.

TABLE 1
Classifier System Variables

	Population I	**Population II**
Length of Full Classifier	24	24
Length of Classifier's Condition	18	18
Length of Classifier's Action	6	6
Number of Classifiers in Population	1000	1000
Probability of choosing # in generating initial population	50%	**40%**
Bid Coefficient	0.1	0.1
Bid Sigma	0.1	**0.15**
Bid Tax	0.01	0.01
Life Tax	0.001	0.001
Bid Coefficient 1	0.1	0.1
Bid Coefficient 2	0.0833	0.0833
Effective Bid Coefficient 1	0.1	0.1
Effective Bid Coefficient 2	0.0833	0.0833
Proportion of Population Selected to Breeding per Epoch	25%	**10%**
Probability of Mutation	2%	**0.1%**
Probability of Crossover	100%	100%
Crowding Factor	100	100
Crowding Sub-Population	100	**150**
Reward Coefficient	2.5	**1.5**
Epoch length (GA period)	100 cycles	**250 cycles**

The results of the learning process are shown in the Figures 4 and 5. The graphs use a mass measure of 1 unit per 1 square inch, therefore the optimal mass is 0.6. For this problem the stress along the modified boundary is constant for the optimal design so the TNSE goes to zero as the component is improved. Figure 4 shows the performance of the system after learning for just 987 learning cycles for Population I. Even though 987 cycles is less than 10 epoches, the system shows that it already has learned to significantly improve the design. Note that the cycles referred to in the figure designate the number of cycles of the classifier system in application mode applied to the initial design. The largest cycle number in the figures is determined by the system's performance, for cycles larger than those shown the system vacillated around its best configuration.

Figure 5 shows the performance of the system after more learning, in this case learning has taken place with population I for 2,162 cycles and separately with population II for 1,527 cycles. Each population started from scratch. After the two populations learned independently, the best 500 classifiers from each were taken to make a population that was used in application mode to generate the results shown in Figure 5. Here the performance is vastly improved over the case shown in Figure 4, the mass is virtually optimal and the TNSE is down from around 0.20 to 0.10.

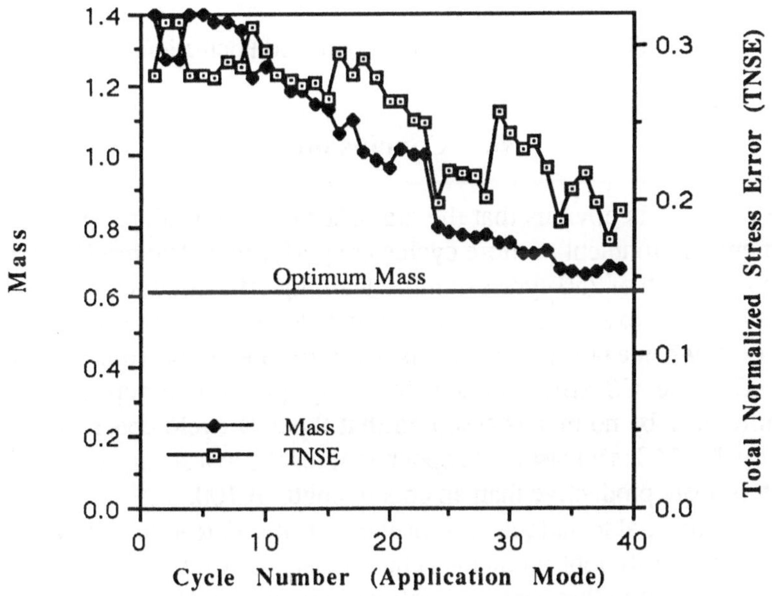

Fig. 4. Performance of Population I after 987 Learning Cycles

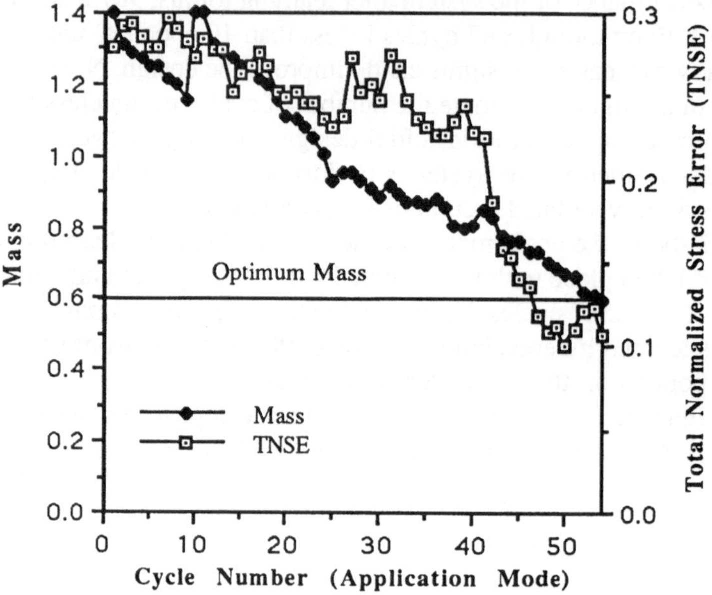

Fig. 5. Performance for Combined Population

6. Conclusions

From the results it is obvious that the classifier system is adapting to the design improvement environment as more cycles are performed. The results also implies that an epoch length of 250 cycles in more learning efficient than the epoch length of 100 cycles. This exemplifies the power and flexibility of the classifier system (CS), for even when a non-best choice is made for one of the variables associated with the CS, the CS still learns effectively just not maximally efficient. Furthermore, it is by no means assumed that the 250 cycle epoch is best epoch length, only that 250 is a reasonable epoch that results in a steady learning rate and one which is more productive than an epoch length of 100.

Another beneficial technique to improve learning is to immerse the classifier system in a sea of different problems. The diversity of stress patterns presented to the classifier system in this paper is relatively myopic in relation to those possible in different designs under various loading. Again, due to the general nature of the CS, the classifiers learned in this problem will provide a much better platform from which to attack more "difficult" problems. The "difficult" refers to a human scale, for the classifier system knows nothing about the environment and may be said to view all two-dimensional structural components as equally difficult.

As mentioned in the introduction, engineering judgement is critical in many aspects of the classifier system application, here are some of those aspects. A critical choice that must be made is the form of a classifier. Some of the questions that must be answered include; how long of a classifier condition (as described in Section 3) is required, should the condition consist of one part or many parts, and what form should the action take? The form of the classifier is tightly coupled with what the condition is going to represent and what the action is going to do. Moving to this next level, an engineer must determine what is important enough to need representation and what actions should be performed. With this in mind, a balance needs to be made between the power of classifier systems to solve problems in a generalized sense and the need to provide the CS with enough information in order to learn. Of course, all the domain dependent information available can be supplied to the CS and it will probably learn faster than without, but then the CS will be trapped in a narrow domain of applicability. This may be suitable for many applications and is not discouraged, however, the thrust of this investigation is generality so as to be able to move forth to tackle problems more difficult than two-dimensional components. The feedback process requires knowledge to judge the merit of changes made to the design among other considerations. There are many other choices that need the engineering eye, all of which can not be covered here and is an area of ongoing research. But just as in other areas of engineering, the choices are made via theoretical background, experience and empirical study.

It has been, and still is, thought by many present engineers in optimization analysis that design sensitivities are a fundamental requirement for shape optimization (Haftka & Grandhi, 1986). The above has demonstrated this is not the case. Design improvement and optimization can be performed without the need for sensitivity information.

It is a further credit to the classifier system that not only did it not need sensitivities but also it could adapt not knowing anything about the initial shape, the representation of the curve under modification, the concept of stress and strain, or optimization.

Therefore, this classifier system has learned to recognize patterns of stress and execute appropriate action. That is, the rules determine what boundary modification is beneficial when a particular stress pattern is present.

7. Current Research and Future Applications

Since this is an embryonic field, there are many divergent paths that one could try to learn more about the application of classifier systems to mechanical engineering component shape improvement. For example, one could try using sensitivity information as a message to the CS, more than three control points could decide the action at a control point, and/or different methods of modification could be tried. We have chosen to ignore the sensitivity information here and in future research because of the large computational overhead required to determine such information once one moves into large two-dimensional or three-dimensional structures and because much of the classifier system's strength is exemplified in its ability to perform and adapt without the benefit of sensitivity information.

After further investigation in the 2-d realm, the CS will be applied to 3-dimensional structural component shape improvement where its potential will be better displayed. For the 3-d case, more information will be fed to the CS per control point modification. However, the structure of the individual classifiers will remain similar. One method of extending the CS from 2-d to 3-d would be as follows; through each control point let pass through it two cords at right angles to each other at their intersection at the control point and follow the surface. Each cord would act analogously to the spline passing through a control point in the 2-d case. For each cord, take the two closest points to the control point at the intersection, and feed each set of three control point stresses into a CS as one did in the 2-d case. Now each cord would come up with a modification for the control point in question, these two suggested modifications would have to be reconciled. The reconciliation would be the modification that would take place. Figure 6 shows the control points used to decide modifications. In Figure 6 the 3-d case shows a subsurface control point that would not be used in the description above. However, it is shown for using stress information below the control point in question will be researched to determine if it provides the CS with exploitable information.

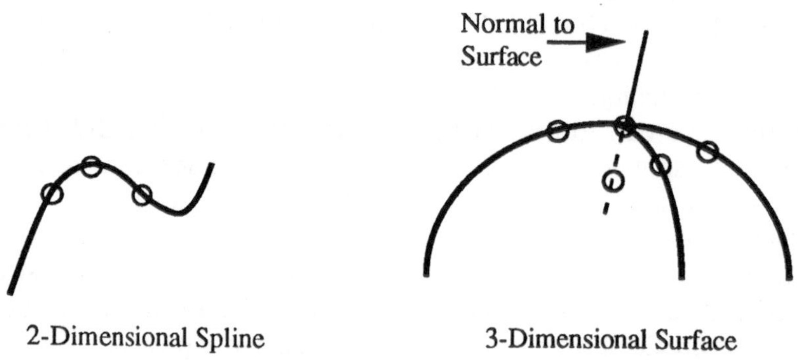

Fig. 6. Control Point Layout on 2-d Boundary and 3-d Surface

The generality of the CS expresses itself with the description of its application to 3 dimensions. One could even think of the 3-d case as two 2-d classifier systems running in parallel, one working on each of the perpendicular cords at each control point (if the technique described above was applied).

The CS computational requirements are drastically less in 3-d than techniques which depend on sensitivity information, since once the stresses are determined, all the information requiring FE analysis for determining the modifications has been determined. For sensitivity based techniques much more FE analysis needs to be performed to generate all the sensitivity information so a modification can be effected.

One could argue that the CS is also computationally expensive due to the massive training necessary. There is high computational costs, but here again the CS benefits greatly from where those computational costs derive. First of all, the training is off-line. That is, training is performed beforehand. In industry on would obtain a pre-trained system, thus avoiding the computational and time costs of the learning phase. Furthermore, there is no reason that the CS would not originally have good rules fed into it by experienced engineers giving it a much higher base from which to learn.

This investigation has shown that the classifier system has a place in the mechanical engineer's toolbox. Due to the power and flexibility of this tool, it is foreseen that the mechanical engineer will find a growing number of applications for which it may be leveraged.

References

Antonoff, M.: 1991, Genetic algorithms, software by natural selection, *Popular Science*, October, 70-74.

Ashley, S.: 1992, Engineous explores the design space, *Mechanical Engineering*, Feburary, 49-52.

Baffes, P. and Wang, L.: 1988, Mobile transporter path planning using a genetic algorithm approach, *SPIE's Cambridge Symposium on Advances in Intelligent Robotics Systems*, Cambridge, MA.

Bramlette, M. F., Bouchard, E. E., Buckman, E. C. and Takacs, L. A.: 1990, Current applications of genetic algorithms to aeronautical systems, *Sixth Conference on Aerospace Applications of Artificial Intelligence*, Dayton SigArt American Computing Machinery, Dayton, OH.

Charniak, E. and McDermott, D.: 1985, *Introduction to Artificial Intelligence*, Addison-Wesley, Reading, MA.

Deb, K.: 1990, Optimal design of a class of welded structures via genetic algorithm, *AIAA-990-1179-CP, Proceedings of the 31st Structures, Structural Dynamics and Materials Conference*, pp. 444-453.

Dorf, R. C.:1983, *Modern Control Systems*, 3rd edn, Addison-Wesley, Reading, MA.

Freeman, L. M., Kumar, K. K., Karr, C. L., and Meredith, D. L.: 1990, Tuning fuzzy logic controllers using genetic algorithms: aerospace applications, *Sixth Conference on Aerospace Applications of Artificial Intelligence*, Dayton SigArt American Computing Machinery, Dayton, OH.

Goldberg, D. D.: 1989, *Genetic Algorithms in Search, Optimization and Machine Learning*, Addison-Wesley, NY.

Goldberg, D. D. and Samtani, M. P.: 1986, Engineering optimization via genetic algorithm, *Electronic Computation*, pp. 471-482.

Haftka, R. T. and Grandhi, R. V.: 1986, Structural shape optimization—a survey, *Computer Methods in Applied Mechanics and Engineering*, Elsevier, pp. 91-106.

Holland, J. H.: 1975, *Adaptation in Natural and Artificial Systems*, University of Michigan Press, Ann Arbor, MI.

Holland, J. H.: 1976, Studies of the spontaneous emergence of self-replicating systems using cellular automata and formal grammars, *in* A. Lindenmayer and G. Rozenberg (eds), *Automata, Languages, Development*, North-Holland, NY, pp. 385-404.

Holland, J. H. and Reitman, J. S.: 1978, Cognitive systems based on adaptive algorithms, *in* D. A. Waterman and F. Hayes-Roth (eds), *Pattern-Directed Inference Systems*, Academic Press, New York.

Melloni, B. J., Eisner, G. M. and Dox, I.: 1979, *Melloni's Illustrated Medical Dictionary*, Williams and Wilkins, Baltimore, MD.

Sandgren, E., Jensen, E. and Welton, J. W.: 1991, Topological design of structural components using genetic optimization methods, *Proceedings of the 1990 Winter Annual Meeting of the American Society of Mechanical Engineers*, Dallas, Texas, pp. 31-43.

Schmidt, L.A.: 1960, Structural design by symmetric synthesis, *Proceedings of the Second ASCE Conference on Electronic Computation*, Pittsburgh, PA, pp. 105-122.

Waterman, D. A.: 1986, *A Guide to Expert Systems*, Addison-Wesley, Reading, MA.

Zhao, Z.: 1991, *Shape design sensitivity analysis and optimization using the boundary element method*, Springer-Verlag, New York.

Zienkiewicz, O. C.: 1980, *The Finite Element Method in Structural and Continuum Mechanics*, McGraw Hill, NY.

GRAMMARS OF FEATURES IN DESIGN

K. N. BROWN, J. H. SIMS WILLIAMS
Department of Engineering Mathematics
University of Bristol
Bristol BS8 1TQ
UK

and

C. A. MCMAHON
Department of Mechanical Engineering
University of Bristol
Bristol BS8 1TQ
UK

Abstract. A method of specifying and generating a set of designed objects known to be assessable is proposed. It is suggested that by generating designs which can be quickly assessed, and through being supplied with advice and assessment as the design proceeds, the designer can improve the design to product cycle. The method is based upon attributed graph grammars which specify valid manipulations of feature models in feature-based design. Semantic functions compute and constrain the feature attributes, and generate a simultaneous assessment as the design progresses. Finally, an example within the domain of stress concentration prediction is presented.

1. Introduction

It is almost universally acknowledged that much of design is dependent upon spatial reasoning, and that mechanical design in particular is dependent upon an understanding of the relationships between geometry, function and behaviour (Matousek, 1963; Dixon, 1986; Smithers, 1989). It is therefore one of the requirements for intelligent design tools to provide advice and information on the consequences of geometry-based design decisions for the function and performance of a designed object.

The nature of the end product of a design depends on many different factors, from meeting the required functionality and spatial constraints to keeping within limits of time and expense. Thus the designer is required to satisfy criteria from many different domains, some of which may be explicitly expressed - e.g function, strength, durability - while others may be implicit production constraints - e.g. machining capabilities or ease of assembly. In many cases, the body of knowledge required can be too great for the designer to consider all factors during the design, and so the design process may involve a repeated cycle of design,

J. S. Gero (ed.), Artificial Intelligence in Design '92, 287–306.

evaluation and redesign. Alternatively, for many factors the designer may not have the experience or knowledge needed to shape the design. The evaluation phase of the design cycle then requires input from analysts and domain experts, and may also involve expensive and time-consuming iterations of prototype manufacture and test. In order to improve this cycle, we believe that design tools should provide a means of giving advice and assessment to the designer across a range of domains. Ideally, this advice should be provided on request while the design is in progress, or, at worst, before the designer enters the more expensive stages of the design process.

Providing such help relating to the geometry of designs involves a trade-off between allowing the designer to design in terms of concepts and structures most useful to the designer, and providing representations which enable computer-based assessment of designs. We maintain that in the domain of mechanical design, the needs of both the designer and the computer are best served by using a feature-based representation. Further, in order to ensure economic design and manufacture, we believe that the designer should be encouraged to design using configurations of elements which are well understood, and should be provided with rapid feedback as to the consequences of design decisions as they are made. We propose, for geometry-based design, an underlying system based on grammars which specify a range of acceptable configurations, and for which we can generate an assessment as the design proceeds. Fitzhorn (1989) proposed a formalization of design as the generation of a string in a language. In his theory, design consists of four elements: the abstract design process, the artifacts of design, design specifications, and the rules for generating artifacts. Although the general design process is here hypothesized to be simple, it is the interactions between the specifications, the generating rules and the design context which provide the apparent complexity. Our work consists of applications within the framework of Fitzhorn's theory, in which we investigate how the specifications and context may be integrated with the generation of the design artifacts.

2. Features

Experience suggests that designers and analysts collect together what they perceive to be important sets of geometric or functional information, which we shall call features, and that their geometric reasoning is based upon those features. We shall define a feature to be

any geometric or functional element or property of an object useful in understanding the function, behaviour or performance of that object.

As examples, consider the objects shown in Figure 1 with attached feature descriptions of various types. Features are conventionally represented by frame or object-like structures, which are instantiations of generic definitions, arranged into

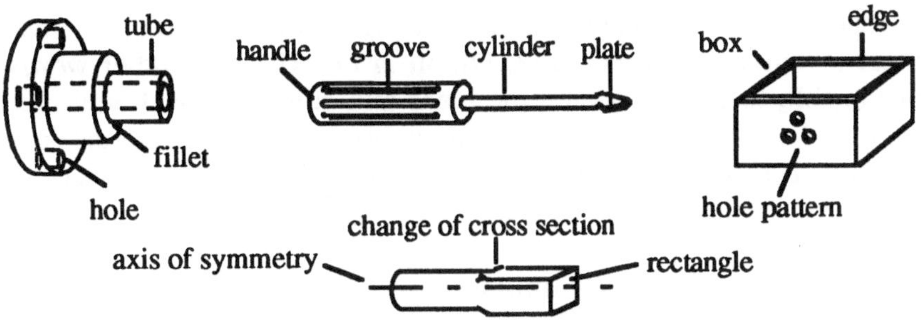

Fig. 1. Features

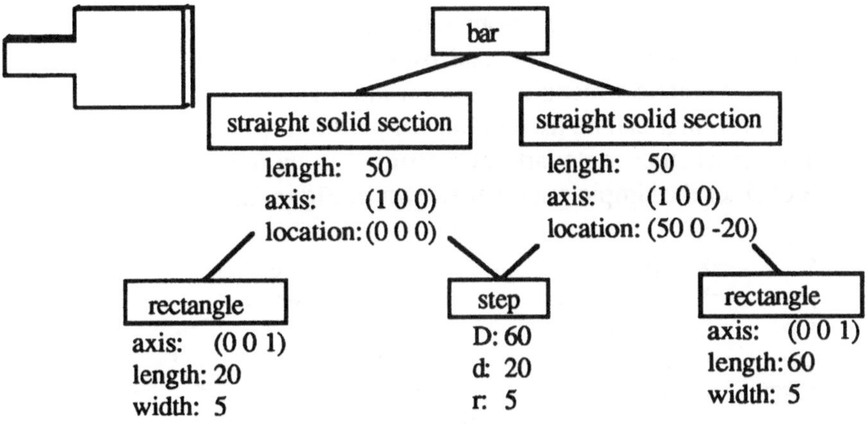

Fig. 2. Feature graph

connectivity graphs. A representation of a simple stepped bar is shown in Figure 2. The use of feature-based representations of geometry has been the subject of much recent research in the application of artificial intelligence to engineering (Henderson, 1984; Dixon *et al.*, 1987; Shah and Rogers, 1988; Juri *et al.*, 1990). For automatic reasoning about geometric objects, feature representations are particularly useful, as the representation can be pitched at a level corresponding to the expected reasoning mechanisms. Features have been used in a wide variety of applications, including stress analysis (Libardi *et al.*, 1986; Brown *et al.*, 1990), group technology classification (Shah and Bhatnagar, 1989), manufacturability evaluation (Luby *et al.*, 1986), machining operations (Juri *et al.*, 1990) and process planning (Gindy *et al.*, 1991).

There are two main methods of obtaining a feature representation. The first is automatic feature recognition, where the designer designs using a conventional solid modeller. The resulting model is then processed and the relevant features are extracted automatically - see, for example, Henderson (1984) or Jared (1984).

The alternative approach involves the designer designing in terms of features from the outset (Shah and Rogers, 1988; Dixon *et al.*, 1987). In other words, the designer is able to concentrate on the ideas and concepts which will meet the required functionality, without worrying about how best to represent the geometry to the computer. From the designer's feature description, the system then creates a solid model automatically.

3. Grammar And Shape

A grammar is a formal method of specifying the members of a set. The grammar defines the syntactic structure of the set by specifying relationships between primitive elements, or transformations which may be applied to those elements. Formal grammars were first proposed by Chomsky (1956), as the means for specifying the syntactic structure of natural language by a set of transformation functions. A grammar can be used either for generating the members of a set, or for determining whether or not arbitrary structures are members of the set. As an example, consider the simple grammar shown in Figure 3.

S	→	NP VP
NP	→	DET N
VP	→	V ADV
VP	→	V
DET	→	the
N	→	man
V	→	runs
ADV	→	quickly
ADV	→	slowly

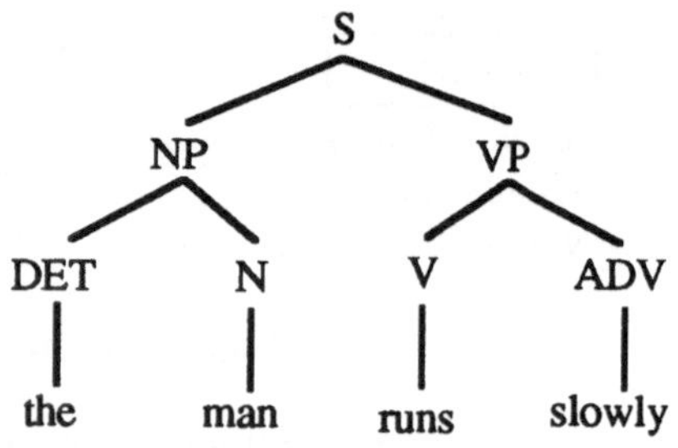

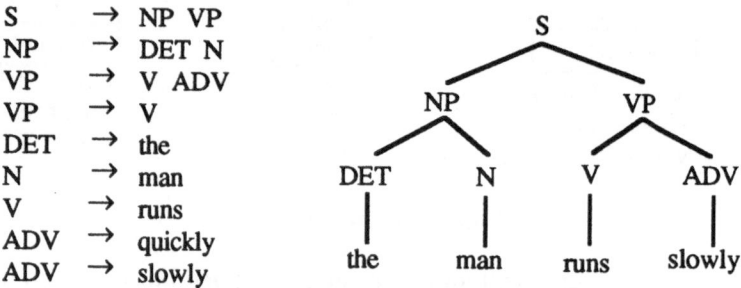

Fig. 3. A simple string grammar

Grammars need not be limited to simple strings of symbols, however. By using shapes as our primitive elements, we can define *shape grammars* (Stiny, 1980), which provide a formal means of specifying sets of shapes. A shape grammar generally consists of a set of transformation rules defined upon a set of shapes plus marker symbols. As a simple example, consider the two-dimensional shape grammar and shape derivations in Figure 4. Shape grammars also serve to characterize the underlying style or structure of constructs within a range of shapes. For example, Stiny and Mitchell (1978) devised a shape grammar which characterized the style of Palladian villas, while Radford and Gero (1985) describe a grammar for the detailed design of house eaves. Both these examples use shape grammars to synthesize designs - starting from an initial symbol, by selectively applying the rules of the grammars, it is possible to generate a villa plan in the style of Palladio, or house eaves in the style of early Australian villas.

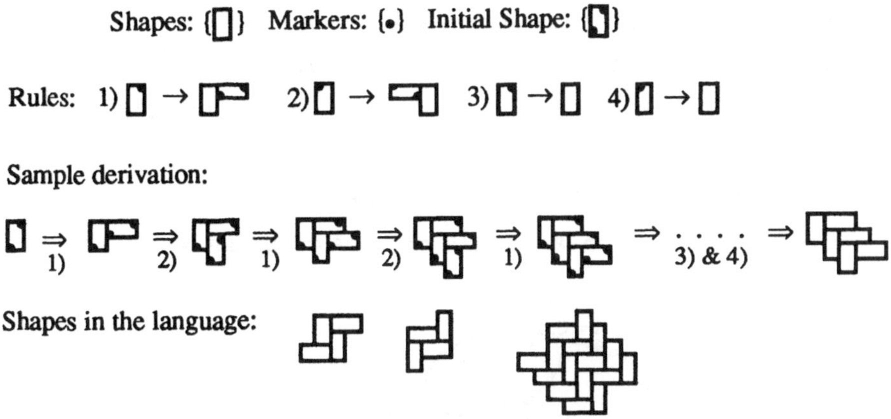

Fig. 4. Shape grammar

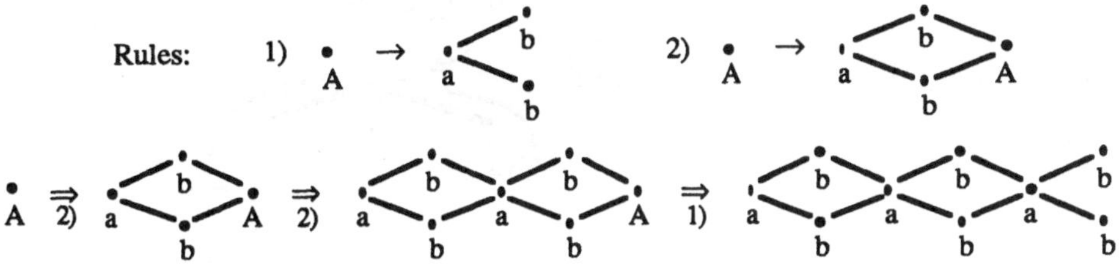

Fig. 5. Graph grammar

Alternatively, shape grammars can be used as an aid to the recognition of complex structures or patterns which are composed of lower level elements. Fu (1974), for example, describes the use of grammars to direct visual pattern recognition.

Two other grammar formalisms with applications in design are graph grammars and attribute grammars. Graph grammars, as their name suggests, operate not on strings or shapes, but on two-dimensional graphs (Ehrig *et al.*, 1987). See Figure 5 for an example of a simple graph grammar, taken from Pfaltz and Rosenfeld (1969).

Knuth (1968) introduced the concept of an attribute grammar (Deransart *et al.*, 1988), in order to specify the semantics of languages. In Knuth's formalism, the meaning of a string of symbols is stored within sets of attributes and values associated with each symbol. Each grammar rule then has some associated attribute relations, which describe the relationships between the attribute values of the symbols in the grammar rule. Attribute values may be inherited (arising from values further up the derivation tree) or synthesized (arising from values lower down the tree). As an example, consider the grammar in Figure 6, which generates arbitrary strings of **a**'s and **b**'s, and records for each **a** the number of **a**'s to the left, and for each **b** the number of **b**'s to the right. Rinderle (1991)

Symbols: { **A, a, b**} Initial symbol: { **A**} A.L = 0

Rules: 1) $\mathbf{A_1 \rightarrow aA_2}$ $\mathbf{a}.n = \mathbf{A_1}.L$, $\mathbf{A_2}.L = \mathbf{A_1}.L+1$, $\mathbf{A_1}.r = \mathbf{A_2}.r$

2) $\mathbf{A_1 \rightarrow bA_2}$ $\mathbf{b}.n = \mathbf{A_2}.r$, $\mathbf{A_2}.L = \mathbf{A_1}.L$, $\mathbf{A_1}.r = \mathbf{A_2}.r+1$

3) $\rightarrow \mathbf{a}$ $\mathbf{a}.n = \mathbf{A}.L$, $\mathbf{A}.r = 0$

4) $\mathbf{A \rightarrow b}$ $\mathbf{b}.n = 0$, $\mathbf{A}.r = 0$

Generation: **A ⇒ aA ⇒ aaA ⇒ aabA ⇒ aabaA ⇒ aabab**

Parse tree:

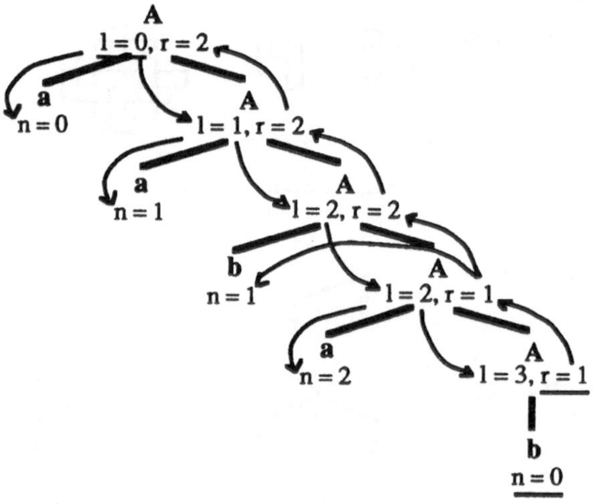

Fig. 6. Attribute grammar

demonstrates how the attributes of grammar symbols may be used to represent engineering parameters to facilitate design, while Brown (1991) introduces the application of attribute grammars of features in feature-based design .

Formal grammars have been applied in a number of cases to engineering design, solid modelling and feature recognition. Fitzhorn (1990) describes how graph-grammars may be used to specify languages of physically realizable shapes using both constructive solid geometry and boundary representations. By representing features as graphs in a solid modeller data structure, graph grammars can be used for automatic feature recognition (Pinilla *et al.*, 1989; Chuang and Henderson, 1991). A grammar (or set of grammar rules) can be defined which specifies a feature class. The task of feature recognition then becomes one of parsing - recognizing which subgraphs in a model can be generated by the grammar. Fu and de Pennington (1991) suggest that graph grammars can be used similarly for design by features, by defining features as graphs in a solid model. A transformational grammar can then be used to generate feature descriptions from a neutral solid modeller, or to create neutral solid model descriptions from the feature graphs.

4. Design, Grammars And Features

Fitzhorn (1990) has shown that, using a graph grammar, we can specify the sets of physically feasible designs generated from solid modelling primitives. As we are attempting to provide advice to the designer as to the consequences of a design, this prompts the question of whether we can specify a language of designs in terms of the more meaningful features discussed earlier. Thus we propose to define grammars of features to specify which conjunctions of features are feasible, and which features can be defined upon others. The language generated by such a grammar will not be equivalent to that generated by Fitzhorn's grammars, as many engineering shapes cannot be represented by current features technology. However, as our interest is in assessment rather than modelling, this is not the drawback it might initially seem. The freedom to design any combination of shape supported by a design tool can result in configurations which cannot be assessed by currently available techniques, or which are known to fail basic criteria. For example, certain combinations of shapes may be too complex to analyse, may be impossible to manufacture with available tools, or may be such that no knowledge exists on suitable assessment methods.

Thus, rather than specify the full range of feasible shapes, we suggest that it may be preferable to use grammars to specify restricted languages of design. In this case, a shape-feature grammar might generate all those configurations which can be analysed, or which meet certain functional or performance criteria, or, as in the architectural examples discussed earlier, conform to an accepted style. By designing within the constraints of a grammar, the designer can then know whether the design can be assessed, or whether it conforms to company standards, for example. In the case of languages of assessable designs, if a number of designs satisfy the required functionality, by selecting those which are generated by the grammar the designer could avoid expensive iterations of the design cycle.

Feature-based descriptions of objects are conventionally represented as graphs or networks, where the nodes of the graph represent feature instances, and the arcs of the graph represent connectivity relationships between features. Thus the specification of a range of objects will be required to be in terms of graph-grammars. The features used to represent objects, however, require a more complicated structure than simply a monadic symbol: the dimensional properties of the features are essential for determining whether or not a given configuration is analysable, or indeed is physically possible. Thus each feature in the grammar will require a number of attributes which will define the dimensions, orientation and location of the feature. As we are proposing the use of grammars for design, the selection of certain attribute values must be under the control of the designer. This is an extension to the concept of a synthesized attribute, in that we cannot specify precisely the value of all attributes through attachments to the grammar rules, but must allow for outside interference. We must relax the restriction which states that an attachment to the grammar rules must define how to compute the values of all attributes: instead we allow an attachment to a rule either to assign values to the attributes of the features occurring in the rule, or to constrain the values of the attributes which may be specified by the user.

5. Graph Grammars

We now present the necessary definitions.

5.1. DEFINITIONS: GRAPHS

Let P be a set of points. A *directed graph (digraph)* G over P is an ordered pair (P_G,E_G) s.t. (such that) $P_G \subseteq P$ and E_G is a set of ordered pairs of points in P_G. Points in P_G are called *nodes* of G, while pairs in E_G are called *arcs* of G. A digraph X is a *subgraph* of G if $P_X \subseteq P_G$ and $E_X \subseteq E_G$. For two digraphs G and H, $G \equiv H$ (G is *isomorphic* to H) if $\exists$ bijections $\psi_P:P_G \rightarrow P_H$ and $\psi_E:E_G \rightarrow E_H$ s.t. for $(x,y) \in E_G$, $\psi_E((x,y)) = (\psi_P(x), \psi_P(y))$.

Graph difference is defined as $G\backslash X = (P_G\backslash P_X, E_G\backslash\{(p,q): p \in P_X \text{ or } q \in P_X\})$.

Let S be a set of symbols s.t. $S = S_N \cup S_A$ and $S_N \cap S_A = \{\}$. S_N and S_A are the *node* and *arc labels* respectively. Let f_G be a function mapping $P_G \rightarrow S_N$ and $E_G \rightarrow S_A$. An *S-labelled digraph* is an ordered pair (G,f_G). S-labelled (X,f_X) is a subgraph of (G,f_G) if X is a subgraph of G, and $f_X = f_G|_X$ (f_G restricted to X). $(G,f_G) \equiv (H,f_H)$ if $G \equiv H$ with bijections ψ_P and ψ_E s.t. $\forall p \in P_G$, $f_G(p) = f_H(\psi_P(p))$ and $\forall (p,q) \in E_G$, $f_G((p,q)) = f_H(\psi_E((p,q)))$; i.e. the bijections ψ_P and ψ_E preserve labels.

$(G,f_G)\backslash(X,f_X) = (G\backslash X, f_G|_{G\backslash X})$.

Finally, we need to define $IN_i(G,X)$ and $OUT_i(G,X)$, the set of i-labelled arcs from $G\backslash X$ to X, and from X to $G\backslash X$ respectively.

$IN_i(G,X) = \{(p,q): p \in P_{G\backslash X}, q \in P_X, (p,q) \in E_G, f_G((p,q)) = i\}$

$OUT_i(G,X) = \{(p,q): p \in P_X, q \in P_{G\backslash X}, (p,q) \in E_G, f_G((p,q) = i\}$

From now on, we will call an S-labelled digraph (G,f_G) over P simply a *graph* G. i-labelled arcs and nodes will be called *i-arcs* and *i-nodes* respectively.

5.2. DEFINITIONS: GRAPH PRODUCTIONS

Graph rewriting rules specify a means of replacing a subgraph X of a parent graph G by a new graph Y. The main problem in defining such rules is stating how the new graph Y is to be embedded in $G\backslash X$. The embedding formulae must not depend upon any particular parent graph. The choice of embedding rules are the main differences between different graph grammar formalisms. Below we present a general method using set functions, similar in style to Nagl's (1987) and Pinilla *et al.*'s (1989) selection operators.

Let Γ be the set of all graphs over P with label set S. For given graphs G and X, the functions will map sets of nodes of G to sets of nodes of G\X. Let π,σ be sets of nodes of G, and let $\rho(P)$ be the power set of P. The symbol $\mapsto$ maps a typical element of the domain onto its image. Formally, we have

(i) $x:\Gamma\times\Gamma\times\rho(P) \rightarrow \rho(P):(G,X,\pi) \mapsto \{z: z \in \pi \;\&\; z \text{ is an x-node in } G\backslash X\}$ $(\forall x \in S_N)$
(ii) $SRC_i:\Gamma\times\Gamma\times\rho(P) \rightarrow \rho(P):(G,X,\pi) \mapsto \{z:z \in P_{G\backslash X} \;\&\; \exists \text{ i-arc } (z,p) \text{ in } G\backslash X \text{ for some } p \in \pi\}$ $(\forall i \in S_A)$
(iii) $TAR_i:\Gamma\times\Gamma\times\rho(P) \rightarrow \rho(P):(G,X,\pi) \mapsto \{z:z \in P_{G\backslash X} \;\&\; \exists \text{ i-arc } (p,z) \text{ in } G\backslash X \text{ for some } p \in \pi\}$ $(\forall i \in S_A)$
(iv) $\sim:\Gamma\times\Gamma\times\rho(P) \rightarrow \rho(P):(G,X,\pi) \mapsto \{z:z \in P_{G\backslash X} \;\&\; z \notin \pi\}$
(v) $\cup:\Gamma\times\Gamma\times\rho(P)\times\rho(P) \rightarrow \rho(P):(G,X,\pi,\sigma) \mapsto \{z:z \in P_{G\backslash X} \;\&\; z \in \pi \cup \sigma\}$
(vi) $\cap:\Gamma\times\Gamma\times\rho(P)\times\rho(P) \rightarrow \rho(P):(G,X,\pi,\sigma) \mapsto \{z:z\in P_{G\backslash X} \;\&\; z \in \pi \cap \sigma\}$
(viii) $\backslash:\Gamma\times\Gamma\times\rho(P)\times\rho(P) \rightarrow \rho(P):(G,X,\pi,\sigma) \mapsto \{z:z \in P_{G\backslash X} \;\&\; z \in \pi\backslash\sigma\}$

Omitting G and X from the notation, we have
(i) $x(\pi)$ is the set of all x-nodes of G\X in π
(ii) $SRC_i(\pi)$ is the set of all nodes in G\X which are the source of i-arcs to nodes in π
(iii) $TAR_i(\pi)$ is the set of all nodes in G\X which are the targets of i-arcs from nodes in π
(iv) $\sim(\pi)$ is the set of all nodes in G\X not in π
(v) $\cup(\pi,\sigma)$ is the set of nodes in G\X in either π or σ
(vi) $\cap(\pi,\sigma)$ is the set of all nodes in G\X in both π and σ
(vii) $\backslash(\pi,\sigma)$ is the set of all nodes in G\X in π but not in σ

Other functions may be defined as necessary, to specify, for examples, nodes in paths of arcs and nodes, or nodes satisfying a given predicate.

A *graph production* is a triple $R = (X,Y,F)$, where F is a set of triples (i,θ,z) or (i,z,θ), where i is an arc label, z is a node of Y, and θ is a combination of the above functions applied to subsets of the nodes of X. When the production is applied to a graph G, θ will specify a set of nodes of G\X. If $(i,z,\theta) \in F$, establish an i-arc from z to every node in θ; if $(i,\theta,z) \in F$, establish an i-arc from every node in θ to z.

If (X,Y,F) is a production, G and H are graphs, and ψ is a bijection s.t. $\psi(X)$ is a subgraph of G, then H is *directly derived* from G by R ($G \Rightarrow_R H$) if
(i) $\psi(Y)$ is a subgraph of H
(ii) $G\backslash\psi(X) = H\backslash\psi(Y)$
(iii) $(p,q) \in IN_i(H,\psi(Y) \Leftrightarrow (i,p,q) \in F \;\&\; p \in \pi$
(iv) $(p,q) \in OUT_i(H,\psi(Y) \Leftrightarrow (i,p,p) \in F \;\&\; q \in \pi$.

5.3. DEFINITIONS: GRAPH GRAMMARS

A graph grammar is a 5-tuple (P,S,T,I,R) s.t. P is a set of points, S is a label set s.t. $S = S_N \cup S_A$, $S_N \cap S_A = \{\}$, $T \subseteq S$ is a set of terminal labels s.t. $T_N \subseteq S_N$, $T_A \subseteq S_A$, $T = T_N \cup T_A$, I is a set of initial graphs, and R is a set of productions of the above form.

G is in the language of graph grammar $\mathcal{G}$ if G contains only terminal labels, $G_0 \in I$ and $\exists$ a sequence of graphs $G_1,\ldots,G_n$ s.t. $G_{i-1} \Rightarrow_R G_i$ and $G_n = G$.

Let S be a set of symbols as before. With each s in S_N, we associate a finite set of attributes A(s). The set of all attributes is $A = \cup_{s \in S} A(s)$. Each attribute a in A has a possibly infinite set of values, V_a. An *attributed graph grammar* is a 5-tuple as above, but with each production rule we associate a set of semantic functions which compute the values of the attrbiutes occurring in X and Y. If $a_1,\ldots,a_t$ are all the attributes occurring in X and Y, then the functions are of the form f: $V_{a1} \times \ldots \times V_{at}$ -> V_i to compute the value of attribute a_i.

The formalism as presented above is not sufficiently flexible to provide a tool for engineering designers. The values of the individual attributes may make individual graph productions inapplicable, even though the subgraph condition on X applies. Similarly, as we are attempting to specify a space in which a designer may operate, and to support design within that space, rather than to synthesize designs automatically, there will be many attributes whose values we will not wish to compute, but would rather constrain to lie within certain limits. Finally, as a matter of notational convenience, we will remove the distinction between terminal and non-terminal labels, but introduce an attribute *marker*: a label with a non-empty marker attribute can be read as a non-terminal, while a label with an empty marker attribute can be read as a terminal label. Formally, we now have that an *attributed graph grammar with constraints* is a 4-tuple (P,S,I,R), where P,S, and I are as above, but the productions are 4-tuples(X,Y,F,C), where C is a set of constraint relations (called the application constraints) on the attributes of the labels occurring in X. The application constraints can be used to restrict rules to the required label and marker attribute combinations. The semantic functions which we attach to the rules are either of the form given above, for computing attribute values, or of the form f:$V_{a1} \times \ldots \times V_{at}$ -> $\rho(V_i)$, for constraining values. A graph is in the language of a grammar if there is a sequence of direct derivations to that graph, and all nodes in the graph have empty marker attributes.

Figure 7 shows an example grammar rule which specifies the addition of a central hole to a flat bar section. For convenience, we write a label X with marker attribute value * as X*.

Application constraint: section.width > 10
section.marker = h

Semantic rules: hole.r ≥ 2
section.d > 2*hole.r-6

Fig. 7. Grammar rule for addition of hole feature

6. Semantics

This proposal is based on the assumption that the assessment of an object may depend on the configuration or structure of the object, and that a grammar can provide the details of that structure. Our aim is to tie in the assessment with the generation of the object, providing immediate feedback to the designer. The attributes of the features can represent information about the object other than simply the parameters required for modelling. For example, weight, manufacturing cost or strength could all be represented as attributes, with their values depending on those of the other parameters. In this way, by computing the attribute values as the design proceeds, information can be supplied to the designer as to the suitability of the design. As attributes may be either synthesized or inherited, it will not always be possible to compute the values of all attributes as their feature symbols are generated - synthesized attributes depend on the values of attributes generated later in the design for their value. This, however, will be an unavoidable feature of design grammars if we want the productions to represent natural design steps.

7. The Example: K_t Prediction In Design

As an example of the approach presented above, we shall present a graph-grammar and attribute system for the design of shafts. The grammar will specify a language of configurations whose stress concentration factors (K_t) can be assessed by a rapid but approximate technique, and the attribute system will predict the K_t values as the design proceeds.

For regular, simple geometries, there exist formulae which relate loading conditions to stress levels. Local changes of geometry modify the simple stress distributions, and produce areas of high stress in the vicinity of the change. There are no simple, generally applicable mathematical characterizations of this stress concentration, however, and beyond the lengthy and computationally expensive technique of finite element analysis the only way of estimating K_t is through the use of a large body of experimental results for standard cases (Peterson, 1974; Lipson and Juvinall, 1963). Real-life objects are generally more complicated than these standard cases, though, and the task of relating the standard results to a given component frequently requires the skill and experience of an analyst. K_t factors are used in the prediction of the strength and durability of components, and knowing their values would help designers to keep stress levels low where required. Some geometry, however, is too complex for this type of assessment, and must be submitted for FEA analysis, while it is possible that there may be simpler structures with the same functionality, but much lower stress levels, which could be used instead. A knowledge-based system for predicting K_t factors based on features has been developed (Brown 1991): below we present some of the knowledge from that system incorporated into an attribute graph-grammar. We shall assume each configuration generated is subjected to a simple bending load. The features required are presented in Figure 8.

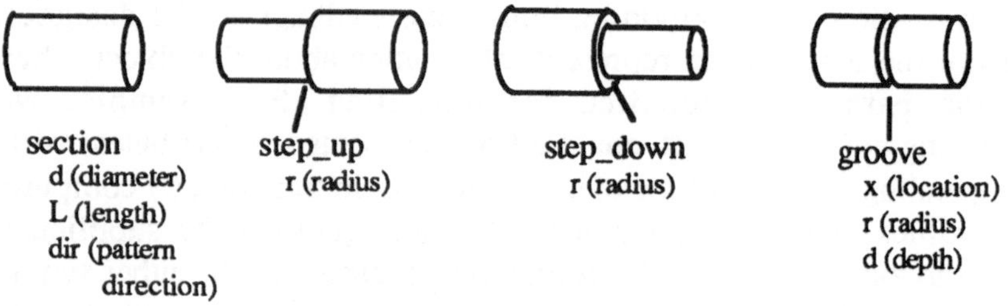

Fig. 8. Feature primitives

Grammar $\mathcal{G}$ (shown in Figure 9) generates, from the left, physically feasible configurations of steps and grooves. The presence of a marker X on a label L on the left-hand side of a production rule is equivalent to the application constraint *L.marker* = *X*. The grooves are constrained to be generated from left to right within each section. The labelled arcs represent an adjency relationship, while the unlabelled arcs represent the feature-subfeature relationship.

A typical generation of a configuration using $\mathcal{G}$ is shown in Figure 10, in both shape and graph form. Examples of some of the other configurations in the language are shown in Figure 11.

8. K_t Semantics

Many of the configurations generated by $\mathcal{G}$ are beyond the capacity of the K_t knowledge base - although they may meet their purpose, and be easily manufacturable, the current state of our knowledge base does not allow us to predict the K_t values. Each feature is assumed to have a range of influence, which

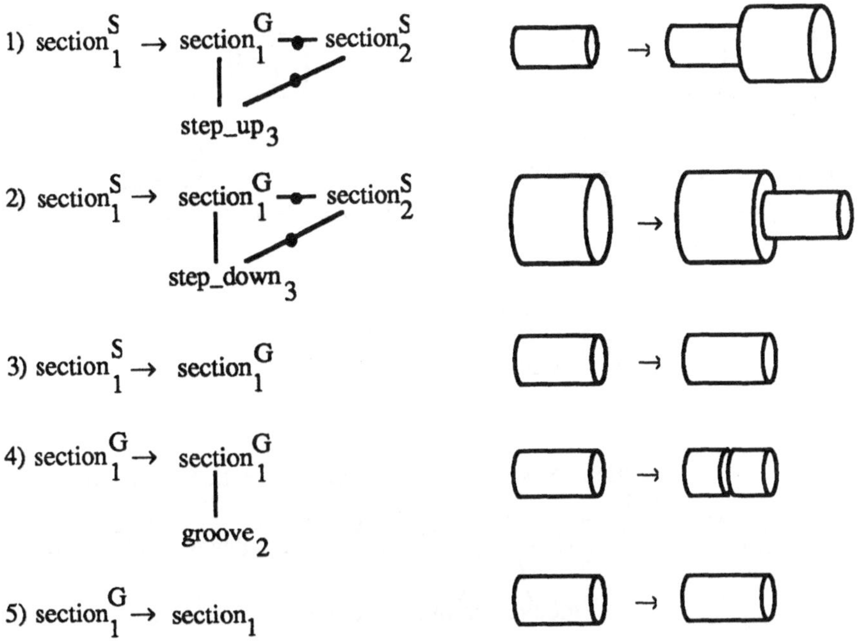

Fig. 9. Grammar $\mathcal{G}$

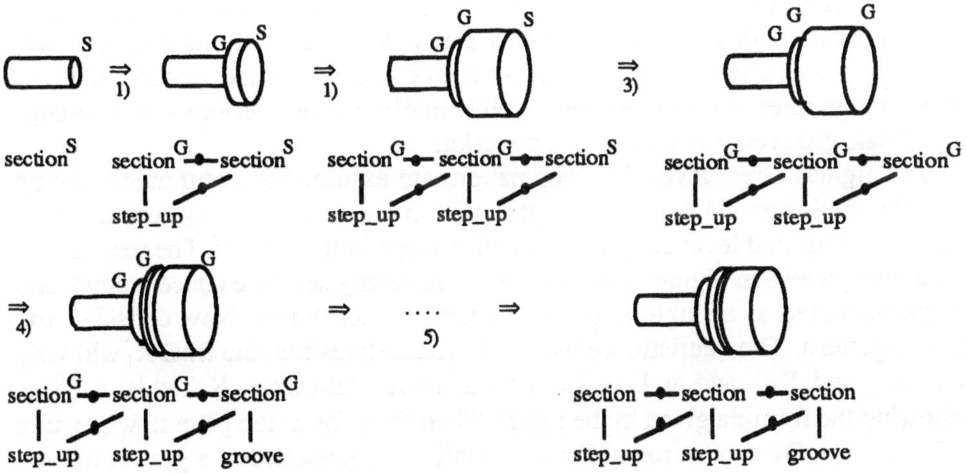

Fig. 10. Generation within grammar $\mathcal{G}$

can be considered to entail a gradual decrease in the concentration induced by the feature. Heuristics are available which compute the range for individual features: for example, the range of a groove is assumed to be three times the depth of the groove, while the range of a step from d_1 to d_2 is assumed to be $1.5*(d_2-d_1)$ in

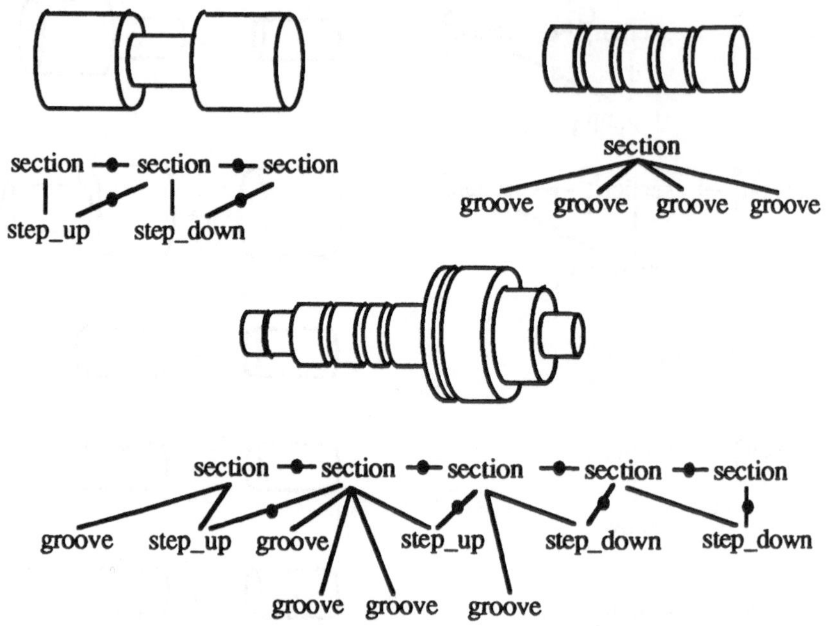

Fig. 11. Configurations in the language of $\mathcal{G}$

the direction of the step. We define a step pattern to be a sequence of steps whose ranges interact. Our knowledge can only assess K_t for monotonic step patterns. Similarly, we cannot predict the effect of a groove on the minor section of a step whose range interacts with the step itself. Finally, we are currently able to cope with single grooves only on any given section.

The highest stress levels in a step pattern are assumed to occur at the step up from the minimum diameter of the pattern, with the precise values depending on the dimensions and location of the remaining steps in the pattern. The second step in the first pattern in Figure 12 is too close to have any separate concentration, and is approximated as a single step with dimensions as shown. Now consider the second pattern. The heuristic we use for patterns states that the final K_t will vary with K_{t1} and K_{t2}, and as L or L_* tend to their maximum, K_t tends to K_{t1}, following the formula given in the figure. Therefore, for a step pattern, we require d, D, r, L, d_*, D_* and L_* to predict K_t. Finally, the presence of a groove in the significant area of a step has the effect of a second step, as shown in Figure 13.

If we wish to constrain the designer to those configurations which can be rapidly assessed (or at least to note when a configuration cannot be assessed), we need to represent the above constraints in the grammar attribute system. To do this, we need to define a number of new attributes. After each step up, we can compute the length the new section must be in order to contribute as a separate step (srf), and the length beyond which the new section starts a new pattern (rf).

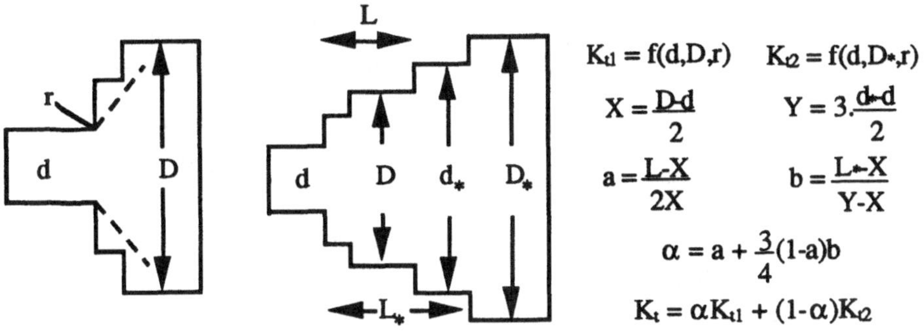

Fig. 12. Step pattern heuristics

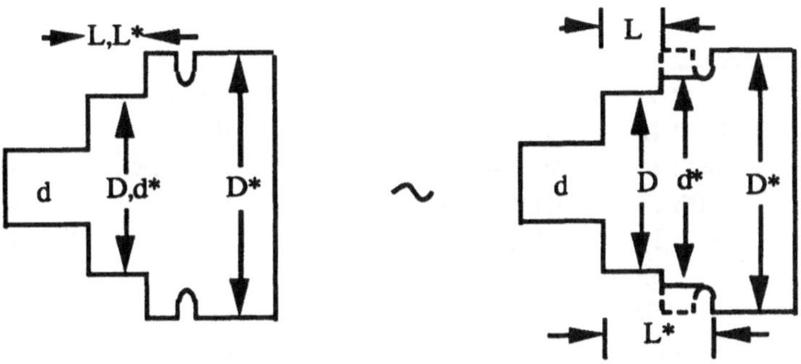

Fig. 13. Groove heuristics

Following this, we can then classify sections according to their contribution to a pattern. The first and last sections in the pattern will be type 0, the first significant step type 2, intermediate steps type 4, and non-contributing steps either type 1 or type 3 (before and after the first significant step respectively). These attributes can be evaluated as the generation proceeds (inherited). We require corresponding attributes for downward step patterns (srb, rb and type), but in this case their values can only be evaluated after the pattern is complete - hence they are synthesized. New downward sections have their dimensions constrained so that the range of their step pattern does not extend back into a previously defined upward pattern, represented by the attribute bc. Finally, to constrain the groove locations, we modify rule 4 (as shown in Figure 14) and use the range attributes of the sections, together with the attributes dir and f, which describe the relative size of the sections to the left and right respectively.

Figure 14 shows a generation of a two-stepped shaft with a groove on the minor section. The middle section is too short to contribute to the K_t value for the step pattern, while the groove is constrained to lie far enough away from the first step to induce an independent stress concentration. The K_t values are obtained

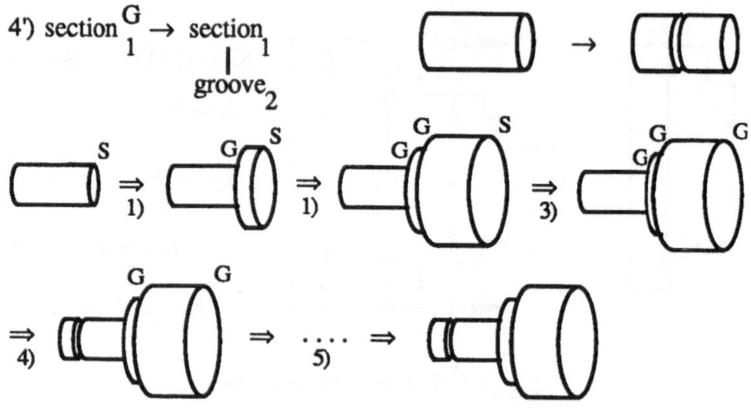

section$_1$	section$_2$	section$_4$	groove$_6$	template$_1$	template$_2$
d : 30	d : 52	d : 60	x : 25	type : step	type : groove
L : 60	L : 10	L : 20	r : 2	d : 30	d : 22
dir : 0	dir : +	dir : +	d : 4	D : ~~52~~ 60	D : 30
type: 0	type: 1	type: ~~2~~ 0	t : 2	r : 3	r : 2
srf :	srf : 11	srf : 5		L : 30	Kt : 2.115
rf :	rf : 33	rf : 35		d* :	
dl : 0	dl : 30	dl : 30		D* :	
dr : 0	dr : 0	dr : 0		L* :	
L-L: 0	L-L: 0	L-L: 0		L' : ~~10~~ 30	
L-R: 58	L-R: 5	L-R: 0		Kt : ~~1.76~~ 1.78	
tl :	tl : 1	tl : 1			
tr : 1	tr :	tr :	step_up$_3$	step_up$_5$	
f : +	f : +	f : 0	r: 5	r: 5	

Fig. 14. Generation and K_t prediction

from a data source as the generation proceeds, and an intermediate value for K_t is produced before the third section is defined. Figure 15 shows the generation of a downward step pattern, with a groove on the intermediate section. As stated above, the attributes representing the range and section classification are synthesized, and can only be generated after the pattern is complete. Once the pattern is complete, the groove can then be added, and its effect on K_t computed.

9. Discussion

The example grammar presented above is too restrictive for real design. The designer is over-constrained, and is unable to generate many obvious and useful configurations. In addition, the semantic functions are relatively complicated for

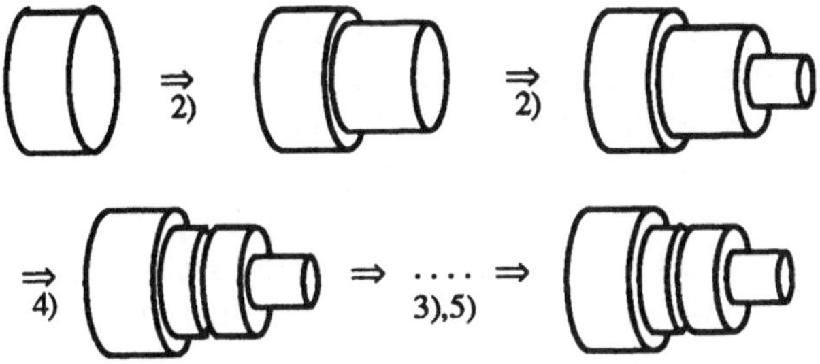

	section$_1$	section$_2$	section$_4$
d	60	50	20
L	20	42	20
dir	0	-	-
type	0	2	~~0~~
srb	5	15	
rb	17	45	
dl	0	0	0
dr	~~0~~ 50	20	0
bc	0	-28	-41.3
L-L	0	5	5
L-R	20	42	0
tl			1
tr	1	1	
f	-	-	0

groove$_6$	
x	23
r	2
d	5
t	1

step_up$_3$	
r	5

step_up$_5$	
r	2

template$_1$	
type	step
d	20
D	~~50~~ 40
r	2
L	~~42~~ 21
d*	50
D*	60
L*	42
L'	62
Kt	~~1.861~~ 1.79

Fig. 15. Generation of a downward step pattern

such a limited language. However, these deficiencies are largely due to a lack of domain knowledge, rather than any inherent deficiency in the approach. Most of our constraints are concerned with prohibiting configurations we cannot assess - if our knowledge base on assessing K_t was more wide ranging, we could dispense with many of the constraints, the semantic functions would be simpler, and the language of the grammar correspondingly larger. The grammar also imposes a directionality on the design - shafts must be generated from left to right. Although this clearly limits its usefulness as a design tool, it has been imposed to keep the semantic functions as simple as possible. There is nothing inherent in the grammar approach which requires such a restriction.

Our main area of interest for future work is whether the approach can be extended to other domains - for example, manufacturing assessment or design for assembly. A specification of the language of designs which can be manufactured by a given process would have obvious benefits. Feature-based assessment for different domains almost definitely requires different, application-specific, feature sets for each domain, however. Current approaches to this problem involve the use of a general design feature set from which application-specific information can be extracted (Shah, 1988). It may be possible to devise a grammar of the design features, for which different semantic attachments compute the relevant parameters for different applications, allowing simultaneous assessment for a number of different domains. Alternatively, it may be preferable to design in terms of a single feature set and to receive a single assessment. The task then would be to translate the finished design to other application-specific sets, and to parse the structures using application specific grammars. The resulting parse trees would then provide the semantic evaluation as before.

One final topic which is suggested by the above work is the possibility of automatic design synthesis using grammars. The grammar rules as we have presented them above are intended to guide the design by explicitly stating which combinations are legal, and predicting the expected effect (within the chosen domain) of rule applications. If we are able to devise a representation for function, then it may be possible to use the semantics to select the most suitable rules from those which can be applied at any given stage. In other words, could we select the configurations which produce the lowest K_t values, for example, or those which are easiest to manufacture?

10. Conclusion

A method has been proposed by which a set of assessable designs may be concisely specified, and made explicit to the designer, using attributed graph grammars which specify valid manipulations of feature models in feature-based design. The formalism has been extended to constrain rather than compute the attributes of the symbols, allowing the designer more control over the generation. By attaching semantic functions to the grammar, it is possible to provide immediate feedback to the designer as to the consequences of design decisions as the design proceeds. An example has been presented from the domain of stress concentration prediction. It is suggested that such a method, if properly developed and extended, may provide a means to improve the design to product cycle.

Acknowledgements

This work has been supported by the ACME directorate of the U.K. Science and Engineering Research Council, initially under grant GR/E/63254, and currently under GR/H/20763: we are grateful for their support. The research has been carried out in collaboration with Rover plc, who are also currently providing additional financial support, for which, again, we are grateful. Finally we would like to thank the anonymous referees for their detailed and comprehensive criticisms of an earlier draft of this paper.

References

Brown, K. N.: 1991, *Conceptual Geometric Reasoning in Artificial Intelligence and Engineering*, PhD Thesis, Dept. of Engineering Mathematics, University of Bristol.

Brown, K. N., Sims Williams, J. H., Devlukia, J. and McMahon, C. A.: 1990, Reasoning With Geometry: Predicting Stress Concentration Factors, *Journal of Artificial Intelligence in Engineering*, 5(4), 182-188.

Chomsky, N.: 1956, Three Models for the Description of Language, *IRE Transactions on Information Theory*, 2, 113-124.

Chuang, S-H. and Henderson, M. R.: 1991, Compound Feature Recognition by Web Grammar Parsing, *Research in Engineering Design* (Special Issue: Formal Languages in Engineering Design, ed. Fitzhorn,P.A.), 2(3), 147-158.

Deransart, P., Jourdan, M. and Lorho, B.: 1988, *Attribute Grammars: Definitions, Systems and Bibliography*, Lecture Notes in Computer Science 323, Springer Verlag.

Dixon, J. R.: 1986, Artificial Intelligence in Design: A Mechanical Engineering View, *AAAI-86*, pp872-877.

Dixon, J. R., Libardi, E. C., Luby, S. C., Vaghul, M. and Simmons, M. K.: 1987, Expert Systems For Mechanical Design: Examples of Symbolic Representations Of Design Geometries, *Engineering with Computers*, 2, 1-10.

Ehrig, H., Nagl, M., Rozenberg, G. and Rosenfeld, A.: 1987, *Graph-Grammars and Their Application to Computer Science*, Lecture Notes in Computer Science 291, Springer Verlag.

Fitzhorn, P. A.: 1989, A Computational Theory of Design, *Preprints: NSF Engineering Design Research Conference*, College of Engineering, University of Massachusetts, pp.221-233.

Fitzhorn, P. A.: 1990, Formal Graph Languages of Shape, *(AI EDAM) Artificial Intelligence for Engineering Design, Analysis and Manufacturing*, 4(3), 151-163.

Fu, K. S.: 1974, *Syntactic Methods in Pattern Recognition*, Academic Press.

Fu, Z. and de Pennington, A.: 1991, Geometric Reasoning Based On Graph Grammar Parsing, *ASME Design Automation Conference*, Miami, USA.

Gindy, N. N. Z., Huang, X. and Ratchev, T. M.: 1991, Feature-based Component Model for Computer Aided Process Planning Systems, *Symposium on Feature-Based Approaches to Design and Process Planning*, Loughborough University of Technology.

Henderson, M. R.: 1984, *Extraction of Feature Information From 3-Dimensional CAD Data*, PhD Thesis, Purdue University.

Jared, G. E. M.: 1984, Shape Features in Geometric Modelling, in *Solid Modelling By Computers: From Theory to Applications* (ed. Pickett, M. S.), Plenum Press.

Juri, A. H., Saia, A. and De Pennington, A.: 1990, Reasoning About Machining Operations Using Feature-Based Models, *International Journal of Production Research*, 28(1), 153-171.

Knuth, D. E.: 1968, Semantics of Context-Free Languages, *Mathematical Systems Theory*, 2(2), 127-145.

Libardi, E. C., Dixon, J. R. and Simmons, M. K.: 1986, Designing with Features: Design and Analysis of Extrusions as an Example, *Mechanical Design Automation Laboratory Technical Report TR 2-86*, Department of Mechanical Engineering, University of Massachusetts at Amherst.

Lipson, C. and Juvinall, R. C.: 1963, *Handbook of Stress and Strength*, MacMillan.

Luby, S. C., Dixon, J. R. and Simmons, M. K.: 1986, Designing with Features: Creating and Using a Features Data Base for Evaluation of Manufacturability of Castings, *Proceedings ASME Computers in Engineering Conference*, Chicago.

Matousek, R.: 1963, *Engineering Design: A Systematic Approach*, Blackie.

Nagl, M.: 1987, Set Theoretic Approaches To Graph-Grammars, in Ehrig *et al.* (eds), *Graph-Grammars and Their Application to Computer Science*, Lecture Notes in Computer Science 291, Springer Verlag, pp41-54.

Peterson, R.: 1974, *Stress Concentration Factors*, Wiley-Interscience.

Pfaltz, J. L. and Rosenfeld, A.: 1969, Web Grammars, *Proceedings of First Joint Conference on Artificial Intelligence*, Washington, DC, pp.609-619

Pinilla, J. M., Finger, S. and Prinz, F. B.: 1989, Shape Feature Description and Recognition Using an Augmented Topology Graph Grammar, in *Preprints: NSF Engineering Design Research Conference*, College of Engineering, University of Massachusetts, pp.285-300.

Radford, A. D. and Gero, J. S.: 1985, Towards Generative Expert Systems For Architectural Detailing, *CAD, Special Issue: Expert Systems*, 17(9), 428-435.

Rinderle, J. R.: 1991, Grammatical Approaches to Engineering Design, Part II: Melding Configuration and Parametric Design Using Attribute Grammars, *Research in Engineering Design* (Special Issue: Formal Languages in Engineering Design), 2(3), 137-146.

Saia, A. and de Pennington, A.: 1991, Feature-based Geometric Reasoning Based on Graph Grammars, *Symposium on Feature-Based Approaches to Design and Process Planning*, Loughborough University of Technology.

Shah, J. J.: 1988, Feature Transformations between Application Specific Feature Spaces, *Computer-Aided Engineering Journal*, December, pp.247-255.

Shah, J. J. and Bhatnagar, A. S.: 1989, Group Technology Classification From Feature-Based Geometric Models, *Manufacturing Review*, 2(3), 204-213.

Shah, J. J. and Rogers, M. T.: 1988, Expert Form Feature Modeling Shell, *CAD*, 20(9), 515-524.

Smithers, T.: 1989, AI-based design versus Geometry-based design, or why design cannot be supported by geomtry alone, *CAD*, 21(3), 141-150.

Stiny, G.: 1980, Introduction to Shape and Shape Grammars, *Environment and Planning B*, 7, 343-351.

Stiny, G. and Mitchell, W. J.: 1978, The Palladian Grammar, *Environment and Planning B*, 5, 5-18.

AN IMPROVED SHAPE ANNEALING ALGORITHM FOR OPTIMALLY DIRECTED SHAPE GENERATION

J. CAGAN and G. REDDY
Department of Mechanical Engineering
Carnegie Mellon University
Pittsburgh PA 15213
USA

Abstract. This paper reviews the new design generation technique called *shape annealing* by Cagan and Mitchell (1992) and proposes modifications to the algorithm to improve design results. Shape annealing is a variation of the simulated annealing stochastic optimization technique. It produces optimally directed shapes within the language specified by a shape grammar by *controlling* the application and selection of shape rules. We present a modification to the basic algorithm which improves performance by 15% while superior shapes evolve.

1. Introduction

Cagan and Mitchell (1992) present the shape generation technique of *shape annealing* which combines concepts of the stochastic optimization method of simulated annealing (Kirkpatrick, *et al.*, 1983) with those of shape grammars (Stiny, 1980). Shape grammars have had great success in the field of architecture in the generation of various types of artifacts. Examples include villas in the style of Palladio (Stiny and Mitchell, 1978), Mughul gardens (Stiny and Mitchell, 1980), prairie houses in the style of Frank Lloyd Wright (Koning and Eizenberg, 1981), Greek meander patterns (Knight, 1986), and suburban Queen Anne Houses (Flemming, 1987). A shape grammar derives designs in the language which it specifies by recursive application of shape transformation rules to some starting shape. In general, at any step in the derivation of a design, there is a choice to be made among different possibilities for rule application. A particular design in the language results from a particular sequence of choices: it is derived by taking a distinctive path through the state-action tree of the grammar, from the starting shape to one of the terminal nodes.

Success of shape grammars has been demonstrated in the architecture domain. For engineering applications, a goal-optimizing approach must be taken; generation of a concept alone is not sufficient. Rather, a design must be optimally driven. Although shape grammars are well able to generate a design once a proper language is created, there has been no discussion of

J. S. Gero (ed.), Artificial Intelligence in Design '92. 307–324.

how to *control* the generation of those shapes. Grammars are routinely demonstrated by exhaustively enumerating possible derivations or by randomly sampling them.

Cagan and Mitchell propose a different approach by specifically using optimization criteria to control the derivation of shapes. The design problem is formulated as an optimization problem with a quantifiable objective function and a set of design constraints. A variation of the stochastic optimization technique known as simulated annealing is then used to control choice among alternative rule applications as shapes are derived. The resulting algorithm, known as *shape annealing*, generates *optimally directed* design solutions. Cagan and Agogino (1991) define optimally directed design as an approach to design which attempts to determine optimal regions of the design space by directing search toward improving the objectives and eliminating suboptimal or dominated regions. The result is not necessarily a globally optimal numerical solution, but rather insight into how the design might be improved relative to the objectives, and a superior (optimally directed) design solution.

Simulated annealing procedures for production of such optimally directed designs generate random perturbations of a given solution state. They absolutely accept the new solution if it is better, or accept it with a certain probability if it is worse. Past applications of simulated annealing have been to determine the optimal solution of continuous or ordered discrete design variables where the form of the solution was known but specific values needed to be determined, or where a top-down refinement was required. Our shape annealing procedure, by contrast, uses a variation of simulated annealing not to determine the values of variables in a given design configuration, but rather to *generate* the configuration itself from a bottom-up approach. Application of this procedure results in evolution of designs: the fittest (most optimal) configurations survive. It requires one or more design criteria (objective functions) with a set of design constraints, a starting shape, and a set of shape transformation rules.

Cagan and Mitchell demonstrated the algorithm by defining a simple shape grammar and using it to pack shapes into a rectangular boundary. Although their results were very good, certain observations were made as to the behavior of the algorithm which resulted in sparse packing toward the start point of the algorithm and denser packings toward the end point. This paper proposes a modification to the shape annealing algorithm which has been experimentally shown to improve performance of the algorithm by 15%. This improvement comes from better control of the distribution of shape removal procedures, and better distribution of the acceptance of inferior designs. Both of these properties of the shape annealing algorithm are discussed in section 4.

The next section summarizes the principles of simulated annealing. Next shape grammars are reviewed. The original and our improved shape annealing procedures are then described and demonstrated. The results are then compared.

2. Simulated Annealing

Simulated annealing is a stochastic optimization technique which has been demonstrated to solve continuous (*e.g.*, Jain, *et al.*, 1990; van Laarhoven and Aarts, 1987, Cagan and Kurfess, 1991) or ordered discrete (*e.g.*, Kirkpatrick, *et al.*, 1983; Jain and Agogino, 1990) optimization problems where the contour of the design space is known. Kirkpatrick, *et al.*, (1983) developed the simulated annealing algorithm based on Metropolis' Monte-Carlo technique (Metropolis, *et al.*, 1953). The idea is analogous to the annealing of metals. A cooling schedule is defined giving a temperature reduction over a certain number of iterations. Temperature is a gradient variable with no physical meaning; the variable is called *temperature* to maintain the analogy with metal annealing. At high temperatures selection of a solution point is quite random while at lower temperatures the solution is more stable; the metal annealing analogy is that at high temperatures the molecules are at a highly random state while at lower temperatures they reach a stable minimum energy state.

With simulated annealing, a feasible state, s_1, is randomly selected and the energy at that state, E_{s_1}, is evaluated. A different feasible state, s_2, is then selected and evaluated to E_{s_2}. For objective minimization, if $E_{s_2} < E_{s_1}$, then s_2 becomes the new solution state. If $E_{s_2} \geq E_{s_1}$, then there is a probability, based on the temperature, that the new state will be accepted anyhow. Acceptance is determined by the probability calculation:

$$\Pr\{E_{s_2}\} = \frac{e^{-\frac{E_{s_2}}{T}}}{Z(T)},$$

where Z(T) is a normalization factor. A random number, r, between 0 and 1 is generated and compared with $\Pr\{E_{s_2}\}$. If $r < \Pr\{E_{s_2}\}$ then the new state is accepted anyhow; otherwise the old state is retained. The temperature is reduced and the process continues until convergence is reached or the temperature reaches 0. Generally, the size of the mutation space is also reduced and asymptotes to zero. In this case, it can be proven (Lundy and Meese, 1986) that if *equilibrium* (*i.e.*, convergence) is reached at each temperature and if the temperature is reduced slowly enough, then the algorithm is guaranteed to determine the global optimum. Because sufficient time cannot be guaranteed for large problems, simulated annealing is used to search only for a good solution and the exact globally optimal solution is often sacrificed for computation time.

Generally, the algorithm is run for several iterations at a given temperature until equilibrium is reached or until a certain number of iterations has occurred. The temperature is then reduced by a pre-determined amount and the algorithm is again run until convergence or an iteration limit is achieved. When there is no accepted new solution at a given temperature the optimum has been found.

Simulated annealing has been used to evaluate a solution for a fixed design configuration or for top down refinement. In this paper we extend the algorithm to actually *generate* the design configuration based on optimization criteria from a bottom up approach.

3. Shape Grammars

Shape grammars were introduced in the architecture literature by Stiny (1980) as a formalism for shape generation. A simple set of grammatical rules are defined which map one shape into a different shape. Only modifications specified by shape rules are permitted. Stiny defines shapes as limited arrangements of straight lines in a Cartesian coordinate system with real axes and an associated euclidean metric. Boolean operations of union and difference are defined on these shapes, as well as the transformation properties of translation, rotation, reflection, scale, and composition. Finally, distinguishing information about an individual shape can be associated through labels. This results in an algebra of shapes and a grammar formalism for algorithms from which languages of shapes are derived.

A shape grammar has four components: A finite set of shapes (S), a finite set of label symbols (L), a finite set of shape rules (R), and an initial shape (I). Thus, from I, R transforms the set (I, L) into a new set (S, L), and, in general, R transforms one set, $(S, L)^-$, of shapes into another, $(S, L)^+$, as:

$$(S, L)^- \overset{R}{\rightarrow} (S, L)^+.$$

Emergent shapes (Mitchell, 1989) can appear through the transformation procedures. Recognition of such shapes with shape annealing is a difficult research issue and is not considered in this paper.

A simple shape grammar is shown in Figure 1 modified from Mitchell (1990). This grammar of half-hexagons can take a single half-hexagon and through rules 1 through 3 convert the shape to a different shape made up of more than one half-hexagon. Rule 4 is a termination rule and the marker in

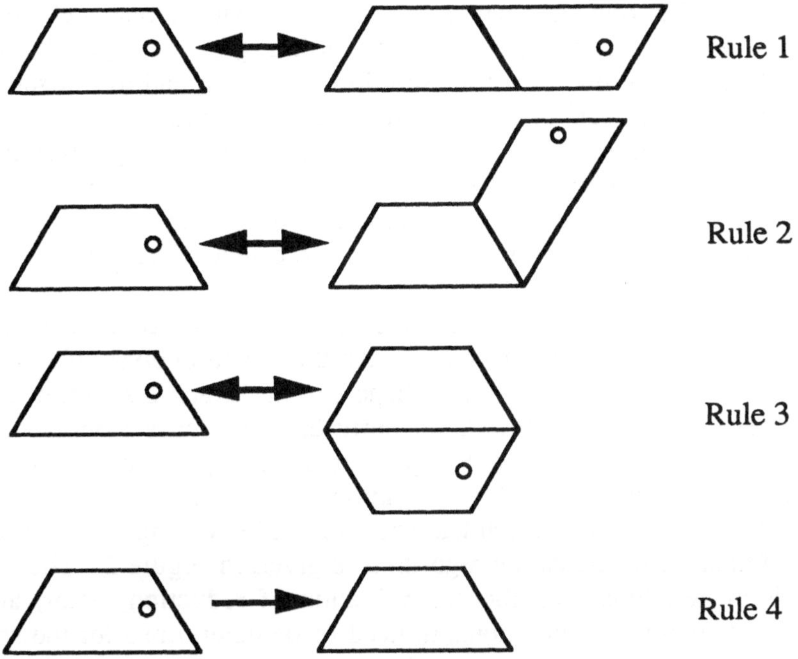

Fig. 1. Example Shape Grammar (from Mitchell, 1990)

the leading shape (indicated by the small circle) is to be removed. The arrows in this grammar point both to the left and the right for rules 1-3. This implies that a single half-hexagon can be converted to one of three shapes of two half-hexagons, or that two half-hexagons in the right-hand configurations can be reduced to a single half-hexagon. Note that successive applications of the grammar can produce intricate shapes built up from half-hexagon sub-shapes.

4. Shape Annealing

Cagan and Mitchell (1992) present the shape annealing algorithm where the concept of simulated annealing is used to determine whether a randomly selected shape rule should be applied at a given design state. Given a current design state, an eligible rule is selected and applied to the design. If the new design does not violate any constraints then it is sent to the Metropolis algorithm which compares the new state to the old state and, based on the temperature profile, determines whether to accept it or not. If the rule violates a constraint then the old state is maintained as the current. Note that a rule may not be eligible for application. The current algorithm actually selects any possible rule and then checks to verify that the rule is eligible.

By only applying additive rules (arrow pointing to the right), the design may rapidly converge on an inferior solution because it can work itself into a state where no rule can be applied without a constraint violation. Our algorithm can back itself out of these local solutions by reversing the previous rule. For every additive rule, there exists a subtractive opposite (arrow pointing to the left); if the subtractive rule is fired immediately following its companion additive rule then the effect of the additive rule is removed.

The result of modeling a subtractive rule for every additive rule is an evolutionary design generation system. As the shape evolves, if it starts to converge on an inferior solution, the shape can backtrack and perturbate to a superior design configuration. This perturbation occurs from generating new design shapes and does not require maintaining a trail of all design decisions; the current state is all that is known about the design rather than a record of all the mutations which had tried and failed during its evolution.

The original shape annealing algorithm is given in Figure 2. The number of outer loop iterations (*i.e.*, the determination of reduction_factor) and the the number of inner loop iterations, n, need to be determined for the specific problem based on convergence of good solutions. The Metropolis subroutine determines acceptance of a rule by applying the probability calculation discussed in section 2. In this algorithm, there is an equal probability that any rule will be selected and an equal probability that rule removal or rule addition will be selected. Also the temperature distribution is formed by multiplying the current temperature by a constant reduction factor over each iteration. These two assumptions will be modified to improve shape generation in the next section.

Note that the size of the mutation space is not reduced at each iteration; this deviates from the normal simulated annealing approach but is required because there is no contour from which to move between neighboring configurations. In actuality, the size of the design space *increases* while the algorithm runs; however, the increase is controlled and the algorithm finds a way to generate an optimally directed design from a countably infinite number of possible configurations. With the shape annealing algorithm, a good solution to this combinatoric problem evolves in polynomial time. It is not guaranteed to be the global optimum; however, by letting the algorithm run a sufficient time, our experience shows the solutions to be quite good.

As an example application of the shape annealing algorithm, consider the shape grammar of half-hexagons shown in Figure 1. Given this language, a countably infinite number of shapes can be generated. As a designer, we desire to place as many half-hexagons as possible into a given space, without overlap, while satisfying the grammar of Figure 2. For purposes of illustration, we consider all spaces to be of a constant area (25 units) and the half hexagon has a long base of one unit and a short base of one half unit. Rule 4 is given for completeness but is actually only applied after the final solution is found. Note that these geometric constraints have limited the number of possible shapes to a finite but very large number.

```
Begin SHAPE ANNEAL
    T = 1.
    Define initial state;
    Evaluate state;
    While T > 0 Do Begin
        success = 0;
        For mutations = 1 to n Do Begin
                                        /* at each temperature mutate  n times
                                        until convergence or limit reached */
           Let temp_state = state;
           Generate rule;
           If rule is applicable
           Then Begin
                Apply rule to temp_state;
                If verify constraints of temp_state
                Then Begin
                     Evaluate temp_state;
                     Test temp_state with Metropolis;
                     If acceptable
                     Then Begin
                          state = temp_state;
                          success = success + 1;
                     End
                End
           End
           If success > limit Then break;
        End
        If success = 0 Then break;              /* no improvement...
                                                solution found */
        T = T*reduction_factor;
    End
End
```

Fig. 2. Original Shape Annealing Algorithm (from Cagan and Mitchell, 1992)

We pose this problem as an optimization problem:

maximize: number of pieces
subject to: pieces are assembled based on shape grammar;
no pieces can overlap;
shape must fit in bounds of defined space.

For this discussion, we limit the defined space to be rectangular and implement the grammar based on the center line of the half-hexagon (the actual rectangle will be slightly larger than 25 square units).

For implementation, the design state is stored by a trace of rule numbers; thus the detailed design is not stored but only a string of rules which create the design is stored. A new rule number is added by storing it at the end of the string, representing selection of the rule with an arrow to the right. Implementation of the rule reversal occurs by randomly generating both positive and negative rule numbers. If a negative rule number is generated (representing an arrow to the left) and the previously applied rule is the analogous positive number, then the positive rule is removed from the rule string. This grammar and resulting shapes could instead be implemented by maximal lines permitting emergent shape properties; emergent properties are not considered in this paper and will be briefly discussed in section 6.

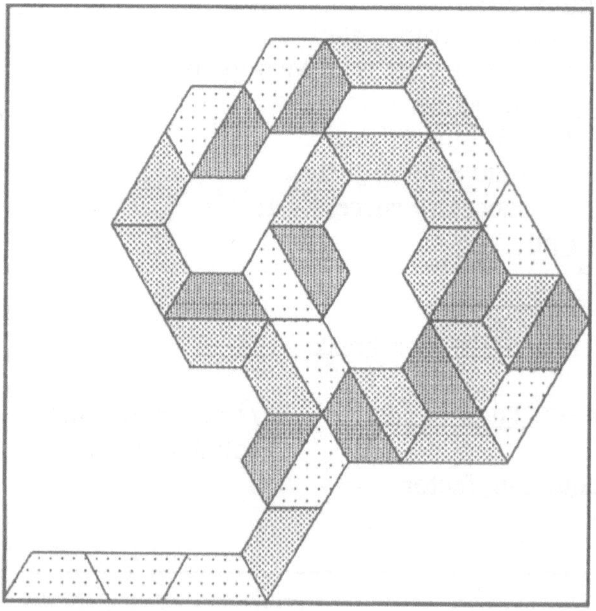

Fig. 3. 5X5 box with 34 half-hexagons

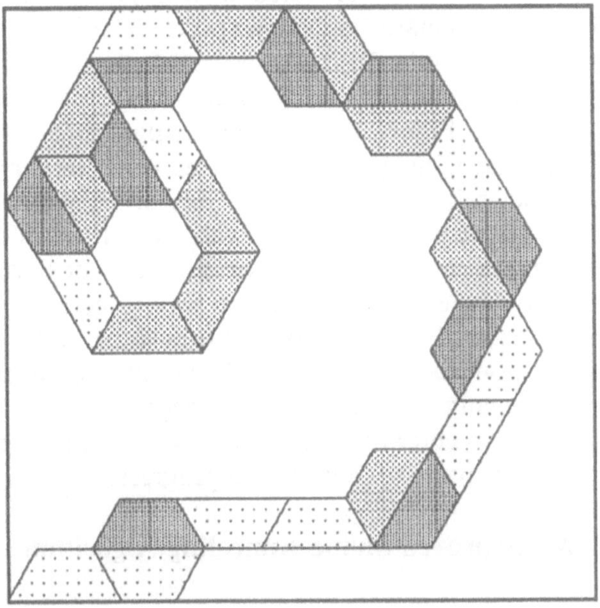

Fig. 4a. Transient solution with 29 half-hexagons evolving to Figure 4b

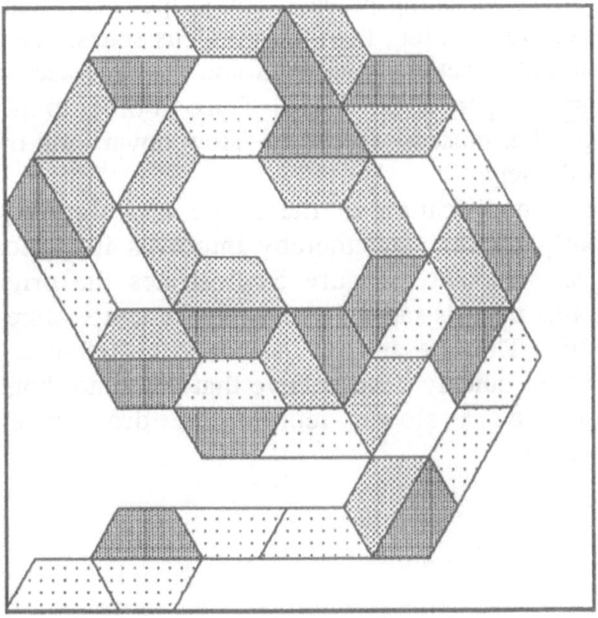

Fig. 4b. 5X5 box with 41 half-hexagons

Figures 3 and 4 show shapes generated with Cagan and Mitchell's shape annealing algorithm for a square boundary space. The fill pattern indicates which rule was applied: the light fill indicates rule 1, the medium fill indicates rule 2, and the dense fill indicates rule 3. Each figure starts with a piece in the bottom left corner. Figure 4 is of particular note. As previously discussed, as a shape is being generated it can work its way into a local solution which, without rule reversal, has no further improvement. Because of the rule reversal, the shape can work its way out of the inferior state and then progress toward a better, optimally directed design. Figure 4a demonstrates a solution state generated on the way to creating the final state of Figure 4b. Note that in Figure 4a there are no other valid additive rules which can be applied from the end state; obviously this solution is an inferior design. As the shape annealing algorithm progresses, the reversal rules become dominant and eventually the shape is able to work out of the "corner" and find a better route with which to progress.

5. An Improved Shape Annealing Algorithm

Observe the shapes generated in Figures 3 and 4. The start point was the lower left corner. The shape is somewhat sparsely packed toward the beginning verses being more densely packed near the end points. At the very beginning of the annealing process, any shapes generated will be better, based on the objective function, than the previous states. The shape quickly goes off in the initial directions generated and has no reason to backtrack to improve the beginning part of the shape. Thus, in order to increase the early packing, the algorithm must be forced to "slow down" and try and pack the early path more densely.

We propose a modification to the shape annealing algorithm which improves the early packing and thereby improves its performance. Two considerations are relevant. Figure 5 illustrates the original annealing schedule. Note that there is rapid decrease in the temperature distribution in the early iterations. Thus the amount of successful high-energy perturbation time decreases rather quickly. We believe that this time should be increased and there should be a slower temperature drop once the gradient significantly changes.

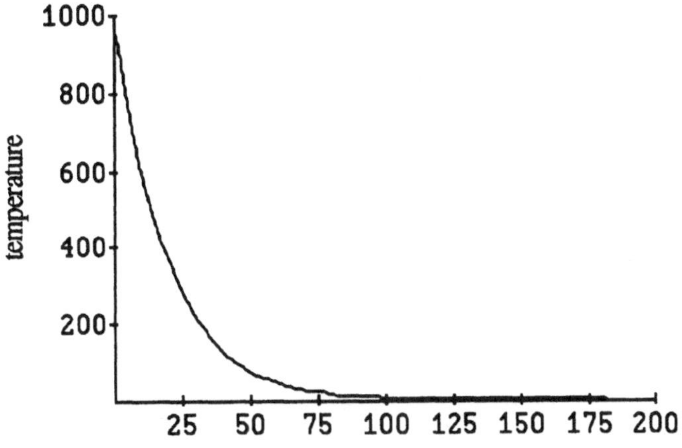

Fig. 5. Original annealing schedule

The second consideration is that there is an equal probability that any rule is chosen and an equal probability that rule removal versus rule addition will be chosen. Thus in our example there is a 1/6 probability that a rule will be removed in the next step after it is added. This probability is constant over the entire annealing schedule. When the shape is densely packed, the number of valid solution paths decreases and thus it is more likely for a rule to be removed. However, toward the beginning of the schedule there are a large number of valid solution paths and thus it is less likely that a rule will be removed (1/6 probability) versus a rule being added (1/2 probability). The shape, then, rapidly progresses and tends to not be so densely packed. If more rule removal operations are selected toward the beginning of the run then the generation will slow down and a denser packing may take place.

Based on these two arguments, an improved shape annealing algorithm is introduced. Instead of using the exponential function:

$$T(i) = (T_{factor}{}^{i})T_{initial},$$

where T_{factor} is the reduction factor, i is the temperature (iteration) step, and $T_{initial}$ is the initial temperature, as shown in Figure 5, we choose the following polynomial function illustrated in Figure 6:

$$T(i) = T_{initial}\left(1 - 10\frac{i}{i_{max}}^{3} + 15\frac{i}{i_{max}}^{4} - 6\frac{i}{i_{max}}^{5}\right),$$

where i_{max} is the total number of iterations. Figure 7a illustrates the original

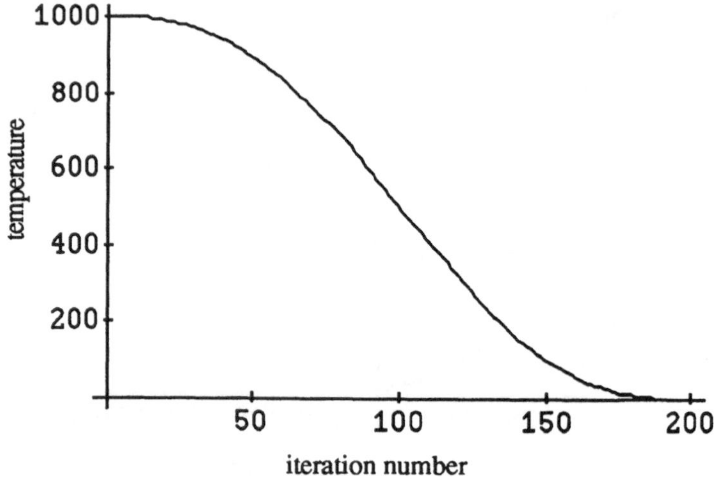

Fig. 6. Improved annealing schedule

probability distribution of accepting a worse solution in a typical run resulting from the Metropolis function, while Figure 7b illustrates the same distribution for a run with the new temperature distribution. Also, instead of maintaining a 1/6 probability of rule removal, the probability is increased to 1/3. Because there is now a greater probability of accepting a worse solution for a longer period of time and more rule removal operations are generated, the algorithm should generate a denser packing toward the start and a better overall performance, according to our arguments.

Figures 8-10 demonstrate results from this modified algorithm, where there is a denser packing in the lower left corner and there is a higher number of pieces in the box. Note in Figure 10 that again the algorithm evolves from an inferior local maximum (10a) to a superior globally directed solution (10b). Note also that many of the shapes from the modified algorithm form a sort of *spiral* in comparison to the shapes from the original algorithm. Based on the shape grammar, the only way for the shape to move in a circular direction is via rule 2. It appears that the global solution might be formed by trying to fill the outside boundary of the square and then work its way toward the center. Thus it would seem that these new shapes are tending toward the global optimum by forming such spirals. As an experimental comparison of the old algorithm to the new algorithm, 40 cases were run for each algorithm, and then results compiled to a mean number. The results illustrate that on average the original algorithm generates 34 pieces while the improved algorithm generates 39 pieces, an increase of 15%. This improvement is significant in any optimizing analysis. We can thus conclude that the modifications to the original algorithm demonstrate significant improvements over the initial results.

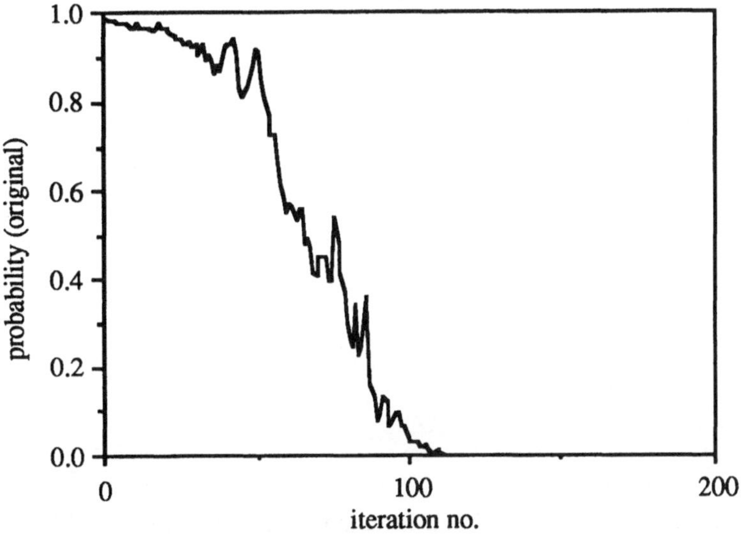

(a)

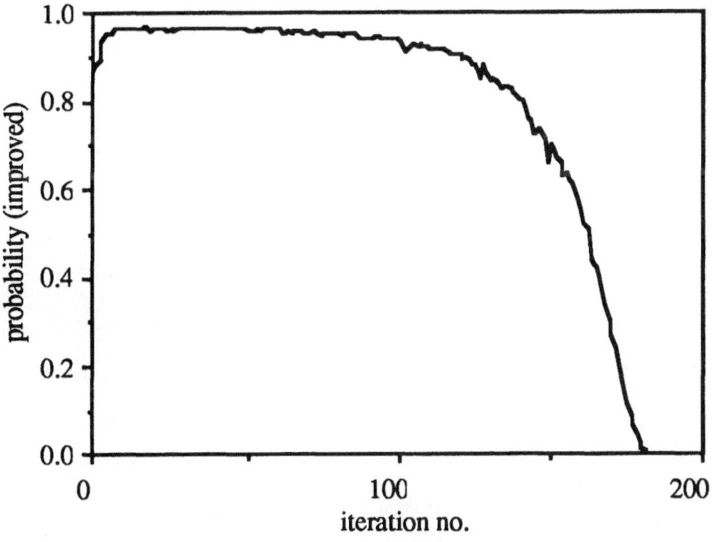

(b)

Fig. 7. Original (7a) and Improved (7b) probability distribution of accepting a worse solution.

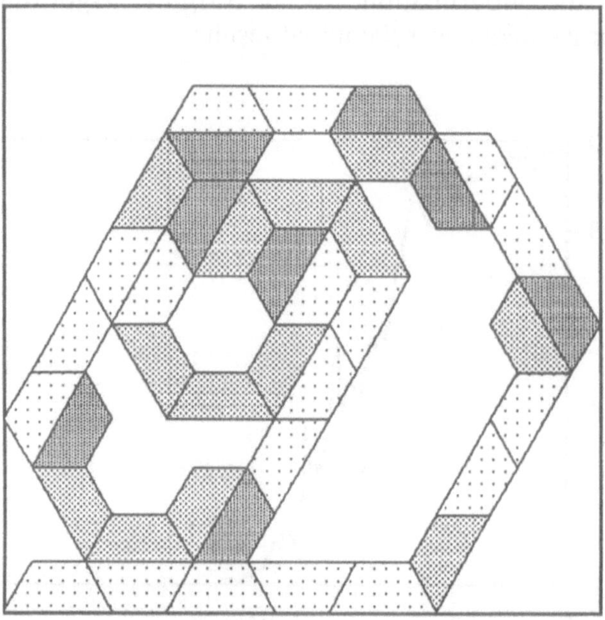

Fig. 8. 5X5 box with 39 half-hexagons from improved algorithm

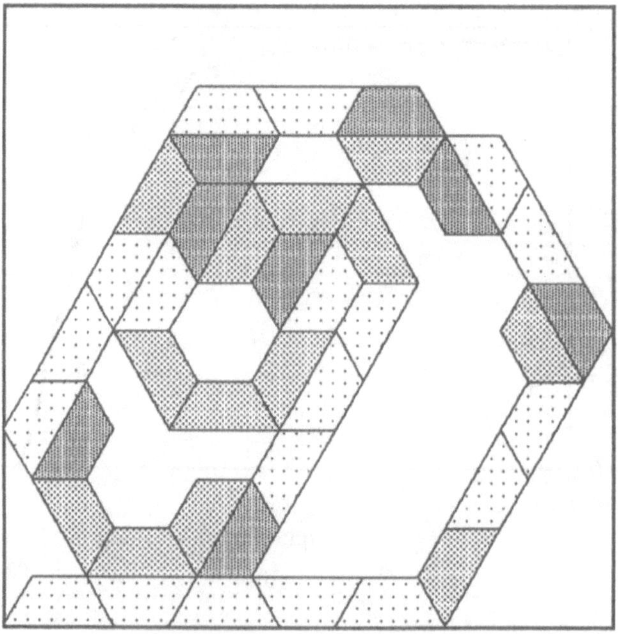

Fig. 9. 5X5 box with 48 half-hexagons from improved algorithm

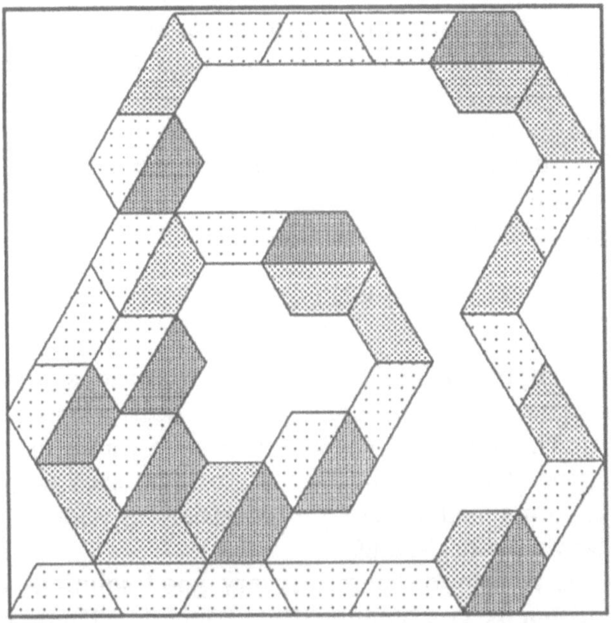

Fig. 10a. Transient solution with 40 half-hexagons evolving to Figure 10b

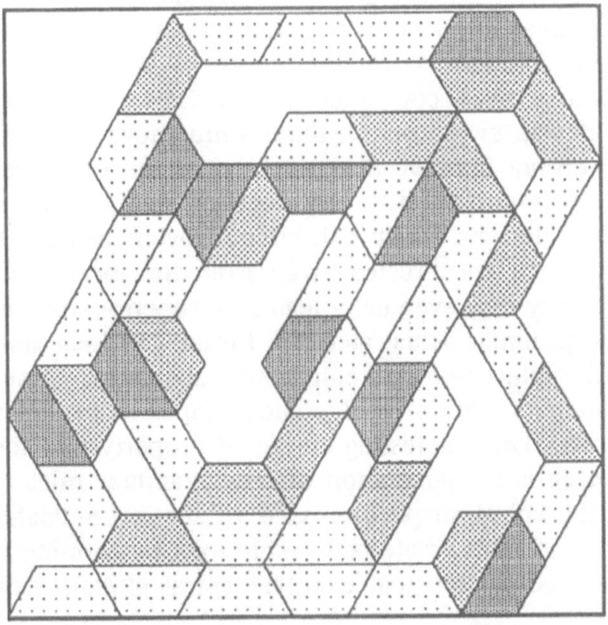

Fig. 10b. 5X5 box with 50 half-hexagons from improved algorithm

The averages computed obviously indicate that some designs are much worse than the ones displayed in the figures. Again, shape annealing is not guaranteed to find the global optimum; although the algorithm is able to jump out of some local optima, it may still converge on an inferior one. In practice, a designer should run shape annealing on a problem more than one time to determine superior solutions; the more time available, the more the algorithm can be run. However, statistically, most designs generated from shape annealing should be reasonable solutions.

6. Discussion

The original and improved shape annealing algorithms present a formal approach to navigate within the combinatoric design space generated by shape grammars. The generated shapes have been found to be sensitive to the initial path generated; the improved algorithm takes advantage of this property and performs better than the original algorithm by creating a denser shape towards the beginning.

Shape annealing is able to determine very good solutions in polynomial time. The algorithm is computationally efficient. There is no need to maintain a trace of design decisions; only the current state and proposed mutation at each step need be stored. The algorithm itself will backtrack out of a configuration if required; however, the designs generated by shape annealing are not guaranteed to be globally optimal. Optimality is sacrificed for generally good solutions by solving an otherwise intractable problem.

Shape annealing, as a controlled Monte-Carlo algorithm, provides a mechanism for design *evolution*. Although mutations may be pursued even if initially they do not improve the design, only the fittest designs survive. This concept is important for engineering design and optimization. Traditionally only designs which exhibit improvement have been pursued; instead designs which are directed away from the local optimum are now considered since they may eventually lead to a superior solution.

The simple grammar illustrated in Figure 1 does not involve the recognition and replacement of emergent subshapes. Extension of the approach to grammars that do involve such shapes is an important direction for future research. One interesting emergent property that could be studied by shape annealing is the generation of new grammar rules. For example, recognition of the spiral shape discussed in section 5 could indicate that certain sequences of rules create better solutions by forming a spiral shape. Rules could then be created that generate spiral characteristics and better solutions.

Shape annealing is used as an intelligent search tool. Domain knowledge can be modeled in the shape grammars. The annealing algorithm then searches the dynamic design space to converge on an optimally directed solution. Thus, in shape annealing, optimization is the intelligent search mechanism, instead of heuristic or other algorithmic approaches. One obvious limitation with this work is that a design objective (or multiple design objectives) must be identified. For engineering design an additional limitation is that the complex functional relationships between shape forms must be understood and computationally represented. These challenging research areas are being pursued as future work. There are many potential applications of this work including knapsack-like problems and various layout problems.

7. Conclusions

We have discussed a powerful method called *shape annealing* to control the generation of shapes with shape grammars based on optimally directed solutions to prescribed needs. We then introduced an improvement to the original algorithm by modifying the annealing schedule and the probability of rule reversal. The application of the shape annealing algorithm to a simple shape grammar has led to the evolution of a large variety of shapes which satisfy the given design requirements. Shape annealing introduces a new way of controlling the application of shape grammars to the design process and stimulates additional research into shape grammars and representations.

Acknowledgements

The authors would like to thank Bill Mitchell for his important discussions about this work and John Corson-Rikert for providing the graphical interface to draw the figures in this paper. This work was sponsored by the Engineering Design Research Center at Carnegie Mellon University, an NSF sponsored research center.

References

Cagan, J. and Agogino, A. M.: 1991, Inducing Constraint Activity in Innovative Design, *AI EDAM: Artificial Intelligence in Engineering, Design, Analysis and Manufacturing*, **5**(1), 47-61.

Cagan, J. and Kurfess, T. R.: 1991, Optimal Design for Tolerance and Manufacturing Allocation, *EDRC Report 24-67-91*, Engineering Design Research Center, Carnegie Mellon University, Pittsburgh, PA .

Cagan, J., and Mitchell, W. J.: 1992, Optimally Directed Shape Generation by Shape Annealing, accepted in: *Environment and Planning B*.

Flemming, U.: 1987, More than the Sum of the Parts: the grammar of Queen Anne Houses, *Environment and Planning B*, **14**, 323-350.

Jain, P., and Agogino, A. M.: 1990, Theory of Design: An Optimization Perspective,

Mech. Mach. Theory, **25**(3), 287-303.
Jain, P., Fenyes, P. and Richter, R.: 1990, Optimal Blank Nesting Using Simulated Annealing, Proceedings of: *ASME Design Automation Conference: Advances in Design Automation - 1988* (Ravani, ed.), **2**, 109-116.
Kirkpatrick, S., Gelatt, C. D. Jr., and Vecchi, M. P.: 1983, Optimization by Simulated Annealing, *Science*, **220**(4598), 671-679.
Knight, T.W.: 1986, Transformations of the Meander Motif on Greek Geometric Pottery, *Design Computing*, **1**, 29-67.
Koning, H. and Eizenberg, J.: 1981, The Language of the Prairie: Frank Lloyd Wright's Prairie Houses, *Environment and Planning B*, **8**, 295-323.
Lundy, M. and Meese, A.: 1986, Convergence of an Annealing Algorithm, *Mathematical Programming*, **34**, 111-124.
Mitchell, W. J.: 1989, A Computational View of Design Creativity, *Preprints of Modeling Creativity and Knowledge-Based Creative Design*, International Round-Table Conference, Heron Island Queensland, December 11-14, pp. 263-285.
Mitchell, W. J.: 1990, *The Logic of Architecture*, MIT Press, Cambridge, MA, p.143.
Metropolis, N., Rosenbluth, A., Rosenbluth, M., Teller, A. and Teller, E.: 1953, *J. Chem Phys.*, **21**, 1087-1091.
Stiny, G.: 1980, Introduction to Shape and Shape Grammars, *Environment and Planning B*, **7**, 343-351.
Stiny, G. and Mitchell, W. J.: 1978, The Palladian Grammar, *Environment and Planning B*, **5**, 5-18.
Stiny, G. and Mitchell, W. J.: 1980, The Grammar of Paradise: on the Generation of Mughul Gardens, *Environment and Planning B*, **7**, 209-226.
van Laarhoven, P.J.M. and Aarts, E. H. L.: 1987, *Simulated Annealing: Theory and Applications*, D. Reidel Publishing Co.

6

INTEGRATED DESIGN ENVIRONMENTS

THE INTEGRATED DESIGN FRAMEWORK

Supporting the design process using a blackboard system

N. R. BALL and F. BAUERT
Engineering Design Centre
Department of Engineering
University of Cambridge
Trumpington Street
Cambridge CB2 1PZ UK

Abstract. Current research into the process of engineering design is extending the use of computers towards the acquisition, representation and application of *design process* knowledge in addition to the existing storage and manipulation of product-based models of *design objects*. This is a difficult task because the design of mechanical systems is a complex, often unpredictable process involving ill-structured problem solving skills and large amounts of knowledge, some of which may be of an incomplete and subjective nature. Design problems require the integration of a variety of modes of working such as numerical, graphical, algorithmic or heuristic and demand flexibility of knowledge representation, reasoning strategies and control in the development of products through synthesis, analysis and evaluation activities. One model of processing that offers a close match to the general characteristics of the design process is the Blackboard model. Blackboard systems permit the effective and efficient co-operation of multiple sources of knowledge through a uniform global database (*the Blackboard*) without constraining these sources to a specific knowledge representation or mode of processing. The development of an *Integrated Design Framework* for the Engineering Design Centre based on a Blackboard architecture will allow the heterogeneous design tools being independently developed within the Centre to be integrated into a single, *flexible design support* system. Two blackboards are required in this system—a *Domain Blackboard* that supports the evolutionary development of design deliverables (e.g. specification, function structure, solution principle, layout) and a *Control Blackboard* that supports the design activities instantiated as part of a specific design strategy (plan of working steps during design). Existing design tools will function as independent *domain knowledge sources* within this environment and the Design Guidelines database developed by the Centre will provide the raw material for the construction of new *control knowledge sources.* This paper presents the architecture of the Integrated Design Framework and an example of how the system can be applied to the design of conrods in a miniature aero engine.

1. Introduction

Commercial industrial design departments are responsible for product quality, production costs, and for planning and forecasting the innovative potential of each product. Organizations try to improve the design process and its resulting products by applying engineering design methodologies and integrating computer-

J. S. Gero (ed.), Artificial Intelligence in Design '92, 327–348.

aided design (CAD) techniques. Commercial CAD systems do not at present adequately support the design thinking and problem solving activities which enable the designer to develop the main goals and steps of a design plan efficiently. To make product design more effective the design process itself should be supported. The purpose of the Integrated Design Framework (IDF) is to provide the designer with a flexible support environment that he can adapt to suit his method of working and the task at hand through the integrated use of the design tools being evolved at the Engineering Design Center (EDC). The development of the IDF is based on research into integrated design methodology, product modelling, design of human computer interfaces and the application of problem solving methods and Artificial Intelligence during design (Pahl and Beitz, 1984; Bauert, 1989; Bauert et al., 1990; Bauert, 1991; Ball et al., 1991).

The EDC is a multi-disciplined research environment bringing together researchers and ideas from the fields of mechanical engineering, materials science, artificial intelligence, machine learning and cognitive science in the development of new design methods. Thus research into the IDF is being conducted from multiple perspectives (cognitive, conceptual, functional, mechanical) involving a variety of terminologies, attitudes and styles (Cebon and Ashby, 1991; Chakrabarti, 1991; Johnson and Thornton, 1991; Stomph-Blessing, 1991; Bauert 1991; Murdoch, 1991). The implementation vehicle for IDF has been chosen so as to induce cohesion across a range of research projects without imposing a technical straitjacket and to reflect to current models about problem solving. To support the levels of human thinking and the communication of knowledge carrying agents with the IDF, the Blackboard model has been chosen as the solution principle (Engelmore and Morgan, 1988; Hayes-Roth, 1985; Nii, 1986a). A blackboard is a communal datastore which every agent can read from and write to and which can be partitioned (into *elements*) to focus on specific areas. Using a *control blackboard* and a *domain blackboard* the strategic control of the design process can be modelled as well as the application knowledge sources and processing design methods in a parallel (competitive) mode.

The paper is structured as follows. Section 2 discusses the theory of blackboard systems. Section 3 describes how the theory has been applied in the development of the IDF. Section 4 presents the structure of blackboard elements. Section 5 outlines the Knowledge Sources being developed at the control level (to model the design process) and at the domain level (to generate the design deliverables). Finally section 6 offers an example design problem solving strategy and a mapping to the IDF prototype.

2. The Blackboard Model of Problem Solving

Blackboard (BB) systems have been demonstrated to be capable of modelling cognitive planning activity across a number of domains (Carter and MacCallum,

1990, Coyne et al., 1990; Dyer et al., 1986; Engelmore and Morgan, 1988; Hayes-Roth et al., 1986a; Nii, 1986b; Sririam, 1986; Tommelein et al., 1987). The key property of a blackboard system is its ability to direct its attention to different tasks according to certain explicit control heuristics. The term 'blackboard' refers to an analogy with the cooperative behaviour of human experts. The blackboard is the autonomous medium through which the experts (or 'knowledge sources') communicate. It is partitioned vertically and / or horizontally to facilitate control, allowing the experts to concentrate on those parts of the facts base that warrant attention.

Blackboard systems can also be viewed as *opportunistic* planning systems. The blackboard is a clearinghouse for suggestions about plan steps which are being made by planning experts. Each expert is expected to make a particular kind of planning decision. Experts do not operate in any particular order—the planning decisions are made only when there is a reason to do so, i.e. in an asynchronous manner. The ordering of steps that characterizes a plan is developed in a piecewise fashion with the plan 'growing' from concrete clusters of expert actions. Opportunistic planning operates in a bottom-up manner when the blackboard focus is directed to a specific partition and in a top-down manner when the focus is directed to a specific expert. This enables parts of the plan to be developed independently to varying levels of detail and offers more flexibility than hierarchical top-down planners.

The blackboard system approach can be mapped onto the design process with respect to the essential task of planning a sequence of design steps. It is well matched to the design problem since both are characterized by:

- a large solution space;
- a variety of input data and a need to integrate diverse information;
- the need for many independent or semi independent pieces of
- knowledge to cooperate in forming a solution;
- the need to use multiple reasoning methods (e.g forward / backward reasoning)
- and multiple heterogeneous knowledge structures;
- an evolutionary solution developed at various levels of abstraction;
- an opportunistic approach to planning.

The power of the blackboard system organization lies in its modularity (of the independent knowledge sources), flexibility (in focussing different knowledge sources on specific aspects of a design task) and robustness (of the distribution of processing amongst independent knowledge sources). A brief overview of the blackboard model is presented here—the reader is referred to the literature for more detail (Engelmore and Morgan, 1988; Hayes-Roth, 1985; Nii, 1986a; Nii, 1986b). The model is composed of two entities—the blackboard data structure and a set of independent knowledge sources. Data on the blackboard is hierarchically organized into *partitions* that are used to create levels of abstraction and temporal substructures. Knowledge sources (KSs) are logically independent, self-selecting modules that are able to monitor and update selected blackboard partitions. Only knowledge sources are permitted to make changes to the blackboard. Based on the latest changes to the information on the blackboard, a control module schedules and executes the next applicable knowledge source. The key aspects of the blackboard system model are:

- *dynamic control* of the self-selecting KSs
- *flexible focus* of attention dictated by blackboard state
- *heterogeneous knowledge structures* within a single blackboard
- *"what's next ?" strategy* which can be data or process driven.

3. The IDF Architecture

The Integrated Design Framework (IDF) architecture is closely based on the BB1 architecture (Hayes-Roth, 1985; Hayes-Roth et al., 1986a). BB1 is a blackboard architecture for control which allows systems to reason about the control of their problem solving behaviour in addition to specific problems and solutions. Two successful applications of BB1 have been the SIGHTPLAN system for designing construction-site layouts and the PROTEAN system for protein structure analysis (Tommelein et al., 1987; Hayes-Roth et al., 1986b). As in the standard blackboard model (Nii, 1986a), functionally independent knowledge sources (KSs) cooperate to solve a problem by applying their knowledge in a globally accessible data structure—the blackboard—that is partitioned along different levels of abstraction. In addition to the domain blackboard on which the domain problem is solved, BB1 has a separate control blackboard for reasoning about its actions.

3.1 OVERVIEW

The top level architecture of the IDF system is shown in figure 1 with two blackboards—control and domain. The control blackboard supports problem solving with its elements of high level abstraction . The designer interfaces to this blackboard via the Design Assistant Graphical User Interface (GUI) which is independent from the blackboard architecture. Its purpose is to provide a friendly, transparent interface that supports a high level dialogue between the designer and all KSs operating within the IDF.

Entries on the control BB (*control elements)* represent what actions are desirable, feasible and actually performed at each point in the *design process.* These control elements can be grouped along dimensions of the blackboard referring to known elements of the design process like design activities or design planning methods and they also can be linked to each other giving paths according to different perspectives of design like cost or manufacturing.

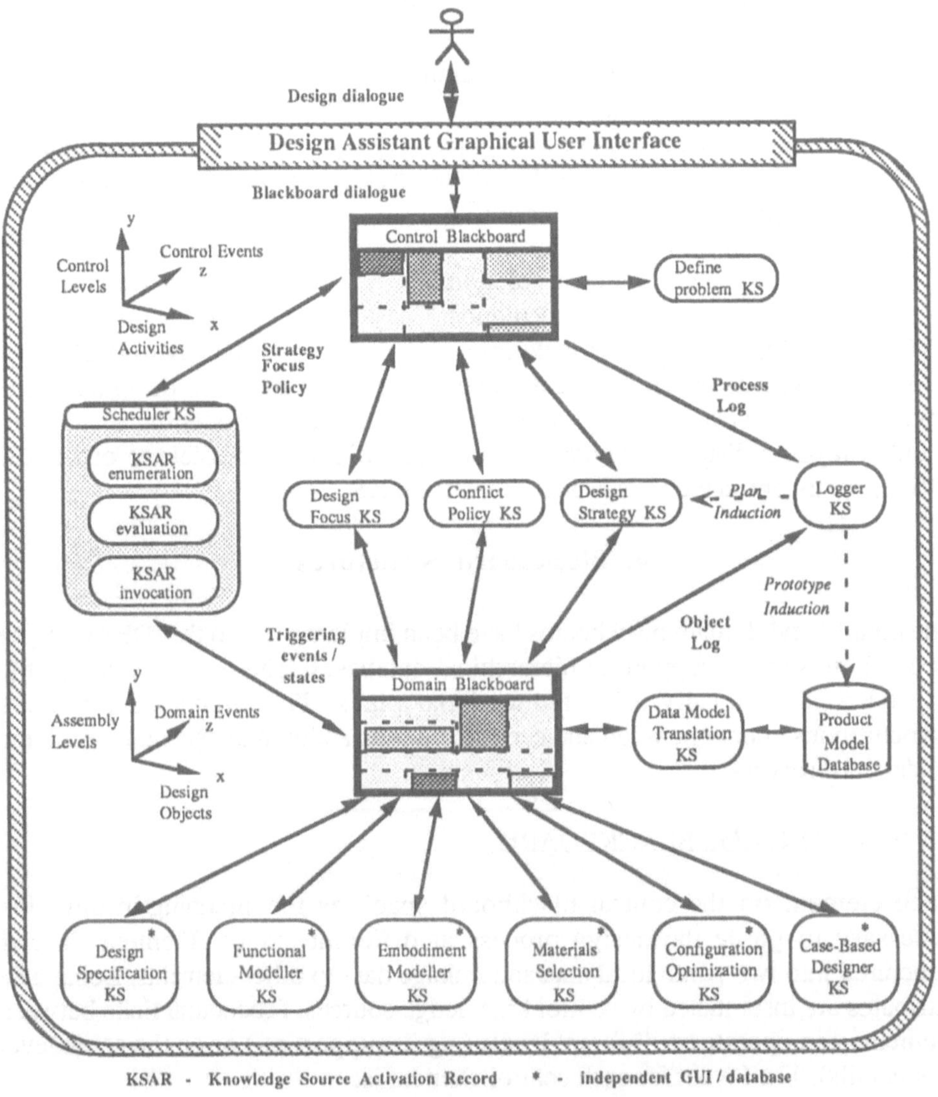

Fig.1 IDF Architecture

Entries on the domain BB (*domain elements*) represent the development of *design objects* (or deliverables) generated during the design process. These domain elements can be grouped into hierarchical assemblies or into development paths showing all deliverables on the domain blackboard.

Knowledge sources are classified by the blackboard on which they operate and the type of knowledge they embody. Three types of knowledge sources (KSs) are required—*System*, *Control* and *Domain*.

System KSs are algorithmic KSs that are domain independent and generate the basic IDF system cycle of:

(1) scheduling and executing control/domain KSs **(Scheduler)**
(2) maintaining the product model **(Data Model Translation)**
(3) logging all blackboard events **(Logger)**

Control KSs are heuristic KSs that model the design process according to the current design problem and designer profile -

(1) definition of the design problem **(Define-Problem)**
(2) sequential plan for problem solution **(Design-Strategy)**
(3) ranking of design objects / activities **(Design-Focus)**
(4) resolution of conflicting rankings **(Conflict-Policy)**

Domain KSs are separate, stand-alone systems that assist the designer in the development of specific design objects from functional modelling through to detailed layouts. Databases owned by domain KSs are accessible to other KSs either directly or via the Data Model Translation KS.

4. Blackboard Structures

The control and domain blackboards have been implemented in the IDF prototype as database servers supporting hierarchical databases. These servers connect to client KSs dynamically using bidirectional pipes. This architecture permits asynchronous client operation and centralized transaction management within the blackboard servers.

4.1 THE CONTROL BLACKBOARD

Each element on the control blackboard specifies the planning information necessary to guide the design process at different stages. Element data is structured into two parts- local data and linkage data to other elements. Local data attributes are instantiated by control knowledge sources. Procedural links between elements can operate on different levels (e.g in support of) or on the same level (e.g entails). The layout for each control element is:

Identifier	{ *Unique (numeric) element identifier*};
Name	{ *Descriptive name* };
Role	{ *Descriptive role in design process* };
BB Coordinates	{ *BB x / y / z dimension indices}*;
Goal	{ *Target action* };
Object	{ *Optional target Design Object* };
Tool	{ *Optional target Knowledge Source* };
Variables	{ *List of Goal variables* };
Termination criteria	{ *Condition for deletion* };
Weighting	{ *Goal Importance* };

Creator	{ *KS that instantiated element* };
Source	{ *Triggering decision/input* };
Status	{ *Current state—active/inactive/completed*};
Creation cycle	{ *Control event when created* }
First-operative cycle	{ *Control event when activated* } ;
Last-operative cycle	{ *Control event when completed* };
Inter element linkage	{ *List of attribute name/value pairs* };

The control blackboard has three axes—control levels, design activities and control events—which define the 3D space of blackboard elements.

The y-axis (control levels) represents different categories of control decision (problem, strategy, focus and policy) which determine which of the system's control procedures operate during particular design activities. This allows the designer to organize his problem solving behaviour in an hierarchical manner in terms of different design strategies and also heuristic design rules and guidelines. The control levels are:

Problem	{ *Overall problem definition* };
Strategy	{ *Sequential plan for solving problem* };
Focus	{ *Weight applied to BB elements / KSs* };
Policy	{ *Global criteria for conflict resolution* };
To-Do-Set	{ *List of triggered and invocable KSs* };
Chosen-Action	{ *KS to be executed at current cycle* }.

A single element on the Problem level defines the overall goal of the design process and the method of control ranging from fully interactive (workbench) to autonomous (expert system). Elements on the Strategy level represent the specific design strategies being adopted at different stages in the process e.g creative design, evolutionary design. Focus level elements are temporary weightings that influence KS scheduling decisions dependent on the current state of the blackboard and availability of knowledge sources. Policy level elements are applied by the Scheduler to resolve any conflicts generated by the current design strategies. To-Do-Set and Chosen-Action decisions are produced by the Scheduler system during the KS evaluation and selection procedure.

The x-axis (design activities) of the control BB represents different generic design steps. They represent the state of development of all models of the product created during the design process in terms of temporal order and degree of model abstraction. Control elements towards the left side of the BB refer to activities planned in the early stages of the design process; those on the right side refer to activities planned later. The design activity axis is broken into a number of steps based on recognized design tasks (Guideline VDI 2221, 1987)). The following activities are proposed:

(1) Clarify and define the task
(2) Determine functions and their structures
(3) Search for solution principles and their combinations

(4) Divide into realizable modules
(5) Develop layouts of key modules
(6) Complete overall layout
(7) Prepare production and operating instructions

Some activities may be clustered into more abstract groupings (e.g. stages), bypassed altogether or reordered depending on the strategies operative for the design problem.

The z-axis (control event) represents the log of all control events recorded during the design process. Each 'time slice' on the z-axis constitutes a 'snapshot' of the design process that is logged (via the logger KS) for later analysis/replay. The capture of log data will allow the designer to reprocess decisions and design activities to see what was important or what happened most frequently.

4.2 THE DOMAIN BLACKBOARD

Each element on the domain blackboard represents an event in the delivery of a design object for a specific part/assembly over time. The element data is structured in two parts—local data and linkage data to other elements. Local data attributes are instantiated by domain knowledge sources. Relational links between elements provide a means whereby elements at one level on the blackboard can inherit information from other elements (normally at a higher level). The layout for each domain element is the same as that of a control element except that the object and tool attributes are not present. The variable list attribute contains pointers to design objects that have been produced by domain KSs during the design process. These objects reside within specific KSs in the prototype system (i.e. are in a KS dependent format). The development of a centralized product model in later versions of the system will enable design objects to be captured in a neutral, KS independent, format. The domain blackboard has three axes—assembly levels, design objects and domain events—which define a 3D element space.

The y-axis (assembly levels) represents the different perspectives of a design object from an abstract conceptual viewpoint to the concrete geometric detail. Increasing levels of abstraction result in the linkage of specific part-based objects into generic assemblies. The assembly levels are:

Product	{ *Overall product level* }
Assembly	{ *Assembly level* }
Part	{ *Part level* }
Feature	{ *Sub-part level* }
Geometry	{ *Primitive level* }

Single product elements represent the development of design objects at the highest level of abstraction, i.e the complete product level. This element will spawn elements at lower levels as the product is decomposed in a top-down fashion.

The x-axis (design objects) of the domain BB is related to the x-axis of the control blackboard and represents design objects produced by interaction of the domain KSs during specific design activities. Note that the definition of domain KS also encompasses the designer. Here the use of part models of the product like function structure, solution principle and form layout is recorded. The design object axis is broken into a number of objects reflecting the deliverables produced during the design process (Guideline VDI 2221, 1987)). These objects are:

(1) Specification
(2) Function structure
(3) Principle solutions
(4) Module structure
(5) Preliminary layouts
(6) Definitive layouts
(7) Product documents

Subsets of objects are required at different levels of the abstraction hierarchy. Details of the objects are not held within the BB elements but referenced by pointers from those elements.

The z-axis (domain events) logs the execution of any domain knowledge source against a specific event for a design object. Each time slice on the z-axis is a snapshot of the design deliverables that is logged for later analysis and prototype extraction. This enables an audit trial of the development of specific objects to be maintained enabling the designer to reason about the evolving structure of the product.

5. Knowledge Source Characteristics

IDF employs three basic of Knowledge Sources—system, control and domain. Each KS is viewed by IDF as a black box whose behaviour is dependent on:

(1)	Triggering criteria	{ *events generating BB elements* }
(2)	Pre-Conditions	{ *specific BB element states* }
(3)	Scheduling criteria	{ *weights applied to elements/KSs* }
(4)	Input	{ *set of input data (inter element)* }
(5)	Output	{ *set of output data (inter element)* }

5.1 THE SYSTEM KNOWLEDGE SOURCES

All system KSs are domain independent procedures whose function within IDF is to generate the basic system cycle (of KS enumeration, evaluation and execution), capture the design objects within a shared product model and log all communication between the designer and IDF KSs. Outline process descriptions are as follows.

Scheduler KS—consists of three procedures that identify which control/domain KS is most applicable given the current blackboard state. The **KSAR** enumerator identifies triggered KSs and creates a 'triggered list' of KSARs (Knowledge Source Activation Records) for contextual and condition matching. Each KSAR that is triggered has its preconditions matched against the relevant BB partitions and dependency linkages. If the match is within limits set by the current strategy decisions then that KSAR is placed on an 'invocable' list for possible execution. The **KSAR evaluator** ranks KSARs on the invocable list (using current Focus/Policy decisions and scheduling criteria) and selects the best KS for execution. The **KSAR invoker** presents the chosen KS to the designer for confirmation before it is executed. If the designer does not agree with the KS chosen, he/she may nominate a different KS for invocation (supplying additional trigger/condition/context data as required).The chosen KS is executed and allowed to update the control/domain BB partitions as required (thus indirectly initiating the logger KS to record the latest events).

Data Model Translation—services domain KS requests for design objects from the IDF Product Model and other standard models. Initially each domain KS will represent parts of the model internally and model data will be disseminated via dialogues between KSs. Ultimately the Product Model will provide a KS independent view of the overall product model in terms of a description and links to its component objects and their underlying design activities.

Logger—functions as an auditor of the system recording BB states after each event and outputing element data to a project log. Initially this log will be a simple sequential file (i.e of a flat format) that will be held internally within the KS and open to interrogation using a simple query language. Ultimately the Logger's output will provide the raw material for an inductive learning process to analyse and extract :

(1) default scripts for specific problems (i.e ,new IDF strategy KSs).

(2) standard prototypes for design objects (Gero, 1990).

Prototype data will be passed to the Case-Based Designer KS (see section 5.3) for assimilation into the "design object experience" of the system. Thus the generation of prototypes will provide a feedback mechanism within IDF whereby prototype information will be used by the Case-Based Designer KS to "seed" the domain blackboard with template objects that best match new object specifications generated during execution of the design strategy.

5.2 THE CONTROL KNOWLEDGE SOURCES

The control KSs aid in solving the design problem by creating, modifying and interpreting decisions on the control blackboard. Four types of heuristic control KS are required in the IDF prototype—define-problem, design-strategy, design-focus, conflict-policy—hat are derived from the existing design guidelines database (Aguirre, 1989). Each KS shares a common structure and implementation code. The outline process descriptions are as follows.

Define-Problem KS—creates the initial control blackboard element at the problem level and defines the entire problem-solving exercise. The creation of a problem element initiates problem solving by triggering at least one domain or control knowledge source. Note that this KS will be configured to reflect the overall behaviour of IDF ranging from an autonomous expert system to interactive workbench environment. Thus there is the facility for the designer to initialize the behaviour of the blackboard according to his task and to have either the control KSs guiding the design process or directly control of the blackboard via the Design Assistant GUI.

Design-Strategy KSs—establish sequential plans for problem solving dependent on problem definition at the strategy level. The concept is based on the belief that systematic planning leads to more optimized engineering procedures. The designer can tell the system which steps of calculation, layout and evaluation he wants to perform. Usually multiple cooperating KSs guide the design process because new decisions may occur at any point in the process and span arbitrary intervals of time. Design plans provide parameters for Design Focus KS to generate Focus level decisions.

Design Focus KS—generates local (temporary) focus level decisions that weight specific types of BB partitions and / or KSs. Focus decisions are used by the Scheduler to rate competing KSs. So we can specify for example that a sizing procedure of a design plan gets a higher priority for the activity "rough layout" while an evaluation activity gets a higher priority during the detailed layout.

Conflict-Policy KSs—generate global (permanent) scheduling criteria at the policy level that are used to resolve conflict between KSs scheduled with equal priority by the focussing process. These KSs are generally fixed knowledge sources like global checklists that aid object creation or methods associated with specific design activities.

5.3 THE DOMAIN KNOWLEDGE SOURCES

The Domain KSs are heterogeneous EDC applications either under development (e.g Embodiment Modeller, Configuration Optimization) or planned (e.g Case-Based Designer). In addition to in-house developments, a number of commercial products will also be integrated into IDF as separate knowledge sources (such as DUCT (Delcam Systems, 1991) and MECHANICA (Rasna Corporation, 1991). Domain KSs are characterized as complete systems (rather than small procedures) with distinct databases and implementation languages whose purpose is to contribute information that will lead to a solution of the design problem

The environment in which the blackboard will operate is similar to that of IBDE (Fenves et al, 1990) in that the knowledge sources are stand-alone systems with diverse knowledge representations viewing the design process from multiple perspectives. Each KS will be linked to the domain blackboard via an interface shell that will allow IDF to view each KS as a 'black box' with clearly defined triggers, pre-conditions, scheduling criteria and I/O (in terms of BB elements). The KS shell will define the contents of each KS and when/how it should be used (via a set of invocation criteria). The KS body may consist of an hierarchical arrangement of subsystems (Carter and MacCallum, 1990)). Once the top level system is invoked, the current process remains within the KS hierarchy until all applicable procedures have been applied.

Eight domain knowledge sources are to be linked to the domain blackboard. Four are currently under development, two are commercial products and the remainder exist only on paper at present. An brief description of each EDC KS follows.

5.3.1 *Design Specification.* The main stages of the design process (Pahl and Beitz, 1984)) are:

- **Clarification of the Task**—establish the specifications/requirements of a design
- **Conceptual Design**—generate broad solutions from the probem statement
- **Embodiment Design**—develop ideas that take concepts to physical forms
- **Detail design**—finalize the details of the design and produce manufacturing instructions and documents.

The *Design Specification* is the key deliverable from the Clarification stage that defines a solution-neutral problem statement in terms of requirements and constraints which can be listed under the following headings:

geometry, kinematics, forces, energy, materials, signals, safety,
ergonomics, aesthetics, economics, manufacture, assembly, quality
assurance, transport, operation, maintenance, timescales, environment.

Each requirement / constraint must be ranked to enable a compromise to be made against available resources through the identification and weighting of mandatory and optional aspects of the problem. The resulting specification forms the basic input to the functional modelling process.

5.3.2 *Functional Modeller*. The conceptual design phase involves the derivation of solution concepts from the knowledge of functional requirements and constraints. The Functional Modeller KS allows the designer to analyse known solution concepts to identify a set of primitive functional elements or building blocks. These can then be combined to form more complicated elements or synthesized into solution concepts to satisfy specified instantaneous functional requirements (Chakrabarti, 1991).

The current system has been developed in LISP using an ART shell on a Symbolics machine. It will be ported to the EDC Sun-based environment but remain LISP based rather than translated to C or C++.

5.3.3 *Embodiment Modeller.* The Embodiment Modeller currently under development in the EDC is the first phase of a new generation CAD system that will allow a designer to capture and process the functional requirements of a design (Johnson and Thornton, 1991). The current system computerizes the embodiment phase of the design process by expanding a functional description into a layout description of generic components and assemblies. Model data is captured in an Object-Oriented database under five classes—components, features, materials, geometry and interfaces. The interactions between components within assemblies is managed via a bi-directional constraint satisfaction system.

The system has been implemented in C++ on the EDC Sun-based workstation environment. The development of a bespoke Object Oriented database has allowed the data attributes and functions associated with specific components to be encapsulated within a single class of object. Permissible relationships between objects are encoded as interface objects.

5.3.4 *KATE (Configuration Optimization).* The Knowledge-based Assistant for Technical merit Evaluation (KATE) (Murdoch, 1991) is a prototype configuration optimization tool developed to identify general mechanical design rules and support conceptual and embodiment design decisions. The evaluation methodology utilized by KATE is based on the concept of *Technical Merit.* This is a 3D design space whose axes are *Duty* , *Cost* and *Confidence.* Duty is defined as the degree to which key technical functions set out in the design specification are fulfilled. Cost represents the degree of difficulty in manufacturing, assembling, maintaining and disposing of the physical embodiment of the technical system. Confidence is a measure of (1) the reliability with which the current systems maintain the Duty and Cost attributes and/or (2) the probability of achieving these. The rationale behind the evaluation of the Technical Merit of a proposed system is to identify whether a change in duty is worth the change in cost .

The system is based on Object-Oriented programming techniques, supporting an interactive graphical object modelling tool and uses a Genetic Algorithm (Goldberg, 1989)) to search the design space.

5.3.5 *Engineering Materials Selector.* The Engineering Materials Selector (EMS) is a PC-based system for the graphical selection of engineering materials from a relational database. The system utilises a flexible high level graphical procedure in which the multi-dimensional material selection problem is broken down into a series of 2D graphical selection stages (Cebon and Ashby, 1991). This enables engineers to use their skills in interpretation of graphical information and avoids the need to evaluate tables of numerical data.

The EMS is a KS which manipulates a large amount of quantitative and qualitative data while leaving the designer free to concentrate on the problems of materials selection and design.

5.3.6 *Cased-Based Designer.* This will be a learning system that will acquire knowledge of design objects through a process of storing and indexing examples of objects in memory. This store of design examples will define the design space for a specific object. Interpretation of designs will involve reasoning from the store by a process of analogy. When the system is presented with a specification of a new design object, it will respond with a *prototype* of that object constructed from a generalization of previous design episodes that generated a similar object.

The analogical process required to map specifications to existing memories is extremely difficult to implement using a symbolic-based system. Therefore a connectionist approach will be taken in the development of the Case-Based Designer similar to that developed by Coyne and Postmus (1990).

6. An Example Design Problem Solving Strategy

6.1 PROCEDURAL DESCRIPTION

As a design case study, a first example "Aero engine" was chosen. The designer presents the problem description to the Design Assistant as : "Design an engine to provide propulsive power for a miniature aircraft". In this first example we want to design a new conrod with a known solution and a new plan whose steps must contain the declaration of the problem and the design constraints. The completed plan constitutes a new specific design strategy consisting of the following steps:

(1) declare statement of problem
(2) declare design constraints (access to design rules)
(3) choose rough form elements out of a layout catalogue
(4) use a calculation concept
(5) size and combine elements according to that concept
(6) choose a material
(7) evaluate against design constraints and approximate costs
(8) choose a manufacturing method (access to design rules)
(9) design to manufacturing needs
(10) evaluate against design constraints and estimate costs
(11) use a simulation concept
(12) vary and optimize dimensions and arrangement of shapes simulating different behaviour
(13) evaluate against design constraints

(14) select final dimensions, tolerances, surface finish
(15) calculate costs

The processing of the design steps starts with the basic requirement : design a connection under given connecting geometry and forces as light as possible. We then choose as rough form elements a torus and prism out of a form element catalogue. The calculation concept is a light and stable design that can withstand buckling. We get a first form and choose aluminium alloy as its material. According to calculated mass, material and form type we can then approximate costs from tables. Design rules for casting layouts guide our decisions about the detailed geometry to be overlaid onto the rough form of the conrod. We can estimate the costs of mouldings according to the size of the planned production run.We use finite element methods as simulation concepts to analyze the conrod's behaviour under load and thermal influences and to optimize its shape. Finally the ground shapes and tolerances are specified, the workpiece model (including the postprocessing for the numerical controlled machine) created and the final costs calculated.

6.2 IDF PROCESS FLOW

The procedural description given in 6.1 can be transformed into a production system format by extracting (from the procedure) independent, self selecting, design activities that generate specific domain objects. Implicit within the procedural description is a two layered description of control and domain KSs. The primary control KS is the design strategy dictating the type and ordering of the design steps. A secondary control KS is the design critic which operates on both control (activity) and domain (object) elements and generates feedback loops within the design strategy. The basic processing of the critic is to (re) focus the system on those activities or design objects that have not been satisfactorily completed according to criteria held globally (within a single evaluation KS) or locally (within the KSs that produced the activity / object under review). This is achieved by resetting focus elements back to their status at the previous control event i.e by rolling back the Control BB status along its z-axis.

A transformation of the procedural description into a series of IDF activity cycles is shown in figure 2(a)-(h). Active KSs are shown as rounded boxes. The text within each box refers to the procedural statement currently represented by an active Control BB element. BB elements are shown in one of three states—inactive, active and completed. Critic KSs, where applicable, are proposed as being local to domain KSs. The System KSs (Scheduler, Data Model Translator, Logger) are not shown on the figures in order to preserve their clarity. At present it is assumed that individual domain KSs will hold parts of the Product Model and global access to these parts will be provided by the Data Model Translator. A brief description of each IDF cycle is offered below and should be read in conjunction with the corresponding figure.

In cycle 1, figure 2(a), the Define Problem KS, initiated by the designer via the Design Assistant, writes an initial element to the Control BB at the Problem level. This element will remain active throughout the design task .

In cycle 2, figure 2(b), a specific Design Strategy KS is triggered, setting elements at the strategy and focus levels of the Control BB and part level of the Domain BB. Active elements are those that reference the first required activity—*clarify and define the task*—and object—*specification.* A Conflict Policy KS is also activated by the presence of the Problem level element and sets two elements at the Policy level effecting layout development. The presence of an active element on the Domain BB triggers activation of the Design Specification KS to generate the specification object. Completion of this KS is dependent on satisfactory evaluation by the Critic.

In cycle 3, figure 2(c), the generation of the specification object leads to a refocussing of the system onto the initial layout of the conrod by creation of a focus element on activity—*develop layouts and key modules*—and object—*preliminary layout.* The presence of an active elements on the feature/layout and part/layout dimensions on the domain BB triggers activation of the Embodiment Modeller KS to generate a generic form of the conrod. Completion of this KS is dependent on satisfactory evaluation by the Critic which is effected by active elements on the Policy level of the Control BB.

In cycle 4, figure 2(d), focus is maintained on the part/layout dimension of the Domain BB to enable the triggering of the Materials Selector KS to choose a suitable material for the conrod. As in cycle 3, completion of this KS is dependent on the Critic and the active Policy elements on the Control BB.

In cycle 5, figure 2(e), the Design Strategy has posted an Control BB element whose goal—*calculate mass and approximate costs*—cannot be serviced by existing EDC domain KSs. In this case, the designer must fulfill the role of an adhoc KS through interaction with the Design Assistant GUI and perform the required calculations externally to IDF. Completion of this task resets the focus element on the Control BB and triggers the Design Strategy to post the next goals in cycle 6. Note that the Conflict Policy elements are completed at this point.

In cycle 6, figure 2(f), the goals—*choose manufacturing method* and *calculate mass/costs*—are posted by the Design Strategy KS and are serviced by the designer interacting directly with the Domain BB via the Design Assistant. The goal—*produce detailed layout*—triggers the DUCT KS to enable the geometry of the part to be developed.

In cycle 7, figure 2(g), generation of the preliminary layout object results in the Design Strategy KS refocussing the system onto the activity—*complete overall layout.* The optimization elements generated by the strategy on the Domain

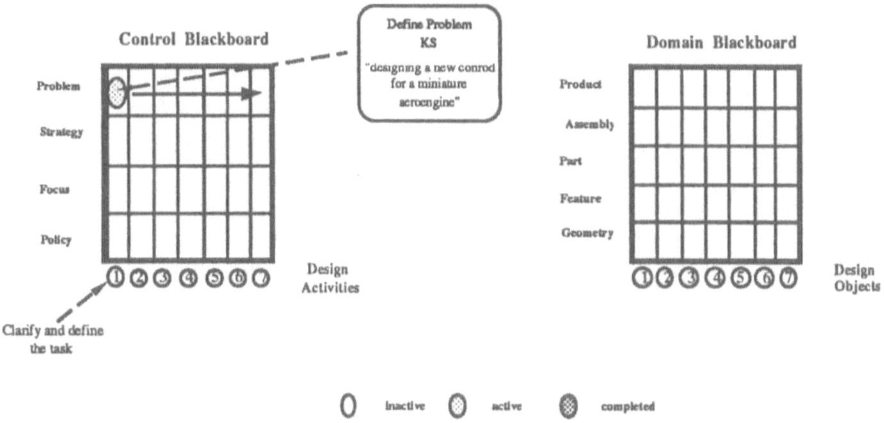

Fig. 2(a).

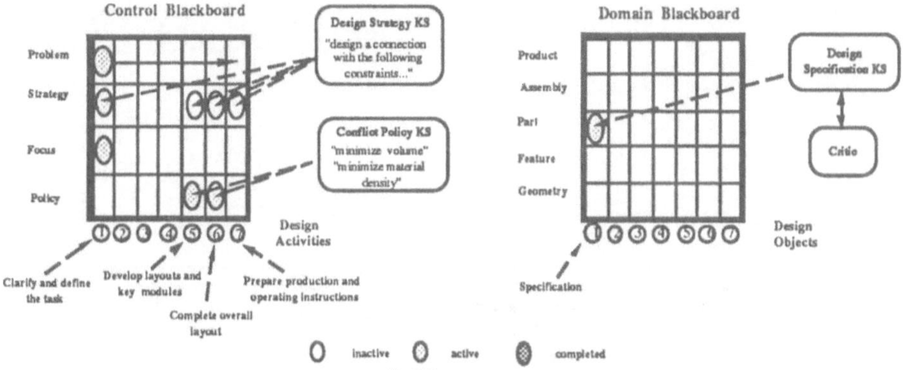

Fig. 2(b).

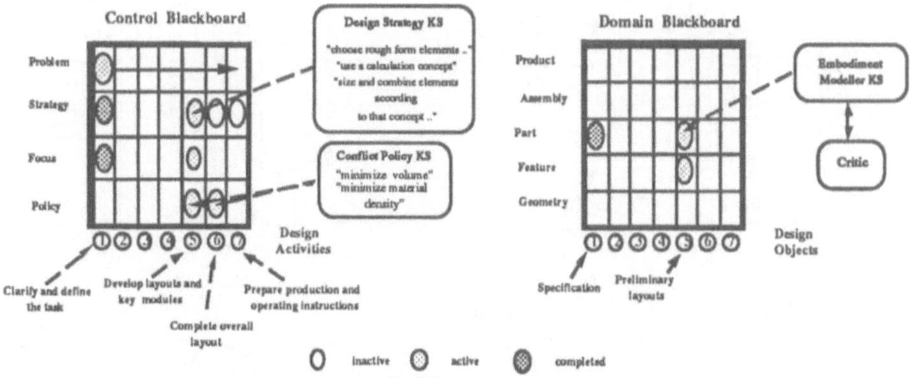

Fig. 2(c).

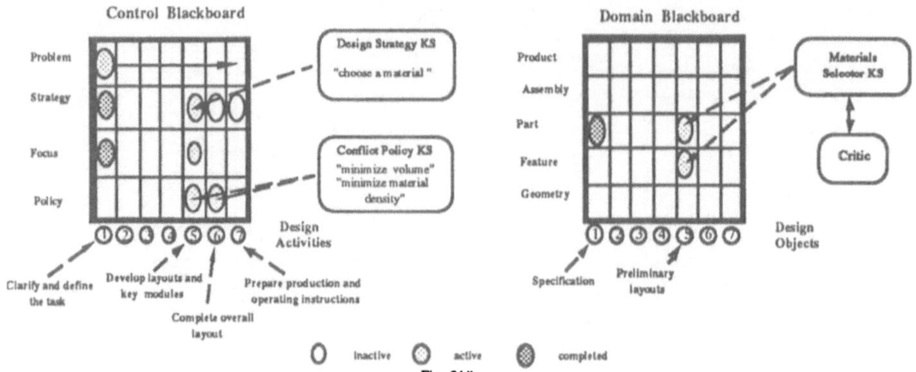

Fig. 2(d).

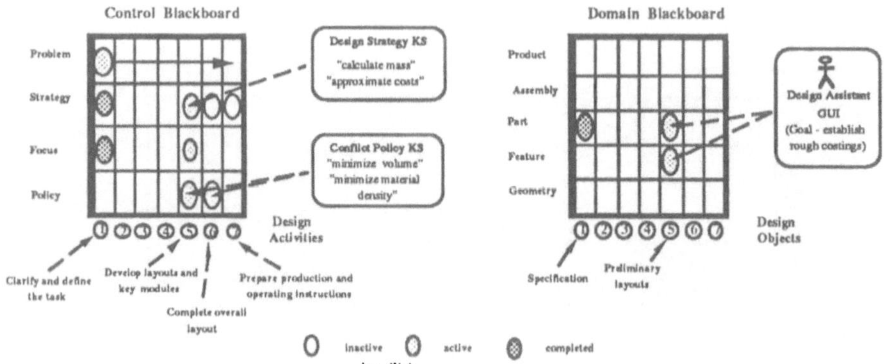

Fig. 2(e).

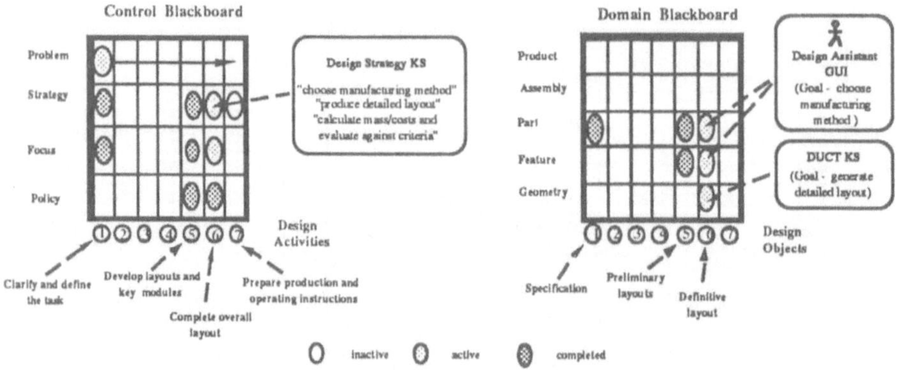

Fig. 2(f).

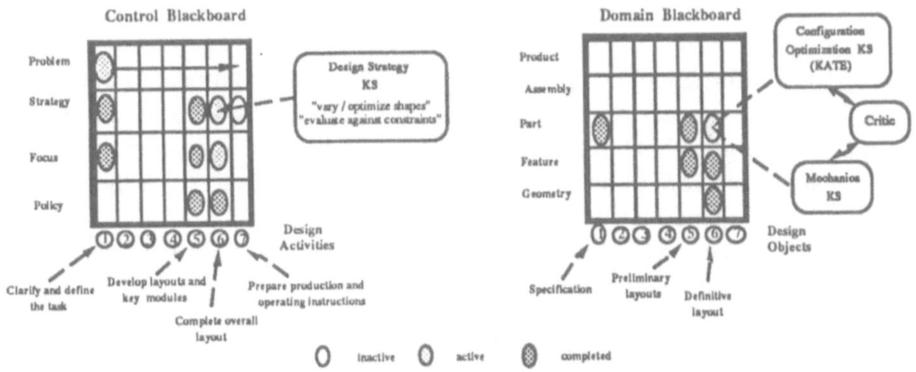

Fig. 2(g).

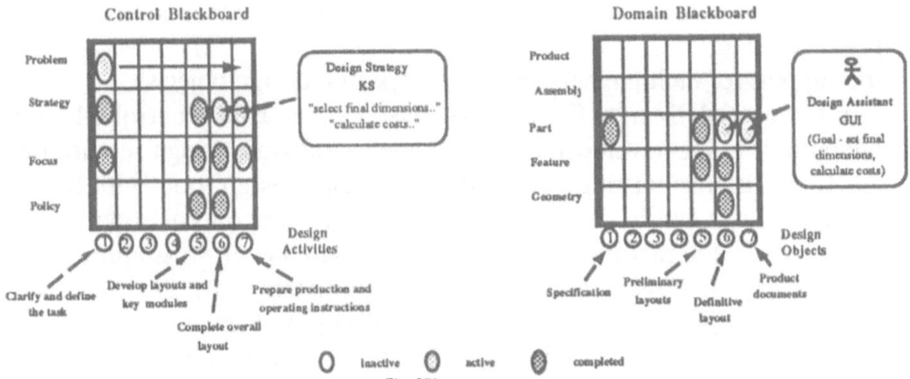

Fig. 2(h).

BB for object—*definitive layout*—trigger optimization KSs which are critiqued according to constraints coded in their local Critic.

In cycle 8, figure 2(h), completion of the optimization activity resets the current focus and triggers the Design Strategy KS to write the final elements to the Control BB for the goals—*select final dimensions* and *calculate costs*. Once again the designer interacts with the Domain BB via the Design Assistant to service these goals. At the completion of cycle 8, all focus elements are reset thus triggering the Define Problem KS to signal completion of the design task.

The above transcript gives an overview of how the IDF system would implement a specific design procedure. Parts of that procedure can be serviced by KSs currently under development within the EDC. Those areas that fall outside the current research must be covered by interaction of the designer with the Design Assistant GUI. This will enable the Logger process to capture the protocol for later replay and analysis.

7. Conclusion

This report has discussed the application of the Blackboard model of cooperative processing to the process of design of mechanical products. This model has been shown to offer a good "fit" to the process of design because it enables:

- a flexible framework to be adapted by the designer so as to tailor the system to his/her working methods and the current design problem;
- the cooperative management of autonomous knowledge sources via BB structures that provide common protocols;
- the co-existence and opportunistic application of several different control strategies ranging from top-down through middle-out to bottom-up;
- an open architecture to be developed that permits the addition of new knowledge sources in an incremental "plug in" fashion at both the control and domain levels;
- the modelling of abstraction levels inherent within the design process and product model.
- the support for multiple perspectives on specific design objects .

An Integrated Design Framework based on a BB1 control blackboard architecture is being developed as the vehicle to integrate design research within the EDC (by linking existing EDC and commercial design tools with high level design guidelines heuristics) to provide a cohesive support environment for the designer. User interaction with this system will be via a graphical, multi-windowed user interface called the Design Assistant. The Integrated Design Framework will provide a platform for EDC design systems to reason in general about their actions as well as about specific problems and solutions and will thus be a testbed for research into general aspects of the design process that are common across a wide range of design methods.

Acknowledgements

This work is supported by funding from the Science and Engineering Research Council.

References

Aguirre G. J.: 1989, *Engineering Design Literature Database*, Department of Engineering, University of Cambridge.

Ball N. R., Bauert F., Stomph-Blessing L. T. M.: 1991, Towards the development of an Integrated Design Framework using the BlackBoard model, *Technical Report No. CUED/C-EDC/TR 7*, Cambridge University Engineering Department, Engineering Design Centre.

Bauert F.: 1989, Composition of product modelling systems based on commercial software and their influence on industrial companies, design methods and education, *Proceedings of the International Conference on Engineering Design ICED89*, Harrogate, UK, Vol. 1, pp. 655–676.

Bauert F., Beitz W., Weise E., Salem N.: 1990, Modelling methods for a flexible computer-aided embodiment design, *Research in Engineering Design*, **2**(1), 15–34.

Bauert F.: 1991, Integration of design methodology, problem solving and design knowledge in product modelling systems, *Proceedings International Conference on Engineering Design ICED91*, Zurich, Switzerland, Vol. 2, 1198–1205.

Carter I. M. and MacCallum K. J.: 1990, *A Software Architecture for Design Coordination*, Engineering Design Research Centre, University of Strathclyde.

Cebon D. and Ashby M. F.: 1991, Computer-based materials selection for mechanical design, *Technical Report No—CUED/C-EDC/TR 3*, Cambridge University, Engineering Design Centre.

Chakrabarti A.: 1991, Functional synthesis of single input single output systems in mechanical conceptual design, *Technical Report No–CUED/C-EDC/TR 8*, Cambridge University, Engineering Design Centre.

Coyne R., Rosenman M., Radford A., Balachandran M. and Gero J. S.: 1990, *Knowledge-Based Design Systems*, Addison Wesley, Reading, Massachusetts.

Coyne R. and Postmus A.: 1990, Spatial applications of neural networks in computer-aided design, *AI in Engineering*, **5**(1), 9–22.

Delcam Systems Ltd: 1991, *CAD-system DUCT5*, Asto Science Park, Birmingham, UK

Dyer M. G., Flowers M. and Hodges J.: 1986, EDISON: An engineering design invention system operating naively, *AI in Engineering*, **1**(1).

Engelmore R., Morgan T. (eds): 1988, *Blackboard systems*, Addison Wesley, Reading.

Fenves S. J., Flemming U., Hendrickson C., Maher M. and Schmitt G.: 1990, Integrated software environment for building design and construction,*Computer-Aided Design*, **22**(1), 27–36.

Gero J. S.: 1990, Design prototypes: a knowledge representation schema for design. *AI Magazine*, Winter, pp. 27–36.

Goldberg D.: 1989, *Genetic Algorithms in Search, Optimization and Machine Learning*, Addison Wesley, Reading Massachusetts.

Guideline VDI 2221: 1987, *Systematic Approach to the Design of Technical Systems and Products* (trans of the German edn 11/86), VDI–Verlag, Düsseldorf.

Hayes-Roth B.: 1985, A blackboard architecture for control, *Artificial Intelligence*, **26**, 251–321.

Hayes-Roth B., Garvey A., Johnson M. V. and Hewett M. A.: 1986a, Application of the BB1 blackboard control architecture to arrangement assembly tasks, *Artificial Intelligence in Engineering*, **1**(2), 85–94.

Hayes-Roth B., Buchanan B.G., Lichtarge O., Hewett M., Altman R., Brinkley J., Cornelius C., Duncan B. and Jardetzky O.: 1986b, PROTEAN: deriving protein structure from constraints, *Proceedings of the AAAI, Volume 2*, Morgan Kaufmann, Los Altos, CA, pp 904–909.

Johnson A. L., Thornton A. C.: 1991, Towards Real CAD, *Proc. ICED-91*, Zurich,Switzerland, Volume 2, pp 896–903.

Murdoch T.: 1991, KATE–an Aid to better Technical Systems, *Technical Report No. CUED/C-EDC/TR 5*, Cambridge University Engineering Department, Engineering Design Centre

Nii H. P.: 1986a, Blackboard systems: the blackboard model of problem solving and the evolution of blackboard architectures, *AI magazine*, 7(2), 39–53.

Nii H. P.: 1986b, Blackboard systems: blackboard application systems, Blackboard systems from a KE perspective, *AI magazine*, 7(3), 82–106.

Pahl G. and Beitz W.: 1984, *Engineering Design*, Springer-Verlag, Berlin.

Rasna Corporation : 1991, *CAE- system MECHANICA*. San Jose, CA, USA.

Sririam D.: 1986, DESTINY: A model for integrated structural design, *AI in Engineering*, 1(2), 109–116.

Stomph-Blessing L.: 1991, The design process of a complex product: selected results on an analysis, *Proceedings of ICED-91*, Zurich, Switzerland, Vol. 1, pp. 342–350.

Tommelein I. D., Vaughan Johnson M., Hayes-Roth B. and Levitt R.: 1987, SIGHTPLAN: A blackboard expert system for construction site layout, *in* J. Gero (ed.), *Expert Systems in Computer-Aided Design*, North-Holland, Amsterdam, pp. 153-167.

IIDE: AN INTEGRATED INTELLIGENT DESIGN ENVIRONMENT FOR AUTOMATION IN DESIGN PROCESS AND MANAGEMENT

J. CHA, W. GUO and Z. SHI
Intelligence Engineering Research Laboratory
Department of Mechanical Engineering
Tianjin University
Tianjin 300072
PR China

Abstract. This paper investigates the integration of design automation by using the methodology of intelligence engineering, which is a newly developed engineering field about knowledge processing techniques. The integration architecture, principles and implementation for an integrated intelligent design environment (IIDE) are presented. A parallel hierarchical structure is proposed for IIDE and a meta-system, serving as IIDE's kernel to manage and control the selection, communication, coordination and operation of the subsystems, is described in detail in its functionality and configuration. An example of redesigning a combustion engine is used to demonstrate the feasibility and capability of the integrated intelligent design environment.

1. Introduction

Computers play an important role in engineering design. Although there have been numerous applications of single domain expert systems and CAD techniques, it is still far from realising the automation of design processes and management in an integrated manner. Therefore human experts have to be heavily involved in the design process with too much manual work. This is not suitable for the knowledge intensive industry in the future (e.g. CIMS).

Design activities can be viewed as a decision-making process. It consists of a series of quantitative computation and qualitative reasoning stages. For example, conceptual design and type synthesis can be categorised as qualitative decision making stages, whereas the detailed design, such as engineering analysis, size synthesis and optimal design, is mainly quantitative computations. Thus, logical, graphical and numerical models should be established to represent various stages of a design for automated processing. Due to the complexity of engineering design, several different knowledge-based systems and various packages of numerical methods and computer graphics may be necessary for problem solving at different stages of a design process, any one of them alone cannot deal with all tasks in design. Therefore we are facing a requirement on integration of numerical calculation with symbolic reasoning, different tasks and functions, as well as many domains and disciplines of knowledge in order to achieve more advanced automation at the decision making level.

From a decision-making automation viewpoint, the traditional CAD techniques are more likely a tool for engineering analysis and drafting than design (Robertson, Ulrich and Fileman, 1991). Even in some design stages that have a strongly numerical nature, such as optimal design and finite element analysis, it is very difficult to

J. S. Gero (ed.), Artificial Intelligence in Design '92, 349–370.

make use of commercial numerical packages alone to solve engineering problems without qualitative decision making (e.g. modelling). The expert system technique has provided a means to cope with symbolic or logical models representing some stages in the design process, but it is only limited within a single domain of heuristic knowledge, so that it has rarely an opportunity to solve real engineering problems of a multidisciplinary nature.

The above viewpoints have been suggested by many scholars, and a great effort has been made to combine artificial intelligence and artificial neural networks with numerical computation (Rao et al., 1987; Cha et al., 1991). Apparently a new methodology is needed to address how to overcome the difficulties in knowledge integration, which cannot be answered by existing techniques for more advanced design automation, where the integration of a large scale knowledge environment is encountered.

In this paper an IIDE is proposed to support the integration of large scale knowledge processing for design automation. The new integration platform is based on the knowledge processing technique of intelligence engineering (IE). IE is a newly developed engineering field to span the gap between artificial intelligence and other branches of computer science and real engineering applications (Rao, 1987; Lu, 1989). Distinguished from AI by the purpose, content and form of study, IE is an application oriented engineering research area which makes use of computers with intelligence to solve real world problems. Different from an expert system technique which can only deal with a single domain of knowledge by symbolic reasoning, IE aims at processing multidomains of knowledge in its various describable forms, such as symbolic, numerical and graphical methods and artificial neural networks. Also IE places more emphasis on knowledge integration, not only knowledge acquisition and representation which are the main tasks of knowledge engineering (Lu, 1989). In summary, knowledge processing techniques, including knowledge acquisition, representation, integration, management and utilisation, and the method for developing integrated intelligent software systems are the main topics of intelligence engineering. Such a methodology can provide engineering design with the capability of integrated automation in the entire design process and management.

With the philosophy of knowledge integration, IE does not exclude other existing techniques which have been developed for years and proven to be useful or even powerful in some respects through intensive applications. In this context, CAD techniques and single domain expert system techniques can serve as basic tools in a broader software environment for the whole design process. CAD techniques can mainly process numeric information and mathematical models and convert data into geometric models using computer graphics. Domain expert systems (rule-based and frame-based) can process symbolic information, induce numeric data into symbolic space and conduct symbolic reasoning to make decisions. The numeric and symbolic techniques complement each other and none of them can completely solve complex engineering design problems alone. It should be pointed out that, to integrate large scale knowledge environments for decision making automation cannot be simply to combine numeric calculations and symbolic reasoning methods; it requires additional theories, methods and techniques to deal with problems raised from knowledge integration, such as coordination, conflict resolution, communication, knowledge sharing and utilisation.

The rest of the paper is organised as follows. Section 2 introduces the architecture of IIDE and describes the key part, namely meta-system, in some detail. Section 3 shows how to implement the IIDE system in general. Finally, applying IIDE to a specific mechanical design is demonstrated with an example in Section 4.

2. The Architecture of IIDE

Engineering design is a complex process requiring knowledge covering many domains so that a number of domain expert systems and numeric programs must be needed to meet this requirement. Therefore the architecture of the environment for integrating these systems has been raised as a key issue.

2.1. The Infrastructure of IIDE

It has been realised that integrating many subsystems into a large scale knowledge environment is often necessary but difficult. Thus, an architecture for distributed intelligent systems has been proposed by Cha (Rao et al., 1989; Cha et al., 1991). This architecture is developed for the same purpose as the blackboard architecture (Hayes-Roth, 1985) but differing from it by more emphasis on intelligent management based on meta-knowledge rather than merely solving communication problems among subsystems.

The architecture of IIDE is characterised as a parallel and hierarchical infrastructure. As a complicated computer integrated intelligent system to imitate a large group of human experts, the IIDE is organised into several levels of a hierarchy and, at each level, there may be one or more software cores residing at the network nodes of computer integrated systems, to control, manage and utilise several subsystems which are attached to it in a parallel manner. The two layers attached to it, core software and subsystems, form an integrated intelligent unit (IIU) which is a building block for a large scale integrated environment (Cha et al., 1991). The system illustrated in Figure 1 is an example of such an IIU. The core software in IIU plays a key role for knowledge integration which will be discussed in detail in the next section as a meta-system.

2.2. The Role of Meta-System in IIDE

As the most important component in IIDE, the meta-system is defined as a control mechanism of meta knowledge, whereas the meta knowledge is referred to as knowledge about knowledge. At present, the concept of meta-level knowledge is

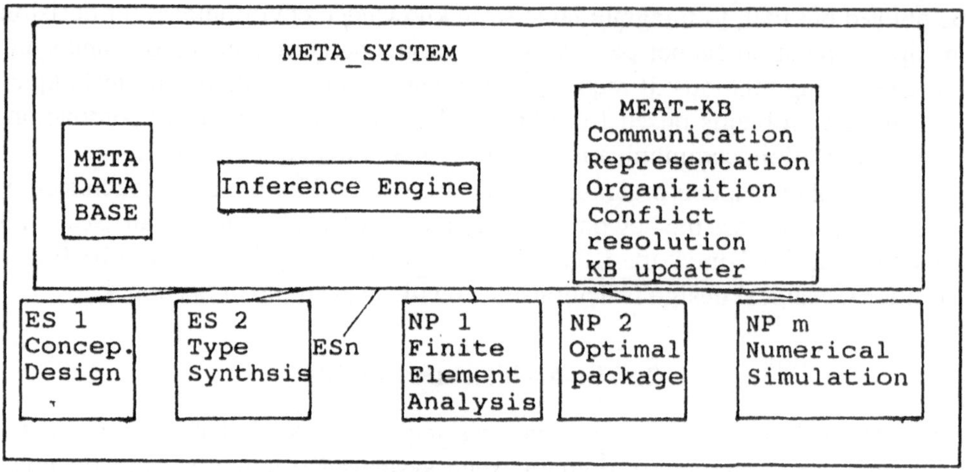

Fig. 1. The architecture of IIDE

widely used in the implementation of meta-rules in order to control the rule-based systems. So far there has appeared no reported meta-level techniques which can solve the integration problems encountered in complex design processes. As a new paradigm, the meta-system approach is proposed to handle the control problem for an integrated intelligent design environment.

The meta-system can be viewed as an expert system to control, supervise, coordinate and utilise the subsystems. Like any common expert system, the meta-system has its own inference engine, database and knowledge base, but it does not provide solutions to any concrete problems associated with specific engineering disciplines. What it should carry out is managerial tasks over the whole system based on its meta knowledge.

2.3. Functionality of a Meta-System

2.3.1. Design process management

Based on the meta-knowledge about the design process in the representation form a of design matrix (described in Section 3), the meta-system can control and manage all the knowledge-based subsystems and numeric programs which are needed in different stages of the design process and integrated into IIDE. This function may include task assignment, resource (software and hardware), allocation (optimal selection), operation instruction, running time monitoring and others. This work is quite complicated for a complex system, especially one that is heterogeneous in nature. To perform this function, the meta-system should know the features and characteristics of each subsystem involved in a specific task and it also should have a better understanding of the design procedure as well as the ability of design process modelling, organising and cooperating.

2.3.2. *Knowledge acquisition*

The meta-system can distribute knowledge into individual expert systems and numeric programs which may be developed for specific tasks and applied at different stages. Such a configuration allows us to write, debut and modify each subsystem separately so that IIDE can be managed efficiently. Another advantage of dividing knowledge into smaller, manageable sizes is to reduce the effort on rule searching and thus to minimise the running time cost. Also, this can make the knowledge base of expert systems easier to change by commercial users as well as their original developers.

2.3.3. *Conflict resolution*

In a complex design process, conflicts may very often arise because controversial decisions can be made on the same issues by different expert systems based on their own domain of knowledge. In such a situation, the meta-system, from a global viewpoint, can provide a near-optimal solution to the conflicts. Some methods, such as priority ranking and rule-based reasoning, have been effectively used in our lab for some cases, which will be described later.

2.3.4. *Concurrent execution*

The meta-system can provide opportunity for two or more individual programs to be run simultaneously. The communication between a software program and the meta-system is subject to one-to-one connection. Therefore, those programs attached to the meta-system can be executed at the same time. The hardware environment suitable for parallel computing can be computer networks or parallel processing computers. This characteristic paves the way for concurrent design in CIMS environments.

2.4. UNIVERSAL DESIGN ARCHITECTURE

With a deep understanding of the design process and managerial ability, the meta-system also can in the acquisition, integration and utilisation of new domains of knowledge so that IIDE can be used for different design applications. Furthermore, the structural advantages of IIDE make all subsystems parallel to each other and to be ordered strictly by the meta-system. Interaction and communication between any two programs must rely on translation by the meta-system. This enables users to readily add or delete programs for different design problems. When a new expert system or numeric package are to be integrated, only the meta-system needs to be modified, whereas the subsystems should remain unchanged. Such features enable any well established software to be incorporated easily into the environment, whenever it is needed.

3. IIDE Implementation

3.1. The Language for IIDE

The selection of a suitable computer language is very important for the IIDE implementation to carry out the mission of knowledge integration. Currently, the meta-system has been implemented with C language for the following reasons. First, C is versatile in numeric computation, graphics and symbolic manipulation. Its capability to handle numeric operation is much more powerful than Lisp, Prolog and it is also superior to FORTRAN and BASIC in terms of symbolic operation. This advantage makes C easier to integrate differently depicted forms of knowledge. Second, C can easily access other language environments through an interface written in a mixture of C and the assembly language. Finally, an extension of C, namely C++, is an object-oriented programming language and is an ideal tool for meta-knowledge representation.

3.2. Implementation of the Meta-System

3.2.1. The implementation method

The meta-system is built essentially with a software, DEST1.0 (Shu et al., 1991), which is an object-oriented knowledge representation and processing tool. DEST1.0 is suitable for the nature of the meta-system in IIDE. A brief description of the main items of the object-oriented structure is given below.

- *Object:* representation of entities in the real world
- *Class:* a group of objects with common characteristics (attributes, operations); class possession hierarchical and inheritance properties
- *Encapsulation:* everything that happens in an object is free from outside interference

With the integration of several knowledge representation means, such as production rules, procedure description and frame description, the object-oriented method is most suitable for representing meta-knowledge of a diverse nature. The hierarchical property of *class* can be used in two ways. One is for knowledge organisation and representation. Combining the inheritance property of *object* and *class*, it can organise knowledge in the same way as it is in the real world and allow the knowledge to be reused. Therefore, knowledge redundancy will be much reduced. Another way is for building the parallel hierarchical structure of IIDE, where independent subsystems are encapsulated by different sub-objects of the meta-system. The functionality of Message and Message passing along with some relevant pattern matching methods also can provide a base for the control mechanism of the meta-system.

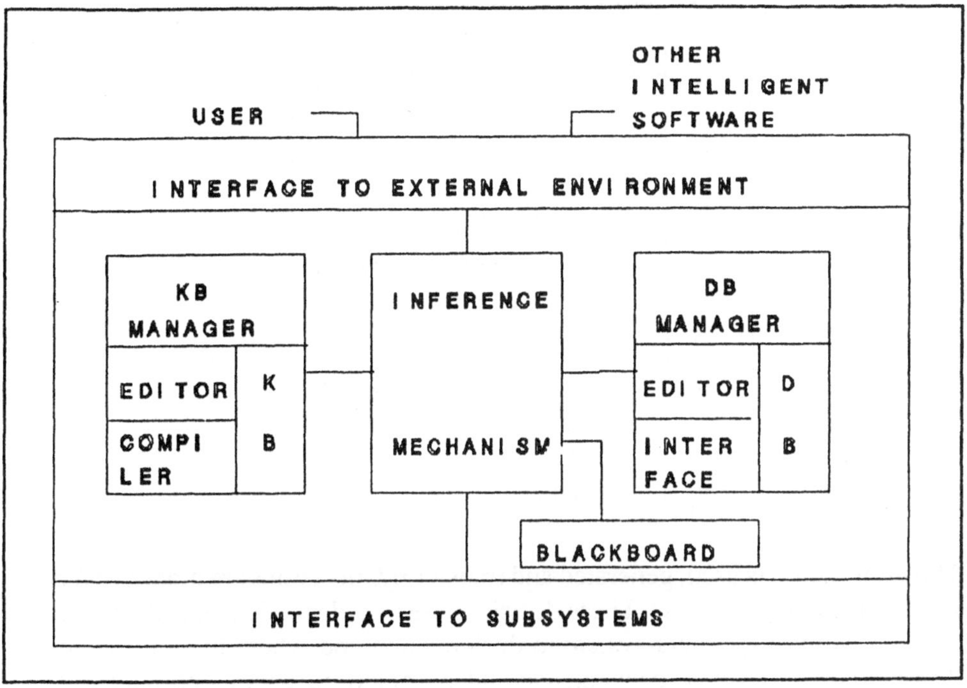

Fig. 2. Structure of a meta-system

3.2.2. Structure of the meta-system

To realise the functions of the meta-system, a structure has been developed to cover all aspects of the integration and management of the system. The meta-system includes the following five main components (Figure 2).

1. *Interface to the external environment.* The interface to the external environment builds communication between the users and the internal software system, as well as among external software systems. The interface includes an icon structured menu that includes windows, dialogue box, security module and editor module. The interface will play a twofold role in an open structured software system: the first is to codify human expertise into the computer system such that it can adopt the most creative intelligence and knowledge in decision making; another is to communicate with other intelligent software systems to extend the system into a much larger scale for more complicated tasks.
2. *Meta-knowledge base.* The meta-knowledge base is the intelligence resource of the meta-system. It serves as the foundation for the meta-system to carry out managerial tasks. As mentioned above, the IIDE is a universal design environment which aims at solving the various engineering design problems. The amount of information associated with the design process is enormous.

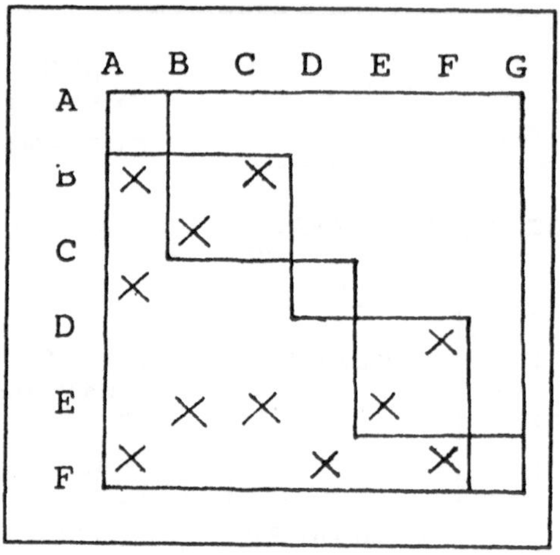

Fig. 3. An example of a design matrix

The information is used, generated and transferred, decisions are made at every stage of the design process. These decisions are interdependent and affect each other. It is up to the meta-system to decide when and where the information flows, when and where decisions should be made, to check and adjust the effectiveness of these decisions on each other, and to find a near-optimal solution to those conflict facts. Therefore a better understanding of the nature of the design process is a most important built-in ingredient of the meta-system. There are several models commonly used to represent design procedures, such as the Direct Graphic, the PERT Chart (Program Evaluation and Review Technique) and the design structure matrices (Steven, 1991). In IIDE, design process modelling is based on a method of design structure matrices, which can overcome the limitation on problem size and complexity found in the Direct Chart and the PERT Chart methods.

The matrix embodies the structure of underlying design activities by mapping relations between tasks in a precise order, which makes interdependence explicit. The matrix reflects such relations in a systematic way so that it is easy to understand, regardless of the size of the matrix. For example, a design process, which had been decomposed into n subtasks, would be represented as a $n*$ square matrix. The task labels are placed down the side of the matrix as row headings and across the top are column headings in the same order. The non-zero element aij of the matrix indicates that task i provides information to task j. Elements below the diagonal represent information transferred to later tasks; elements above the diagonal indicate information feedback to earlier tasks. Figure 3 shows a design matrix with seven tasks.

The meta knowledge base consists of a compiler and a structured frame knowledge representation facility. The compiler can convert the external knowledge representation that is obtained through the editor, and is easily understood by users, into an internal representation form in order to be processed by the inference mechanism. Characterised by diversity and variety, the meta-knowledge is represented in the object-oriented frame structures of DEST1.0. There should be several modules in the frame to represent different parts of the meta-knowledge base. For example, the communication standardisation module is for heterogeneous subsystems to communicate with each other, and the control module and conflict resolution module can serve for general management purposes, whereas the module of knowledge about subsystems and the task assignment module must be built according to each specific application. The advantage of an open structured meta-knowledge base allows the performance of more intelligent functions in decision making, which was previously solved by human experts.

3. *Database.* The database in the meta-system functions as a global database for the whole integrated intelligent system, whereas those databases attached to the subsystems can only be used by individual subsystems locally. The database contains an editor, an interface to the inference engine, a management system and a physical storage structure. The interface converts the external data representation form into an internal form. The control of data flow in the database is provided either by the inference engine through reasoning on corresponding modules in the meta-knowledge base, or by classified users with certain authority. A database management system performs data processing functions.

4. *Inference mechanisms.* Due to the diversity of the meta-knowledge and the variety of its representation forms, the inference mechanism in the meta-system adopts various inference methods, such as forward chaining, backward chaining, exact reasoning, inexact reasoning, and their combinations. The inference mechanism conducts reasoning on meta-knowledge. Additionally, it also carries out various actions according to the outcome of reasoning, e.g. passing data between any two subsystems and storing new data in the database. Therefore there are some functional modules in the mechanism which further extend the functionality of the inference mechanism.

5. *Interface to internal subsystems.* This component of the meta-system should be established based on each specific application. The interface connects every individual subsystem which is used in IIDE and under the control and management of the meta-system. Each module of the interface can convert a nonstandard data format from a specific subsystem into a standard format in IIDE. The conversion between the standard formats of any two different languages can be carried out by the inference engine based on a module of the meta-knowledge base.

4. Application Case Study

The IIDE approach has been applied to several real engineering design problems in the Intelligence Engineering Research Laboratory at Tianjin University. One of them is described below to illustrate the implementation of the new methodology for engineering design problems.

4.1. Background of the Problem

There is a great need to redesign combustion engines based on a new technique, namely the inject combustion theory. For various types of existing engines, a combustion chamber design according to the new theory can reduce fuel consumption and increase power capacity by 15-20proposed and further developed by Professor Liu in Tianjin University and has been patented in China, the United States and four major developed countries in Europe. The redesign process making use of the new technique involves symbolic reasoning based on the know-how type of knowledge of the combustion experts, numeric computations, geometric structure arrangement and engineering drawing. The impediment to apply this new technique (combustion engine design theory, principles and criteria) into engineering applications is that only few human experts on the new technique are available to many customer specified design tasks, and the redesign work is too time consuming if the conventional design approaches (on board drafting, numeric optimal design) are employed. In order to raise the design automation level for improving design efficiency and quality, we have developed an integrated intelligent system with the IIDE strategy for helping the design experts in engine redesign.

4.2. IIDE Configuration for Engine Design

The IIDE serves as a consultant and associate of the human design expert in problem solving rather than a replacement. In the integrated environment, design activities are mainly conducted by the meta-system and several subsystems, which play different roles with their domains of knowledge. The system is configured with a meta-system as its core component and several subsystems parallel to the meta-system (Figure 4).

4.3. Engine Design Process Management

4.3.1. Design process representation

The decomposition of the design process into structured subtasks is the first step toward the automated management of the process. The target is to decompose the complex design process into smaller segments, which can be easily handled by existing design techniques.

From the decomposition structure of an engine system (Figure 5) and the information view of the design process (Figure 6), the design matrix can be easily

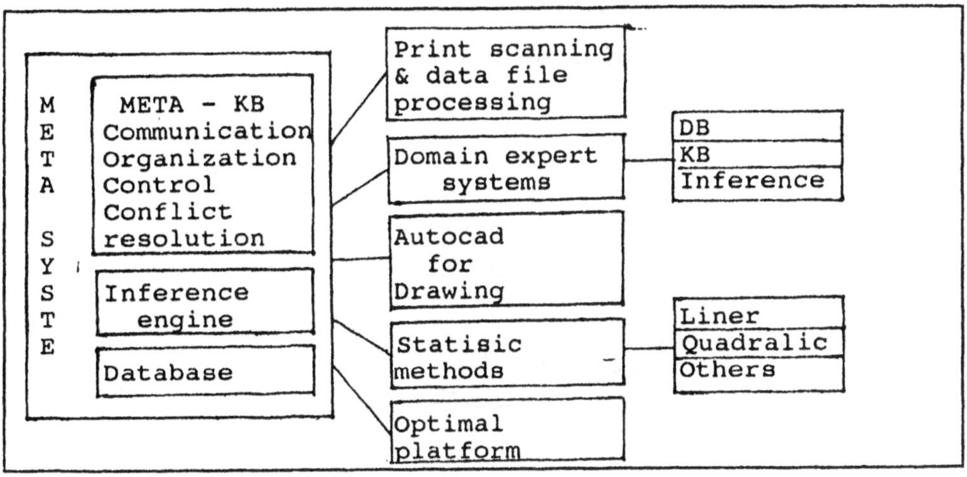

Fig. 4. IIDE configuration for engine design

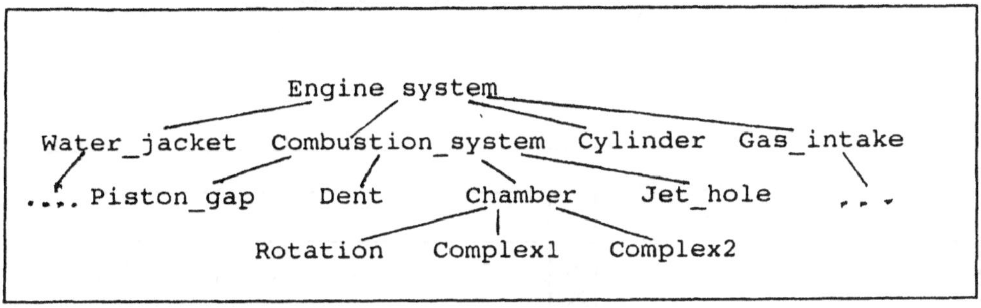

Fig. 5. Decomposition of the engine system

established (Figure 7), in other words, the design process can be mapped into the matrix structure.

4.3.2. Design process reorganisation

Due to the tremendous amount of information involved in the process and the complexity of this information flow, one of the main characteristics of the design process is design iteration, which reflects the information circulation and is displayed as task loops in the design matrix. The design iteration, indicated by the revision of decisions that had been made at previous stages of design, is the result of using incomplete and imperfect information or making conflicts among these decisions.

The meta-system has adopted the partitioning and tearing algorithms (Robertson et al., 1991) to rearrange the design tasks so that the availability of information required at each stage of the design can be maximised and an initial ordering of the tasks can be set up to start the first iteration. In addition to existing methods, the

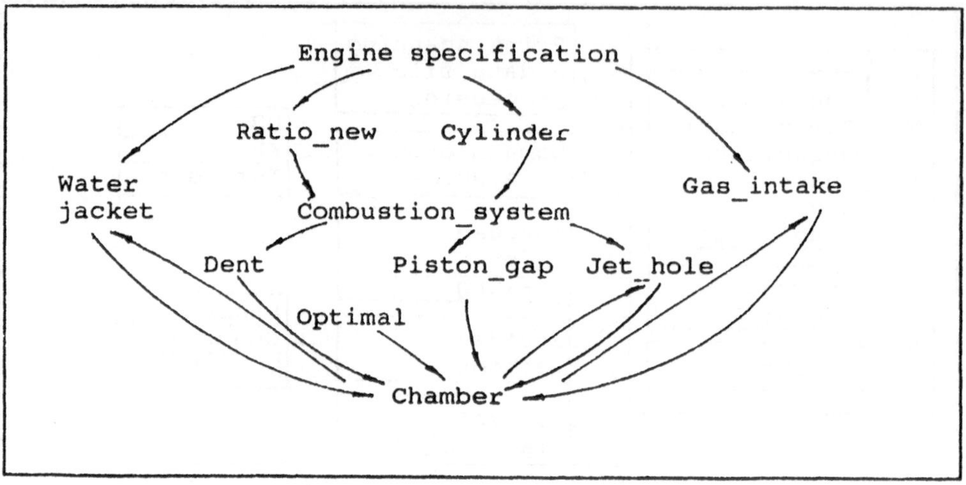

Fig. 6. Engine design information chart

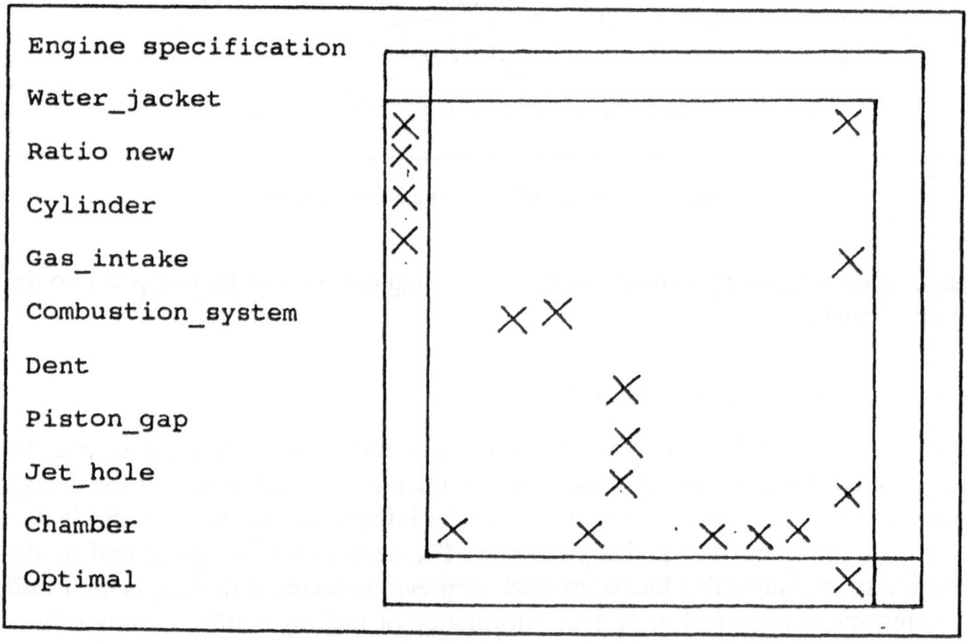

Fig. 7. Engine design matrix

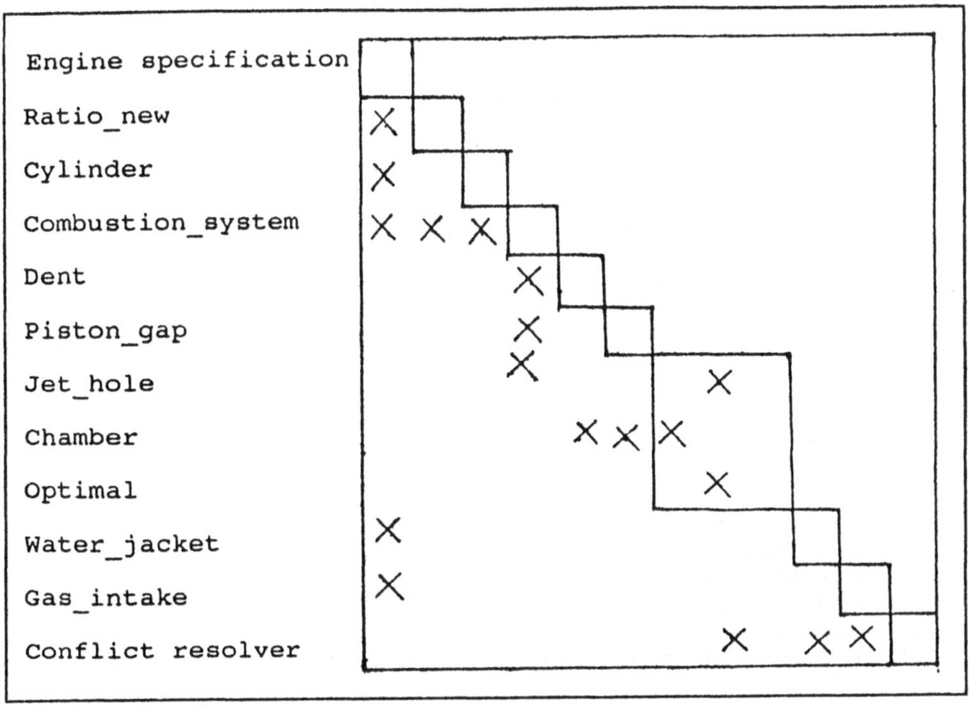

Fig. 8. Reorganised design matrix

meta-system is endowed with the capability of conflict resolution by adding one more task in the design process to decouple the loosely coupled tasks (see Steven et al., 1990). The conflict resolution is detailed in Section 4.3.4. All these efforts can improve design management in automation. Figure 8 demonstrates the reorganised design matrix.

4.3.3. Design process control

Since all subsystems are encapsulated by the interfaces, the meta-system adopts a feed back control mechanism with which the subsystems can be invoked by receiving certain messages from the meta-system, and in turn each subsystem sends back the execution results to the meta-system. The meta-system may pass the feed back information further to other subsystems as needed. The control is realised by means of message passing and pattern matching, which are borrowed from the blackboard control architecture. They can play a mediator between heuristic reasoning and algorithmic computing procedures, thus serving as a synchronisation mechanism between two processes.

The meta-system has two specific agents for its control function, namely `planner` and `monitor`, which are coded in the object `control`. The `planner`

TABLE I
The control mechanism

```
    Unit: control in_KB met.kbs
    Superclass: meta_control
    Memberslot: planner from control
    Inheritance: METHOD
    Valueclass: METHODS
    Cardinality.Min: 1
    Cardinality.Max: 1
    Values: planner
METHOD planner(design_plan:keywords)
BEGIN
    send (design) to "engine_specification";
    send (design) to "ratio_new";
    .
    send (design) to "conflict_resolver";
END.
Memberslot: monitor from control
    Inheritance: METHOD
    Valueclass: METHODS
    Cardinality.Min: 1
    Carnality.Max: 1
    Values: monitor
METHOD monitor(design_monitor:keyword,source:string)
BEGIN
    reason(_FRAME,"Monitor_rules");
    .
    send (design) to source;
END.
```

is responsible for message passing. All messages and their sequence are produced through the design process organisation, represented by the resequenced design matrix. Therefore the `planner` should be dynamically set up for each specific design task. The `monitor` is responsible for receiving and analysing the information returned from subsystems. It can figure out the causes for running time breakdown occurring in subsystems, and then deal with it automatically or to notify and advise the user to fix it. The knowledge used in this process is task independent and inherited from the general management modules. The way in which the meta-system control mechanism works is shown in Table I.

The two agents correspond to two methods in the object `control`. The messages in the `planner` method are organised in the same sequence as in the reorganised design matrix. The subsystems will be invoked according to the sequence of tasks, so `planner` should be built for specific design projects. The agent `monitor` is established by using the inheritance feature of `class`. When `monitor` is activated, it first reasons with `Monitor_rules` to identify the causes of breakdown occurring in subsystems. `Monitor_rules` can be used for general purposes as it inherits properties from the superclass `meta_control`. Then `monitor` may use troubleshooting routines to fix any problems in the subsystems. After that, a message will be sent to `source` to resume the design process or to notify the user through the interface in case the problem can not be fixed by routines.

4.3.4. Design process coordination

All activities involved in the engine design are filled by the above subsystems. When they work together for a common goal, it is not as simple as exchanging information among them automatically. Each subsystem concentrates on a particular domain of knowledge so that they rarely possess the knowledge in different domains. There must be some conflicts coming up when decisions are made based on different domains of knowledge.

In engine design processes, the expert systems for `chamber` design, `gas_intake` design and `water_jacket` are built based on different disciplines. The main concern for `chamber` design is the quality of combustion process, which can be affected by speed and form of fire propagation, the temperature distribution, etc. These important facts are all dependent on the structure, shape and relevant parameters of `chamber`, `jet_hole` and other facts. The main concern for `gas_intake` design is the gas intake efficiency, including the quantity (the volume intake in a stroke) and the quality (the pressure and the ratio of gas mixed with air). The concern of `water_jacket` design is to provide a better cooling environment, such as high heat transmission efficiency, to avoid bad working conditions (high temperature, high pressure) in the engine.

Due to space limitation, the decisions made by these subsystems may often conflict. In an engine design process, the way to solve potential conflicts is for the meta-system to identify conflicts and then compromise by rearranging each part of the design according to its importance from a global point of view. The following segment of codes, Table II, shows how the meta-system works.

The program `part_assemble` is used to assemble local structures into a global coordinate. The two predefined rule sets in the object `conflict`, namely `ID_rules` and `Global_rules`, can be used for identifying or solving conflicts. The `conflict_type` indicates where the conflict occurred, e.g. the `conflict_type = 1` means that at section `Y_0_Z_25`, the `chamber` design, the `water_jacket` design and the `gas_intake` design interfere with each other. After reasoning with the rule set of `Global_rules` (e.g. rule 20, 21 in Table II), a compromise can be reached. Due to the current design specification, the `water_jacket` design has the highest priority so that the other two (the `chamber` design and the `gas_intake` design) should conform to it and be modified according to some constraints (represented by `parameterlist` in Table II).

TABLE II
Conflict resolution

```
   Unit: conflict in_KB meta.kbs
   Memberslot: conflict_type from conflict
   Inheritance: Override.Values
   Valueclass: integer
   Cardinality.Min: 1
   Cardinality.Max: 1
   Values: unknown
   Memberslot: ID_rules from conflict
   Inheritance: Override.Values
   Valueclass: RULES
   Cardinality.Min: 1
   Cardinality.Max: 1
   Values: {
   rule 10
   fact "section" = "Y_0_Z_25" and
        chamber.w/2+water_jacket.w+gas_intake.Max_w>=50
   then _FRAME.conflict_type = 1
   .
}
   Memberslot: Global_rules from conflict
   Inheritance: Override.Values
   Valueclass: RULES
   Cardinality.Min: 1
   Carnality.Max: 1
   Values: {
   rule 20
      fact Engine_speed >= 4500 and
           fuel_mon <= 80 and
           ratio_new >= 9.5
      then cooling efficiency is prior
   rule 21
      fact cooling efficiency is prior
      then "part1" = "chamber" and
           "part2" = "gas_intake"
   .
}
   Memberslot: resolution from conflict
   Inheritance: METHOD
   Valueclass: METHODS
   Cardinality.Min: 1
   Cardinality.Max: 1
   Values: resolver
METHOD resolver (resolution:keyword)
BEGIN
   part_assemble();
   reason(_FRAME,"Global_rules");
   .
   send(modify,parameterlist) to "part1";
   send(modify,parameterlist) to "part2";
   .
   end;
END.
```

4.4. Subsystems Description

4.4.1. Expert system for engine design

The expert systems (the `chamber` design, the `gas_intake` design and the `water_jacket` design) make use of the inject combustion theory and other knowledge about cylinder head design (for example, aerodynamics for the process of gas intake, compression and exhaust, combustion for process combusting and cooling), which is represented by object-oriented knowledge representation methods. The values of the structure parameters are determined through interpolation algorithms and some empirical formulas which can nest in the object-oriented structure, but also can be obtained from external numerical packages. The expert systems are written in C and share a common forward chaining inference engine. The knowledge about compression ratio is organised into an object-oriented frame manner, which is demonstrated as an example in Table III.

As shown in the program, `ratio_new` is the name of the frame, representing the object of ratio_new calculation. `_FRAME` indicates current object. `r_n`, `r_f` and `r_m`, etc. denote the components of the object. The value of these components can be obtained either from reasoning with heuristic knowledge rules, such as `r_m`, or with numeric computations, e.g. `r_f` and `r_dp`, or by inheriting from other objects, such as `r_o`. In such a structure, all methods, algorithms and knowledge are organised in the frame and the process can be carried out automatically by the frame itself as the task is assigned.

4.4.2. Automatic optimization modelling

The expert system for optimization modelling is developed to provide mathematical models to the parameter optimal design package. Depending on the knowledge about the relationship and constraints of different parts and parameters of a chamber with respect to functionalities, the inference mechanism can produce mathematical models for the optimal design subsystem (Figure 9). The knowledge is also represented in an object-oriented frame structure.

4.4.3. 2D Graphic Scheme Display

The subsystem for 2D graphic display receives the numeric results in the form of object-oriented data structure from the subsystems of structure design, parameter optimal design and conducts assemble operations on previously designed parametric graphic components according to the information received. The design results, shown in Figure 10, can be displayed graphically to assist users judging the relevance of the design, if the design is acceptable then they will be stored for future use in engineering drawing.

TABLE III
ratio_new O-O representation

```
Unit: ratio_new in_KB engine.kbs
Memberslot: r_n from ratio_new
Inheritance: Override.Values
Valueclass: real
Cardinality.Min: 1
Carnality.Max: 1
Values: Unknown
Memberslot: r_m from ratio_new
   Inheritance: Override.Values
   Valueclass: real
   Cardinality.Min: 1
   Cardinality>Max: 1
   Values: Unknown
Memberslot: r_m rules from engine_sys
   Inheritance: Override.Values
   Valueclass: RULES
   Cardinality.Min: 1
   Cardinality.Max: 1
   Values: {
   rule 053
      fact T_requir.material = "aluminium"
      then _FRAME.r_m=0
   rule 054
      fact T_requir.material = "castiron"
      then _FRAME.r_m = -0.7
}
Memberslot: ratio_M from ratio_new
   Inheritance: METHOD
   Valueclass: METHODS
   Cardinality.Min: 1
   Cardinality.Max: 1
   Values: ratio_M
METHOD ratio_M(design:keyword)
VAR
   design:keyword;
BEGIN
   reason(_FRAME,"r_m_rules");
   _FRAME.r_f:=0.5*(T_requir.mon_no-75)/10;
   _FRAME.r_dp:=-0.5*(%_requir.d_piston-92)/10;
   _FRAME.r_o:=T_requir.oldcomp_ro;
   .
   ratio_n:=_FRAME.r_n;
   END.
```

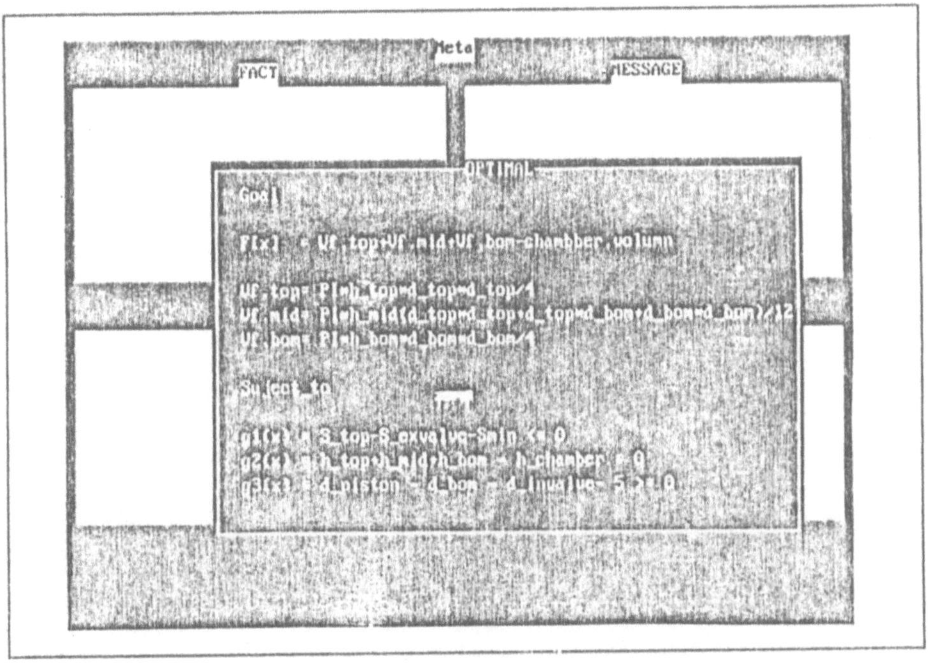

Fig. 9. Chamber optimal modelling

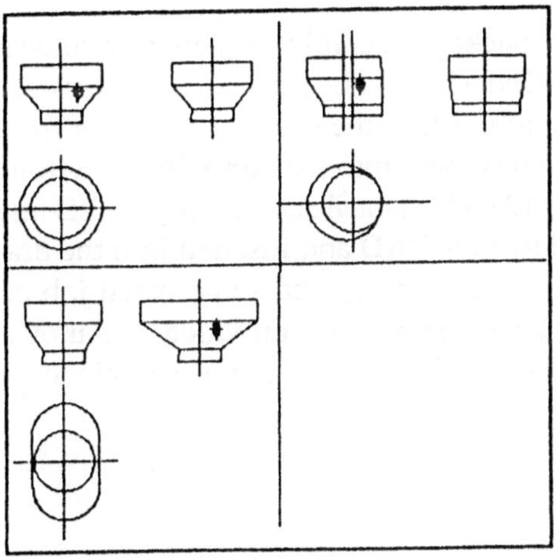

Fig. 10. Multiple chamber schemes

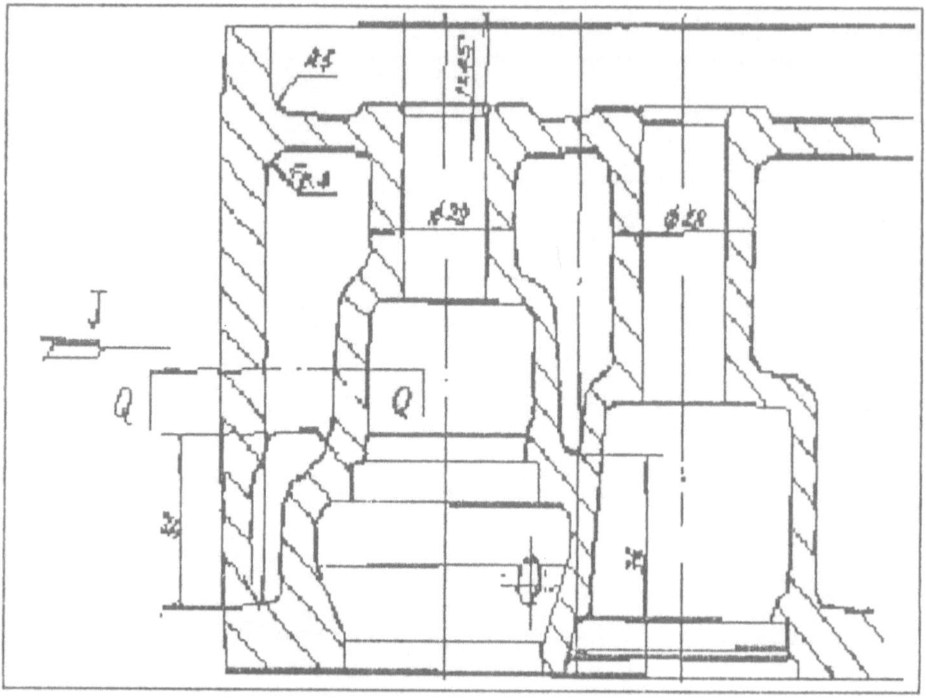

Fig. 11. Engine blueprints scanning

4.4.4. Engineering drawing

The AutoCAD software package is employed here as a subsystem for engineering drawing of the whole cylinder head. The subsystem makes use of a scanning device and a relevant software to produce the graphic data files from existing blueprints of engine heads, then inputs them into the AutoCAD environment for editing and revising (Figures 11 and 12). The data file containing the 2D image of the redesigned chamber is then passed to AutoCAD and inserted into the drawing (Figure 11). A parametric graphic component library and some batch job commands have been developed in AutoLisp language which is provided by AutoCAD. This component library and command fields will help users to raise efficiency in revising drawings.

4.4.5. Optimal design

Optimal design subsystem (optimal platform) obtains mathematical models from the expert system of optimization algorithms. The optimum results are passed to the graphic display subsystem through the meta-system. Some other numeric computation programs are also stored in optimal design subsystems and are available to other subsystems.

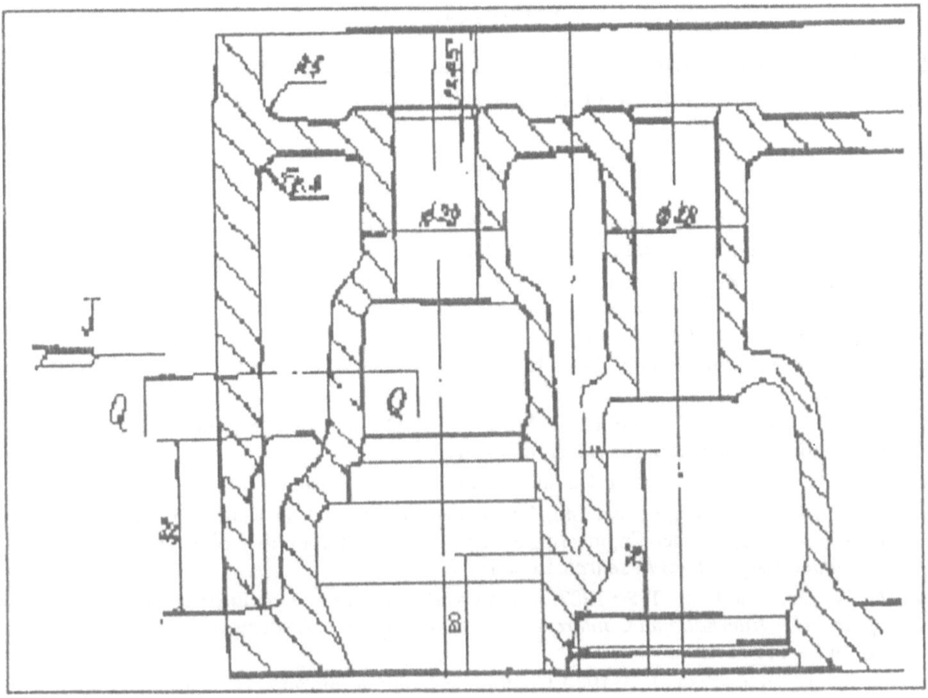

Fig. 12. Revised engine prints

5. Conclusions

An integrated intelligent design environment is proposed for enhancing automation in the design process and management. Building on the top of current CAD techniques and expert system methods, IIDE can help realise decision making automation based on knowledge processing techniques to meet the challenge of the upcoming knowledge-intensive industry. The parallel hierarchical structure of IIDE is developed. In IIDE, the different expert systems and numeric computation routines are coordinated by a meta-system, which is endowed with a better understanding of the design process. These expert systems and numeric routines are written in different languages or programming tools and used separately in different design processes. In this way, we can easily add new programs and reduce the scope of rule search to enhance efficiency. Moreover, the near optimal solution can be selected among the conflicting decisions, and parallel processing becomes feasible in the integrated intelligent system. The IIDE has been used for a mechanical design and shown great potential in assisting human experts in engineering design. Future work will have more emphasis on increasing the intelligence of the meta-system for the needs of concurrent engineering.

Acknowledgements

The authors would like to acknowledge financial support from the National Science foundation of China and the chinese State Commission of Education

References

Cha, J. Z., Rao, M., Zhou, J. and Shi, Z. C.: 1990, Integrated intelligent software system and its applications in mechanical engineering, *Journal of Tianjin University*, **2**.

Cha, J. Z., Rao, M., Zhou, J., Zhao, Z. Y. and Guo, W.: 1991, New process on integrated environment for intelligent manufacturing automation, *Proceedings of the 1991 IEEE International Symposium on Intelligent Control*.

Gebala, D. A. and Eppinger, S. D.: 1992, Methods for analysing design procedures, *ASME DTM'92*, pp.227.

Robertson, D., Ulrich, K. T. and Fileman, M.: 1991, CAD system and cognitive complexity: beyond the drafting board metaphor, *ASME DTM'91*.

Hayes-Roth, B.: 1985, A blackboard architecture for control, *Artificial Intelligence* **26**, 251–321.

Lu, S. C.-Y.: 1989, Knowledge processing for engineering automation: a summary of current research in the Knowledge-Based Engineering Systems Research Laboratory, *NSF Grantees Conference, Manufacturing systems Research*, Berkeley, CA.

Rao, M., Tsai, J. P. and Jiang, T. S.: 1987, A framework of integrated intelligent systems, *Proceedings of the IEEE International Conference on System, Man and Cybernetics*, Alexandria, Virginia, pp. 1133–1137.

Rao, M., Jiang, T. S. and Tsai, J. P.: 1989, Combining symbolic and numerical processing for real-time intelligent control, *Eng. Applic. AI*, **2**, 19–27.

Shu, Y. Q., Yang, H. B. and Zhou, J.: 1991, *DEST1.0 User Menu*.

Steven, D. E., Whitney, D. E., Smith, R. P. and Gebala, A.: 1990, Organizing the tasks in complex design projects, *ASME, DTM'90*.

Su, S. Y. W. and Lam: 1990, Object-oriented knowledge base management technology for improving productivity and competitiveness in manufacturing, *NSF Grantees Conference on Design and Manufacturing systems*, Tempe, AZ.

KASE: AN INTEGRATED ENVIRONMENT FOR SOFTWARE DESIGN

S. BHANSALI and H. P. NII
Knowledge Systems Laboratory
Stanford University
701 Welch Road, Bldg C
Palo Alto CA 94304
USA

Abstract. Software design consists of determining a high-level organization of a system that meets a given problem specification. We present a prototype system called KASE (Knowledge Assisted Software Engineering) that helps system analysts and designers design and redesign software systems. The KASE environment provides knowledge representation and reasoning tools to integrate knowledge about general software design principles, prototypical software architectures, and application domain. Unlike CASE tools and module interconnection languages that represent the structure of a software system without its semantics, the goal of KASE is to integrate both the structure and the semantics of software modules and to provide *active* assistance in the design of systems. We illustrate the design process in KASE through an example, focusing on the ability of KASE to provide a co-operative man-machine design environment.

1. Introduction

Software system design is a high-level organization of system components that meets a given problem specification. The design process consists primarily of decomposing problem specifications into functional modules or subsystems, defining the interfaces and the dependencies between the modules, allocating high-level tasks to computing agents, and determining representation schemes for data. A good software design should enable the implementation of the software to proceed in a modular and largely routine manner.

Like all design activities, designing software systems is a knowledge-intensive task. Several studies, e.g. (Adelson and Soloway, 1985; Guindon, Krasner, and Curtis, 1987) have found that the predominant cause of failures among system designers is lack of knowledge—knowledge about the task domain, knowledge about design schemas, knowledge about design processes.

• One major difficulty in software design is the need to integrate *knowledge from multiple domains*. As one moves towards abstract description of software design, the description becomes increasingly tied to the description of the

J. S. Gero (ed.), Artificial Intelligence in Design '92, 371–389.

application domain. That is, the concepts and constraints of the task domain are integral parts of software design. In addition, Jeffries *et al.* reported that a designer's knowledge becomes more domain-specific and detailed as he gains expertise (Jeffries, Turner, Polson, and Atwood, 1981). Thus, in addition to knowledge about software—architectures, algorithms, data structures, programming languages—creating software design requires *domain-specific knowledge*.

• Because requirements are ambiguous and incomplete (Swartout and Balzer, 1982; Parnas and Clements, 1986), and because both the requirements and the environment in which the software operates are in a constant state of flux, there is a need for problem structuring during software design (Guindon, 1990). *Problem structuring* is the process of discovering missing information and redefining the problem space based on the new information. Missing information often includes unstated problem statements which are implicit aspects of the application domain.

• With few exceptions, current tools for software design, i.e. CASE tools, are little more than graphical editors that can help in creating high-level diagrams (e.g., a data flow diagram) and in ensuring the syntactic consistency of these diagrams. These tools serve to document a software design, but they do not posses knowledge about the task domain, and hence cannot provide *active support* in the initial design process, or in the subsequent modifications to the design. Tools like gIBIS (Conklin and Begeman, 1989) provide an environment where the rationale for a design process can be captured and referenced. However, like CASE tools, they too provide only *passive* assistance to a designer (that is, the initiative for the design process lies only with the user with the tools simply providing a medium for documentation).

What is needed are software design tools that *explicitly model domain knowledge* and facilitate the *reuse* of this knowledge by (1) delivering relevant knowledge in the appropriate context and (2) ensuring that design errors are not introduced during design modifications. An environment containing such tools should also record the decisions made in the design process. The record can be used to determine dependencies between various design steps, to retract design decisions, and to provide a context for design modifications when requirements change.

Finally, in order to aid understanding, the environment needs to provide tools that help the designer visualize both the structure and the behavior of the designed artifact. As Brooks notes, software is inherently a very complex artifact and is difficult to visualize in its entirety (Brooks, 1987). What is needed are ways to "see" software from different *viewpoints*. The LaSSIE system (Devanbu, Brachman, Selfridge, and Ballard, 1991) uses four views: the domain model view, the architectural view, feature view, and code view. These views correspond to different knowledge domains apparent in the design. In addition to these views a designer should be able to "see" software using pictorial notations - e.g. an integrated view of control flow, data flow, and functional decomposition of modules (Guindon, 1992).

2. KASE (Knowledge Assisted Software Engineering) Environment

We describe a software design environment called KASE that provides *active* assistance in the design of a software system. Some of the basic ideas underlying KASE are:

- identify the knowledge domains used in software design,
- encode the knowledge separately, and
- provide tools to instantiate and integrate the knowledge in the service of designing a system that meets the problem specifications.

The current prototype KASE contains in its knowledge base a prototypical software architecture (called *generic architecture*) that can be used as a basis for designing solutions to a class of problems (called *generic problem)*, a *domain model* that can be used to interpret the problem specifications, and heuristics for designing applications using the generic architecture. As shown in Figure 1, the objective is to expand a given problem statement in terms of the concepts and operations in the domain model, instantiate the generic architecture, and map the domain concepts and operations onto the architecture. We call this process *customization* – customize a generic architecture to fit the application.

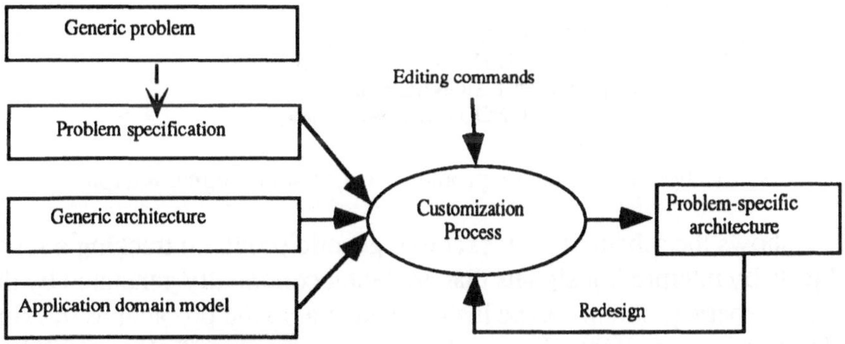

Fig. 1 Customizing an architecture

2.1 GENERIC PROBLEMS

A generic problem represents a class of problems; individual problems are specified as instances of the class. By identifying problem classes, one can design knowledge representation schemes, architectures, and reasoning processes which are appropriate for the general problem, and reuse them for several different problem instances. Therefore, the identification and specification of such a class of problems is a task that precedes, and is a prerequisite of, the subsequent design activity using KASE[1].

The specification of a generic problem results in the creation of a problem schema which is analogous to the notion of *cliches* in the Programmer's Apprentice (Waters, 1985). A problem schema specifies the high-level structure of a problem specification and has certain *roles*, which represent the parameters of the problem, and *constraints* on the values of the roles. Instantiating these roles with specific values results in the creation of a specific problem specification.

```
Continuous-Signal-Interpretation   :Generic-problem
    Signal-Inputs: )<var> : (SEQ :FROM <int> :TO <int> (<fields>)
                              <field-description>)]+
    Body: WHILE <formula> DO <statements> ENDWHILE
    Task  Assumptions:  <task-assumptions>
    where
    <fields> ::= <identifier> | <identifier> <fields>
    <field-description> ::= EXIST <objects> SUCH-THAT <condition>
    <statements> ::= <statement> ; <statements> | <statement>
    <statement> ::=  (IF <formula> THEN-DO <statement>) |
                     (FORALL <vars>  <formula> DO <statement>) |
                     (PRINT <terms>)
    <formula> ::=  (FORALL <vars> <formula>) |
                   (EXIST <vars> <formula>) |
                   (AND <formulae>) |
                   (OR <formulae>) |
                   (IMPLIES <formula> <formula>) |
                   (NOT <formula>)
                   (Predicate <terms>)
    <task-assumptions> ::=  (UNRELIABLE-SIGNAL  <var>) |
                            (INCONSISTENT-SIGNAL S <var> <var>) |
                            (REDUNDANT-SIGNAL <var>) |
                            (ASYNCHRONOUS-SIGNAL  <var>) |
```

Fig. 2. Specification of the generic problem of continuous signal interpretation.

Figure 2 shows the schema for an example generic problem: tracking a set of mobile objects by interpreting signals that are being continually generated by the objects. (This generic problem can be instantiated, e.g. to the problem of tracking aircrafts from radar and voice signals (Brown, Schoen and Delagi, 1986) or tracking ships from sonar data (Nii, Feigenbaum, Anton and Rockmore, 1982). This problem has three parameters: (i) the specification of the input signal(s); (ii) the main body or functional description of the problem in the form of an extremely high-level program; and (iii) certain characteristics of the domain and the

[1]It is assumed that a library of generic problems, associated architectures, and the domain model are an integral part of the KASE environment.

environment. The constraints on the schema roles are specified by specifying a grammar for instantiating the roles.

2.2 GENERIC ARCHITECTURES

Associated with each generic problem is a set of (possibly one) generic architectures, which can be used to create a system for solving instances of the generic problem. A generic architecture is a collection of *parameterized modules* and intermodular dependencies. A parameterized module is a logical collection of software entities like procedures, types, etc. in which some of the entities are abstracted as parameters. A parameter can be, among other things, an algorithm, a representation scheme, or a sub-module. The design process is viewed as an instantiation of the various parameters comprising a generic architecture. However, even though our approach to design can be considered as an instance of parameterized design, the parameters can be fairly complex entities and the design task is non-trivial (see Section 3 for examples).

The structure of the generic architecture determines the basic solution strategy for solving the problem. For example, the continuous signal interpretation problem given earlier can be solved using a symbolic, knowledge based approach, or by statistical analysis of the data and the two solutions would have radically different architectures. The degree of detail in specifying an architecture depends on the generality of the associated generic problem. Thus, one could have a very general architecture (for example, a generic blackboard architecture), which could be used on a large number of problems sharing certain properties, or one could have a very specific architecture (for example, an architecture for a payroll system) with detailed specifications of all the functional modules.

A module description includes information about the input and output data flows of the module, the submodules/supermodules structural relations, the services it requires from other modules, the services it provides to an external module, the precondition and postconditions for each service provided by the module, and/or a program template that implements each service. Some of this information is useful in visualizing the organization of a software system from different views. For example, the input and output data flow and the submodule relationship are used to construct a data flow diagram; the program templates and the data flow are used to construct a modified Action diagram (Martin and McClure, 1988) which shows both the flow of control and data within a module. Thus, KASE provides an environment which is a natural extension of the environment provided by commercial CASE tools.

The most interesting aspect of the module description is that some of its attributes are viewed as parameters of the module. Associated with each parameter attribute is a method which can be used to determine the value of the parameter. The complexity of the method depends on the type of the parameter.

For example, it may be a simple process of selecting between a pre-determined list of alternatives, or it may involve sophisticated reasoning using domain knowledge and heuristic rules.

Figure 3 shows the functional decomposition view of a generic architecture for the continuous-signal-interpretation problem class and Figure 4 shows a partial description of the signal-interpreter module of the generic architecture.

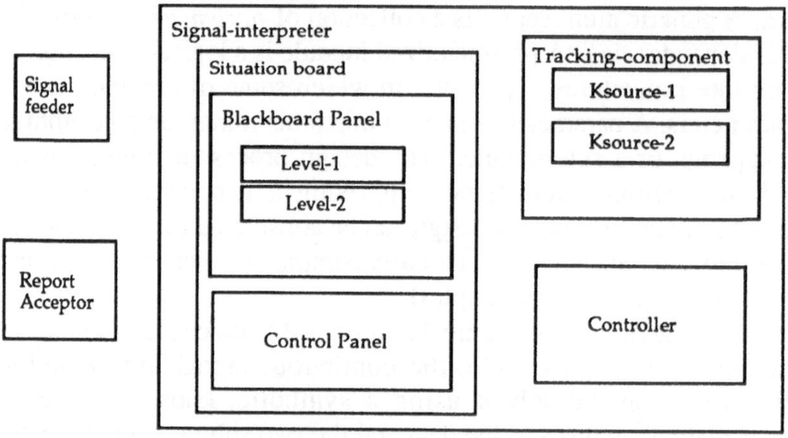

Fig. 3. A functional decomposition of the generic architecture for the continuous-signal interpretation problem. The architecture shows the main modules comprising the architecture.

2.3 DOMAIN MODELS

The third and final knowledge encoding in the KASE environment consists of a model of the application domain. This process has been called domain modeling (Neighbors, 1984). The domain model provides the ontology of terms and operations used to describe an application domain independent of a specific problem specification; several different problem specifications can then be stated in a high-level language using this ontology. This provides a powerful form of reuse that is one of the keys to large scale improvement in software productivity.

Signal-Interpreter **isa** module
submodules Situation-board, Tracking-component, Controller
supermodule CSI-system
inputs ?s : SEQ(signal)
outputs ?r : SEQ(report)
requires (print-report ?r), (read-next-signal) :signal, (start-execution)
provides (main)
calls report-acceptor, signal-feeder
called-by nil
parameters
1) Controller

```
    2) SituationBoard
constraints
    1) Controller is instantiated to an EventDriven-Controller iff SituationBoard is
         instantiated to an EventDriven-SituationBoard.
    2) Only the TrackingComponent should have a dataflow into the SituationBoard.
    3) Only the Controller module can call the Tracking-Component.
         .....
service-definitions
main()
  begin
    start-execution()                  % It simply calls the service start-execution
  end                                  % provided by the Controller module.
```

Fig. 4. Representation of the Signal-interpreter module in the generic architecture.

We are developing an object-oriented modeling environment that creates a model from three different perspectives - static or object model, behavior model, and functional model (Rumbaugh, Blaha, Premerlani, Eddy, and Lorensen, 1991). For the purposes of this paper we limit ourselves to the object and the functional models.

The primary components of the static domain model are *objects* and *relations* between the objects. An object is an abstraction of some entity in the application domain, e.g. an aircraft or a signal. Associated with each object is a set of *attributes* which are properties that describe an instance of an object. We distinguish between *primitive* attributes and *derived* attributes. Primitive attributes are those that must be explicitly specified for each instance of an object; derived attributes are those that can be computed from primitive attributes of the object and/or other data. With each derived attribute is a set of *definitions* that describe how the value is to be computed.

Besides attributes, there are a set of *methods* associated with each object. Methods are operations that change the state of an object. The description of a method includes the pre- and post-condition stating when the method is to be applied. Examples of how objects, relations, and the operations on them are modeled are given in Section 3.

2.4 CUSTOMIZATION PROCESS

The customization process consists of refining a selected generic architecture into a detailed architectural specification based on the model of the domain and the problem specification. In KASE, the customization process is performed in an interactive and mixed-initiative setting. The role of KASE in the design process is that of an intelligent design associate that provides suggestions on how to refine the architecture, carries out the commands invoked by the user, informs the designer of constraint violations in the design, keeps a record of the design steps and the dependencies between the steps so that incremental modifications to the design can be done efficiently.

The knowledge used by KASE in providing these kinds of assistance includes general, domain independent knowledge about software design, architecture-specific knowledge for the instantiation of various architectural parameters, as well as specific heuristic knowledge about design related to a particular domain. Most of the domain independent design knowledge is represented in the form of constraints (e.g. those relating different levels of a data flow diagram), and KASE contains mechanisms which automatically keep track of these constraints as well as heuristics for resolving constraint violations (Nii, Aiello, Bhansali, Guindon, and Peyton, 1991). The architecture specific knowledge includes a set of constraints governing the relationships between different components of the architecture, a library of reusable modules and schemas which can be used to instantiate the architectural parameters, and a collection of heuristic rules and algorithms which can be invoked by a designer to instantiate certain parameters and optimize the design.

2.5 REDESIGN

We subscribe to the view that most software design is redesign. A new design is created by modifying parts of an extant design. The modification may involve changing a design either due to a design error or as a result of a change in the problem requirements or the computing environment. KASE uses different mechanisms to support these two kinds of modifications.

2.5.1 Redesign due to error in original design. KASE automatically checks for violations of several kinds of constraints and helps the designer modify the architecture to resolve the inconsistencies. The constraints in KASE are currently divided into three categories.

(1) General architectural constraints. These are constraints that have to be satisfied by all architectural designs. Examples of such constraints are: "Every data link must have a consumer and a producer" and "If a module X requires a service F, then there must be some module Y which is called by X and module Y provides service F".

(2) Specific architectural constraints. These are constraints that are specific to all instances of a particular generic architecture. For example, an architecture based on the blackboard model might have constraints: "There must be no data flow or control flow between two knowledge source modules" and "If the BlackboardPanel is instantiated to a Data-driven Blackboard Panel then the Controller module should be instantiated to a Data-driven controller module".

(3) Stylistic constraints. Stylistic constraints are those that are derived from design principles that are considered 'good', for example, "A module must not be

decomposed into more than *n* submodules at any level of abstraction." Stylistic constrains may be divided into general stylistic constraints and constraints specific to an architecture. However, currently all stylistic constraints in KASE are general constraints.

Each constraint in KASE is associated with a *trigger*, a *predicate*, and an optional *resolving-action*. A trigger is a set of actions that can potentially cause the constraint to be violated, for example, the constraint, "Every data link must have a consumer and a producer," has a trigger consisting of two actions Delete-module and Add-datalink. A *predicate* is a Lisp expression that checks to see whether the constraint is actually violated. *Resolving-action* is a set of actions that may be taken to remedy the constraint violation.

KASE monitors the design activity and flags each constraint that is triggered by a user action. When a user indicates the completion of a design session, KASE checks the predicates for each flagged constraint to see whether the constraint is actually violated. Quite often, a constraint that gets violated by a design action is resolved by a later action, and such constraint violations should be, and are, transparent to the designer.

When KASE reports a constraint violation, the designer can ask KASE for a list of suggestions on how to resolve the error. Depending on the nature of the constraint, KASE presents a list of different actions that may be taken to remove the constraint violation. (KASE currently does not consider the effects of any of the actions on the rest of the architecture; thus, it is possible that some of the suggestions offered by KASE remove the current constraint violation but may introduce other constraint violations.) The user can then choose either one of the actions suggested by KASE or take some other action.

2.5.2 Redesign due to change in requirements. KASE provides tools that can help a designer in modifying parts of a design to meet new requirements without having to start from scratch. First, KASE maintains a history of all the design steps and allows the user to go back to any previous state of the design. It does this by replaying the design history from the initial state to the desired state.

A second redesign support provided by KASE is in localizing the effects of a design change. KASE uses dependencies between design steps to structure a linear design history into a lattice. When the user wants to undo the effect of a particular design step, KASE uses the position of that design step in the derivation history to determine what other design steps are affected by it (Bhansali, 1992).

3. An Example System Design

In order to illustrate the capabilities of the KASE environment, we will step through the design of a system called ELINT (Brown, Schoen, and Delagi, 1986) that was developed several years ago as part of a multi-sensor information fusion system called TRICERO. The objective of the ELINT subsystem was to analyze processed, passively acquired, real-time radar data from radar emitting devices (called *emitters)* in aircrafts. The primary input to ELINT was time-ordered streams of processed observations from multiple data collection sites. In addition, ELINT received periodic feedback from another system in TRICERO about specific aircraft activities. The outputs from ELINT consisted of periodic status reports about the activities of clusters of radar-emitting objects in the area under surveillance. (A cluster of emitters could be a single aircraft with more than one radar-emitting devices or more than one aircraft.) The report included such information as the position and heading of clusters, description of possible aircrafts comprising the cluster, and an indication of the cluster's current activity.

The architecture of the original ELINT system was based on the blackboard model (Nii, 1989). The basic components of a blackboard-based system are a shared, global data structure called the *blackboard* which represents the current problem-solving state, a set of components called *knowledge sources* that contribute information to the blackboard leading to a solution of the problem, and a *control* component that monitors the information on the blackboard and decides what knowledge source to execute next. The design of a blackboard-based system essentially consists of designing these three components.

3.1 GENERIC PROBLEM, GENERIC ARCHITECTURE AND DOMAIN MODEL

We assume that a generic problem that abstracts the ELINT problem specification and a generic architecture that abstracts a class of blackboard-based architectures have already been encoded in KASE (Figs. 2 and 3). The design of the ELINT system begins by first building a model of ELINT's domain. Figure 5a shows examples of two objects and a relation in this domain. Figures 5b and 5c show the definition of a derived attribute (emitter.position) and an operation (create-new-cluster).

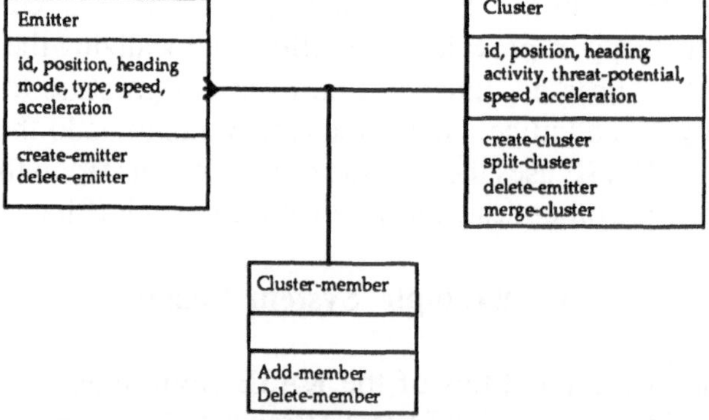

(a)

```
emitter.position(e:emitter, t:time) = p:position
definitions
(1) From-two-positions-and-bearings(s1,s2: site)
      precondition: s1 <> s2
      definition: (= p (direction-intersection
                            (line-of-bearing s1 e t)
                            (line-of-bearing s1 e t)
                            (site-position s1 t)
                            (site-position s2 t))).
```

(b)

```
Method: create-new-cluster(e:emitter)
precondition: (AND (= C (SET (c: cluster)))
                   (FORALL (c : cluster) (or (not (co-located c e))
                                             (not (co-headed c e)))))
postcondition: (EXIST (c:cluster)
                  (and (= (SET (c:cluster) (UNION C {c}))
                       (cluster-member c e)
                       (= (cluster.position c) (emitter.position e) ...)))
```

(c)

Fig. 5. (a) Examples of objects (emitter and cluster) and relations (cluster-member) in the domain model.The second level of each box shows attributes and the third level shows methods on the object. (b) Definition of a derived attribute. (c) A method definition.

3.2 DESIGNING THE ELINT SYSTEM

Step 1 Specifying the ELINT problem. The design of a system using KASE begins by first specifying a problem as an instance of a known generic problem (Figure 6). This is an instance of the generic problem shown in Fig.1.

```
Problem: Aircraft-monitoring-system
isa: Continuous-Signal-Interpretation
Inputs:
  1) observation : SEQ (:FROM 1 :TO EOF (eid sid eloc sloc lob etype emod qual t)
                   :ST (EXIST (e s)
                          (and (emitter e) (site s) (= eid (emitter-id e)) (= sid (site-id s))
                               (= sloc (site-position s t)) (= lob (line-of-bearing s e t))
                               (= emode (emitter-mode e t))(= etype (emitter-type e))
                               (data-quality qual)))))
  2) feedback : SEQ(:FROM 1 :TO EOF (cid es t)
                :ST (EXIST (c) (and (cluster c t) (= cid (cluster-id c))
                                    (= es (cluster-emitters c t)))))
Program:
  (WHILE true DO
    (IF (= (div (current-time) 100) 0) THEN-DO
        (FORALL c (cluster c) DO
          (PRINT (current-time)
                 (cluster-id c) (cluster-position c) (cluster-heading c)
```

(cluster-activity c)(cluster-aircraft-types c)
(cluster-threat-potential c)))))

Task Assumptions:
(Unreliable-signal observation)
(Redundant-signal observation)
(Asynchronous-signal feedback)
(Asynchronous-signal observation)
(Inconsistent-signals observation feedback)

Fig. 6. The specification of the ELINT problem.

Step 2 Selecting the generic architecture. After the problem has been specified as an instance of one of the generic problems, the system presents the user with a set of different generic architectures associated with the generic problem. The user must select one of these architectures to serve as a starting point for his design. Currently, there is only one architecture associated with the continuous-signal-interpretation problem, and it is automatically selected by the system.

Step 3 Customizing the architecture. When a generic architecture is selected, KASE creates a customization menu which lists all applicable commands that can be used to refine the architecture. These commands represent methods that can be used to instantiate the various module parameters.

The modules of a generic architecture are stored in a reusable-module library. As shown in Figure 7 some of these modules represent a class of modules. When a generic architecture is selected, it must be instantiated by replacing a module class by an instance of that module. For the current architecture, there are four module classes—Controller, SituationBoard, ControlPanel, and the Ksources. Figure 7 shows the choices for the Controller, SituationBoard, and the Ksource modules.

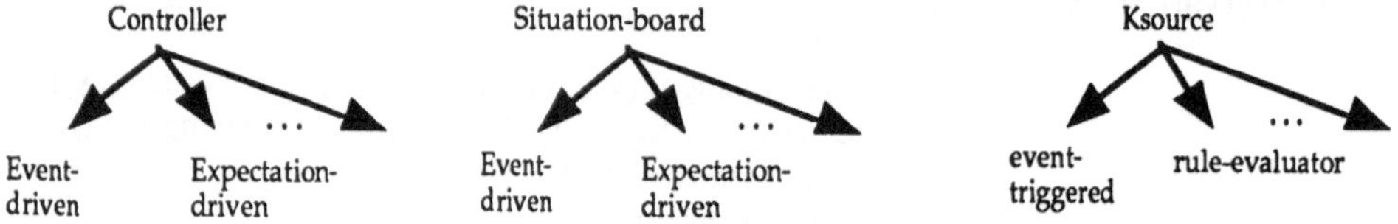

Fig. 7. Possible instantiations of some generic modules.

The user selects one of the choices from the customization menu. Note that the choices for instantiating different modules are not independent. For example, if the SituationBoard is chosen to be Event-driven then it constrains the Controller to be the Event-driven Controller module. When a user selects a particular instance for a module, KASE uses architectural constraints to automatically prune the set of available choices for the remaining modules.

The KASE environment does not prescribe a pre-determined order in which to compose the architecture from modules, nor does it force the user to complete the instantiation of all the modules before proceeding to the next step. Rather, the initiative lies with the designer most of the time, and he/she can build parts of the architecture in any order. Furthermore, the designer can specify only a subset of the modules comprising the architecture. Thus, for the example architecture, the user may choose to instantiate the SituationBoard and the Ksources, but leave the Controller uninstantiated.

Step 4 Determining the blackboard levels. When a particular instance of a module class has been selected, KASE uses the parameters of that module to update the customization menu by commands that can instantiate the parameter. Figure 8 shows the customization menu after the modules SituationBoard and Ksource have been instantiated.

It can be seen that the customization menu contains commands that represent some of the fundamental issues in designing a blackboard system: How should the blackboard be structured? How should the problem-solving expertise be divided into knowledge sources? What are the specific operations to be performed by those knowledge sources? Other commands that are added later deal with control issues: What should be the overall control paradigm—data-driven or model-driven or some other hybrid technique? What should be the conflict resolution mechanism in the scheduler?

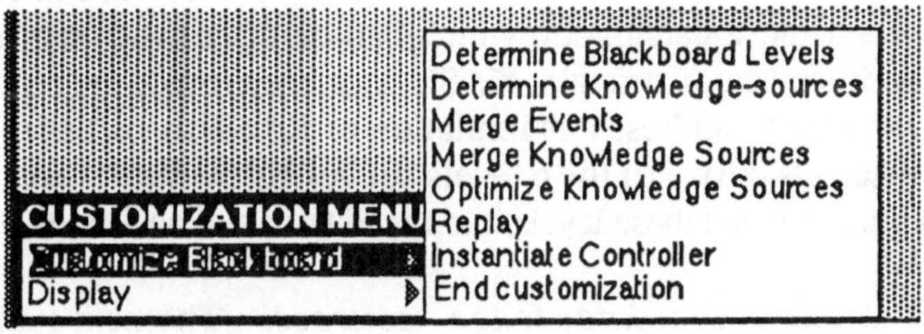

Fig. 8. The customization menu.

One of the first activity in designing a blackboard system is to specify the data structure of the blackboard. The KASE environment provides low-level editing commands which can be used to create and describe modules. However, it is expected that a designer would first use KASE's inferencing mechanism to obtain initial values for the parameters and then modify them by subsequent refinements and optimization commands.

KASE employs various kinds of general inference mechanisms and heuristics for instantiating parameters. For example, it uses several different heuristics to determine the set of levels into which the BlackboardPanel should be divided. One of these heuristics is based on a general goal-directed reasoning strategy. It first constructs a functional dependency graph showing the relationships between the output objects that the system must produce and its input signals. From the dependency graph the system deduces a set of objects and relationships and their attributes which are needed to compute the values of the output objects. Figure 9 shows the set of blackboard levels that are suggested by KASE when the user selects Determine Blackboard Levels from the customization menu.

Level	Attributes
Cluster	id aircraft-types cluster-emitters position activity heading threat-potential
Emitter	id type position heading
Site	id position
Observation	eid sloc emod sid lob qual eloc etype t
Feedback	cid es t

Fig. 9. Blackboard levels suggested by KASE

In this example, the user may decide that the representation of sites is not needed, since the information about collection sites is only used to calculate the position of an emitter, a calculation which can be done when the input signal (observation) is processed. (Currently, KASE does not provide the rationale for its suggestions; we plan to address this limitation of KASE as part of future work). Thus, the user selects only 5 of the 6 levels suggested by the system. KASE then automatically incorporates these levels into the BlackboardPanel module.

Step 5 Determining the knowledge sources. The next typical step in designing blackboard systems is to determine the knowledge sources that operate on the blackboard. The user indicates this by selecting Determine Knowledge-sources from the customization menu.

The specific (event-triggered) knowledge source that is chosen for this module can be represented abstractly as a triple $<P, O, E>$ where P represents the precondition events that must occur in order for the knowledge source to be invoked, O is a set of operations performed by the knowledge source that modify the data on the blackboard, and E is the set of events (or control data) that can be posted by the knowledge source to signal the specific changes on the blackboard brought about by the operations.

The set of knowledge sources are determined using a three-step process. First, the system determines all relevant operations based on the structure of the blackboard. Next, it generates a list of potential events that can be used to signal the occurrence of each of the operations. Finally, the system uses the functional dependency graph created earlier to generate the sets of preconditions for each operation. At the end of this step the system has a list of (14, for this example) knowledge source operations.

Next, the system groups some of these operations together into a single knowledge source. Several different heuristics can be used for grouping the operations. The heuristic used by the system is based on the presence of common preconditions: *group all operations that have the same set of preconditions into one knowledge source*. This heuristic results in the creation of 10 knowledge sources shown in Figure 10.

	PRECONDITONS	OPERATIONS	EVENTS POSTED
1	(emitter-mode updated) (cluster-emitters updated)	[COMPUTE CLUSTER-ACTIVITY]	(cluster-activity updated)
2	(processed obs) (cluster-emitters updated)	[COMPUTE CLUSTER-AIRCRAFT-TYPES]	(cluster-aircraft-types updated)
3	(emitter-position updated) (cluster-emitters updated)	[COMPUTE CLUSTER-POSITION]	(cluster-position updated)
4	(cluster-position updated) (emitter-mode updated) (cluster-emitters updated)	[COMPUTE CLUSTER-THREAT-POTENTIAL]	(cluster-threat-potential updated)
5	(emitter-heading updated) (emitter-position updated) (processed feedback)	[COMPUTE CLUSTER-EMITTERS]	(cluster-emitters updated)
6	(processed obs)	[COMPUTE EMITTER-POSITION] [COMPUTE EMITTER-HEADING]	(emitter-position updated) (emitter-heading updated)
7	(split-cluster occurred) (delete-cluster occurred) (merge-cluster occurred) (emitter-heading updated)	[CREATE CLUSTER]	(create-cluster occurred)
8	(cluster-position updated)	[DELETE CLUSTER]	(delete-cluster occurred)
9	(emitter-heading updated) (cluster-emitters updated)	[COMPUTE CLUSTER-HEADING] [SPLIT CLUSTER]	(cluster-heading updated) (split-cluster occurred)
10	(clock-event ?t)	[GENERATE OUTPUTS] [PROCESS-INPUT OBS] [PROCESS-INPUT FEEDBACK]	(generated-outputs) (processed obs) (processed feedback)

Fig. 10. Knowledge sources suggested by the system for the example.

3.3 REFINING THE DESIGN

Step 6 Optimizing the design. As with blackboard levels, the final decision on how to structure the knowledge sources is up to the user. Typically, the user would keep all the knowledge sources suggested by KASE and refine the set by adding, deleting, or modifying the knowledge sources. However, a common design phenomenon is the desire and/or the necessity to optimize the design. KASE provides a set of optimization commands.

One of these commands merges events. It may be the case that whenever a particular event occurs it is usually accompanied by another event, for example, whenever the emitter-heading changes, so does the emitter-position, and the designer may not want to treat the second event as a distinct event. KASE provides a command for merging a set of events and automatically replaces each occurrence of the merged events by the new event in each of the affected knowledge sources. Similarly, KASE provides a command to merge knowledge-sources using one of the pre-compiled optimization heuristics.

This example illustrates one of the guiding themes of our approach: *Divide the design task between a human and KASE in a way that exploits the unique skills of each.* In general, the human is better equipped to decide when to apply an optimization technique and what heuristics/technique to use for the optimization, whereas the machine is better equipped to carry out the optimization task, propagate the effects of those changes to other parts of the program, remember the optimization task, and if necessary, undo the effects of the optimization operations later.

Step 7 Checking for consistency. The initial generic architecture selected by a designer is consistent (i.e., the intermodular constraints are satisfied), but is incomplete since some of the modules and the parameters of some modules have yet to be determined. The customization task seeks to complete the architectural design, but in the process a designer may introduce inconsistencies in the design. In the current example, when the user indicates the end of the customization session, KASE detects one architectural constraint violation: "The following events are posted by knowledge sources, but do not trigger any knowledge source: ..." This is an example of a design error that can be easily introduced and may be difficult to detect. In this case, it actually exposes a flaw in KASE's own event-generation heuristic, which is based on the simplistic assumption that each action that is performed by a knowledge source should be reported as an event on the blackboard.

Step 8 Resolving errors. When KASE reports a constraint violation, the designer can ask KASE for a list of suggestions on how to resolve the error.

Depending on the nature of the constraint, KASE presents a list of different actions that may be taken to remove the constraint violation. For the above example, KASE offers just one suggestion: Delete the events that are not preconditions of any knowledge source. If the user accepts this suggestion, KASE automatically updates the knowledge source by deleting the offending events.

3.4 REDESIGN

Step 9 Redesigning. We illustrate KASE's redesign support very briefly using an example. Suppose the user changes the problem specification slightly so that instead of simply producing reports, ELINT also accepts queries about the expected positions of clusters and emitters based on their past activities. The user decides to incorporate this change by treating queries as expectations (instead of events) and processing them separately on the SituationBoard. One of the design steps the user takes is to decompose the ControlPanel in the SituationBoard into two panels, EventPanel and ExpectationPanel using the Copy-module command provided by KASE. KASE will check and report all constraint violations to the user. An example of one of the constraints violated in this case is that the data link into the ExpectationPanel has no producer.

4. Related Work

Much of our approach shares ideas as well as terminology of task-oriented methodologies for design (e.g. Brown and Chandrasekaran, 1989; Chandrasekaran, 1986), although our work evolved independently of them. Our approach of viewing the design task as an incremental refinement of a generic architecture using a series of knowledge-based editing command is largely inspired by research in knowledge-based software engineering (e.g. Graves, 1991; Johnson and Feather, 1991; Lubars and Harandi, 1989; Waters, 1985). Research in the reuse of generic architectures for software design is recently receiving increased attention following a DARPA-initiated Domain-specific Reusable Architecture project.

5. Conclusion

The KASE system represents our initial attempt in building a prototype environment that can offer varying degrees of assistance to a software designer by employing diverse sources of knowledge. Our current work is focusing on extending the domain modeling representation to capture the dynamic behavior of a system by modeling states, transitions, events, and actions, and integrating it with the object and functional view of the system.

The maximum payoff from using the KASE environment is achieved when one has reusable architectures that can be used to design systems for a family of related problems. We need to identify such architectures and problem classes and use KASE for designing software systems for problems belonging to such problem classes.

An issue that we are interested in is to see how sensitive the domain representation is to the choice of an architecture. In other words, we would like to explore how much of the application domain knowledge encapsulated in a domain model can be reused to design software systems based on completely different architectures.

In parallel with the above, we are exploring the issue of design rationale capture and its reuse during redesign. KASE's current redesign capabilities were mentioned briefly in this paper. We are interested in extending these capabilities so that KASE can automatically incorporate certain changes in problem requirements into the design by using analogical or case-based reasoning (Bhansali and Harandi, 1991).

Acknowledgements

The KASE system is a result of several people's work. We gratefully acknowledge the contributions made by Nelleke Aiello, Raymonde Guindon, Liam Peyton and Go Nakano who wrote most of the code for KASE.

References

Adelson, B. and Soloway, E.: 1985, The role of domain experience in software design, *IEEE Transaction on Software Engineering*, **SE-11**(11), 1351-1360.

Bhansali, S.: 1992, Generic software architecture based redesign, *AAAI Spring Symposium on Computational Considerations in Supporting Incremental Modification and Reuse*, Stanford, CA.

Bhansali, S. and Harandi, M. T.: 1991, Synthesizing UNIX shell scripts using derivational analogy: an empirical assessment, *Ninth National Conference on Artificial Intelligence*, Anaheim, CA, AAAI Press/MIT Press, pp. 521-526. .

Brooks, F. P., Jr.: 1987, No silver bullet: essence and accidents of software engineering. *IEEE Computer*, April

Brown, D. C. and Chandrasekaran, B.: 1989, *Design Problem Solving: Knowledge Structures and Control Strategies*, Morgan Kaufmann, San Mateo, CA:.

Brown, H. D., Schön, E. and Delagi, B. A.: 1986, An experiment in knowledge-based signal understanding using parallel architectures, *Technical Report STAN-CS-86-1136*, Department of Computer Science, Stanford University.

Chandrasekaran, B.: 1986, Generic tasks in knowledge-based reasoning: high-level building blocks for expert system design, *IEEE Expert*, **1**(3), 23-30.

Conklin, J. and Begeman, M.: 1989, gIBIS: A tool for all reasons, *Journal of the American Society for Information Science*, **40**, 200-213.

Devanbu, P., Brachman, R. J., Selfridge, P. G. and Ballard, B. W.: 1991, LaSSIE: A knowledge-based software information system, *Communications of the ACM*, **34**(5), 34-49.

Graves, H.: 1991, Lockheed environment for automatic programming, *6th Annual Knowledge-Based Software Engineering Conference*, Syracuse, NY, pp. 78-89.

Guindon, R.: 1990, Knowledge exploited by experts during software system design. *International Journal of Man-Machine Studies*, 33(3), 279-304.

Guindon, R.: 1992, Requirements and design of DesignVision, an object-oriented graphical interface to an intelligent software design assistant, *ACM Proceedings of CHI'92*, Monterrey, CA.

Guindon, R., Krasner, H., and Curtis, B. (eds): 1987, *Breakdowns and Processes During The Early Activities Of Software Design By Professionals*, Ablex.

Jeffries, R., Turner, A., Polson, P., and Atwood, M. E. (eds): 1981, *The Processes involved in designing software*, Lawrence Erlbaum, Hillsdale, NJ.

Johnson, W. L. and Feather, M. S.: 1991, Using evolution transformations to construct specifications, *in* M. Lowry and R. McCartney (eds), *Automating Software Design*. AAAI Press, Cambridge, MA.

Lubars, M. D. and Harandi, M. T.: 1989, Addressing software reuse through knowledge-based design, *in* T. J. Biggerstaff and A. J. Perlis (eds), *Software Reusability*, ACM Press, New York, NY.

Martin, J. and McClure, C.: 1988, *Structured Techniques: The Basis for CASE*, Prentice Hall, NJ.

Neighbors, J.: 1984, The DRACO approach to constructing software from reusable components, *IEEE Transactions on Software Engineering*, **10**(9), 564-573.

Nii, H. P., Aiello, N., Bhansali, S., Guindon, R., and Peyton, L.: 1991, Knowledge Assisted Software Engineering (KASE): An introduction and status June 1991, *Technical Report KSL-91-28*, Knowledge Systems Laboratory, Computer Science Department, Stanford University.

Nii, H. P., Feigenbaum, E. A., Anton, J. J. and Rockmore, A. J.: 1982, Signal-to-symbol transformation: HASP/SIAP Case Study, *AI Magazine*, Spring, 23-36.

Nii, P.: 1989, Blackboard systems, *in* A. Barr, P. Cohen, and E. Feigenbaum (eds), *Handbook of Artificial Intelligence*, Addison-Wesley, New York, NY.

Parnas, D. L. and Clements, P. C.: 1986, A rational design process: How and why to fake it, *IEEE Transactions on Software Engineering*, 12, 251-257.

Rumbaugh, J., Blaha, M., Premerlani, W., Eddy, F. and Lorensen, W.: 1991, *Object-Oriented Modeling and Design*, Prentice Hall, Englewood Cliffs, NJ.

Swartout, W. and Balzer, R.: 1982, On the inevitable intertwining of specification and implementation, *Communications of the ACM*, **25**(7), 438-440.

Waters, R. C.: 1985, The Programmer's Apprentice: a session with KBEmacs, *IEEE Transactions on Software Engineering*, **11**(11), 1296-1320.

7

DESIGN RATIONALE

DRCS: AN INTEGRATED SYSTEM FOR CAPTURE OF DESIGNS AND THEIR RATIONALE

M. KLEIN
Boeing Computer Services
PO Box 24346, MS 7L-64
Seattle WA 98124-0346
USA

Abstract. In current design practice, the rationale for design decisions is captured, if at all, as a collection of paper documents, project and personal notebook entries as well as designer's recollections, and is maintained distinctly from the description of the design itself. Increasing demands for higher product quality and lower cost, the growing complexity of the artifacts we design as well as the highly distributed nature of modern manufacturing enterprises is making it increasingly critical that design rationale be captured in a highly usable form, and in particular one that allows us to harness the power of computers to support our activities. Existing rationale capture systems, however, have important limitations: they are either not oriented specifically towards the design process or use a representation not easily generalizable to the full range of potential design problems. This document presents a rationale capture language and system that transcends these limitations by being able to capture designs and their rationale in an integrated way.

1. The Need for Design Rationale Capture

When an artifact is designed nowadays the typical output of this activity is a blueprint, CAD file or some other kind of document that describes the final result of a long series of deliberations and tradeoffs. The original design requirements, the different alternatives pursued as well as the underlying intent and logical support for each design choice are usually lost, or are represented at best as a scattered collection of paper documents, project and personal notebook entries as well as the recollections of the artifact's designers. This design rationale information can be very difficult to access by human agents and is represented such that computers can provide little support for managing and utilizing it.

The challenges of intensified global competition and the growing complexity of the artifacts we design are making it increasingly critical that design rationale be captured in a highly usable form. The potential benefits of such capture are manifold. Explicitly represented design rationale can help individual designers clarify their own thinking about the design (Yakemovic, 1990), e.g. by keeping track of the relationships and differences between the options explored (Kleer, 1985; Lee, 1991), ensuring that all relevant issues and requirements have been addressed (McCall, 1987), detecting flaws in one's reasoning (MacLean, 1989), tracking the consequences of changes in

J. S. Gero (ed.), Artificial Intelligence in Design '92, 393–412.

requirements and design decisions (MacLean, 1989) and so on. The benefits become even greater when one considers that design is typically undertaken by multiple agents over time. The reasoning used to make a decision becomes available for others (e.g. representatives from different aspects of the artifact life cycle such as production and maintenance) to critique and augment from their own perspectives (Lee, 1990). The participants whose contributions are affected by a change in the requirements or design can be identified readily (MacLean, 1989). Designs developed by others which addressed similar requirements and difficulties can be retrieved, understood and then modified to meet current needs (Mostow, 1987; Steinberg, 1984). Documentation production is simplified (Balzer, 1984). Design rationale can help new members in design teams quickly familiarize themselves with a project, and can serve as a basis for training new designers (Boose, 1991).

In order to achieve these benefits, however, significant challenges must be met. The representation we use must allow designers to express their design reasoning in a natural way while at the same time be formal enough to support useful computational services. In addition, the process of describing rationale should impose minimum overhead on the design process. Most existing approaches (Lee, 1991; Yakemovic, 1990; McCall, 1987; MacLean, 1991; Potts, 1988; Newman, 1990; Boose, 1991), however, represent the rationale for decision-making in general but not design in particular, reducing their generality, expressiveness and computational usefulness. The corresponding rationale capture systems are not integrated with the tools used to describe designs, resulting in spotty capture of design rationale, the potential for inconsistency between the rationale and design descriptions as well as the tendency to waste time on issues that later prove unimportant (because the designers cannot focus their discussion on issues revealed by actual inspection of the evolving design description) (Fischer, 1991). While systems that integrate design and rationale representations do exist (Fischer, 1989; Thompson, 1989), they use design representations that can not be easily generalized to domains other than the ones (e.g. kitchen design) they were developed for. The fundamental problem in both cases is that these rationale languages and systems do not provide an *integrated* and *generic* framework for describing designs and their rationale.

In this paper a design rationale capture tool called DRCS (the Design Rationale Capture System) that avoids the limitations of previous work in this area is described. The rationale language underlying DRCS is built upon previous work on decision rationale capture as well as a generic model of design reasoning and is designed to capture in a natural way all important aspects of design decisions, including descriptions of the artifact and the plans used to design and produce it. The DRCS user interface supports rationale capture as an integral part of the design description process throughout the design life cycle. The DRCS language, the DRCS interface and possible directions for future work will be discussed below.

2. The DRCS Rationale Language

Rationale is essentially a record of the reasoning process an individual used to reach certain conclusions. If a rationale description language is expressed in terms that accurately mirror the individuals' reasoning process, it will be easier to use (Gruber, 1987). A rationale language for the medical domain, for example, would be much less useful if it did not include terms like "hypothesis", "evidence", "symptom" and so on, since these are entities used in medical reasoning. Underlying the DRCS rationale language, then, is a model of how designers *think*. We shall begin by outlining this model and then provide an overview of the language that attempts to capture such reasoning.

2.1. THE DESIGN REASONING MODEL

A central aspect of a design reasoning model is, of course, a commitment to how the design itself should be represented and refined. A design, considered over the entire life cycle, includes descriptions of both physical artifacts (the physical entity produced) and temporal artifacts (the plans followed to define and actually produce the artifact). We consider each in turn below.

In DRCS, physical artifacts are viewed as collections of modules, which can represent entire systems, subsystems or their components, each with characteristic attributes, whose interfaces (with their own attributes) are connected by connections of a given type. The resources used by a module (e.g. cost, weight) are represented as a particular kind of attribute (Figure 1).

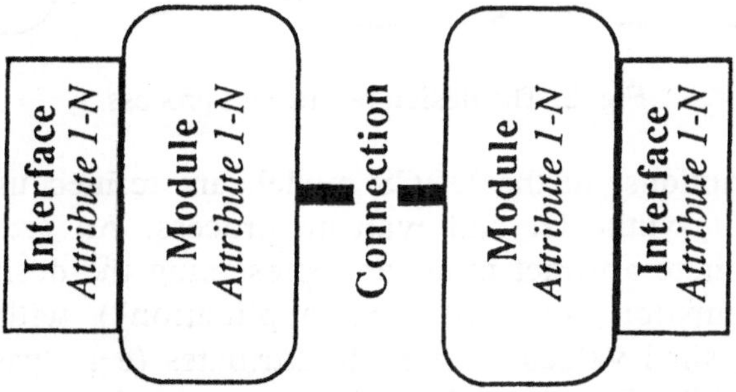

Fig. 1. The design description scheme.

A computer, for example, can be described as a set of VLSI chip modules with attributes describing their functionality, power consumption and so on. The modules' interfaces (pins) are connected to each other via (electrical) connections realized as deposited wires. At another level, we can view an entire board as a module connected via a connector to the bus and indirectly to other boards. At yet a higher level, a computer can be viewed as a set of interacting subsystems including disk I/O, memory and the CPU. The interfaces and connections at this level describe the data and control protocols connecting these sysems. Hydraulic systems, similarly, can be viewed as collections of pipe, switch, tank and pump modules inter-connected via hydraulic connections (e.g. threaded pipe).

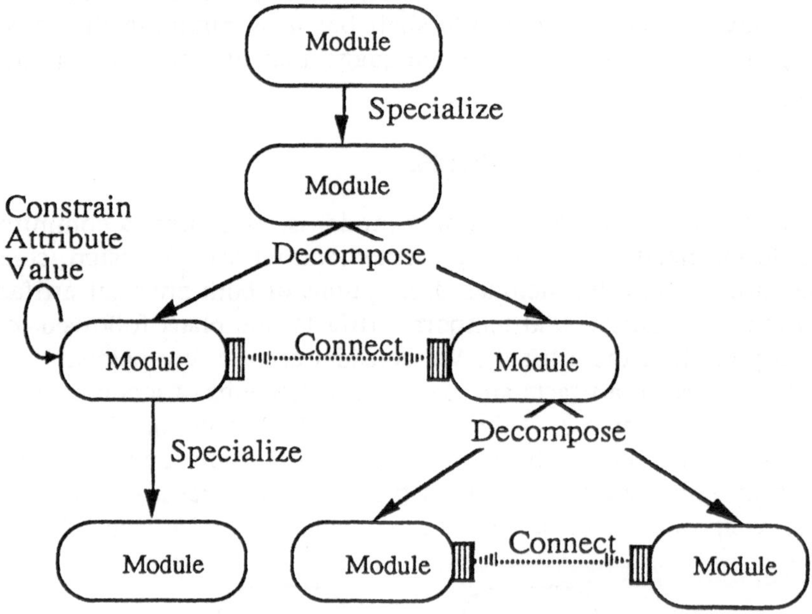

Fig. 2. The design refinement process.

Artifact descriptions, in the DRCS model, are refined using an iterative least-commitment synthesize-and-evaluate process. An artifact description starts as one or more abstract modules representing the desired artifact (e.g. "airplane", "computer" or "software application") with specifications represented as desired values on module attributes (e.g. "passenger capacity should be > 350"). This is refined into a more detailed description by constraining the value of module attributes, connecting module interfaces (to represent module interactions), decomposing modules into sub-modules and specializing modules by refining their class (Figure 2).

If we were designing an airplane, for example, we might decompose the top-level "airplane" module into wing, tail and body section modules as well as electrical, hydraulic and mechanical subsystem modules. The interactions between modules (e.g. the physically connected wing and body sections) are represented as connections between module interfaces.

Plans are viewed as (perhaps partially) temporally ordered collections of tasks with associated attribute constraints (Figure 3). An artifact production plan would thus be represented as a sequence of tasks corresponding to operations such as machining, inspection and the like. Every task includes one or more primitive actions that actually implement the task.

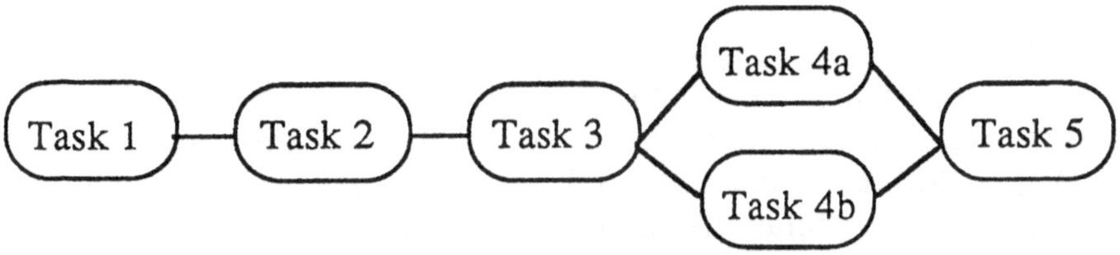

Fig. 3. A plan description.

Plans, like artifacts, are defined in an iterative least-commitment manner. The essential difference is that the basic entity is a task rather than a module, and tasks are temporally ordered rather than connected via interfaces. For both physical and temporal artifacts, DRCS provides a constraint language that allows indefinite descriptions and thus least-commitment design, a commonly-used approach that supports conflict avoidance and early conflict detection (Stefik, 1981; Klein, 1991).

In parallel with the iterative refinement of the design description is evaluation of the design with respect to how well it achieves the specifications. Based on this analysis we may choose to select one design option over another or modify a given option in order to address an identified deficiency. The stages of specification identification, design option definition, evaluation and selection/modification can be interleaved arbitrarily and thus often occur in an iterative fashion in the design process.

Designers, in addition to reasoning about the design itself (i.e. at the domain level) , also reason about the *process* they use to define the design (at the *meta*-level) (Stefik, 1981). A designer may have a plan for how the design itself will be created, made up of tasks like "collect requirements", "develop options", "perform trade study" and so on. If several design options are available, a designer may reflect on which option to select. If a conflict between two or more design goals and actions occurs, a fix for the conflict needs to be found. The design reasoning process is generally goal-driven, in the sense that actions are taken as part of strategies intended to achieve goals such as meeting a specification, refining a design option, making a control choice, resolving a design conflict and so on.

This generic model of design reasoning is based on classical Systems Engineering work (Blanchard, 1981) as well as AI models of artifact design ((Mcdermott, 1982), (Marcus, 1987; Brown, 1985; Mittal, 1986; Tong, 1987; Brown, 1984; Klein, 1991)) and planning ((Stefik, 1981) (Chapman, 1985; Faragher-Horwell, 1990)). These models have been applied successfully to a wide variety of domains including electrical, electronic, hydraulic and mechanical systems as well as software.

2.2. THE RATIONALE LANGUAGE

The DRCS rationale language captures design reasoning using a vocabulary of assertions consisting of *entities* like modules, tasks, specifications and versions as well as *claims* made about these entities. Claims come in two main types. A pre-defined vocabulary of *relation* claims is provided for the purpose of describing relationships between assertions; they can describe the design (e.g. module-1 *has-submodule* module-2, module-3 *has-attribute* attribute-1, attribute-1 *has-value* (> 350)) as well as the rationale for design decisions (e.g. value-1 *is-derived-from* procedure-1, claim-1 *is-supported-by* claim-2). In addition there is an all-purpose *text* claim that users can use to enter natural language text if what they wish to say can't conveniently be expressed using the existing entitiy and relation claim types. Any claim can serve as part of the rationale for another claim, so we can make claims about the design, claims describing rationale for design decisions, claims concerning why we should believe this rationale (or not) and so on recursively. The net result of describing design rationale in this way is a graph of entity and text claim instances connected by relational claims.

In the sections below, the vocabulary of claims and entities that make up the DRCS rationale language will be described. This discussion will be divided into five sections. As we have seen, design reasoning consists of defining/tracking design requirements (i.e. *evaluation*) as well as *synthesis*. Designers usually have some *intent* underlying a given action, as well as supporting *arguments* for believing that the action is the correct one to take. Since there are often multiple options for any decision, designers generally explore a space of multiple alternatives, or *versions*. Each of these five aspects will be considered separately. This division is made, however, purely for expository purposes. In actual use, of course, all aspects of the language can be used as appropriate at any time.

2.2.1. *Synthesis*. The DRCS language synthesis component is responsible for capturing the actions used to define artifacts and plans. Let us consider artifacts first (Figure 4):

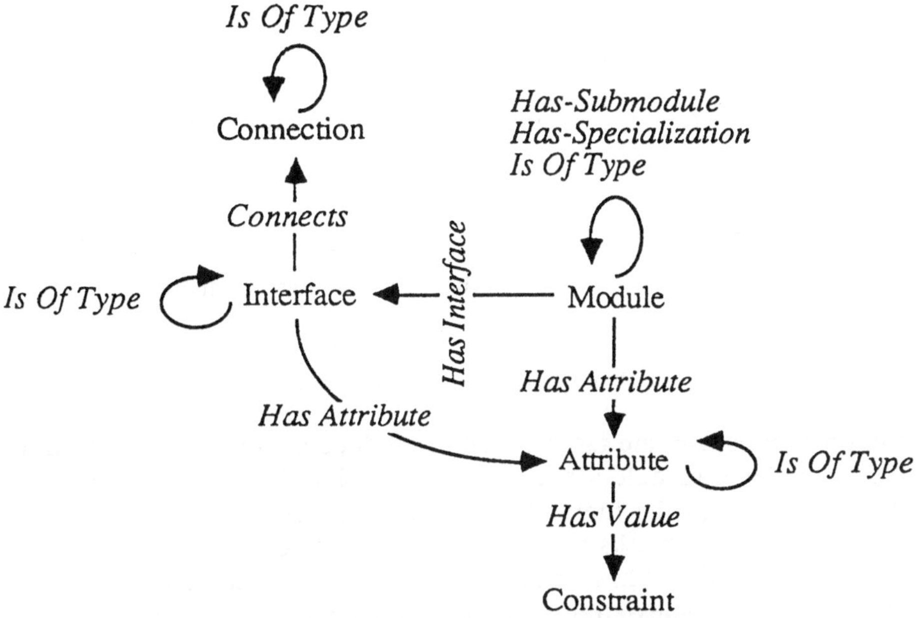

Fig. 4. Artifact synthesis entities and relationships.

In this figure, entities appear in plain font, while relation claims appear as directed arcs with italic labels; the latter signify that the given relationship can hold between the arc source and target. All entities that allow the "is-of-type" relation have an associated type taxonomy that a user can select from to define the entity type. This type taxonomy will in general include both generic and domain-specific elements and can be used to support useful computational services. This will be discussed in more detail below.

The basic entities for artifact description include modules, attributes, interfaces and connections. Modules can have submodules or specializations. Attributes can have values, expressed using a constraint language. Constraint languages are used to express indefinite descriptions, and have been used extensively in numerous design and planning systems (Stefik, 1981; Tong, 1987; Mitchell, 1985; Sussman, 1980). The DRCS constraint language provides a wide range of constructs including "absolute" constraints like inequalities, ranges and sets as well as "relational" constraints such as boolean and mathematical equations.

Plan descriptions are captured as follows (Figure 5):

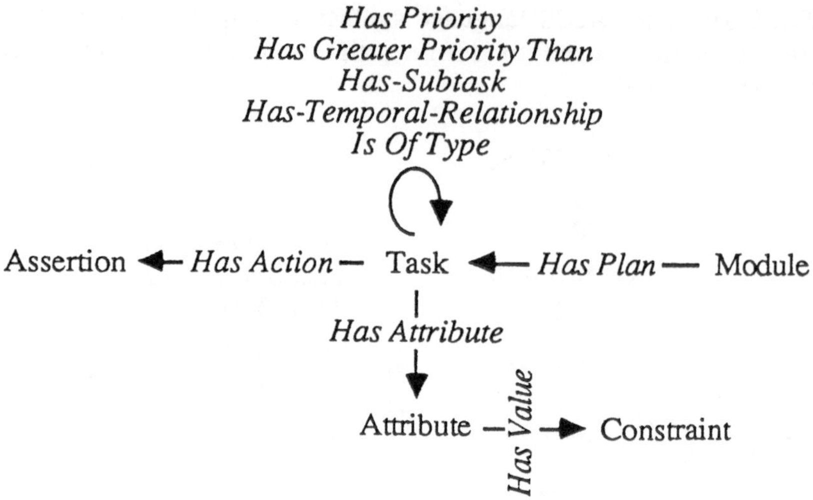

Fig. 5. Plan synthesis entities and relationships.

The basic entities here are tasks. Plans to produce an artifact are related to the artifact's top-level module via a "has plan" claim. Every plan is represented, as we have seen, as the hierarchical decomposition of a top-level task into temporally ordered subtasks with associated primitive actions. This is captured using "has-subtask", "has-action" and "has-temporal-relationship" (e.g. "comes-before", "comes-after") claims. Task actions can be any addition to the design/rationale description, i.e. any assertion. Plans can have priorities. For both artifacts and plans,"uses-resource" attributes are important; they describe use of a given resource type (e.g. time, weight, money, tools, people).

2.2.2. *Evaluation.* The evaluation component of the DRCS rationale language is used to capture design specifications as well as how well they have been achieved (Figure 6). Design and plan specifications are defined as desired values (using the constraint language) for design and plan attributes. Attributes and specifications can have different types, priorities and subsumption relationships. Attributes and specifications can have types. Specification types include objectives, requirements and preferences, and differ in how critical they are and thus by implication how willing we are to relax them. A design version can be said to achieve a given specification; precisely how is described in the rationale for that claim (see below).

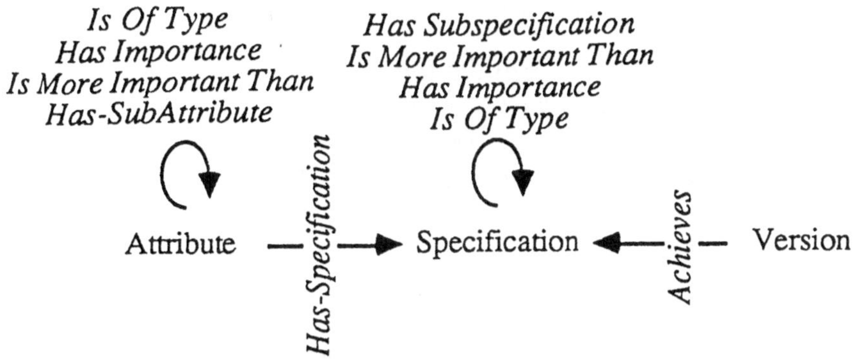

Fig. 6. Evaluation entities and relationships.

2.2.3. *Intent Model.* Whenever a designer takes some kind of design action, he or she is usually pursuing a strategy to find an answer for some problem; i.e. has some *intent* when taking that action (Figure 7):

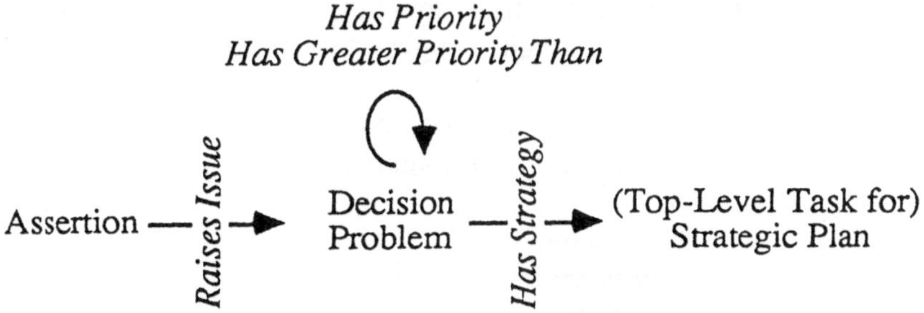

Fig. 7. Intent model entities and relationships.

Any assertion in a design description can raise a decision problem. There is, in fact, a pre-enumerated set of decision problem types; one for every relation defined for a given assertion type. There is, for example, a decision problem for the "Has Submodule" relation on modules (where the problem is determining how to decompose the module), a decision problem for the "Has Value" relation on attributes (where the problem is determining what the attribute value should be) and so on. Some decision problems can have greater priority than others. The strategy used to try to resolve a decision problem is represented as a "has-strategy" link to the top-level task of a plan.

2.2.4. *Versions Model.* The versions model is used to capture how the designer creates and explores the space of design alternatives (Figure 8).

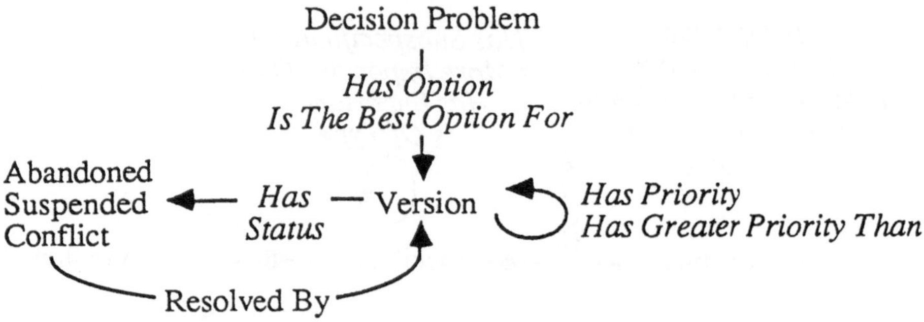

Fig. 8. Versions model entities and relationships.

New versions are created whenever a situation that calls for tentative decisions and/or exploration of multiple alternatives, i.e. when defining options for a decision problem. Every option for a given decision problem is asserted in a different version. The versions storing the options can have differing priorities as well as statuses. If a given version has the status "conflict", we can indicate which alternate version resolves that conflict. The preferred option for a decision problem can be represented using an "is the best option for" claim.

2.2.5. *Argumentation Model.* The purpose of the argumentation model is to allow one to describe the reasons for and against believing claims (Figure 9):

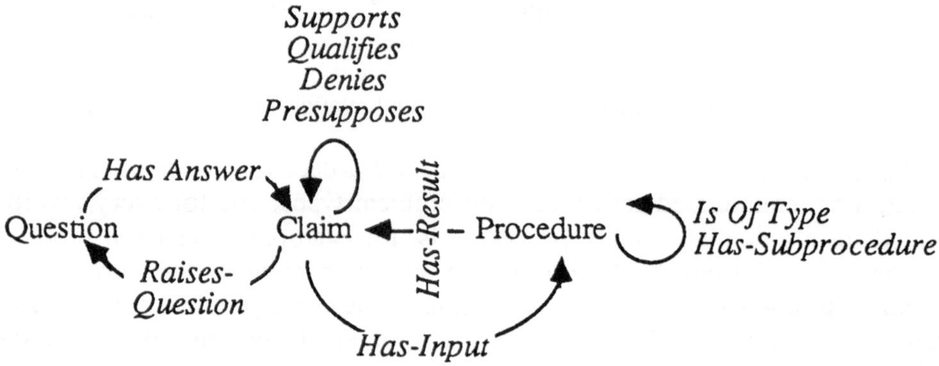

Fig. 9. Argumentation model entities and relationships.

The basic entities include claims (both relation and text claims) as well as procedures and questions. Claims can support, qualify, deny or presuppose one another. One can use the "has-result" and "has-input" claims to link claims to the procedures used to derive them and the inputs to those procedures; procedures can be mathematical equations or more unstructured information such as textual reference sources, handbooks, catalogs, standard engineering tables and so on. An individual can raise "questions" about the validity of a claim and assert that given claims answer these questions. Any synthesis, evaluation, intent, versions or argumentation claim can itself be the subject of argumentation claims.

Consider the following example of this language in use (Figure 10):

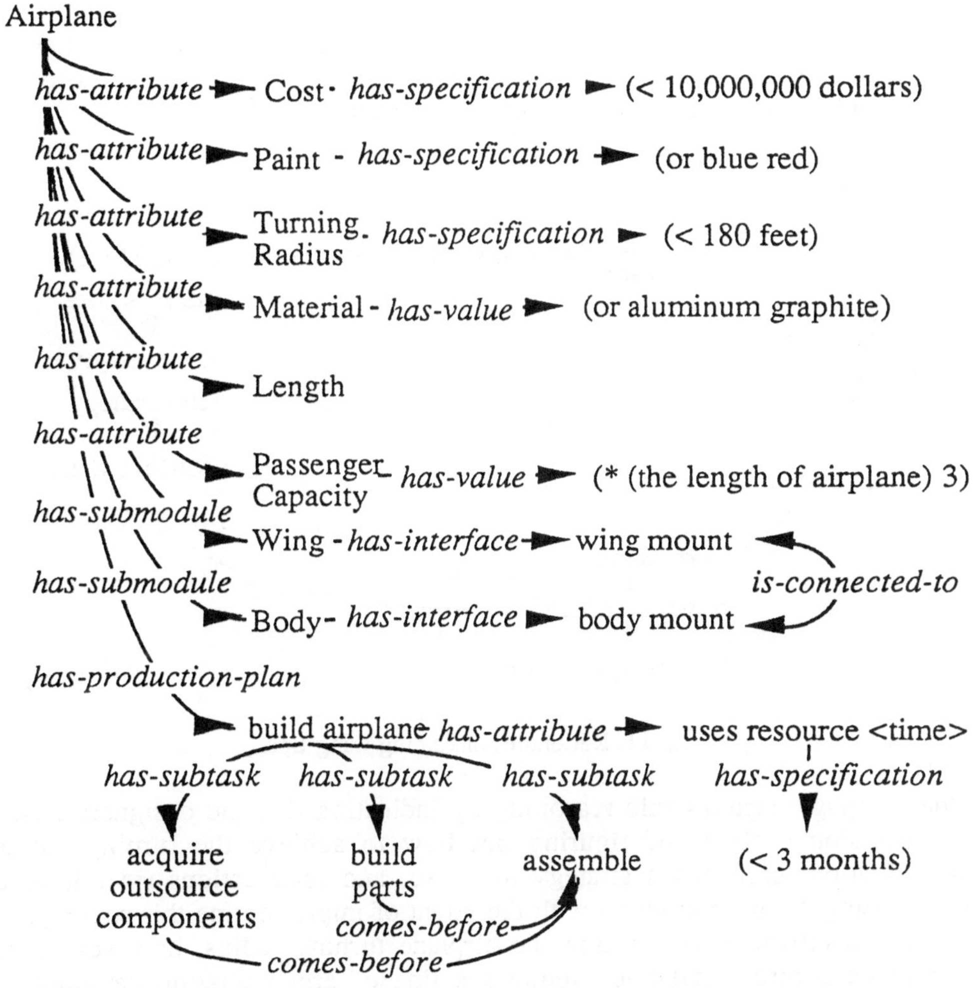

Fig. 10. Example of initial design specs and commitments.

Imagine some designers are working on the preliminary design for a new airplane. The designers begin by establishing the initial set of requirements and then perform some preliminary design. In this example, the designers want the final airplane to cost less than 10 million dollars, have a turning radius of less than 180 feet and so on. They have made some initial commitments; the plane will be made out of either aluminum or graphite, have a passenger capacity that is a particular function of its length, and consists of inter-connected wing and body submodules. The plane will be manufactured by a process that should take no more than 3 months per plane and consists of partially ordered subtasks including "acquire outsource components" etc.

As the design process continues the designers begin to address achieving the turning radius specification. One possible strategy for doing so is to use folding wings to reduce the airplane's turning radius (Figure 11).

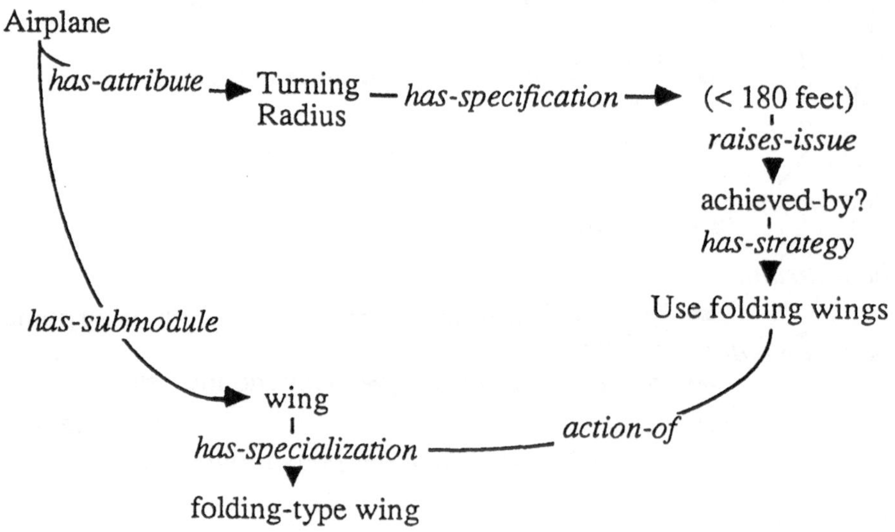

Fig. 11. Rationale for folding wing tips.

The language captures this reasoning by indicating that the designers raised the decision problem of figuring out how to achieve the turning radius specification, proposed a strategy to do so, and took actions (in this case specializing the wing module) with the intent of implementing this strategy.

The specification concerning the airplane turning radius, however, turns out to be controversial and produces a line of argumentation captured as follows (Figure 12):

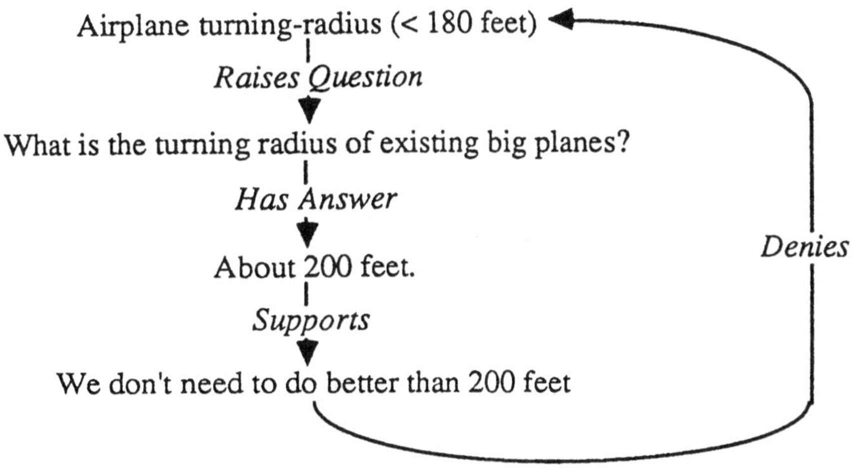

Fig. 12. Argumentation concerning the turning radius specification.

A designer wondered what the turning radius of existing big planes was; he goes on to claim that since existing big planes have as much as a 200 feet turning radius, this new plane need not do any better. Should this specification be replaced by a less stringent one, the designers can use the rationale graph to determine what derived decisions, such as the choice of folding wings, potentially need to be re-considered.

Design decisions can sometimes conflict with each other and lead to actions taken to resolve the conflict (Figure 13). At one point in the design process, designers are exploring the choice of aluminum as the airplane material and blue as the airplane skin paint. A designer asserts, based on reference to a published table of viable paint/material combinations, that this option is not viable, i.e. that these choices represent a conflict. Inspection of the rationale reveals that the choice of aluminum helps satisfy the requirement for a quick manufacturing process, but the paint choice has no particular support and another paint option is consistent with the initial specifications. They therefore choose to try a different paint color.

In the figure, rectangles represent versions; assertions believed in a version appear inside the corresponding rectangle. Options for decision problems (like finding the color for the airplane) are represented using "has-option" links to the version storing the assertions describing the option. An "achieved-by" claim links the specification for a quick manufacturing plan to the version that achieves it; this claim in turns provides support for the choice of the material aluminum. The logical support for the conflict claim is represented using a procedure that describes the inputs (choices of paint and material) as well as the way they support the conclusion (in this case via a table of viable paint/material combinations).

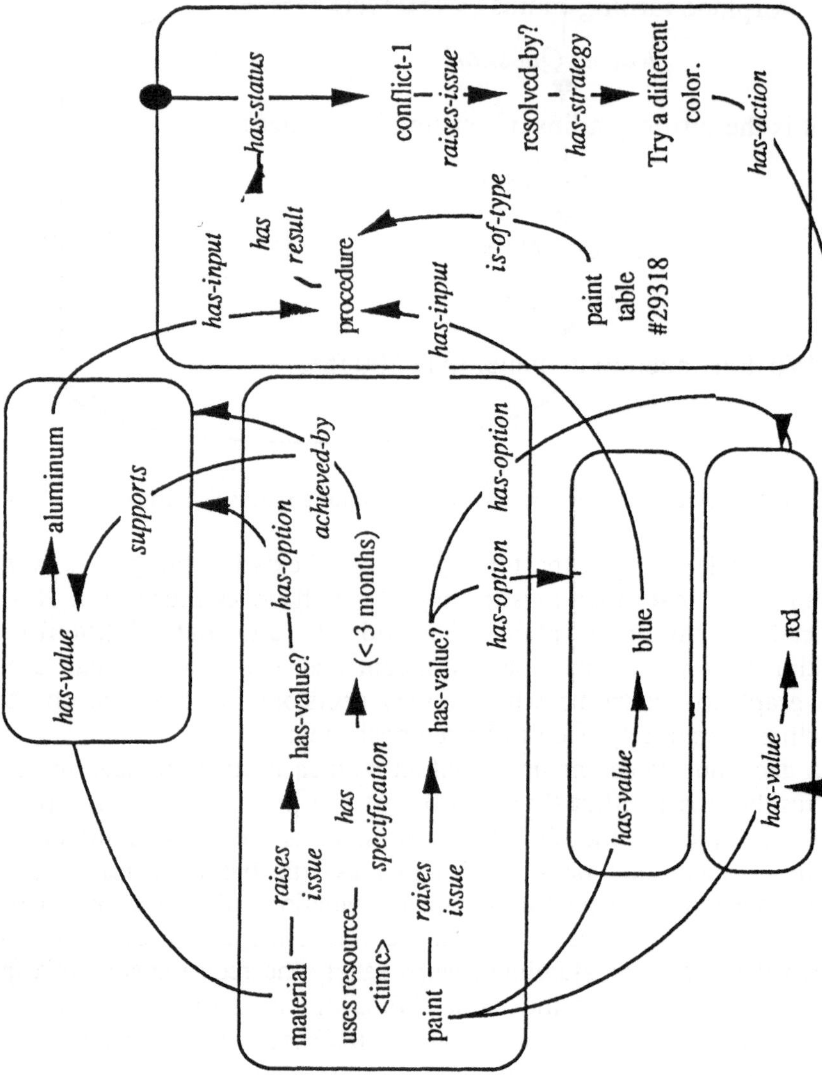

Fig. 13. Rationale for a conflict and its resolution.

These examples should help clarify how the DRCS rationale language can support useful services. Using this language one can, for example, search for all controversial design decisions underlying a given decision (by looking for underlying claims with many supports and denies claims), determine the consequences of withdrawing a design choice (by deleting all derived decisions without independent support), review the options explored for a given problem (by checking all the has-option claims stemming from the decision problem) and so on.

The DRCS design rationale language is essentially an extension (with substantial modification) of research to date in decision rationale capture, and is most closely related to DRL (Lee, 1991) and JANUS (Fischer, 1989). The essential difference between this and previous work is that DRCS integrates a rationale language with a generic design description language applicable to a very wide range of design domains. This integration allows *greater expressiveness* than that supported by previous work. In generic decision rationale languages the requirements, decision problems and options are described as natural language text that (in the absence of robust natural language understanding capability) is essentially opaque and allows few computational services. In DRCS, by contrast, these are captured using a structured language with explicit, computer-interpretable semantics. A version is described in terms of inter-connected modules and tasks. Specifications are described as a desired value (expressed using a constraint language) for an attribute. Decision problems are selected from a pre-enumerated set with known semantics. The information-theoretic content is thus sigificantly higher, but not at the cost of using a domain-idiosyncratic design representation such as that used in JANUS.

The more explicit semantics of DRCS makes *greater computational support* possible. Since decision problem and specification semantics are known, for example, it makes it much easier to fetch previous design cases which dealt with similar challenges. Since alternatives and their rationale are both represented in a structured way, we can more easily find the differences and similarities between candidate design versions; in addition to simply listing the differences in components, connections and so on we can uncover the points where the designs diverged and what the rationale for each choice was. Knowledge about entity types can be used to support conflict detection, classification and resolution (Klein, 1991). Acquiring new design knowledge by generalization from design rationale (Mitchell, 1985) is also likely to be facilitated.

In addition to making greater computational support possible, DRCS appears to be in some ways *more natural* than previous languages for describing design rationale. Rationale can be attached directly to the design aspect (module decomposition, attribute value etc) it refers to, rather than to a piece of text. DRL, for example, has to resort to the expedient of representing specifications as subgoals of decision problems rather than as desired values for module attributes. Rationale for design decisions based on "programmatic" concerns (e.g. keeping the design definition or manufacturing process from being too resource-intensive) can be captured as links between claims concerning the resource limits of a plan and claims describing the design artifact. DRCS incorporates a model of intent (based on (Wilensky, 1983)) absent from most decision rationale work; while DRL for example includes a "goal" entity it provides no way of linking goals to the strategies used to achieve them and the actions that implement the strategies.

3. The DRCS Interface

The DRCS system provides a graphical interface that allows users to interactively enter and view designs and their rationale using the language described above. The system is currently implemented as a stand-alone tool with simple built-in artifact and plan editors. For this rationale capture technology to be truly useful, however, it needs to be integrated into the tools designers actually use; as noted above, designers tend not to use rationale capture facilities that aren't an integral part of their usual design activity. The purpose of the DRCS system is thus to improve our understanding of how to write effective rationale capture interfaces; the lessons from this work will eventually be incorporated into rationale capture systems implemented as enhancements of design tools in actual or planned use. DRCS is currently implemented in Common Lisp on a Symbolics workstation.

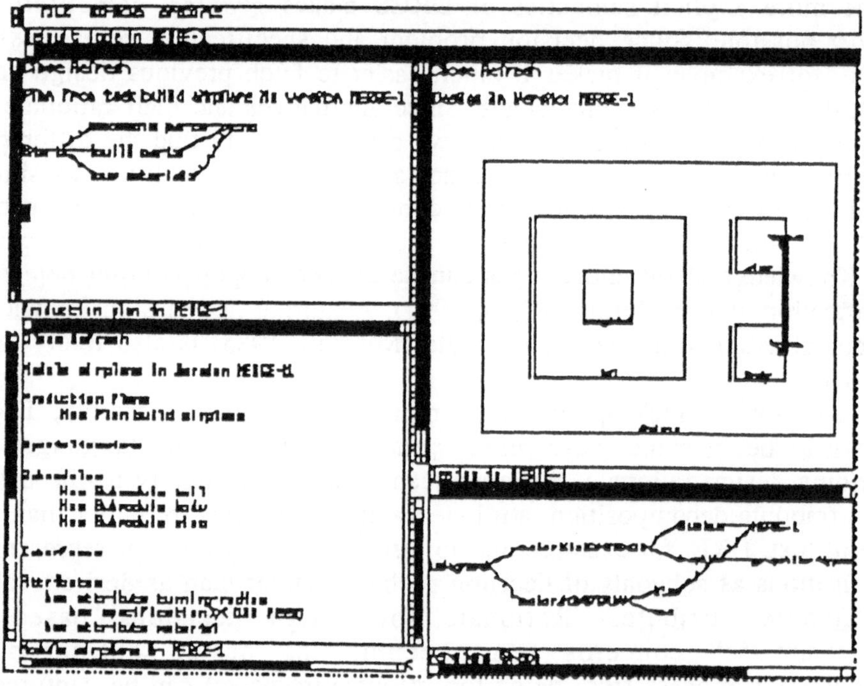

Fig. 14. Example of the DRCS interface in use.

The DRCS architecture consists of a design rationale database plus a direct-manipulation pointer, window and menu-based graphical interface like that familiar to users of Macintosh and Windows systems (Figure 14). The graph-based rationale representation underlying DRCS is expressive but not always the best form in which to view and edit designs. The interface accordingly allows the user to create windows that present the desired subset of the database using many different perspectives, each designed to highlight different aspects of the design/rationale description. One can create windows, for example, that display the current artifact design as rectangular modules with lines representing connections, show the ordering of tasks in a given plan as PERT charts, reveal the argument structure affecting a given claim as a graph, present the current set of versions as a lattice and so on. These windows dynamically update themselves whenever the subset of the rationale database they view changes, so they are continuously up to date. In addition to simply displaying the database contents in some pre-defined format, one can create instances of "analysis" windows which present useful analyses, such as pointers to circular or incomplete argument structures or questionable decision choices, that help the designer produce better designs.

The topmost window in the display contains the menubar; each item in that window creates a menu of options when selected. "File" menu options include saving the current state of the rationale database and reading in a previously saved example. The "Windows" menu allows one to create new windows or cycle through existing ones, while the "Special" menu allows one to create analysis windows. In this example, the user is currently viewing the versions graph (lower right), the artifact design in one of the versions (upper right), a description of one component in that design (lower left) and a description of the plan used to produce that component (upper left).

The printed representation of every entity and claim is mouse-sensitive; a menu of options that make sense in the current context for that assertion appears when it is clicked on using the pointing device. There are two classes of options for any assertion; perspective creation and editing. Perspective creation options create windows that view the assertion from a given perspective. When clicking on the top-level task of a plan, for example, one can create windows that view it either as temporally-ordered leaf tasks or as a task decomposition hierarchy. Editing options allow one to update the design rationale database; for any assertion the menu will include a list of all the types of claims one can make about that assertion. To add an attribute to a module, for example, one clicks on the module and selects the "Define Attribute" option; the user is then prompted for an attribute name and an instance of that attribute connected to the module via a "has-attribute" claim is created. To connect two module interfaces, one selects the "Make Connection" option for one interface and then selects the interface to connect to. To add support for a claim one selects the "Is Supported By" option for that claim and then either selects an existing claim or creates a new one; the appropriate "supports" relation between these claims is automatically created.

The DRCS interface builds upon previous work in this area such as gIBIS (Conklin, 1987) and SIBYL (Lee, 1990). The key difference between DRCS and these systems is that DRCS integrates design capture of both physical and temporal (plan) artifacts with rationale capture in a single tool. Users thus have no need to switch tools when describing the design as opposed to its rationale, can attach rationale directly to the design claims of interest, and can focus their efforts on describing rationale that the evolving design description reveals as critical.

4. Future Directions

DRCS has been used successfully to describe design rationale for a variety of simple new designs as well as re-represent information for more complicated ones; based on this experience we feel that the basic viability of this approach has been substantially validated.

There is, however, a rich range of possibilities for future growth. The rationale capture language needs to be augmented, e.g. to capture geometric information (by incorporating a feature-based geometric representation) and tentative or "fuzzy" argumentation (Lee, 1990; Pearl, 1982; Fedrizzi, 1988). DRCS currently is realized as a stand-alone single-user prototype; to be truly useful it needs to be realized as a multi-user augmentation of tools used by designers in their daily work.

The critical issue of high rationale capture overhead has to be addressed. Two basic approaches have been taken; reducing the burden of expressing rationale (Lakin, 1989; Boose, 1991; Baudin, 1989; Klein, 1991; Mark, 1990; Gruber, 1991) and "rewarding" the users for their trouble by providing useful services (Lee, 1991; Mark, 1990). We are considering providing natural language understanding capabilities to allow users to describe rationale as English text. This is less daunting than one might imagine because (1) only a limited subset of English is likely to be needed to express design rationale, and (2) the current context can help reduce the semantic ambiguity of natural language text; if a user has clicked on an attribute value to describe it's rationale, for example, the referent of the phrase "this value" or "this attribute" is easy to determine. In general, the attempt will be made to maximize the ability of the rationale capture tool to "fill in the gaps" and provide useful services to its users.

Acknowledgements

I would like to gratefully acknowledge the contributions of Ted Kitzmiller, Mike Anderson and Ahmed Zayan in the development of the design rationalelanguage and rationale capture interface described herein. Ted provided valuable feedback on this paper. I also benefited from the comments of many of my other colleagues at Boeing Computer Services.

References

Balzer, R.: 1984, Capturing the design process in the machine, *Rutgers Workshop on Knowledge-Based Design Aids.*

Baudin, C., Sivard, C. and Zweben, M. A.: 1989, Model-based approach to design rationale conservation, *IJCAI-89 Workshop on Model-Based Reasoning.*

Blanchard, B. S. and Fabrycky, W. J.: 1981, *Systems Engineering and Analysis,* Prentice-Hall, Englewood Cliffs, NJ.

Boose, J. H., , J. M. B. and Shema, D. B.: 1991, A decision support system for automating engineering trade studies, *Hawaii International Conference on System Sciences.*

Brown, D. C. and Chandrasekaran, B.: 1984, Expert systems for a class of mechanical design activity, *in* J. S. Gero (ed.), *Knowledge Engineering in Computer-Aided Design,* North-Holland, Amsterdam, pp.259-282.

Brown, D. C.: 1985, *Failure Handling In A Design Expert System,* Butterworth.

Chapman, D.: 1985, Planning for conjunctive goals, *Technical Report*, Massachusetts Institute of Technology.

Conklin, J. and Begeman, M. L.: 1987, gIBIS: A hypertext tool for team design deliberation, *Proceedings of Hypertext 87,* pp. 247-251.

Faragher-Horwell, R., Murphy, A. R., Nguyen, T. P., Purdon, D. J., Small, D. H. and Sharma, K. J.: 1990, Automated Process Planning (APP) domain independent shell design issues: Year End Report, *Technical Report BCS-G2010-98*, Boeing Proprietary, The Boeing Company, March.

Fedrizzi, M., Kacpryzk, J. and Zadrozny, S.: 1988, *An Interactive Multi-User Decision Support System For Consensus Reaching Process Using Fuzzy Logic With Linguistic Quantifiers,* Elsevier Science Publishers, Vol. 4, pp. 313-327.

Fischer, G., McCall, R. and Morch, A.: 1989, Design environments for constructive and argumentative design, *Proceedings of CHI.*

Fischer, G., Lemke, A. C., McCall, R. and Morch, A. I.: 1991, Making argumentation serve design, *Journal of Human Computer Interaction.*

Gruber, T. R. and Cohen, P. R.: 1987, Design for acquisition: Principles of knowledge-system design to facilitate knowledge acquisition, *IJMSS 26*, 143-159.

Gruber, T. R.: 1991, Learning why by being told what, *IEEE Expert.*

Kleer, J. D.: 1985, Choices without backtracking, *AAAI-85*, pp.79-85.

Klein, M. and Lu, S. C. Y.: 1991, Insights into cooperative group design: experience with the LAN designer system, *in* G. Rzevski and R. A. Adey (eds), *Proceedings of the Sixth International Conference on Applications of Artificial Intelligence in Engineering (AIENG '91),* University of Oxford, UK, pp.143-162.

Klein, M.: 1991: Supporting conflict resolution in cooperative design systems, *IEEE Systems Man and Cybernetics* **21**(6).

Lakin, F., Wambaugh, J., Leifer, L., Cannon, D. and Sivard, C.: 1989, The electronic design notebook: Performing medium and processing medium, *Visual Computer* **5**(4), 214-226.

Lee, J.: 1990, Sibyl: A tool for managing group decision rationale, *Proceedings of CSCW 90*, pp. 79-92.

Lee, J. and Lai, K. Y.: 1991, What's in design rationale?. *Human-Computer Interaction.*

MacLean, A., Young, R. M., Bellotti, V. and Moran, T.: 1989, Design rationale: The argument behind the artifact, *Proceedings of CHI*, Austin TX.

MacLean, A., Young, R., Bellotti, V. and Moran, T.: 1991, Questions, options and criteria: elements of a design rationale for user interfaces, *Journal of Human Computer Interaction: Special Issue on Design Rationale.*

Marcus, S., Stout, J. and McDermott, J.: 1987, VT: An expert elevator designer, *Artificial Intelligence Magazine* **8**(4), 39-58.

Mark, W. and Schlossberg, J.: 1990, Design memory, *Proceedings of the Knowledge Acquisition for Knowledge-Based Systems Workshop.*

McCall, R.: 1987, PHIBIS: Procedurally heirarchical issue-based information systems, *Proceedings of Conference on Planning and Design in Architecture.*

Mcdermott, J.: 1982, R1: A rule-based configurer of computer systems, *Artificial Intelligence* **19**, 39-88.

Mitchell, T. M., Steinberg, L. I. and Shulman, J. S.: 1985, A knowledge-based approach to design, *IEEE Transactions on Pattern Analysis and Machine Intelligence PAMI-7*, **5**, 502-510.

Mitchell, T. M., Mahadevan, S., and Steinberg, L. I.: 1985, LEAP: A learning apprentice for VLSI design, *Proceedings of IJCAI*, IJCAI, pp. 573-580.

Mittal, S. and Araya, A.: 1986, A knowledge-based framework for design, *American Assocation of Artificial Intelligence*, pp. 856-865.

Mostow, J. and Barley, M.: 1987, Automated reuse of design plans, *Proceedings ICED*, IEEE, August, pp. 632-647.

Newman, S. and Marshall, C.: 1990, Pushing Toulim Too Far: Learning from an argument representation scheme, *Technical Report*, Xeroc PARC.

Pearl, J.: 1982, Reverend Bayes On Inference Engines: A distributed hierarchical approach, *AAAI-82*, pp. 133-136.

Potts, C. and Bruns, G: 1988, Recording the reasons for design decisions, *Proceedings of the 10th International Conference on Software Engineering*, pp. 418-427.

Stefik, M. J.: 1981, Planning with constraints (Molgen: Part 1 & 2), *Artificial Intelligence* **16**(2), 111-170.

Steinberg, L. and Mitchell, T.: 1984, A knowledge-based approach to VLSI CAD: The redesign system, *21st Design Automation Conference, IEEE.*

Sussman, G. J. and Steele, G. L.: 1980, Constraints—A language for expressing almost-hierachical descriptions, *Artificial Intelligence* **14**, 1-40.

Thompson, J. B. and Lu, S. C. Y.: 1989, Representing and using design rationale in concurrent product and process design, *Proceedings of the Symposium on Concurrent Product and Process Design, ASME Winter Annual Meeting*, pp. 109-115.

Tong, C.: 1987, AI in engineering design, *Artificial Intelligence in Engineering* **2**(3), 130-166.

Wilensky, R.: 1983, *Planning And Understanding*, Addison-Wesley, Reading.

Yakemovic, K. C. B. and Conklin, E. J.: 1990, Report on a development project use of an issue-based information system, *CSCW 90 Proceedings*, pp. 105-118.

TOWARD A KNOWLEDGE MEDIUM FOR COLLABORATIVE PRODUCT DEVELOPMENT*

T. R. GRUBER
Stanford Knowledge Systems Laboratory
701 Welch Road, Building C
Palo Alto CA 94304 USA

J. M. TENENBAUM
EIT, Incorporated
459 Hamilton Ave, Suite #100
Palo Alto CA 94301 USA

and

J. C. WEBER
Lockheed AI Center
3251 Hanover St.
Palo Alto CA 94304 USA

Abstract. Information sharing and decision coordination are central problems for large-scale product development. This paper proposes a framework for supporting a *knowledge medium* (Stefik 1986): a computational environment in which explicitly represented knowledge serves as a communication medium among people and their programs. The framework is designed to support information sharing and coordinated communication among members of a product development organization, particularly for the tasks of design knowledge capture, dynamic notification of design changes, and active management of design dependencies. The proposed technology consists of a shared knowledge representation (language and vocabulary), protocols for foreign data encapsulation and posting to the shared environment, and mechanisms for content-directed routing of posted information to interested parties via subscription and notification services. A range of possible applications can be explored in this framework, depending on the degree of commitment to a shared representation by participating tools. A number of research issues, fundamental to building such a knowledge medium, are introduced in the paper.

1. Introduction

1.1. The Need for Knowledge Sharing and Communication Coordination in Cooperative Product Development

Product development is a knowledge- and communication-intensive process. For collaborative product development, team members need to have access to the knowledge underlying decisions by other members. They also need to assess the impact of their decisions on each other, and notify the affected parties in an appropriate way. Effective communication is especially important whenever anything affecting an existing design decision changes. In product development, something is always changing — perhaps a design requirement, an unanticipated simulation or

* This work was supported by DARPA prime contract DAAA15-19-C-0104 through Lockheed subcontract SQ70A3030R, monitored by the U. S. Army Ballistic Research Laboratory.

J. S. Gero (ed.), Artificial Intelligence in Design '92, 413–432.

test result, the availability of a component, or an improvement to the manufacturing process. Reacting quickly to such changes is essential for quality and productivity.

While computers are used extensively in product development, existing tools do little to facilitate information sharing and coordination. If anything, they aggravate the problem by isolating information at tool boundaries. Most computer tools support particular tasks in engineering (e.g., geometric modeling, analysis), manufacturing (e.g., process planning, scheduling) or business (e.g., cash flow analysis). Often, the principal output of one tool is a piece of paper that is mailed or faxed to team members in other departments. Those individuals must then re-enter the relevant information in the format required by their tools, resulting in team members being cut off by their tools. They make decisions on the basis of inconsistent or out of date information. Moreover, only information concerning the artifact is generally available; critical information about the decisions leading to design choices and their underlying rationale is seldom ever captured much less communicated.

1.2. THE ROLE OF A KNOWLEDGE MEDIUM

Imagine a computational environment in which explicitly represented knowledge serves as a communication medium among people and their programs. Such an environment is called a knowledge medium (Stefik 1986). Information exchange and decision making of team members is mediated by a shared knowledge base and interaction protocols. On-line tools, such as CAD/CAM, document editors, and electronic mail interfaces interact by posting information about different aspects of the design to a shared environment, in a common representation. The interactions among shared objects are represented with new knowledge structures that span the applications and engineering disciplines. Formal behavioral models are related to simulation demonstrations; design decisions to requirements; text and audio annotations to formal schematics. The relationships among shared objects range from unspecified dependency ("object X is in some way dependent on object Y") to precise analytic constraints ("$Y = 2x + 3$") and computer-generated explanations of causality ("X caused Y to fail").

When something in a product knowledge base changes, automatic analysis determines how the change relates to other product data, either by following explicit links or by inference. The change in a parameter triggers a spreadsheet-like re-calculation of dependent parameter values in an equation model. The change in a different parameter invokes a rule-based constraint checker to verify that the value is within the allowable range. When dependencies are not captured precisely, the shared environment can notify all designers that the changes potentially affect and the designers make the determination of if and how to respond.

1.3. SHADE: A Representational Framework for Design Knowledge

As a start toward this vision, we are developing a representational framework called SHADE[1] to support the sharing of knowledge about designs and the coordination of information about design changes to interested parties. By *representational framework* we mean a common knowledge representation (language and vocabulary) and software interaction protocols designed to support a set of knowledge-based services. SHADE will provide the infrastructure in which the information created and manipulated by heterogeneous engineering tools can be interrelated and coordinated. Such tools include current and future word processors, CAD drawing tools, model editors, simulators, constraint checkers, and electronic mail. The target services include design knowledge capture, dynamic notification of design changes, and active management of design dependencies.

SHADE is at the core of next generation concurrent engineering systems under development at Stanford and Lockheed. The framework is expected to lead to a new generation of tools such as electronic design notebooks for recording and retrieving engineering knowledge, selective dissemination of design decisions and changes, and explanation and validation of design requirements.

The project to develop the SHADE representation and associated computational environment has only recently begun. In this paper we illustrate the intended functionality in the context of engineering redesign (Section 2), then specify the requirements (Section 3), describe the technical approach (Section 4), characterize the important research issues (Section 5), and briefly discuss the research methodology (Section 6).

2. A Scenario of Use

This section will describe a scenario in which an engineer is working in the proposed SHADE environment supporting cooperative product development. The first part of the scenario is a series of interactions between a designer and current, state of the art, engineering tools. The second part describes a hypothetical extension in which SHADE representations and services provide assistance in a collaborative setting.

2.1. Today: Eclectic Use of Isolated Design Tools

The setting is an engineer redesigning a planar manipulator, a two-fingered robot manipulator that moves a payload around in a planar workspace. The engineer is working with editors for two design descriptions: the textual requirements document and a component/connection graph of the power supplies, motors, and linkages comprising the electromechanical subsystem. The requirements document contains a section marked with a change bar. The marked text prescribes an

[1] for SHAred Dependency Engineering

increase in the maximum payload weight over the previous design. The engineer is tasked with modifying the design to accommodate the changed requirement.

The engineer starts by simulating the manipulator. She calls up a case library of simulation scripts configured for the previous design. Looking at the titles, she selects one labelled "demonstration of planar manipulator under maximum loading conditions," and runs it. The simulation system shows a graph of the position and velocity of a series of payloads under a standard test trajectory. The manipulator works fine. Then the designer adjusts the mass parameter of the payload and runs the simulation again. The simulation shows the manipulator slowing down and then failing to complete the test trajectory with the heavier payload. This is the behavior that the redesign is to fix.

To accommodate the heavier payload, she decides to replace the present motor in the planar manipulator with a bigger one. She calls up the new motor from a library, and using her component/connection editor, replaces the old one. She selects the component in her graphical interface, and then gives it an annotation: "bigger motor needed for heavier payload requirement." She then links the annotation to the previously circled requirement in the document using a mouse gesture.

She now runs the simulation using the new design, and verifies that the new motor behavior achieves the required behavior. Using a menu she saves the script and labels it with "new motor handles heavier payload." Using the mouse she graphically links this annotation to the circled requirements text and the motor icon in the component/connection editor.

This scenario is not typical of current practice, but it is completely plausible to implement with existing technology. None of the individual tools are new: the document editor, the component-centered CAD editor, the simulation system with model library. The user interface technology also exists: selecting arbitrary objects with the mouse, drawing hyperlinks between objects, attaching textual annotations to links. What is missing is the infrastructure.

2.2. TOMORROW: LOOSELY-COUPLED TOOLS IN A COLLABORATIVE ENVIRONMENT

Consider an extension to the scenario in which the designer is not working alone, but with others in an organization. When the designer selects objects in editors, runs simulation programs, and links objects and data with annotated links, the environment records these design activities in a permanent record called a design history. The system indexes the history by the selected objects and data. This history is accessible by other engineers later in the product lifecycle, and it persists after the original designers become unavailable.

One of these downstream engineers is trying to plan the manufacture of the artifact and discovers that the motor specified in the design is no longer available from the suppliers. Since she will have to substitute another motor, she needs to know why the specified motor was chosen. So she looks into the design history, keying the search on the motor part number. She is shown the information manipulated by the original designer: the changed requirement, the component/connection graph, the manufacturer's data sheet, and the related simulation results. This helps her find a compatible replacement by telling her something about the role of the previous motor in the overall design.

Consider a further extension to the scenario, in which engineers work *concurrently* and exchange data and share models as they are created and refined. Now the design knowledge environment is an active medium, *notifying* interested parties of relevant changes to the distributed design knowledge. In this scenario, when the designer replaces the motor, the change triggers a set of notifications sent to tools used by other engineers. One is for the manufacturing engineer responsible for milling the frame for the manipulator. After being notified of the motor change, she notices that the new motor has a larger shaft diameter. She acts on this information by increasing the size of the hole in the CAD specification of the mounting bracket. Then she invokes her process planning tool to verify that the specified bore hole is within manufacturability limits, and produces a new plan for machining the bracket. If the process planning tool uncovers some manufacturability problems, then appropriate designers can be notified immediately.

3. Requirements for a Representational Framework

Support for the kinds of services suggested in the scenario imposes an interesting set of requirements for a representational framework.

First, SHADE must provide a representation for shared design knowledge. The representation must be shared in the sense that participating tools can assert and exchange information within a common conceptualization. For example, a service for capturing relationships among a requirement, a motor selection, and a simulation result needs a representation for these classes of information, and a mapping to the data available from tools. The representation must be expressive enough to capture domain-specific knowledge — not just data – used by participating tools. Simulation models, manufacturability constraints, and data dependency graphs are examples of knowledge to be represented.

Although the shared representation must be able to incorporate many kinds of engineering knowledge, this doesn't imply that the representation must be capable of expressing the union of all distinctions made by the participating engineering tools. In fact, the challenge is to support degrees of shared knowledge from minimal sharing to strong common models.

The second requirement is closely related. SHADE must provide a means to represent the relationships among "foreign" data in heterogeneous formats from very different tools. In the scenario, the designer related a variety of sources of information. The design requirements were selected in text editors as strings, the model for the motor was selected from a component library and inserted as an icon into an editor, then incorporated as a behavior model in a simulation. The framework needs some way to link these data in a globally consistent way.

One option is a common data format, standardized down to the data structures. This demands a huge commitment from the developers of the participating tools, and is difficult to achieve outside of tightly integrated product lines. Instead, we reject any requirement that the fine-grained structure of all design elements be "standardized" into a common data model. All that is required is the ability to *encapsulate* design elements with a identifier and some characterization of what the element denotes. For example, a single design move might be reflected in a complex series of transactions on a CAD model, resulting in a new shape for a part. For the purpose of representing the *relationship* between the change in the geometry and other elements such as the dynamical behavior and manufacturing cost, only an identifier of the design move need be represented in the shared knowledge base. The geometric details are left to the geometry model supported by the CAD editor. The contents of the encapsulated design element is said to be *opaque* to the shared representation.

Third, as a knowledge medium for human members of product development teams, the representation must support both structured information that is interpretable by computers and unstructured information that may only be meaningful to people. Examples of the former include behavior equations and netlist descriptions of circuits. Examples of the latter include E-mail discussions, pages from a handbook containing text and diagrams, or animated displays produced by a simulation program. In the scenario, the justification of the design decision included a mix of machine-interpretable data (e.g. replayable simulations) and human-only information (e.g., annotations, informally specified requirements). A representation for such as mix of objects is called *semi-structured* (Malone, Yu and Lee 1989).

Fourth, the framework must provide mechanisms for content-based information routing and automatic invocation of demons in response to changes in the data. In content-based routing, when a piece of information is made available to the shared environment, it is delivered to interested consumers on the basis of the information content. The mapping of content to addressee is based on a specification of interest called a *subscription.* This is the kind of routing provided by electronic "news clipping" services, typically on the basis of keyword indexing. The framework must provide the language for specifying interest, the mechanisms for implementing the mappings, and the protocols by which participants post information for distribution. Similarly, it must establish a protocol for associating changes in the data with the activation of "demons" (programs).

4. Elements of the SHADE Technology

SHADE is designed to achieve the aforementioned requirements and to enable the kinds of computational support illustrated in the scenario. There are three basic parts to the SHADE approach. First, design knowledge in the shared environment is specified and exchanged using a *shared knowledge representation.* This is a standard syntax and basic domain vocabulary, in a reusable form, along with a knowledge base that is accessible by all of the participating engineering tools in the framework.

Second, knowledge is exported from engineering tools into the shared environment via a protocol for encapsulation and publication. Encapsulation transforms tool-specific data into a form that is in conformance to the common knowledge representation. Publication is the process of posting new information, such as changes to a design, in way that can be monitored by demon mechanisms.

Third, the content-based routing of information among participating tools is achieved with a protocol for *subscription* (specifying interests) using the shared knowledge representation as a specification language and the *propagation* of new information to interested parties based on their subscriptions. A SHADE subscription looks like a query to a deductive database; it is an expression with some free variables to be instantiated by the desired information. The path of information flow from producer to consumer is determined according to the contents of the information packets (i.e., what the expressions mean) rather than by pre-arranged addressing.

These are the essential commitments of the approach. Let us now examine some important details.

4.1. A Shared Knowledge Representation for Specifying and Exchanging Design Knowledge

At the heart of SHADE is a representation for knowledge about a designed product. The knowledge may come from any of several sources, including human engineers and varieties of software tools. The representation formalism must be very expressive to capture the kinds of intra-tool relationships motivated by the scenario (e.g., design decisions involving requirements, structural modifications, simulation results, and textual annotations). Analyses of information demands and usage of designers in several domains support this need to represent many kinds of design knowledge from many sources (Gruber and Russell 1992, Kuffner and Ullman 1990). In addition, statements in the representation will be interpreted by humans and by tools that do not use the representation internally. For these reasons we have chosen to base the SHADE representation on a declarative language for first-order logic called KIF (Genesereth, Fikes and et al. 1992). KIF is a proposed specification for a standard Knowledge Interchange Format, or Interlingua. It has the desired properties of interpreter-independence (it is based on well-defined semantics which do not assume any particular inference procedure) and expressiveness (it is sufficient to represent anything that can be stated in first-order logic, and also allows statements to refer to other statements as terms). KIF, however, only provides the basic syntax and formal semantics.

The bulk of the shared representation is a *common ontology* of design knowledge. In philosophy, an ontology is a systematic account of existence. In the context of AI systems, we identify the ontology with the set of formal terms with which one represents knowledge (which determine what can exist for the program). Practically, then, a common ontology is the vocabulary for representing the knowledge needed for some purposes in some domain. It is a dictionary of classes, relations, functions, and object constants, along with their definitions in human-readable text and machine-interpretable KIF sentences. In SHADE we are using a system called Ontolingua (Gruber 1992) for developing and maintaining the ontologies.

One can think of the design knowledge ontology as an evolving theory of distinctions that are worth representing in a *shared* design knowledge base (not an arbitrary design tool). The SHADE project is tasked with producing an ontology for design knowledge for the purposes described earlier: design knowledge capture, dynamic notification of design changes, and active management of design dependencies. Much of the intellectual foundation has been laid, and we are drawing from representation work in design automation (Williams 1991), decision support (Boose, Shema and Bradshaw 1991), design rationale (Lee and Lai 1991a), dependency management (Lubars 1991, Petrie 1991), applied mathematics (Givan, McAllester and Zalondek 1991, Wolfram 1991), behavior modelling (Falkenhainer and Forbus 1991, Genesereth 1990, Palmer and Cramer 1991, Ziegler, Elzas and Oren 1989), and the representation of everyday human experience (Lenat, Guha, Pittman, Pratt and Shepherd 1990).

We plan to develop a basic ontological foundation and then extend it in the directions indicated by how it is used in practice by engineers. The initial SHADE ontology (based on (Gruber and Russell 1990) and an informal study of group ontology development) will include classes for behavioral descriptions, constraint expressions, design decisions, dependencies, design criteria, design moves, design parameters, functional descriptions, requirements, and structural descriptions. These classes reflect distinctions that have some utility for the design rationale capture and change notification tasks, and which can be naturally mapped onto existing engineering tools. With experience and the addition of participating tools, SHADE's ontology of design knowledge will be extended to integrate simulation and synthesis models, engineering data from on-line databases and libraries, and models of engineering processes throughout the product lifecycle.

The language and vocabulary are used to create a *shared knowledge base* (SKB) of information about the relationships among changing data in the participating tools. The domain of the SKB is *not* all possible information about the design (i.e., the union of all the tools' domains), but how the data produced and consumed by tools are related. The shared knowledge base serves as a common carrier for information communicated by participating tools, conceptually like a shared memory or bus. However, the SKB need not be centrally located or permanently stored; it may actually be distributed among the knowledge bases of the participating tools. This is the subject of ongoing work beyond the scope of this paper.

What is in the shared knowledge base? Consider, for example, the first scenario. When the designer graphically links the requirement text, the new motor, and the simulation, she is actually instantiating a template for recording design decision information. The relationship between these elements would be asserted into the SKB using sentences like the following. The sentences formally relate objects representing the requirement text, motor, and simulation scenario, using a design-decision object.[2]

```
(instance-of dd-1 design-decision)
(instance-of rt-1 requirement-text)
(mentioned-requirements dd-1 rt-1)
(instance-of sc-1 structural-component)
(relevant-structures dd-1 sc-1)
(instance-of ss-1 simulation-scenario)
(relevant-behaviors dd-1 ss-1)
```

That is, the design decision `dd-1` is like a "frame" that captures a systematic relationship among the requirement `rt-1`, the device component `sc-1`, and the results of a simulation, `ss-1`.

All tools that participate in the SHADE environment have access to the contents of the SKB, and can assert to it. The mapping from tool representations to the shared representation is the topic of the next subsection.

4.2. A Protocol for Encapsulation and Publication

The second component of the SHADE framework is the protocol by which loosely coupled, heterogeneous tools can export (publish) information to the SKB. The protocol is constrained by an interesting set of requirements:

- Each tool has its own internal representation and data model.
- Only some of the information processed by the tools is machine- interpretable (e.g., equations are; natural language text and bitmap images are not).
- Only some of the data can even be *represented* in the shared ontology (because the vocabulary is always incomplete).

The SHADE approach to publication is based on a few simple ideas. First, "foreign" data from the tools can be organized in terms of objects and sentences about the objects. For instance, in structure editors the objects might be components, and in a word processor, the objects might be text fragments selected by the user. Properties or attributes of objects, and their relationships to other objects, can be stated as sentences referring to the objects. Second, although there is a finite set of persistent objects in an design environment, the number of sentences one can think up about these objects is unlimited. Therefore, in the context of an arbitrary collection of independently designed tools, it is possible to share a database of objects, but not to share a complete knowledge base of sentences about them. We say that such objects are *opaque*, because the shared KB can represent the identity

[2] This example is only meant to convey the style of representation. The actual ontology for representing design decisions is under development.

of the object, but not its internal details. The fourth idea unifies the problem of representing foreign data with the problem of incorporating semi-structured data. Whether the object is a formal geometric entity, fully interpreted by its originating CAD tool, or a semi-structured videotape of a simulation, interpretable only by human observers, the same opaque-object abstraction can be applied. In other words, SHADE can encapsulate an entire video as if it were a foreign object coming from a CAD tool.

The SHADE publication protocol follows naturally from this analysis. The details are under design, but the basic policy for encapsulation and publication mechanism is as follows.

- Each tool exports facts about objects it manages. The objects are encapsulated as *handles* to the shared environment (i.e., identifiers that the tool associates with persistent internal data structures). The publication protocol does not specify *which* objects and facts to export (see Section 4.3).
- Opaque objects are denoted by unique names in the shared representation. SHADE is responsible for assigning identifiers from the global namespace to object handles given by exporting tools. The unique names are formal object constants in KIF.
- The unique identifiers in the shared knowledge representation denote the objects in the world, not their binary representations in tools. The only meaning imparted to an opaque object in the shared knowledge base is its unique identity; it can be tested against other opaque objects for equality. This is sufficient to index and search for these identifiers, using simple pattern matching (no deep inference). Everything else known about an opaque object is represented with KIF sentences. The existence of an object is implied by it being asserted as an instance of a class. A permanent property of an object is asserted with a binary relation or unary function.
- SHADE applications can assert relationships among opaque objects, such as the constituents of design decisions. These relationships have a well-defined interpretation (specified in the common ontology) even if the contents of the opaque objects are only interpretable by humans or specific tools. This is guaranteed by requiring that all relation terms used in sentences be defined in the shared ontology.
- Exporting tools are responsible for the user interface that presents exported objects to humans. Some user gestures will indicate that an object is selected, and therefore party to some relationship (e.g., a design decision in the scenario). SHADE programs may call on tools to display previously exported objects. This may require the tool to reconfigure to a previous state (see Section 5.3 on issues of representing versions and time).

In the the design decision scenario of Section 2.1, when the user selects a requirement in the requirements editor, the editor associates the text string with a locally unique handle (which could be the actual bits encoding the data). It exports this handle with the assertion that the text is of type requirement-text. SHADE maps the tool-specific handle to a unique name in the SKB (`rt-1`), producing the assertion

```
(instance-of rt-1 requirement-text)
```

The same process applies to the export of the motor component from the structure editor and the simulation scenario from the simulation tool. If other members of the project team were interested in the simulation data, they could ask SHADE to request that the scenario `ss-1` be presented. The term `ss-1` denotes the scenario. On the simulation tool end the scenario is stored as a script specifying behavior models, input conditions, and other input parameters. Recreating the scenario is achieved by rerunning the simulation using the script.

4.3. Subscription and Notification

As stated earlier, a major service enabled by SHADE is content-based routing of information. As facts about opaque objects are published, the information is made available to interested parties. Conceptually, any SHADE participant can query the shared knowledge base. However, as a practical matter the virtual knowledge base may be distributed over many local knowledge bases/tools. Furthermore, it is impractical for every program to continually monitor all sources of shared information on the network. This is why it is important for the representational framework to provide the basic services for coordinating the flow of information among tools based on specifications of information needs and information generated.

The essential elements of the SHADE mechanism for content-based information routing are subscription, notification, and determining relevance. A subscription is a query using the using the shared knowledge representation. Notification is the delivery of the answer, also in the shared knowledge representation. When information is exported to the shared knowledge base, those parties who subscribed to that information are notified of the new information. Determining relevance is the task of deciding which parties would be interested in a given piece of information. There are many open research questions concerning how the messages can be routed, and how parties to the information negotiate the exchange (see (Genesereth and Tenenbaum 1991) for details).

In the scenario of Section 2.2, the manufacturing engineer wanted to be notified when any component is added to the artifact structure model. She might specify this as

```
(interested-in agent-9 '(instance-of ?x structural-component))
```

This says that `agent-9` should be informed when any new structural components are posted to the shared environment. Agents are programs obeying the protocol, in this case agent-9 is acting on behalf of the manufacturing engineer. When the new motor is added to the design, the agent might be informed of the fact as it was posted:

```
(instance-of sc-1 structural-component)
```

The human who owns the monitoring agent might be notified with an email message. Upon notification that *something* is happening with regard to a component, she could ask the program for details, which might turn up the link between the new motor and the design decision:

```
(relevant-structures dd-1 sc-1)
```

Note the subscription and notification rely on the shared definition of the term `structural-component` in the shared ontology. The engineer's understanding of this term needs to be in agreement with the tool that posted the new information (practically, the agreement is with its programmer). Similarly, the design history capture tool commits to terms in the ontology for representing decisions and relating them to requirements, structures, validation test results, etc. But in this kind of subscription, the ontological commitment is very circumscribed. Relevance can be determined syntactically, with a simple pattern matching procedure. Nothing else matters about the `?x` but that it is a structural-component. The opaque object for the motor might be represented in the originating tool as a bitmap image, a geometric model, a table of manufacturer data, or a set of equations. *For the purposes of notification and decision management,* these distinctions may not matter.

Even with agreement on terms by all parties, it is difficult to anticipate the actual form in which a fact may be asserted. For the same reason, it is also difficult to formulate many queries as simple facts that will match syntactically with assertions made by information producers. The long-term power of content-based routing as an interoperability principle comes when one allows inference between subscription and notification. The knowledge used to infer the connection between new information and the relevant subscriptions is represented in *relevance theories.* Relevance theories are also represented in the shared representation.

4.4. Reasoning About Dependencies and Change with Relevance Theories

With a declarative representation one can write theories about the relevance of changes in the shared KB to specified interests. A theory might be formulated as a set of rules that describe how a change in one design element is related to a change in another. For example, one theory specifies that if a component is replaced, then the features of the replacement part might potentially change. This theory is just a compact way of stating that the weight of a replaced component may change, the cost of a replaced component may change, and so forth. Theories allow computer programs to infer relationships rather than just look them up. They let humans specify a generalization rather than enumerate a set of cases. Our manufacturing engineer can now write a subscription to be notified if anything changes in the

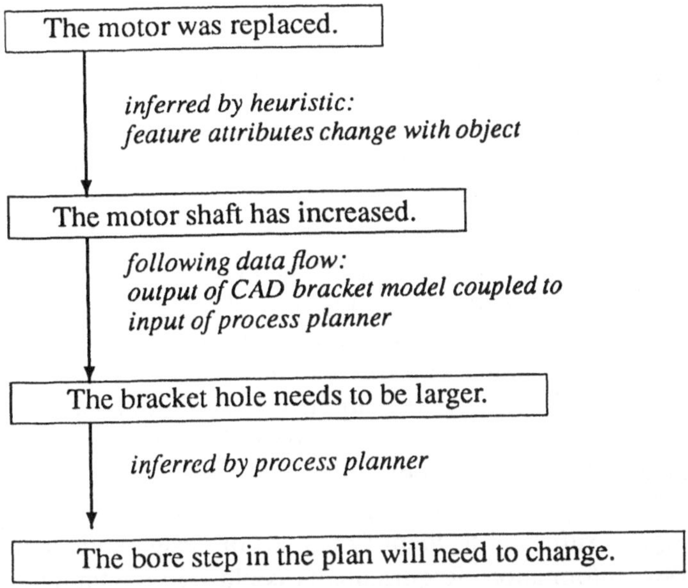

Fig. 1. Path of relevance reasoning and notification

properties of the structural components that might effect the milling process on the brackets.

In SHADE, relevance theories will be used to infer paths of notification in a design knowledge base. Figure 1 illustrates such a path from the planar manipulator scenario described in Section 2.2. The nodes represent changes to design elements. The arcs correspond to inferences about the relevance of changes in one element to changes in another. The first inference could be made by the simple heuristic described above: when the motor was replaced, the motor shaft diameter increased. The second inference relies on a similar heuristic, that associates paired features of mating parts; when the motor shaft diameter changes, the corresponding bracket hole must also be changed. The application of this heuristic is complicated in our scenario because the parts involved are modelled by two different CAD tools (one for simulating the motor assembly and another for creating the process plan for the mounting bracket). The relevance of shafts to holes was determined by a human engineer, and reflected in a model of the data flow between these two tools. In the third inference, a change in the input specification to the process planner leads to a change in the output plan for the milling machine. The fact that changed inputs can lead to changed output might be inferred by a general heuristic. Exactly how the change to the mounting hole will impact the process plan is inferred procedurally by the process planner itself.

5. Research Issues

5.1. The Relative Utility of Ontological Commitment

The SHADE framework has the desirable property that it can provide some level of service at almost any level of knowledge sharing. An absolutely minimal ontology would be a single class of opaque objects and a single kind of relation or link. This is equivalent to conventional hypermedia, where nodes are undifferentiated, unstructured data objects (text, sound, video bites) and the links are untyped relations. Even this minimal ontology allows the linking of design information across engineering tools. The shared knowledge base, object encapsulation, publishing, and subscription technology would all contribute to improved information access. Subscription would be primitive; one could subscribe to specific objects or everything (the latter case is equivalent to hypermedia browsing). This is the level of ontological commitment of design notebooks used to support collaboration.

With a slight enrichment in the ontology, functionality improves rapidly. A weak theory of design is sufficient to identify classes of design knowledge into which to classify new information. Requirements could be differentiated from structure descriptions and the output of simulations could be related to decisions. This kind of ontology could support a design history. A design history is a log of design moves, abstract operations on design elements such as requirements and structures. Capturing design histories in a SHADE-like tool environment could result in a record with a lot of context: the objects, the actors, the tools, and the state of the documents being operated on by the tools. These can be used to selectively retrieve design elements (e.g., "retrieve all structural changes produced by Fred last week using the Cadence schematic editor"). Moreover, if the data were classified and organized into general categories corresponding to their role in the design (e.g., a component selection identified as an "alternative" with respect to a design decision), the resulting record is much easier to index and search. Importantly, the simple ontology works when the instances are interpretable by humans, as with textual and graphical annotation. This is the level of ontological commitment of collaboration tools such as gIBIS (Conklin and Begeman 1988, Conklin and Yakemovic 1991) and SYBIL (Lee 1990, Lee and Lai 1991b), which support documentation of group deliberation and decision making based on a simple ontology.

With a commitment to a rich ontology of formally represented design knowledge, the relevance of new information to expressions of interest can be inferred, rather than just matched. In addition, if the vocabulary of design knowledge permits a chain of reasoning from source to effect, then the answers – the information delivered to subscribers – can be structured as well. In particular, systems can be built to explain *how* something is relevant. For example, the torque output by a motor might be computed by an equation linking it to input current and a motor constant. In this case, the relevance of motors to torque is inferred via the equation. The resulting explanation is that the motor replacement changed the motor constant, which resulted in a new torque via the equation. This way, simulators that generate quality explanations for humans (Forbus and Falkenhainer 1990, Iwasaki and Low 1991) can be part of the knowledge medium. For example, explainable simulation can be used to capture specifications of intended function and context of use that are hard to represent formally (Gruber 1990, Gruber 1991). In general, if formal reasoning is used to infer relevance, then the resulting proof can serve as an explanation of how something is relevant to the object of interest. This is another way in which the knowledge medium is active; the *invocation* of simulation tools on formal data produces information which can then be published into the shared information spaces.

These potential applications suggest that increasing power results from an increasingly rich shared representation. This is a central hypothesis being explored in the SHADE project: that the utility of the services enabled is a function of the level of commitment to a shared ontology. A family of research questions emerge. What is the minimal ontology sufficient to give useful functionality? What services are enabled by what kinds of shared representation? If we view subscription/notification as information retrieval, do the hit rate and accuracy improve with increasingly rich shared ontology?

5.2. PARTIAL META-MODELS

The SHADE approach to knowledge sharing occupies an interesting niche between document and object linking protocols (OLE, Publish and Subscribe, OMG) and complete standardized formal data models used for engineering tool integration (e.g., PDES, CAD Framework Initiative). Both integration approaches play a role in supporting collaborative engineering. We have described a representational framework that achieves the capabilities of both. By encapsulating foreign data, it supports hypertext-like links and browsing among formal objects maintained by heterogeneous engineering tools as well as informal (unstructured) data contained in engineering documents. Since the representation is based on an expressively comprehensive language, SHADE also allows complete specification of shared data models. These specifications are sometime called meta-models (Fulton 1991, Kiriyama, Yamamoto, Tomiyama and Yoshikawa 1989), since they describe the information content of base-level databases.

The SHADE framework allows us to explore the continuum of integration from content-independent linking to full meta-models. The degree of integration is determined by how much of the information manipulated by participating tools can be represented in the shared representation. We are entertaining the notion that there is a useful middle ground, a partial meta-model, in which some of the data can be represented completely, and the rest can remain tool specific or only interpretable by humans. The hypothesis is that the domain of the shared ontology should be the relationships between data exchanged by tools, and not the knowledge manipulated by the tools internally to produce their results.

5.3. Representing Change in Design Information

We have deliberately been silent on how a SHADE participating tool determines what information is worth exporting. Since the shared ontology is incapable of representing all the distinctions used internally by tools, there will always be a need to determine what is potentially relevant to the world. For the same reason, a subscription cannot tell a tool precisely what aspects of an object to be informed of, since one cannot anticipate all such properties.

For the sake of simplicity we will focus this issue on the case of design change, and assume that there is only only level of granularity of "objects" in a tool that might be mentioned in published facts. In many varieties of object-oriented representation, there is a clear enumeration of those facts about an object which are its properties. One could ask to be notified if any of those facts were stated or retracted, using free variables as needed to generalize the query. For example, to subscribe to changes in a slot of a given object, one says

```
(interested-in me '(particular-slot object ?any-value)).
```

To subscribe to a change in *any* slot of a given object, one might write:

```
(interested-in me '(?any-slot object ?any-value))
```

These are just two of many possible generalizations of change over an object. If we wanted to know about the *time-varying* change in a parameter of any model of an object, the sentence might look like this:

```
(val (temperature (has-model motor-2 ?model)) ?time ?value)
```

The point is that what is persistent and immutable – what constitutes a design change or even an object – is tied up with how we conceptualize change and represent facts about changes. If objects are persistent, what are versions? Are facts tagged with version numbers, or associated with versions of objects? How can representations of ordinal time stamps or versions be coordinated across tools?

5.4. Inferring Relevance of New Information to Specified Interests

Inferring relevance in general is intractable (Levy 1992). However, specialized theories can implement certain kinds of relevance reasoning efficiently. In particular, when the representation is based on opaque, semi-structured objects, and the shared representation of relationships among these objects forms a semantic network in the SKB, relevance can be approximated by a spreading-activation traversal of these relations. For example, if the designer records the relations among the design elements such as requirements, component choice, and simulation results, the relevance of one design element to other shared objects can be viewed in terms of the paths between them in the network. Unfortunately, almost any object can be connected to another over a path of links. One kind of relevance theory theory would specify *which* links are plausibly relevant for a given query, constraining propagation to traverse only these links.[3]

This is only one of many heuristic approaches to determining relevance. There are also several ways one might represent the relevance theory. Framing the relevance question in the SHADE context of a potentially distributed, shared, semi-structured information space raises more interesting questions. Where is the relevance reasoning performed (i.e., at the producer, consumer, or a mediator agent)? Where is the relevance theory kept? How can end users write their own, as an extension of specifying their interests? Results could be applied to the larger contexts of information search over wide area networks. SHADE can provide an excellent test bed for experimentation in this area.

6. Discussion and Plans

We have laid the foundation for a knowledge medium that could support the information and communication intensive activities that dominate cooperative product development. These activities include engineering knowledge capture and retrieval, change notification and active management of design dependencies. The approach places all product knowledge in a coherent framework, relating, for example, requirements, artifact and process models, design decisions, test results, and engineering knowledge. Designed to integrate human organizations as well as design tools, it accommodates information in formats ranging from unstructured multi-media documents to formal engineering models.

Conventional approaches to integrating engineering knowledge rely on standardized data structures or unified meta-models, both of which require substantial commitments from tool designers. SHADE departs from such approaches in two fundamental ways. First, tools and data are encapsulated rather than unified. Second, tools communicate through a shared representation that is developed incrementally with use.

We are seeding the ontology with general primitives for relating the information produced by existing tools (e.g., text editors, email, on-line engineering handbooks, feature-based CAD, simulators with model libraries) for tasks such as design rationale capture and the management of design dependencies. From the start, users will be able to link objects from their tools in a hypertext style. Even this minimalist node-and-link ontology can support basic information sharing functions such as retrieval and notification. Initially, only a small amount of information used by human engineers will be machine interpretable. Similarly, only a fraction of the information embedded in existing systems will be available externally. As some of those relationships prove useful, they can be formalized and added to the shared

[3] This technique has been used successfully in information retrieval tasks (Cohen and Kjeldsen 1987).

ontology. The additional ontological commitments enhance the selectivity of retrieval and notification operations and enable the relevance of indirectly linked information to be inferred. This novel, pragmatic approach to ontology development is motivated by the heterogeneity of existing software and the adaptability of humans.

The ontology and information routing services are being developed in coordination with two other research projects concerned with knowledge sharing. DARPA's Knowledge Sharing Effort is developing mechanisms and conventions for the sharing and reuse of heterogeneous knowledge bases and knowledge-based systems (Neches, Fikes, Finin, Gruber, Patil, Senator and Swartout 1991). This project is developing the KIF interlingua, intelligent agent communication protocols, specifications for common representation systems, and mechanisms for developing and maintaining shared ontologies. The Palo Alto Collaborative Testbed (PACT) is an integration of four experimental engineering environments from Stanford and Lockheed, covering such diverse engineering disciplines as process planning (Cutkosky and Tenenbaum 1991), software engineering (Weber, Livezey, McGuire and Pelavin 1992), circuit layout (Genesereth 1989), and device modeling (Iwasaki and Low 1991). PACT provides a test bed for experiments in propagating design changes across human, tool, and even tool-framework boundaries.

Acknowledgements

The local AI and Engineering research community has significantly influenced the ideas in the paper. We wish specifically to acknowledge the contributions of Mark Cutkosky, Richard Fikes, Mike Genesereth, Jay Glicksman, Bruce Hitson, Larry Leifer, Bill Mark, Jim McGuire, Rich Pelavin, and Dan Russell. Participants of the DARPA knowledge sharing effort, particularly Steve Cross, Bob Neches, Mark Stefik, and Gio Wiederhold, have also helped shape our thoughts. Finally, thanks to Morton A. Hirschberg, US Army Ballistic Research Laboratory, for last-minute technical editing.

References

Boose, J. H., Shema, D. B. and Bradshaw, J. M.: 1991, Knowledge-based design rationale capture: Automating engineering trade studies, *in* M. Green (ed.), *Knowledge Aided Design*, Academic Press, London.

Cohen, P. R. and Kjeldsen, R.: 1987, Information retrieval by constrained spreading activation in semantic networks, *Information Processing and Management* **23**(4), 255–268.

Conklin, J. and Begeman, M. L.: 1988, gIBIS: A hypertext tool for exploratory policy discussion, *Proceedings of the 1988 Conference on Computer Supported Cooperative Work (CSCW-88)*, ACM, Portland, Oregon, pp. 140–152.

Conklin, J. and Yakemovic, K. B.: 1991, A process-oriented paradigm for design rationale, *Human Computer Interaction* **6**(3/4).

Cutkosky, M. R. and Tenenbaum, J. M.: 1991, Providing computational support for concurrent engineering, *Systems Automation, Research and Applications* **1**(3).

Falkenhainer, B. and Forbus, K.: 1991, Compositional modeling: Finding the right model for the job, *Artificial Intelligence* **51**, 95–143.

Forbus, K. D. and Falkenhainer, B.: 1990, Self-explanatory simulations: An integration of qualitative and quantitative knowledge, *AAAI-90*, Boston, pp. 380–387.

Fulton, J. A.: 1991, The semantic unification meta-model: technical approach, *Standards working document ISO TC184/SC4/* WG3 N 81 (P 0)*, IGES/PDES Organization, Dictionary/Methodology Committee. Contact James Fulton, Boeing Computer Services, P. O. Box 24346, MS 7L-64, Seattle, WA 98124-0346.

Genesereth, M. R. and Tenenbaum, J. M.: 1991, An agent-based approach to software, *Technical Report Logic-91-6*, Stanford University Logic Group. Under revision.

Genesereth, M. R., Fikes, R. E., et al.: 1992, Knowledge Interchange Format, Version 3.0 Reference Manual, *Technical Report Logic-92-1*, Computer Science Department, Stanford University.

Genesereth, M. R.: 1989, Designworld, *Technical Report Logic-89-3*, Computer Science Department, Stanford University.

Genesereth, M. R.: 1990, DSL reference manual, *Technical Report Logic-90-3*, Computer Science Department, Stanford University.

Givan, R., McAllester, D. and Zalondek, K.: 1991, Ontic91: Language specification and user's manual, *Technical report*, MIT.

Gruber, T. R. and Russell, D. M.: 1990, Design knowledge and design rationale: A framework for representation, capture, and use, *Technical Report KSL 90-45*, Knowledge Systems Laboratory, Stanford University.

Gruber, T. R. and Russell, D. M.: 1992, Derivation and use of design rationale information as expressed by designers, *in* T. Moran and J. H. Carroll (eds), *Design Rationale*, Erlbaum. In preparation.

Gruber, T. R.: 1990, Model-based explanation of design rationale, *Technical Report KSL 90-33*, Knowledge Systems Laboratory, Stanford University. Also appears in Proceedings of the AAAI-90 Workshop on Explanation.

Gruber, T. R.: 1991, Interactive acquisition of justifications: Learning "why" by being told "what", *IEEE Expert* **6**(4), 65–75.

Gruber, T. R.: 1992, Ontolingua: A mechanism to support portable ontologies, *Technical Report KSL 91-66*, Knowledge Systems Laboratory, Stanford University.

Iwasaki, Y. and Low, C. M.: 1991, Model generation and simulation of device behavior with continuous and discrete changes, *Technical Report KSL 91-69*, Knowledge Systems Laboratory, Stanford University.

Kiriyama, T., Yamamoto, F., Tomiyama, T. and Yoshikawa, H.: 1989, Metamodel: An integrated modeling framework for intelligent CAD, *in* J. S. Gero (ed.), *Artificial Intelligence in Design*, Computational Mechanics Publications, Southampton, UK, pp. 429–449.

Kuffner, T. A. and Ullman, D. G.: 1990, The information requests of mechanical design engineers, *Design Studies* **12**(1), 42–50.

Lee, J. and Lai, K.-Y.: 1991a, A comparative analysis of design rationale representations, *Technical Report 91-121*, MIT Center for Coordination Science.

Lee, J. and Lai, K.-Y.: 1991b, What's in design rationale, *Human Computer Interaction* **6**(3/4).

Lee, J.: 1990, Sibyl: A qualitative decision management system, *in* P. Winston and S. Shellard (eds), *Artificial Intelligence at MIT: Expanding Frontiers*, MIT Press.

Lenat, D. B., Guha, R. V., Pittman, K., Pratt, D. and Shepherd, M.: 1990, Cyc: Toward programs with common sense, *Communications of the ACM* **33**(8), 30–49.

Levy, A.: 1992, Irrelevance in problem solving, *Technical report*, Stanford University, Knowledge Systems Laboratory.

Lubars, M. D.: 1991, Representing design dependencies in an issue-based style, *IEEE Software* pp. 81–89.

Malone, T., Yu, K. and Lee, J.: 1989, What good are semi-structured objects? adding semiformal structure to hypertext, *Sloan Working Paper Tech. Report 3064-89-MS*, MIT.

Neches, R., Fikes, R., Finin, T., Gruber, T., Patil, R., Senator, T. and Swartout, W. R.: 1991, Enabling technology for knowledge sharing, *AI Magazine* **12**(3), 16–36.

Palmer, R. S. and Cramer, J. F.: 1991, SIMLAB: Automatically creating physical systems simulators, *Technical Report TR 91-1246*, Department of Computer Science, Cornell University.

Petrie, C.: 1991, Planning and replanning with reason maintenance, *Technical report*, MCC.

Stefik, M.: 1986, The next knowledge medium, *The AI Magazine* **7**(1), 34–46.

Weber, J., Livezey, B., McGuire, J. and Pelavin, R.: 1992, Spreadsheet-like design through knowledge-based tool integration, *International Journal of Expert Systems: Research and Applications*. To appear.

Williams, B. C.: 1991, A theory of interactions: unifying qualitative and quantitative algebraic reasoning, *Artificial Intelligence* **51**(1-3), 39–94.

Wolfram, S.: 1991, *Mathematica: A System for doing Mathematics by Computer*, Addison-Wesley.

Ziegler, B. P., Elzas, M. and Oren, T. (eds): 1989, *Modelling and Simulation Methodology: Knowledge Systems Paradigms*, Elsevier North Holland.

ENDURING SUPPORT

On defeasible reasoning in design support systems

B. S. LOGAN, D. W. CORNE
Department of Artificial Intelligence
University of Edinburgh
Edinburgh, UK

and

T. SMITHERS*
Artificial Intelligence Laboratory
Vrije Universiteit Brussel
Brussels, Belgium

Abstract. The task of a design support system is conventionally conceived as one of providing the designer with solutions to specific parts of a design problem. In this paper we argue that this approach is fundamentally flawed. We identify two main modalities of support: the production of *necessary* consequences and the production of *possible* consequences of the current design description, and discuss the problem of devising an architecture capable of providing such support using the Edinburgh Designer System (EDS) as an example. We describe the difficulties inherent in integrating the derivation of possible consequences into the architecture of EDS and argue that while in principle such difficulties can be overcome, in practice the goal of providing globally consistent solutions to particular parts of the design problem is unattainable. We propose a different approach in which the design support system *explores* the consequences of various design decisions. The results of this exploration are represented as counterfactual conditionals which we believe more closely approximates the information required by a designer.

1. Introduction

We take designing things to be a particular kind of intelligent behaviour. We further presume that the nature of the process underlying this kind of behaviour has a general form common to all instances and types of designing. For us, then, what human designers do when they design things is to be taken as a way of implementing this process—possibly the only way at the moment—but not as a way of defining it. In taking this stance we are able to identify the design process in the abstract, independent of its human implementations, and thus to ask how it might be understood in computational terms.

The long term aim of our AI in Design research is a computational theory of design process which can be used to both explain design behaviour and to predict design performance in all its forms; natural or artificial. In pursuit of this aim we have developed what we call the exploration-based model of design process (Smithers & Troxell, 1990), and have been attempting to verify and develop it by building experimental design support systems (Buck et al., 1991; Smithers et al., 1990).

* SWIFT Professor of AI on leave from the Department of Artificial Intelligence, University of Edinburgh.

J. S. Gero (ed.), Artificial Intelligence in Design '92, 433–454.

Our research method involves investigating the design process by trying to build design support systems that actively participate in design tasks with human designers. In this way we are able to control how much and which parts of the overall design process we try to directly investigate at any time. It also emphasises the problems of human-computer interaction which we inevitably have to face up to. This approach contrasts with other methodologies found in AI in Design research which attempt to build automated design systems. The role of of the design support systems that we build and experiment with can be identified within the following framework used to organise our research programme:

1. *Exploration-based model of design process*, in which we are seeking to develop a general model of the design process which attempts to identify and relate all its essential characteristics and properties.
2. *Domain models of design process*, in which we are seeking to understand how the general model is to be specialized to model the design process in a particular domain. This typically involves identifying the nature, type, and amounts of knowledge required to pursue design tasks in the domain, and identifying the nature, type, and scope of the activities engaged in when solving design tasks in the domain. In this way, what are called routine, innovative, and original design tasks can be identified as different configurations of the general model.
3. *Design support models*, in which we are trying to identify those of the activities identified by the domain models that can be supported in an integrated and coherent way using a computer-based support system, and to identify that part of the knowledge required to design in the domain which could be represented within such a support system as part of its support activities.
4. *Design support system architectures*, in which we are attempting to identify the functional components, organisational structure, and control capabilities required of a computer-based design support system that can meet the needs of human designers operating in a particular domain.
5. *Design support system implementation techniques*, in which we are seeking to identify and develop computational techniques and implementation approaches suitable for the building of effective experimental design support systems.

In this paper we discuss the problem of devising and implementing an architecture for a design support system capable of meeting the needs of human designers operating in a particular domain. We focus on a particular kind of support commonly advocated in the literature—the derivation of *possible consequences* from a partial design description, i.e., providing advice or assistance in solving parts of a design problem. We describe the difficulties inherent in integrating the derivation of possible consequences into the architecture of a particular design support system, the Edinburgh Designer System, and argue that while in principle such difficulties can be overcome, in practice the goal of providing globally consistent solutions to particular parts of the design problem is unattainable. More generally, our approach to this problem will hopefully serve to illustrate our research methodology and the kinds of understanding it produces. To set the context and to motivate the particular problem we will be concerned with, we first briefly reiterate some essential aspects of the exploration-based model of design and two characteristics of design support models that we take to be domain independent to a large extent.

2. The Exploration-based Model of Design

The exploration-based model of design process sets out to address the inherently ill-structured nature of design tasks head on. It characterises the design process as one that starts with an initial requirement description that is incomplete and possibly inconsistent. In many cases the stated objectives are in direct conflict with one another and the designer cannot simply optimise one requirement without suffering losses elsewhere. Different trade-offs between the requirements result in a whole range of acceptable solutions, each likely to prove more or less satisfactory under different interpretations of the requirements. It is the very inter-relatedness of all these factors which is the essence of design problems rather than the isolated factors themselves, and it is the identification and structuring of relationships between these criteria that forms the basis for the design process (Lawson, 1980). The design process cannot therefore be adequately characterised as a *search* process in which the task is essentially one of selection or optimisation over a completely defined space of alternatives.

The initial requirement description is analysed into a final complete and consistent requirement description via the synthesis of possible design descriptions that satisfy some or all of the identified requirements. These descriptions are typically not serially ordered and are intended to span the space of possible designs thought relevant to the task in hand. This process of analysis-through-synthesis necessarily involves the discovery of the structure of the problem to be solved. As possible solutions are constructed and developed they provide an increasingly detailed context against which to test the designer's hypotheses, and the evaluation of a proposal can result in the discovery of previously unrecognised relationships and criteria. Solutions to particular sub-problems are apt to be disturbed or undone at a later stage when new aspects are attended to and the considerations leading to the original solution are forgotten or not noticed. As a result, while the final solution may satisfy all the requirements that are evoked when it is tested, it may violate some of the requirements that were imposed (and temporarily satisfied) at an earlier stage in the design. These explorations help the designer appreciate which requirements may be most readily achieved and those that may be neglected without loss. As part of this process, the designer learns which criterion values will achieve the design goals and how much variation of these values can be tolerated while still achieving acceptable performance. The designer also discovers the implications of achieving the current goal, and any other decisions required to make the attainment of these goals consistent with the existing solution. A large part of the design process is thus devoted to discovering the nature and scope of the problem set by the requirement description which itself changes as exploration takes place. See (Smithers & Troxell, 1990) for more details.

3. A Model of Design Support

The design process outlined above involves a continual cycle of generating alternative design solutions or part solutions and the derivation of the consequences of these solutions in an attempt to understand their implications for other design criteria. These alternatives may embody radically different approaches to the problem or they may be variations on a common theme or both. Each of these solutions may in turn be broken down into a set of simpler sub-problems, (for example, the design of an assembly may be reduced to the design of its constituent components) together with the problem of integrating the resulting part solutions, or it may be broken down into a set of associated functions. Of particular concern are inconsistencies which arise in attempting to satisfy multiple goals, for example when the derived performance fails to meet the design requirements or when the proposed solution implies two or more values for a given design parameter.

We can therefore identify two important aspects of the support required when carrying out design tasks:

1. *Strategic support* for the overall control of the design process. A major task of any design support system is one of complexity management: controlling the generation of alternative solutions to particular sub-problems; comparing the alternatives; selecting the most promising candidates for further development; and revising and refining the requirement description.
2. *Tactical support* for the construction and development of alternative design solutions. The system should provide support for context specific inference based on design decisions, constraints and the requirements defining the design problem, together with physical laws, heuristics and other domain knowledge relating the parameters of the design. In addition the system should, where possible, provide assistance in solving particular design problems, drawing on the large amounts of knowledge encoded in design handbooks, codes of practice and in the expertise of individual designers.

In an previous paper (Logan, Millington & Smithers, 1991) we discussed some of the problems of providing strategic support. In this paper we concentrate on the problems of providing tactical support within the blackboard-based system architecture that we have been developing for the past seven years, see (Buck et al., 1991; Smithers, 1987; Smithers et al., 1990). This system is called the Edinburgh Designer System or EDS for short.

4. System Architecture

EDS consists of four principal subsystems which together embody the model of design support outlined above: *knowledge representation*; *inference*; *consistency maintenance*; and *context management*. In this section we briefly describe the architecture of EDS and how it supports the exploration of the space of possible designs. In the following sections we concentrate primarily on the consistency maintenance and context management sub-systems of EDS and their role in providing tactical support. Context management, knowledge representation and the overall design of the system are discussed in more detail elsewhere (Logan, Millington & Smithers, 1991; Logan & Smithers, 1992; Smithers et al., 1990).

EDS represents domain knowledge in a structured knowledge-based called the Domain Knowledge Base (DKB). This contains definitions of domain objects, called *module classes*, related by *part_of* and *kind_of* relations. Each module class declares a set of parameters, variables and constraints which define a particular class or type of object. This knowledge of the domain is used by a series of inference engines or support systems to infer the consequences of the designer's decisions. There are two main kinds of support system: *general purpose support systems* (GPSSs) and *special purpose support systems* (SPSSs). General purpose support systems use constraint propagation techniques to infer the *necessary* consequences of the designer's decisions. In general these sub-systems are invoked automatically and can be allowed to proceed relatively unhampered by the designer with the sole constraint that they should refrain from re-deriving information already known about the design. Special purpose support systems, on the other hand, advise on how to solve particular design problems. In general these sub-systems offer *possible* consequences, often in the form of one or more design decisions which extend the current design description. They tend to be goal directed and are usually invoked explicitly by the user who then interacts with the support system to reduce the number of options until a single solution is found.

Control of the interactions between the support systems is in the style of a blackboard system (Hayes-Roth, 1985). In the blackboard model the knowledge required to perform a particular task is partitioned into a series of knowledge sources (KSs) each of which performs a particular sub-task. The current state of the task and the results produced by the knowledge sources are recorded in a global database or blackboard. The knowledge sources use information on the blackboard to derive new information using algorithmic procedures or heuristic rules. The system maintains an agenda of knowledge source activation records (KSARs) which identify possible inferences the system could perform. The decision as to which knowledge source to apply in any particular situation is determined dynamically based on the current solution state (and in particular the latest additions to the blackboard) and on the ability of the knowledge sources to contribute to the developing solution. In EDS each support system is implemented as one or more knowledge sources which derive consequences of the current design description represented on the blackboard.

The system pays particular attention to any inconsistencies derived by the knowledge sources as these are often indicative of inconsistencies in the problem requirements or problems with the proposed solution. A design description must be consistent if it is to refer to anything. At the same time we have to recognise that inconsistencies are inevitable—a design description is typically inconsistent for much of its history as the designer explores the space of possible designs attempting to meet the various design requirements. However all a contradiction between two design decisions indicates is that any inference which depends on both decisions is of no value; it is still important to draw inferences from each of the decisions independently.

The desire to derive as much information as possible from inconsistent design descriptions led to the adoption of an assumption-based truth maintenance system (ATMS) for EDS (de Kleer, 1986a). The ATMS builds and maintains a dynamic datastructure, the Design Description Document (DDD), and provides an interface between the contents of the DDD and the other sub-systems, passing out relevant pieces of information to them as required and incorporating new information which it receives from them into the dependency structure. Each piece of information on the blackboard is associated with with an ATMS *node*. A datum p is said to have a *justification* if there exists a set $q_1, \ldots, q_n$ of nodes from which it can be derived. The set $q_1, \ldots, q_n$ are termed the *antecedents* of the justification and p is termed the consequent. A subset of the data, called *assumptions*, are taken as primitive and everything else is derived from them. In EDS these represent the basic design decisions made by the user. By tracing backwards through the supporting justifications, the ATMS identifies the set(s) of assumptions (i.e. design decisions) on which a datum ultimately depends. Such a set of assumptions is called an *environment*. The set of environments in which a datum is known to be derivable is is called its *label*. A datum which has a non-empty label is said to have support, i.e., it can be consistently derived from the assumptions forming each of the environment(s) in its label. If the designer's decisions subsequently turn out to be mutually inconsistent, the ATMS restores consistency by deleting from the label of each datum node all of the environments which contain the inconsistent assumptions. Data which are only derivable in such an environment (i.e., they can only be derived from an inconsistent set of assumptions) become unsupported and hence cannot form the basis of further inferences.

The central role of the ATMS in the system architecture has resulted in a number of extensions to the conventional ATMS model. For example, the ATMS truth maintains the knowledge source activation records within the DDD on an equal footing with existing data and its justifications. This automatically removes KSARs whose antecedents are discovered to be inconsistent. Similarly, blackboard systems are usually designed and implemented as 'single context' problem-solving systems in which the knowledge sources work together to construct one consistent solution. Truth maintaining the blackboard has therefore resulted in a number of departures from the conventional practice in blackboard systems, notably the absence of deletions (except for KSARs) or amendments as it is unclear how these fit into an assumption-based truth maintenance scheme.

Design proceeds by creating instances of module classes and assigning values to their parameters to define one or more possible solutions.[1] When the user makes an assumption one (or more) datum nodes are created to hold the new information. This information is examined by the knowledge sources to see if it, together with any information already assumed or derived, can be used to make further inferences. If a knowledge source is able to make an inference, it generates a bid in the form of a knowledge source activation record which is scored and merged into the agenda. When the KSAR reaches the front of the agenda, it is executed and the results are claimed into the ATMS. This information may in turn form the basis for a new round of bids and this cycle continues until no executable KSARs remain in the agenda. As new values for parameters or bounds on them are assumed or derived, consistency checks are performed between constraints and values by the *valueConflict* KS. Conflicts result in the creation of a justification for the distinguished node ⟨*false*⟩ recording the fact that the assumptions involved are mutually inconsistent and causes the ATMS to partition the assumptions into mutually consistent sets. If there is no conflict, EDS marks this by justifying the datum ⟨*consistent*⟩ and proceeding as usual. The user is viewed as a knowledge source whose 'bids' are always processed first. This allows the system to follow several lines of reasoning as it attempts to infer the consequences of the user's design decisions, while still giving priority to user input.

5. General Purpose Support Systems

A general purpose support system (GPSS) is a reasoning system which performs inferences common to many of the sub-tasks which together constitute the overall design process. Note that 'general' in this sense does not mean that such inferences are necessarily useful in other design domains, although they may be. There are currently four support systems (or 'engines') implementing general purpose support capabilities within EDS (Smithers et al., 1990):

1. the *Evaluation Engine* which handles value propagation, constraint satisfaction and expression simplification;
2. the *Algebraic Manipulation Engine* (AME) which takes an arbitrary set of equations and solves them for any number of variables in which they are linear;
3. the *Relation Manipulation Engine* (RME) which performs value interpolation and relational operations on tabular data; and
4. the *Spatial Reasoning Engine* (SRE) which performs spatial inferencing.

[1] This is an oversimplification—the user can also define new parameters and constraints and assemble novel designs from existing modules.

These are implemented as 35 knowledge sources.[2] There are also two knowledge sources which do not form part of any support system: the *Direct Inference System*, a simple rule interpreter, which can be used to implement user-defined knowledge sources and support systems; and the *USER-KS* which handles user input.

The general purpose support systems can be subdivided into two main types: those which are *user-invoked* and those which are *system-invoked*, i.e., they are triggered automatically by information being claimed into the DDD. This distinction is based largely on pragmatic considerations, and in particular the computational costs associated with the support system. Ideally, all the support systems would be invoked automatically. Our objective is to assist the designer in exploring the space of possible designs by providing information useful in determining the future development of the current design solution, rather than (as conventionally happens) by providing the means to obtain such information. However our understanding of the design process at the level of an individual designer solving a particular design problem is insufficient to determine which of the many inferences the system could make are most appropriate at any given point in the problem solving process. This basic difficulty is compounded by two additional problems:

1. the design description is constantly changing, both as a consequence of assumptions made by the designer and information derived by the system from the current design description; and
2. any inferences made by the system may subsequently be invalidated if the underlying assumptions are discovered to be inconsistent.

One strategy would be to try to derive everything we can about the design. While this may result in the derivation of some useless information, it allows us to have confidence in the information we do produce, as any inconsistencies implicit in the design description which the system is capable of finding will be discovered. This is the rationale underlying the choice of the blackboard and its opportunistic control strategy. At each cycle, the system applies the knowledge sources to the current design description, reviewing any as yet uncompleted work in the light of what was discovered at the last cycle—inconsistencies, 'more interesting' facts etc.—and adjusts its inference priorities accordingly. However this approach is not practical in its pure form. While the KSs are selected with a view to producing useful consequences, not all of them will be equally useful in a given situation, and at any point there will be many more inferences we can make than we have the resources to make. We therefore compromise. If the computational costs associated with a particular support system are high, the decision to invoke it is typically left to the user due to the difficulty of determining *a priori* the relevance of the information produced to the user's current interests and objectives. On the other hand, if the computational costs are low, the support system is typically invoked automatically,

[2] The definition of a knowledge source in the blackboard literature is rather vague. In another interpretation, we could say there are 4 knowledge sources, one for each support system, each of which consists of a number of rules or 'methods'. However such an interpretation obscures the relationship between the KSs and the justifications they produce.

even if the results may not be of immediate interest to the user. Hopefully this will also reveal any inconsistencies before the computationally expensive support systems are invoked by the user. The difficulties associated with this heuristic are part of a larger problem involving the determination of the context of design tasks and the control of inference which is considered in (Logan, Millington & Smithers, 1991). In EDS, the Evaluation Engine and the user-defined support systems built using the Direct Inference System are automatically invoked by the system, while the AME, RME and SRE are all user-invoked.

6. Special Purpose Support Systems

Unlike a general purpose support system, a special purpose support system or *specialist* is a reasoning system which provides support for a particular design task, such as design for manufacture, design for maintenance or cost analysis. This may involve advising on particular aspects of design solutions with respect to, for example, their cost or manufacturing implications, or suggesting general strategies for resolving identified and well-understood conflicts. In general, this requires specialist knowledge which is unique to both the task and domain. Such support systems tend to be goal directed and would typically operate under the close supervision of the user.

In EDS special purpose support systems are implemented using a Prolog meta-language or *shell*. The EDS shell is a backward-chaining rule-based inference system which provides a framework for building specialised design support systems based on design heuristics in the form of production rules. This approach is natural given the architecture of EDS and is in accordance with much of the existing work on 'support systems' for design. The production rules are constructed using, a library of Prolog clauses which allow examination of the DDD, DKB map, ATMS justification structure and the Prolog versions of any loaded modules. These clauses act like normal Prolog clauses, they succeed or fail and, if appropriate, instantiate variables.

However there are a number of problems in integrating these SPSSs into the EDS architecture.[3] While the EDS shell provides a flexible environment for the definition of new support systems, such systems may violate some of the implicit assumption on which the ATMS-blackboard model is based. Two main problems can be identified: determining what advice to give; and truth maintaining the results. The former involves finding out which design alternatives are consistent with the current solution, while the latter involves ensuring that any inconsistencies which arise in the future are detected.

The task of a special purpose support system is typically to find a solution to

[3] There are currently no special purpose support systems in EDS. A single SPSS which advised on the selection of appropriate subclasses for the transmission module on the basis of the values of the transmission parameters was implemented as a demonstration. However this system does not form part of the current version of EDS and no other special purpose support systems have been developed.

a particular design problem. To be useful such proposals must be selective; it is rarely possible or even desirable to enumerate all possible ways of solving a design problem. Presumably such a solution should be consistent with the current design context, otherwise the task of the specialist is trivial.[4] However the consistency or otherwise of a particular datum with the current state of the DDD cannot be determined by the KS that derived it. In the current implementation, the blackboard is suspended when a knowledge source is running. As a result, the consequences of any claims made by the KS are not derived until the next blackboard cycle. There is no general mechanism to request the proof of a particular goal. If a KS wants a piece of work done by another KS it must know how to claim an appropriate KSAR. Even if it can generate a KSAR, the calling KS cannot wait while the sub-goal is proved. If it cannot proceed without the information, it must fail and rely on being re-invoked by its trigger when the new information becomes available.

Even if it were possible for an SPSS to invoke the blackboard and have any derived results checked for consistency without losing overall control of the system, such an interaction model is incompatible with the 'opportunistic' control strategy of the ATMS-blackboard. Furthermore, if an inconsistency is detected, the system must backtrack, deleting any claims and/or assumptions made prior to the discovery of the inconsistency. At present while assumptions can be deleted claims cannot. Any label propagation performed prior to the detection of the inconsistency must also be undone and the new nogood set(s) deleted from the label of the false node. It is not clear how this could be achieved.

Clearly the advice offered by a specialist may have considerable influence on the design process and should therefore form part of the design history. However even if the specialist can determine what advice to give the system cannot simply extend the design description by making additional assumptions which achieve the goal. Even if the additional assumptions are consistent with the rest of the design description this is typically not acceptable as the system cannot explain *why* the solution is what it is. It simply records the fact that if the design description is extended in a certain way, a particular consequence follows. To *justify* such an assumption, it would be necessary to record why it was preferred to other possible extensions to the design description. This reflects the basic asymmetry between the system and the user; while the system is expected to be able to explain (and hence justify) everything it does, the user is often incapable of explaining why they investigated a particular alternative or enumerating their reasons for preferring one alternative to another. Without knowledge of the assumptions justifying the output of a specialist we cannot detect potential inconsistencies between the user's and the specialist's assumptions. Clearly what can be derived is limited by the system's knowledge of potential inconsistencies (in the form of constraints relating parameter values) and what it has been told by the user. While we cannot trap all

[4] There is also the problem of what the solution is to be consistent with if the current design description is itself inconsistent.

inconsistencies in the user's assumptions, it seems prudent to avoid compounding these problems by allowing the system to make unsupported assumptions.

If the specialist simply offers 'advice' in the form output presented to the user or claimed into the the ATMS (e.g. "use material x") rather than attempting to extend the design description the position is even worse. If the advice is not claimed into the ATMS then it will not form part of the design history and it will be impossible to determine if the advice should ever be withdrawn as a consequence of changes to the current design context. Even if the advice is truth maintained and the specialist is explicit about the supporting assumptions on which the advice is based, if the user subsequently acts on the advice by making the corresponding assumption(s) (i.e., uses material x in the current design description), any inconsistencies which arise later may not be detected because the supporting assumptions justify the the result claimed by the specialist, not the user's assumption(s). Extending the design description therefore seems preferable, even though this will typically result in other KSs being invoked to derive the necessary consequences of the specialist's advice. Indeed this could be seen as a positive advantage since it means that the implications of the new decision are derived automatically.

Typically such decisions depend on the presence or absence of certain design characteristics. If all the assumptions required by the specialist to derive its advice form part of the current design description there is no difficulty. Such inferences can be viewed as necessary consequences of the current design solution and the decision to use the specialist. However if the problem cannot be solved using only the facts currently known about the design, the specialist musk ask the user for the information it requires or make *additional* assumptions about the future development of the design. Alternatively, the specialist can use the *absence* of certain features in the current design description to justify its conclusions. Note that this still involves the system 'completing' the design; it assumes that because P is currently not derivable, it never will be derivable.[5] For example, if an SPSS is capable of deriving one of a set of disjunctive choices based on the absence of information from the DDD, the dependency on this lack of information must be recorded. In a rule-based SPSS (defined using the EDS shell) the ordering of the rules imply additional constraints on any resulting derivation. Given a left-to-right depth-first execution strategy, the rules:

$$\begin{aligned} &A \wedge B \rightarrow D \\ &A \wedge C \rightarrow E \end{aligned} \tag{1}$$

are equivalent to:

$$\begin{aligned} &A \wedge B \rightarrow D \\ &A \wedge C \wedge \neg Pr(B) \rightarrow E \end{aligned} \tag{2}$$

[5] This is called the 'contingent nature of absence' in (Tsang, 1991). Tsang states: "decision making may not be based on absence of constituents since any absent constituent may be present at a later stage in the process."

where Pr denotes the object-level provability relation. In Equation (2) the dependence of E on the inability to prove B is made explicit. This requires a meta-level statement about the derivability or otherwise of B to express the implicit reliance on negation as failure at the object-level. These constraints form a set of additional antecedents to the justification supporting any derived result.

This dependency cannot be expressed directly using logical negation at the object-level. The formula:

$$A \wedge C \wedge \neg B \rightarrow E \tag{3}$$

states that B must be *false* for E to be derivable, whereas Equation (2) states only that B should not be *derivable*. If we are willing to interpret negation as failure at the meta-level, we can obtain the same effect at the object-level by using the *closed world assumption* (or equivalently by using logical equivalence in definitions) (Clark, 1977).[6] While the *valueConflict* KS effectively enforces a (strict) form of the closed world assumption for instantiated variables allowing the derivation of $\langle false \rangle$ from $colour(block1, blue)$ and $colour(block1, red)$, the DKB syntax lacks the power of full first-order logic. As a result it is impossible to express universally quantified constraints such as $(\forall x)(\neg B(x))$. The problem is not one of detecting an inconsistency between two values, but recording the dependence of the result on the absence of *any* value for $B(x)$.

Such inferences effectively require a default logic; whether D or E is derivable depends on current state the DDD and what can be inferred from it. For example, if B is subsequently assumed or becomes derivable in the current environment, E will become inconsistent and the environment must be partitioned. However, as we have seen, even if we could represent the fact that B cannot be proved, we cannot establish whether it can in fact be proved without abandoning or seriously compromising the blackboard model of control, unless the KS can establish its own antecedents without assistance from the rest of the system.

Soliciting information from the user is also somewhat problematic. If the information supplied by the user leads to the derivation of an inconsistency which undermines the current state of the KS, what should the KS do? Should the inconsistency be recorded—after all, it involves a user assumption—and the KS fail? Alternatively, should the inconsistency be viewed as a 'possible inconsistency'—since a SPSS by definition only infers possible consequences—and the inference discarded? The answer to this question seems to depend on whether information solicited from the user 'counts as' an assumption (would the user have made the same assumption if the specialist had not asked for the information), and more generally whether negative information in the form of an inconsistent partial solution is regarded as useful information which should form part of the design history. If it is, then even if a consistent solution can be found, any inconsistent alternatives

[6] Note that this changes the semantics; $\neg Pr(B)$ becomes equivalent to $\neg B$. If this is to be consistent, Pr must accurately represent the object-level proof relation, i.e., $\neg Pr(B) \leftrightarrow Pr(\neg B)$ (Kowalski, 1979).

tried before the consistent solution was derived should also form part of the design history as a record of what didn't work and a partial justification of why the derived result was produced. Keeping a record of inconsistent alternatives may be the only principled approach, as any attempt to prove the consistency of a particular result is relative to some resource bound and subsequent processing may result in the derivation of an inconsistency.

7. The ATMS and Blackboard Models of Inference

Superficially, many of these difficulties appear to be the result of a conflict between the ATMS and blackboard models of inference. In the ATMS model described by de Kleer (1984), a problem-solver deduces new data from old through the application of rules or 'consumers'. Associated with each consumer is a set of data which form the antecedents of the consumer. A consumer is invoked when all of its antecedents are believed in the current context. Every inference resulting from a consumer invocation is recorded as a justification, which explicitly represents the dependency of the derived result on the antecedents of the consumer. In contrast, the blackboard model is much less specific about what constitutes an inference. In the blackboard model, the knowledge needed to solve the problem is partitioned into a number of independent chunks roughly corresponding to areas of specialisation within the task, and which divide the overall problem into a series of loosely coupled sub-tasks. These tasks or areas of expertise are implemented as 'knowledge sources'; inference systems or procedures capable of solving problems in their particular domain. Given such a broad characterisation of inference, a precise definition of what constitutes a knowledge source is difficult. Nii (1986) defines a a knowledge source as anything which "makes change(s) to blackboard object(s)".

There are obvious differences between the single rule application of the ATMS and the less easily characterised inferences performed by knowledge sources in the blackboard model. Whereas the ATMS model is monotonic with respect to proof, i.e., once a proposition has been derived it may not be deleted or changed, the blackboard model permits arbitrary modifications and deletions of blackboard objects. As a result it has been necessary to place a number of restrictions on the operation of the blackboard to allow truth maintenance. The most important is probably the absence of deletion and amendment. In the ATMS model, problem-solving is always carried out relative to a particular context. However the deletion of an item may have consequences for contexts other than the context in which the decision to delete the item was made. Amending an item suffers from the same problem (in that it implicitly involves deletion), unless the amendment entails the item in all the contexts in which the item is currently believed. Even with these restrictions a number of problems remain, particularly with respect to the integration of special purpose support systems, and further restrictions are necessary if the ATMS-blackboard model is to be coherent.

SPSSs infer possible rather than necessary consequences, which may involve relying on the absence of information in a particular context to draw a conclusion. To characterise this kind of inference requires a default logic. Even if we had some way of establishing the consistency of a particular datum, in the current implementation it is impossible to record the dependency of derived results on the continued absence of a particular set of assumptions. To do so would require a non-monotonic ATMS. A 'non-monotonic' ATMS is a monotonic ATMS augmented with 'out-nodes'; nodes representing propositions for which proofs cannot be found. An out-node *out-p* is defined to hold in the context of an environment E if and only if p is not a member of the context of E. We can view this definition as a general principle by which environments are completed with sets of out-nodes. The extension-base (the set of out-assumptions which hold for a given environment) and hence the context of an environment is non-monotonic with respect to assumptions and justifications. Since *out-p* holds in an environment E only when p is not a member of the context of E, the derivation or assumption of p results in the retraction of *out-p* and everything which depends on it. The ATMS used by EDS is monotonic. While it is possible to represent normal defaults using the approach proposed by de Kleer (1986c) (given a suitable extension to the DKB syntax to handle quantified negation) it computes incorrect labels for non-normal defaults (Junker, 1989).

As a result, unless care is taken, knowledge sources may violate the assumptions on which the ATMS-blackboard is based. If the inferences performed by the system are to be truth maintained, a number of limitations have to be placed on the design and implementation of knowledge sources:[7]

1. all assumptions used in deriving a result must form part of its justification;
2. a knowledge source may not use the absence of information to justify a conclusion; and
3. a knowledge source may not invoke another knowledge source which might render its antecedents inconsistent.

Condition (1) requires that a knowledge source be *functional*, i.e., the consequent of any resulting justification must be a function of its antecedents and nothing but its antecedents. Condition (2) follows immediately from Condition (1) and the fact that the ATMS-blackboard is monotonic. Condition (3) is a consequence of the inability of the ATMS-blackboard to backtrack. Since the ATMS-blackboard is monotonic, the only way to implement defeasible reasoning is to explicitly manipulate the contents of the DDD to reflect what is currently believed. Leaving aside the difficulties of backtracking over label propagation, such an approach violates the basic ATMS model.

8. The Role of Defeasible Reasoning in Design

If the analysis presented above is correct, we seem to be faced with a choice: either we accept the constraints on support systems presented in the previous section or

[7] These limitations are similar to the restrictions on consumers identified by de Kleer (1986b).

we have to extend the current system in a number of ways. Neither of these options is very attractive. If we stay within the bounds of what can be truth-maintained by the ATMS-blackboard the types of inferences that can be performed by special purpose support systems are very limited. On the other hand, implementing SPSSs as originally envisaged would require fundamental changes to the architecture of the system.

However we believe that these difficulties are indicative of more fundamental problems with our knowledge level analysis (the model of design support). We feel that extending the current implementation in the ways outlined above would in fact run counter to our exploration-based model of the design process, in placing total responsibility for a solution on a single sub-system. In effect it amounts to design without exploration. We are trying to graft a search/problem-solving based approach to design onto a system which is based on the idea that design is exploration, not search. We have argued that design problems cannot be solved this way; they are too complex and there would be too many exceptions to any rules we might formulate. Rather it is necessary to explore the space of possible designs; to try things and observe their consequences.

If we look at what happens when the designer asks a 'specialist' for advice about some aspect of a design (typically "how do I solve this problem" where the 'problem' may be the selection of a material or a dimension, or some more general goal such as what type of structure to use or whether a building should be single storey or multi-storey), we find that the specialist uses the characteristics of the of the design description, often together with information solicited from the user, to select one of a number of possible materials or to compute an appropriate size for the component. However it does not (and cannot) investigate the wider implications of such a decision.

Assuming the specialist makes some attempt to ensure that the advice it offers is consistent with the current design solution there are two problems:

1. the number of possible sources of inconsistency the specialist can check is limited; and
2. even if it were possible for a specialist to ensure that its advice is *currently* consistent, if an inconsistency is subsequently detected what is the user to do?

These problems arise because the capabilities of specialists are finite. They are constrained to produce their advice based on the local problem context and cannot have knowledge of the consequences of their proposals for all design criteria in all situations.[8] Moreover, as we have seen, the resulting information cannot easily be truth maintained. We are therefore left with advice based on limited knowledge

[8] Newell (1990) terms this a *trap-state mechanism*: i.e., a mechanism which is both local (contained within a small region of the system) and whose result must be accepted. Such a mechanism "can itself (by assumption) only be a source of a finite amount of knowledge; and when that knowledge fails (as it must if the tasks are diverse enough) there is no recourse (also by assumption). Thus the system is trapped in a state of knowledge, *even if the knowledge exists elsewhere in the system.*" (p. 221, emphasis added) The point is that we are more likely to discover possible conflicts by utilising all the resources of the system rather than by relying on a single sub-system.

which may be invalidated if the designer violates any of the assumptions made by the specialist about the future development of the design solution. Even if a KS could detect and respond to the source of the inconsistency by backtracking to another solution, it is ultimately bound to fail. The problem is not solvable by search by definition, so no consistent solution can be found. The assumptions the specialist makes to justify its conclusions about the presence or absence of certain design characteristics either mark a failure to perform exploration (i.e., if a consistency test was performed they would be inconsistent) or if they are consistent now (an exhaustive consistency check was possible) they will become inconsistent.

When an inconsistency is detected what should the specialist do? The design description is simply a collection of constraints which are jointly inconsistent. How is the specialist to know which constraints it can change to restore consistency? The initial specification of a design problem is typically incomplete and inconsistent. These characteristics may or may not be apparent to the designer when design begins. More often the problem is discovered to be incomplete and inconsistent as the designer explores the space of possible designs. Even if the problem is initially consistent, it may become inconsistent as the designer completes the problems description by adding 'missing' requirements or as the designer makes decisions about the nature of the design solution. There is no reason to suppose that designers abandon such requirements or decisions when they discover them to be inconsistent (see, for example, (Rowe, 1987)). Nor is there any reason to suppose they should; it may be possible to achieve a much 'better' overall design solution, albeit to a slightly different problem, by retaining the designer's (inconsistent) requirement or decision and relaxing one of the original problem requirements.

Even if we were to assume the 'original' or given set of (inconsistent) requirements were to take priority over the constraints defining a partial solution, i.e., those added by the designer, how does the specialist know which changes to the solution would be acceptable and which would not? To make such a decision would require at least a knowledge of the relative importance of the constraints and the possibility of their relaxation or the relative merits of various trade-offs between design the criteria. In short, *each* 'specialist' would have to know as much as the human designer. In fact it would have to know more, since much of this information is not available until the designer enters into a process of negotiation with the client or regulatory authorities responsible for the constraints. This does not seem to be a very promising basis for the design of 'modular' support systems.[9]

[9] Meyers, Pohl and Chapman (1991) in their work on the ICADS system have attempted to overcome this problem by centralising knowledge of possible conflicts between domain experts (i.e. specialists) and how these can be resolved in a single 'Conflict Resolver' module. While this allows domain experts to be developed independently of one another without worrying about inconsistencies in the advice they produce, it simply moves the problem elsewhere. With admirable candour they conclude that: "the assumption that the conflict resolution set required for the coordination of a representative number of domain experts can be largely predefined . . . is obviously not valid." and "the reasoning capabilities of the current ICADS working model . . . cannot be extrapolated to a real world working model involving two or three dozen domain experts and a comparable number

If the specialist is sufficiently clever it will produce nothing since none of the alternatives it can propose are consistent. If we were to imagine an extremely sophisticated specialist that knew everything about the domain, i.e., it effectively completes the design in all possible ways in an attempt to find a consistent solution to the design problem it has been given, then if the requirements are inconsistent—as they typically will be—all it can do is fail. The useful information produced by such systems concerning the consequences of decisions and their inconsistencies with the problem requirements and other decisions is produced by accident. The answers they produce are useless *as answers to the designer's original question* as they are bound to be inconsistent, but the side-effects of trying to find the answers (i.e the exploration) is useful. Paradoxically the more limited the specialist, the more useful its results. As it becomes more limited, it makes more assumptions which result in more consequences being derived and hence more inconsistencies being discovered. Would it therefore not be better to produce this information in an orderly way—looking for the inconsistencies and recording them—rather than looking for something which doesn't exist?

If this argument is correct there is no point in the specialist even trying to be consistent since it can't know all the possible sources of inconsistency; and if it did it couldn't produce a solution. From a logical point of view all the solutions it can produce are equivalent; the specialist may as well claim the first thing that comes into its head. Resolving the problem requires either changing it or finding a new way to solve it.

9. The Exploration Knowledge Source

We are therefore investigating a third approach which we believe offers a more appropriate solution to these problems which involves allowing the system itself to explore the space of possible designs. We envisage a number of 'exploration' KSs either user-invoked or triggered by aspects of the current design description, whose function is to propose extensions to the current design description *without first attempting to determine if the extension is consistent.* Clearly the majority of such extensions will be inconsistent with the current design description. However this is not the point—they serve to highlight the implications of plausible design options. We are not suggesting that such exploration KSs should produce parameter values at random, rather they will investigate the consequences of appropriate values given the current design and the priority ordering on goals defined by the user. In fact such knowledge sources will be very like specialists, except that they simply serve to initiate a line of exploration rather than offering considered 'advice'. As such they can afford to be rather more permissive in their proposals.

of knowledge bases. Any attempt to predefine the necessary conflict resolution set, even if this were possible, would serve to constrain the design space and restrict rather than enhance the creative ambiance of the design environment." (p. 915)

We represent the results of such exploration as conditionals or counterfactuals:

$$conditionals : \text{if } x = 5 \text{ then } y = 10 \quad \text{when } x \text{ is undefined}$$
$$counterfactuals : \text{if } x = 5 \text{ then } y = 10 \quad \text{when, for example, } x = 2$$

At present, while we can represent propositional conditionals, we do not derive them. The dependency information expressed by the conditional remains implicit in the justification structure and labels of the ATMS.

We can either allow the system to make assumptions or we can justify the hypothesis directly by the partial design description which triggered the 'exploration' KS. This greatly simplifies the representation of possible consequences, in that they are now a *monotonic* consequence of a partial design description. Conditionals summarise the justification structure, collapsing long chains of dependency relationships into single statements which summarise the structure of the problem. The (possibly counterfactual) conditional 'if $x = 5$ then $y = 10$' can be interpreted as stating that if the assumption '$x = 5$' were added to the current context (replacing any existing value for x) then '$y = 10$' would become derivable. By expressing such dependency information in this way we can highlight the implications of adding the assumption '$x = 5$' to the current context and suppress the fact that the conclusion '$y = 10$' depends on additional assumptions which also form part of the current context. The conditional is therefore only true in those contexts containing the additional assumptions on which it depends. However this is true of all derived information. The fact that the conditional depends on additional assumptions is recorded by the ATMS in the justification structure and the label of the conditional. If the underlying assumptions on which the conditional relies are subsequently discovered to be inconsistent, the conditional will lose support. Similarly it will cease to be true in any inconsistent extension of the current context.

By using the other knowledge sources to derive the consequences of possible design decisions we escape the problems of 'compartmentalisation' identified above and it allows us to use our existing blackboard architecture to derive the consequences of possible extensions to the current design description. Furthermore the 'exploration' knowledge sources seem to approximate more closely human design knowledge expressed as heuristics or rules of thumb about what might be a good idea to try in the current situation. An heuristic only says that if the condition is true the inference *may* be sound. The condition of an heuristic does rely on characteristics of the design description to identify those situations in which it might be useful from those in which they may not, but these do not attempt to be exhaustive. If the condition of an heuristic is true in some situation it does not mean that the heuristic will work; rather the condition only has to be sufficiently discriminating that the heuristic works enough of the time to be useful. If the heuristic doesn't work, simply try something else.

By relieving the SPSSs of the responsibility of knowing all the likely consequences of their suggestions (and in particular their unanticipated interaction with other parts of the design) we make it easier to engineer such capabilities in our systems. The combination of generator and test (together with the ability to summarise the resulting justifications structure as a conditional) works in the same way as a human designer, drawing on the full capabilities of the system in the form of KSs and module classes to assess the implications of proposals suggested by purely local considerations. To put it another way, a specialist no longer has to know everything. Rather than the specialist having to make explicit all the assumptions on which its advice is based—an impossible task as the set of such assumptions is infinite—it simply has to produce reasonable suggestions. For example, knowledge of statutory controls need no longer be encoded within the specialist (although it may be prudent to do so). While changes in legislation may invalidate a particular proposal made by the specialist, so long as this fact is detected by another KS the result will still be sound. The conditionals produced by the 'exploration' KSs cannot be used to answer precise questions about how a particular design problem *should* be solved, however we would argue such questions are unanswerable. The result of such exploration is design knowledge—knowledge of the type the designer is seeking about the structure of the problem.[10] One can imagine a persuasive front-end to the system responsible for presenting the results of the system's exploration to the user: "look, you can halve your costs if you use material x". We believe that such statements better capture the intuition underlying the notion of possible inference as it appears in the model of design support.

10. Conclusion

The task of a design support system has conventionally been conceived as one of providing the designer with solutions to specific parts of a design problem. The objective of such systems is to produce, either directly by elaborating the design description or indirectly in the form of 'advice' to the designer, consistent extensions to the current design proposal.

In this paper, we have argued that this goal is unattainable. It is important to stress that the problems identified above are not confined to EDS. In particular, they are not artefacts of the architecture of EDS. Rather the ATMS-blackboard makes explicit two important assumptions common to all design support systems, namely:

1. *Soundness*: that the advice given by the system should be 'good' advice. At a minimum it should be consistent with any existing partial solution.
2. *Constancy*: that the advice given should continue to be sound in the face of extensions to the design description.

While no system is infallible, we shall argue that a system is useful to the extent it approximates these goals. This can be seen most clearly if we examine what happens when a system violates one or other of these constraints. If a system fails,

[10] This is hardly surprising given that we have argued that the purpose of exploration is to produce such knowledge and (up to a point) it doesn't matter who does the exploration.

it is because the extension it proposes is (or becomes) inconsistent.[11] In many situations, although useless, this is relatively benign behaviour—the failure will be immediately obvious to the designer. However if we consider the designer's reasons for wanting to use such a system the failure of the system becomes more serious. If the designer is simply delegating a task to the system which the designer understands there is not much of a problem (although there is always the risk of an oversight on the part of the designer). However if the designer is relying on the system to augment his or her own skills and the advice produced by the system is inconsistent with the rest of the design the designer is stuck—they must adapt their design to the system's limitations or abandon the system and attempt to solve the problem on their own. On the other hand if the problem is more subtle and no obvious inconsistency results (for example if the designer and the system base their decisions on inconsistent assumptions which are never made fully explicit or if the system's advice subsequently becomes inconsistent) the consequence may be design failure.

To date these conditions have not been problematic because research efforts have mainly focussed on small, independent systems. The scope of such systems is typically fairly limited, so that it is not possible for the system to guarantee soundness. It is up to the designer to resolve any conflicts which arise between the advice offered by the system and that offered by other specialists or between the system and the existing design solution. Similarly, no attempt is made to ensure constancy. While the designer is free to re-run the system if an inconsistency arises or if the designer suspects the assumptions on which the system's advice rests have materially changed, the system itself cannot detect the revisions to the design description which would result in its advice being undermined. Independently, such systems are incapable of overcoming these limitations. Indeed it is only by concentrating on a single aspect of the problem that any progress has been possible in the development of such systems.

It is these assumptions of soundness and constancy which make the goal of modular 'assistants' capable of producing globally consistent solutions to parts of the design problem unattainable. The capabilities of such systems do not matter so long as their *purpose* is to propose consistent extensions of the design description. To solve any part of the design problem, a specialist effectively has to solve all of it, which eliminates all reasonable implementations. Design is essentially a holistic activity which cannot be broken down into sub-problems which are then solved independently.

We have proposed a solution to the problem of possible inference in the context of EDS based on the notion of 'exploration' knowledge sources. While this proposal is unique to the EDS architecture, we believe our general approach and in particular the objective of generating knowledge of the structure of the problem could be

[11] A system which claims to produce 'optimal' solutions can also fail if there exist better alternatives to the solution it proposes.

profitably applied to systems whose architecture is radically different from that of EDS. This is not just a change in emphasis, but a change in objectives which reflects an alternative view of the nature of the design process. We believe this to be a more fruitful approach than further attempts to develop existing search-based architectures.

Acknowledgements

The development of the Edinburgh Designer System was partially funded by the UK Science and Engineering Research Council as part of the Alvey large scale demonstrator project Design to Product (DtoP), other parts of which are funded by GEC plc and Lucas Diesel Systems (a division of Lucas Automotive Ltd.) whose active involvement in our research through the DtoP project we gratefully acknowledge, together with that of the other DtoP collaborators at Leeds University Department of Mechanical Engineering and at Loughborough University Department of Computing Studies, Department of Engineering Production, and the Human Factors in Advanced Technology research centre. Work on EDS is currently supported under SERC grant No. GR/F/6200.1.

Amaia Bernaras, Alistair Conkie, Ming Xi Tang and Nils Tomes read an earlier draft of this paper and made many helpful comments.

References

Buck, P., Clarke, B., Lloyd, G., Poulter, K., Smithers, T., Tang, M. X., Tomes, N., Floyd, C. & Hodgkin, E.: 1991, The Castlemaine project: development of an AI-based design support system, in *Artificial Intelligence in Design 91*, J. Gero, ed., Butterworth-Heinemann, 583–601.

Clark, K. L.: 1977, Negation as failure, in *Logic and Data Bases*, H. Gallaire & J. Minker, eds., Plenum Press, 293–322.

Dressler, O.: 1990, Problem solving with the NM-ATMS, in *Proceedings of the 9th European Conference on Artificial Intelligence*, L. C. Aiello, ed., Pitman Publishing, London, England, 253–258.

Gregory, S.: 1987, *Parallel Logic Programming in PARLOG*, Addison-Wesley.

Hayes-Roth, B.: 1985, A blackboard architecture for control, *Journal of Artificial Intelligence* 26, 251–321.

Junker, U.: 1989, A correct non-monotonic ATMS, in *Proceedings of the International Joint Conference on Artificial Intelligence*, Morgan Kaufman, 1049–1054.

de Kleer, J.: 1984, Choices without backtracking, in *Proceedings of the National Conference on Artifical Intelligence*, Austin, Texas, 79–85.

de Kleer, J.: 1986a, An assumption-based TMS, *Artificial Intelligence* 28, 127–162.

de Kleer, J.: 1986b, Problem-solving with the ATMS, *Artificial Intelligence* 28, 197–224.

de Kleer, J.: 1986c, Extending the ATMS, *Artificial Intelligence* 28, 163–196.

Kowalski, R.: 1979, *Logic for Problem Solving*, North Holland.

Lawson, B.: 1980, *How Designers Think*, Architectural Press, London.

Logan, B. S., Millington, K. & Smithers, T.: 1991, Being economical with the truth: assumption-based context management in the Edinburgh Designer System, in *Artificial Intelligence in Design 91*, J. Gero, ed., Butterworth-Heinemann, 423–446.

Logan, B. S. & Smithers, T.: 1992, Creativity and design as exploration, in *Modeling Creativity and Knowledge-Based Creative Design*, J. S. Gero & M. L. Maher, eds., Lawrence Erlbaum, Hillsdale, New Jersey, 149–188, (in press).

Meyers, L., Pohl, J. & Chapman, A.: 1991, The ICADS expert design advisor: concepts and directions, in *Artificial Intelligence in Design 91*, J. Gero, ed., Butterworth-Heinemann, 897–920.

Newell, A.: 1990, *Unified Theories of Cognition*, Harvard University Press, Cambridge, Mass..

Nii, H. P.: Summer 1986, The blackboard model of problem solving and the evolution of blackboard architectures , *AI Magazine*.

Reiter, R.: 1980, A logic for default reasoning, *Artificial Intelligance* 13, 81–132.

Rowe, P. G.: 1987, *Design Thinking*, MIT Press, Cambridge, Mass..

Smithers, T.: 1987, The Alvey large scale demonstrator project Design to Product, in *Artificial Intelligence in Manufacturing, Key to Integration*, T. Bernhold, ed., North Holland, Amsterdam, 251–261.

Smithers, T., Conkie, A., Doheny, J., Logan, B., Millington, K. & Tang, M. X.: 1990, Design as intelligent behaviour: an AI in design research programme, *Artificial Intelligence in Engineering* 5, 78–109.

Smithers, T. & Troxell, W. O.: 1990, Design is intelligent behaviour, but what's the formalism, *Artificial Intelligence in Engineering Design, Analysis and Manufacturing* 2, 89–98.

Tsang, J. P.: 1991, A combined generative and patching approach to automate design by assembly, in *Artificial Intelligence in Design 91*, J. Gero, ed., Butterworth-Heinemann, 485–502.

8

CASE-BASED REASONING IN DESIGN—1

ASKJEF: INTEGRATION OF CASE-BASED AND MULTIMEDIA TECHNOLOGIES FOR INTERFACE DESIGN SUPPORT

J. BARBER*
S. BHATTA**
A. GOEL**
M. JACOBSON*
M. PEARCE**
L. PENBERTHY*
M. SHANKAR**
R. SIMPSON*
and E. STROULIA**

**Human Interface Technology Center*
NCR Corporation
500 Tech Parkway, NW
Atlanta GA 30313 USA

and

***College of Computing*
Georgia Institute of Technology
801 Atlantic Drive
Atlanta GA 30332-0280 USA

Abstract. AskJef is a prototype AI system that helps software engineers in designing human-machine interfaces. It provides a memory of interface design examples, primitive domain objects, and design principles, guidelines, errors and stories. The design examples are represented graphically and decomposed temporally. The different types of knowledge are cross-indexed to enable the designer to navigate through the system's memory. AskJef helps software engineers in (1) understanding interface design problems by illustrating and explaining solutions to similar examples, and (2) comprehending the domain of interface design by illustrating and explaining the use of design guidelines. It uses text, graphics, animation and voice to present relevant information to the designer.

1. Motivations and Goals

That we need better human interfaces is axiomatic among designers. The ease of using an artifact, and of learning how to use it, depends in large part on the interface to the artifact. In the words of Jef Raskin, "the interface is the product."[1] Unfortunately, as Norman (1988) has argued, the interfaces to even many everyday artifacts leave much to be desired.

Perhaps because software engineering is a relatively young discipline, many software interfaces need improvement. By software interfaces, we mean not just

[1] Jef Raskin, a leading authority in interface design, was the domain expert on our project. The AskJef system is named after him.

J. S. Gero (ed.), Artificial Intelligence in Design '92, 457–475.

interfaces to computers, but, more generally, human interfaces to all information-processing machines including cash registers and automatic teller machines. These interfaces typically are designed by software engineers who have little or no formal training in interface design. In addition, interface design generally does not receive enough time or attention in most software engineering projects. What makes the design of software interfaces especially hard, however, is the *ill-structured* nature of the domain. Software engineers generally do not have access to adequate guidelines that they can productively use in designing interfaces. Although guidelines such as Common User Access (CUA) (IBM, 1991a; 1991b) exist, there is no standard set of comprehensive design guidelines in the industry. Further, the guidelines are often vague, inconsistent, incomplete, inaccessible, poorly documented, hard to comprehend and even harder to use (Tetzlaff and Schwartz, 1991).

Supporting software engineers in interface design thus presents a real-world challenge to Artificial Intelligence (AI). Given the creative nature of the task and the ill-structured nature of the domain, it seems unlikely that present-day AI technologies can be used for building autonomous interface design systems. The issue then becomes whether and how AI technologies can be used to *help* software engineers. The goal of this paper is to examine a specific hypothesis to this end, namely *case-based reasoning and multimedia provide a useful combination of technologies for supporting design in ill-structured domains, and, more specifically, for helping software engineers in designing human-machine interfaces.* By case-based reasoning, we mean reasoning from previous experiences.

This hypothesis is based on two considerations. First, some studies of interface design, for example (Tetzlaff and Schwartz, 1991), show that software engineers typically prefer specific examples to general guidelines as advice on their tasks. Previous experiences with interface design are a major source for such examples. Second, recent AI research, for example (Barletta and Hennessy, 1989; Dyer, Flowers and Hodges, 1987; Goel, 1991; Hinrichs and Kolodner, 1991; Huhns and Acosta, 1987; Maher and Fenves, 1987; Navinchandra, 1991; Sycara and Navinchandra, 1989), shows that case-based reasoning is a useful approach to design.

To elaborate and evaluate the above hypothesis, we constructed a prototype AI system called AskJef. AskJef has two main operational goals. The first is to help the interface designer understand a given problem and to indicate the structure of possible solutions, thus shortening the time needed to produce an acceptable initial interface design. This is done by presenting examples of past solutions to similar problems. Adaptation of these examples to the new problem specification is left to the designer. The second goal of AskJef is to help the designer understand the interface design domain. This is done by illustrating and explaining design guidelines pertaining to the current problem, thus operationalizing the interface design knowledge in the context of specific problems. These goals are based on a survey of NCR interface designers that we conducted at the start of the AskJef project.

The design, construction, and evaluation of AskJef has raised a number of new practical, theoretical, and methodological issues. For example, how to present information to the software engineer (AskJef uses text, graphics, animation and voice), how to cross-index design examples and guidelines (we developed an indexing vocabulary), how to represent the content of an example (AskJef uses both graphical and symbolic representations), how to decompose complex examples (AskJef decomposes them temporally), and how to navigate the system memory. The remainder of this paper discusses these issues in detail.

2. System Architecture

AskJef contains two cooperating modules: memory and interface. The memory module manages different types of knowledge such as design examples and guidelines. Given a problem specification, the memory manager accesses relevant knowledge from memory. The interface module interprets the designer's actions, allows him to specify a problem, and uses several modalities to present information that the memory manager retrieves.

AskJef runs in the Microsoft Windows 3.0 environment. Its memory manager is implemented in the ART-IM knowledge tool from Inference Corporation. Its interface manager is implemented in ToolBook, an interface programming environment from Asymmetrix Corporation. Communication between the two modules is through Windows Dynamic Data Exchange (DDE).

3. Knowledge Representation

The AskJef system integrates several types of knowledge: interface design examples, primitive domain objects, design principles, guidelines, error prototypes and stories. All these types of knowledge can be broadly categorized into *experiential* knowledge, consisting of interface design examples and stories, and *domain* knowledge consisting of the interface objects, principles, guidelines, and prototypical errors. The organization of these types of knowledge in AskJef's memory is shown in Figure 1.

3.1. Interface Design Examples

Each interface design example is a previous interface developed for a specific task. The example of an interface developed for customer-activated terminals (CATs) in fast-food restaurants, which customers use to order food items without human assistance, is shown in Figure 2. Examples are used in AskJef to help the designer understand the given problem and to suggest the structure of possible solutions to the problem. They provide the designer with a basis for the design of an interface for the current problem. Examples are also used for the

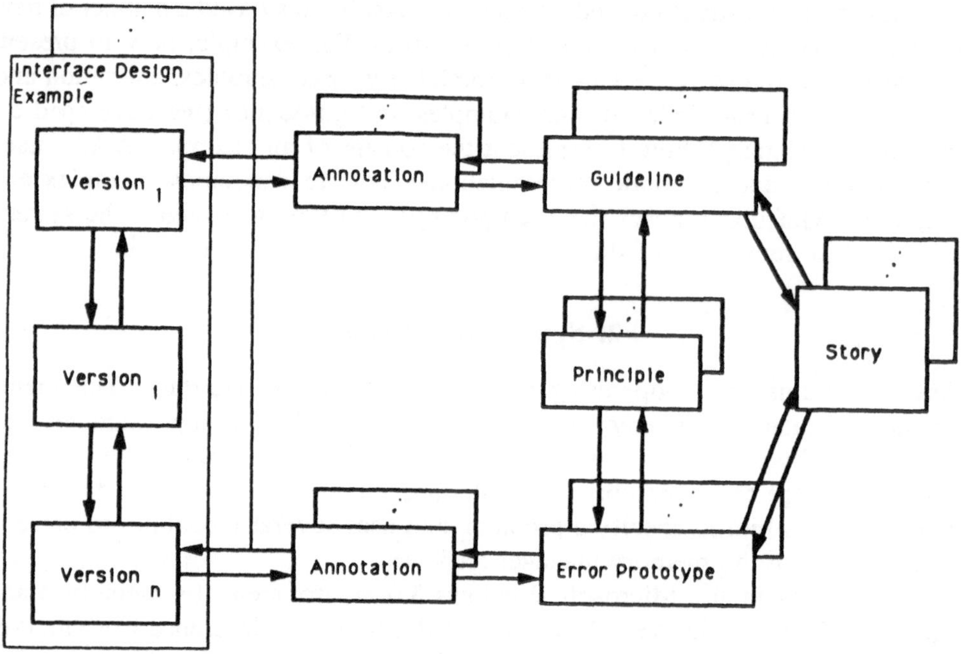

Fig. 1. Knowledge Organization in AskJef

operationalization of abstract principles and guidelines. Guidelines are used to evaluate the examples. This evaluation is stored as an annotation associated with each relevant example.

Interface design in general is an iterative process that often leads to the design of multiple versions of the same interface. Each version typically is a complete interface that attempts to satisfy the given problem specification in a better way than previous versions. Figure 3 illustrates a second version of the interface for CAT shown in Figure 2. Each version also has associated annotations. Each annotation discusses how an interface object satisfies or violates a particular guideline. These annotations are helpful for illustrating, explaining, justifying and critiquing various aspects of the interface. AskJef contains 5 design examples in its memory, with 9 versions and 38 annotations. The examples and their versions in AskJef are examples of actual interfaces, previously designed by NCR software engineers at Human Interface Technology Center.

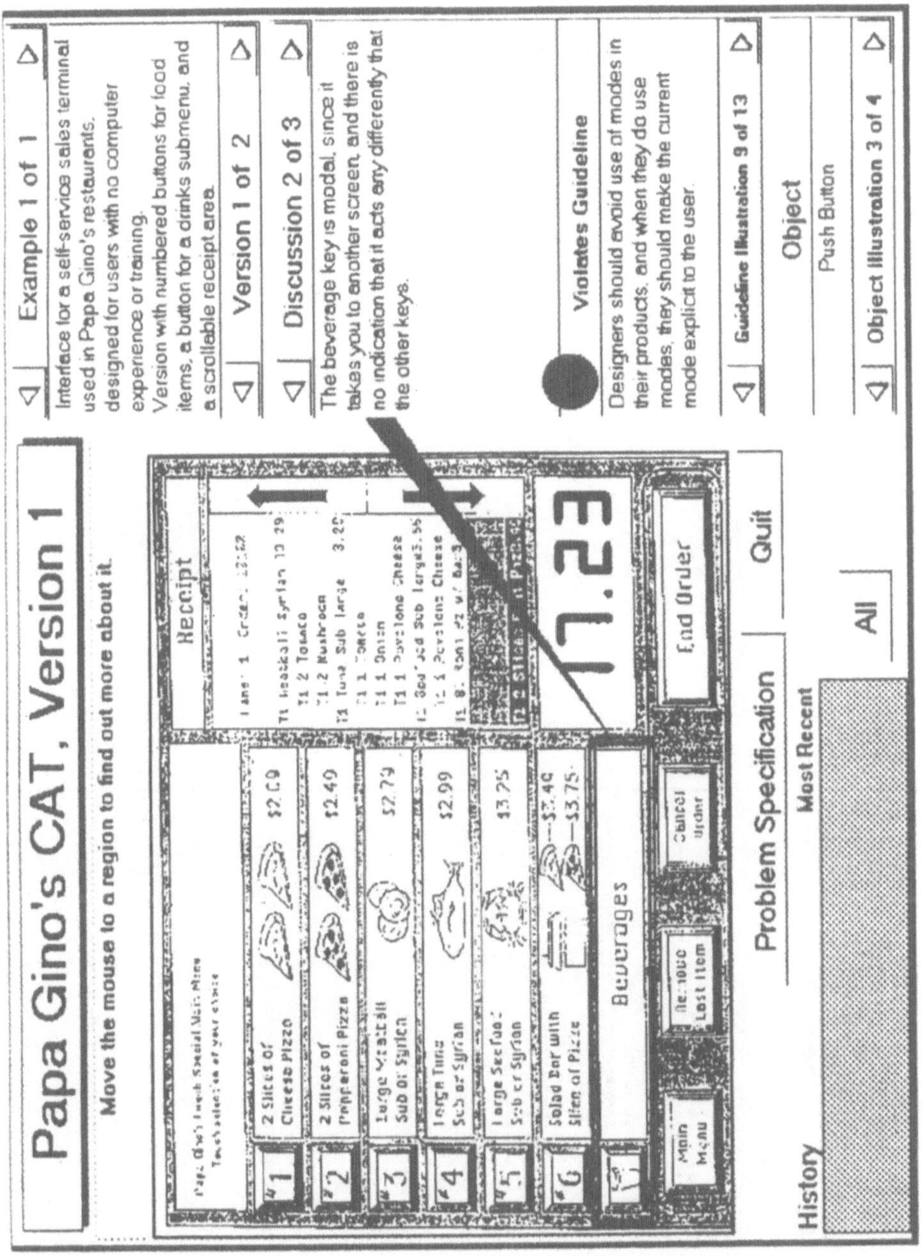

Fig. 2. Interface Example Screen (Papa Gino's CAT, Version 1)

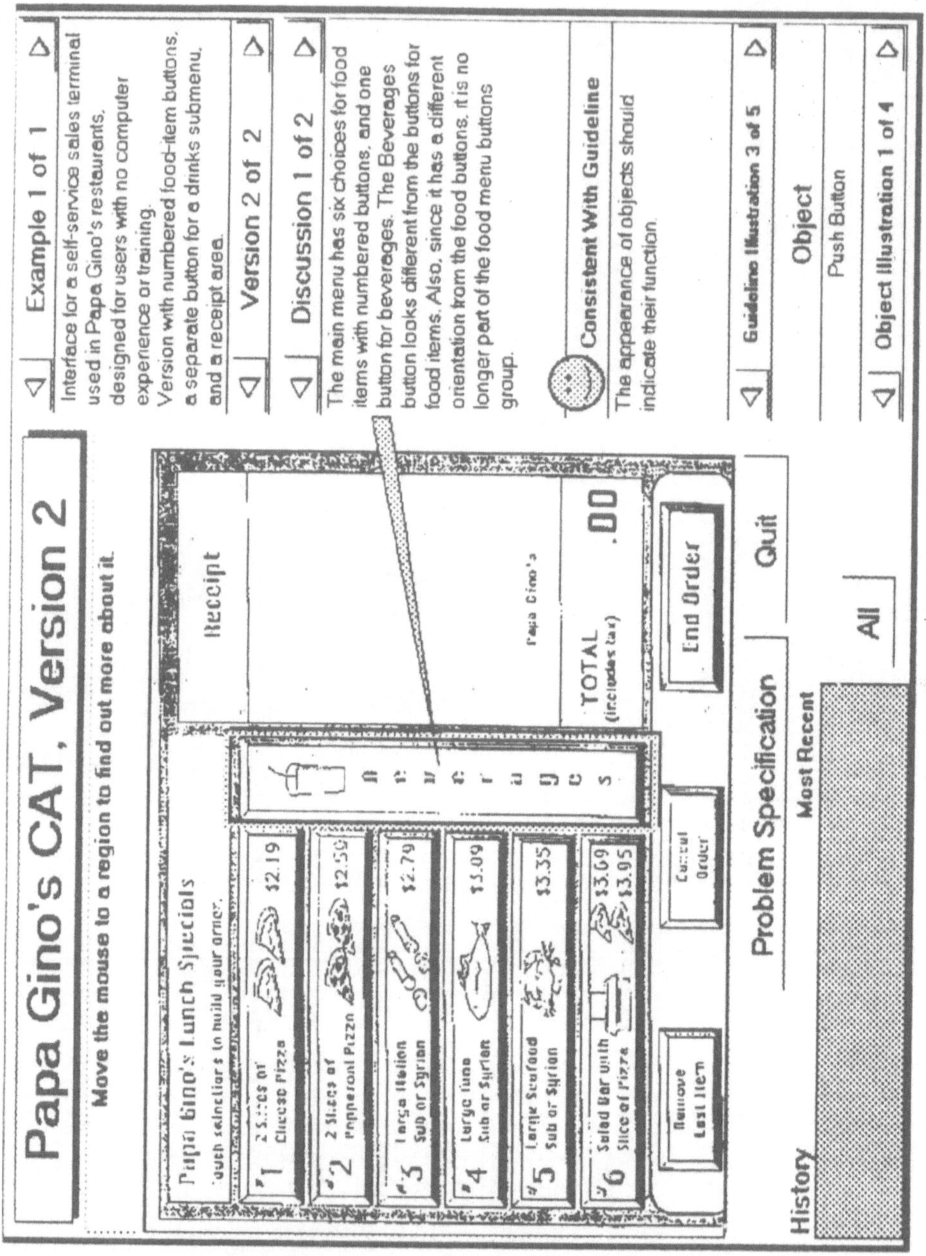

Fig. 3. Interface Example Screen (Papa Gino's CAT, Version 2)

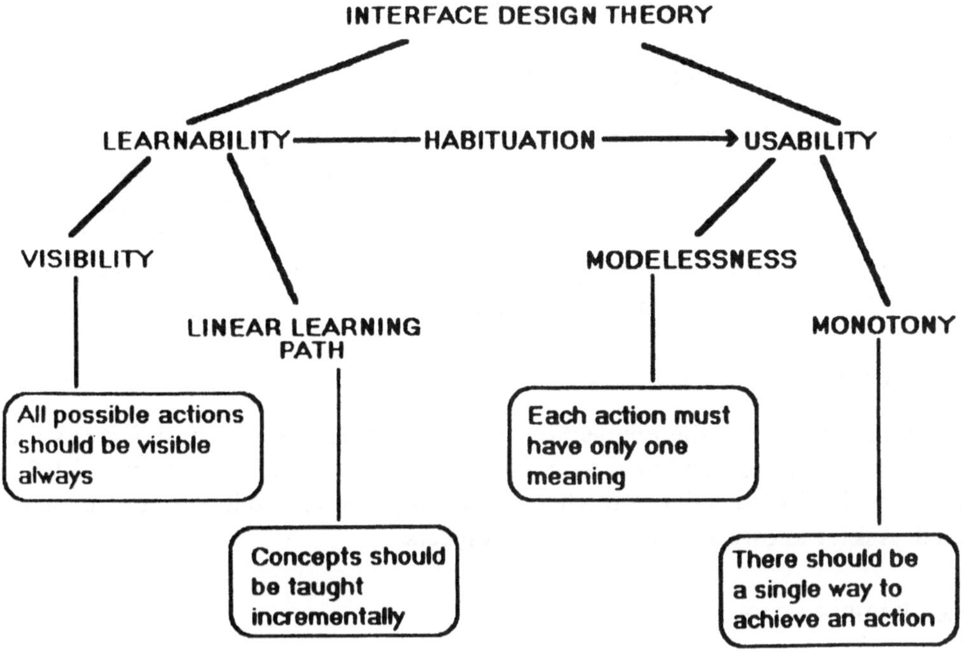

Fig. 4. AskJef's Principles of Interface Design

3.2. Principles of Interface Design

The principles in AskJef are derived from the interface design theory of Jef Raskin. These principles explain the properties of good interfaces at a very general level and help to make the designers aware of important design issues and techniques. The principles illustrated in Figure 4 are often domain independent. For example, the principle `Modal interfaces should be avoided` is valid both for software and mechanical interfaces. These principles are used in AskJef to organize and motivate guidelines and prototypical errors. AskJef contains 7 principles.

3.3. Guidelines for Interface Design

A guideline in AskJef is a domain-specific heuristic that pertains to specific interface design subtasks and objects. An example of a guideline is `Make things that behave the same, look the same`. Guidelines in AskJef, while not taken from any one source such as IBM's Common User Access (CUA) guidelines (IBM, 1991a; 1991b), they are representative of those commonly used in the computer industry. These guidelines provide the interface designer with prescriptive heuristics and are presented in the context of the current problem.

AskJef currently contains 47 guidelines.

3.4. Prototypical Errors in Interface Design

In AskJef, prototypical errors are known errors committed by novice interface designers. They are empirical in nature and usually result from the violation of a principle. An example of a prototypical error is `Recalcitrant interfaces divert the attention of the user from the task to the interface.` In AskJef these prototypical errors are used to warn designers of potential pitfalls. AskJef currently contains 11 error prototypes.

3.5. Primitive Interface Objects

Interface objects are the elementary building blocks such as *push buttons* and *menu bars* that are combined to construct interfaces. Each object in AskJef is accompanied by guidelines on when it is appropriate to use it and how it can be used. AskJef knows about 15 interface objects.

3.6. Illustrative Stories

Stories in AskJef are cross-domain examples, gleaned from knowledge engineering sessions with Jef Raskin, that illustrate specific principles, guidelines, or errors. For example, one story in AskJef explains how some aircraft cockpit designs often violate the principle of modelessness as illustrated in Figure 5. In AskJef, stories are primarily used as alternative illustrations of principles or guidelines or errors within the context of specific interface design examples. They help convince the designer that these suggestions should be heeded to. AskJef currently knows 47 stories.

4. Modes of Representation

Knowledge in AskJef is represented in three modalities: symbolic, graphical and audio. Since software interfaces are generally visual in nature, the interface design examples are represented graphically. However, the graphically represented examples are indexed symbolically so that they can be retrieved given a problem specification in symbolic form. Each interface example points to several annotations. The representation of an annotation also consists of symbolic and graphical parts. The graphical part points to a specific feature in the interface example; the symbolic part points to the guideline or error that the feature illustrates. Principles, guidelines and errors are represented symbolically. Primitive interface objects are represented symbolically, with pointers to the graphically represented examples and symbolically represented guidelines. The representation of the stories consists of three parts: symbolic, graphical and audio. The text of a story is represented symbolically, the animation portion is represented

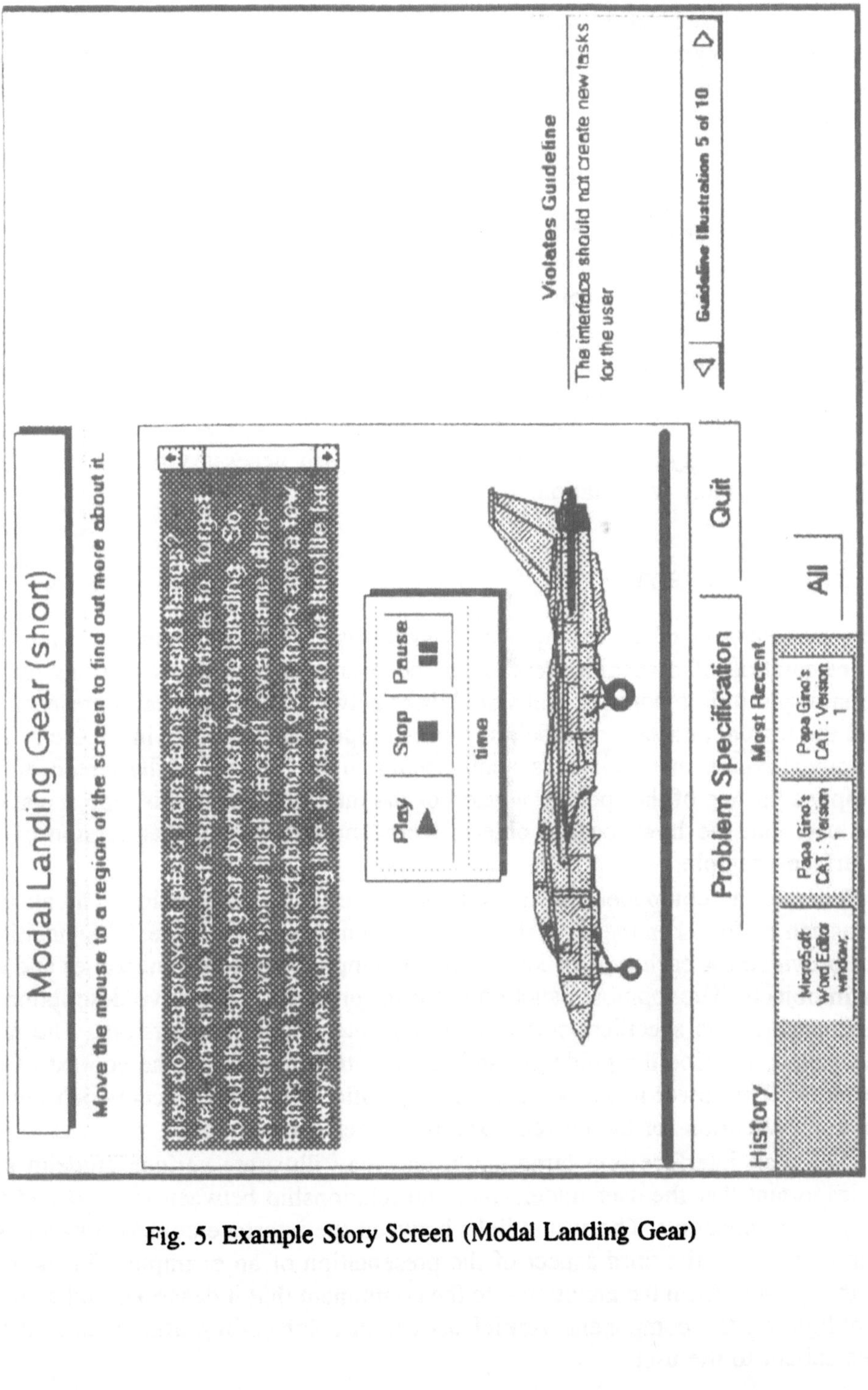

Fig. 5. Example Story Screen (Modal Landing Gear)

graphically, and the narration of the story is in audio form. This representation of AskJef's knowledge in multiple modes results in a useful memory organization in which symbolic, graphical and audio representations are combined and cross-indexed. We will return to the issue of memory organization a little later.

5. Presentation of Information

One of the major functions of the AskJef interface manager is to communicate useful information to the designer. In general, the presentation modality may depend on the intended user, the task, the domain, and the type of information that has to be presented. AskJef uses several techniques to present information: text, graphics, animation, and sound. Principles, guidelines, and error prototypes are presented in text. Interface design examples are presented graphically since they consist of screen layouts. Stories generally incorporate text and graphics but the complexity of some of the stories makes it necessary to use sound and animation in their presentation.

5.1. Example Presentation

The presentation of an example in AskJef has three components: a graphical representation of a specific version of the example; textual annotations on the version and its components; and visual links between the graphical representation and the textual annotations. Each of these components is shown in Figure 2. The graphical representation is consistently found on the left side of the screen, and is simply a bitmap of the specific version of the interface. From this representation, the user can see how specific objects were implemented in that version of the interface example.

The second component of an example is the textual annotation. The annotations can be found on the right side of the screen shown in Figure 2. Annotations are provided for each type of knowledge: examples, versions, guidelines and domain objects. The topmost panel on the right provides a narrative description of the interface, its specification, background, and its various versions. The middle panels describe the guideline and its operationalization in the context of the example. The lower panel describes the specific interface objects which are the focus of attention for the current example under consideration.

Since an interface is a large entity and may illustrate various guidelines, it is important that the user understands the relationship between the text and the graphic components. The visual links between the interface components and the annotations are the third aspect of the presentation of an example. This is done with a pointer from the annotation to the component that it describes and a "box" highlight on the component. AskJef uses both color coding and icons to make this salient to the user.

5.2. Story Presentation

The layout for the stories is similar to that for the examples. For each story, AskJef displays the text of the story and the principle, guideline or error that the story illustrates. In some stories a voice narrating the story and an animated description of the story are also presented as indicated in Figure 5. The combination of audio narration and synchronized animation is an especially effective means of communicating the point of the story.

6. An Example of AskJef in Action

The interface designer begins a session with AskJef at the problem specification screen, shown in Figure 6. Help for all screens is always available in a message area located at the top of the screen, directly below the title bar. The designer selects the appropriate categories relevant to the current problem. For example, if the designer is trying to get help on designing a customer-activated terminal (CAT) for a fast-food restaurant, he might select the categories indicated with a "X" in Figure 6.

The designer then selects the "Example" push-button. In response AskJef retrieves interface examples from its memory. For the above problem specification AskJef retrieves the interface example shown in Figure 2. The designer may browse though various annotations on the displayed version and see different versions of the current example. The designer may also browse the annotations pertaining to a guideline or an object. For instance, the Papa Gino's CAT example shown in Figure 2 contains a menu of pushbuttons that violates one of AskJef's guidelines. The button labeled "Beverages" is used by Papa Gino's customers to get a submenu of available drinks. It looks like the other buttons, which can be confusing. This violates the guideline `Things that act different should look different.`

By navigating along the "Guideline Illustrations" in Figure 2, the designer reaches the second version of Papa Gino's CAT, shown in Figure 3. One of the annotations in this version explains how the above problem was corrected: The "Beverage" button is now set apart to reflect the fact that it acts differently from the other buttons.

When the designer has browsed through the current example he might realize that the current problem needs to be specified differently. To do this, the designer selects the "Problem Specification" button. The system responds by saving the current context of the selected problem and creating an icon in the "History" area that will allow the user to revisit this specification if needed, and returning the user to the problem specification screen.

The problem specification limits the set of examples the designer can view. However, he may decide at some point that he wishes to view all available examples, whether they match the problem specification or not. The designer

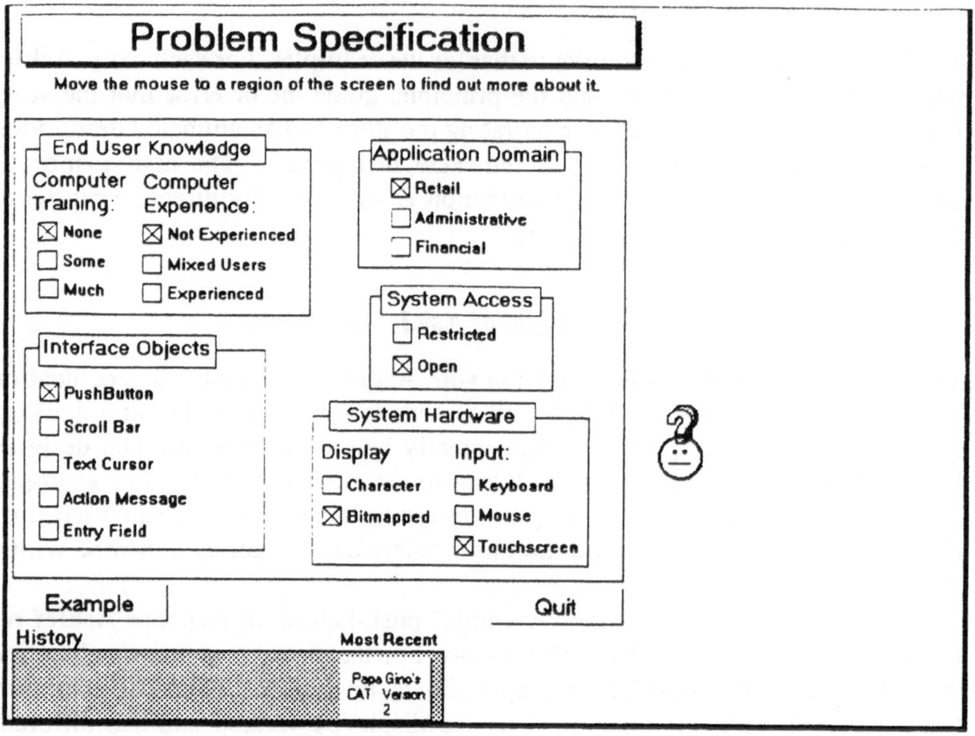

Fig. 6. The Problem Specification Screen

can do this by clicking on a button on the example-viewing screen labeled "All". This makes the entire set of examples available for browsing. It also creates an icon in the "History" area, which can be clicked on to go back to a restricted set of examples.

7. Memory Organization

We now return to the issues of organization of AskJef's memory and the retrieval of information from it.

7.1. Example Decomposition

In practice, the design specifications for an interface are often incomplete and are sometimes internally inconsistent. Software engineers usually design interfaces iteratively, in a "generate and test" fashion. Each iteration produces a new version of the interface design. As each version of the interface is developed, it is evaluated, redesign decisions are made and then implemented in the next version. The various versions collectively constitute a history of design decisions that lead to the final design. Each design example in AskJef contains a temporal evolution of an interface design via its intermediate versions. This knowledge is useful both for advising designers on specific problems and in tutoring them on issues of design. It provides an account of how interface designs evolve, the kinds of errors designers make, and how they can be fixed.

7.2. Principle, Guideline and Error Hierarchies

Principles in AskJef are organized in a generalization-specialization hierarchy. Guidelines and errors are also organized in their own generalization-specialization hierarchies. Guidelines are linked to principles through *enables* links. Similarly, errors are linked to principles via *disables* links. Each guideline or error contains information about its *positive* and *negative illustrations* in the form of examples and stories. A positive illustration of a guideline is an example that conforms to it and a negative illustration is one that violates it. For error prototypes, the situation is reversed.

7.3. Object Categories

The guidelines at the lowest level of the guideline abstraction hierarchy are associated with the interface design objects, such as *push buttons*, *menu bars*, and *scroll bars*. Our usability studies showed that when the designer is at the implementation stage of interface design, it is useful for AskJef to provide access to information about interface design objects, the associated guidelines and the illustrations of how these objects can be used. Following this study, we represented the interface-design object descriptions in AskJef and organized them into hierarchies around the categories that are prevalent in the domain. For example, menu bars and scroll bars are "control objects." Each object description refers to the guidelines of two types – *when-to-use* and *how-to-use* – and also refers to the *positive* and *negative illustrations* of interface designs that use this object.

7.4. Retrieval and Navigation

When interface designers ask questions about problems, they usually ask questions about specific problems they have with their own design rather than abstract ones. For this reason, a designer's access to AskJef's memory is primarily through the interface design examples. Once an interface design example is retrieved from memory, AskJef's memory organization provides direct access to the other types of knowledge such as guidelines and stories.

The retrieval process begins when the designer specifies a design problem. AskJef uses the problem specification as a probe to retrieve appropriate examples. Example retrieval is based on elimination of examples that fail to meet the specification and not on the selection of examples that match the problem specification. Consequently, the fewer the features in the specification, the larger is the set of examples retrieved from the system memory. AskJef presents the closest matching example to the designer and also informs him of other examples, rank-ordered by their similarity to the problem specification. The similarity of an example to a problem is measured by the number of matching features in their specifications.

The designer may then browse through the examples presented by AskJef. If he wishes, he can also navigate through the larger set of all examples contained in AskJef's memory. The dimensions along which the designer can navigate are: examples, versions, discussions, guideline illustrations, and object illustrations. That is, he can view additional examples that match the current problem specification, additional versions of the same example, additional discussions for the current version, additional illustrations of the current guideline or error prototype, or additional illustrations of the current interface object. Note that the problem specification limits the scope of memory that the designer can navigate through. However, the designer can explicitly change the scope of his navigation to the whole memory by specifying that he is interested in all the examples. AskJef keeps track of all the points in its memory the designer visited and creates a "landmark" that saves the current context and the location within that context each time he moves to change the problem specification. The landmark can be the name of an example, version, or annotation which the designer was visiting. These "landmarks" are displayed to the designer as a "history" of his visits, and allow him to quickly return to any landmark at any time simply by selecting the landmark.

8. Evaluation

Before designing AskJef, we conducted a survey of NCR software engineers who design interfaces to determine what characteristics of a system like AskJef would be most useful. The design of AskJef followed the points made by the surveyed interface designers.

In our experiments with AskJef, we found that the features shown in Figure 6 are discriminative of the different examples stored in AskJef's memory. Of course, the above list of features is not complete but is intended only to be representative of the types of features that might be useful in a large system. Additional work is necessary to produce a comprehensive set of features for problem specification.

During and after its implementation, we evaluated AskJef using both evaluation by experts and usability testing. Expert evaluation was used because it typically identifies many more problems than other techniques do (Jeffries, Miller, Wharton and Uyeda, 1991). Usability testing was used because it is good at identifying serious problems and has the added advantage of identifying recurring problems (Jeffries, Miller, Wharton and Uyeda, 1991).

8.1. METHOD

Expert evaluation went on throughout the development of AskJef, which allowed us to rapidly create several prototypes of the system. The evaluators were experts in the fields of interface design, human-computer interaction, artificial intelligence and software engineering. The evaluators recorded and discussed their notions of what areas of AskJef's knowledge a designer might want to explore and how he might navigate to these areas. They also worked with AskJef looking for grammatical errors, wording of messages, terms used for commands and so forth. These "clerical" problems were also recorded and corrected. We collected a series of questions from the expert evaluations.

Later in the development of AskJef, two formal usability tests were conducted using eight NCR employees not involved with the development of AskJef. Two interface designers and two software engineers participated in the first test, and one interface designer and three software engineers participated in the second. Both tests were videotaped so they could be reviewed later in greater detail. In the first test, the users worked through a brief tutorial of AskJef and then used the system for two tasks: design of an interface, and redesign of an interface. These tasks allowed us to assess the usability of AskJef's guidelines as well as the usability of AskJef itself (Molich and Nielson, 1990; Thovtrup and Nielson, 1991). After completing the two tasks, the users answered the set of questions initially raised by expert evaluations. Their answers led us to a second set of questions.

In the second test, users worked through an extensive tutorial that required a great deal of interaction with AskJef, and then used the system to solve an interface design task. Again, after completing the design task, the users answered the revised set of questions raised by the first test.

8.2. RESULTS

Results from the two usability tests came from comments users made during testing and from answers to the test questions. Results are divided into those that pertain to the problem specification and those that pertain to the interface examples. Designers were generally satisfied with how a problem was specified in AskJef. The problem specification screen (Figure 6) was effective in allowing them to obtain relevant design information for their design or redesign task. Having categories of specification items helped them break down their tasks. However, they did note a number of additional categories and items that they would like to see in the problem specification screen. Designers also noted that some items should be mutually exclusive. For example, it is unlikely that access to the system being designed would ever be both open and restricted.

Designers found that annotated examples were an effective means for advising them on interface design. Examples allowed them to quickly see effective and ineffective ways of designing interfaces that were relevant to their problem, and to understand how specific guidelines should be applied. However, they felt that interactive rather than static examples would be even better. Designers would also like to directly select the aspect of a particular interface example presented.

AskJef's extensive navigation controls initially overwhelmed some users but they quickly learned the different navigational routes. After working through AskJef, users typically wanted additional methods of navigating. According to many users, moving from an example with a violation of a particular guideline to an example with a correction of the violation of that guideline would be a useful addition. Users would also like to directly move from one guideline to the next. Finally, having a history of problem specifications was a very useful feature that some users likened to a bookmark to which they could go back.

9. Related Research

Using stories and cases for automated advising and training of humans has been strongly advocated by Schank and his colleagues (Schank, 1990). They have constructed ASK-TOM, an advisory system in the domain of trust banking (Schank *et al.*, 1991). ASK-TOM provides an ever-present expert for query by novice consultants. It contains video clips of a domain expert talking about typical problems in trust banking. One of the primary assumptions in ASK-TOM is that novice's questions can be anticipated and answered. Memory navigation in the system is completely user-directed along pre-established associations.

Using cases specifically for helping designers has been advocated recently by Kolodner and her colleagues (Kolodner, 1991; Goel *et al.*, 1991a). They have constructed ARCHIE, an advisory system in the domain of architectural design. Like interface design, the domain of architecture design is ill-structured.

Our work on AskJef is partly inspired by ASK-TOM and ARCHIE, and shares many assumptions with them. However, AskJef differs from ASK-TOM and ARCHIE in several significant ways. First, while all three system use examples, only AskJef provides a temporal decomposition of examples into different versions of the examples. Second, while ASK-TOM relies almost exclusively on stories and ARCHIE similarly relies exclusively on cases, AskJef contains not only examples and stories but also design guidelines, principles, and objects. Thus the types of knowledge in AskJef are quite different from those in these previous systems. Third, while ASK-TOM's and ARCHIE's knowledge is almost entirely domain-specific, AskJef contains domain-independent design principles and cross-domain stories in addition to domain-specific design examples and guidelines. Fourth, while all three systems address the issue of indexing examples in memory, only AskJef addresses the issue of cross-indexing different types of knowledge so that a designer can navigate from one type of knowledge to another. Thus the indexing problem in AskJef is quite different. Finally, while both ASK-TOM and ARCHIE use examples primarily for suggesting solutions to problems, AskJef also uses examples for illustrating design guidelines.

The AskJef system is also related to recent work on use of hyper-media to build design environments, for example (Fischer, McCall and Morch, 1989). The two lines of research appear to be complementary. Case-based reasoning focuses on the content and organization of experiences while hypermedia appear to provide a useful scheme for navigating through the memory.

9.1. Further Research

We are now constructing a new version of AskJef that builds on our experience with the current version. Some of the key considerations in the design of the new system are (i) extending the vocabulary for the specification of design problems, (ii) describing and indexing examples at multiple levels of decomposition, (iii) defining the semantics of design principles, guidelines and errors more carefully, and (iv) helping the designer form a mental model of the system's memory organization.

10. Conclusions

The AskJef experiment leads us to draw several conclusions.

Ill-structured domains: The domain of interface design is ill-structured. While some design principles and guidelines do exist, they are generally difficult to understand or to apply to specific problems. One method for overcoming this difficulty is to use examples to explain the principles and guidelines and to illustrate their usage. AskJef demonstrates the usefulness of this approach.

Case-Based reasoning: Examples provide a useful approach for structuring knowledge, especially in ill-structured domains such as interface design. Examples are very effective in communicating the point about abstract guidelines and principles. Previous cases of interface design are a major source of examples. AskJef demonstrates the usefulness of examples in supporting interface design.

Integrating multiple technologies: AI as a stand-alone technology is insufficient for real world problems. In order to engineer successful usable systems, AI technologies must be combined with other appropriate technologies so that users *can* use them and will *want* to use them. AskJef shows how a combination of AI technology with conventional software and multimedia technology leads to synergetically expanding capabilities.

Multimedia technology: Multimedia technology provides tremendous flexibility for presenting information. Combinations of multiple modalities of presentation, such as animation with audio narration, can be used to maximize the communication value of presentations. AskJef shows how the use of multimedia can enhance the effectiveness of a design support system.

Real world problems: Finally, AskJef is a demonstration of the application of known technology to a real-world problem. AskJef is also a demonstration of the use of real-world problems to identify new theoretical issues and constructs, such as graphical representation of examples, temporal decomposition of examples, and cross-indexing between different types of knowledge represented in multiple modalities.

Acknowledgements

We are grateful to Jef Raskin for allowing us to interpret and use his principles of interface design and stories that illustrate the principles. We are also thankful to Alana Anoskey, Albert Badre, Richard Catrambone, Bart Elias, Dick Henneman, Mike Inderrieden, Alex Kirlik, Janet Kolodner, Beth Meyer, Lynn Miller, David Rubini, and Ali Vassigh for their critiques of the AskJef system, and Adam Lewis, Antai Peng, Khun So, Kim Stout, and Sam Wagner for participating in usability testing. We thank Kavi Mahesh and anonymous reviewers for their comments on an earlier draft.

References

Barletta, M. and Hennessy, D.: 1989, Case Adaptation in Autoclave Layout Design, *Proc. DARPA Case-Based Reasoning Workshop.*

Dyer, M., Flowers, M. and Hodges, J.: 1986, EDISON: An Engineering Design System Operating Naively, *Proc. First International Conference on AI Applications in Engineering.*

Fischer, G., McCall, R. and Morch, A.: 1989, Design Environments for Constructive and Argumentative Design, *Proc. CHI'89,* ACM, New York, pp. 269-275.

Goel, A.: 1991, Representation of Design Functions in Experience-Based Design, *Proc. IFIP WG 5.2 Conference on Intelligent Computer-Aided Design,* Columbus, Ohio, pp. 269-292.

Goel, A., Kolodner, J., Pearce, M., Billington, R. and Zimring, C.: 1991, A Case-Based Tool for Conceptual Design Problem Solving, *Proc. Third DARPA Workshop on Case-Based Reasoning,* Washington D.C., Morgan Kaufmann, Los Altos, CA, pp. 109-120.

Hinrichs, T. and Kolodner, J.: 1991, The Roles of Adaptation in Case-Based Design, *Proc. Ninth National Conference in Artificial Intelligence,* pp. 28-33.

Huhns, M. and Acosta, E.: 1988, Argo: A System for Design by Analogy, *IEEE Expert,* Fall.

IBM: 1991a, *Systems Application Architecture Common User Access Guide to User Interface Design* (SC34-4289-00, June 4, 1991 Draft), IBM, Armonk, New York.

IBM: 1991b, *Systems Application Architecture Common User Access Advanced Interface Design Reference* (SC34-4290-00, June 4, 1991 Draft), IBM, Armonk, New York.

Jeffries, R., Miller, J.R., Wharton, C. and Uyeda, K.M.: 1991, User Interface Evaluation in the Real World: A Comparison of Four Techniques, *Proc. CHI'91,* ACM, New York, pp. 119-124.

Kolodner, J.: 1991, Improving Human Decision Making through Case Based Decision Aiding, *AI Magazine,* 12(2), pp. 52-68.

Maher, M. and Fenves, S.: 1985, HI-RISE: An Expert System for the Preliminary Structural Design of High Rise Buildings, *in* J. Gero (ed) *Knowledge Engineering in Computer-Aided Design,* North Holland, Amsterdam, Netherlands.

Molich, R. and Nielson, J.: 1990, Improving a Human-Computer Dialogue, *Communications of the ACM*, **33**, pp. 338-348.

Navinchandra, D.: 1991, *Exploration and Innovation in Design: Towards a Computational Model*, Springer-Verlag, New York.

Norman, D.: 1988, *The Psychology of Everyday Things*, Basic Books Inc., New York.

Schank, R.C.: 1990, *Tell me a Story*, MacMillan, New York.

Schank, R.C., Ferguson, W., Birnbaum, L., Barger, J. and Griesing, M.: 1991, ASK TOM: An Experimental Interface for Video Case Libraries, *Proc. Thirteenth Annual Conference of the Cognitive Science Society*, Lawrence Erlbaum, Hillsdale, NJ, pp. 143-148.

Sycara, K. and Navinchandra, D.: 1989, A Process Model of Case-Based Design, *Proc. Eleventh Cognitive Science Society Conference.*

Tetzlaff, T. and Schwartz, D.R.: 1991, The Use of Guidelines in Interface Design, *Proc. CHI*, ACM, New York, pp. 329-333.

Thovtrup, H. and Nielson, J.: 1991, Assessing the Usability of a User Interface Standard, *Proc. CHI'91*, ACM, New York, pp. 335-341.

DEJAVU:

A CASE-BASED REASONING DESIGNER'S ASSISTANT SHELL

T. BARDASZ
Computervision
14 Crosby Drive
Bedford MA 01730 USA

and

I. ZEID
Department of Mechanical Engineering
Northeastern University
Boston MA 02115 USA

Abstract. A case-based reasoning (CBR) designer's assistant shell called DEJAVU is presented. The shell is the implementation of a methodology and a computational model for mechanical design automation. The methodology is based on analogical reasoning and the computational model uses CBR. DEJAVU provides a flexible and cognitively intuitive approach to acquiring and utilizing design knowledge. DEJAVU is a domain independent shell that can incrementally acquire design knowledge in the domain of the user. It, therefore, provides a design environment that can learn from the designer until it can begin to perform design tasks autonomously or semi-autonomously. The main components of DEJAVU are a knowledge base of design plans, a design plan system, and an opportunistic problem solver. DEJAVU is written in Lisp using CLOS. DEJAVU is the first step in developing a robust designer's assistant shell for mechanical design problems. One of the major contributions of DEJAVU is in developing a clean architecture that can be extended to become more robust over time. The details and merits of the DEJAVU shell are discussed. An example is included to illustrate the use of DEJAVU.

1. INTRODUCTION

The field of mechanical design methodology has been experiencing a surge of activity over the past few years. This is due, in large part, to a concentrated effort toward developing mechanical design into a science in the hope that clear, founded approaches toward design automation can be developed. Texts have been generated on the process of design (Hubka, 1982; Ostrofsky, 1977; Pahl and Beitz, 1984) but most of the models are prescriptive in nature. That is, they describe how design should be performed and not how it is currently being executed. Recently, protocol analysis has proven to be a useful tool for understanding the design process and generating descriptive models (Adelson, 1989; Guindon, 1989). Protocol analysis is performed by studying a recorded session of a designer solving a design problem.

There is yet to be a complete model of the design process. Coupled with the fact that there are many different methodologies being successfully used, it becomes ex-

J. S. Gero (ed.), Artificial Intelligence in Design '92, 477–496.

tremely difficult for a design automation system to experience wide use if it relies on one or more of the existing design methodologies. The selection of only one methodology handcuffs the system into solving only a select set of problems. The adoption of all techniques forces the system to have the ability to reason about design problems and the best method for solution, in addition to actually solving the problem.

An interesting view of design is achieved by considering design as a problem space. In a problem space definition of design, the initial state is the design problem statement. The final state is the resultant mechanical artifact that satisfies the problem statement. The operators are any actions that the designer can take to progress the design from its initial state to the final state (e.g. decompositions, attribute assignment, component instantiation, analysis invocation, etc.). The intermediate states of the design are partial or incomplete designs that satisfy some set of the original problem criteria. The sequence of operators that are used to transform the initial state to the final state is called the transformation sequence.

Using this view of mechanical design, the complexities involved in developing a mechanical design automation system are evident. First, the search space is enormous. It encompasses virtually every incomplete or intermediate design that satisfies some set of the initial specifications. A weak method such as Means Ends Analysis would become unmanageable quickly given the combinatorics involved. Therefore, a mechanism for choosing a search strategy through the problem space must be employed to reduce the search space. Secondly, the final state is not a state at all. It is actually a space in itself. It is the space of all candidate designs that satisfy the original problem specification. So there is no concrete final state to work backwards from.

The ideal solution to the problem space problem would be to eliminate the search, or at least constrain it in some fashion through the use of some search strategy. We have already identified the fact that the selection of one strategy would limit the applicability of the design system. The ideal system would select a strategy based on the problem specification. The set of heuristics necessary to select a strategy based on the problem definition is ambiguous. Designers select a strategy not only based on the design problem statement, but also on their own level of knowledge and the set of tools that are available for application to the design problem.

The problem of developing a widely applicable design system then becomes:

1 How do we acquire a designer's heuristics for selecting a problem solving strategy based on the problem specification?
2 How do we integrate this strategy into a computational environment that can utilize this knowledge?
3 How do we allow the system to "grow" by continuously incorporating these strategies, and allowing them to be reused?

This is the set of problems that are addressed in this paper.

2. The Design Methodology

The solution to the design problem described in the previous section is the development of a design methodology and a computational model that allows the participation of past design knowledge to solve new design problems. Such a methodology and a model are very appealing given the fact that designers usually reason from past experience to solve new design problems. More specifically, the reuse of design solutions or approaches from past designs make up a large part of a designer's activities for design problem solving. The methodology and the model provide a design environment which can acquire design knowledge incrementally from designers and utilize it later to solved design problems.

The design methodology developed in this paper is based on derivational analogy. Derivational analogy in design is focused at learning and reusing idealized design plans to solve new problems. Derivational Analogy is a subset of the Analogical Problem Solving research in Machine Learning. Analogical problem solving is based on an approach that utilizes past or model situations to solve new problems. Many varied systems have been developed that use analogical problem solving as its base (Hall, 1989). Derivational analogy is differentiated from typical analogical reasoning in that what is transferred are the lines of reasoning, decision sequences, accompanying justifications, etc., as opposed to just a description of a previous situation. Derivational analogy is focused at building a new solution from an idealized past solution process description.

The use of derivational analogy for design has been described and/or attempted previously (Bardasz and Zeid, 1991a; Huhns and Acosta, 1987; Mostow and Barley, 1987; Steinberg and Mitchell, 1985; Wile, 1983). A majority of systems were developed in the electrical design domain and only one (Bardasz and Zeid, 1991a) describes analogical problem solving for mechanical design. The basis for the use of derivational analogy in design is that it is much easier to modify aspects of the design process (to fit a new design situation), than it is to modify the designed artifact itself. Therefore, these design systems focus on the use of design process plans as their working knowledge base elements.

In design, reasoning from the design process is highly desirable for two reasons. The first is that a representation of a designed object itself would have no notion of the constraints or behavior that the original designer intended. A retrieved design plan would represent the constraints and intent of the designer, in the design plan, thus ensuring a valid design in more cases. The second effect is that it unburdens the knowledge base from keeping variant forms of the same solution. A single design plan can be responsible for generating a whole family of design solutions in a particular class (e.g. there could be one plan for a screw that could represent all of the variations; self–tapping, different heads, etc.).

In using derivational analogy for design, there are three main issues: 1) knowledge base organization, 2) design plan form, and 3) the modification of old plans to fit new situations. Derivational Analogy has been mostly concerned with the last two issues. In a review of systems that utilize derivational analogy (Mostow, 1989), only one of the systems wasn't hand fed a relevant design process to start from. In Bardasz and Zeid (1991a), a knowledge base of design plans is described with reference to the work in Case–based Reasoning. In Bardasz and Zeid (1991b), that model was furthered by a complete description and implementation.

Problems with the use of Derivational Analogy for design are described in (Mostow, 1986). The two main reasons for failure in replaying design derivations are described as "missing preconditions" and "context sensitive references". A precondition is a description of the state that the design must be in for the next step of the design derivation to be valid. If a precondition is not given, then failure may occur for several reasons: 1) The step cannot be executed, 2) The result of the step is invalid, 3) the step does not achieve its goal or 4) a different step should be executed. In a "context sensitive reference" a step may refer to some component or attribute of the design plan that is specific to its original context. Since the design plan is being replayed in a possibly different context, a reference could fail for several reasons: 1) any reference that cannot be clearly resolved, 2) a reference to something that is not in the current context, or 3) a reference that becomes resolved incorrectly.

The use of derivational analogy for design is relatively new. The research is driving to better models of the design process (Mostow, 1985), representations of knowledge bases for design problems (Bardasz and Zeid, 1991b), and more elegant solutions to the referencing problems described above.

3. The Computational Model

A computational model of the design methodology described in Section 2 is developed using CBR. Case–based reasoning is very similar to derivational analogy (Seifert, 1989). Both approaches are based on the use of past solutions to solve new problems. The difference between the research communities comes in their focus on which tasks they are primarily interested in. The derivational analogy research is focused primarily on the process of mapping a previous solution to the current problem. The case–based reasoning research community is focused, in large, on efficient memory models to support the retrieval of relevant cases. In order to realize a complete system, both facets of the process must be given equal attention.

Case–based reasoning (CBR) is based on the principle that problem solving can benefit from solutions to past problems. The knowledge base of plans could be by learned from experience, example or from being told. The major advantage of this approach over expert systems is that when presented with a similar problem, CBR systems do not re–reason from an initial set of facts and rules. Instead, it uses a solution which embodies the reasoning applied in solving a similar problem.

This approach holds a key benefits with respect to solving mechanical design problems. The application of case–based reasoning does not require the application of a single, specific problem solving strategy for its solutions. In fact, if implemented correctly, CBR should be able to store, retrieve and assemble solutions regardless of the problem solving strategies embodied in the individual solutions. The only requirement that the solutions must adhere to is the adoption of a general mechanism which allows them to share information and represent constraints between themselves. This is because there is no guarantee that the CBR system will only assemble solutions of like types.

Case–based reasoning is intuitive, and can be viewed in everyday life. For example, a person does not reason from general principles when faced with the problem of how to get to work. This problem has been solved several times before and the plan associated with the goal is enacted. The plan could look something like going to the car, starting it up, driving it to work, parking, and walking to one's office. The reasoning aspects of this task are left at the wayside and the plan is enacted as a script (Schank and Abelson, 1977).

Although the approach is intuitive, it is not necessarily easy. For instance, how about if the person is a consultant and the place of work is different every day? What happens when the car is in the workshop? If it won't start? These problems can be associated with case–based reasoning aspects such as mapping and plan failure.

Case–based Reasoning has been employed on a multitude of problems (Hammond, 1989; Kolodner, 1988); caterers, recipe makers, dispute mediation, criminal sentencing, etc. There are three systems that have been documented as design systems in the CBR literature. The term design is used rather loosely, and is misleading if one is from the mechanical design research community. One system called Julia (Hinrichs, 1989; Shinn, 1988) uses case–based reasoning and constraint propagation to design a catered meal. Another system called CYCLOPS: A Design Problem Solver (Navinchandra, 1989) solves design problems in the domain of landscaping. It is focused around finding potential problems when designing a new neighborhood. The last system is called FIRST, a Case–Based Mechanical Redesign System (Daube and Hayes–Roth, 1990). The FIRST system redesigns structural beams. The FIRST system is focused at repairing design failures.

In applying case–based reasoning to mechanical design, there are three issues that are fundamental to the success of the approach. The first issue is in identifying what will be stored in the knowledge base (or memory). The development of the structure of a design solution needs to be addressed in order to determine the working components. The second issue is in defining the memory model. The most widely used scheme for memory organization is the EMOP–based memory. In this model, events are stored in a "discrimination net"–like structure indexed by features that describe the event. The first most critical task is in identifying the set of features that will allow the retrieval of relevant past solutions to the current problem. The feature definition will have a dramatic effect both on the size of the resultant model as well as on the usefulness of the model to return "good" solutions. The third issue is in providing a coherent scheme for the Mapping problem. Mapping is the process of making a retrieved solution applicable to the current problem. Since design is in many cases composed of sub–problems, it is feasible that design solutions to these sub–problems can come from completely different design sessions. It is the responsibility of the mapping component to modify the solution structure to work together toward the solution of the current problem without failure.

The computational model is implemented in a system called DEJAVU. DEJAVU is written in Lucid Common Lisp with CLOS on a SUN4. The remainder of the paper describes DEJAVU, its main components, and how it can be used to solve mechanical design problems.

4. DEJAVU Shell and Architecture

The DEJAVU shell and architecture is the embodiment of the computational model (case–based reasoning) described in Section 3. Its primary goal is to assist designers in solving design problems by leveraging solutions developed for previous design problems. This corresponds to the design methodology described in Section 2 (analogical reasoning).

The form of the design solution for DEJAVU is the design rationale or design plan. The basis for the design plan is a transformation sequence (as in problem space definitions (Newell, 1980)). This has several advantages. The two most important being; a) independence from a particular design problem solving methodology (Bardasz and Zeid, 1991) and b) the ability to capture and represent the transformation sequence directly from a human–computer interaction in a CAD environment (Takala, 1989). A transformation sequence can be associated with a goal, or goals, and information to help in its replay (e.g. determine if it is appropriate, determine if it achieved its intended goal(s), etc.). In this way, a transformation sequence is developed into a design plan, and is the building block from which the case memory of DEJAVU is devised.

The ability to build and retrieve design plans that corresponds exactly or near–exactly to a prescribed design problem is a powerful concept. An even more powerful concept is the ability to compose the solution to a brand new design problem by assembling the solutions to its subproblems into a **design plan structure**. In fact, this design plan structure can be simultaneously composed of both retrieved and taught design plans. Once the design plan structure is complete and is deemed to work correctly, it can be converted to a design plan and placed in the case memory. This allows the designer to interact with the system freely, where the system retrieves relevant design plans for the "known" subproblems and the designer generates the design plans for the "new" design subproblems. All the while, these learned and retrieved design plans are assembled into a design plan structure that represents the solution to the original design problem.

The goal of the DEJAVU shell is to aid the designer in building a design plan structure, composed of learned and retrieved design plans, that solves the current design problems. Once this design plan structure is built, the shell should help in integrating the constituent design plans into a cohesive whole. In order for DEJAVU to satisfy these requirements, it must have the following capabilities:

1. A method for teaching the system a design plan.
2. A mechanism for the designer to retrieve a relevant design plan.

3. A mechanism to build a total design solution from taught and retrieved design plans.
4. A mechanism for integrating the taught and retrieved design plans into a working "whole"
5. A mechanism for evaluating design plans.

The remainder of this section describes the interactions between the designer and the DEJAVU shell.

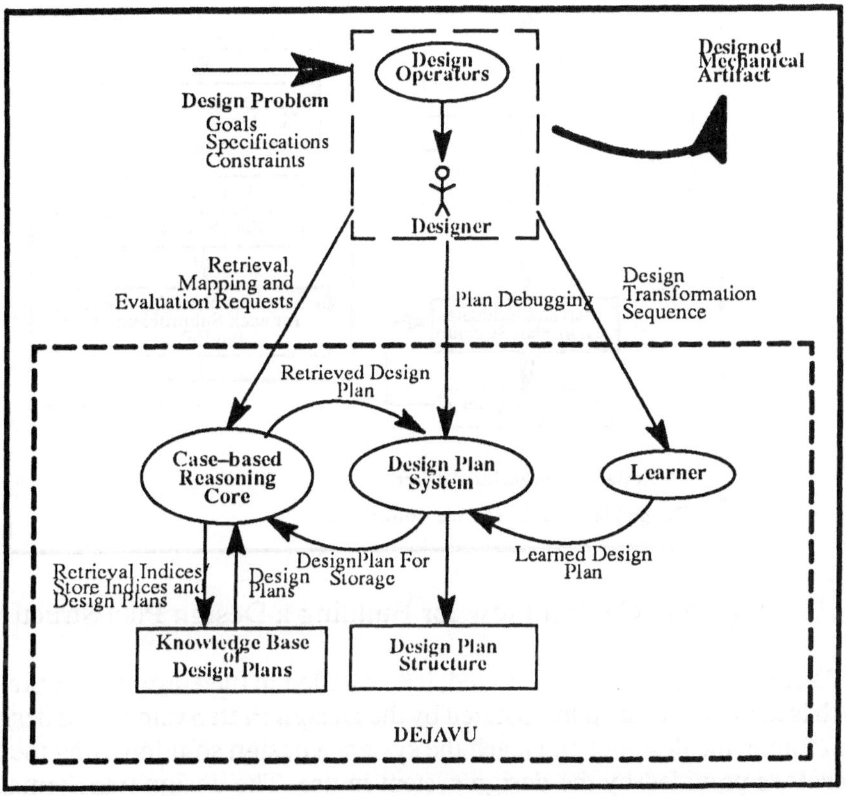

Figure 1. Flow of Information in DEJAVU

4.1. User Interaction

The flow of information between the designer, DEJAVU and the components of DEJAVU is shown in Figure 1. The designer is given a design problem in terms of its goals, constraints and specifications. The designer then queries the **Memory Model** which is part of the **Case-based Reasoning Core** for a relevant design plan. The case-based reasoning core will try to retrieve the most relevant design plan.

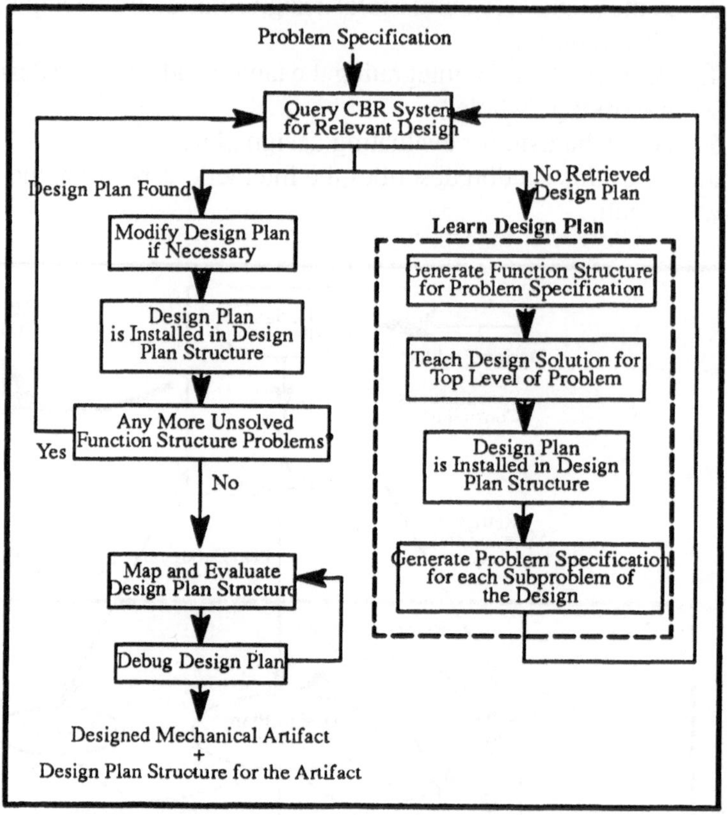

Figure 2. Control Flow for Building a Design Plan Structure

If a candidate design plan is found, it is installed in the **Design Plan Structure**, which is maintained and administered by the **Design Plan System**. If a design plan is not found, the designer can teach the system a design solution using the **Design Operators** provided by the design system in use. The design transformation sequence, which is a result of solving the design problem (Takala, 1989), is converted into a design plan by the **Learner** (Bardasz, 1991; Bardasz and Zeid 1991a). This learned design plan is then installed in the design plan structure. This interactive building of the design plan structure from learned and retrieved design plans is depicted in the Figure 2.

There is no guarantee that the learned and retrieved design plans that compose the design plan structure will work together effortlessly. The process of transforming retrieved design plans from their original context of creation to the current context is the responsibility of the **Mapping Process**. The mapping process is also part of the Case-based Reasoning Core. Its goal is to automatically identify and rectify any referencing problems problems between components of the design plan structure.

Once the design plan structure is complete, mapped and debugged, it is stored in the knowledge base of design plans through the Memory Model.

Over time, the shell would continuously gain knowledge in the domain of the designer. In the beginning, teaching the system design solutions would be the predominant activity. This will not be burdensome as long as the teaching is performed in the same environment that would normally be used for design problem solving. As time goes on, the system will be able to assist the designer more by utilizing the stored design plans from previous sessions. Long term, the system should completely alleviate the designer from performing the common, routine design problems associated with the domain of deployment.

4.2. Architecture

Figure 3 is a block diagram of the architecture of the DEJAVU Designer's Assistant Shell. In accordance with the capabilities described earlier, there are clear modules that map to the identified capabilities.

The **Learner** is responsible for providing a method to allow the designer to teach the system the solution to a design problem. It takes in a transformation sequence and converts it into a design plan. It allows the designer the ability to elaborate the plan with goals, preconditions, postconditions, etc. The resultant design plan is loaded into the design plan system and becomes part of the design plan structure.

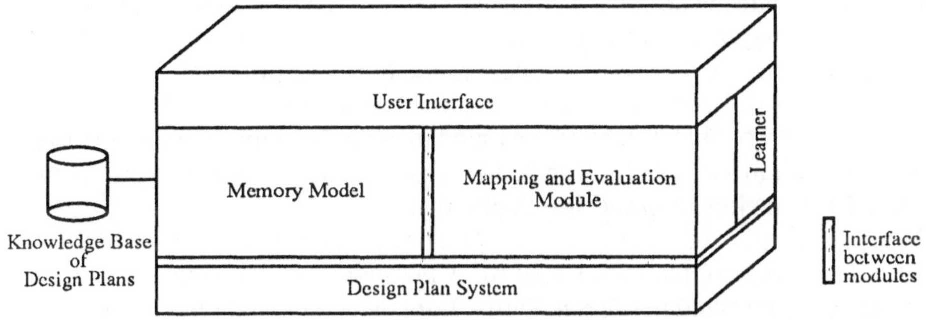

Figure 3. Block Diagram of the DEJAVU Architecture

The **Memory Model** is the module that allows the user to store and retrieve design plans from the **Knowledge Base Of Design Plans.** If a design plan is retrieved, it is installed in the design plan structure through the design plan system interfaces.

The **Design Plan System** provides the capability to build design plan structures whose constituents are learned and/or retrieved design plans. It has the capability to evaluate a design plan structure. It also contains interfaces that allow the other modules to query the design plan structure and invoke its evaluation.

The **Mapping and Evaluation Module** integrates the constituent design plans of the design plan structure into a cohesive whole. It interrogates the design plan structure, through interfaces to the design plan system, to determine where problems exist in the solution. It then invokes a set of heuristics to rectify the problems. If a solution to the problem is found, the offending design plan is edited in place to correct the problem. The Mapping and Evaluation Module also interacts with the Memory Model of Design Plans. It is possible in the DEJAVU system to have memory elements participate in the design plan system (described more completely in Section 5.). The Mapping and Evaluation Module can evaluate these memory elements during its execution, and edit the design plan structure to record the change.

The following three sections describe the Memory Model, Design Plan System and the Mapping and Evaluation Module of the DEJAVU shell. A more detailed account as to their goals and implementation is given in the specific sections.

5. Memory Model

The knowledge base of design plans is based on a model described in (Kolodner, 1984), which is EMOP–based. It is composed of four hierarchical contexts for the storage and retrieval of 1) Product design plans, 2) Assembly design plans, 3) Mechanical component design plans and 4) Recurring engineering problem plans. The indices for retrieval are derived from the user's description of the design problem. In order to alleviate the designer(s) from being restricted to a consistent set of terms and/or terminology, a semantic network is interfaced into the memory model to provide "elaborations" and cross–references to the traversal process.

The contexts devised for the mechanical design memory model follow an extended taxonomy of objects that can be found in assembly representations (Libardi, Dixon, and Simmons, 1988). The contexts are **Products, Assemblies, Components** and **Recurring Engineering Problems.**

Many of the plan re–use systems to–date have focused on the goals of the plan to index it in memory (Hammond, 1989b). A taxonomy of mechanical design problems has been presented in (Dixon, Duffy, Irani, Meunier, and Orelup, 1988). The information was expanded to serve as the set of indices for each of the different event contexts for the mechanical design memory model. In Figure 4, the indices for each of the event–types are shown. The definitions of the indices follows:

1. **Class** – A description of the artifact type (e.g. shaft, bearing, gear, electric_motor, etc).
2. **Goals** – The goal that the artifact should satisfy (e.g. transmission_of_torque, load_support, vibration_damper, etc.)
3. **Goal Situation** – A descriptive list of terms used to describe the situations under which the object best serves its goal (e.g. Helical Gear – high load, high rpm, parallel shafts, etc).
4. **Input** – A list of the attribute names that represent the known information at the initial state of the design of the artifact (hp, load, geometric parameters, etc.).

5. **Output** – A list of the attribute names that represent information derived from the execution of the design plan (max–stress, geometric parameters, etc.)
6. **Assessment** – The criteria by which the design of the artifact is/was to be judged (min_weight, min_cost, max_speed, max_efficiency, etc.).
7. **Phenomena** – The physical phenomena used to satisfy the goals (e.g. friction, transverse stress, heat_transfer, etc.).
8. **Participants** – A list of the other types of mechanical artifacts that are required for the design artifact to exist (e.g., hole – surface, finite_element_preparation – mechanical component, etc.)
9. **Shape** – The rough shape of the mechanical component (e.g. cylindrical, rectangular, amorphous, etc.).

These indices will be used to differentiate one design problem from another. They also represent the content frame information that is found in the EMOPs of each of the memory levels. In that regard, they are used to describe the common aspects of the design problems stored under an EMOP.

Class Model Goals Goal Situation Input Output Assessment **Product Indices**	Class Goals Goal Situation Input Output Phenomena Assessment **Assembly Indices**
Class Goals Goal Situation Input Output Assessment Phenomena Shape **Component Indices**	Class Goals Goal Situation Input Output Participants Phenomena **Recurring Engineering Problem Indices**

Figure 4. Indices for Mechanical Design Event Types

The features used to define the goals, assessment criteria, phenomena, shape, etc., are flat and symbolic. It is a necessity for these features to match the indices in the memory structure, if it is to be traversed. It would be naive to assume that a group of designers (or even one designer) would use the same vocabulary consistently all the time to describe these aspects of the design. It would also be overly restrictive to impose a fixed vocabulary to describe these features (never mind the indomitable size of the task to identify a complete vocabulary for mechanical design).

To alleviate this "common vocabulary" problem, the memory model is interfaced to a semantic network. The semantic network is capable of representing relations between the vocabulary terms that are presented to the memory model. The relations that are most important are **a–kind–of** and **synonymous**. These relations are used during elaboration as index fitting alternatives. Index fitting is a technique that can be used when the traversal mechanism is at an EMOP with a set of features, but none of the features match any of the available indices.

A "weighted feature" scheme is used to select the most applicable or "best case" event from a set of events that are returned after traversing the memory model. Each type of index in the memory model has a weight associated with it. The result of the traversal is a list of events that satisfy some or all of the input features. For each event, the set of matching features between the source features and the features of the retrieved event are determined. For each feature type in the set, its weight is added to the score of the event. The event with the highest point total wins. In this manner, the user can assign priority to the matching of certain conditions. If it is absolutely essential, above all else, that the event returned satisfy the *assessment* criteria, then the designer can give that index the highest weight. The only feature that represents a special case is the **goal** feature. When the designer has specified a goal that the artifact must fulfill, then that goal must match the goal of the returned event.

In the above knowledge base, there are four possible results as to what type of result is returned to the designer:

1 **Complete Match:** The features of the design problem are matched completely and a design plan is returned "as is" with the intention that it be executed directly without modification.

2 **Partial Match:** Only a subset of the input features are matched. A modifiable design plan is returned with the intention that the designer adapts it to fit the new case completely without extension.

3 **No Match:** None of the features of the design problem are matched. No design plan is returned.

4 **Generalization Match:** The features match a generalized concept in the knowledge base (e.g. a mechanical object that transmits rotary motion). A more specific instantiation cannot be found from the input. In this case, the memory object (EMOP) that represents the generalization is returned to participate in the design problem solving.

Each of these different types of results has different consequences to the designer. A Complete Match means the design problem is satisfied completely. A Partial Match means that some of the design problem features were not addressed and the designer should modify the design plan to fit the problem completely. A No Match requires the designer to "teach" the system the solution. A Generalization Match may require a more specific problem specification or it can be left as is, to be automatically expanded by the mapping and evaluation process, if possible.

6. Design Plan System

The design plan system provides the basic knowledge representation and reasoning mechanism for the design solution. The design plan is the atomic unit of knowledge in the knowledge base. It is an evaluable representation of the solution to the design problem. The design plan system provides the backbone to the DEJAVU system in that it is the focal point of the solution composition and evaluation.

The design plan system used by DEJAVU is the Concept Modeler by Wisdom Systems®. The Concept Modeler is an object–oriented system for defining mechanical design plans. The Concept Modeler is programmed in Lisp using CLOS. With the Concept Modeler, a designer can represent the design of a class of mechanical products/assemblies/components in one "part" specification. The evaluation of the part specification may yield different results depending upon the values of the input variables of the design plan. For a complete description of the Concept Modeler the reader is referred to the Reference Manual (The Concept Modeler Reference Manual, 1990).

The Concept Modeler employs a demand–driven or lazy evaluation scheme with dependency–directed backtracking. Virtually any property of a part definition can be a variable. The variable definition can be composed of an equation, function, or method that references other variables in the design plan. When a design plan is loaded into the system, no computation of variable values is performed. In fact, a variable is not evaluated unless if it is explicitly requested by the user or by some other operation that references it, hence the demand–driven evaluation. A dependency network is generated dynamically and maintained for the variables in the design plan. If a variable's value changes, all of the variables that depend on it become unevaluated. In order for the variable to obtain a value, it must be re–evaluated. This combination of demand–driven evaluation with dependency–directed backtracking maintains the consistency of the part definition.

An operator called "the" provides a facility that allows the designer to reference variables defined anywhere in the design plan structure. The "the" operator allows a part's properties to have an expression for its value that can be dependent on any other property in the design plan.

It is the "the" operator that allows the demand–driven evaluation of a design plan structure. When the the specification of an attribute or any component of a subpart specification includes the "the" operator, the situation for demand–driven evaluation has been developed. The algorithm is quite easy to visualize, in evaluating the value of a variable definition, if it references another variable through the "the" operator, 1) find the variable, 2) if it has a value, return it, else evaluate that variable and use its resultant value in the computation. This recursively evaluates/demands all variables that are required for the original variable value to be computed.

7. MAPPING AND EVALUATION MODULE

The Mapping and Evaluation (MAPE) Module is responsible for mapping the design plans, of the design plan structure, from their original context into the current context. It is possible that the design plan structure is composed of learned design plans and design plans that are retrieved from previously unrelated design problem solutions. Each design plan may have references that are sensitive to the context in which it is created. The MAPE Module will identify these context–sensitive references, and invoke a knowledge source (KS) that contains heuristics to resolve them.

Another major responsibility of the MAPE Module is in resolving generalizations (EMOPs), retrieved from the memory model, into specific instantiations of design plans. A generalization may exist in the design plan structure that refers to a class of mechanical artifact. The MAPE Module monitors the design problem solving process to determine if any information is generated that might help in transforming the generalization into a particular mechanical design plan. If so, a KS is invoked to further evaluate the information and expand the generalization, if possible.

The basis for an effective approach to the mapping problem is to perform mapping and evaluation simultaneously. This capitalizes on the fact that the information made available to the mapping process is increasing as the design is becoming progressively more evaluated. With this basis in mind, the design of the MAPE module centers around the use of a blackboard architecture in conjunction with the demand–driven nature of the Design Plan System. A blackboard architecture inherently supports the participation of multiple activities (e.g. mapping and evaluation) in an integrated environment. The demand–driven nature of the Design Plan System enables the specific identification of missing/incorrect references per evaluation request.

The top–level design for the Mapping and Evaluation (MAPE) Module is shown in Figure 5. The MAPE Module is composed of the MAPE Design Plan System interfaces, a Design Plan Interrogator, and the Generic Blackboard Development (GBB) system (Corkill, Gallagher and Murray, 1986). The Design Plan Interrogator interrogates the Design Plan Structure through the MAPE Design Plan Interfaces to dynamically define the problem solving environment for the GBB system. The GBB system is then invoked in an opportunistic problem solving scheme to simultaneously map and evaluate the Design Plan Structure.

The implementation of the MAPE Module complements DEJAVU's approach of aiding the designer in building a dynamic design plan structure. The Design Plan Structure is developed completely by the designer. It is influenced by the knowledge base of the system, and the method of design problem solving used by the designer. Therefore, the Design Plan Structure is dynamically defined and composed of possibly both learned and retrieved design plans. In concert with this approach, the problem solving environment of the MAPE Module is dynamically defined to provide a better correspondence to the Design Plan Structure.

The goal of the MAPE is to progress the design plan structure toward a complete design solution. A "complete" design solution is achieved when the design plan structure can be expanded into all of its constituent design plans and if all of the attributes defined for each design plan can be evaluated without error.

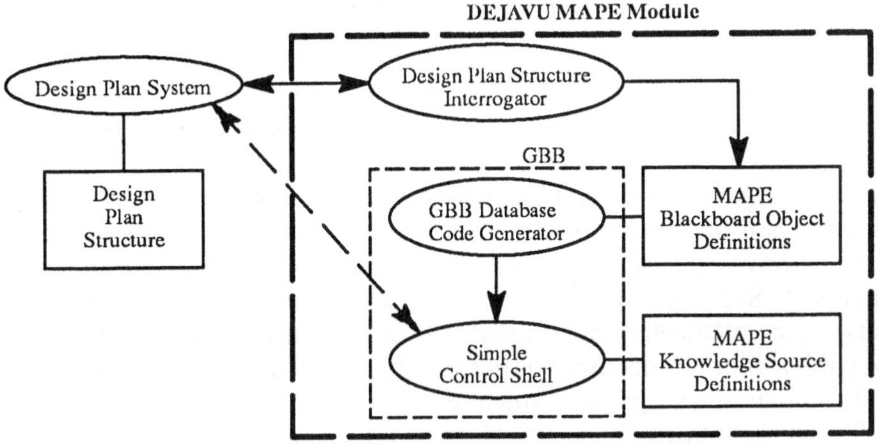

Figure 5. MAPE Module Top–level Design

Each design plan in the design plan structure has a space corresponding to it somewhere in the blackboard structure. The spaces are used to represent the state of the evaluation of the design plan. All attribute values that are derived during the evaluation of the design plan will be posted to the design plan's space. In addition, if the evaluation of an attribute of a design plan is halted because of a reference that cannot be evaluated, the offending evaluation is posted to the blackboard as a precondition for the continuation of the design plan's evaluation. These preconditions, if unresolved, actually identify the referencing problems in the design plan structure.

When the user invokes Mapping and Evaluation Module, the following actions occur:

- The Design Plan Structure is interrogated and the Blackboard Structure corresponding to the design plan structure is generated.
- An initial KS is invoked to trigger the invocation of KS's that expand each design plan and evaluate each and every one of its attributes.
- If a failure occurs in the evaluation of a design plan, a precondition is created that identifies the invalid reference.
- When an attribute unit is added to the blackboard structure, all preconditions are checked to determine if they are still valid. If a precondition is

satisfied, a unit is created that re–starts the evaluation of the design plan that was held up.

- Generalizations/EMOPs in the design plan are processed to determine if they can be resolved.
- If there are preconditions remaining after all of the initial KSs have evaluated, heuristics are applied to further help the mapping.
- If preconditions still exist after the execution of the heuristics, a report is generated for the designer that explicitly states where the problems are in the design plan structure, and what the offending reference is that has generated it.

This design for the mapping process has several desirable characteristics. First, the opportunistic approach guarantees the generation of the maximum amount of information of the design plan structure is generated before the mapping process is attempted. Second, it cleanly integrates heuristics that can resolve mapping problems in a simultaneous mapping and evaluation scheme. Lastly, the design allows the participation of memory nodes from the knowledge base to participate in the evaluation of the design plan structure.

These capabilities are provided by an architecture that uses opportunistic problem solving in a blackboard environment to simultaneously map and evaluate the design plan structure. The nature of the blackboard system provides a clean design for the "evaluation" knowledge sources, mapping problem heuristics and generalization/EMOP expansion heuristics to participate together. The goal being to generate a flawless, complete design plan structure that solves the intended design problem.

8. An Example

The design of a car is discussed in this section to illustrate how the DEJAVU system is used in solving mechanical design problems. Figure 6. shows the car assembly. The assembly is kept to three levels deep for simplicity reasons. The designer requires a six–cylinder, fuel–injected engine to meet the high horsepower (HP) and speed design requirements of the car.

The designer begins with the interaction of DEJAVU by describing the car design problem. A retrieval request is then made to DEJAVU (memory model) using the features that describe the design problem. DEJAVU responds with a "no match" indicating that it has no design plan that corresponds to the specification. The designer, therefore, decomposes the car into a body, an engine, and a suspension as shown in Figure 6. The designer now attempts to utilize DEJAVU to solve each of these subproblems. To illustrate the different retrieval cases, let us assume that the system has no body design plans but does have engine and suspension design plans.

Upon the designer request for a body design plan, DEJAVU returns "no match". In this case, the designer teaches DEJAVU how to design a car body. All of the body requirements and specifications are used to create the body design plan and it is installed in the design plan structure.

When the designer inputs the engine specification, DEJAVU returns a "partial match". The retrieved engine design plan has four cylinders and a carburetor. Through use of the Design Plan System in DEJAVU, the design modifies the existing engine plan design plan to correspond to a six cylinder engine with fuel injection.

The suspension retrieval request returns an "exact match" because a suspension existed in the memory model that corresponded exactly to the design problem specification.

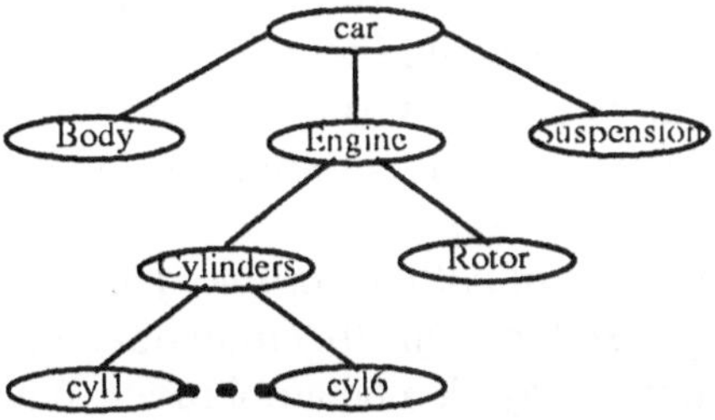

Figure 6. Example Car Assembly

Thus far, the car design plan structure has three subplans; a new body design plan, a modified engine plan, and an existing suspension plan. Will these subplans work together correctly and cohesively? Since the engine and suspension design plans are old, they may contain references to elements of their previous designs. For example, the suspension may reference the previous design's wheel well clearance. The engine may reference its previous design's motor mount locations, or type of transmission. The Mapping and Evaluation Module will identify and attempt to repair the invalid references utilizing the heuristics of the MAPE module and the semantic network of DEJAVU.

Once the mapping and evaluation process is complete, the new car design plan is stored in DEJAVU's knowledge base according to its original design specification and any new specifications the design would like to add. DEJAVU has learned the design solution for this particular car design problem. This car design plan may be retrieved and used "as is" or modified for future designs that have the same or similar design problem specification.

9. Discussion

The issues identified for a case–based reasoning designer's assistant shell had to be resolved, at one level or another, by DEJAVU. By developing a system, and addressing these issues, several gains were made: 1) a better understanding of these issues and the repercussions of various approaches, 2) the development of a clean, extendible CBR architecture with well defined interface boundaries, and 3) the identification of semantics as what appears to be the most significant, pervasive issue in the way of a robust CBR designer's assistant shell. It is worthwhile to discuss each of the issues in terms of how it was addressed in DEJAVU.

Case acquisition was accomplished by facilities provided by the design plan system. The design plan system allows the dynamic creation of a design plan. The dynamic creation is performed in terms of defining the inheritance, properties, property expressions, and subparts of a new design plan object. The mechanism and information required to support the recollection of a relevant design case is fundamental to the success of the system. For DEJAVU, an EMOP based memory model with four contexts was used to store design plans.

The representation for a case in the DEJAVU system is a Concept Modeler "part". It is a declarative definition defining the components of a design and their relationships. The components could include geometry in their definition.

Semantics was a main issue that surfaced when implementing DEJAVU. Semantics is so critical to a case–based reasoning system because the organization of the knowledge base and its contents (the design plans) are driven by the designer's understanding and expression of both the problem and the design. The knowledge base and design are expressed in terms that may not be consistent across the users of the system. These terms may also be ambiguous, having different or multiple meanings. In DEJAVU, flat strings are used to build the semantic network. These clearly lack the expressive nature required for describing design concepts.

10. CONCLUSIONS

A case–based reasoning mechanical designer's assistant shell called DEJAVU is implemented and described. DEJAVU directly satisfies needs identified from the evaluation of design protocols. Its main components are: a model of memory for mechanical design plans, a design plan system, and a mapping module that integrates previously unrelated design plans toward a common design goal. The issue of semantics is identified as critical to the success of the approach. Overall, the approach is seen as viable and achievable with several tough issues remaining before a complete implementation can be realized.

Future work on DEJAVU includes various areas. It will be integrated with existing CAD systems to facilitate the interactive creation of design plans. Its design plan system will be extended to include, in addition to geometric definition and relationships, information related to its goal structure, preconditions, postconditions, alternative strategies, etc. The use of flat strings in interfacing with the memory model and the semantic network will be enhanced to encompass objects more closely related to design. In addition, a large portion of the heuristics involved in the Mapping and Evaluation Module dealt with referencing problems. These heuristics will be enhanced to deal with all aspects of mapping problems that occur when using a case–based reasoning system for design.

References

Adelson, B.: 1989, Cognitive Research: Uncovering How Designers Design, Cognitive Modelling: Explaining and Predicting How Designers Design, Journal of Engineering Design, **1**(1)

Bardasz, T.: 1991, *Mechanical Design Automation via Case–based Reasoning: An Incrementally Intelligent Approach,* Masters Thesis, Northeastern University, June 1991.

Bardasz, T., and Zeid, I.: 1991a, Applying Analogical Problem Solving to Mechanical Design, CAD Journal, **23**(3), pp. 202–212.

Bardasz, T., and Zeid, I.: 1991b, Analogical Problem Solving in Mechanical Design: The Knowledge Base Model, 1991 ASME Computers in Engineering Conference, Vol. **1**, pp. 169–176.

Bardasz, T., and Zeid, I.: 1990, Proposing Analogical Problem Solving to Mechanical Design, 1990 ASME Computers in Engineering Conference, Vol. **1**, pp. 181–186.

Brownston, L., Farell, R., Kant, E., and Martin, N.: 1985, *Programming Expert Systems in OPS5: An Introduction to Rule–based Programming*, Addison–Wesley Pub. Co, Menlo Park, CA, 1985.

Corkill, D. D., Gallagher, K. Q., and Murray, K. E.: 1986, GBB: A generic blackboard development system, Proceedings of the National Conference on Artifical Intelligence, Philadephia, PA, pp. 1008–1014.

Daube, F., and Hayes–Roth, B.: 1990, A Case–Based Mechanical Redesign System, 1990 AAAI Proceedings, pp. 1402–1407.

Dixon, J. R., Duffey, M. R., Irani, R. K., Meunier, K. L., and Orelup, M. F.: 1988. A Proposed Taxonomy of Mechanical Design Expertise, 1988 ASME Computers in Engineering Conference, Vol. 1, pp. 41–46.

Gallager, K. Q., Corkill, D. D., and Johnson, P. M.: 1988, GBB Reference Manual: GBB Version 1.2, Technical Report 88–66, Department of Computer and Information Science, University of Massachusetts, Amherst, MA

Guindon, R.: 1989, A Cognitive Study of High–Level System Design:Implications of Opportunistic Behaviors For A Theory Of Design, in Proceedings of the 1989 NSF Engineering Design Research Conference, University of Massachusetts, Amherst.

Hall, R. P.: 1989, Computational Approaches to Analogical Reasoning: A Comparative Analysis, Artificial Intelligence, Vol. 39, pp 39–120.

Hammond, K., (ed.): 1989, *Proceedings: 1989 Case Based Reasoning Workshop*, Morgan Kaufman, San Mateo, CA.

Hammond, K.: 1989b, *Case–Based Planning: Viewing Planning as a Memory Task*, Academic Press, San Diego, CA.

Hinrichs, T. R.: 1989, Towards an Architecture for Open World Problem Solving, *in* K. Hammond (ed.), *Proceedings: 1989 Case–based Reasoning Workshop*, Morgan Kaufmann Publishers, Inc, San Mateo, CA, pp. 115–118.

Hubka, V., 1982, *Principles of Engineering Design*, Butterworth, London.

Huhns, M. N., and Acosta, R. D.: 1987, Argo: An Analogical Reasoning System for Solving Design Problems, Technical Report AI/CAD–092–87, MCC, Microelectronics and Computer Technology Corporation, AI/KBS and VLSI CAD Programs, 3500 West Balcones Center Drive, Austin, Texas 78759.

Kolodner, J. L. (ed.): 1988, *Proceedings: Case Based Reasoning Workshop*, Morgan Kauffman, San Matteo, CA.

Kolodner, J. L.: 1984, *Retrieval and Organizational Strategies in Conceptual Memory: A Computer Model*, Lawrence Erlbaum Assoc., Hillsdale, NJ.

Libardi, E. C., Dixon, J. R., and Simmons M. K.: 1988. Computer Environments for the Design of Mechanical Assemblies: A Research Review, Engineering with Computers, Vol. 3, pp 121–136.

Mostow, J.: 1989, Design by Derivational Analogy: Issues in the Automated Replay of Design Plans, Artificial Intelligence, **40**(1–3), pp. 119–184.

Mostow, J.: 1986, Why are Design Derivations Hard to Replay? *in* T. Mitchell, J. Carbonell and R. Michalski (eds.), *Machine Learning: A Guide to Current Research*, Kluwer, Hingham, MA, pp. 213–218,

Mostow, J.: 1985, Towards a Better Model of the Design Process, AI Magazine, 6(1), pp. 44–57.

Mostow, J., and Barley, M.: 1987, Automated Re–use of Design Plans, *in* W. E. Eder (ed.), *Proceedings of the 1987 ASME International Conference on Engineering Design*, Boston, MA, pp. 632–647.

Navinchandra, D.: 1989, Case–Based Reasoning in CYCLOPS, A Design Problem Solver, *in* J. Kolodoner (ed.), *Proceedings: 1988 Case Based Reasoning Workshop*, Morgan Kauffman, San Matteo, CA, 1989, pp. 286–301.

Ostrofsky, B.: 1977, *Design, Planning, and Development Methodology*, Prentice–Hall, Englewood Cliffs, NJ.

Pahl, G. and Beitz, W.: 1984, *Engineering Design*, Springer–Verlag, London.

Schank, R. C. and Abelson, R. P.: 1977, *Scripts, Plans, Goals and Understanding*, Lawrence Erlebaum, Hillsdale, NJ.

Seifert, C. M.: 1989, Analogy and Case–Based Reasoning, *in* K. Hammond (ed.), *Proceedings: 1989 Case–Based Reasoning Workshop*, Morgan Kaufmann Publishers, San Matteo, CA, pp. 125–129.

Shinn, H. S.: 1988, Abstractional Analogy: A Model of Analogical Reasoning, *in* J. Kolodoner (ed.), *Proceedings: 1988 Case–Based Reasoning Workshop*, Morgan Kaufmann Publishers, San Matteo, CA, 1988, pp. 370–387.

Steinberg, L. I., and Mitchell, T. M.: 1985, The Redesign Systems: A Knowledge Based Approach to VLSI CAD, IEEE Design and Test, pp. 45–54.

Takala, T.: 1989, Design Transactions and Retrospective Planning, Tools for Conceptual Design, *in* V. Akman, P. J. W. ten Hagen and P. J. Veerkamp (eds.), *Intelligent CAD Systems II: Implementational Issues*, Springer–Verlag, pp. 262–272.

Wile, D. S.: 1983, Program Developments: Formal Explanations of Implementations, Communications of the ACM (CACM), 26(11).

The Concept Modeler, Release 1.2, Reference Manual, Wisdom Systems®, Pepper Pike, Ohio, 1990.

A CASE-BASED DESIGN AID FOR ARCHITECTURE

E. A. DOMESHEK and J. L. KOLODNER
College of Computing
Georgia Institute of Technology
Atlanta GA 30332-0280
USA

Abstract. This paper summarizes the current status of a project to construct a design aiding system for architects. The **Archie-II** system is an application of case-based reasoning techniques to the task of assisting human designers. The focus on *design aiding*, the choice of *case-based techniques*, and the resulting specification of a *case browsing system* are reviewed and justified in the first section. The balance of the paper then focuses on the ways in which design cases can be carved up for presentation to designers and how the resulting pieces can be indexed and organized so as to make them available at appropriate times in the design process.

1. Conceptual Design of a Conceptual Design Assistant

The value of rapid, effective, and creative design is becoming ever more apparent in today's global economy, hence the growing emphasis on applying computers to design. Unfortunately, much as we might wish otherwise, we do not yet have an algorithm for design. Researchers in the AI and Design community can take that lack either as a challenge or as a constraint. Those interested in improving the design process in the short run had better take it as a constraint. As such, it suggests that any useful design system of the near future will fall at the low end of the intelligence spectrum — it will more likely be a tool supporting current practitioners in their existing practices, than a design agent autonomously carrying out pieces of the design task on its own.

Taking this view, the question we want to ask is not "how can a computer generate designs?" but rather, "how can a computer help a designer?" That, in turn, leads us to ask a series of questions: What do designers currently do? Which of their tasks are currently problematic? Which might be done better with computer assistance? Which tasks provide a base onto which new processes might be grafted? What we are looking for is places where computers' abilities can address designers' needs and complement designers' abilities.

1.1. Conceptual Design

Conceptual design, as the first step in the design process, is in some sense the most influential. Commitments made there — determinations of what the problem is and what the broad solution should look like — have the largest effects and are the hardest to undo later. A mediocre or lousy concept can be carried through

J. S. Gero (ed.), Artificial Intelligence in Design '92, 497–516.

skillfully, but it will never equal a similarly executed design based on a clear and clever concept. If we want to improve design, then conceptual design seems like an area with the potential for a high payoff.

But surely, if we lack an algorithm for design, then conceptual design must be one of the more problematic aspects. Conceptual design is a squishy sort of task, mired hopelessly in the open-ended real world — a task without a clear specification because part of the problem to be solved is identifying the problem. Conceptual design seems a realm for mysterious qualities like insight, inspiration, and creativity. This probably accounts for why so little work in AI and Design has been addressed to the problem of conceptual design.

On the other hand, if we are not trying to automate design, but rather to help designers, perhaps there is reason for hope. A system that supports designers' creativity need not itself be creative. We must first consider what sort of assistance would be useful, and for that, we can look at how conceptual design is currently done. While we want a model of the conceptual design process to guide us, as long as we are not proposing that the system *do* design, it need not be a completely specified operational model; a loose qualitative model may tell us enough to get on the right track.

Looking at designers in the early, conceptual, phase of design, one thing is clear: instead of trying to work in a vacuum, designers spend much of their time thinking about existing designs, reviewing the literature, pouring over formal and informal documentation of earlier works. This is true in a field like architecture where architects delight in touring interesting buildings, browse magazines like *Architectural Record*, and often pull old blueprints from the company files. It is also true in fields where designers are, in fact, working on something radically new; for example, the next generation of nuclear reactors – lessons learned from past experiences help in focusing the designer on issues of importance.

One way we might hope to assist conceptual design is to make it easier for designers to find, browse, and learn from records of relevant old designs. The combination of a *cognitive model* describing how people can benefit from past experiences and a *technology* for finding and presenting such experiences could form the basis for computer systems that offer designers a new and valuable sort of assistance with their conceptual design problems.

1.2. Case-Based Reasoning

The rationale for the research program reported here is that we *have* such a combination of cognitive model and technology. Case-based reasoning (CBR) (Schank, 1982; Kolodner, Simpson & Sycara, 1985; Hammond, 1989) is a paradigm within Artificial Intelligence based on a memory-centered cognitive model; the basic idea is that people are good at figuring out what to do in new situations largely because they are able to remember and adapt things they did (or saw others do) in similar previous situations. In addition to applications emphasizing reasoning tasks like planning (Hammond, 1989; Alterman, 1988), scheduling (Mark, 1989), decision-making (Bain, 1986), diagnosis (Hunter, 1989; Bareiss, 1989; Koton, 1988), explanation (Kass, et al., 1986; Ram, 1989), argumentation (Ashley & Rissland, 1987; Simpson, 1985; Sycara, 1987), and even design (Hinrichs, 1991, Goel, 1989), CBR is being developed into a technology for building systems that assist human users by presenting them with useful information chosen from organized memories of past experiences (Kolodner, 1991; Domeshek, 1992; Bareiss, Ferguson & Fano, 1991).

In the course of developing the many experimental systems cited above, a wide range of roles have been identified for memories of particular past experiences: a past case might suggest one course of action or warn against another; it might contribute to the evaluation or detailed adaptation of a possible response; while analyzing a situation, it might focus attention on particular questions or suggest specific answers. The problem is that getting a system to apply a case in any of these varied ways requires a large amount of idiosyncratic real world knowledge. While this remains a major roadblock preventing widespread development of autonomous reasoning systems, it does not prevent cases from being usefully applied by a human user who can be counted on to possess and apply sufficient common sense. This is the realization that has driven some researchers to focus on the technology of case retrieval and presentation, building systems to aid humans in their decision making (Kolodner, 1991; Barber, J., Bhatta, S., Goel, A. Jacobson, M., Pearce, M., Penberthy, L., Shankar, M., Simpson, R., and Stroulia, E., 1992).

A case-based decision aid works roughly like this. The user describes his/her problem to the system. The system recalls cases that are similar and presents them to the user. The user makes decisions about the solution in the new situation based on those cases, updates the problem and solution specifications, and repeats the process. Cases recalled by the system can help the user by pointing out suggestions for solutions, mistakes made in old cases, and results of carrying out old solutions in the world. These systems tend to look like clever hypermedia systems; the cleverness lies in how the information is organized and how it is presented. The aim is to reflect the cognitive needs and abilities of human users engaged in a particular task.

1.3. ARCHIE-II: A CASE-BASED DESIGN AID FOR ARCHITECTURE

The AI lab at Georgia Tech's College of Computing has been at the forefront of research in CBR in general, and the application of CBR to design in particular. Work on three case-based design aiding systems is currently in progress. Two of those projects are attacking the problem of design in the domain of architecture. **Ask-Archie** and **Archie-II** are both decended from an earlier system known as **Archie** (Goel, Kolodner, Pearce, Billington, and Zimring, 1991). This paper describes work on the **Archie-II** system.

Figures 1 and 2 give some idea of what the current version of **Archie-II** looks

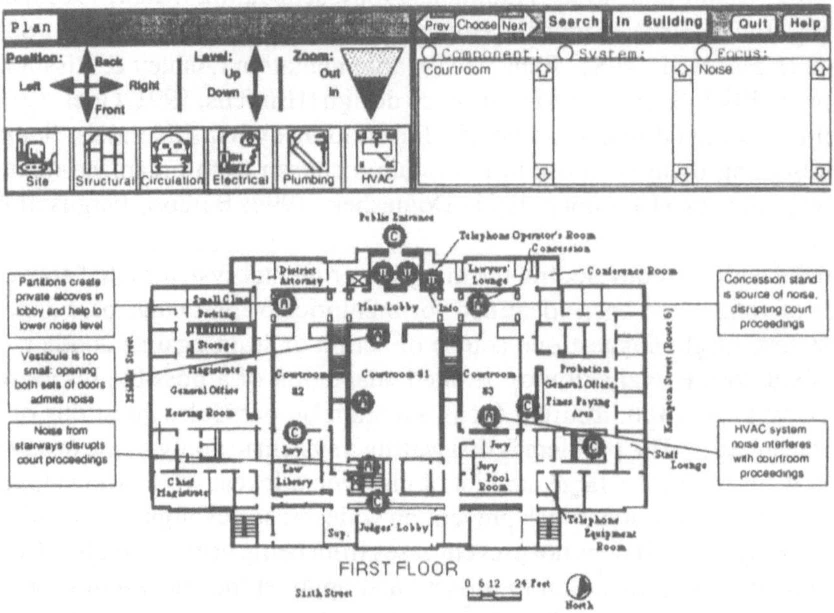

Fig. 1. Archie-II Sample Browsing Screen

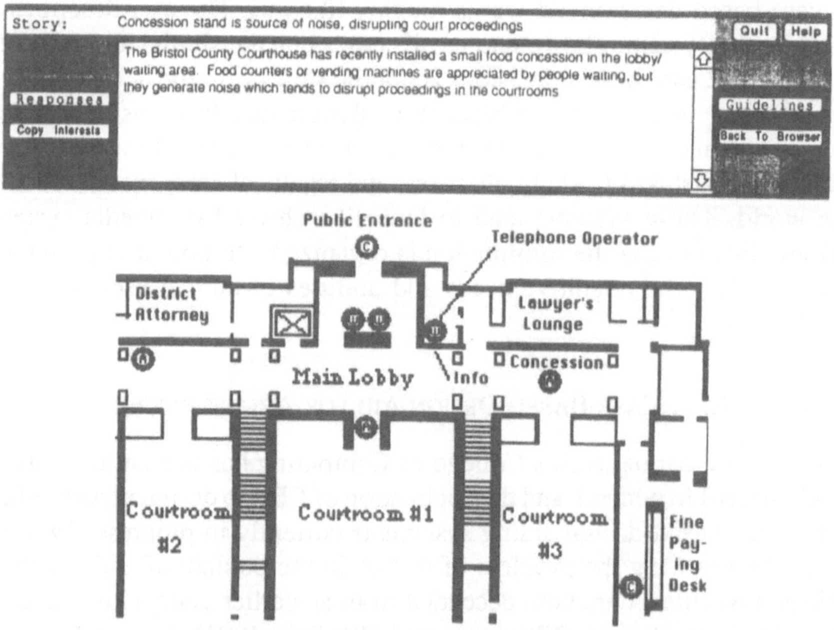

Fig. 2. Archie-II Sample Story Screen

like; in these figures the system is displaying information about the Bristol County Courthouse in New Bedford MA (drawn from, Building Diagnostics Inc., 1988). One key thing to notice in figure 1 is that the system makes it possible for users to browse over architectural graphics (in this case building plans) just as they normally would in current practice; we have found this mimicing of familiar media to be a powerful means of communication with architects. In addition, note that, just as on standard blueprints, the graphic images may be supplemented by textual annotations offering more detailed explanations of aspects of the figure; in **Archie-II** these annotation appear in boxes around the edge of the graphic, and also serve as mouseable buttons providing access to still more information about the plan.

Figure 2 shows what the screen looks like after clicking on the annotation box *"Concession stand is source of noise, disrupting court proceedings."* The browsing controls filling the upper third of figure 1 have been replaced, in figure 2, by a box of text telling in more detail the story behind the chosen annotation. Associated with the story box are a new set of buttons offering the user a chance to look at still more information about this aspect of the building's design. By choosing to examine the design **responses** related to this story, users can learn about possible fixes for the problem introduced by the concession. By choosing to examine the design **guidelines** related to this story, users can learn about the goals operative in this situation and plans for achieving those goals; from there, they may eventually be led to consider other stories (perhaps drawn from other buildings), that relate to the same guidelines, illustrating other successes and failures related to the initial story shown here.

1.4. CONCEPTUAL DESIGN OF A PAPER ON CONCEPTUAL DESIGN AIDING

This first section has been concerned with characterizing the design problem, focusing on the conceptual design phase, justifying a case-based approach to assisting designers with this phase, and giving some idea of what such a system would look like. The rest of this paper is broken into two major sections. In Section 2 we focus on how design cases can be carved up into useful chunks; while in the context of most AI systems this would be considered as a problem in *representation*, in the context of a browsing system like **Archie-II** figuring out what is important in a case and what clusters of facts are related becomes largely a problem of *presentation*. Section 3 discusses how the presentation chunks identified in Section 2 are organized for browsing and search; this discussion includes some memory organization and some representational content. Section 4 closes the paper with a short summary of where we currently stand in the implementation of **Archie-II** and places this system in the context of a larger research program on design involving the development of other case-based aiding systems.

2. Presentations

Given that **Archie-II** is to serve as a case browser, emphasizing *presentation* of information to users over *adaptation* or *application* of past solutions by machine, the system's memory organization and contents are determined by what would make sense to present to a user. Our notion of browsing assumes there is a lot of information in the system and that the user will need to sort through it quickly, skipping much that is irrelevant to their current concerns. To be useful, a case library ought to contain records of many different designs; furthermore, each design case can potentially contain a large amount of information. The need to skim will hold both at the level of cases (which for **Archie-II** means building projects) and *within* cases.

In fact, **Archie-II**'s system of presentations is primarily an attempt to facilitate browsing at a grain size smaller than entire design cases. Whole building designs are too large and complex to directly help an architect with most design decisions. To figure out what sorts of case chunks *would* be useful to designers, we have to ask and answer several questions: What design tasks are case materials supposed to help with? How is old case information supposed to aid in those tasks? What sorts of information make useful contributions to necessary tasks?

We have set our sights on supporting designers in the following four tasks:

- problem comprehension;
- solution brainstorming;
- goal-directed composition of high-level solutions;
- goal-directed filling in of details.

Together these account for much of what must be accomplished at the level of conceptual design. As for the question of how case materials are supposed to help with these tasks, we turn to prior CBR research on design which suggests that memories of past experience can make the following sorts of contributions:

- proposing solutions;
- identifying pitfalls and opportunities;
- suggesting evaluation criteria.

Information chunks that can support these tasks and provide these services cluster design features with an evaluation of how they perform with respect to goals that commonly arise in the design process. In other words, useful chunks present how some design goal was or could be implemented and the effect of that implementation. In presenting these chunks to users, the presentation needs to mention those facts about a piece of a design that affect some goal (or small set of related goals) along with the goal (or goal set) that is being addressed and that renders those facts relevant.

Based on this analysis, we have identified three classes of chunks worth presenting to designers; two kinds come directly from cases, the third kind generalizes across cases. *Stories* are exactly the sorts of goal-focused evaluative case descriptions sketched above; design stories are intended to teach lessons by example, and the lessons they teach can be about attempted solutions that worked or failed, situations where goals may or may not be relevant, ways of judging how a design is likely to fare with respect to some goals. The second class of chunks, loosely titled *documentation*, presents information organized by means other than goals; in particular, design documentation clusters information according to decompositions in terms of structural components and functional systems. Note that stories also often focus on particular structural or functional parts of designs, but documentation is distinguished by its lack of a particular evaluative "point;" it is provided for those times when the designer really wants access to as many details of a past design as the system can provide. Finally, *guidelines*, the third class of presentation in **Archie-II**, are the abstractions behind story lessons; design guidelines generalize across cases and provide a way of relating parts of cases to one another.

We propose carving cases into chunks of these sorts and allowing users to browse those chunks in some controlled but natural way as they engage in design. The three presentation classes — stories, documentation, and guidelines — are discussed in this section. For each, we will define the class (describing and giving examples of interesting sub-classes) and justify the inclusion of these classes in our system. We shall show that documentation helps in organizing the presentation of the other sorts of chunks so that they can be examined and navigated easily. In Section 3, we discuss the situations in which it makes sense to present this information to a user and how individual chunks are indexed.

2.1. STORIES

Complete records of existing complex designs are usually not immediately useful guides to new design. When architects look to old designs for help on new designs, they are usually looking for the sorts of help that we have already suggested cases can provide: ideas for solutions to general problems or specific issues, hints about pitfalls or opportunities, or reminders of evaluation criteria for parts of their design. Furthermore, complete records of the design process (including goals, constraints, options, decisions, evaluations, and revisions) and its outcomes (including interactions, mistakes, and fixes) are exceedingly rare. The fact is, not every aspect of every building is equally interesting (either absolutely, or in some particular design context). What is really wanted from experience is not complete exhaustive records of past designs, but a few good stories that make points relevent to the current phase of the current design.

The basic idea of a *story* is that it is a selective presentation about some particular design case that has some *lesson* to teach (Schank, 1991) — a lesson more focused than "this is how to design a building." Stories pick out related facts from the mass of information about a design, and marshal the selected facts into a narrative or argument, highlighting the underlying causation and the critical relationships that account for some outcome. Stories make good presentation chunks because they are focused enough to aid decision making; we know the kinds of decisions we are trying to help a designer make, and we can pick stories out of our cases that support that kind of decision making.

Stories are an effective means of communicating for several reasons. Instead of simply presenting facts, stories organize facts to make a point, thereby making the facts (as well as the point) more memorable. Stories can illustrate generalizations, giving a better idea of when a rule is applicable (and when it is not), showing how it can be carried out, helping the hearer follow the underlying causality, and exemplifying typical or atypical outcomes. Factual stories can also outdo generalizations as convincing arguments because they claim that the observed outcome truly followed from the stated facts in at least one real situation.

We have identified three different classes of stories that we believe ought to appear in our system[1]: (1) *point stories* discuss some features of a design, describing how they contribute towards, or undermine, some particular goal or plan for some goals; (2) *interaction stories* discuss how some features of a design case can be interpreted with respect to several design goals and plans, perhaps advancing some while frustrating others; (3) *cluster stories* briefly mention the effects of some set of related features in a case without implying any necessary relationship between their various effects. These three classes of story represent different ways of slicing up a design case, suitable for users with different goals. Each is discussed in turn.

2.1.1. Point Stories

Point stories are the simplest sort of story. They deal with the interpretation of design features with respect to a single design goal or plan. The story in figure 2 is an example of a point story. It discusses how the presence of the concession stand in a particular courthouse creates problems for the goal of maintaining a quiet work environment in the courtrooms. As another example, this time of a *positive* effect, consider the following story:

> The Bristol County Courthouse is noteworthy for its successful entrance designs. The building features a separate entrance for juvenile cases to help keep children out of the public eye and away from serious offenders. The entry for accused persons is separate from both the public and the juvenile entrance, and its security is enhanced by use of a sally port. A separate entrance for witnesses was not considered in the design of this building.

In this story, the multiplicity of entrances is analyzed as contributing to the goal that various user groups remain segregated for reasons of security, comfort, and fairness. This story might interest designers considering how to maintain segregation in other circumstances, or those considering the issue of entrance design and wanting to evaluate their decisions' potential effects. In the section on

[1] Analysis due to Anna Zacherl

indexing, we will see how the system can provide access to this story under such conditions. In addition, this story might be of interest as an example of successfully adhering to a particular guideline (discussed in the next subsection), offering the kind of specifics that make it easier to interpret an otherwise abstract rule.

Note that the story does not include all the details of how these entrances were constructed; it only mentions those specifics that support the analysis with respect to the goal of segregation. If, as system builders, we had access to details such as exact entrance dimensions, materials, and construction we would want to make them available in the system, but not as part of this story. Instead, an entrance could be a represented component of the design with such specifications attached; the details of components like entrances are then accessible from within stories that mention them, allowing users who become intrigued by the story to easily see more design details.

2.1.2. Interaction Stories

Analyses describing how design features affect single goals, while often useful idealizations for making a point, rarely capture the full complexity of any particular situation. In many cases, when a designer "makes a mistake" — introduces a design feature that has some negative effect — it turns out that there was some *positive* reason for introducing that feature. The actual mistake might have been adopting peculiar priorities for goals or failing to realize that a goal was going to be affected by a decision made in the service of another goal. To offer appropriate aid to a user, it is important to understand when such tradeoffs are being made (perhaps without explicit foreknowledge), or when designs exhibit synergies or compound mistakes (Domeshek, 1992).

For example, consider a situation where a designer has chosen black and white signs for a public lobby faced in black and white marble: the problem is that the signs end up blending into the background and visitors to the building fail to receive the guidance they need; the original reason for the design choice was to have the signs fit into the aesthetic scheme of the lobby. A lesson from this situation might be that aesthetic considerations should not be allowed to subvert basic functional performance, and the facts of this case suggest conditions under which this conflict is likely to arise. An interaction story carries the lesson along with details about one way a sign's function can be subverted (low contrast with background) and suggestion of when to watch out for such interactions (when making decisions about color coordination).

In **Archie-II**, interaction stories provide access to underlying point stories (and point stories allow the user access to covering interaction stories). The following interaction story from **Archie-II**'s current corpus discusses several implications of the Bristol Courthouse's prisoner entrance design, cited in the earlier point story about multiple entrances serving the goal of segregation:

> The Bristol County Courthouse has designated entrances for the public, juveniles,

> and the accused with the holding area placed immediately inside the accused entrance. The separate accused entrance optimizes security at the entrances and is far enough from the courtrooms to provide good acoustic isolation (and thus a quiet environment for trials). However, security and acoustic isolation of work areas are both compromised during circulation of the accused from the holding area to the courtrooms because the necessarily long path runs adjacent to, and even crosses through, unsecured work areas.

This interaction story discusses the pros and cons of the prisoner entrance design; the point story presented earlier offered an account focused on the success of the design with respect to entrance segregation. Point stories focus on how certain features affect a certain goal, and linked interaction stories make it possible for the user to understand this effect in a larger context. A detailed accounting of the linkages between presentations will be given in Section 3.2.

2.1.3. Cluster Stories

The final class of stories currently included in the design for **Archie-II** are cluster stories. To a large extent, cluster stories are a browsing convenience. As will be discussed in more detail in Section 3, **Archie-II** provides several different ways of searching for information, and any of these browsing techniques may initially yield a large set of candidates. The system needs to provide some organized way of viewing the space of candidate presentations to keep the user from feeling overwhelmed. Cluster stories are one way of providing an overview that helps ease the load of browsing such candidate sets.

For example, placing annotations on a blueprint, as in Figure 1, is a way to impose a simple spatial organization on stories. But this is not the only sort of organization we might want, and this simple technique will sometimes break down. When there is a lot to say about one small region of a building or when there is some relationship between several stories located close to one another, a cluster story can stand in for a set of stories, serving as a kind of table of contents organizing the several things there are to say about the region. For example, **Archie-II** attaches the following cluster story to one particular room:

> The court law library is adequate in the Bristol County Courthouse, however, the shelving could be more efficient. The library is also used for showing videos of drivers arrested for Driving Under the Influence of Liquor.

Segments of this text serve as mouseable buttons allowing access to the point stories being referenced: design of an adequate law library; poor design of shelving units; use of the library as a video screening room.

Stories are not limited to clustering in space or on graphics. For example, the following cluster story organizes a set of point stories all related to a particular *component* of a particular *functional system*, but one that happens to be spread throughout the building:

In the Bristol County Courthouse, the surveillance cameras throughout the building effectively supplement the work of the court officers responsible for courthouse security.

2.2. GUIDELINES

Stories describe *instances* of design, and often these can be thought of as illustrations of more abstract *rules* of design. These rules are rarely firm, so we call them *design guidelines*. Looked at the other way, guidelines generalize over stories. Like stories, guidelines talk about ways of coping with design issues manifesting in the context of various structural and functional parts of buildings. Unlike stories, however, guidelines do not talk about specific outcomes; they talk mainly about goals, plans, and constraints.

Given this notion of guideline, we can now interpret the "point" of a point story as some commentary on, or interpretation of, a design guideline; a story tells of success or failure in meeting a goal or following a plan, and its concrete details point out pitfalls and opportunities, perhaps speaking to the applicability of the guideline. An interaction story most often reports clashes or synergies that result from the relevance of several different guidelines.

There are several different types of guideline. Some guidelines simply prescribe goals to be achieved, while others map from goals to general mechanisms for achieving those goals or to detailed plans. Guidelines can deal with issues at any grain size, ranging from large strategic decisions to more focused tactical decisions to specification of detailed constraints. Guidelines vary considerably in their force; some describe requirements (perhaps derived from building codes), others describe policies, or simple recommendations. Guidelines may focus on single issues, or may discuss tradeoffs required by goal interactions.

Our first example guideline illustrates how the general issue of safety is associated with a plan calling for a particular type of segregation with specific implication for building layout:[2]

> Judges and jurors should not be subjected to the possibility of prisoner threats, confrontations, or violence. Prisoner circulation should be distinct from that of the court staff.

This guideline maps from some high level goals to a relatively specific suggestion about building layout: keep certain circulation paths distinct. Other guidelines, such as the one below, relate guidelines to each other. This one points out the prevalence of a particular wayfinding problem and thereby simply raises other goals:

> Many persons who visit the district courts do so in order to pay fines. It should be clear and obvious where they are to go. As these people enter the building they should be able to find the clerk's office quickly and easily.

[2] The content of this example is taken from a study of courthouses in the state of Michigan, (King, J., Moore, E.O., Johnson, R.E., and Guregian, S.A., 1981).

Here we are told that accessibility of the clerk's office to fine payers is an issue, but we are simply told to make the office easy to find. Having a space be easy to find is a rather high level goal in itself and we must refer to other guidelines that address this issue. **Archie-II** contains guidelines telling us that wayfinding to any function of the building by any unfamiliar user of the building can be eased by proper use of signs and by proper location of a function's space. Part of the more specific guideline about signage runs as follows:

> There are many factors to consider when making signs for a building. For instance, disabled users will have special needs that must be considered; problems with vision, stature, and comprehension are particularly important for sign design and placement. Signage both inside and outside such buildings should be accompanied by easily understood symbols. Symbols require neither the ability to read nor knowledge of a particular language; also, when properly designed, symbols can reiterate the message, thus aiding individuals in their task of interpreting sign messages. Signs should be large and should use bold contrasting colors. If signs are to project from the plane of the wall or hang from the ceiling, then they should be mounted with a bottom edge at least 7 feet above the floor. Avoid sharp edges or exposed fasteners on signs. Design lighting that does not cause glare on nearby signs. Place signs to provide maximum visual exposure, particularly along routes of travel.

Finally, after several steps, we have again mapped from high level goals down to relatively specific recommendations for how part of the building should be designed. The point is that the goal/plan decomposition behind guidelines serves as an important means of generalizing across situations and connecting stories with related lessons.

The organization, representation, and presentation of guidelines is an area of **Archie-II** under active development as this paper is being written. The current approach to organizing this data is to start with a goal/plan hierarchy, rooted at the highest level in an analysis of the functions of particular building spaces over their life cycle from the perspective of particular building user populations. Starting from these high level goals, we can then elaborate a space of guidelines varying in grain size and specificity, in source and in force. We are currently concentrating on fleshing out the part of the guideline space covering accessibility and usability issues, primarily as they arise in courthouses and office buildings for users and staff. We have a long way to go before even this small corner of the larger architectural design guideline space is filled in.

2.3. Facts and Context

So far, with stories and guidelines, we have discussed case materials comprising analyses and evaluations of parts of designs. **Archie-II** also has to organize and present whatever is available in the way of basic documentation for designs. This documentation includes facts and figures, and it includes graphics of many different kinds; it includes summaries that provide an overview of the entire design, and it includes detailed specifications for individual parts; it includes the original requirements for the design, evolving requirements, and the eventual design solution that resulted. All this additional information might simply exacerbate the problem of managing efficient browsing; the trick of effective memory organization is to use some of the additional required presentations as organizational tools that actually simplify browsing.

Supporting rapid selective browsing of large case libraries requires the system to provide users with a feel for how much information is available, roughly what its content might be, and where they stand in surveying it. A system should give an overview of the information available — perhaps shaping it into a space, locating the user in the space, and marking off regions that have already been visited; a system should standardize its presentation formats so that users can learn to expect where certain types of information can be found, allowing them to navigate and skim more effectively; a system should, to the extent possible, use familiar presentation formats that users are already adept at comprehending. The presentations used in **Archie-II** to document basic facts about designs meet most of these goals and thus provide a helpful context for lesson browsing.

2.3.1. Design Graphics

In the domain of architecture, many of the basic facts about a design can be recorded and presented in various graphic forms. It is standard practice for architects to pour over sketches, diagrams, detailed plans, renderings, and photographs of existing or planned buildings; these are presentations we can expect potential users to understand and accept. In fact, use of graphics seems to be standard practice in most design disciplines that deal primarily with the design of physical artifacts, and may be quite common even in the design of processes (where time lines and information flows are often charted).

Graphics are a great example of presentations that serve not only to communicate specific information about a design, but also to help manage the complexity of browsing a design's content; graphics can organize other presentations and help users maintain a sense of what parts of the building they have been studying. Note that even though **Archie-II** primarily relies on two-dimensional depictions of buildings, graphics are not restricted to representing purely physical context; the parts they depict may also reflect functional systems (such as plumbing, electrical, or HVAC systems) and possibly even design goals (such as zones with particular requirements for privacy, acoustic isolation, or security). Linking graphic displays together in a network that reflects physical contiguity or other logical relationships between parts helps users maintain a sense of location in a space of possible presentations. Locating a story on a graphic (as in Figures 1 and 2) can also serve as a way of identifying the story, helping maintain a sense of what information is available and what has been seen.

Graphics also serve well as context for *interpreting* other presentations. The stories brought up by clicking on annotations are not just organized by the graphics, the graphics illustrate the story as well. The story of how noise from the concession stand disrupts court proceedings really only makes sense when you can see the plan of the building showing the concession's location relative to the entrances to the courtrooms.

2.3.2. Design Descriptions

Much design data takes the form of facts and figures, possibly boring but certainly useful in the right circumstances. One of those circumstances is when an architect is just trying to find their way into a browsing system like **Archie-II**. An architect given a new commission might want to assess the program he or she has been given; scanning existing buildings with similar purposes might allow useful comparison of issues like budgets, aggregate size, and level of finishes. In specific niches architects are often concerned with idiosyncratic descriptive statistics; for example, when designing speculative office space, important features include percentage of leasable space and maximum lease depth. These are among the kinds of information that ought to be available in design descriptions.

Design descriptions, when serving as a means of finding direction in the browsing process, are probably best presented in groups forming *library overviews* that summarize some set of available buildings for the user. Such overviews, generated in response to users' initial partial description of their situations, can give an idea of how many relevant cases the system knows about, as well as a feel for the range, variability, and typical values for any selected descriptive dimensions. These overviews ought to serve as the jumping-off point for inspection of more detailed design descriptions, design graphics, and stories.

Detailed design descriptions of the various structural and functional parts of a building are also useful. They can, for instance, provide additional data not included in any story. They can organize data that would otherwise be spread through the text of several stories. While stories are an effective way to organize design information, we cannot expect to anticipate every lesson that might be learned from every building. Sometimes designers will become curious about details beyond those that seemed immediately relevant to a story. As suggested earlier, there ought to be a way to get from a story about entrances to detailed information about those entrances. It also ought to be possible to get from a presentation of the information about an entrance to the stories in which that entrance figures.

3. Indexing

The problem of browsing in **Archie-II** is not so much a problem of figuring out what sort of chunk to examine or what kind of presentation to look at, as it is a problem of selecting *which* particular chunk to study. A useful full scale version of the system will have a huge amount of information in it and more chunks than any one user could browse (especially since a user's main concern should be designing a building, and not just browsing past cases). The problem of selecting useful items from out of a large memory is a common one in all of CBR; it has been labeled the *indexing problem.*

The indexing problem is so named because of it's similarity to the common problem of finding what you want in a library or in a large book, and because the initial conception of a solution to the problem in CBR is similar to the approach taken to the more familiar problem. To find a topic in a book, you make a guess at a good description of the topic in which you are interested and you look through the index in the back of the book for a listing that matches; if the contents of the book are well indexed and if you have guessed a good probe, then the index will tell you where to find relevant material. In CBR, the approach to finding a case (or a chunk of case) in a case library is similar, though the nomenclature is different: in CBR, the equivalent of a book's index is called an indexing system, while the description you use as a probe and the individual entries in the indexing system are called indexes. Nonetheless, the problem of access to appropriate chunks is the same, and the dependence on arriving at good descriptions of those chunks is similar.

Archie-II uses two different styles of indexes. The kind we have been talking about so far are called *descriptive* indexes. These indexes are intended to support reasonably flexible forms of search through the corpus. The issue raised by such indexes is how to design a simple but comprehensive language for describing the conditions under which chunks are worth retrieving. The other kind of indexes are *relationship* indexes. Relationship indexes are best thought of as being more like the kind of pointers authors embed directly in their texts — for example, "see Section 3.2 below for more discussion of relationship indexes." Relationship indexes only make sense when it is possible to make a limited set of reliable predictions about what sorts of information a user should want to see next when in a particular context.

3.1. DESCRIPTIONS

Stories are the chunks we would most like to index by descriptions; they most directly achieve the purposes we defined for memory chunks: proposing solutions, identifying pitfalls and opportunities, and suggesting evaluation criteria. To settle on a way of describing stories, we need to figure out what stories can be about. The major thing that distinguishes a story from any other chunk of a design is that it selects and organizes design features in order to account for an outcome with respect to some instance of a design issue. That means we have to worry about describing design issues and the contexts in which they give rise to more specific goals. A story can be about safety in the courtroom (a design issue and a structural component). A story can be about the efficiency of the air conditioning system (a design issue and a piece of a functional system). A story can be about privacy in the holding cells as affected by the surveillance cameras (a design issue in the context of both a structural component and a functional system). To complete the mapping from design issue to design goal, we often need to pay attention to *who* cares about the issue, and *when* it matters. Safety means different things to a judge and a prisoner; it means different things to people using the building than to those building the building.

The need to map from design issue to design goals generates our sketch of **Archie-II**'s descriptive indexes. The indexes are composed of descriptors specifying a design issue and some of the other four components, listed here (including at least a structural or functional building part):

- a design issue;
- a structural component;
- a functional system;
- a stakeholder perspective;
- a phase in the lifecycle.

Design issues are the highest level concerns that apply in particular and sometimes idiosyncratic ways to every building and to many parts of buildings. Examples issues include *efficiency, accessibility, usability, safety, comfort, privacy, symbology*, and *aesthetics*. The structural component descriptor may refer to either a generic class of structure (floor, wing, room), or a specific structural component (the main floor, the north wing, the staff lounge). The functional system descriptor may refer to either a generic system (plumbing, electrical, HVAC), or a specific component from such a system (a pipe, valve, switch, duct, blower, thermostat). These are the sorts of building parts about which architects must make design decisions, so these are the kinds of parts about which the system must be able to tell stories. Stakeholders are the various classes of individuals and institutions that have concerns about the building. Stakeholders include the *designer, builder, owner, community, residents, users, mission-staff*, and *support-staff*. The lifecycle phases during which these stakeholders are likely to have particular goals for the building include *construction, use, maintenance*, and *renovation*.

We can use this system to describe some of the stories used earlier as examples. For example, the original point story about the value of several entrances can be described in the following two ways.

Design Issue	Privacy	Safety
Structural Component	Juvenile Entrance	Prisoner Entrance
Functional System		
Stakeholder	Juveniles	Community
Lifecycle Phase	Use	Use

As a more complex example, consider the interaction story built around the courthouse entrance design; that story might be described as a tradeoff with design features that successfully ensure security at the prisoner entrance undermining security within other working parts of the building. To build this description we need to introduce one additional descriptor to the format. Since an interaction story is a discussion of how several issues were differentially affected by some design features, descriptions of interaction stories will need to be able to describe more than one issue and its outcome. In this example, the two columns are to be read not as two different descriptions, but as a tradeoff description, the first column a positive outcome, the second negative. We expect to limit such descriptions to pairwise tradeoffs each mentioning only two goals and outcomes.

Design Issue	Safety	Safety
Structural Component	Prisoner Entrance	Prisoner Path
Functional System		
Stakeholder	Community	Users
Lifecycle Phase	Use	Use
Outcome	Positive	Negative

The information described here as potentially belonging in a story description is intended to support three major uses of stories: proposing solutions, identifying pitfalls and opportunities, and suggesting evaluation criteria. Given a goal, it is possible to search for descriptions that have that goal and a positive outcome to find stories that tell of solutions for the goal. Given information about a part of a design, it is possible to search for descriptions mentioning that part and having either positive or negative outcomes to find stories that identify pitfalls and opportunities. Given both a goal and a part of a design, it is possible to search for descriptions with either positive or negative outcomes to gather stories that between them should suggest additional features that determine difference in outcome and hence make for good evaluation criteria.

3.2. Relationships

Relationship indexes are simpler than descriptive indexes. They are essentially hard links between presentations intended to direct the user's attention to related material. In any situation, **Archie-II** can make some reasonable predictions about what chunks would be worth seeing next and can make those recommendations visible to the user as browsing options.

The number of relationship types is small and fixed. In the original discussion of the various chunks, mention was made of the sorts of transitions between presentations that we intended to support. Figure 3 gathers all of that information in one place. To summarize briefly:

- point stories link to interaction stories in which they participate, and interaction stories link to their underlying point stories;
- point and interaction stories link to guidelines whose application or failure they illustrate, and guidelines in turn link back to the stories over which they generalize;

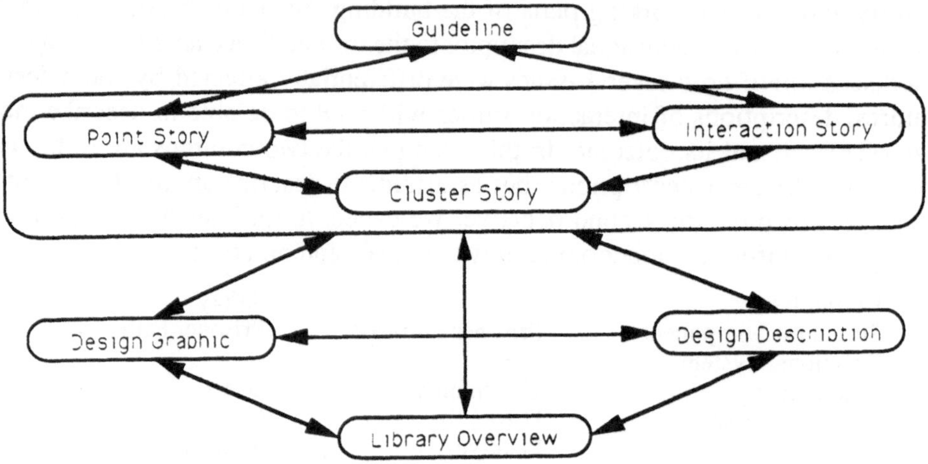

Fig. 3. Presentation Linkages in Archie-II

- cluster stories can link to either point or interaction stories which link back to their clusters;
- library overviews link to design graphics, design descriptions, and stories, all of which link back to overviews, and to one another.

One way to think of these links is as part of the system's interface, but they can also be thought of part of the system's memory network. The catch, of course, is that this memory is primarily intended for perusal by a human user, rather than for autonomous search and manipulation by the system itself. The point is that the content described above and the organization pictured here are useful in supporting the design process.

4. Project Status

Archie-II currently exists as a set of twenty analyzed stories and their accompanying guidelines, a preliminary vocabulary for describing those stories, and an interface prototype developed in Supercard™ on an Apple Macintosh™. The system's stories are drawn from a set of Post-Occupancy Evaluation reports prepared by architectural consulting firms as part of procurement review processes of various government agencies. The guidelines come from existing compilations of guidelines, codes, and expert analysis.

The current interface supports presentation of the several types of case chunks discussed above. While it does not currently implement all of the different presentation modes, it gives a good feel for the sort of interaction we have in mind. We are in the process of using this interface prototype to solicit feedback on the design from human/computer interaction and visualization specialists, and from architects who constitute the system's potential user community. We are asking them to evaluate proposed functionality, the organization of that functionality on screen, and the vocabulary available to users for indexed searches. So far we have received much interest and encouragement from architects who have seen the system.

By the time this paper appears, we will have been through several iterations of the interface design and should have a system with several dozen stories organized in its memory. Current areas of active development include the accumulation of more stories and guidelines, and the further elaboration of indexing formats and vocabulary. These activities are closely interrelated, since in the case-based paradigm, it is the existence of data to be indexed that drives the development of indexing formalisms.

Conceptual design is both intellectually fascinating and economically important. We believe that by taking a *case-based aiding* approach to conceptual design, we will be able to learn something about the process of design, and also have a positive effect on the product of design. It is this mixture of scientific and technological thrusts, each with short term and long term payoffs, that make us optimistic about the future of **Archie-II** and systems like it.

Acknowledgements

Many people have worked on the development and conception of the systems described in this paper. Craig Zimring, our colleague in architecture, and Richard Billington, our research programmer, have been constant collaborators throughout. The **Archie-II** team also includes Joshua Bleier, Anna Zacherl, and Kadayam Vijaya. An earlier version of the system, **Archie**, was developed with the help of Ashok Goel, Michael Pearce, and Lucas Sentosa; lessons learned from that effort have substantially influenced our current direction. This work has been supported in part by the Defense Advanced Research Projects Agency, monitored by ONR under contract N00014-91-J-4092. All views expressed are those of the authors.

References

Alterman, R.: 1988, 'Adaptive Planning', *Cognitive Science* **12**, 393-422

Ashley, K. and Rissland, E. L.: 1987, Compare and Contrast: A Test of Expertise in *AAAI-87*, Morgan Kaufmann Publishers: San Mateo, CA.

Bain, W.: 1986, *Case-Based Reasoning: A Computer Model of Subjective Assessment*. Ph.D. Thesis, Dept. of Computer Science, Yale University, New Haven, CT.

Barber, J., Bhatta, S., Goel, A. Jacobson, M., Pearce, M., Penberthy, L., Shankar, M., Simpson, R., and Stroulia, E.: 1992, AskJef: Integration of Case-Based Reasoning and Multimedia Technologies for Interface Design Advising, in this volume.

Bareiss, E.R.: 1989, *Exemplar-Based Knowledge Acquisition: A Unified Approach to Concept Representation, Classification, and Learning*, Academic Press: Boston, MA

Bareiss, R., Ferguson, W., and Fano, A.: 1991, The Story Archive: A Memory for Case-Based Tutoring in *The Proceedings of the 1991 DARPA Workshop on Case Based Reasoning*, Morgan-Kaufmann Publishers: San Mateo, CA.

Building Diagnostics Inc.: 1988, *Post-Occupancy Evaluation: New Bedford County Courthouse.*, Office of Programming: Boston, Mass., Project No. CBR 85-1 STU, Division of Capital Planning and Operations, June, 1988

Domeshek, E.A.: 1992, *Do the Right Thing: A Component Theory for Indexing Stories as Social Advice*. Ph.D. Thesis, Yale University, New Haven, CT.

Goel, A.: 1989, *Integration of Case-Based Reasoning and Model-Based Reasoning for Adaptive Design Problem Solving*. Ph.D. Thesis. Department of Computer and Information Science. The Ohio State University.

Goel, A., Kolodner, J. L., Pearce, M., Billington, R., and Zimring, C.: 1991, *Archie: A Case-Based Architectural Design System*. Technical Report No. GIT-CC-91/18. College of Computing, Georgia Institute of Technology, Atlanta, GA.

Hammond, K. J.: 1989, *Case-Based Planning: Viewing Planning as a Memory Task*, Academic Press: Boston, MA.

Hinrichs, T. R.: 1991, *Problem Solving in Open Worlds: A Case Study in Design*. Ph.D. Thesis. Technical Report No. GIT-CC-91/36. College of Computing, Georgia Institute of Technology, Atlanta, GA.

Hunter, L. E.: 1989, *Knowledge Acquisition Planning: Gaining Expertise Through Experience*. Ph.D. Thesis. Yale University, New Haven, CT.

Kass, Alex M. & Leake, David B.: 1988, Case-Based Reasoning Applied to Constructing Explanations in *Proceedings of the Workshop on Case-Based Reasoning (DARPA)*, Morgan-Kaufmann Publishers: San Mateo, CA.

King, J., Moore, E.O., Johnson, R.E., and Guregian, S.A.: 1981, *The Michigan Courthouse Study Vol 1*, Thomson-Shore, Inc: Dexter Michigan

Kolodner, J. L.: 1991, 'Improving human decision making through case-based decision aiding', *AI Magazine* **12:2**, 52–68

Kolodner, J. L. & Simpson. R. L.: 1989, 'The MEDIATOR: Analysis of an Early Case-Based Problem Solver', *Cognitive Science* **13:4**, 507–549

Kolodner, J.L., Simpson, R.L., and Sycara, K.: 1985, A Process Model of Case-Based Reasoning in Problem Solving in *Proceedings of IJCAI-85*,

Koton, P.: 1988, Integrating Case-Based and Causal Reasoning in *Proceedings of the Tenth Annual Conference of the Cognitive Science Society*, Lawrence Erlbaum Associates, 167–173

Mark, W.: 1989, Case-based Reasoning for Autoclave Management in *Proceedings of the Second Workshop on Case-Based Reasoning*, Morgan Kaufmann Publishers: San Mateo, CA.

Ram, A.: 1989, *Question-driven understanding: An integrated theory of story understanding, memory and learning*. Ph.D. Thesis. Yale University, New Haven, CT.

Schank, R.C.: 1982, *Dynamic Memory: A theory of reminding and learning in computers and people*, Cambridge University Press: New York, NY

Schank, R.C.: 1991, *Tell Me a story*, Charles Scribner and Sons: New York, NY.

Simpson, R.L.: 1985, *A Computer Model of Case-Based Reasoning in Problem Solving: An Investigation in the Domain of Dispute Mediation*. Ph.D. Thesis. Technical Report No. GIT-ICS-85/18, School of Information and Computer Science, Georgia Institute of Technology, Atlanta, GA.

Stanfill, C., and Waltz, D.: 1986, 'Toward Memory-Based Reasoning', *Communications of the ACM* **29:12**

Sycara, E.P.: 1987, *Resolving Adversarial Conflicts: An Approach to Integrating Case-Based and Analytic Methods*. Ph.D. Thesis. Technical Report No. GIT-ICS-87/26, School of Information and Computer Science, Georgia Institute of Technology, Atlanta, GA.

Turner, R. M.: 1989, *A Schema-Based Model of Adaptive Problem Solving*. Ph.D. Thesis. Technical Report No. GIT-ICS-89/42. School of Information and Computer Science, Georgia Institute of Technology. Atlanta, GA.

9

CASE-BASED REASONING IN DESIGN—2

A MODEL-BASED APPROACH TO BLAME-ASSIGNMENT IN DESIGN

E. STROULIA, M. SHANKAR, A. GOEL
Artificial Intelligence Group
College of Computing
Georgia Institute of Technology
801 Atlantic Drive
Atlanta GA 30332-0280 USA

and

L. PENBERTHY
Intelligent Systems Group
Human Interface Technology Center
NCR Corporation
500 Tech Parkway NW
Atlanta GA 30313 USA

Abstract. We analyze the blame-assignment task in the context of experience-based design and redesign of physical devices. We identify three types of blame-assignment tasks that differ in the types of information they take as input: the design does not achieve a desired behavior of the device, the design results in an undesirable behavior, a specific structural element in the design misbehaves. We then describe a model-based approach for solving the blame-assignment task. This approach uses structure-behavior-function models that capture a designer's comprehension of the way a device works in terms of causal explanations of how its structure results in its behaviors. We also address the issue of indexing the models in memory. We discuss how the three types of blame-assignment tasks require different types of indices for accessing the models. Finally we describe the KRITIK2 system that implements and evaluates this model-based approach to blame assignment.

1. Introduction

Design is a very common and wide-ranging information-processing task. The general design task takes as input the specification of a set of constraints on the design of an artifact. It has the goal of giving as output the specification of a structure for the artifact that satisfies the given constraints. In this paper we are primarily concerned with conceptual (or preliminary) design of physical devices. A typical design task in the domain of physical devices is to design a device that achieves a given set of behaviors. The conceptual phase of this design task takes as input a specification of the behavioral and structural constraints on a device, and has the goal of giving as output a high-level specification of a structure for the device that can deliver the desired behaviors.

Blame (or credit) assignment is a major subtask of the design task. The blame-assignment task occurs, for example, in redesign. Given a device that fails to achieve some desired function, the blame-assignment subtask of redesign is to identify the structural faults responsible for the device failure. The blame-assignment task also occurs in experience-based (or case-based) design. In experience-based design, new

J. S. Gero (ed.), Artificial Intelligence in Design '92, 519–537.

design problems are solved by adapting the solutions to old ones. Modification of the structure of existing designs to meet a new behavioral specification is thus a critical task in experience-based design. Identification of the structural "fault" responsible for the "failure" of the old design to deliver the new behaviors is another instance of blame assignment in design.

This paper has three main goals. First, we analyze the blame-assignment task in the context of experience-based design (and redesign) of physical devices. We identify three types of blame-assignment tasks that differ in the types of information they take as input: the design does not achieve a desired behavior of the device, the design results in an undesirable behavior, or a specific structural element in the design misbehaves. Second, we describe a model-based approach for solving the blame-assignment task. This approach uses structure-behavior-function (SBF) models of physical devices. The models capture a designer's comprehension of the way devices work in terms of a causal explanation of how their designs result in their behaviors. The models are based on a component-substance-field (CSF) ontology and represented in a behavioral representation language (BRL). This language provides a vocabulary for capturing the causal processes underlying the functioning of devices, including cyclic causal processes and interactions between multiple causal processes. Third, we address the issue of indexing knowledge of causal processes in memory. We discuss how the three types of the blame-assignment task require different types of indices for accessing knowledge of the causal processes.

The KRITIK2 system implements and evaluates this model-based approach to blame assignment. In addition to retrieving SBF models and using them for blame assignment, KRITIK2 also repairs designs and revises models. A discussion of design repair and model revision is beyond the scope of this paper and we will not discuss it any further (see Goel 1991a for design repair and Goel 1991b for model revision). Instead, we focus on showing how KRITIK2 validates the model-based approach for two of the three types of blame-assignment tasks in three different domains: electrical circuits, heat exchangers, and angular momentum controllers.

2. Types of Blame Assignment

Blame assignment is a classical problem in artificial intelligence (Minsky 1963; Samuel 1967). Given a system (e.g., a plan, a program, a problem solver) that fails to deliver the behaviors desired of it, the general blame-assignment task takes as input a specification of the structure of the system, the desired behaviors of the system, and the behaviors delivered by the system. It gives as output a specification of the structural faults responsible for the failure of the system to deliver the desired behaviors.

Blame assignment is a very common subtask of the design task. To see how blame assignment occurs in design, suppose that a user supplies a design specification such as: "Design an electrical circuit for a special kind of flashlight that produces light of intensity X, uses batteries of type Y, and does not produce more than Z amount of heat". [1] Note that the above design specification contains (i) the function desired of the design, viz. "produces light of intensity X", (ii) a structural constraint on the design, viz. "use batteries of type Y", and (iii) a behavioral constraint on the design, viz. "does not produce more than Z amount of heat". Now suppose that the designer retrieves from her memory a previously designed circuit that satisfies similar constraints and attempts to adapt the old design to satisfy the new constraints. To decide on modifications to the old design, she would need to identify the causes for the old design's "failure" to satisfy the new constraints. This is an instance of the blame-assignment task.

To take the example one step further, suppose that the designer generates an electrical circuit for the new flashlight but an evaluation of the design shows that the bulb in the circuit dies when the switch is pressed. Note that this specification of the design failure refers to a specific structural component "the bulb". Again, to decide on modifications to the circuit design, she would need to identify the causes for the design's failure. This too is an instance of the blame-assignment task.

We have identified three different blame-assignment tasks that occur in design of physical devices. The three tasks differ in the type of information they take as input: (i) the design does not achieve a desired function of the device, for instance, the retrieved design in the above example may not deliver the function of producing "light of intensity X"; (ii) the design results in an undesirable behavior, for example, the retrieved design may produce "more than Z amount of heat"; and (iii) a specific structural component in the design misbehaves, for example, the bulb in the circuit dies when the switch is pressed. While the three blame-assignment tasks differ in the type of information they take as input, the information they give as output is the same, viz. the structural causes for the design failure. In the first case, the retrieved design may not deliver the function of producing "light of intensity X" because the battery in the circuit supplies too high an electrical voltage. In the second case; the design may result in the undesirable behavior of producing "more than Z amount of heat", again because the battery supplies too high an electrical voltage. In the third case, the structural component (bulb) in the design may misbehave (die when the switch is pressed) yet again because the battery supplies too high an electrical voltage. As we discuss below, solving these different types of blame-assignment tasks calls for different types of knowledge.

[1] In this paper we distinguish between several different types of behaviors of a system. First, we distinguish between two types of behaviors of a system: (i) output behaviors and (ii) the *internal causal behaviors* that result in the output behaviors. Next, we make a distinction between two types of output behaviors: *intended output behaviors*, which we call *functions*, and unintended output behaviors that may arise as side effects. Finally, we make a distinction between two types of unintended output behaviors: *undesirable output behaviors* and unintended output behaviors about which neither the designer nor the user cares. The blame-assignment task refers to the output behaviors of a system, which include its functions.

3. Model-Based Blame-Assignment

In principle, a number of methods such as associative reasoning, case-based reasoning and model-based reasoning, are potentially applicable to the blame-assignment task. The choice of a particular method, in general may depend on factors such as the nature of the task (what type of blame-assignment task is it), the nature of the domain (what types of knowledge are available in it), the properties of the method (what types of knowledge and computational resources it requires) and the properties of the desired solution (what measures of correctness and optimality it should satisfies) (Goel and Chandrasekaran 1990).

In the KRITIK project we are investigating the use of qualitative structure-behavior-function (SBF) models for solving blame-assignment tasks in the domain of physical devices.[2] The content of these models is design-specific (though the underlying ontology and language are design-independent). The SBF model of a device captures a designer's comprehension of the causal behaviors that explain how the structure of the device produces its output behaviors (including its functions) (Goel 1991a). The hypothesis underlying the use of SBF models is that they are sufficient for solving blame-assignment tasks. The rationale for their use is that SBF models of physical devices are readily available and that model-based reasoning is a robust mechanism. Other methods, such as associative reasoning and case-based reasoning, while applicable, use highly situation-specific knowledge; model-based reasoning makes use of more general knowledge.

The SBF model of a design captures only the "declarative" part of the knowledge needed for blame assignment. In addition to these models, KRITIK2 uses abstract plans that capture the "procedural" portion of its blame-assignment knowledge. A blame-assignment plan specifies a specialized search procedure in the form of a sequence of abstract operations (Goel 1991a). When instantiated in the context of a design failure, it searches the SBF model of the design to locate the structural causes of the design failure. These abstract blame-assignment plans are design-independent.

3.1. Accessing Blame-Assignment Knowledge

The next issue is how to organize the SBF model of a design, i.e., how to index the causal behaviors that explain how the design structure produces the output behaviors of the design. The other half of this issue is how to organize the plan memory, i.e., how to index the abstract plans that search the SBF models. These issues are important because both the success and the efficiency of the model-based approach to blame-assignment depends on its ability to access the right causal behavior in the SBF model to search and the right plan with which to search the model.

[2] "Kritik" is a Sanskrit word that roughly translates to "the designer".

KRITIK2 uses *multiple indices* into the causal behaviors: desired functions, undesirable output behaviors, and structural components. The rationale for using these indices is that they directly correspond to the three types of blame-assignment tasks we described in the previous section: the design does not achieve a desired function of the device; the design results in an undesirable behavior; and a structural component in the design misbehaves. The advantage of this indexing scheme is that it provides rapid access to the causal behavior(s) in the SBF model that is useful for solving a given blame-assignment task.

The high-level organization of abstract plans in KRITIK2 is *task-specific*. That is, the plans are indexed by the type of blame-assignment task to which they are applicable. For example, the plans for searching the SBF model when the design does not achieve a desired function of the device are indexed by the types of functional differences between the desired and the delivered functions (Goel 1991a). The rationale and advantage of this indexing scheme is that it provides rapid access to the abstract plan that is useful for solving a given blame-assignment task.

4. Structure-Behavior-Function Models

KRITIK2's SBF models are based on a component-substance-field ontology. In this section we limit the discussion to components and substances. We discuss fields in a later section.

In KRITIK2, each device is described in terms of its function, its structure and its internal causal behaviors that allow its structure to achieve its function. The schema used for the representation of a design is shown in Figure 1(a).

Structure

KRITIK2 views a physical device as composed of a set of components and substances. The components are connected to one another and the substances flow through the components. Knowledge of the structural components of the device is organized in a structure-substructure hierarchy. Each device can be viewed as a set of connected components, and each component can be viewed as a device that can be further decomposed in terms of its components. The structure schema is shown in Figure 1(b).

Function

The function of a device is also represented as a schema. In this paper we focus on device functions that can be viewed as transformations from the input state to the output state of the device. The slots and the types of fillers in the schema shown in Figure 1(c). [3] The *given* state specifies the input state of the device; the *makes* state specifies the output state of the device. The *stimulus* slot specifies the interaction of the device with its external environment. The representation of a function of a device also specifies knowledge of the internal causal behavior

[3] In Figure 1, the slots enclosed in {} are optional and those that are followed by a * can exist zero or more times in the schema

design : (function : functional specification
causal-model : set of internal causal behaviors
structure : partonomic hierarchy of device components
history : acquisition history)

1(a) **Design Schema**

structure : (is-a-part-of : structural-part
consists-of : set of structural parts
affects-transitions : set of transitions where it plays a role
affected-by-transitions : set of transitions by which it is affected)

1(b) **Structure Schema**

functional specification : (makes: final behavioral state
{given: initial behavioral state }
by: causal behavioral-state sequence
stimulus: input from the external environment
{provided: secondary function })*

1(c) **Functional Specification Schema**

behavioral state : (previous: previous state
next: next state
enabled-by: preceding state-transition
enabling: succeeding state-transition
substance-schema: substance description at current state :
location
(property value)*
{contained substances description })

1(d) **Behavioral State Schema**

state-transition : (previous: preceding state
next: succeeding state
using-function: component's primitive function in mode
{structural-relation }
{qualitative-relation }*
{enabling-condition }*)

1(e) **State Transition Schema**

Fig. 1. **Structure-Behavior-Function model**

of the device that results in the function. In Figure 1(c), the causal behavior is specified by the *by-behavior* slot. Thus, the function schema contains knowledge of both a function and knowledge of the index to the causal behavior that results in the function. The *provided* slot specifies the conditions under which the behavior achieves the function.

Causal Behavior

As substances flow through the device components, their properties may change and also the mode of operation of the components may change. The sequence of these substance state and component state changes constitute *causal behaviors*. Knowledge of causal behaviors is represented as directed acyclic graphs (DAGS). The nodes in this graph represent the behavioral states and the edges represent state transitions. The states (i.e., the nodes) and the state-transitions (i.e., the edges) are themselves represented as schemas.

Behavioral State

A behavioral state is represented in the form of either a substance schema, a component schema, or a field schema. The *substance-schema* specifies the location, property, and values of properties of a substance. Substances can be either substances such as water and Nitric Acid or abstract substances such as heat or electricity. A substance in a particular device may also contain other substances, which are also described in the substance-schema shown in Figure 1(d).

State Transition

A state transition is also represented as a schema. The slots point to both the preceding and succeeding states of the transition. A transition may be caused by primitive types of behavioral interactions between substances and components such as *allow* and *pump* (Goel 1991). The slot *using-function* is a pointer to the specific function of a specific element in the device structure. A state transition may also be annotated by the enabling conditions under which the behavioral interactions result in the transition. One type of enabling condition, denoted by *under-condition-transition*, may act as an index to another behavior. The state-transition schema is described in Figure 1(e).

5. Blame Assignment in Experienced-Based Design

In experienced-based design, given a new problem specification, the design with the closest function to the new desired function F_{new} is retrieved from the design memory. The structure of the retrieved design S_{old} is adapted to transform the function it delivers from F_{old} to F_{new}. Once a new structure S_{new} is produced, it is tested. If it delivers the desired function, it is stored in the design memory for further reuse. Often, the testing reveals undesirable behaviors. In these cases, redesign is needed.

5.1. Blame Assignment in Design Adaptation

The specific task of blame assignment that we address in this section takes as input the specifications of the function F_{new} desired of the design and the function F_{old} delivered by the design, where the desired function F_{new} is similar to but different from the delivered F_{old}. It has the goal of giving as output a specification of the structural element(s) in S_{old} which is(are) responsible for the failure of the structure S_{old} to deliver the desired function F_{new}.[4] For blame-assignment tasks of this type, KRITIK2 uses the function F_{old} as an index to the causal behaviors that result in it.

An Example : The design of a Nitric Acid cooler

[4] The term "structural element" of a structure of a device entails both the components and the substances constituting the structure of the device.

Let us consider as an example the task of designing a Nitric Acid cooler (NAC_{new}) to reduce the temperature of some quantity of Nitric Acid from some initial temperature $T1$ to some final temperature $T2_{new}$ Let us also suppose that the design-retrieval task returns the structure of the design of a Nitric Acid cooler (NAC_{old}) which reduces the temperature of the same quantity of Nitric Acid from temperature $T1$ to temperature $T2_{old}$, where $T1 - T2_{new} >> T1 - T2_{old}$. Clearly, the desired function of cooling Nitric Acid from $T1$ to $T2_{new}$ is similar to but different from the delivered function of cooling Nitric Acid from $T1$ to $T2_{old}$.

Informally, the functioning of this device can be described as follows: Hot Nitric Acid is pumped in the device by a pump. Then it flows through a pipe into a heat-exchange chamber. There it goes through another pipe which is enclosed in a container of cold Water. Heat is exchanged between hot Nitric Acid and cold Water, as a result of which the temperature of out-flowing Nitric Acid is lower than that of in-flowing Nitric Acid; the temperature of cold Water increases correspondingly. The cold Water is pumped into the heat-exchange chamber by a water-pump. The temperature of the out-flowing Nitric Acid depends on (i) the temperature of the in-flowing Nitric Acid, (ii) the temperature of the in-flowing Water and (iii) the flow rate of Nitric Acid and the flow rate of Water. The functional specification of the old Nitric Acid cooler is shown in Figure 2(a). The internal causal behavior that achieves the cooling of Nitric Acid is shown in Figure 2(b).

For this example, the task of blame assignment takes as input the specification of the structure of NAC_{old}, the specifications of the desired function of cooling Nitric Acid from $T1$ to $T2_{new}$ and the delivered function of cooling Nitric Acid from $T1$ to $T2_{old}$ It has the goal of giving as output that structural element S_{old} that is responsible for the failure of the structure of NAC_{old} to cool Nitric Acid from $T1$ to $T2_{new}$.

KRITIK2's model-based method in this example works as follows: The function F_{old} of NAC_{old} is used as an index to retrieve the causal behavior $CoolNitricAcid$ shown in Figure 2(b). Since the value of the Nitric Acid temperature $T2$ needs to be changed, KRITIK2 traces the behavior $CoolNitricAcid$ which explains the Nitric Acid flow through the device, from $state4$ and backwards. The Nitric Acid temperature is the same in both $state4$ and $state3$ but changes in $state2$. The transition $transition2\_3 : state2 \rightarrow state3$ explains the temperature change. The initial Nitric Acid temperature $T1$, the quantity of heat exchanged between Nitric Acid and Water $Q2 - Q1$, and the Nitric Acid flow rate R, are added to the set of optional changes, since they affect the final temperature $T2$. Also, because $transition2\_3$ points to $transition6\_7$ of the behavior $HeatWater$, as necessary for $transition2\_3$ to occur, KRITIK2 also traces behavior $HeatWater$, from $state7$ and backwards.

The algorithm is described in more detail in Figure 3. The computational advantages of the algorithm arise from three sources. First, since F_{old} is used as an index into B_{old} the algorithm quickly localizes the start of search in B_{old}. Second, it traces only parts of B_{old} and the behaviors it indexes because the cros-indexing

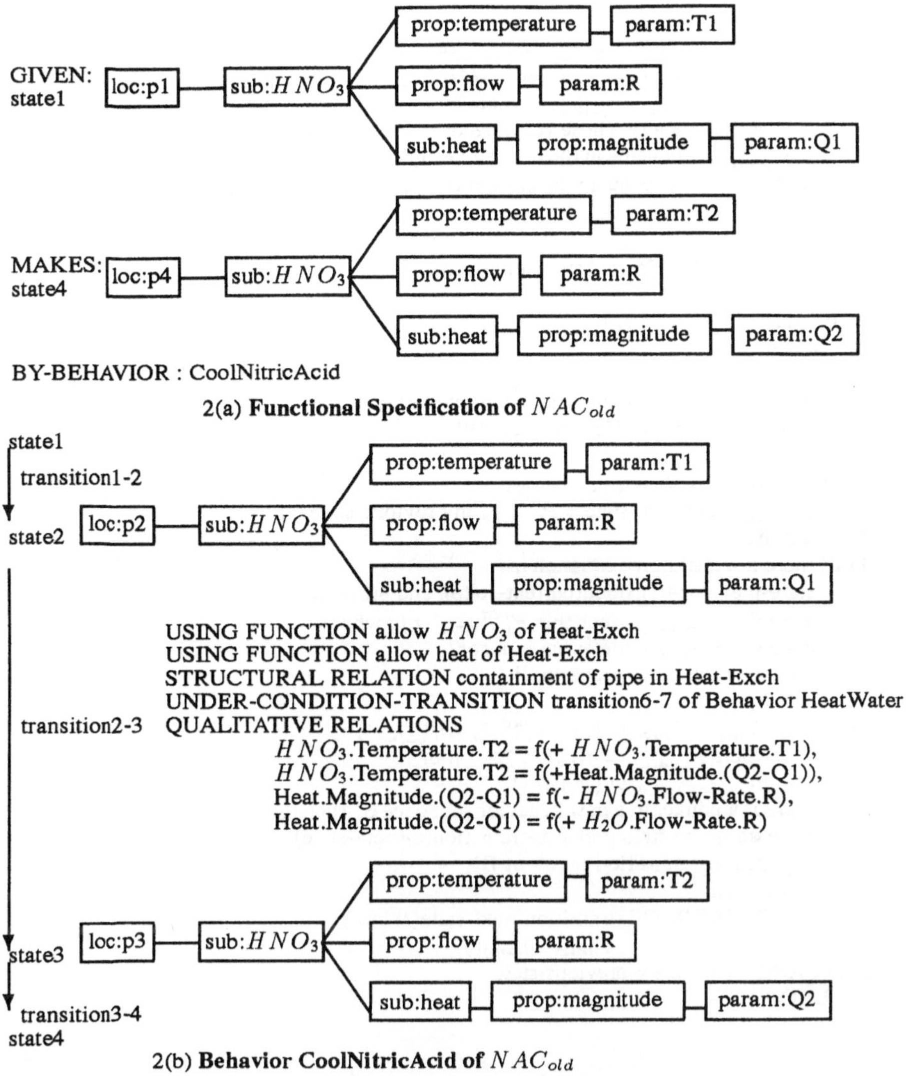

2(a) **Functional Specification of** NAC_{old}

2(b) **Behavior CoolNitricAcid of** NAC_{old}

Fig. 2. **The** NAC_{old} **design**

among internal behaviors allows it to focus the search.

5.2. Blame Assignment in Redesign

The specific task of blame assignment that we address in this section takes as input (i) the specification of the undesirable behavior $B_{undesirable}$ that the design exhibits, and (ii) the structural element in which the undesirable behavior is observed. It has the goal of giving as output a specification of that structural element of S_{new} which is responsible for the side-effect $B_{undesirable}$. For blame-assignment tasks of this type KRITIK2 uses structural elements of the device as indices into the causal

```
BLAMETRACE( F_old, F_new )
  F_new - F_old = ( Initial_State_new - Initial_State_old,
                    Final_State_new - Final_State_old)
  S_must-change = {(P_i, P^old_i.value, P^new_i.value), where
                    P_i is the property whose value needs to be changed,
                    P^new_i.value is its value in the desired design
                    P^old_i.value is the value in the current design
                    such that P^old_i.value ≠ P^new_i.value,
  B_old = function-by-behavior(F_old)
  S_must-change = BACKTRACE (behavior-final-state(B_old), S_must-change)

BACKTRACE (state, S_must-change)
  current-state = state
  LOOP
        previous-state = state-previous-state (current-state);
        IF previous-state = NIL, THEN Exit LOOP
        CASE :
        (0) IF for every P_i in S_optional-change
                P^old_i.value in current-state = P^old_i.value in previous-state
            THEN NIL
        (1) IF there is a qualitative equation in
                state-previous-transition(current-state), and property
                P_i in S_must-change such that P^old_i = f(c) , where
                c a component parameter, param, or
                  another substance property, P_l
            THEN S_must-change =
                    S_must-change ∪ {(P_l, P^old_l.value, P^new_l.value)}
                    where P^new_l.value = f_inv(P^new_i.value), or
                    S_must-change ∪ {param}
        (2) IF there is a pointer to a new behavior sequence B' such that
                a the transition state-previous-transition(current-state)
                depends on a transition trans of B'
            THEN spawn
                    BACKTRACE (transition-next-state(trans),
                                         S_must-change)
        END-CASE current-state = previous-state
        goto LOOP
  END-LOOP
```

Fig. 3. **Algorithm 1**

behaviors in which the elements play a role.

An Example : The Hubble Space Telescope

In order to make the present discussion more concrete, let us consider the specific problem of redesigning the reaction wheel assembly (RWA) aboard the Hubble Space Telescope (Keller, Manago, Nayak and Rua 1988). The desired function of RWA is to make the telescope point at a chosen area of the sky. The structure of the RWA consists of a rapidly spinning rotor mounted on a shaft. The rotating shaft is connected to a stator at both ends via assemblies of anti-friction

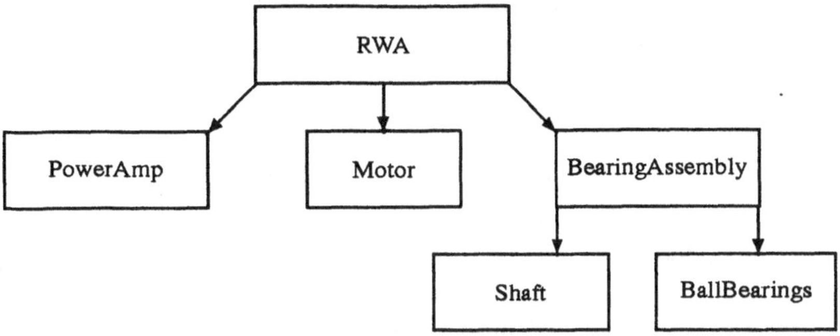

Fig. 4. **Part of the Reaction Wheel Assembly (RWA) structure**

```
BearingAssembly :
    is-a-part-of : RWA
    consists-of : (Shaft, BallBearings)
    affects-transitions :
        ((BearingAssembly.Heat-Affects-BearingAssembly.Temperature))
    affected-by-transitions : ()
```

Fig. 5. **The BearingAssembly Structure Schema**

ball bearings. The power that drives the rotor comes from a motor that is remotely controlled from earth. The stator itself is mounted on the walls of the telescope bay. The structure of the RWA, and its hierarchical organization, is shown in Figure 4. The structural schema that KRITIK2 uses to represent of structure of the BearingAssembly is shown in Figure 5.

The functioning of RWA is based on the law of conservation of angular momentum. When the telescope is to be oriented towards a specific direction, a signal from the earth is sent to the motor that results in a change in the power supplied by the poweramp to the motor. This causes a change in the motor's angular momentum which in turn affects the angular momentum and angular velocity of the shaft. Due to the conservation of angular momentum, the angular momentum of the telescope as a whole changes in the opposite direction. When the telescope nears its desired orientation, a change in the angular momentum of the telescope in the opposite direction is achieved in a similar manner, and the telescope angular velocity is reduced to zero. A common problem in the operation of RWA arises due to friction in the bearing assemblies. The load on the bearings due to the rapid spin of the rotor causes deformation of the bearing balls which results in increased frictional forces in the bearing assembly. This causes generation of heat in the bearing assembly. Since the increase in temperature depends on the load on the bearings, a typical redesign solution to this problem is to increase the load capacity of the bearings by increasing the size of the balls. This increased temperature in the bearing assembly is an example of an unintended and undesirable behavior.

Figure 6 shows the internal causal behavior of the RWA as represented in KRITIK2. Note that the change in the Shaft.AngularVelocity has both desired and undesired effects. The desired effects are depicted by the empty dashed boxes in the left of the Figure. The side-effect which is of interest for the particular example, is shown by the rest of the dashed boxes in the middle and right of the Figure. Note that each dashed box in Figure 6 represents a state-transition, each vector represents an under-condition-transition between the state-transitions it connects, and the solid boxes represent component-states.

The input to the blame-assignment task in the RWA example is an undesirable behavior (high temperature) localized to a specific structural element in RWA (the bearing assembly). The output is the specific structural element(s) in RWA responsible for the high temperature in the bearing assembly. KRITIK2's algorithm for blame assignment in the context of redesign is shown in Figure 7. This algorithm assumes that there are no loops in the causal dependencies in the model.

In the RWA example, the algorithm uses the BearingAssembly (whose specification is shown in Figure 5) as an index into the causal model as indicated in the bottom right of Figure 6. The index points to the behavior in the model in which the temperature of the bearing assembly is affected, namely the $BearingAssembly.Heat - Affects - BearingAssembly.Temperature$. The *under-condition-transition* pointer in this behavior is traced to the "preceding" behavior $Shaft.FrictionalForce - Affects - BearingAssembly.Heat$. The behaviors in the model are traced backwards until the algorithm reaches the behavior $Shaft.AngularVelocity - Affects - BearingAssembly.FrictionalForce$. The transition in this behavior is annotated with a qualitative relation that specifies that the frictional force is inversely proportional to the diameter of the bearing balls. Since the diameter of the bearing balls is an independent variable in the model (it does not depend upon the preceding behavior) the algorithm concludes that this is a structural element responsible for the undesirable behavior of RWA. The final output of the blame-assignment process is a list of such structural elements.

5.3. Cycles and Fields

The examples described above are limited in two very important ways. First, they are devices whose internal causal behaviors can be represented as DAGs (directed acyclic graphs). Their internal causal behaviors do not contain cyclic behaviors, such as feedback for example. Second, they contain only components and substances, where a substance can flow from one component to another only if the two components are interconnected. They do not contain fields that can influence at a distance.

Although the class of physical devices consistent with the two assumptions stated above is quite large, there are many other devices that contain cycles or fields or both. Consider for example the electromagnetic household buzzer, shown

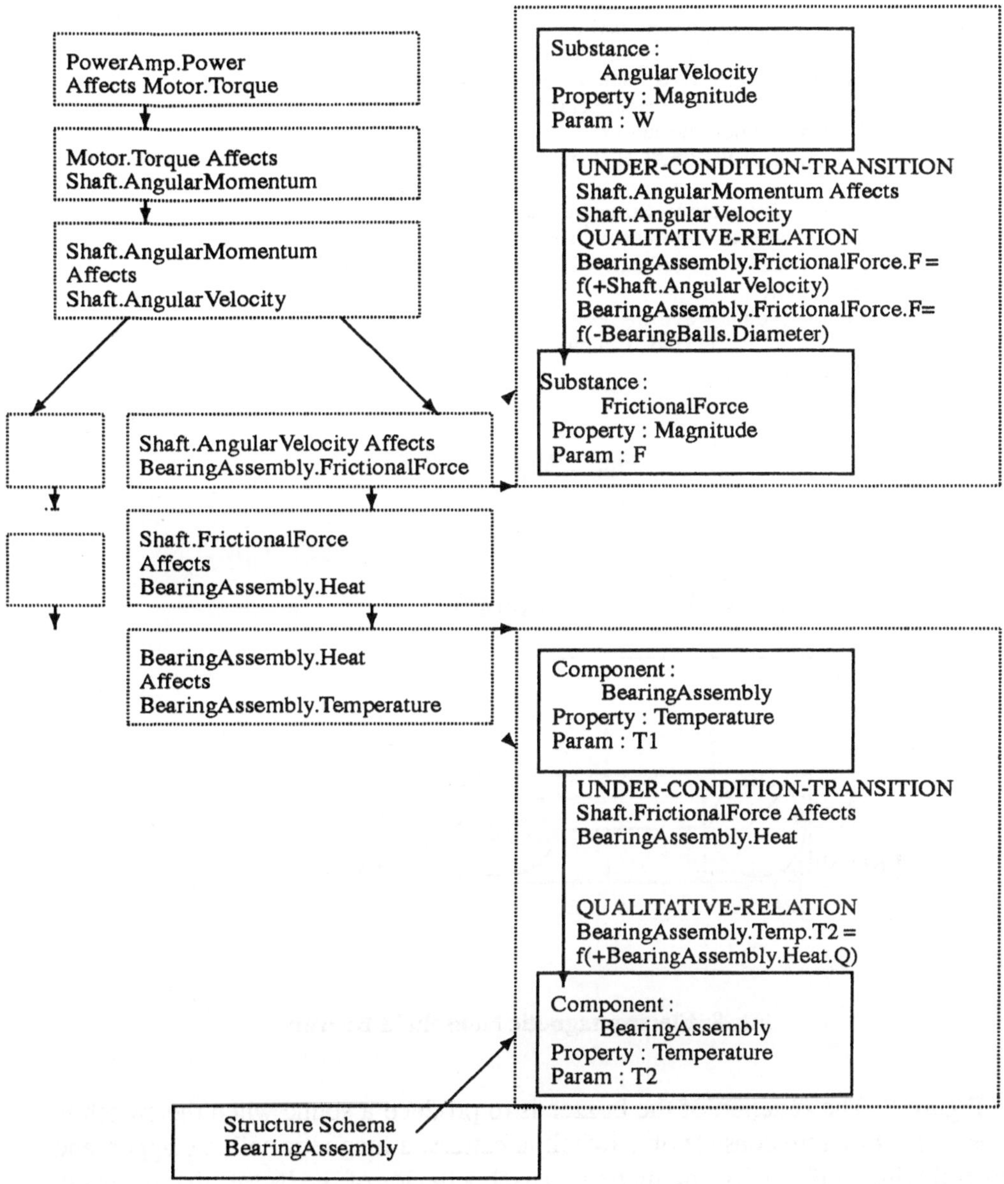

Fig. 6. **Part of the Internal Causal Behavior of RWA**

```
BLAMETRACE(Pointer-to-Structure, B_undesirable)
  Starting_Points = affected-by-transitions(Pointer-to-Structure)
        FOR EACH element IN Starting_Points
              Structural_Causes = Structural-Elements
                                        that affect this transition
              UnderConditionTransitions =
                    List of FOR EACH element IN Structural_Causes
                                        affected-by-transitions(element)
              FOR EACH cause IN Structural_Causes
                    IF affected-by-transitions(cause) = ∅
                    THEN PossibleFaults = PossibleFaults ∪ {cause }
              END-FOR
              Starting_Points = Starting_Points
                                   ∪ UnderConditionTransitions
        END-FOR
```

Fig. 7. **Algorithm 2**

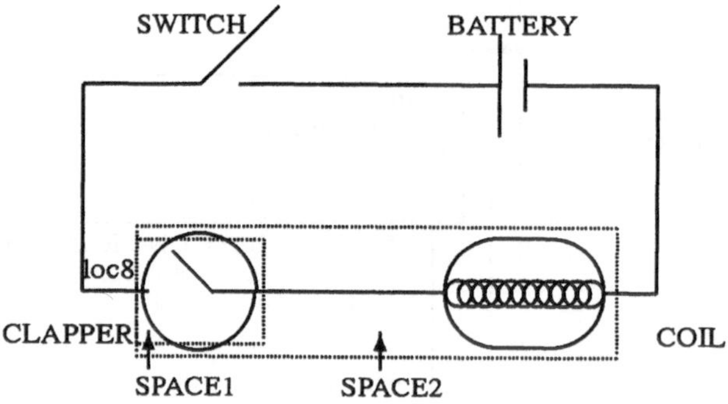

Fig. 8. **Electromagnetic Household Buzzer**

in Figure 8. The function of the buzzer is to produce a sound when the switch is closed. Its structure consists of a switch, a battery, a magnetic coil, a clapper, and several wires. Informally, the internal causal behavior of the buzzer that results in its function is the following: When electricity flows through the coil, it creates a magnetic field in the space around it. The clapper is within this space. In resting state, the clapper is stationary and the circuit is closed at point $loc8$, as shown in Figure 8. When the switch is pressed, the circuit is closed, electricity flows from the battery in the coil, and a magnetic field is created. The magnetic field attracts the clapper towards the coil, and opens the circuit at $loc8$. The moment the circuit opens, electricity flow stops, thus destroying the magnetic field in the coil, and releasing the clapper to its resting state. When the clapper returns to its resting

field-schema: field-source : a point in the device space
field-range : the range within which the field influences
intensity : a property of the field
influence-function : qualitative relation that describes the intensity change with the distance from the field.

Fig. 9. **Field Schema**

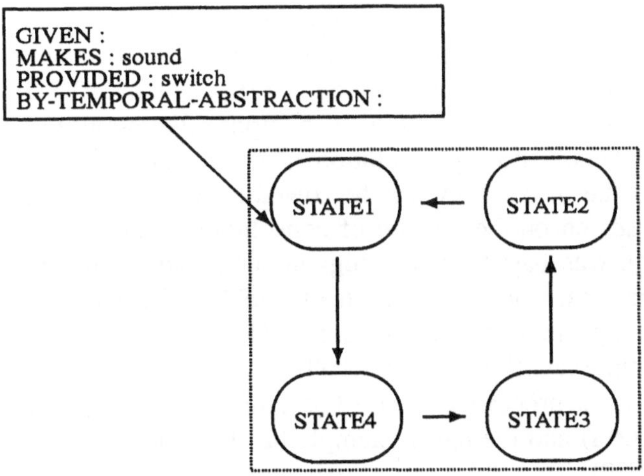

Fig. 10. **The** ***by-temporal-abstraction*** **primitive**

state, the circuit closes again, and the behavior is repeated.

The behavior described above depends critically on the repeated creation and destruction of the magnetic field in the coil. The magnetic field differs from substances such as water in that the field can influence the device components at a distance, as long as they are within the field space. Properties of components that happen to be within the field space might change. A field can be characterized by the location of the field source, the range of its influence, its intensity, and the function that describes how its intensity changes with the distance from the field source. KRITIK2's representation of a field is shown in Figure 9.

The function of the device is to produce a continuous sound. This sound is produced as the result of a sequence of events, namely, the clapping of the clapper against the circuit when it returns to its resting position. The device function can be viewed as a temporal abstraction of the repetitive causal behavior. KRITIK2 uses the primitive *by-temporal-abstraction* to represent this relation between the function and the causal behavior of the device as illustrated in Figure 10. Since the internal causal behavior occurs repeatedly, the pointer from the *by-temporal-abstraction* slot to the cyclic behavior can point to any of the states as initial state. A simple extension is also necessary to the blame assignment algorithm (i) to start the blame assignment process in the initial state to which *by-temporal-abstraction* points, and (ii) to terminate the trace of the behavior when it reaches the initial behavior state the second time around.

6. Related Research

KRITIK2 evolves from our earlier work on the KRITIK system. The original KRITIK used case-based reasoning for designing physical devices and model-based reasoning for adapting design cases (Goel 1991a, 1991b). KRITIK used structure-behavior-function models based on a component-substance ontology and represented in the behavioral representation language for design modification. Its component-substance ontology was borrowed from Bylander and Chandrasekaran's (1985) research on the method of consolidation, and its behavioral representation language was based on Sembugamoorthy and Chandrasekaran's (1986) work on the functional representation scheme. KRITIK2 extends KRITIK's ontology, model, and language to include fields. KRITIK could handle simple interactions between multiple causal processes; KRITIK2 extends it to accommodate causal side-effects and cyclic processes. The NAC example described here has been adapted from (Goel 1991a) and the RWA example has been adapted from (Goel and Chandrasekaran 1989).

Our decomposition of the task of design modification into the subtasks of blame assignment and repair generalizes Goel and Chandrasekaran's (1989) decomposition of the redesign task into the subtasks of diagnosis and repair. Similarly, our proposal for using device functions, undesirable output behaviors, and structural elements as indices to causal processes generalizes their proposal for using functions and structural components as indices. More importantly, KRITIK2 for the first time integrates these ideas into a broader framework, implements the framework in a computer program, and validates its sufficiency in several domains.

KRITIK2 also builds on other work in artificial intelligence on design. Its use of abstract plans, for example, is based on the method of plan selection and instantiation developed by Sacerdoti (1976), Friedland and Iwasaki (1985), and Brown and Chandrasekaran (1989) in their work on design problem solving. However, KRITIK2's plans differ in their indexing, which is specific to the task of blame assignment, and execution, which results in backward search of the causal behaviors embedded in the SBF models.

The idea of using causal models for analyzing designs of physical devices dates at least as far back as Rieger and Grinberg's (1978) work on commonsense algorithms. They did not, however, relate the causal model of a device with its functions or provide an account of how the causal model can be used for blame assignment.

Gero, Lee and Tham (1991) have recently proposed the use of functions as indices to causal models in a manner very similar to the functional representation scheme (Sembugamoorthy and Chandrasekaran 1986) from which KRITIK2's organizational scheme evolved. Umeda *et al.* (1990) have integrated the component-substance ontology of consolidation (Bylander and Chandrasekaran 1985) with the functional indexing scheme in a representation that closely resembles the SBF models used by the KRITIK2 system. Neither Gero, Lee and Tham nor Umeda *et al.* describe how their SBF models can be used in design modification.

Stallman and Sussman (1977) introduced dependency-directed backtracking in their work on design modification to decide what component to modify when a design fails to achieve the desired functions. KRITIK2's SBF models serve a similar purpose: they express the causal dependencies between the device states which enables the designer to trace the structural cause for the device's failure to achieve a desired function. Steinberg and Mitchell's (1985) REDESIGN system's use of the purposes of structural components in a design and KRITIK's use of functional abstractions of structural components are similar. REDESIGN's design knowledge, however, is largely associative rather than in the form of SBF models. Murthy and Addanki's (1987) PROMPT system uses graphs of models to decide on the applicability of modification heuristics. KRITIK2 uses task-specific organizations of general SBF models and blame-assignment plans for design modification. Further, REDESIGN and PROMPT are concerned with design modification in the context of redesign only; KRITIK2 can handle design modification in the context of experience-based design as well as in redesign problem solving. The CADET system, (Navinchandra, Sycara and Narasimhan 1991) uses a functional organization of causal behaviors for design modification in the context of case-based design. Their approach is similar to that of the original KRITIK system. CADET, however, neither uses multiple indices into causal behaviors, nor handles cyclic processes or other complex interactions, nor does it solve different kinds of blame assignment tasks. KRITIK2 does all this and several domains.

6.1. Further Research

KRITIK2's model-based method for blame-assignment raises a number of new issues. Current work on KRITIK2 focuses on representation of and reasoning about (i) interactions among causal behaviors, (ii) cycles including feedback and feedforward, (iii) fields, and (iv) large devices in complex engineering domains.

7. Conclusions

Our experiments with KRITIK2 lead us to three main conclusions. First, blame-assignment tasks in design modification can be of three types: (i) the design does not achieve a desired function of the device, (ii) the design exhibits an undesirable behavior, and (iii) a specific structural element in the design misbehaves. Second, each of the three blame-assignment tasks requires different indices into the causal behaviors of the design. In particular, task (i) requires design functions as indices into the causal behaviors that result in the functions, and task (iii) requires structural elements of the design as indices into the causal behaviors in which they play a role. Third, KRITIK2's model-based method is sufficient for blame-assignment tasks of types (i) and (iii). This model-based method uses structure-behavior-function models. The range of physical devices and causal processes covered by KRITIK2 provides some evidence for the generality of its representations and methods.

Acknowledgements

This research has benefited from discussions with several colleagues in the Cognitive Science Program at the Georgia Institute of Technology and at the Laboratory for Artificial Intelligence Research at the Ohio State University. We are especially thankful to Sambasiva Bhatta, B. Chandrasekaran, and S. Prabhakar. We thank Kavi Mahesh and anonymous reviewers for their comments on an earlier draft. This work has been partially supported by a grant from the NCR Corp. and an IBM Fellowship.

References

Brown, D. and Chandrasekaran, B.: 1989, *Design Problem Solving: Knowledge Structures and Control Strategies*, London, UK, Pitman, 1989.

Bylander, T. and Chandrasekaran B.: 1985, Understanding Behavior Using Consolidation, *Ninth International Joint Conference on Artificial Intelligence*, pp. 450-454.

Friedland, P. and Iwasaki, Y.: 1985, The Concept and Implementation of Skeletal Plans, *Automated Reasoning*, **1**, pp. 161-208.

Gero, J.S., Lee, H.S., and Tham, K.W.: 1991, Behaviour: A Link between Function and Structure in Design, *Proceedings IFIP WG5.2 Working Conference on Intelligent CAD*, pp. 201-230.

Goel, A.: 1991a, A Model-Based Approach to Case Adaptation, *Proceedings of the Thirteenth Annual Conference of the Cognitive Science Society*, pp. 143-148.

Goel, A.: 1991b, Model Revision: A Theory of Incremental Model Learning, *Proceedings of the Eighth International Workshop on Machine Learning*, pp. 605-609.

Goel, A. and Chandrasekaran, B.: 1989, Functional Representation of Designs and Redesign Problem Solving, *Proceedings of the Eleventh International Joint Conference on Artificial Intelligence*, pp. 1388-1394.

Goel, A. and Chandrasekaran, B. : 1990, A Task Structure for Case-Based Design, *Proceedings of the 1990 IEEE International Conference on Systems, Man, and Cybernetics*, pp. 587-592.

Keller, R., Manago, C., Nayak, P., Rua, M.: 1988, Internal Memo, Knowledge Systems Laboratory, Stanford University.

Minsky, M.: 1963, Steps Towards Artificial Intelligence, *Computers and Thought*, Feigenbaum and Feldman (editors), McGraw-Hill, New York.

Murthy, S. and Addanki, S.: 1987, PROMPT: An Innovative Design Tool, *Proceedings of the Sixth National Conference on Artificial Intelligence*, pp. 637:642.

Navinchandra, D., Sycara, K. and Narasimhan, S.: 1991, Behavioral Synthesis in CADET, a Case-Based Design Tool, *Proceedings of the 7th Conference on Artificial Intelligence Applications*, pp. 217-221.

Rieger, C. and Grinberg, M.: 1978, A System for Cause-Effect Representation and Simulation for Computer-Aided Design, *Artificial Intelligence and Pattern Recognition in Computer-Aided Design*, J. Latombe (editor), Amsterdam, Netherlands, North Holland, pp. 299-334.

Sacerdoti, E.: 1977, *A Structure for Plans and Behavior*, Lawrence Erlbaum, Hillsdale, NJ.

Samuel, A.: 1967, Some Studies in Machine Learning Using the Game of Checkers - II, *IBM Journal*, **11**(11),601.

Sembugamoorthy, V. and Chandrasekaran, B.: 1986, Functional Representation of Devices and Compilation of Diagnostic Problem-Solving Systems, *Experience, Memory and Reasoning*, J. Kolodner and C. Riesbeck(editors), Lawrence Erlbaum, Hillsdale, NJ, pp. 47-73.

Stallman, R. and Sussman, G. : 1977, Forward Reasoning and Dependency-Directed Backtracking in a System for Computer-Aided Circuit Analysis, *Artificial Intelligence*, **9**, pp. 135.

Steinberg, L.I. and Mitchell, T.M.: 1985, The REDESIGN System: A Knowledge-based Approach to VLSI CAD, *IEEE Design and Test of Computers*, **2**(1),45.

Umeda, Y., Takeda, H., Tomiyama, T., Yoshikawa, H.: 1990, Function, Behaviour, and Structure, *Applications of Artificial Intelligence in Engineering, vol 1, Design, Proceedings of the Fifth International Conference*, Gero (editor), Springer-Verlag, Berlin, pp. 177-193.

A STUDY ON STRUCTURAL OPTIMUM DESIGN BASED ON QUALITATIVE SENSITIVITIES

M. ARAKAWA and H. YAMAKAWA
Department of Mechanical Engineering
Waseda University
3-4-1 Ohkubo Shinjuku
Tokyo 169
Japan

Abstract. In the process of designs, designers sometimes are needed to set up mathematical models in order to solve those design problems. And a design optimization is often carried out based on those models. If you observe that the design is one of processes of finding the best or better solution under design constraints, optimization would become one of the most important parts of the design process to obtain a sophisticated design. However, it is not always a simple task for the designers to make models quantitatively, especially in such upper flow of design process as a conceptual design or a preliminary design. Even in the problems in upper flow of design, it is often much easier to estimate difference between two design candidates qualitatively. Thus, optimization with aid of qualitative grasp would be very helpful and important. In this paper, we first propose a basic idea of qualitative sensitivity and qualitative optimality, and present a method of qualitative optimum oriented design. There, we make use of fuzzy reasoning system for qualitative grasp. We compare the results obtained by the proposed method with those by conventional quantitative method. And then we develop Qualitative Sensitivity Aided Optimum Oriented Design System by Referring "Model Case Base", which combine the merits of both model base reasoning and case base reasoning in order to indicate qualitative sensitivities effectively and was named by the authors. We apply this system to the layout optimization problem of parabolic antenna. Through these results, we examine the efficiencies of our method.

1. Introduction

Design is a process of finding the best or better solution through many decisions and compromises by consideration of plenty of factors of design objects. And when you want to carry out design with the aid of computer, you need to model those objects, and treat them in mathematical way. In many cases, these models are required to express completely in quantitative manner. And once you have achieved a quantitative model, you can carry out optimization with those mathematical models. Optimization is a process of finding the most reasonable solution (optimum solution) by giving mathematical models, and is a method to obtain the result which is desirable to coincide with that of well trained designers, so called experts. If you recognize that design is one of processes to find better solution, optimization would consider to be one of the most important part of the design process. However, in design process, there exist many factors which are hard to

J. S. Gero (ed.), Artificial Intelligence in Design '92, 539–558.

express quantitatively. Figure 1 is an illustrative flow of a design process which shows a domain of a task of human beings and that of computer. It shows that lower the process goes, more domain computer should deal with. Thus, it suggests that necessity of quantitative modeling become larger as the design process goes lower. So that if you want to treat upper part of the design process, it would not be easy and effective to carry out optimization in quantitative manner. However, there would be some models which can not be expressed completely in a quantitative manner but in a qualitative or semantic manner. For example, if you consider a manufacturing of a structure, it is clear that the design with less numbers of welding is better. In this case, the function of "manufacturing" have sensitivity "increasing" with respect to the variable "the numbers of welding". From such a standpoint, we will show a basic idea of a "qualitative sensitivity", which can be expressed qualitatively. And one of the purposes of this paper is to introduce optimization with qualitative sensitivities.

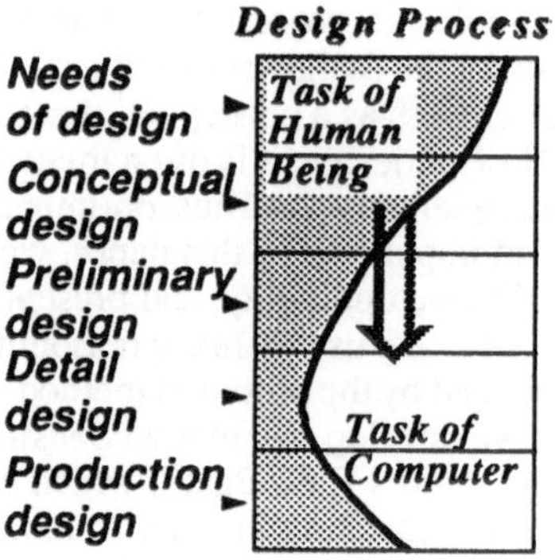

Figure 1 Design process and domains of tasks

Referring the past investigations, there exist the studies by Cagan J. & Agogino A.M.,(1987,1991) which examined optimization with qualitative grasp by using both Monotonicity Analysis and symbolic computation. And in the studies of Arakawa M. & Yamakawa H.(1989,1990,1991a,b), qualitative optimization methods were proposed, examined and made it clear that optimization can be carried out even with qualitative grasps. In those studies, we have done optimization interactively by using qualitative reasoning with sign logic. However, owing to sign logic, there occurs lots of ambiguity, thus we need to carry out a trial optimization for the insurance of optimality. In Arakawa M. & Yamakawa H.(1991b), fuzzy language and extended fuzzy rules have been developed to each operation, and been made a better inferring system. Since it was based on sign logic, it still have a problem of ambiguity. In this study, we have extended fuzzy reasoning and achieved better inference rules than the conventional method. According to these modification, we don't need to carry out a trial optimization, and make it possible to carry out optimization with more qualitative models. We first demonstrate the proposed method to a simple optimization problem, and examined difference between the obtained result and the quantitative optimization result and check the efficiency of the method. Then we have developed "Qualitative Sensitivity Aided Optimum Oriented Design System by Referring Model Case Bases", which aids the designer in deciding qualitative sensitivities during the interactive optimization process with indicating qualitative sensitivities through comparing two model case. In this study, we newly call "Model Case Bases", which means that the proposed system is made by combining merits of both the model base reasoning and the case base reasoning to indicate qualitative sensitivity effectively. Then we use this system and carry out optimization without using of quantitative sensitivities. Through these demonstrations, we have examined the efficiency of the proposed method and also the proposed system.

2. General Definition of Qualitative Optimization Problem

In this section, we will explain a formulation of optimization problem and an algorithm of the proposed method. And also, we will show an idea of the qualitative optimality which we will make use of as the criterion of optimality in qualitative optimization.

2.1 FORMULATION OF OPTIMIZATION PROBLEM

Generally, an optimization problem can be expressed as Eq.(1)

$$\min_{\mathbf{x}\in X^N} f(\mathbf{x}) \tag{1}$$

subject to

$$g_j(\mathbf{x}) \leq 0 \quad :(j=1,2,\ldots,M)$$

where

$$\mathbf{x}=\{x_1,x_2,\ldots,x_N\}^T$$

If X is a set of real numbers, then it will be called as an optimization with continuous variables, and if X is a set of integer numbers, then it will be called as an integer optimization. In practical design problems, the domain of design variables are not always real numbers. For example, thinking of a truss problem; the number of nodes is an integer, in most of the design cross sectional areas may be chosen from some industrial standards, and considering of material characteristics, a material may be chosen among the definite number of materials and these values may have specified values and those may be considered as discrete numbers. In such a case, design variables might be chosen from given values or data base. In this study, we assumed that design variables should be chosen from given specified values. According to this formulation, it may be thought as a kind of combinational optimization problem or a searching problem of an optimum point among discreet values. To treat these kinds of problem we do not adopt the searching techniques in the field of artificial intelligence but the qualitative sensitivities. By the way, there still exists a problem whether the obtained solution of this formulation could be called "optimum solution" in global sense. But from practical points of view, we relax the meaning of the word "optimum" to "optimum oriented". Then the obtained solution will surely be close to the best one among the solutions of a satisfying problem in Simon (1969) which is expressed in Eq.(2)

$$
\begin{aligned}
&\text{satisfy} \\
&\quad f_i(\mathbf{x}) - \hat{f}_i \le 0 \quad :(i=1,2,\ldots,L) \\
&\text{subject to} \\
&\quad g_j(\mathbf{x}) \le 0 \quad :(j=1,2,\ldots,M) \\
&\text{where} \\
&\quad \mathbf{x} = \{x_1, x_2, \ldots, x_N\}^T \\
&\quad \hat{f}_i : \text{desired value of } f_i(\mathbf{x})
\end{aligned}
\tag{2}
$$

2.2 GENERAL ALGORITHM OF PROPOSED METHOD

Here, we show a general algorithm of qualitative optimization in the following:

(1) Select initial conditions which will satisfy Eq.(1)
(2) Determine qualitative sensitivities
(3) If there is no active constraint, select a state which may have the most improving (decreasing or increasing) of objective function. And go to (5)
(4) If there are any active constraints, find a state which will satisfy constraints and have higher expectation for improving objective function.
(5) Judge whether the state will satisfy the qualitative optimality. If it doesn't satisfy the qualitative optimality, go to (2).

In Fig.2, we will show the general algorithm of the proposed method. In section 3, we will explain how to infer the design in detail. One of big differences between proposed method and previous works in Arakawa M. & Yamakawa H.(1991b) is that we don't need to carry out trial optimization. According to this improvement, it become possible to carry out optimization with qualitative sensitivities, even without a definite modeling. We will start to explain qualitative optimality which we use in algorithm (5).

2.3 IDEA OF QUALITATIVE OPTIMALITY

The quantitative optimality can easily conclude as following

Quantitative Optimality

> If there is no possibility to improve objective function or if it violates at least one constraint with updating design variables $\mathbf{x}_{opt}$ in any directions which seem to improve objective function, then $\mathbf{x}_{opt}$ is local optimum.

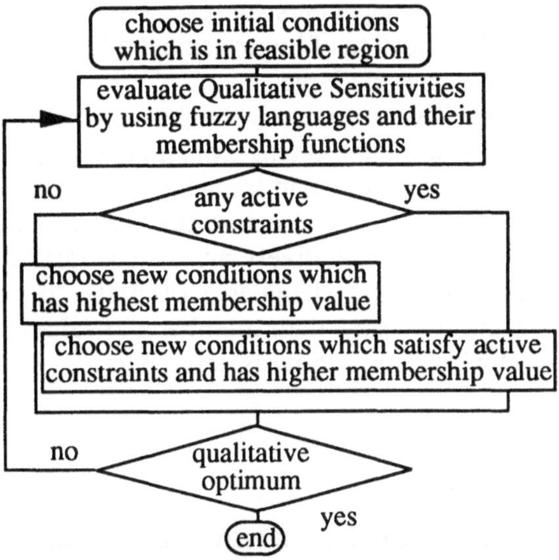

Figure 2 Algorithm of proposed method

An analogous to this optimality, the qualitative optimality to be proposed follows;

Qualitative Optimality

Consider an improvement to the next state from the current state. If we cannot find any other states which will not violate at least one constraint qualitatively, then the current state is qualitative local optimum.

3. Inference Algorithm by Using Fuzzy Reasoning

In this section, we will propose an algorithm of inferring the next set of design variables which is a main part of the proposed method.

3.1 DEFINITION OF QUALITATIVE SENSITIVITY AND ITS MEANGING

In iterative process of ordinary quantitative optimization with continuous design variables by the mathematical programming methods, they usually calculate sensitivities of objective function and constraints, and then calculate amounts of improvement for design variables by using these values. So that in each step, the relation between current design variables and the new design variables can be expressed as following in general,

$$x_{new}=x_{current}+P(x_{current},\nabla f,\nabla g)\cdot\Delta \qquad (3)$$

Where P is a vector of which components are functions of $\mathbf{x}_{current}$, $\nabla f, \nabla g$ and based on which we determine the direction of improvement of design variables. Δ works to improve design variables not too large.

In this study, we consider a case that new design variables may be chosen from the given values. Then it is not so important problem to calculate ∇f so accurately or quantitatively. Main concerning lies in knowing the amounts of effects of the changes in the design variable x_k on objective function and constraints qualitatively. These effects evaluated qualitatively are called "qualitative sensitivities" and assigned by three kinds of fuzzy languages ("small", "normal" and "large"). These languages show the effects by their membership values. Qualitative sensitivities of function h(x) with respect to x_k is defined as

$$QS(h(\mathbf{x}),x_k)=\mathrm{Sign}\left(\frac{\partial h}{\partial x_k}\right)\cdot\begin{Bmatrix}\mu_{small}\\ \mu_{normal}\\ \mu_{large}\end{Bmatrix} \tag{4}$$

where $\mathrm{Sign}(\partial h(\mathbf{x})/\partial x_k)$ is the sign of quantitative value of $\partial h(\mathbf{x})/\partial x_k$. It means that we don't need to calculate the exact values of $\partial h(\mathbf{x})/\partial x_k$. We allocate meaning to each language as following.

small :It means that h(x) will not increase so much even we update x_k in the direction of $\mathrm{Sign}(\partial h(\mathbf{x})/\partial x_k)$

normal :It means that h(x) will increase when we update x_k in the direction of $\mathrm{Sign}(\partial h(\mathbf{x})/\partial x_k)$

large :It means that h(x) will increase greatly when we update x_k in the direction of $\mathrm{Sign}(\partial h(\mathbf{x})/\partial x_k)$

Allocating these meanings, you can decide membership values in some quantitative ways. If you are well trained designer, you can determine these values adequately from experience or intuition. But if you are not so, you need some quantitative guidance or histories of past same kinds of design cases; for which we can have model or case base database.

3.2 MEMBERSHIP FUNCTIONS TO FUZZY SETS OF QUALITATIVE SENSITIVITIES AND DESIGN VARIABLES

We use two kinds of fuzzy sets; one is a fuzzy set of qualitative sensitivities and the other is a fuzzy set of design variables. In this section, we will explain the membership functions to both kinds of the fuzzy sets. In fuzzy languages and fuzzy numbers, their meanings greatly depend on their membership functions. And also they have important roles in fuzzy reasoning. We will set expediently a membership function to each fuzzy set as follows. As a type of membership function, we can use any function if it is simply non-decreasing in its left hand side and non-increasing in its right hand side by its definition, but here we use the function in Eq.(5) for its simplicity.

$$M(x)=\begin{cases} 1-\left|\frac{x-m}{w}\right| & \text{for } M(x)>0 \\ 0 & \text{otherwise} \end{cases} \tag{5}$$

According to Eq.(5), we may have to decide the value m (representing number) which makes the membership 1 and width w. In this study, we set membership function to each qualitative sensitivity in Eq.(6a,b,c).

small	$:m=x_k, w=1.5\text{width}$	(6a)
normal	$:m=x_k\pm \text{width}, w=2\text{width}$	(6b)
large	$:m=x_k\pm\text{width}, w=2.5\text{width}$	(6c)

where x_k represents the current value of design variables and "width" represents a standard unit increment among its database. As for design variables, we set membership function of each discreet given value x_i in Eq.(7).

$$x_i \qquad :m=x_i, w= 2\ \text{width} \tag{7}$$

3.3 PROPOSED INFERENCE ALGORITHM BY USING FUZZY REASONING

Figure 3 shows schematically the outline of the proposed inference algorithm by comparing with that of the gradient projection method (GPM) in a case with two design variables. Where current design variables are **x** and constraint g_j is an active constraint. In GPM, we determine new design variables $\mathbf{x}_{new}$ by projecting ∇f to the active constraint g_j by using ∇g_j. In the proposed method, new design variables can be obtained as following. Let's consider the inference of new design variable of x_k in general. And where we assumed qualitative sensitivity of objective function f with respect to design variable x_k as $QS(f,x_k)$ (Suppose $\text{Sign}(\partial f/\partial x_k)$ is "minus"). We then think intersection of two fuzzy sets; fuzzy sets by qualitative sensitivities and those by design variable. Then superior of this intersection set will be a degree of influence of fuzzy language to the design variable. Then we add multiple of a degree of influence of each fuzzy language to the design variable and qualitative sensitivity, and set this value as an reasoning (Ex) of the design variable.

$$Ex_f(x_{kcandidate})=\sum_{i\in\{\text{small},\text{normal},\text{large}\}} \mu_i\times\sup\left\{\min_{x\in R}\{M_{i,f}(x), M_{kcandidate}(x)\}\right\} \tag{8}$$

For example, in Fig. 3, a reasoning of next state ("dominant") will be adding membership value in point A times μ_{small} and membership value in point B times μ_{normal}. In the general algorithm of part (3) and first half of part (4), which have explained in the section 2.2, we choose each new design variable as to have the

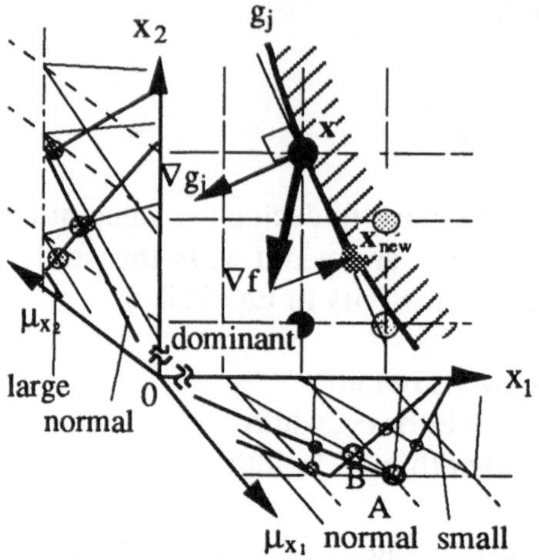

Figure 3 Fuzzy Reasoning Method

maximum of reasoning in this study. And we call the obtained design variables as $\mathbf{x}_{dominant}$. Last part of general algorithm of part (4), we choose values as following: We assumed that constraint g_B violates a constraint. Then we can find a direction that constraint g_B to correct back within the constraint after we set a qualitative sensitivity of constraint g_B to design variable $\mathbf{x}_{dominant}$, we can calculate reasoning to each discreet design variables under consideration of the influence of constraint g_B.

$$Ex_g(x_{kcandidate}) = \sum_{i\in\{small,normal,large\}} \mu_i \times \sup\left\{\min_{x\in R}\{M_{i.gB}(x), M_{kcandidate}(x)\}\right\} \tag{9}$$

We thought that Eq. (8) gives a reasoning of decreasing objective function and Eq. (9) is a reasoning of correct back from the violated constraint. And we will think of infimum of these two reasoning as following.

$$Ex_{fg}(x_{kcandidate}) = \inf\{Ex_f(x_{kcandidate}), Ex_g(x_{kcandidate})\} \tag{10}$$

And we try to consider a next state of design variables from higher order of these reasoning until it become satisfying the violated constraint.

4. Numerical Example (By Using the Proposed Qualitative Inference)

Let's treat a simple optimization of 3-bar truss (in Fig. 4), as an example, in order to compare with the results obtained by the proposed method with quantitative optimization.

Where optimization problem can be expressed as following.

$$\text{Minimize} \quad M(\mathbf{x}) = 2\sqrt{2}\,\rho x_1 + \rho x_2 \tag{11a}$$

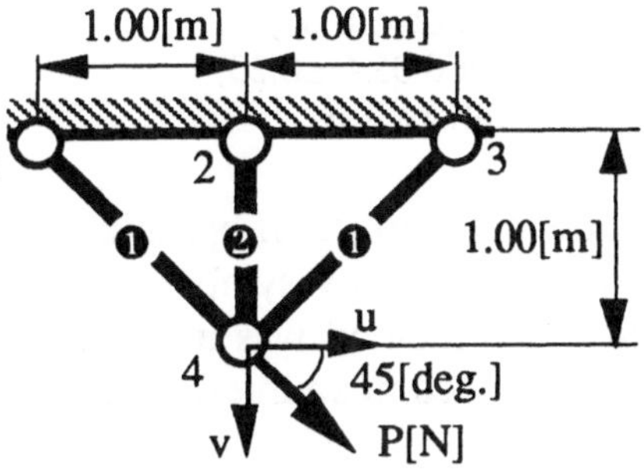

Figure4 3-bar truss

subject to constraints:

$$g1:\ u(\mathbf{x}) = \frac{P}{E x_1} \leq u_a \tag{11b}$$

$$g2:\ v(\mathbf{x}) = \frac{P}{E(x_1+\sqrt{2}x_2)} \leq v_a \tag{11c}$$

$$g3:\ \sigma_{ten}(\mathbf{x}) = \max\left\{ \frac{P(\sqrt{2}x_1+x_2)}{x_1(\sqrt{2}x_1+2x_2)}, \frac{P}{x_1+\sqrt{2}x_2} \right\} \leq \sigma_{tena} \tag{11d}$$

$$g4:\ \sigma_{com}(\mathbf{x}) = \frac{Px_2}{x_1(x_1+\sqrt{2}x_2)} \leq \sigma_{coma} \tag{11e}$$

where

- P :load(=1,00x10^3[N])
- E :Young's Modulus(=2.05x10^{11}[Pa])
- ρ :density(=7.86x10^3[Kg/m^3])
- u_a :allowance displacement for x direction (=2.50x10^{-6}[m])
- v_a :allowance displacement for y direction (=1.00x10^{-6}[m])
- σ_{tena} :allowance tensile stress (=3.00x10^5[Pa])
- σ_{coma} :allowance compressive stress (=5.00x10^5[Pa])
- x_1,x_2 :design variables (Cross sectional areas of ❶ and ❷ respectively) [m^2]

And we use following equations in order to evaluate the qualitative sensitivities from the obtained quantitative sensitivities.

$$y = \left| \partial h/\partial x_k \times \text{width} \big/ h(\mathbf{x}) \right| \tag{12a}$$

$$\mu_{small}(y) = \begin{cases} 1-20y & (\text{for } 0 \leq y \leq 0.05) \\ 0 & (\text{otherwise}) \end{cases} \tag{12b}$$

$$\mu_{normal}(y) = \begin{cases} 20y & (\text{for } 0 \leq y \leq 0.05) \\ 1 - 20/3\,(y-0.20) & (\text{for } 0.05 \leq y \leq 0.20) \\ 0 & (\text{otherwise}) \end{cases} \tag{12c}$$

$$\mu_{large}(y)=\begin{cases} 0 & (\text{for } 0\leq y\leq 0.05) \\ 20/3\,(y-0.05) & (\text{for } 0.05\leq y\leq 0.20) \\ 1 & (\text{otherwise}) \end{cases} \tag{12d}$$

Table 1 and Fig. 5 show comparison of optimization results. The quantitative result of optimization was obtained by symbolic computation by Mathematica, and the result of the proposed method was implemented by Macintosh Allegro Common Lisp and it was obtained in an interactive way.

Table 1 Results of Optimization

contents	quantitative method	initial condition	proposed method
x_1 [m^2]	2.53×10^{-3}	9.800×10^{-3}	2.516×10^{-3}
x_2 [m^2]	1.66×10^{-3}	9.800×10^{-3}	1.707×10^{-3}
Mass [kg]	6.93×10	2.95×10^{2}	6.94×10
g_1 [m]	1.93×10^{-6}	5.00×10^{-7}	1.94×10^{-6}
g_2 [m]	1.00×10^{-6}	2.06×10^{-7}	9.89×10^{-7}
g_3 [Pa]	3.00×10^{5}	7.22×10^{4}	3.00×10^{5}
g_4 [Pa]	1.34×10^{5}	4.23×10^{4}	1.38×10^{5}

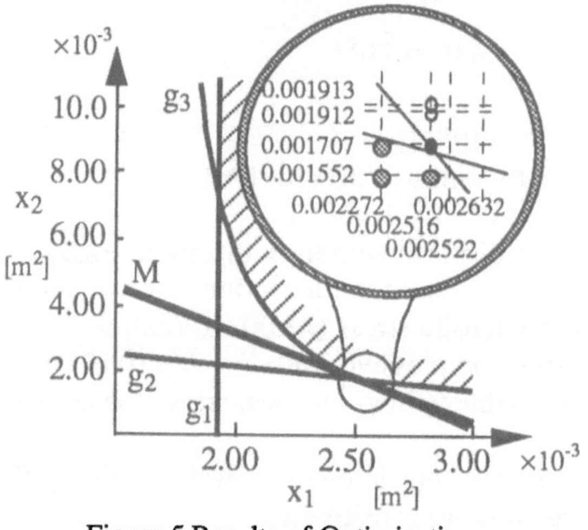

Figure 5 Results of Optimization

Figure 5 shows the vicinity of optimum solution, and a magnifying view in the circle shows the relationships between discreet values and optimum solution of quantitative method in detail. The black dot shows the current state and it became also the optimum state mentioned below. The masked dots show the states without increasing objective function. From this figure, you can see that it will be impossible to improve without violating at least one constraint. Thus, the current state was the optimum state from the qualitative optimal standpoint.

From the results explained preceding, though the formulations of Eq. (6) and

Eq. (12) which will play important roles in fuzzy reasoning are settled in expedient ways, we could obtained the optimum solution very close to the quantitative optimum solution by the proposed method. It shows that it would be possible to carry out optimization based on qualitative sensitivities, when we could determine sensitivities even in qualitatively grasps. In another words, it shows one of the possibilities of qualitative optimizations.

5. Qualitative Sensitivities Aided Optimum Oriented Design System By Referring Model Case Base

In this section, we will introduce a general concept of the developed system to determine qualitative sensitivities by referring model case base. The system is made use of both model base reasoning and case base reasoning. We use models to indicate qualitative sensitivities, not by moving parameter of the model and deciding discrepancies of the behaviors between modeled system and real system as done in ordinary model base reasoning process, but by comparing two model cases. Choice of two model cases is carried out by indexing techniques of case base reasoning. Thus, in this study, we call these data base "model case base".

We just refer to case base reasoning system simply to compare with the developed system.The case base reasoning is a system which solve a problem by using the data base of the past cases. The typical set up of case base reasoning system is in Fig. 6. Case base reasoning is the system which finds a solution of given problem by repeating following processes until it succeed in composing a new satisfying case for the problem; the main processes are decomposing the past restored cases, retrieving their characters and composing a new case.

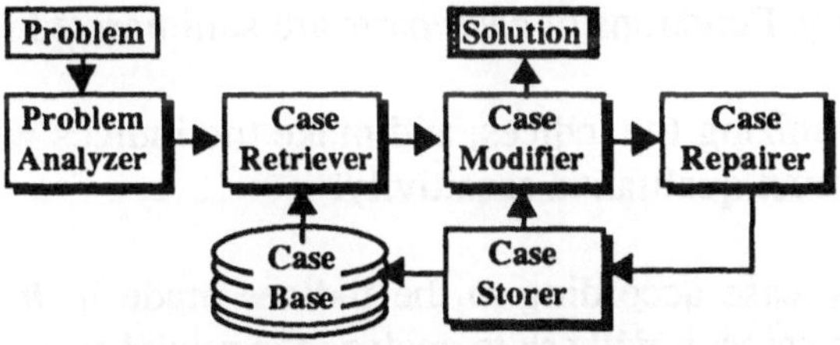

Figure 6 Set-up of Case Base Reasoning

In the proposed system, we do not seem to evaluate qualitative sensitivities directly from model case base. What we want to do is to retrieve two model cases which have same indices besides nominal design variables and to analogize model cases whose indices are required to retrieve two model cases explained just before. And after retrieving more than two model cases, we compare one another and indicate an example of qualitative sensitivities to the designer. Thus, what is called "Problem Analyzer" in case base reasoning is still important in the proposed system. It makes indices to the design case and also to its environments. Then we can search for the cases such that only nominal design variable changes and all the other environments unchanged in order to indicate qualitative sensitivity for that design variable. If there are no model case base for such indices, we can analogize the model case with the other model cases which have almost same indices besides nominal design variable and the other index. After then, it will be possible to indicate the qualitative sensitivity. In this system, we don't need all of the components in case base reasoning. Because we don't need to solve the problem. Thus, the proposed system is much simpler and more reliable in indicating qualitative sensitivities than the case base reasoning, because the reasoning sensitivities is much more difficult than the reasoning their values. The set-up of the proposed system is in Fig. 7.

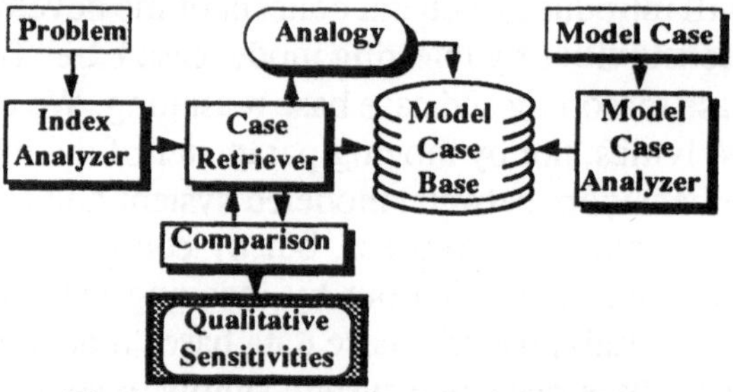

Figure 7 Set-up of Developed System.

This system can be divided into two major parts. One is a right hand side in Fig.7, and they are the parts of indexing a model case and storing them to the model case base, in another word, preprocessing part. The other is a left hand side in Fig.7, the parts of analyzing what kind of indices to search, searching or analyzing the case from the given indices, comparing two model case and instructing the qualitative sensitivity. Functions of each parts are summarized in the following.

1) Index Analyzer

A part of organizing the indices, and make the indices which seems to be compared to instruct qualitative sensitivity.

2) Case Retriever

Searching the case according to the indices made in Index Analyzer. If there are no such cases, it will ask to analyze the model case or analogize from the existed model cases. If you choose to analogize, it will retrieve the more than two indices which seems to be able to analogize from the existed model cases.

3) Model Case Analyzer

A part to index the given model case.

4) Analogy

A part to analogize the model case which is required to compare two model cases. The required indices are already given in "Case Retriever".

5) Comparison

Comparing two model case, and show the difference between them and indicate the sample of qualitative sensitivity in order to support the decision of the designer.

The reasons why we use the model cases, not the cases, are that we want all the other design conditions to remain same, and that we want to be avoid a dependency of the results on the cases

6. Example (By The Proposed System)

In this section, we try to use the qualitative sensitivity aided optimum oriented design system to solve the optimum design problem. And in this example, we will treat an optimum layout problem of supporting structure of a big parabolic antenna. These kinds of problems are dominated in preliminary design in Fig. 1. Layout optimization problem is one of the important topics in engineering design, and recently some kinds of special methods, such as COC Method and Homogenization Method and so on, were examined in Echenauer et. al 1990. But generally speaking, these kinds of problems are tedious and difficult for the optimization.

6.1 DEFINITION OF OPTIMUM LAYOUT PROBLEM ON PARABOLIC ANTENNA

We are going to treat an example according to following definition.

$$\text{Minimize } M(\mathbf{x}) \tag{13a}$$

subject to

$$\delta(\mathbf{x}) \lesssim 1.25\ \delta_{initial} \tag{13b}$$

where

$\mathbf{x}=\{x_1,x_2,x_3,x_4\}^T$

M :mass of the structure

δ :r.m.s. deviation of the surface

x_1 :link type ($\in$ (Type 1, Type 2))

x_2 :lib type ($\in$ (Type 1, Type 2))

x_3 :link number ($\in$ Even integer (≥ 4))

x_4 :lib number ($\in$ Integer (≥ 2))

And "$\lesssim$ " in Eq.(13b) is a fuzzy inequality and which means that left hand side is about larger than right hand side. So that Eq.(13b) means that we will allow r.m.s deviation up to about 25% larger that the initial condition, and it is only a measure. In Fig. 8 and Table 2, we show constant conditions for model case analysis. These are sample conditions only for model case expediently and they have no relationships to the decision making or constraints in detail design which will be done after preliminary design. Figure 9 shows the basic types of Link and Lib in an antenna supporting structure.

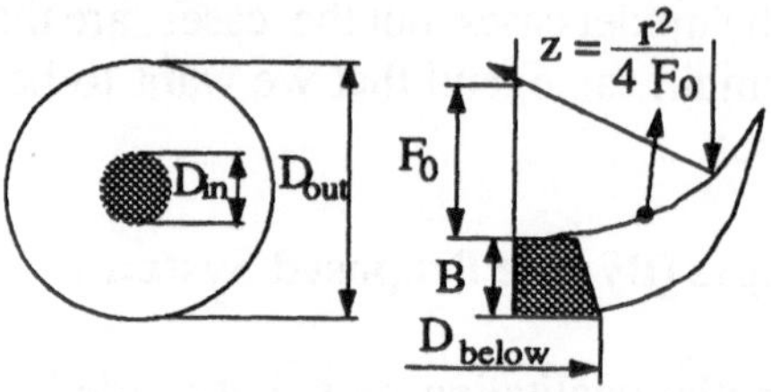

Figure 8 Instruction of Model of Parabolic Antenna

Table 2 Modeled Constraints

contents		*values*
Young's Modulus	Pa	2.05×10^{11}
Density	kg/m3	7.86×10^{3}
D_{out}	m	10.0
D_{in}	m	2.00
D_{below}	m	2.40
F_0	m	5.00
B	m	1.00
A_{link}	m2	3.479×10^{-3}
A_{lib}	m2	6.913×10^{-3}

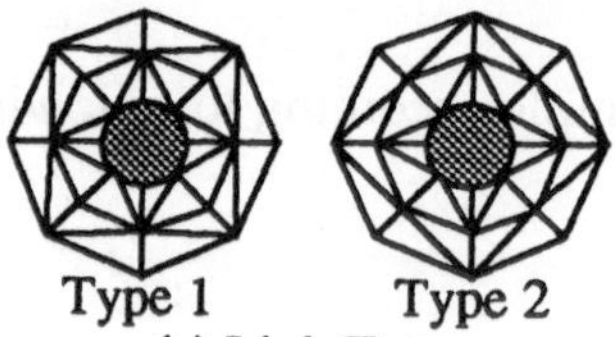

(a) Link Type

(b) Lib Type

Figure 9 Types of Layout

6.2 QUALITATIVE SENSITIVITY AIDED OPTIMUM ORIENTED DESIGN FOR THE EXAMPLE PROBLEM

In this example, we prepare the following indices, where we relates each meaning of index to the type of the design variables.

•Case Number	(CN)	Integer
•Analysis Method	(AM)	FEM/Analogy
•Mass	(M)	Real
•r.m.s. deviation	(R)	Real
•Link Type	(KT)	Type 1 / Type 2

•Lib Type (BT) Type 1 / Type 2
•Link Number (KN) Integer
•Lib Number (BN) Integer

Table 3 shows initial model case base expediently.

Table 3 Initial Model Case Base

CN	AM	M	R	KT	BT	KN	BN
1	F	17.2 X10³	1.54X10⁻⁴	1	1	4	10
2	F	14.1	1.62	1	1	3	10
3	F	12.3	1.76	1	1	3	8
4	F	9.26	2.23	1	1	2	8
5	F	7.70	2.49	1	1	2	6
6	F	14.4	2.34	1	2	3	10
7	F	14.1	1.53	2	1	3	10
8	F	15.2	1.56	2	1	4	8
9	F	10.5	1.83	2	1	3	6
10	F	11.0	2.91	2	2	2	10

Now, we are going to show how the system give the information of the qualitative sensitivities briefly. Initial Stage will be in Case Number 1. Let's try to consider qualitative sensitivities of M and R to KN. Then it tries to search the indices which have the same indices without KN, that means indices (KT=1, BT=1, BN=10). Then it finds CN=2. In this case, Analysis Method (AM) is the finite element method (F), so that the results of analysis could be reliable. As they are $M=14.1\times10^3$, $R=1.62\times10^{-4}$ it shows the information that

$$QS(M,x_3) \rightarrow \frac{14.1-17.2}{17.2} \times \frac{1}{3-4} = 0.180$$

$$QS(R,x_3) \rightarrow \frac{1.62-1.54}{1.54} \times \frac{1}{3-4} = -0.0519$$

And also, it shows some standards of translating these values to qualitative sensitivities or some times history of the decision of the designer. Showing some standards and the history of the decision would be good support for the designer. Next, let's think of qualitative sensitivities to BN. Trying to search indices (KT=1, BT=1, KN=4), it can't find such stages at all. So that we need to analyze either by FEM or analogizing based on model case base. In the analogy, we try to move one design variable, in this case KT, BT or KN. Now, if we move KN to 3, then it will be possible to show information of sensitivities to BN. It means that from CN=2 and CN=3, it can analogize the stage (KT=1, BT=1, KN=4, BN=8). In this case,

$$M=\left(\frac{12.3-14.1}{14.1} \times \frac{-2}{8-10} +1\right) \times 17.2\times10^3 = 15.0\times10^3$$

$$R=\left(\frac{1.76-1.62}{1.62} \times \frac{-2}{8-10} +1\right) \times 1.54\times10^{-4} = 1.67\times10^{-4}$$

By using these values, we can give information of qualitative sensitivities as preceding. And also, it would be possible to store the result of analogy, until there will be special need for accurate values. Because, in this system, we don't need accurate solution of analysis, what we need is to show the difference between two stages. And reason why we do not analogize the information of difference but the result of analysis, is that we want to avoid strange behavior caused by non-linearity, and also we want to grow the model case base. In this examples, we treat the functions which have accurate quantitative values, but we can also use the functions without accurate quantitative values. For example, the function "difficulty in manufacturing" will be clearly better with less number of the elements, but we never know how much it would be quantitatively. Even though, it might be not so difficult to determine some approximate number to the function. Then the proposed system could instruct qualitative sensitivities and help the decisions of the designer. In the following example, we use Eq. (12b,c,d) as standards of qualitative sensitivity.

6.2 RESULTS OF EXAMPLE

We construct the qualitative sensitivity aided optimum oriented design system explained in the previous section, and solve the optimum layout problem of supporting structure of parabolic antenna by using the proposed method. The process of optimization is shown in Fig. 10 and Table 3, and final model case base is shown in Table 4.

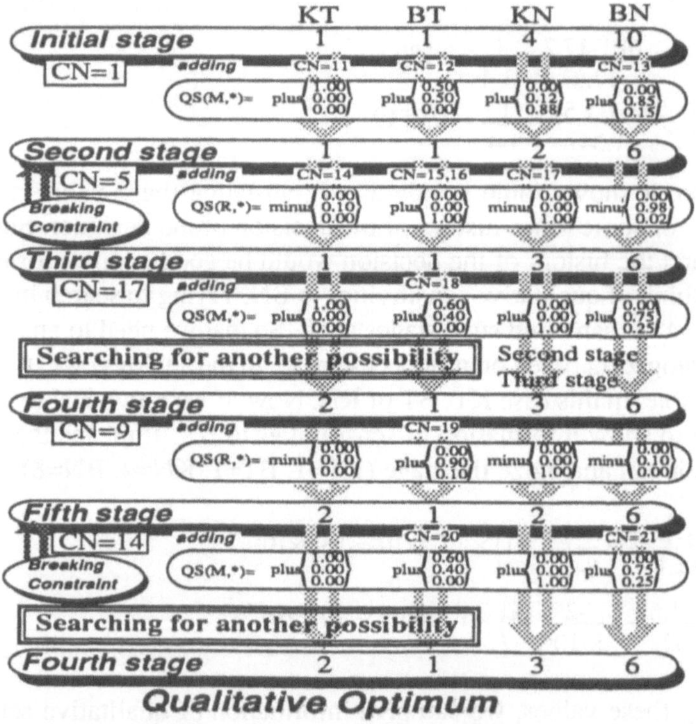

Figure 10 Process of Qualitative Optimization

Table 3 Process of Qualitative Optimization

Stage	M	R	KT	BT	KN	BN
Initial	17.2 $\times 10^3$	1.54$\times 10^{-4}$	1	1	4	10
Second	7.70	2.49	1	1	2	6
Third	10.5	1.93	1	1	3	6
Fourth	10.5	1.83	2	1	3	6
Fifth	7.70	2.37	2	1	2	6

Table 4 Added Model Case Base

CN	AM	M	R	KT	BT	KN	BN
11	A	17.2 $\times 10^3$	1.45$\times 10^{-4}$	2	1	4	10
12	A	17.6	2.02	1	2	4	10
13	A	14.7	1.66	1	1	4	8
14	F	7.70	2.37	2	1	2	6
15	F	9.46	3.20	1	2	2	8
16	A	7.87	3.25	1	2	2	6
17	A	10.5	1.93	1	1	3	6
18	A	10.8	2.37	1	2	3	6
19	A	10.7	2.17	2	2	3	6
20	A	7.86	2.81	2	2	2	6
21	A	9.26	2.12	2	1	2	8

In Fig. 10 and Table 3, we can see that the mass decreased to about 61% of initial one and the r.m.s deviation increased only 19%, thus the result satisfy constraints of the problem. As we cannot calculate quantitative sensitivities in this problem, so that we cannot compare the result with quantitative optimization, but at least looking all model case base, you may see that this result would be the most suitable one. Thus this result might be said optimum solution even in quantitative manner. By the way, let's state the process briefly. In the initial stage, as there are no active nor violating constraints, we improve the design variables by the qualitative sensitivity of mass. In the second stage, the r.m.s deviation was violating the constraint, thus we improve the design variables by considering the qualitative sensitivity of r.m.s. deviation and previous qualitative sensitivity of mass. When we chose the highest order of reasoning, we reached to the third stage. As this stage satisfy constraints, we try to improve by considering the qualitative sensitivity of mass. Then we reached to the stage in the same way as in the second stage. So we tried to consider both qualitative sensitivity of r.m.s. deviation in the second stage and qualitative sensitivity of mass in the third stage, and choosing the highest order of reasoning, we reached to the state same as the third stage again. So we chose the second highest reasoning, then we reached to the fourth stage.This state satisfies the constraints, so that we tried to improve design variables by the qualitative sensitivities of mass, to reach to the fifth stage. In this stage, the constraints of r.m.s deviation was violating, so that we tried to improve by considering both qualitative sensitivities of r.m.s deviation in the fifth stage and that of mass in the fourth stage. Then we choose the highest order of reasoning and reached to the fourth stage again. So that we tried to find the next stage by choosing the design variables according to the higher order of reasoning, but we couldn't find any state without violating some constraints nor increasing mass. So that we decide that the fourth stage would satisfy the qualitative optimality, and decide the design variables as optimum solutions in qualitative way.

And examples in analogy with the model case and giving information of the qualitative sensitivities are shown in Fig. 11 and Fig 12, respectively. We want to add here that we stressed only on analyzing the behavior of the system and we didn't make any efforts to make the better interface in interactive process. In Fig. 11 and Fig.12, bold letters show input of the designer. Where "cs" means the current state, and by Index Analyzer, the system chose "qb11" and "qb12" as the base of analogy. By these two model case, "pra" shows that the system could analogize the case with (KT, BT, KN, BN)=(2,1,4,10). By function (analogy), the system shows the result of analysis. In Fig.15, you can see that the system indicated the qualitative sensitivities by using Eq.(14b,c,d). We don't have to pay so much attention to those values which are given by analogy with model case in the sense of quantitative manner. However, indicated values are not so important than in their qualitative amount. They are to give just some kinds of information.

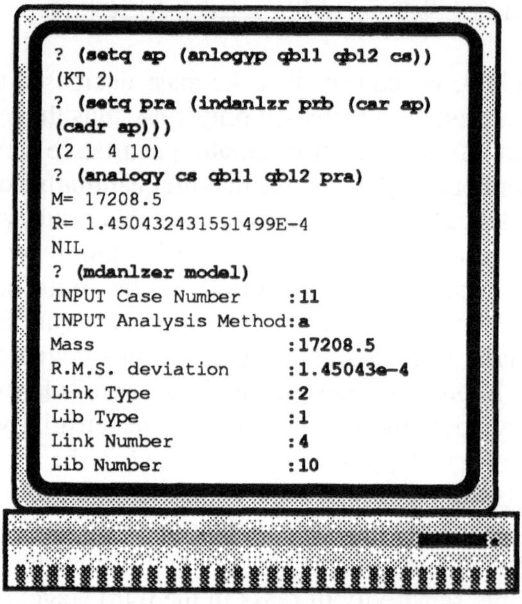

Figure 11 Example of Analogy of Model Case

On closing this section, we would like to summarize the obtained consideration from this example.

(1) We develop the qualitative sensitivity aided optimum oriented design system by referencing model case, and used it during the optimization process in deciding qualitative sensitivity. As the system can analogize the model case from existing model case base, we can carry out this system even with small

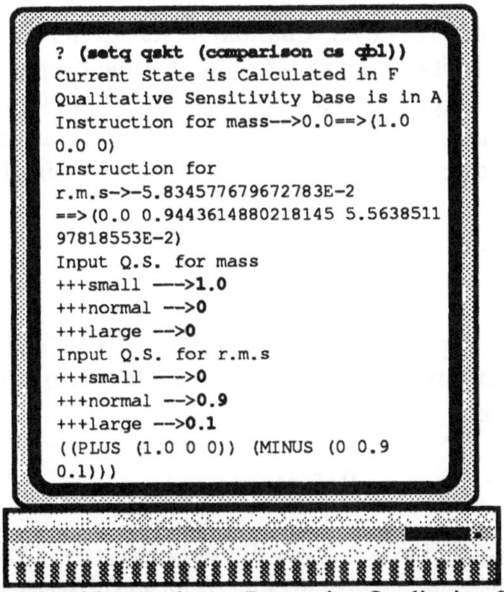
? (setq qskt (comparison cs qbl))
Current State is Calculated in F
Qualitative Sensitivity base is in A
Instruction for mass-->0.0==>(1.0
0.0 0)
Instruction for
r.m.s->-5.834577679672783E-2
==>(0.0 0.9443614880218145 5.5638511
97818553E-2)
Input Q.S. for mass
+++small --->1.0
+++normal -->0
+++large -->0
Input Q.S. for r.m.s
+++small --->0
+++normal -->0.9
+++large -->0.1
((PLUS (1.0 0 0)) (MINUS (0 0.9
0.1)))

Figure 12 Example of Interaction to Determine Qualitative Sensitivities

number of initial model case base, and also we can have qualitative sensitivity easily. Although obtained sensitivities had no assurance mathematically in exact mean, we can still carry out optimization with those approximate sensitivities. (qualitative sensitivities)

(2) By using determined qualitative sensitivities, we carried out the layout optimum problem of parabolic antenna. We could carry out optimization even with the model that each sensitivities could not be quantitatively calculated in advance. Thus, we showed one possibility to carry out optimization in preliminary design process by using proposed method.

7. Conclusions

1) We proposed an optimization method with qualitative sensitivities making use of fuzzy language and qualitative optimality. We applied the proposed method to the simple optimization problem with quantitatively calculated sensitivities. The result of optimization was nearly equal to the result of quantitative optimum one. Thus we may say that the proposed method had efficiency once we can determine qualitative sensitivity.

2) We developed the qualitative sensitivity aided optimum oriented design system by referencing model case. And applied it to the optimum layout problem of parabolic antenna. Through this example, it turned out that it is possible to use the proposed method to the preliminary design process. From that the proposed method might be used even in the rough modeling without any quantitative or mathematical informations.

3) We applied the proposed method only to the problems which have definite mathematical modeling. However, qualitative sensitivity doesn't need such a mathematical models in itself. Thus, we think that the proposed method can be used in upper part of design process than detail design. In future, we would like to revise the fuzzy reasoning inference system and also extend application field to the upper part of the design process and make the domain of application field for optimization wider.

Acknowledgements

This is a part of studies supported by Fellowships of the Japan Society for the Promotion of Science for Japanese Junior Scientists. We would like to thank them for their financial supports.

References

Arakawa,M. & Yamakawa,H.(1989),A Study on Multi-Objective Optimum Design Applying Qualitative Reasoning,Proc. of the Int. Symp. on Advanced Computers for Dynamics and Design'89,JSME:p.267-272

Arakawa,M. & Yamakawa,H.(1990),A Study on Optimum Design Applying Qualitative Reasoning,Trans. of J.S.M.E,Vol.56CNo.522:p.267-272(in Japanese)

Arakawa,M. & Yamakawa,H.(1991a),A Study on Interactive Multicriteria Optimization Method Using Qualitative Reasoning,Trans.of J.S.M.E,Vol57CNo.534:p.413-419(in Japanese)

Arakawa,M. & Yamakawa,H.(1991b),A Study on Multicriteria Structural Optimum Design Using Qualitative Reasoning, Artificial Intelligence in Design'91(edited by John Gero),Butterworth&Heinemann:p.839-855

Cagan.J & Agogino,A.M.(1987),Innovative Design of Mechanical Structures From First Principles,AI EDAM,1(3): p.169-189

Cagan.J & Agogino,A.M.(1991),Inducing Constraint Activity in Innovative Design,Artificial Intelligence in Engineering Design,Analysis and Manufacturing,5(1)

Kobayashi,S.(1991),Perspective of Case-Based Reasoning,SIG-KBS-9102-4,JASI:p.29-38(in Japanese)

Eschenauer,H.,Mattheck,N.,Olhoff,N.(Edited)(1990),Engineering Optimization in Design Processes,Springer-Verlag:p.27-90

Simon,H.A.(1969),Science of Artificial,MIT Press

Walsh,G.R.(1975),Methods of Optimization,John Wiley & Sons

ADAPTATION OF SPATIAL DESIGN CASES*

K. HUA, I. SMITH, B. FALTINGS
Artificial Intelligence Laboratory, LIA
Federal Institute of Technology (EPFL)
1015 Lausanne, Switzerland

and

S. SHIH and G. SCHMITT
Computer Aided Architectural Design, CAAD
Federal Institute of Technology (ETH)
8093 Zurich, Switzerland

Abstract. Any design process involves two kinds of knowledge: *domain* knowledge and *design* knowledge. In this paper, we focus on the formulation of design knowledge within the building domain. Representations which formulate building design knowledge include production rules, shape grammars, prototypes and cases. To avoid blind search, design knowledge must be indexed by function according to different aspects. We argue that search for a solution that accommodates several aspects is best carried out through iterative refinement of *cases*, and that a precise *geometrical* model of the case is required to link different aspects. This leads us to employ cases as a design knowledge representation and *adaptation* as a reasoning methodology for design. We describe a procedure for adaptation of building structures to new environments using interleaved processes of dimensional and topological modifications. We show results of a prototype which implements this procedure.

1. Introduction

In this paper, we assume that design can be seen as a problem-solving activity involving two kinds of knowledge: *design* knowledge for generating structures, and *domain* knowledge which is used to evaluate them and guide the search process (Faltings, 1989). If design processes do not include extensive blind search while still covering a large space of possibilities, the generation of structures must be directed by desired functions. Ideally, enough design knowledge results in systems which generate structures that satisfy these functions.

Functions refer to various aspects of a design; each function is associated with their own abstraction of the final artifact. For example, for structural safety and serviceability a building is a collection of beams, columns and loads; for energy efficiency the building is a set of heat sources, sinks and conductors; and for circulation patterns it is a set of spaces and doors (Figure 1).

Design knowledge must be formulated in terms of these different abstractions. For example, design knowledge about functional spaces can be formulated by modeling a building as a set of rectangles (Flemming, 1978), while for aspects of aesthetics and architectural form, shape grammars may be useful (Gips, 1975; Stiny, 1975; Mitchell, 1977). Each of these representations allows formalization of

* This paper describes research carried out as part of a project sponsored by the Swiss National Research Program in Artificial Intelligence (NFP 23).

J. S. Gero (ed.), Artificial Intelligence in Design '92, 559–575.

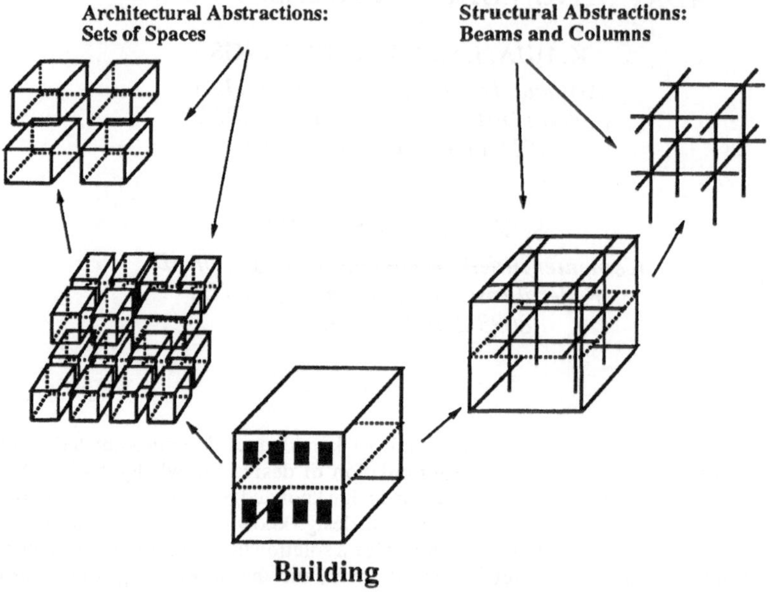

Fig. 1. *Functions of a design are defined according to different abstractions. Abstractions can be linked by reference to a common building model.*

important design knowledge, but only with respect to each individual aspect. This brings up the *integration problem*: models obtained in different abstractions must be *integrated* into a single building model in order to form a basis for constructing the artifact.

While it is straightforward to interpret a building model in terms of different abstractions, the abstracted models by themselves usually do not determine a unique model. In fact, the integration of different abstractions leads to another search process to find a building model which is compatible with all abstractions. The original problem is just translated into another design problem. Hence, the first major argument of our paper is as follows:

> Design knowledge for all abstractions in a design system must be formulated in terms of a single building model to avoid the integration problem.

This observation has been implied by (Gross, 1989). However, using a single building model also makes solution by top-down search intractable. When knowledge in different abstractions is linked to a common analogical representation, search processes in different abstractions are no longer independent. Generating a solution now implies a *single* search process based on building models. The complexity of this search process becomes the *product* of the complexities of search in the individual abstractions, thereby making the problem intractable in practice.

Problems associated with the complexity of design problems have motivated the use of *prototypes*(Gero, 1990) or *case-based reasoning*(Schank, 1982; Kolodner,

1984; Hammond, 1986). Case-based reasoning offers the advantage that design starts with a completely specified, feasible solution. *Adaptation* of such a solution to a new problem is expected to take significantly less effort than complete *generation* of a solution from initial criteria, thus reducing complexity. In the context of a computer-based design system, additional advantages of case-based reasoning are realized because it is not necessary to reason about all aspects of the design. A case might have desirable features which the design system does not know about, but there is a good chance that the adaptation of the case to a new problem preserves such features.

The paradigms of prototype and case-based reasoning have been the subject of much research into intelligent CAD. For example, (Gero, 1990) has proposed prototypes as generalized schemes of design knowledge which is categorized further according to function, behaviour and structure. Other work in building design includes ARCHIE (Goel et al, 1991), IBDE (Fenves, 1989), CADSYN (Maher and Zhang, 1991), DDIS (Wang and Howard, 1991), as well as work by Rosenman et al (Roseman, et al 1991) and Oxman (Oxman 1991).

Existing research into case-based reasoning for architectural design does not address the integration problem. For example, the ARCHIE system contains a set of independent modification procedures to eliminate discrepancies between cases and new problems, but it cannot deal with situations where these methods generate conflicting descriptions in different abstractions. On the other hand, in our research we do not consider the issue of case *selection*, since we believe that designers wish to maintain control of this process themselves.

In this paper, we describe a method of case adaptation using an interleaved processes of *dimensional* and *topological* adaptation. For dimensional adaptation, the integration problem is addressed by mapping different abstractions directly to a model of the case and through using a *dimensionality reduction* process. This process resolves conflicts between abstractions before the actual adaptation takes place, while at the same time limiting the modification to those that actually address the discrepancies. For topological adaptation, the system uses problem-specific transformation rules. We are replacing problem-specific rules with a process of *syntax-directed translation* which is applied to models represented according to different aspects. Syntax-directed translation is a process based on fundamental principles, thereby ensuring the existence of a mapping of the modifications applied within abstractions to the building model.

This paper is structured as follows. In Section 2, we elaborate upon the relationship between aspect models and a common building model and define the processes of dimensional and topological adaptation. In Section 3, we show how the two processes are interleaved to carry out case adaptation in a consistent manner. Finally, Section 4 contains a description of the prototype implemented so far and includes some results.

2. Design Knowledge and Cases

A case is a precise model of an actual building, and satisfies functional criteria in many different aspects. The idea of using a case to solve a new problem is to apply the way that these criteria have been achieved for the case to a new problem. This requires a problem-specific generalization, which we call *adaptation* to distinguish it from the problem-independent generalizations which are the focus of machine learning.

We consider two forms of adaptation:

- *dimensional* adaptation for modifying dimensions while maintaining presence of the components and their connections.
- *topological* adaptation for changing the components and their connections themselves while maintaining basic functions.

Every adaptation can violate functional features of a case. Dimensional changes may destroy features of architectural form and structural stability. Topological changes may change circulation patterns and other functions. When adaptation is performed using explicit models of these functions such violations are avoided provided that the scope of the models is not exceeded. Even when functions exceed this scope, there is a good chance that functions are maintained if the case is appropriate.

Another important characteristic of adaptation is that of *minimality*: the adaptation should vary only those aspects of the case in which the new problem is different from its original environment. For example, there is no reason to reconsider the layout of rooms if external constraints related to them have not changed. This means that an adaptation does not attempt to *improve* upon the case, but assumes that it already represents a good solution. Only explicit *discrepancies* between the case and the new context are treated during adaptation. This is an important difference between the case-based knowledge representation and parametric design, where each new problem initiates a new global tradeoff.

2.1. Dimensional Adaptation

Dimensional adaptation changes only numerical dimensions of the building model that represents the case, we call these *base parameters*. Changes must be *constrained* to maintain the functional characteristics. These constraints are expressed by algebraic equalities and inequalities in terms of parameters of the model. We distinguish the following types of constraints:

- *geometrical* constraints: equalities which maintain coherence of the geometric structure.
- *structural* constraints: inequalities which ensure structural safety and serviceability.
- *normative* constraints: inequalities or equalities which ensure compliance with building codes.

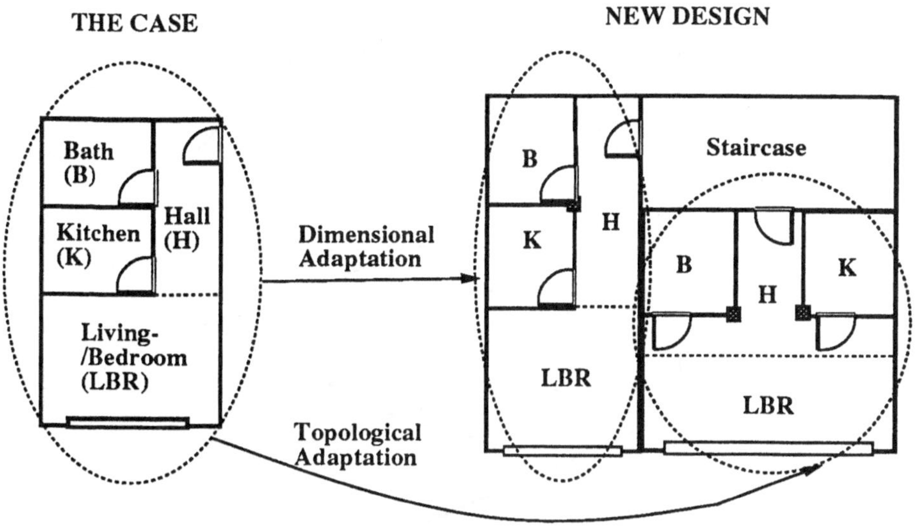

Fig. 2. *An illustration of topological and dimensional adaptation. Dimensional adaptation occurs in response to* discrepancies, *violations of constraints that occur when the design is placed in a new environment. Topological adaptation occurs when dimensional adaptation cannot achieve a feasible result.*

– *functional* constraints: inequalities or equalities which maintain functional or form characteristics which are judged to be favourable.

Many of the inequality constraints in fact participate in *tradeoffs* of different functions. For example, the size of different rooms is a tradeoff to make best use of the total available space. An important tenet of case-based reasoning is that such tradeoffs are made in a near-optimal way. This means that many of the inequalities should be replaced with equalities which reflect the way that tradeoffs have been achieved.

Discrepancies are the motivation for adaptation; they manifest themselves as constraints which are violated. The purpose of dimensional adaptation is to correct discrepancies by changing values of parameters involved in the corresponding constraints. The complete set of base parameters involved in the conflicting constraints is called the *raw* adaptation parameterization.

As an example, consider the case shown in Figure 2. By unification of the original environment of the case and the new environment, an initial placement of the case is obtained. However, values of the base parameters representing the new environment conflict with the constraints which represent the fact that all elements of the case must fit into the site. The base parameters involved in these violations are the positions of those elements that fall outside the site. They form the *raw* adaptation parameterization.

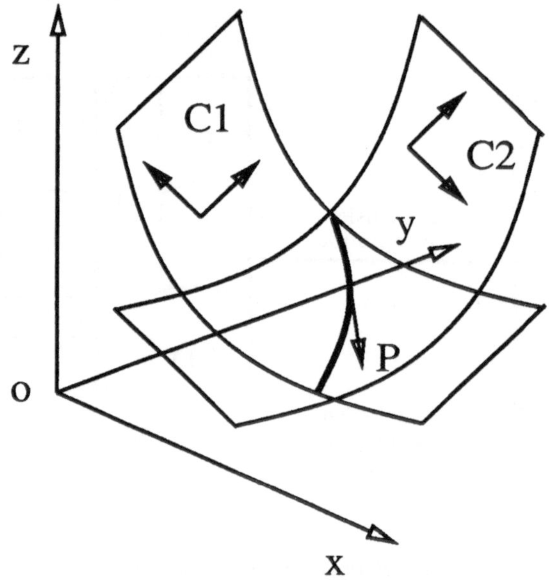

Fig. 3. *Equality constraints on a set of parameters correspond to surfaces, and varying parameters means navigating on the intersection of the surfaces. Through explicitly solving these constraints* - dimensionality reduction - *this intersection can be reparameterized.*

While the raw adaptation parameterization could be used to adapt the case, this poses severe problems of control. Since there is a large number of constraints, every change of a parameter immediately propagates to many others, which in turn might cause conflicts with the original change. A solution to these problems is found through explicitly *solving* the constraints on the case instead of simply propagating changes through them. This corresponds to a *dimensionality reduction* of the adaptation parameterization, as shown schematically in Figure 3.

Projection of the raw adaptation parameters into the subspace of base parameters leads to the final *adaptation parameterization* which implicitly satisfies all constraints. This global solution takes into account *all* discrepancies and *all* constraints in one single step. Therefore problems of control, arising when parameter changes are propagated through a network of constraints, are avoided. It solves the integration problem defined earlier for constrained dimensional adaptation of cases.

2.2. Topological Adaptation

The adaptation parameterization is valid locally for small variations of the building. When large variations are attempted, the system may run up against limitations: for example, it is not possible to reduce the height of a floor to less than 2.4 meters.

When such constraints make adaptation by purely dimensional means impossible, topology of the building must be changed.

As topological changes modify the basic elements of the design, the constraints to maintain the quality of the building can no longer be represented by dimensional constraints. They must be defined in terms of the abstractions which make the functional characteristics apparent. Topological adaptation can be carried out with problem-specific transformation rules. A more general way is to use syntax which defines the category of a given case according to particular abstraction that make the functional characteristics apparent. Based on this syntax, we can find new topologies for case adaptation.

Syntax-directed translations are based on an abstraction in which only the interrelationships among design elements are of concern. Such an abstraction can be described with a graph, where the vertices of the graph are used to represent design elements and the edges are used to describe relationships. Modifications of these abstractions must respect topological constraints which correspond to features of concern to the designer. An important observation underlying syntax-directed translation is the following. Consider a graph grammar which can generate the topology of the case, and a trace of this generation, the parse tree, the grammar itself implicitly contains a number of constraints on topology, and additional constraints can be explicitly defined. The space of topological transformations which are consistent with implicit and explicit constraints is obtained by substituting substructures of the parse tree with other applicable productions of the grammar.

Grammars were invented to define the syntax of languages. Because of its usefulness for computation, grammars are also extended to describe other structures such as graphs (Pfaltz, et al. 1969; Feder 1971) and shapes (Stiny, 1975; Gips, 1975). Computer applications of grammars regarding pictorial forms include the recognition of hand writing, wave forms, fingerprints, traffic systems, and chromosomes (Fu, 1982) as well as the generation of biological forms (Lindenmayer, 1968). The applicability of grammar in design domains has also been shown in the analysis or the synthesis of designs (Koning, 1981; Knight, 1981; Flemming, 1986).

More generally, formal language theory defines syntax-directed translation as a process linking two grammatical derivations, an input and an output structure, which are not necessarily based on the same grammar. The two grammars have to share the same set of non-terminals, and their productions must be in exact correspondence, i.e. containing the same non-terminals in different permutations. The translation process starts with the syntax analysis of the input expression and uses it to guide the generation of the output expression. Applying this concept to designs, we use the input grammar to analyze the design context and the output grammar to generate design compositions. This gives us a general and compact way to define sets of candidate topological transformations.

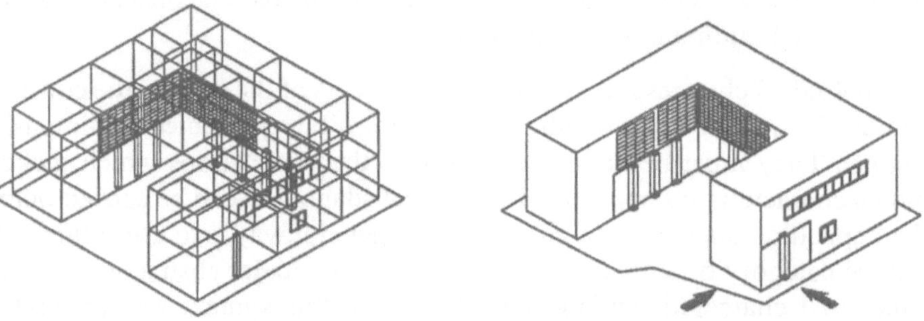

Fig. 4. *An example of an adaptation. The original case, shown on the left, is parameterized as shown by the arrows on the right. The parameterization is specifically formulated according to differences in shape between the original and new building site. The drawing on the left shows the complete building model, the right only shows the external shape.*

3. The Case Adaptation Procedure

Case representation and problem specification

A case is represented by placing the following items into frames which are organized into a hierarchical structure.

- A CAD model of the building, including a *base parameterization* of the building geometry and detailed dimensions as well as geometrical, structural, normative and functional constraints.
- Environmental features (natural boundaries, slope, access roads, orientation, shading objects in the environment) and their relation to building coordinates.
- A topological model which can be generated automatically according to the base parameterization and relations between spaces.
- Domain dependent transformation rules for possible topological discrepancies or grammars for syntax-directed translations.

Cases having this representation form *design knowledge* within the system. A new design problem is represented through different information for environmental features, topological and dimensional constraints relating to the new environment.

The example of Figure 4 has 162 base parameters and 101 geometrical constraints. Geometrical constraints are part of the CAD model and express the connectivity of the walls of the building. Limitations for dimensional generalization are constraints on size and height of each element in the building. The environment is defined by a polygon formed by surrounding building elements or natural boundaries. For example, features concerning the opening of the "u" are: facing south, direction of the slope, and the direction of the access road. Topological relationships are established according to the adjacencies of spaces in the building.

Stages of Case-based Adaptation

Reasoning with case knowledge proceeds according to the following steps when problem-specific topological-transformation rules are employed.

1. *Evaluation* of the existing case in the original and new environment in order to find topological and dimensional discrepancies.
2. If there are topological discrepancies, call corresponding transformation rules to transform the topological graph of the case to meet needs of new site while preserving the functions. Rebuild the dimensional constraints after topological transformations.
3. If there are dimensional discrepancies, identify the violated dimensional constraints and define a set of *constraints* on the *local* generalization of the design. Add these constraints to the constraint list stored in the case. Normative constraints and functional constraints which contain inequalities are kept separate.
4. Using constraints which are not satisfied in the new environment and constraints in *local* generalization of the design, define all the variables in this local constraint network with *adaptation parameters* through a domain independent *dimensionality reduction* process.
5. Vary the adaptation parameters in order to ensure that no constraints are violated. Propagate changes to related variables inside relevant areas of the building.
6. Check the validity of adaptation by verifying according to the constraints that are not included in the dimensionality reduction, such as those constraints which are expressed in terms of inequalities. If no constraints are violated, stop with success. If some constraints are violated, proceed to the next step.
7. If constraints are not maintained in the new design problem, trigger topological transformation rules which relax constraints in the related constraint set. If there is a transformation which preserves design features of the case, go back to step 1, otherwise the case is not suitable.

Computation of Case Adaptation

Case adaptation begins with insertion of the case into a new design environment. This is done through a functional evaluation process which finds an insertion such that the number of discrepancies is minimized. Insertion of a case into a new context takes place at two levels: topological and dimensional. Topological insertion begins first. In this step, two topological graphs of the form of the case and the site of new design problem are connected. This is done by comparing the site of case and the site of new design, including specifications of functional requirements. This process finds an insertion such that most of the functional requirements are satisfied. Once topological insertion is finished, dimensional constraints are built according to the topology created. For example, Figure 5 shows how the site of Felder house and a new site are connected which decides the insertion of form of the case into

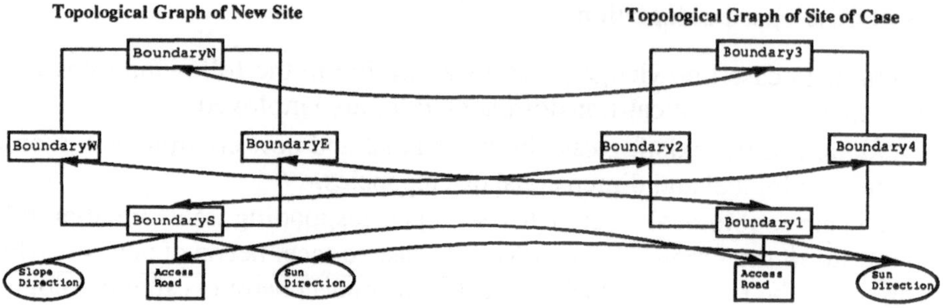

Matches of topological graphs: new site and site of the case

Fig. 5. *Matching of a new site and site of Felder house.*

the new site. In this example, the only functional requirement is the orientation of the house(facing south). The left part of Figure 6 explains how dimensional insertion is carried out. After insertion, only a small number of discrepancies are found. In the figure, triangles indicate positions which are maintained. The hatched area outside the boundary of site defines dimensional discrepancies. The middle part of Figure 6 shows parameters considered for adaptation. Upper-case names are variables. Names in lower case are constants. Points with a cross are positions that can be changed in both X and Y directions during the adaptation. Points with a circle are positions that can only be changed in X direction. The right part of Figure 6 gives constraints on the considered variables. This is a simplified example for the first floor of the Felder house example in Figure 4. When considering the complete case, there are many more constraints and parameters.

Violated dimensional constraints are identified through dimensional discrepancies. Generalization of the case is carried out by transforming constants in the case into variables. Architectural cases are large-scale design problems which involve thousands of design parameters. Dimensionality reduction is useful for transforming the problem into a manageable form. Dimensionality reduction is done without explicit domain knowledge. For rectangular representations, constraints are mostly linear, and dimensionality reduction can be implemented through a Gaussian elimination based process. Recently, more general constraint solving techniques have been developed within the context of constraint logic programming. For most problems, only a few parameters remain after dimensionality reduction. For example, in Figure 6, after the insertion of the Felder house into a new environment, only two parameters require consideration, as shown in Figure 7.

Parameters are modified one after another along the directions indicated. Modifications are propagated directly to variables in affected areas of the building. Each step of modification is verified using normative constraints such as building codes for size, height of spaces and building elements. The left part of Figure 7 is the result after dimensional adaptation. Middle and right parts of the figure show de-

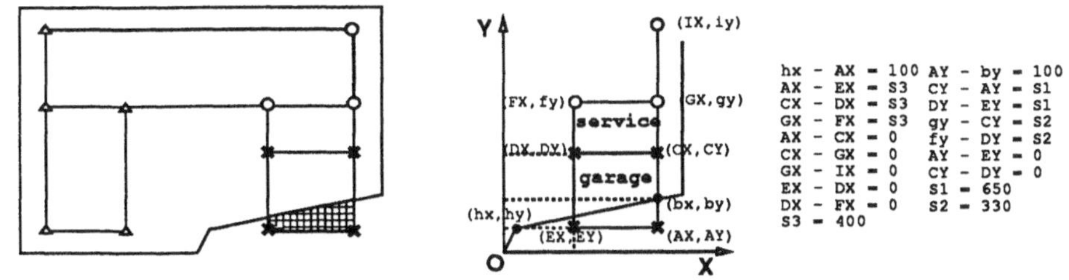

Insertion with minimum dimensional discrepancies **Considered variables** **Relevant constraints**

Fig. 6. *Simplified dimensional adaptation of part of the Felder house.*

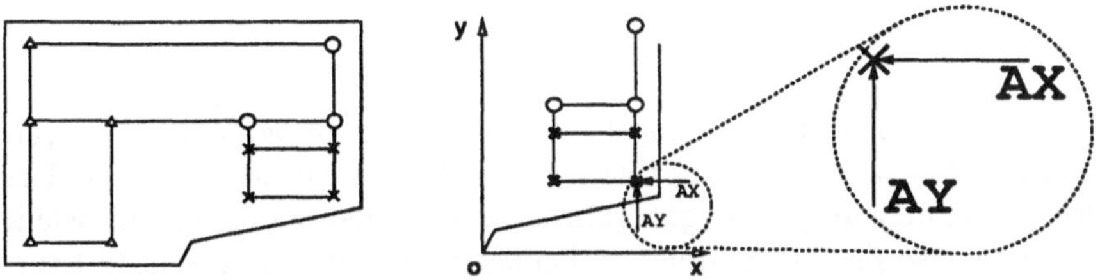

After dimensional adaptation **Local details of dimensional adaptation** **After dimensionality reduction two parameters remain: AX, AY**

Fig. 7. *Simplified dimensional adaptation of part of the Felder house continued.*

tails of dimensional adaptation and two remained parameters after dimensionality reduction on the generalization of the case.

If violation of constraints cannot be avoided, dimensional adaptation fails. Topological transformation is triggered to displace, to add and/or to delete building elements according to violated constraints. Topological adaptation is carried out through application of transformation rules. These rules ensure that design features of the case are preserved during transformation.

Figure 8 shows an example of topological transformation of the Felder house in Figure 4. The left part of the Figure gives the transformation in terms of a topological graph. The right part is the final result of building after the transformation. Topological transformation does not ensure successful adaptation. If one topological transformation fails, new topologies are tried. If all possible topologies fail, the case is not suitable.

However, problem-specific rules are not an ideal approach for topological transformations, since it is hard to define rules that preserve design features and it is difficult to maintain design features when many rules perform topological transformations on the case. We are investigating a new way to do this through a syntax-directed transformation. This will provide us with a method that transforms the case topologically at a more global level while preserving important features of the case.

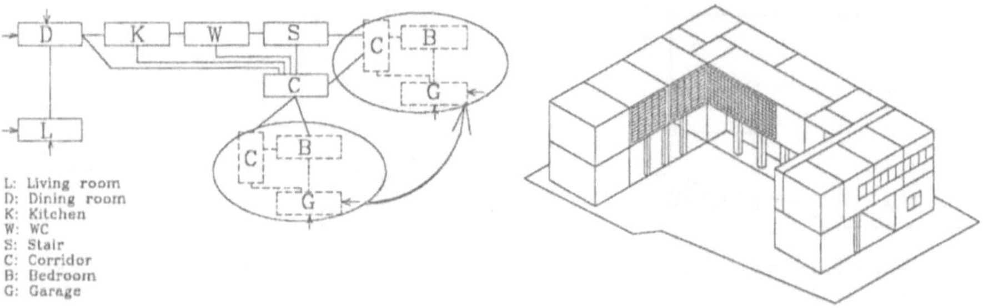

Fig. 8. *The highest levels of the topological graph for the example, and a variation of the shape by rules based on the topological graph.*

Grammars for defining topological changes

A case is viewed as the result of a translation from a design problem to a specific design solution. Our assumption is that this translation can be generalized and applied to different contexts. The generalization is carried out using a translation grammar which describes the correlation between context and design. In the example of the Felder house, the context is simplified as a linear configuration of squares, which are represented as a string composed of letters "m" and "r", where "m" means placing a square then moving forward and "r" means placing a square then turning right, see Figure 9. For the output of the translation, a set of design elements, represented with letters "a" to "f", is used to synthesize the design. In each of these elements, circles represent insertion points, and arrows indicate new directions and positions for the next element to be inserted. Thick lines are external walls and thin lines are interior partitions.

In the second stage, the design case is compared to the complete output of the assumed translation process in order to choose similar alternatives. This process reduces computation when the case does provide a sound basis for adaptation, and in addition this process minimizes the possibility of destroying the quality of the case. A grammar which describes acceptable configurations of functional spaces is then used to check the feasibility of the chosen candidates and to arrange functional spaces.

In Figure 9 and Figure 10, we use the Felder house as an example. The case is represented as "daebbcbf" according to the six design elements. A U-shape, represented as "mmrmmmrm", is given as an imperative configuration with which the design has to comply. In the translation process, the input grammar is used to analyze the input configuration - to parse the input string "mmrmmmrm", while the output grammar is used to generate the output string with the guidance of the parse tree. The "+" sign in the output string means "or". The generated output string "$d(a+b)(c+e)(a+b)(a+b)(a+b)(c+e)f$" can be expanded into sixty four variations. Using an algorithm described in (Fu, 1982), our program compares the case to the sixty four variations and selects six most similar ones as candidates.

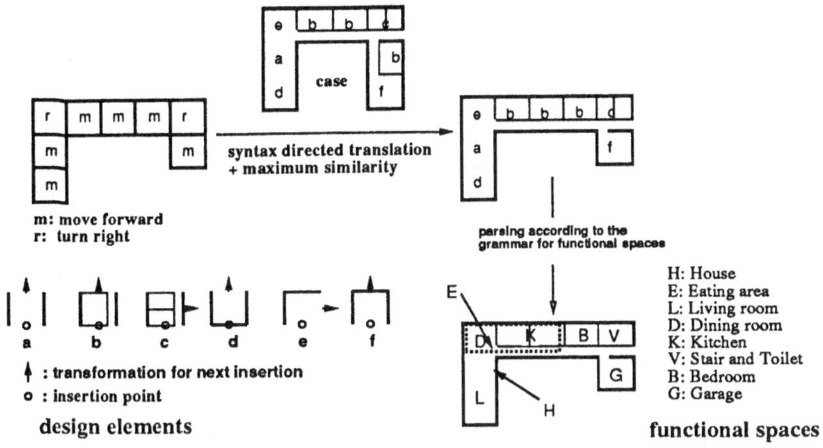

Fig. 9. *The process of topological adaptation.*

```
Grammar for syntax directed translation
Initial state: S, nonterminals: (S, F, T)

          input        output
1   S -> mF    :    dF
2   F -> m     :    f
3   F -> T     :    T
4   F -> mF    :    (a+b)F
5   T -> r     :    (c+e)
6   T -> rm    :    (c+e)F
7   T -> rmF   :    (c+e)(a+b)F

Grammar for functional spaces
initial state:H, non-terminals:(H,E,L,V,B,K,D,G)

H -> (LE+EL)(VB+BV)G
E -> KD+DK
L -> (d+a+e)(a+f)
V -> c
B -> Bb+b
K -> bb+b
D -> a+d+e+f
G -> d+f
```

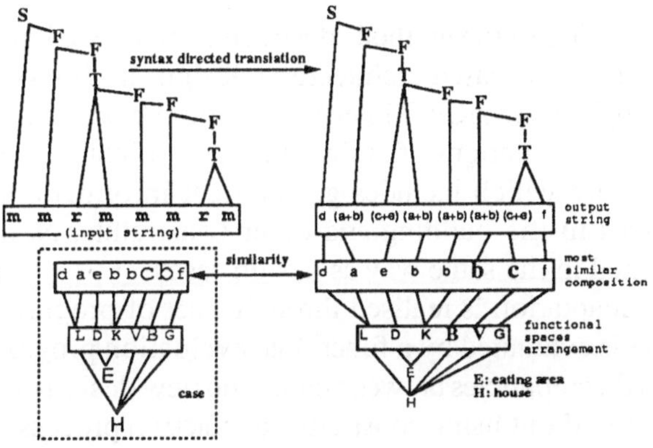

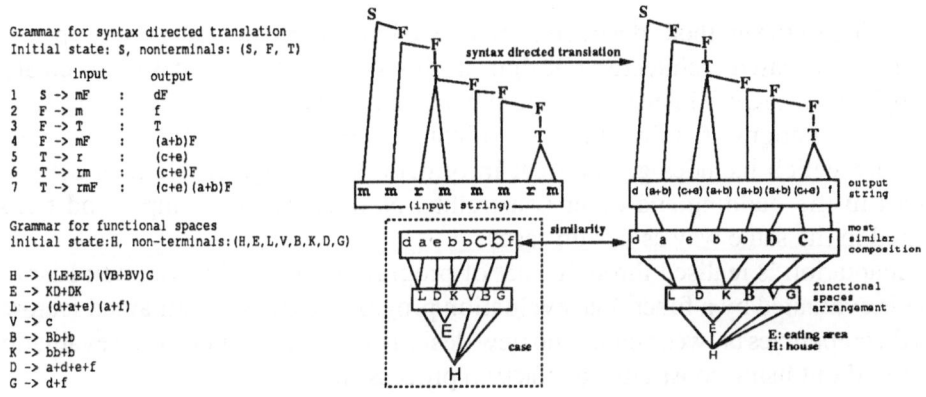

Fig. 10. *Grammars and the derivations of the input and the output strings.*

Using the grammar for functional spaces to verify the six candidates, we found one which is compatible to the functional requirements. The arrangement of spaces is also realized in this process. Figure 11 shows the case, the six candidates, and result of the adaptation. As a summary, the process adapts the given case to the required configuration "mmrmmmrm", and derives a design "daebbbcf", which accommodates functional spaces "LDKBVG".

This example demonstrates the potential of using grammars as a platform for recording design knowledge necessary for case adaptation. This method improves upon using case dependent rules for topological adaptation. Using syntax directed translations, topological transformations are carried out in a more general manner within the scope of a set of predefined design elements.

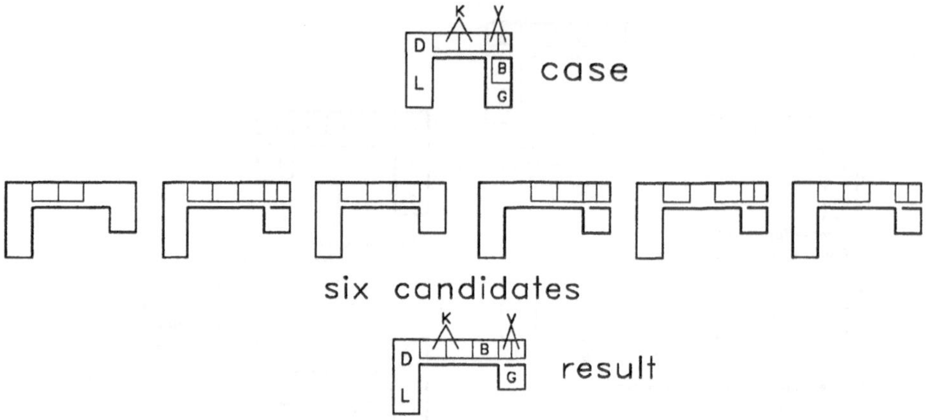

Fig. 11. *The case, six similar candidates, and the result of adaptation.*

4. Implementation of a prototype system

According to the methods described above, we have implemented a prototype system for case-based architectural design. In this system, original cases are created through an AutoCAD interfacing program and stored as AutoCAD drawings. Internal representations of building in form of entities are translated into a frame-based structure which includes geometrical information, topological relations between spaces in the building and other constraints. New contexts are input and transformed in the same way as design cases.

Adaptation is realised through a set of program packages in Common Lisp. A case is evaluated by a functional evaluation program to detect maintained features and discrepancies between cases and new design problems. Dimensional evaluation is carried out using constraint satisfaction processes.

Currently, we are investigating rectangle shaped building spaces and elements. These cases contain linear equations and inequalities. Dimensionality reduction is implemented with a Gaussian elimination based algorithms for equalities. Since most cases involve redundantly specified constraints that are unacceptable for most mathematical algorithms, we implemented a general linear constraint satisfaction program which can resolve redundantly specified, under-constrainted and/or over-constrainted linear constraint systems for detailed design of elements such as windows, doors, opens and other elements associated with rooms. For the linear constraints which are encountered in rectangular spaces, Gaussian elimination is sufficient. For more general constraints, languages developed in the field of constraint-logic programming are necessary.

When changes of adaptation parameters do not generate a solution, topological transformations are triggered according to violated constraints. Geometric constraints are built according to the modifications of topological structure. The process goes back to the beginning of dimensional adaptation after each successful

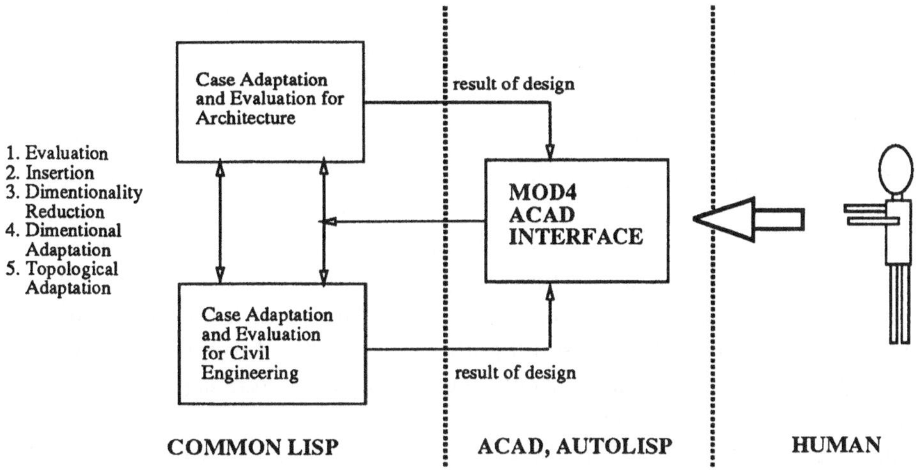

Fig. 12. *System structures of the prototype system.*

topological transformation.

Several full-scale design cases have been tested for different design sites and different design specifications. In addition, a set of well defined small cases are used to verify the correctness and efficiencies of computation.

Civil engineering structures are represented in the same way and are processed in a separate package. This package is used to adapt and test the structure after architectural adaptations to the case. A feedback is given to the main control when such a process is called. We are implementing a general mechanism which will consider integration of architectural and civil engineering abstractions with the ideas presented at the beginning of the paper. Figure 12 gives an outline of the prototype. We have implemented a complete prototype that consists of creation of cases and new design problems from AutoCAD to drawing the design result back to AutoCAD for final structural and architectural evaluation by domain experts.

Current implementation of the prototype system provides us with experience related to investigation into computational issues of case-based design. Research into *dimensionality reduction* for a broader range of constraints and their relation to propagation and verification is under way. We also expect that a better way to combine topological adaptation with dimensional adaptation may appear after syntax-directed topological adaptation described in Section 3 is fully implemented and tested.

5. Conclusions and Final Remarks

The research described in this paper has been motivated by two main issues. The first is that case-based design is more efficient than top-down development for realistic problems. The second is that case-based design methods can solve the problem of integrating different aspects of a design.

The prototype we developed has shown that adaptation of complete buildings does not require excessive amounts of knowledge. It would be extremely difficult to develop a formalism of this size which could produce the complete designs that result from using our case adaptation process. On the other hand, one must be aware that this performance is achieved at the expense of limited *coverage*: adaptation is successful only if the starting case is sufficiently close to a feasible solution.

Our prototype has also shown how it is possible to improve integration. Aspects of structural engineering, common sense, and architectural form are included in a *precise geometrical model*. Such integration is possible through creation of specific adaptation parameterizations; conflicts between different aspects are resolved before adaptation even begins.

Our research has indicated that the adaptation of cases is a useful methodology for intelligent design systems, and we argue that the case-based knowledge representation is fundamentally more powerful than grammar or rule-based formalisms alone. This is in contrast to most other work in case-based reasoning, where introduction of cases is seen only as an efficient method for knowledge-acquisition. We emphasize that generalization of cases *during execution and according to the problem* is important for the solution of integration problems inherent in design.

So far, our prototype is only capable of adapting single cases in response to geometric discrepancies. The next step of our research will be to consider more functions and to use the idea of *dimensionality reduction* to improve integration. When these problems are solved, combination of several cases will become possible at a detailed geometrical level and this should create a substrate for innovative designs.

Acknowledgements

This work is a result of collaborative research between the authors and ICOM(Steel Structures), EPF Lausanne. We would like to thank Simon Bailey for his work on implementation of some of the ideas described herein.

References

B. Faltings: "An Architecture for Design in Large Domains," The Third IFIP WG 5.2 Workshop on ICAD 1989

J. Feder: "Plex languages," Inf. Sci. 3, 225-41 (1971).

S. J. Fenves, U. Flemming, C. Hendrickson, M.L. Maher and G. Schmitt: "An Integrated Software Environment for Building Design and Construction," Symposium Proceedings for CIFE, Stanford University, March, 1989

U. Flemming: "Wall Representations of Rectangular Dissections and Their Use in Automated Space Allocation," *Environment & Planning B5* 1978

U. Flemming: "More Than the Sum of Parts: The Grammar of Queen Anne Houses," Environment and Planning B: Planning and Design 14, no. 3: 323-350, 1986

K. S. Fu: "Syntactic Pattern Recognition and Applications," Prentice-Hall, Inc., Englewood Cliffs, N.J. 07632, 1982

John S. Gero: "Design Prototypes: A Knowledge Representation Schema for Design," AI Magazine, Vol. 11, No. 4, 1990

J. Gips: "Shape Grammars and Their Uses," Birkhuser Verlag, Basel Switzerland, 1975

A. K. Goel, J. L. Kolodner: "Towards a Case-based Tool for Aiding Conceptual Design Problem Solving," DARPA Case-based Reasoning Workshop, 1991, pp. 109-120

M. Gross: "Relational Modeling: a basis for computer-assisted design," The Electronic Design Studio, MIT Press, 1989

K.J. Hammond: "Case-based Planning: An Integrated Theory of Planning, Learning and Memory," Ph.D Thesis, Yale, 1986

T. W. Knight: "The Forty-One Steps," Environment and Planning B 8, no. 1: 97-114, 1981

J. L. Kolodner: "Retrieval and Organization Strategies in Conceptual Memory: A Computer Model," Lawrence Erlbaum Associates, Hillsdale, NJ, 1984

H. Koning, and J. Eizenberg: "The Language of the Prairie: Frank Lloyd Wright Prairie Houses," Environment and Planning B 8, no. 3: 295-323 1981

A. Lindenmayer: "Mathematical Models for Cellular Interactions in Development," Parts I and II, Journal of Theoretical Biology 18, pp 280-315, 1968

M. L. Maher, D. M. Zhang: "Case-based Reasoning in Design," Artificial Intelligence in Design, Butterworth Heinemann 1991, pp. 137-150

W. J. Mitchell: "The Computer's Role in Design," Chapter 2 in: Computer-Aided Architectural Design, Van Nostrand Reinhold Company, New York, 1977

R. E. Oxman and O. M. Oxman: "Refinement and Adaptation: Two Paradigms of Form Generation in CAAD," Proceedings of CAAD Future 1991, Zurich, Switzerland, pp. 291-306

J.L. Pfaltz and A. Rosenfeld: "Web Grammars," Proc. First Int. Joint Conf. Artif. Intell. May 1969, Washington, D.C., pp 609-19

M. A. Roseman, J. S. Gero, and R. E. Oxman: "What's in a Case: the Use of Case Bases, Knowledge Bases and Databases in Design," Proceedings of CAAD Future 1991, Zurich, Switzerland, pp. 263-278

D. J. Rosenkrantz: "Programmed Grammars: a New Device for Generating Formal Languages," 8th IEEE Annu. Symp. Switching Automata Theory, Austin, Tex., 1967, Conf. Rec.

R. C. Schank: "Dynamic Memory: A Theory of Reminding and Learning in Computers and People," Cambridge University Press, London, 1982

G. Stiny: "Pictorial and Formal Aspects of Shape and Shape Grammars," Birkhuser Verlag, Basel Switzerland, 1975

J. Wang and H.C. Howard: "A Design-dependent Approach to Integrated Structural Design," Artificial Intelligence in Design, Butterworth Heinemann 1991, pp. 151-170

10

DESIGN ANALYSIS

AN INTELLIGENT MODELLING ASSISTANT FOR PRELIMINARY ANALYSIS IN DESIGN

D. P. FINN
Hitachi Dublin Laboratory
Trinity College
University of Dublin

and

J. B. GRIMSON and N. M. HARTY
Trinity College
University of Dublin
Dublin 2, Ireland

Abstract. This paper describes work in progress aimed at developing an intelligent system for assisting engineers with the task of preliminary analysis of design problems. Preliminary analysis precedes detailed numerical analysis and involves formulating, evaluating and assessing engineering problems with the objective of proposing higher level intermediate analysis models. The approach taken is based on Chandrasekaran's propose-critique-modify method which is adapted for preliminary analysis. The use of this method is justified by viewing preliminary analysis as a knowledge-based modelling activity based on successive proposing, evaluation and refinement of candidate analysis models. The system architecture is based on exploiting a number of artificial intelligence techniques including model based reasoning, case based reasoning and rule based reasoning. A modelling options case base assists engineers in proposing candidate analysis models. Engineering 1st principles and formulae are utilised within an artificial intelligence framework to provide a means of evaluating and critiquing candidate analysis models. The system is integrated with an existing CAD system. The problem domain covered is application independent but will initially focus on analysis of domains associated with heat transfer problems.

1. Introduction

Analysis of design problems using numerical techniques constitutes an important task in engineering practise. Although analysis techniques may vary from problem to problem, the process of analysis is generally considered to consist of a number of interrelated phases (Babuska, 1990; Bathe, 1990). These phases are illustrated in Figure 1 and include: obtaining a behavioural understanding of the design problem or physical system to be analysed[1], specifying a suitable mathematical model which is adequately representative of the physical system, carrying out numerical simulation using a numerical technique such as the finite element method, visualising

[1] In this work, the term *Physical System* is used to describe the complete design problem that is being analysed.

J. S. Gero (ed.), Artificial Intelligence in Design '92, 579–596.

simulation data using graphical techniques and interpreting simulation results with reference to the physical system and mathematical model. The phases of obtaining a behavioural understanding and specifying a mathematical model are often called early or preliminary analysis (Finger, 1989), and are essential if later numerical analysis is to be successful. In recent years, advances in

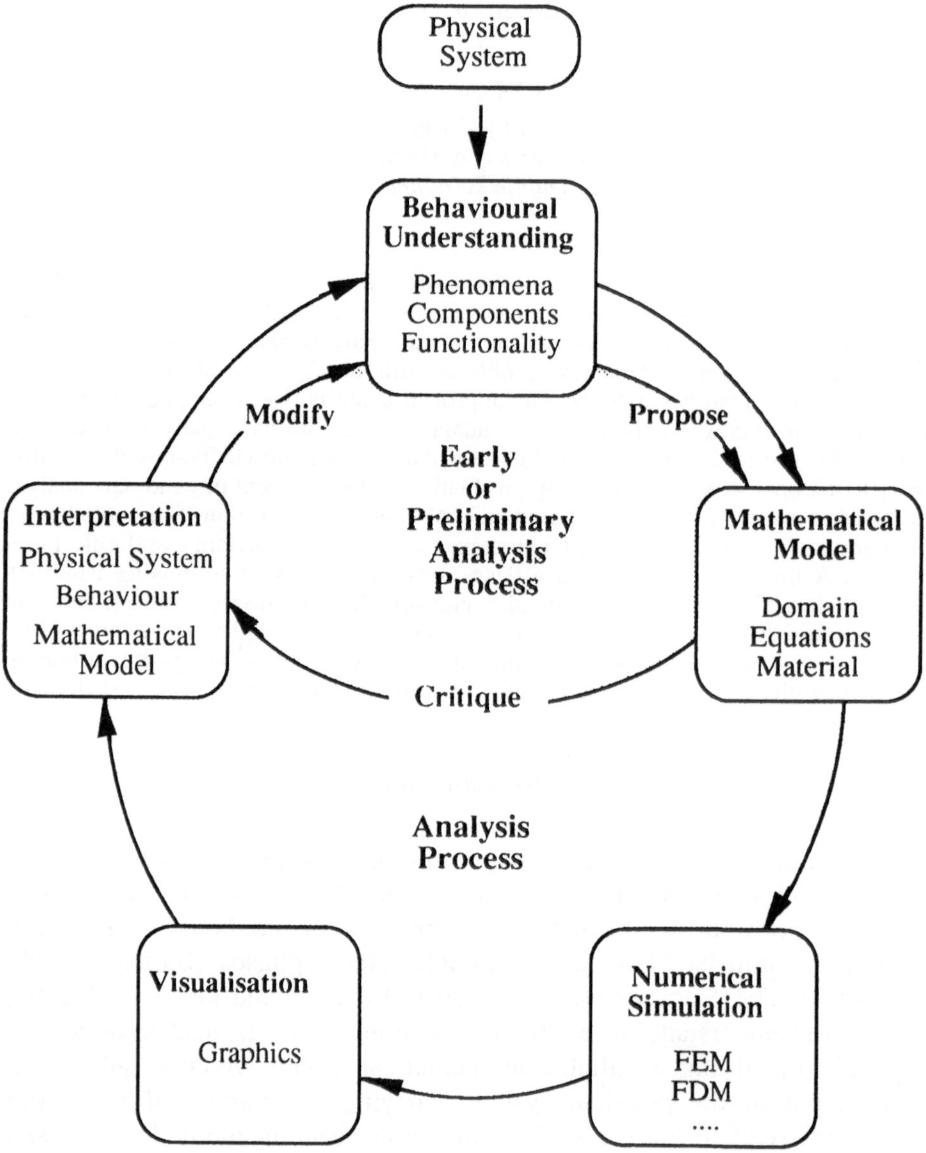

Fig. 1. Preliminary Analysis and the Analysis Processes in Design

numerical research have led to significant developments in the numerical simulation phase of analysis, while parallel advances in post-processing techniques have permitted sophisticated visual presentation of simulation data. Although, these developments have made significant inroads into analysis as a whole, preliminary analysis has not as yet been incorporated within the computer-aided engineering analysis agenda. This important phase is primarily cognitive and experiential in nature and until recently was best considered left to the experience and skill of the engineer. However, recent research opinion has indicated that the use of Artificial Intelligence techniques may provide a promising approach for developing software tools that deal with these 'non-numerical' aspects of design analysis (Finger, 1989; Oden, 1990; DeKleer, 1990).

This paper focuses on the mathematical modelling process of preliminary design analysis. An intelligent, interactive modelling assistant that aims to assist engineers in formulating, evaluating and assessing candidate mathematical models in preparation for numerical simulation of physical systems is proposed. This approach is argued in favour of automated modelling and is based on Chandrasekaran's propose-critique-modify method which is adapted for the modelling task in preliminary analysis (Chandrasekaran, 1990). A modelling options case base assists engineers in proposing candidate mathematical models. Engineering 1st principles and formulae are utilised within an artificial intelligence framework to provide a means of evaluating and critiquing the candidate mathematical models. The system architecture is based on exploiting a number of artificial intelligence techniques including model based reasoning, case based reasoning and rule based reasoning. The system is integrated with an existing CAD system.

The paper is laid out as follows: Section 2 discusses the various issues and options involved in mathematical modelling. Section 3 examines previous work from the perspectives of qualitative reasoning and engineering analysis. Section 4 outlines the approach followed in this work. Section 5 details the overall system architecture and system components. Finally, a brief conclusion and future directions are given.

2. Mathematical Modelling in Preliminary Analysis

Preliminary analysis can be considered as a cyclic or iterative activity that involves assessment of a given physical system, with the objective of proposing and evaluating candidate mathematical models in preparation for later detailed analysis such as numerical simulation (Babuska; 1990, Bathe ;1990). In Figure 1, mathematical modelling of physical systems is shown as a phase of preliminary analysis. As can be expected, understanding the physical system in terms of its behaviour and function is an essential pre-requisite to the mathematical modelling process. Understanding of physical systems is generally a mental activity that is based on reasoning about the underlying system behaviour and functionality and this phase has been a subject of considerable research interest using qualitative reasoning techniques (Forbus,1990). The research work described in this paper does not address this agenda, but focuses on the follow-up tasks of deriving and evaluating candidate mathematical models.

2.1 MODELLING TO DERIVE A MATHEMATICAL MODEL

Except for simple cases, it is usually not feasible to analyse in detail all aspects of a physical system because of its inherent complexities. Furthermore, perceived redundancies of certain aspects of the physical system may prompt that such aspects be ignored altogether. Therefore it is usual to apply various modelling techniques that assist by reducing the system complexity. Modelling techniques can be applied to any of the following aspects of a physical system: geometry domain, phenomena, boundary conditions, initial conditions, material properties or mathematical equations. Decisions regarding the application of modelling techniques is the key issue associated with mathematical modelling. The major challenge to the engineer; is identifying the importance of different system aspects, applying an appropriate modelling technique and finally assessing the suitability of the resulting model for analysis.

Modelling techniques can either be in the form of *simplifications* or *idealisations* and these terms are discussed in the following sections.

2.1.1. *Simplifications in Mathematical Modelling.* Simplifications are modifications applied to either the geometry domain or phenomena so that the resulting physical system is more amenable to detailed analysis. These modifications may change a system component thereby reducing complexity, or remove altogether a system component from consideration of detailed analysis. Figure 2 illustrates the various geometric and phenomena simplifications and they are explained in more detail in the following paragraphs:

Geometric Four basic types of geometric simplifications can be applied to a physical system domain, namely: dimensional reduction, geometric symmetry, feature removal and domain alteration.

Dimensional reduction involves reducing the degree of spatial analysis or

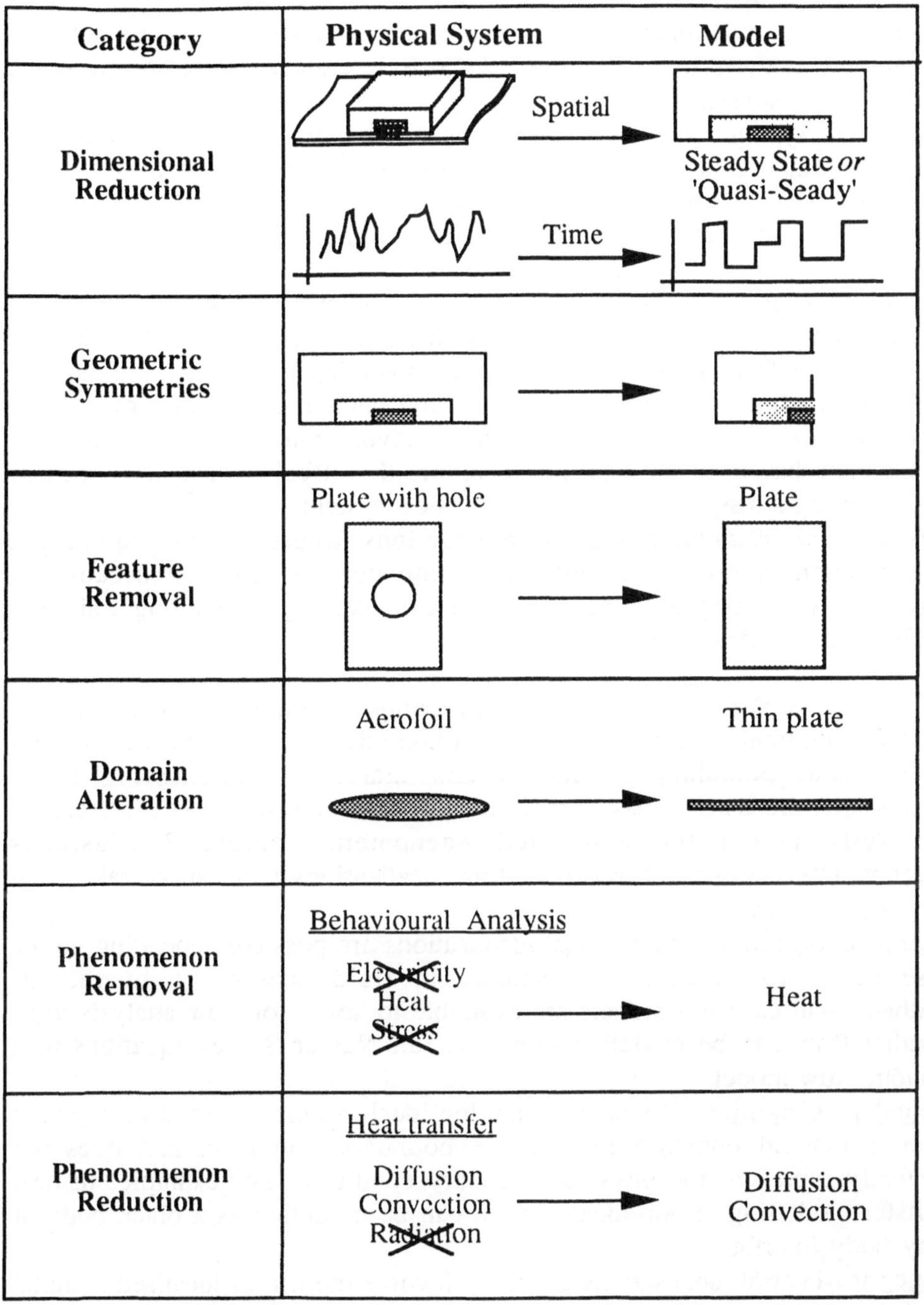

Fig. 2. Simplifications in Mathematical Modelling

time analysis. Spatial analysis can involve reduction from 3-dimensional to 2-dimensional or 1-dimensional analysis. Time analysis can involve reducing a transient analysis to a 'quasi-analysis' (discrete steady steps in time) or a steady state analysis.

Geometric symmetries involve removing redundant domains by identifying spatial symmetries and applying compensatory boundary conditions.

Feature removal involves removing some engineering feature that is not expected to contribute significantly to the overall analysis result, e.g., a hole or fin.

Domain alteration involves changing some aspect of the spatial domain so that the analysis is simplified, e.g., modelling a thin aerofoil as a thin plate.

Phenomenon Two types of simplifications involving phenomena are possible and these include: phenomenon removal and phenomenon reduction.

Phenomenon removal involves the removal from analysis a complete phenomena based on the decision to ignore the effect of that phenomenon, e.g., ignoring stress effects within the physical system.

Phenomenon reduction applies to situations where a multi-component phenomenon exists and a particular component is removed because its significance is judged to be minor importance, e.g., removing radiation analysis from a heat transfer problem.

2.1.2. *Idealisations in Mathematical Modelling*. Idealisations involve the use of mathematical expressions that mathematically describe the system phenomena, boundary conditions and material constituency. These expressions are usually based on behavioural assumptions and therefore do not fully model the associated phenomena. Figure 3 illustrates diagrammatically the various modelling idealisations that can be taken and they are as follows:

Phenomenon Many mathematical idealisations are possible depending on the case being considered. For example in fluid analysis, a number of mathematical equation models are available to solve for flow analysis, e.g., parallel flow can be modelled using the full Navier Stokes equations or a Couette flow model.

Boundary Conditions Boundary condition idealisations can involve applying a mathematical equation to model a boundary condition that does not perfectly represent the physical boundary condition. For example, in heat transfer modelling, a non-ideal surface can be modelled as a black body or gray body interface.

Material Material idealisations generally involve the use of idealised material laws to model some complex material behaviour. For example modelling an expected non-linear material response using a linear approximation function.

3. Related Work

Similar work that deals with modelling in preliminary analysis can be observed to come from two distinct research backgrounds: artificial intelligence researchers involved in qualitative reasoning and engineering researchers involved in knowledge-based analysis systems. Currently, research is generally at an early stage and varies from innovative qualitative reasoning techniques applied to simple physical systems to expert system approaches applied to specific engineering design domains. In the following sections, work from the qualitative physics and engineering agendas are discussed separately.

3.1 QUALITATIVE PHYSICS

The most relevant work from qualitative physics relates to the issues of multiple ontologies and automated modelling associated with high level analysis of physical systems (Weld, 1990). Addanki (1990) deals with automated modelling of large complex systems using a paradigm called *the graph of models*. The basic concept of this approach is that a physical system

Category	Physical System	Model
Phenomenon	Fluid Flow →	Full Navier Stokes Model *or* Stokes Model *or* Couette Flow Model
Boundary Condition	Decreasing Temperature Boundary →	Constant Temperature *or* Constant Flux
Material Law	Non - Linear Material Response →	Piecewise Linear *or* Linear

Fig. 3. Idealisations in Mathematical Modelling

can be represented by a number of small, interlinked knowledge-based analysis models where each model is abstracted on the basis of its underlying assumptions. Modelling progresses by automatic selection and changing of analysis models until the initial model requirement is satisfied. This approach is demonstrated for analysis and design of gear systems. Iwasaki (1989) proposes a generalised approach that aims to generate an equation model from an initial structural model. An intermediate process model and engineering conservation laws are used to assist in the specification of the equation model. Hobbs (1985) presents a framework for linking multiple models at different abstraction levels. The approach assumes a detailed global theory and discusses how simpler, idealised theories can be related to the global theory using articulation axioms. Falkenhainer and Forbus (1991) propose a compositional modelling approach that uses explicit modelling assumptions which permit domain knowledge to be decomposed into semi-independent fragments, each describing various aspects of the physical system. A model composition algorithm is used to compose a suitable model that represents the physical system and is based on information about the general domain theory, a structural description and system behaviour.

3.2 THE ENGINEERING PERSPECTIVE

In a general overview of engineering modelling and analysis, Zienkiewicz (1990) argues that because of the complex nature of the mathematical modelling phase of preliminary analysis, it is unlikely that an automated modelling approach is a feasible objective. Finger and Dixon (1989) in a wide ranging survey discuss the state of research in mechanical engineering analysis and design. They argue that a need exists for work that addresses the early stages of the analysis process and conclude that the results of such research could act as an interface to numerical analysis systems or a tool in the preliminary stages of analysis. Shephard (1990) discusses the various modelling options that are possible when specifying a mathematical model for analysis. An overall integrated framework for modelling of physical systems is presented. Development work on this project is currently on-going and an application based on aircraft modelling is presented.

3.3 DISCUSSION

Work from the qualitative reasoning field has focused on the goal of automated modelling in preparation for numerical modelling. However, most of the problems tackled to date have been simple problems and are far removed from the reality of the engineering world (Forbus, 1990). In another report, Falkenhainer and Forbus (1991) conclude that the necessary knowledge to model complex systems and capture the breadth and depth of an engineer's knowledge will be of orders of magnitude larger than todays qualitative physics. Engineering researchers argue that the complexity associated with modelling makes this task very difficult to tackle using an automated modelling approach (Zienkiewicz, 1990). Therefore, in this work, an intelligent interactive modelling assistant that aims to help engineers in the modelling aspects of preliminary analysis is proposed as a more achievable target.

4. Research Approach

The purpose of this section is to describe the coverage of the project within the overall context of preliminary analysis. This is achieved by discussing first the reasoning process associated with the mathematical modelling aspect of preliminary analysis and by then outlining the precise aims and coverage of the current work.

4.1 THE REASONING PROCESS IN MATHEMATICAL MODELLING

In Section 2, the range of available modelling options were outlined. This section discusses the reasoning basis by which these options are chosen.

Consider the initial stages of modelling when an engineer formulates a conceptual understanding of the physical system. This involves reasoning about possible phenomena, geometric features, component functionality and system behaviour. Using this information, the importance of different aspects of the physical system can be assessed. The next objective is to create and assess candidate mathematical models for analysis. It is proposed that there are two main reasoning stages involved in this process, namely, *model construction* and *model evaluation*. *Model construction* involves choosing and applying different modelling options to specify a candidate mathematical model. This stage can be considered a function of the motivation and purpose of the analysis, the modelling options available and any constraints on applying these options to the problem domain. *Model evaluation* involves assessing the sufficiency of a candidate mathematical model for representing the physical system. Evaluation can be carried out using engineering estimations or a simple trial numerical simulation. Using the results of the evaluation, the suitability of the candidate mathematical model can be assessed. By reconsidering model construction, other modelling options can be applied, candidate mathematical models evaluated and finally the most suitable model selected. Model construction and evaluation are different in their focus: construction is essentially a subjective knowledge-based search process based on engineering and modelling experience, evaluation is a knowledge-based assessment process based on applying engineering analysis techniques.

4.2 APPROACH

Currently the approach favoured by many researchers focuses on developing an automated modelling facility for preliminary analysis. The fundamental tenet of the automated modelling agenda aims to develop a system that automatically reasons about some physical system with the objective of carrying out model construction and model evaluation.

In this work an interactive modelling assistant is proposed which participates with the engineer in the task of mathematical modelling. The following points summarises the main features of this approach:

- The system is based on an intelligent interactive modelling assistant rather than an automated modelling system.
- The aim of the system is to provide an environment that supports the engineer in selecting, applying and evaluating the various mathematical modelling options.
- A modelling options case-base is provided to advise the engineer on the various modelling options available and the context in which they are often used. However, the decision of which modelling option to choose and where to apply it remains ultimately with the engineer.
- Engineering principles and formulae are used as the basis for evaluating modelling decisions. The approach followed automatically reasons about each candidate mathematical model, selects and applies the appropriate engineering formulae and solves the formulae. This technique can be viewed as a knowledge-based engineering estimation process and is similar to the "back-of-an-envelope" calculations often used by engineers as an evaluation method[2]. Use of engineering formulae for evaluation is favoured over simple trial numerical simulation because of associated difficulties with the numerical issues of stability, convergence and computer resources.
- The information provided includes estimations of phenomena and predictions of required computer resources. By comparing different candidate models, the effect of a particular modelling decision can be assessed by the engineer. The absolute accuracy of the estimations themselves is not a critical issue because the major emphasis is based on the

---

[2] The term "back of the envelope" analysis describes a widely used informal engineering technique that combines engineering knowledge, principles and formulae to assess and evaluate engineering problems.

comparison of different mathematical models. This is based on the argument that candidate models will vary quite significantly and therefore the focus is on the relative relations between models.

• The importance of the concepts of *assistant* and *interaction* are very important in this work. In this way, the approach can be viewed as an interpretation of the propose-critique-modify method defined by Chandrasekaran (1990) for design tasks. In this case, Chandrasekaran's approach is adapted to the modelling task, where the engineer is assisted in *proposing* (constructing) a candidate mathematical model, the system automatically *critiques* the proposal and based on this information the engineer *modifies* the initial proposal.

• The work will initially concentrate on the domain of modelling associated with general heat transfer analysis.

5. System Architecture

This section presents an overview of the system architecture and discusses in detail the various system components. Initial implementation concentrates on the domain of engineering heat transfer and aims to assist users with modelling heat conduction, convection and radiation problems for two dimensional, steady state problems.

5.1 SYSTEM OVERVIEW

Figure 4 represents an overview of the system architecture. This architecture reflects the adoption of the propose-critique-modify method adapted for the task of an interactive modelling assistant. Modelling consists of a two stage process, namely, model construction and model evaluation. Construction is carried out by the engineer by proposing and modifying candidate mathematical models using a CAD interface with guidance from the modelling case base and information provided by the engineering estimations. Once a particular mathematical model is proposed, a knowledge-based CAD representation is passed for model evaluation. Evaluation consists of verification and critiquing. Verification is automated and the associated reasoning and evaluation processes are fully documented and explained by the system. Information provided includes estimations of the phenomena and required computer resources. Critiquing is carried out by the engineer by comparing these estimations with (a) the original guidance provided by the modelling case base and, (b) other proposed mathematical models. Using this information the engineer can modify the initial model, thereby proposing a new candidate model for further evaluation. In this fashion, modelling evolves as a process of successive refinement.

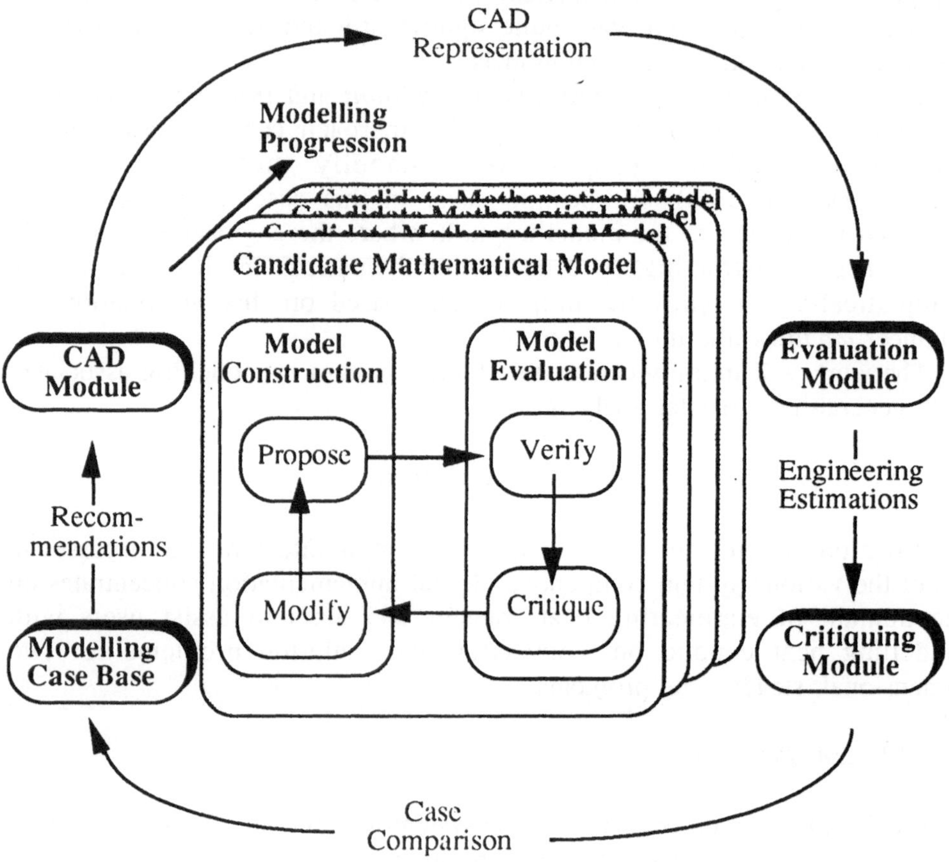

Fig. 4. System Image

5.2 SYSTEM COMPONENTS

System components can be divided according to the tasks of model construction and model evaluation.

5.2.1 Model Construction: Proposing and Modifying. Proposing of a candidate mathematical model is carried out using an interactive knowledge-based CAD editor and the modelling case-base. Figure 5 illustrates the system components associated with mathematical model construction. Model construction is carried out as follows: the problem geometry is specified using the CAD editor, geometric features are incorporated and phenomena and boundary condition models are chosen with guidance from the modelling case base. New candidate models can be assessed by adding or

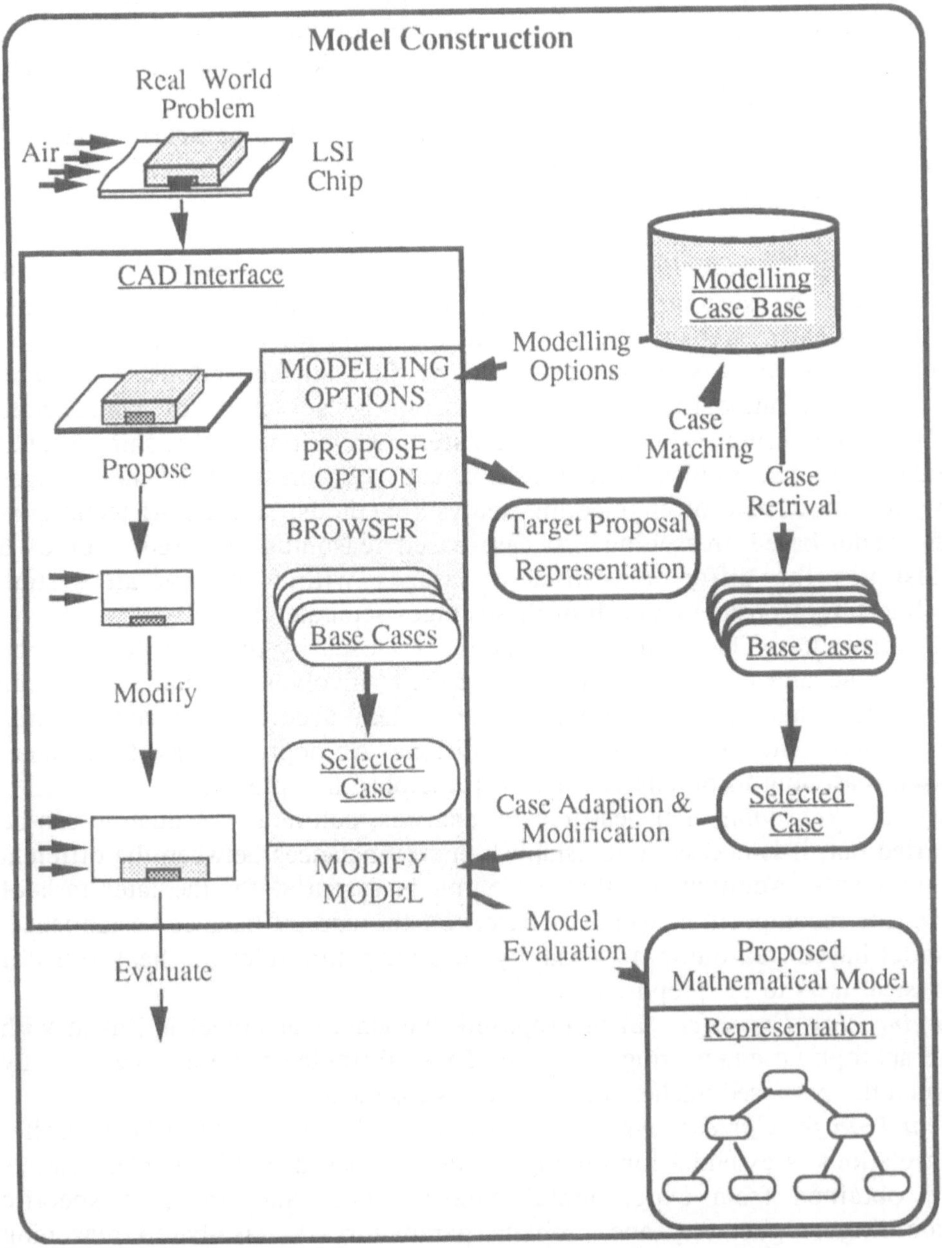

Fig. 5. Mathematical Model Construction: Propose - Modify

removing features, specifying alternate phenomena or boundary condition models, reducing dimensions, taking symmetries or substituting material models. Once the user proposes a particular modelling option, the CAD interface creates a target case which forms the basis for matching and retrieving suitable base cases. Each base case provides information such as the context for applying the modelling option, expected modelling advantages and disadvantages and a generalised technical evaluation of the modelling option. Once the engineer is satisfied with a particular modelling option, model evaluation can proceed on the basis of the proposed mathematical model.

5.2.2 Model Evaluation: Verifying and Critiquing. Verification is the process by which the system provides engineering estimates of the proposed mathematical model thereby providing the basis for critiquing the model. Verification consists of a number of distinct stages and these include, problem decomposition, problem reasoning, selection of appropriate engineering formulae, use of heat transfer correlations, establishing spatial nodal relations, problem recomposition and solution. Each of these stages involve knowledge of engineering analysis methods, and the AI techniques of model-based reasoning and case-based reasoning are used. Figure 6 illustrates the different verification stages within the overall system architecture. Considering each of these stages separately:

Decomposition The mathematical model proposed by the engineer must be interpreted and prepared for evaluation. This involves reasoning about the physical system and decomposing the system according to components, phenomena modes and boundary conditions. An approach based on model based reasoning using engineering 1st principles is exploited.

Physical Reasoning For heat transfer analysis, before any evaluation can be carried out, it is necessary to establish energy balances between the different components. Additionally, this reasoning is essential for the later task of problem recomposition which is based on the energy balances established. Model-based reasoning is used by applying the relevant heat transfer physical laws to the proposed model.

Engineering Formulae Each proposed mathematical model is linked with the appropriate engineering formulae. These formulae represent the basis by which the proposed mathematical model is assessed.

Heat Transfer Correlations In the heat analysis domain, use of heat transfer correlations is essential for solving convection based problems. Correlations are obtained from experimental measurements and are case specific according to geometry and problem parameters. A case-based reasoning approach is used for matching and retrieving suitable correlations.

Spatial Nodes Once the engineering models, formulae and correlations have been established, it is necessary to establish spatial nodes for each component. This is essential for linking the engineering formulae to the physical components. Figure 7 illustrates a number of generalised cases for a

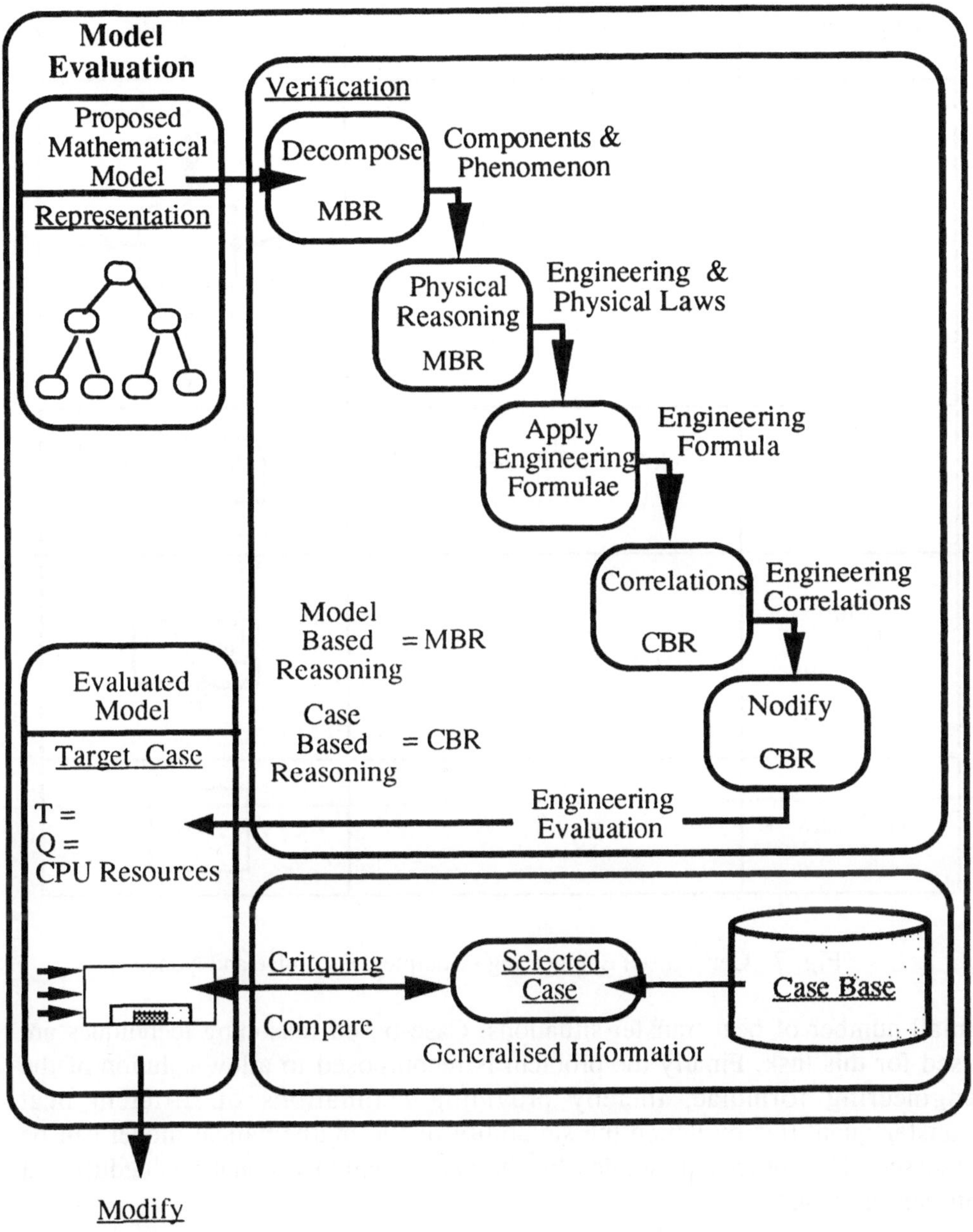

Fig. 6 Mathematical Model Construction: Verify - Critique

Engineering Principle	**Engineering Formulae**	**Spatial Node Case**
Conservation of Energy	$\Sigma f + q = 0$	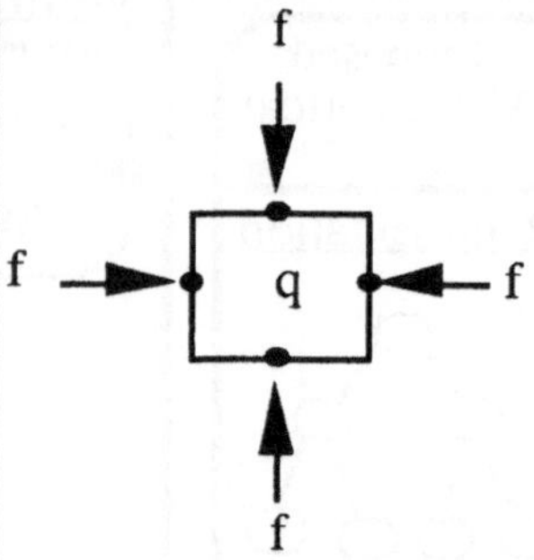
Fourier's Law	$f = - kA \cdot (T1-T2)/d + V \cdot \rho AcT$	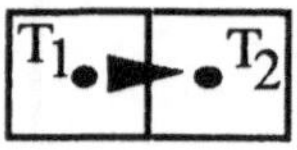
Newton's Law of Cooling	$f = f(v)A\ (Tav - T)$	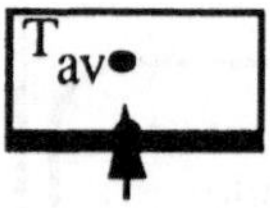
Watt's Law	$q = - i^2 d/\sigma A$	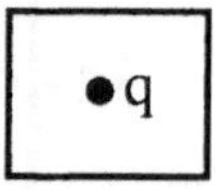

Fig. 7 Generalised Engineering Principles, Formulae and Nodes

small number of heat transfer situations. Case-based reasoning techniques are used for this task. Finally the problem is recomposed to allow solution of the engineering formulae, thereby providing estimations of different heat transfer quantities by which the suitability of the mathematical model can be assessed. This process is guided by the energy balance equations deduced at an earlier stage.

Critiquing is carried out by the user by comparing the engineering estimations provided during the verification process with earlier proposed models as well as the generalised information provided in the base case. A number of issues must be borne in mind when critiquing:

• Critiquing is based on comparing a range of mathematical models and assessing the importance of different features, phenomena modes or equation models.
• The accuracy and validity of engineering estimations must be considered in relative terms rather than absolute terms. For example, if evaluation of a LSI chip with fins indicates a cooling rate of 10 mWatts and without fins a cooling rate of 4 mWatts, then a possible conclusion is that the inclusion of fins in the mathematical model is important. On the other hand, if an evaluation indicated that the radiation component constituted only 5% of total heat transfer and required a third of total simulation resources, then a possible decision may be not to carry out a numerical analysis for radiation.

5.3 IMPLEMENTATION

Implementation is currently at an early stage, an initial CAD interface has been developed and work is now proceeding on the specification of a preliminary modelling case base.

6. Conclusions

An intelligent mathematical modelling assistant for preliminary design analysis has been described. The modelling process is argued to consist of two tasks: model construction and model evaluation. The approach adopted is based on Chandrasekaran's generalised propose-critique-modify method which is adapted for an interactive modelling system. The system aims to aid the engineer with model construction and automate the task of model evaluation. First principle engineering formulae are integrated in a knowledge-based evaluation environment and provide the basis for evaluating and critiquing candidate mathematical models. The artificial intelligence techniques of model-based reasoning, case-based reasoning and rule-based reasoning are exploited. The system focuses on engineering analysis and design problems described by partial differential equations and will initially be implemented in the heat transfer domain. In this way independence of application domain is achieved. Additional research issues under consideration include; the use of a blackboard architecture to augment the system interface and providing guidance on the reliability of the engineering estimations.

Acknowledgements

The authors would like to express their appreciation to N. Hurley, E. Cahill, N. Hataoka and N. Sagawa for their helpful comments.

References

Addanki, S. Cremonini, R and Penberthy, J.S.: 1990, Reasoning about Assumptions in Graph of Models, *Readings in Qualitative Reasoning about Physical Systems,* Morgan Kaufmann, CA, pp. 546-552.

Babuska, I.: 1990, The Problem of Modelling the Elastomechanics in Engineering, *Computer Methods in Applied Mechanics and Engineering,* North Holland, Amsterdam, **82**(1-3), 155-182.

Bathe, K.J., Lee, N. and Bucalem, M.L.: 1990, Use of Hierarchical Models in Engineering Analysis, *Computer Methods in Applied Mechanics and Engineering,* North Holland, Amsterdam, **82**(1-3), 5-26.

Chandrasekaran, B.: 1990, Design Problem Solving: A Task Analysis, *AI Magazine,* **11**(4), 59 - 71.

DeKleer, J.: 1990, Qualitative Physics: A Personal View, *Readings in Qualitative Reasoning about Physical Systems*, Morgan Kaufmann, CA, pp. 1-8.

Falkenhainer, B. and Forbus, K.D.: 1991, Compositional Modelling: finding the right model for the job , *AI Journal*, Elsevier Pub., Amsterdam, **51**(1-3), pp. 95-143.

Finger, S. and Dixon J.R.: 1989, A Review of Research in Mechanical Engineering Design. Part II: Representations, Analysis, And Design for the Life Cycle, *Research in Engineering Design,* **1,** pp. 121-137, Springer-Verlag, Berlin.

Forbus, K.D.: 1990, Qualitative Physics: Past, Present, and Future, *Readings in Qualitative Reasoning about Physical Systems*, Morgan Kaufmann, CA, pp. 11-39.

Hobbs, J.R.: 1985, Granularity, *Proc. of IJCAI '85,* Morgan Kaufmann, pp. 432-435.

Iwasaki, Y., Doshi., T., Gruber, R., Keller, R. and Low, C.M.: 1989, Equation Model Generation: Where do the Equations Come From? *Proceedings on 1989 Workshop on Model Based Reasoning,* AAAI, pp. 120-123.

Oden, J.T.: 1990, Reliability in Computational Mechanics, *Computer Methods in Applied Mechanics and Engineering*, North Holland, Amsterdam, **82**(1-3),1-3.

Shephard, M.S., Baehmann, P.L., Georges, M.K. and Korngold, E.V.: 1990, Framework for Reliable Generation and Control of Analysis Idealisations, *Computer Methods in Applied Mechanics & Eng.,* North Holland, **82**(1-3), pp. 257-280.

Zienkiewicz, O.C. and Zhu, J.Z.: 1990, The Three R's of Engineering Analysis and Error Estimation and Adaptivity Computer Methods in Applied Mechanics and Engineering, North Holland, Amsterdam, **82**(1-3), pp. 95-113.

Weld, D.S. and DeKleer, J.: 1990, Multiple Ontologies and Automated Modelling *Readings in Qualitative Reasoning about Physical Systems*, Morgan Kaufmann, CA, pp. 481-483.

AN INTELLIGENT MODELLING ASSISTANT FOR PRELIMINARY ANALYSIS IN DESIGN

D. P. FINN
Hitachi Dublin Laboratory
Trinity College
University of Dublin

and

J. B. GRIMSON and N. M. HARTY
Trinity College
University of Dublin
Dublin 2, Ireland

Abstract. This paper describes work in progress aimed at developing an intelligent system for assisting engineers with the task of preliminary analysis of design problems. Preliminary analysis precedes detailed numerical analysis and involves formulating, evaluating and assessing engineering problems with the objective of proposing higher level intermediate analysis models. The approach taken is based on Chandrasekaran's propose-critique-modify method which is adapted for preliminary analysis. The use of this method is justified by viewing preliminary analysis as a knowledge-based modelling activity based on successive proposing, evaluation and refinement of candidate analysis models. The system architecture is based on exploiting a number of artificial intelligence techniques including model based reasoning, case based reasoning and rule based reasoning. A modelling options case base assists engineers in proposing candidate analysis models. Engineering 1st principles and formulae are utilised within an artificial intelligence framework to provide a means of evaluating and critiquing candidate analysis models. The system is integrated with an existing CAD system. The problem domain covered is application independent but will initially focus on analysis of domains associated with heat transfer problems.

1. Introduction

Analysis of design problems using numerical techniques constitutes an important task in engineering practise. Although analysis techniques may vary from problem to problem, the process of analysis is generally considered to consist of a number of interrelated phases (Babuska, 1990; Bathe, 1990). These phases are illustrated in Figure 1 and include: obtaining a behavioural understanding of the design problem or physical system to be analysed[1], specifying a suitable mathematical model which is adequately representative of the physical system, carrying out numerical simulation using a numerical technique such as the finite element method, visualising

[1] In this work, the term *Physical System* is used to describe the complete design problem that is being analysed.

J. S. Gero (ed.), Artificial Intelligence in Design '92, 579–596.

to provide a useful feedback about the design of a device based on the result of simulation, the system must be able to evaluate the predicted behavior with respect to the knowledge of the function.

In this paper, we focus on the task of design verification using both knowledge of the structure of a device and its intended functions. In particular, we address the question of when one can say a behavior predicted by a prediction system achieves the desired function in the manner intended by the designer. We use Functional Representation (Sembugamoorthy & Chandrasekaran 1986) to represent the function of a device and the expected causal process for achieving the function. We will present a formal definition of matching between an expected behavior and a predicted behavior. Finally, we will demonstrate behavior verification based on the definition, using an actual example of behavior predicted by a simulation system.

1.1 BEHAVIOR-ORIENTED AND FUNCTION-ORIENTED APPROACHES TO MODELING

Research in model-based reasoning about physical systems has emphasized representation of structures and reasoning about behavior from the knowledge of their structures and physical principles. Several model-based reasoning systems have been built (Falkenhainer & Forbus 1991, Crawford et al. 1990, Iwasaki & Low 1991) that formulates a model of a device based on its structure and predict its behavior. An important requirement in the approach taken in these systems, which we shall call the *behavior-oriented* approach, is that the knowledge is stored in small pieces, each representing a conceptually independent physical phenomenon such as a physical process or an aspect of the behavior of a component. For the pieces to be composable, each of them must be defined in a context-independent manner as much as possible in the sense that there is no unstated assumption about the surroundings of a component or the function of the whole device. These systems predict a behavior in terms of a sequence or a graph of states, each of which is characterized by the set of applicable knowledge pieces, implied constraints, and variable values.

This type of model-based reasoning capability is useful for a system aimed to help in design, since it allows the system to formulate a behavior model automatically and to simulate its behavior so that the designer can discover behavioral implications of design decisions easily. However, an account of behavior in the form of a sequence of states must be evaluated to be useful for further development of the design. Does the predicted behavior achieve the desired function? Does it do so in the way the designer intended? These are crucial questions in providing a useful feedback to the designer. In order for a model-based reasoning system to answer such questions, it must have knowledge of the function of the device -- WHAT it is supposed to do -- and the expected behavior -- HOW it is supposed to achieve the function.

Functional Representation (FR) is a representational scheme for the functions and expected behavior of a device. FR represents knowledge about devices in terms of the functions that the entire device is supposed to achieve and also of the sequence of causal interactions among components that lead to achievement of the functions. FR takes a top-down approach to representing a device in contrast to the bottom-up approach of behavior-oriented knowledge representation and reasoning schemes. In Functional Representation, the function of the overall device is described first and the behavior of each component is described in terms of how it contributes to the function, while in a behavior-oriented approach, the behavior of the entire device is inferred from those of individual components.

In order to evaluate a design, one must be able to predict the possible behavior of the design, as well as to determine whether the predicted behavior achieves the expected functionality. Verification that a behavior of a designed artifact achieves the desired goal must ascertain the following:

(1) the overall function of the device is achieved,
(2) the expected chain of events happen in the predicted behavior, and
(3) the causal connections expected between events exist in the predicted behavior.

The purpose of this paper is to investigate this concept of behavior verification and provide a formal definition of behavior verification of a design with respect to its intended functions and the expected causal processes for achieving the functions. As an example of a model-based reasoning system, we use DME (Device Modeling Environment) developed at Stanford University (Iwasaki & Low 1991). Given a design of a device, DME formulates a computational model and predicts its behavior.

This paper is organized as follows: In Section 1.2, we will briefly describe DME. In Section 2, we formally define functions, expected behavior, and what it means for a predicted behavior to match an expected behavior. Section 3 presents an example of behavior verification. We conclude by discussing future work and related work in Section 4.

1.2 DEVICE MODELING ENVIRONMENT

DME is a program developed by How Things Work project (Fikes et al. 1991). The goal of the project is to provide a computational environment for design of electromechanical devices, and DME is the device modeling program which forms the core of the environment. Given the topological description of a device and initial conditions, DME formulates a mathematical model and simulates its behavior.

In DME, knowledge about physical phenomena is organized into *model fragments* in the knowledge base. Each model fragment represents knowledge of a conceptually distinct physical phenomenon such as a physical process, component behavior characteristics, etc. DME takes an input description of the initial state, including the topological model of the device, and searches the knowledge base for model fragments that are applicable to the given situation. Equations to describe the behavior of the device are formulated from the constraints associated with the set of model fragments thus found. The equations are used to predict the behavior of the device qualitatively using QSIM (Kuipers 1986) or numerically. During prediction, if there are any changes in the set of applicable model fragments, the set of equations is updated accordingly and prediction continues with the new equation model.

Some model fragments represent instantaneous changes, which are phenomena that take place too quickly to model as continuous phenomena. Such model fragments do not have constraints but they have consequences, which are facts to be asserted. When an instantaneous model fragment becomes active, a new state is generated immediately to follow the current state, and the consequences are asserted in the new state.

A model fragment m can be interpreted as one large implication of the form $P_m \Rightarrow E_m$ or $P_m \Rightarrow R_m$, where P_m, E_m, and R_m denote respectively the conditions for the applicability of the model fragment, the behavior constraints (equations in the case of a continuous phenomenon), and the consequences of m (in the case of a discontinuous phenomenon).

Definition 1. A device state is represented as a set of state variables $\{V_S\}$ consisting of values of all the variables of interest in the description of the device. State variables can be either continuous or discrete.

Definition 2. A device trajectory, T_r, represents the course of behavior of the device over time. It is a linear sequence of states.

2. What does it mean to verify that a design achieves an expected function?

In this section, we define what it means for such a simulated behavior to achieve the expected behavior represented in FR. This requires introduction of the notions of a *causal process description* (CPD) and a *function* in FR. A CPD is a causal explanation of how certain states of interest come about by exhibiting a sequence of causal transitions. The transitions are annotated by different types of causal explanation.

Definition 3. A Causal Process Description (CPD) is defined as a pair *{C, G}*, where *C* is the applicability condition and *G* is a directed graph *G* = *{N, L}*. *C* specifies the condition under which the device is expected to behave as specified by *G*. *C* is a necessary condition for applicability of CPD but not a sufficient condition. *N* is a set of nodes and *L* is a set of directed links among nodes. Each node represents a partial description of a state. There are two distinguished nodes in *N*, the initial node, N_{init}, and the final node, N_{fin}. Each link represents a causal connection between nodes. The graph may be cyclic, but there must be a directed path from N_{init} to N_{fin}.

A link may have an attached *qualifier, By-function-<f>-of(c)*, where *c* is a component, to indicate the conditions under which the transition will take place. A link can also have *annotations* of the types, *Provide(p)*, *If(p)* and *Trigger(p)*, where *p* is a wff, to indicate the type of the causal explanation to account for the transition.

In order for us to be able to relate a *Tr* and a CPD, we require that *each node in a CPD must be given a definition in the form of a wff about objects and predicates defined in terms of model fragments attributes.* We let *def(n)* denote such a definition of a node *n*. For example, the node "Battery charging" is defined as $dC/dt > 0$, where *C* is the variable, charge-level of the battery. With such a definition, a node in a CPD becomes a partial description of a state in *Tr* using attributes defined in the model fragment library.

Definition 3 mentions qualifiers and annotations that can be attached to the links between the nodes in CPD. The full list of proposed annotations can be found in (Sembugamoorthy & Chandrasekaran 1986) and (Keuneke 1991). For a link from n_i to n_j, an annotation *By-function-<f>-of(c)* means the causal interactions going from n_i to n_j must involve *c* achieving its junction. The purpose of a qualification is to allow a causal transition to be explained in further detail by CPD's of component *c*. In contrast, qualifiers allows one to specify further conditions on the causal transition. *Provided(p)* means that condition *p* must hold during the causal transition. *If(p)* means that the condition *p* must hold at state n_i. *Trigger(p)* means that *p* must not hold before n_i, but must hold at some point after n_i (inclusive).

In summary, a CPD describes a causal process from some perspective at the device level, and the conditions on the causal transitions and explanations of them are given as part of the description. Figures 1 and 2 show examples of CPD's. They are for the electrical power system (EPS) aboard a satellite, which we will use in Section 3 as an example. N_{int} and N_{fin} in each CPD are indicated by a box in dashed lines and a box in thick lines respectively.

Definition 4: A function F is defined as a quintuple $\{Type_F, P_F, Dev_F, C_F, G_F\}$, where

- $Type_F$: One of {*ToMake, ToMaintain, ToPrevent, ToControl*}.
- P_F: The functional goal, i.e. the wff that the function is to make true.
- Dev_F: The device that this function is a function of. This has to be a model fragment in the DME's knowledge base.
- C_F: The condition which specifies when the function must be achieved.
- G_F: The set of CPD's describing the causal mechanism to achieve the function.

Figure 3 shows the function of EPS. We consider four types of functions; *ToMake, ToMaintain, ToPrevent,* and *ToControl* (Keuneke 1991). Note that a device can have multiple functions, in which case each function will be represented separately. In case of a function of type *ToControl*, P_F must be of the following form:

$(= v_0\ f(v_1 \ldots v_n))$,

where v_i's are variables and f is some function of its arguments.

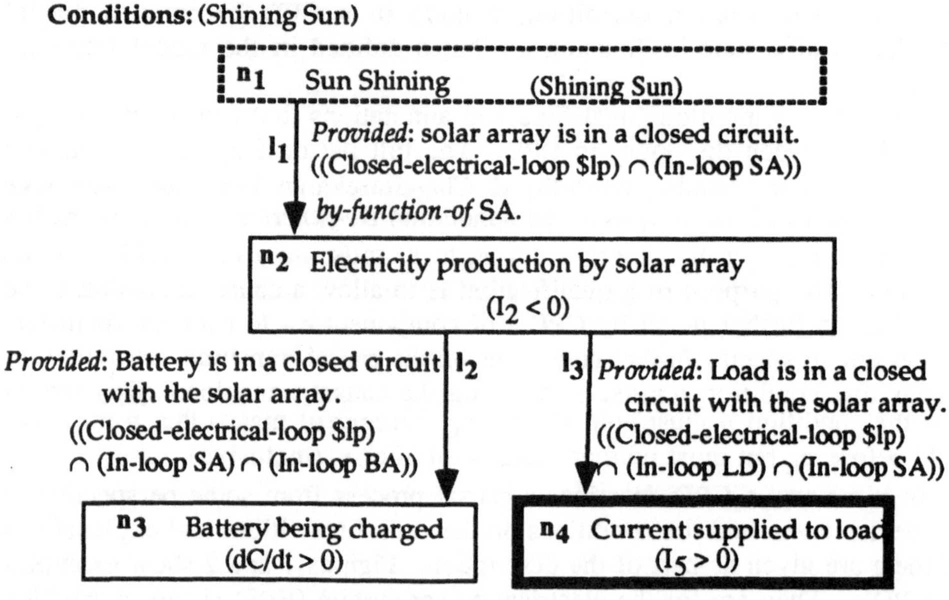

Figure 1: CPD_1 of EPS

Conditions: ~(Shining Sun) ∪ (Active Battery-over-charged)

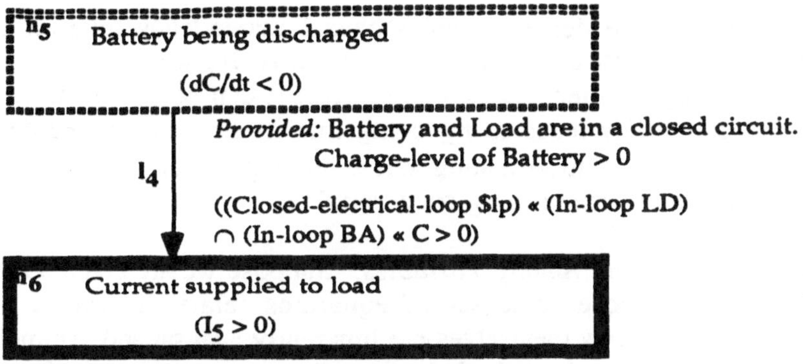

Figure 2: CPD 2 of EPS

Given a device description and initial conditions, we can generate a *Tr*. Suppose we also have an intended function for the device and associated CPD's. Intuitively, we would like to say that the device achieves the function in *Tr* if (1) the functional goal is achieved, (2) there are states in *Tr* matching all the nodes in the CPD in the specified temporal order, and (3) for each causal link in the CPD, there is a causal path in *Tr* that connects the cause to the effect. In order to make these conditions more precise, we must define the concept of a causal path in a trajectory. Then, we will define what it means for a trajectory to match a CPD.

For the rest of this paper, we use the following notation: We will attach *[s]* to wff's, model fragments and variable to denote the following axioms:

p[s]: A wff *p* holds in the state *s*.

m[s]: The phenomenon represented by *m* is active in *s* .

v[s]: A wff that asserts the value of *v* holds in *s*.

We will use notations such as <, >, ≤, and ≥ to express ordering among nodes in a CPD and states in a trajectory. We write "$n_1 < n_2$" where n_1 and n_2 are nodes in a CPD to indicate that n_1 is strictly causally upstream of n_2. For states s_1 and s_2 in a trajectory, "$s_1 < s_2$" means that s_1 strictly precedes s_2 in time. Note that the ordering is partial for nodes because a node can have multiple incoming and outgoing nodes. Ordering is total for states because a trajectory is a linear sequence of states.

Function

Type$_F$: *ToMaintain* *P$_F$*: (Powered Load) *C$_F$* : T

Dev$_F$: EPS *G$_F$*: CPD_1, CPD_2

Figure 3: The function of EPS

2.1 CAUSAL DEPENDENCY IN A TRAJECTORY

We now present the definition of a causal dependency relation between axioms $p1$ and $p2$ in a trajectory, *Tr*. Intuitively, we say $p2$ is causally dependent on $p1$, written "$p1 \Rightarrow_c p2$", when it can be shown in *Tr* that $p1$ being true eventually leads to $p2$ being true in *Tr*. Before we define the causal dependency relation among wff's more precisely, we introduce the notion of *causal ordering* among state variable.

2.1.1 *Causal Ordering Among Variables.* Suppose we are given a system of variables, and suppose some set of equations relate the values of these variables. Equations by themselves are inherently acausal and symmetric, but even when people represent the behavior of a system by a set of equations, they often perceive directed causal relations among variables. Causal ordering theory (Iwasaki & Simon 86) is used to reveal causal dependencies among variables in a set of equations and produce a graph structure that encodes these relations. In order to apply the procedure, one must have a set of independent equations, each of which represents a conceptually distinct mechanism in the situation. One must also know the variables which are externally controlled. Such variables are called *exogenous* variables.

Given a set of N equations which satisfy these requirements, the first step of the causal ordering procedure is to isolate all the subsets of variables whose values can be determined independently of the remaining variables. Such a subset of variables can be found by identifying a set of n equations which contains exactly n variables but which itself does not include a proper subset containing the same number of equations as variables. Such subset is called a *minimal complete subset.* The variables in any minimal complete subset are the "uncaused causes" of the system, and they are causally independent of other variables. Each exogenous variable constitutes a minimal complete subset.

Next, the equations in all minimal complete subsets are removed from the original set of equations and their variables are also removed from the remaining equations, producing a reduced set of $N - m$ equations in $N - m$ variables, where m is the total number of equations (and variables) in all the minimal complete subsets. Then, a new independent subset of variables is determined in the reduced set. This process repeats until the set can no longer be reduced. For each equation in the original set, the variable that was reduced last is said to be *causally dependent upon* all the other variables in the equation, and a directed graph can be generated to depict the causal dependency structure of the entire set, with nodes representing variables and links representing causal dependency relations among them. Also, for each variable v in a minimal complete subset, we define $D(v)$ to be the set of all equations in the set. In other words, $D(v)$ is the set of all the equations that directly determine the value of v.

We will write $v_1 \rightarrow_C v_2$ when v_2 is causally dependent upon v_1 according to the definition of causal ordering. When a minimal complete subset consists of more than one variable, the causal ordering procedure does not impose ordering among them since such a situation indicates the existence of a feedback loop among them. In such a case, we say the variables are *interdependent* and write $v_1 \leftrightarrow_C v_2$.

Even though the above description is given in terms of variables and equations, which imply domains of continuous variables and function, the concept applies to domains of continuous as well as discrete variables. The "equations" in the case of discrete variables can be any axiom that can be used to determine the value of one variable depending on other variables, as long as such an axiom represents some conceptual mechanism in the situation. For example, the control of a sprinkler system that turns on between 1 and 2 am every day can be represented by an axiom,

$1 \leq time \leq 2 \Rightarrow$ (On Sprinkler).

2.1.2 *Causal Dependency Relations.* Given the causal ordering procedure we can now proceed to defining causal dependencies among wff's.

Definition 5. The causal dependency relation, denoted $a_i \Rightarrow_C a_j$, is defined between two wff's, a_i and a_j, in the descriptions of states in a trajectory Tr. We write "$a_i \Rightarrow_C a_j$" and say "a_j *depends on* a_i" or "a_i *causes* a_j." The relation $\Rightarrow_C$ is transitive.

The following conditions specify when a wff can be said to be causally dependent on another in Tr:

(a) If $p[s_0], p[s_1] \ldots p[s]$ for all states from s_0 up to s, (in other words, p was part of the initial conditions and never changed), we say that $p[s]$ is exogenous, and write $\phi \Rightarrow_C p[s]$.

(b) If there exists a state $s_j < s$ such that, $\sim p[s_j]$, and $p[s_{j+1}]$, where s_{j+1} is the immediate successor of s_j in Tr, and there exists $m[s_j]$, such that $p \in R_m$, and $p[s_i]$ for all s_i between s_{j+1} and s inclusive (in other words, p becomes true at some point before s as a consequence of the phenomenon represented by m.), we say $m[s_j] \Rightarrow_C p[s]$.

(c) For each $p \in P_m$, we say $p[s] \Rightarrow_C m[s]$. In other words, for each phenomenon active in s, we say that the phenomenon being active is dependent on its precondition being satisfied.

(d) If $v_1 \rightarrow_C v_2$ according to the definition in Section 2.1.1, we say $v_1[s] \Rightarrow_C v_2[s]$.

(e) For each equation e in $D(v)$ for a variable v and a phenomenon m such that $e \in E_m$, we say $m[s] \Rightarrow_C v2[s]$. In other words, we say v

depends on the phenomenon giving rise to the causal relation between v and whatever other variables v depends on.

(f) $v2[s] \Rightarrow_C v1[s]$ and $v1[s] \Rightarrow_C v2[s]$ if $v2 \leftrightarrow_C v1$ in the causal ordering in s.

(g) $v'[s1] \Rightarrow_C v[s_2]$, where s_2 is the state immediately following $s1$, and v' the time-derivative of v in $s1$.

2.2 When is a function achieved?

We listed in Definition 4 different types of functions of devices and components. In this subsection, we spell out the conditions under which a device is said to achieve each type of function in a trajectory.

Definition 6: Let s_Z denote the final state in *Tr*, and *Dev* denote either Dev_F or one of its components. A function F is said to be achieved in a trajectory *Tr* in any of the following cases depending on $Type_F$. In all the cases C_F must hold in the initial state of *Tr*. In the following,

Case 1: When $Type_F$ = *ToMake*, F is achieved by *Tr* if

1.1 P_F, the functional goal, holds in the final state s_Z. We denote this by $P_F[s_Z]$. And,

1.2 There is some device variable v, and some state s in *Tr*, such that $v(s) \Rightarrow_C P_F[s_Z]$ (*i.e.* this fact causally depends on the operation of the device).

Case 2: When $Type_F$ = *ToMaintain*, F is achieved in *Tr* if in all states s_i in *Tr* , the following is true: For some s_j such that $s_j \leq s_i$ in *Tr*,

2.1 $P_F[s_i]$, and

2.2 There is some device variable v such that $v[s_j] \Rightarrow_C P_F[s_i]$.

Case 3: When $Type_F$ = *ToControl*, F is achieved in *Tr* if

3.1 $v_0[s_Z] = f(v_1[s_Z], \ldots v_n[s_Z])$ (*i.e.* the functional relation holds between the value of the controlled variable and the values of the controlling variables in the final state),

3.2 $v_i[s_Z] \Rightarrow_C v_0[s_Z]$ for $1 \leq i \leq n$ (*i.e.* the value of the controlled variable in the final state causally depends on the controlling variables), and

3.3 There is some device variable v and some state s in *Tr*, such that $v(s) \Rightarrow_C v_0[s_Z]$

Case 4: When $Type_F$ = *ToPrevent*, F is achieved in *Tr* if $\sim P_F[s]$ for any state s in *Tr* (*i.e.* F is achieved in *Tr* if the functional goal of F does not hold in any state). We make the closed world assumption that $\sim p$ unless p is explicitly known to hold.

Definition 7: A trajectory *Tr* is said to match a CPD if there is a mapping *st* from nodes in the CPD to the states in *Tr* that satisfies the following conditions:

1. for each node *n* in the CPD there is a state *st(n)* in *Tr* where *def(n)* holds, and
2. for any nodes n_1 and n_2 in the CPD, $st(n_1) \leq st(n_2)$ iff $n_1 < n_2$, and
3. for each causal link *l* from n_1 to n_2, there is a causal path $def(n_1)[st(n_1)] \Rightarrow_c def(n_2)[st(n_2)]$ in *Tr*, where *def(n)[st(n)]* denotes that *def(n)* holds in the state *st(n)*. Furthermore, if *l* has an attached qualifier, *Provided(p)*, *p* must hold for all states between $st(n_1)$ and $st(n_2)$ inclusive. If *l* has an attached annotation, *By-function-of*, which points to a component *o*, there must be a causal path $o[s] \Rightarrow_c def(n_2)[st(n_2)]$ for some state *s* such that $st(n_1) \leq s \leq st(n_2)$.

Clause 1 of the above definition ensures that for each node in the CPD, there is a state in *Tr* that matches it. Clause 2 makes sure that the temporal ordering of causes and effects in the CPD is preserved in the temporal ordering of their corresponding states in *Tr*. Finally, Clause 3 ensures that the causal paths exist in *Tr* that correspond to the causal links in the CPD.

Armed with the Definitions 1 through 7, we are now ready to state precisely what we mean by verification that a predicted behavior achieves the expected behavior.

Definition 8: We say that a trajectory *Tr* of a device achieves the expected behavior with respect to a function *F* when the following conditions are met. Tr_i denotes the subsequence of *Tr* from the initial state up to and including state s_i:

1. *F* is satisfied in *Tr* according to Definition 6, and
2. F is achieved in the expected manner, which is verified as
   - Case 1: if $Type_F$ is not *ToMaintain*, *Tr* matches one of the CPD's of *F* according to Definition 7,
   - Case 2: if $Type_F$ is *ToMaintain*, for each state *s* in *Tr*, a match between Tr_i and one of the CPD's exists such that $s_i = st(N_{fin})$ for the final node N_{fin} of the CPD.

Clause 1 of this definition makes sure that the function is achieved in the trajectory. Clauses 2 ensures that the function is achieved in the way the designer intended. We must distinguish the cases where the type of the function is *ToMaintain* and others, because if the function is to make or prevent some condition, we need only to show that the condition is brought about (or prevented) in the intended manner. However, if the function is to maintain some condition throughout the trajectory, we must show that the condition is in fact brought about in the intended manner for every state.

3. Example: EPS behavior

In this section, we demonstrate behavior verification with an example of the electrical power system (EPS) aboard a satellite orbiting the earth (LMSC 1985). A simplified schematic diagram of the EPS is shown in Figure 4. The components of the EPS are a solar array (SA in Figure 4), a rechargeable nickel-cadmium battery (BA), a load representing all the electrical loads on board (LD), a relay (K1), and a device called a charge current controller (CCC) for controlling the relay. The solar array generates electricity when the satellite is in the sun, supplying power to the load and recharging the battery. The battery is a constant voltage source when it is charged between 6 and 30 ampere-hours. When the charge level is below 6 or above 30 ampere-hours, the electromagnetic force produced increases or decreases as it is charged or discharged.

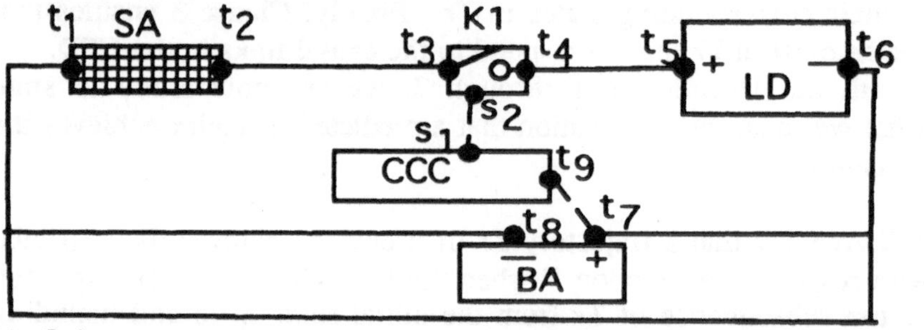

SA: Solar array
LD: Electrical load on board
BA: Rechargeable battery
CCC: Charge current controller
t_1 through t_9 : Electrical terminals
s_1, s_2: Signal terminals
K1 : Relay
—— Signal connection
- - - - - Sensor data connection
—— Electrical connection

Figure 4: Electrical Power System

Since the battery can be damaged when it is charged beyond its capacity, the charge current controller opens the relays when the voltage reaches 33.8 volts to prevent the battery from being over-charged. The charge current controller (CCC) has a sensor connected to the the positive terminal (t_7) of the battery to sense the voltage. When it reaches 33.8 volts during a sun-light period, it turns on the relay K1. When the relay is energized, it opens and breaks the electrical connection, preventing further charging of the battery and switching the current source for the load from the solar array to the battery. When the relay is open or when an eclipse period begins, the charge-level starts to decrease. When the charge-level decreases to 6.0, the voltage will start to decrease. At 31.0 volts the CCC turns K1 off to close if it has been opened.

The main purpose of the EPS is to supply electricity to the load constantly. This function of the EPS was shown in Figure 3, and the CPD's for achieving the function were shown in Figures 1 and 2.

2.1 EPS MODEL FRAGMENTS

The structure of EPS as shown in Figure 4 is given to DME as a collection of model fragments each representing a components and their connections. These model fragments represent the static aspect of the situation, and they are always active. In addition, there are model fragments representing various behavioral aspects of the components. They are activated/deactivated during the simulation according to the state of the world. Some of them are shown below with their conditions *(P_m)*, behavior constraints *(E_m)*, and results *(R_m)*. Voltage and current are measured at terminals. The sign convention for current is that the current at a terminal is positive into the component owning the terminal. For the rest of the example, we will use the abbreviations shown below in parentheses to refer to the model fragments.

Model fragments concerning behaviors of battery

Battery-normal-operating-range (BN)

P_m: (Rechargeable-battery \$b) ∩ 6.0 amp-hours < (Charge-level \$b) < 30.0 amp-hours

E_m: (EMF \$b) = 33.0 volts

Battery-over-charged (BO)

P_m: (Rechargeable-battery \$b) ∩ (Charge-level \$b) > 30.0 amp-hours

E_m: (EMF \$b) = M^+(Charge-level \$b)

Battery-under-charged (BU)

P_m: (Rechargeable-battery \$b) ∩ (Charge-level \$b) < 6.0 amp-hours

E_m: (EMF \$b) = M^+(Charge-level \$b)

Behaviors of the solar array

Solar-array-generating (SG)

P_m: (Sun \$s) ∩ (Shining \$s) ∩ (Solar-array \$a) ∩ (In-closed-circuit \$a)

E_m: (Current-thru-terminal (Plus-terminal \$a)) < 0

Solar-array-in-eclipse (SE)

P_m: (Sun \$s) ∩ ~(Shining \$s) ∩ (Solar-array \$a)

E_m: (Current-thru-terminal (Plus-terminal \$a)) = 0

Solar-array-in-open-circuit (SO)
P_m: (Sun $s) ∩ (Shining $s) ∩ (Solar-array $a) ∩ ~(In-closed-circuit $a)
E_m: (Current-thru-terminal (Plus-terminal $a)) = 0

Behaviors of the charge current controller
Turn-k1-on (ON)
P_m: (Charge-current-controller $ccc) ∩ (Signal (Signal-terminal-1 $ccc)) = off ∩ (Voltage-at-terminal (Voltage-sensing-terminal $ccc)) ≥ 33.8 volts
R_m: (Signal (Signal-terminal-1 $ccc)) = on
Turn-k1-off (OFF)
P_m: (Charge-current-controller $ccc) ∩ (Signal (Signal-terminal-1 $ccc)) = on ∩ (Voltage-at-terminal (Voltage-sensing-terminal $ccc)) ≤ 31.0 volts
R_m: (Signal (Signal-terminal-1 $ccc)) = off

Behaviors of the relay
Relay-closed (RC)
P_m: (Relay $r) ∩ (Relay-closed-p $r)
E_m: (voltage-at-terminal (electrical-terminal-one $r)) =
(voltage-at-terminal (electrical-terminal-two $r))
(current-thru-terminal (electrical-terminal-one $r)) =
- (current-thru-terminal (electrical-terminal-two $r))
Relay-open (RO)
P_m: (Relay $r) ∩ ~ (Relay-closed-p $r)
E_m: (current-thru-terminal (electrical-terminal-one $r)) =
- (current-thru-terminal (electrical-terminal-two $r))
Relay-closing (CL)
P_m: (Relay $r) ∩ ~(Relay-closed-p $r) ∩ (Signal (Signal-terminal $r) = off
R_m: (Relay-closed-p $r)
Relay-opening (OP)
P_m: (Relay $r) ∩ (Relay-closed-p $r) ∩ (Signal (Signal-terminal $r) = on
R_m: ~(Relay-closed-p $r)

In addition, there are three model fragments used to model the sun, **Sun-rise** (RISE), **Sun-set** (ST) and **Reset-orbit-time** (RST). As it takes approximately 100 minutes for the satellite to go around the earth once, Orbit-time is a 100-minute clock, which is reset to 0 when it reaches 100. We model the sun as rising and setting when Orbit-time = 0 and 60 respectively instead of modeling the satellite as revolving around the earth.

We will use the following notations for quantities.

I_i	Current through terminal t_i into the component owning t_i
V_i	Voltage measured at terminal t_i
C	The charge level of the battery
R_{ld}	The resistance of the load
R_{ba}	The internal resistance of the battery
EMF	The electromotive force of the battery
Time	Orbit-time

Each node in the behaviors of EPS can now be defined precisely in terms of these model fragments and their attributes. Using this vocabulary, the function of EPS shown in Figure 3 translates to the following condition:

EPS Function: $I_5[s] > 0$ for all $s \in Tr$.

The precise definition of each node in CPD_1 and CPD_2 using these model fragments and their attributes are shown in parentheses in Figures 1 and 2.

3.1 SIMULATED BEHAVIOR.

We simulated the behavior of EPS on DME. In the initial state, *Time* is between 0 and 60, sun is up, the relay is closed, and the charge level is between 6 and 30.0 amp-hours. From this initial state, a number of behaviors are possible. In qualitative simulation mode, because of the ambiguity of qualitative simulation, there are multiple possible trajectories. Tables 1 presents one of the possible trajectories of EPS generated by DME. The variable values are shown with their magnitude and the sign of their derivative. In a state where the derivative is undefined, the sign is shown as x. The right most column of the table shows the set of active models in each state. The set of all active model fragments in each state is actually much larger, but since most of them represent components, terminals, junctions, etc. and are active throughout the simulation, we show only the ones that change their activation status at some point. A model fragment that becomes activated in a given state is shown in bold. A × over a model fragment indicates that it becomes deactivated in the given state.

3.1.1 *Trajectory* Tr_1. The behavior we consider is summarized in Table 1. In the initial state s_0, since the sun is up and the relays are closed, the charge-level is increasing. When it eventually reaches 30 amp-hours (s_1), the battery enters the over-charged state and the voltage level starts to rise. When it reaches 33.8 (s_3), CCC changes the signal to on (s_4), and K1 opens (s_5). At this point, the solar array stops generating current, and the battery starts to discharge. The charge level and the voltage starts to decrease. Soon, the sun sets (s_7). As the charge level continues to decrease, the battery returns to the normal operating range (s_9) It eventually becomes under-charged (s_{11}), and the voltage starts to decrease below 33.0 volts. When it reaches 31.0 volts (s_{12}), the signal to K1 is turned off (s_{13}), and K1 closes (s_{14}). Soon, the sun rises again (s_{17}), and charging resumes.

Following are equations that are generated by DME during the simulation. On the right of each equation, we indicate the model fragment that gives rise to the equation. "Junction" indicates that the equation was generated as a behavior constraint of electrical junction model fragments (not shown in Figure 4 to simplify the figure.) Note that e_5' is applicable when e_5 is not, and vice versa.

$e_1: I_2 = -I_1$	*SG*	$e_7: V_5 = R_{ld} I_5$	*LD*
$e_2: I_2 + I_3 = 0$	*junction*	$e_8: V_7 = V_5$	*junction*
$e_3: I_3 + I_4 = 0$	*RC*	$e_9: V_5 = V_4$	*junction*
$e_4: I_4 + I_5 + I_7 = 0$	*junction*	$e_{10}: V_4 = V_3$	*RC*
$e_5: EMF = M+(C)$	*BO or BU*	$e_{11}: V_3 = V_2$	*junction*
$e_5': EMF = 33.0$	*BN*	$e_{12}: dC/dt = I_7$	*BA*
$e_6: V_7 = EMF + R_{ba} I_7$	*BA*		

R_{ld} and R_{ba} are exogenous.

The causal ordering among variables in the states where the relay is closed is shown in Figure 5. The causal ordering when the relay is open is mostly similar except that the links from I_2 to I_3 and from I_3 to I_4 are missing[1]. In the figure, the variable at the head of an arrow is causally dependent on the variable at the tail. The causal links are labeled with equations that are responsible for the link. The link labeled *i* is an integration link to a variable from its derivative. Variables I_5, V_5, I_7, and V_7 are inter-dependent. The equations responsible for their inter-dependence are e_4, e_6, e_7 and e_8.

We prove that this trajectory satisfies the expected behavior of EPS. Due to the limitation of space, we omit the proof that the qualifiers, *Provided*, and *By-function-of*, on causal links are satisfied. The proof of these conditions is straightforward.

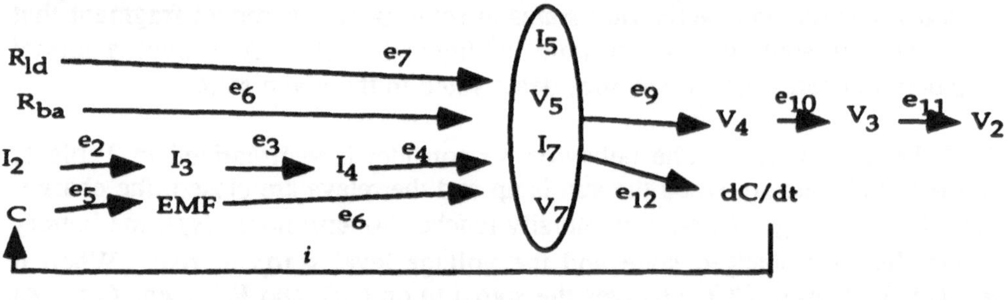

Figure 5: Causal ordering when K1 is closed

1 When the battery is in its normal operating range, the causal ordering is slightly different since e_5' instead of e_5 is applicable, eliminating the link from *C* to *EMF*.

Table 1: Trajectory, Tr_1, of EPS

state	Sun	Relay	Signal	I_2	I_5	I_7	V_7	C	Time	Active Models
s_0	shine	close	off	- std	+ std	+ std	33.0 std	6-30 inc	0-60 inc	BN, SG, RC
s_1								30 inc		~~BN~~, SG, RC
s_2							33.0-33.8 inc	30-inf inc		BO, SG, RC
s_3							33.8 x			BO, SG, RC, ON
s_4			on							BO, SG, RC, ~~ON~~, OP
s_5		open		0 stc		- std	33.0-33.8 dec	30-inf dec		BO, ~~SG~~, ~~RC~~, SO, RO
s_6									60 inc	BO, ST , SO, RO
s_7	~shine								60-100 inc	BO, ~~ST~~, SE, ~~SO~~, RO
s_8							33.0 dec	30 dec		~~BO~~, SE, RO
s_9							33.0 std	6-30 dec		BN, SE, RO
s_{10}								6 dec		~~BN~~, SE, RO
s_{11}							31.0-33.0 dec	0-6 dec		BU, SE, RO
s_{12}							31.0 dec			BU, SE, RO, OFF,
s_{13}			off				0-31.0 dec			BU, SE, RO, ~~OFF~~, CL
s_{14}		close								BU, SE, ~~RO~~, ~~CL~~, RC
s_{15}									100 x	BU, SE, RST , RC
s_{16}									0 x	BU, SE, ~~RST~~, RISE , RC
s_{17}	shine			- std		+ std	0-31.0 inc	0-6 inc	0 inc	BU, ~~SE~~, ~~RISE~~, SG , RC
s_{18}									0-60 inc	BU, SG, RC
s_{19}							31.0 inc			BU, SG, RC
s_{20}							31.0-33.0 inc			BU, SG, RC
s_{21}							33.0 inc	6 inc		BU, ~~SG~~, RC
s_{22}							33.0 std	6-30 inc		BN, SG, RC

(1) That the function of EPS is satisfied by behavior Tr_1 is clear from Table 1 since $I_5 > 0$ in all states. That EPS or its components takes part in bringing about the fulfillment of the goal is subsumed by the proof in part (2) and (3).

(2) The following proof applies for s being one of states s_0 to s_4, and s_{17} through s_{22}, where the condition of CPD_1, *(Shining Sun)*, holds.

Let $st(n_1) = st(n_2) = st(n_3) = st(n_4) = s$. It follows that $st(n_1) \leq st(n_2) \leq st(s_3) \leq st(s_4)$. Since $(Shining\ Sun)[s] \cap (< I_2\ \ 0)[s] \cap (> dC/dt\ \ 0)[s] \cap (> I_5\ \ 0)[s]$, we have $def(n)[s]$ for all n in CPD_1.

Proof of l_1: $(Shining\ Sun)[s] \Rightarrow_c I_2[s]$.

$(Shining\ Sun)[s] \Rightarrow_c SG[s]$ because of Definition 5.c.

$SG[s] \Rightarrow_c I_2[s]$ because of Definition 5.e.

It follows that $(Shining\ Sun)[s] \Rightarrow_c I_2[s]$.

Proof of l_2: $I_2[s] \Rightarrow_c dC/dt[s]$

$I_2 \rightarrow_c dC/dt$ in s as shown in Figure 5.

It follows that $I_2[s] \Rightarrow_c dC/dt[s]$ because of Definition 5.d.

Proof of l_3: $I_2[s] \Rightarrow_c I_5[s]$

$I_2 \rightarrow_c I_5$ in s as shown in Figure 5.

It follows that $I_2[s] \Rightarrow_c I_5$ because of Definition 5.d.

Therefore, CPD_1 of EPS is achieved in states s_0, s_1, and s_{17} through s_{23} of Tr_1.

(3) The following is true for s being one of states s_5 through s_{16}, where the condition of CPD_2, $\sim(Shining\ Sun) \cup (Active\ BO)$, holds.

Let $st(n_5) = st(n_6) = s$. It follows that $st(n_5) \leq st(n_6)$.

Since $(< dC/dt\ \ 0)[s] \cap (> I_5\ \ 0)[s]$, we have $def(n)[s]$ for all nodes n in CPD_2.

Proof of l_4: $I_5[s] \Rightarrow_c dC/dt[s] \cap dC/dt[s] \Rightarrow_c I_5[s]$

$I_5 \rightarrow_c dC/dt$ in s as shown in Figure 5. It follows that $I_5[s] \Rightarrow_c dC/dt[s]$ because of Definition 5.d.

Therefore, CPD_2 of EPS is achieved in states s_2 through s_{16} of Tr_1.

3. Discussion

In this paper, we formalized a number of notions that were relatively informally specified in the Functional Representation language and defined matching between an expected behavior represented in Functional Representation and a predicted behavior. We also demonstrated its use in deciding whether a particular trajectory of a device achieves an expected behavior.

Our primary goal is to use the knowledge of functions and expected behavior for the purpose of design verification. It is important that the definition of behavior verification we have presented is not biased towards any particular perspective about what are more important than others as a causal factor. In other words, it does not require that the function or the expected behavior be described from a particular point of view. This definition of verification of a predicted behavior with respect to a function and an expected behavior is inclusive enough to allow a trajectory to match many representations of functions or expected behaviors. Likewise, there can be any number of trajectories that can be shown to match a given expected behavior as there can be any number of designs that accomplish the same functionality. Thus, the mapping between trajectories and an expected behavior is many to many. However, if the goal is to verify that a predicted behavior achieves a given expected behavior, this non-uniqueness of a match is not a problem. Our definition does not establish that the given expected behavior is the only correct causal story for a given trajectory, nor that the trajectory is the only correct way to achieve the function. However, the definition does establish that a given design achieves the function in an expected manner, which is what is needed for our purpose of design verification.

Our next step is to implement a program that takes a functional representation and a trajectory and automatically proves whether or not the expected behavior is realized in the trajectory.

3.1 RELATED WORK

Bradshaw and Young (1991) and Franke (1991) have also proposed representations of the knowledge of a purpose and their use in design. They represent the intended function in a manner that is similar to the way functions are represented in Functional Representation. Bradshaw and Young built a system called Doris, which uses knowledge of purpose for evaluating behaviors generated by qualitative simulation as well as for diagnosis and explanation.

The focus of Franke's work on representing functions is slightly different from ours or Bradshaw and Young in that he represents the purpose of a design modification and not that of a whole device. He developed a representation scheme, called TED, in which he expresses the purpose for making a modification δ in a structure using the same function types as those in Functional Representation. Thus, in order to prove that a function is achieved by a modification δ, he must compare the behavior of structure M and that of M', which is M with the modification δ. Another important characteristic of TED's representation of functions is that it can be a sequence (not necessarily a linear) of partial descriptions. The representation of a function in TED typically says "δ guarantees σ," where σ is a sequence, called scenario, of partial descriptions. The sequence of partial descriptions is matched against states in a sequence of qualitative states generated by QSIM.

The most important difference between our work and the works by Franke's or by Bradshaw and Young's is that we take not only the functions but also the causal interactions into account in evaluating behavior. We feel that it is important to test whether it is in fact the causal processes intended by the designer that are responsible for bringing about the achievement of the functional goal, since the satisfaction of the functional goal does not necessarily indicate that the design is functioning as intended. We believe that evaluating a trajectory with respect to the causal process as well as the function allows one to uncover hidden flaws in a design which may otherwise go undetected.

References

Bradshaw, J. A. and Young, R. M.: 1991, Evaluating Design Using Knowledge of Purpose and Knowledge of Structure. *IEEE Expert*, April.

Crawford, J., Farquhar, A., and Kuipers B.: 1990, QPC: A Compiler from Physical Models into Qualitative Differential Equations. *Proceedings of the Eighth National Conference on Artificial Intelligence.*

Falkenhainer, B. and Forbus, K.: 1988, Setting up Large-Scale Qualitative Models. *Proceedings of the Seventh National Conference on Artificial Intelligence.*

Fikes, R., Gruber, T., Iwasaki, Y., Levy, A. and Nayak, P.: 1991, How Things Work Project Overview. Technical Report, KSL 91-70, Knowledge Systems Laboratory, Stanford University.

Franke, D. W.: 1991, Deriving and Using Descriptions of Purpose. *IEEE Expert*, April.

Iwasaki, Y. and Low, C. M.: 1991, Model Generation and Simulation of Device Behavior with Continuous and Discrete Changes. Technical Report KSL-91-69, Knowledge Systems Laboratory, Stanford University.

Iwasaki, Y. and Simon, H.A.:1986, Causality in Device Behavior. *Artificial Intelligence* 29.

Keuneke, A.: 1991, Device Representation: The Significance of Functional Knowledge. *IEEE Expert*, April.

Kuipers, B.: 1986, Qualitative Simulation. *Artificial Intelligence* 29.

LMSC: 1984, *Support Systems Module System Procedure for Pointing and Control Subsystem (SE-23, Vol. V)*, Lockheed Missiles and Space Company document # D889545A.

Sembugamoorthy, V. and Chandrasekaran, B.: 1986, Functional Representation of Devices and Compilation of Diagnostic Problem-Solving Systems, *in* Kolodner, J.L. and Riesbeck, C.K. (eds), *Experience, Memory, and Reasoning*, Lawrence Erlbaum Associates, Hillsdale, NJ.

ISSUES IN INCREMENTAL ANALYSIS OF ASSEMBLIES FOR CONCURRENT DESIGN

J-C. TSAI, R. KONKAR and M. R. CUTKOSKY
Center for Design Research
Department of Mechanical Engineering
Building 02-530, Duena Street
Stanford University
Stanford, CA 94305-4026
USA

Abstract. Incremental analysis methods permit fast response when assessing the effects of changes made to an evolving design. The objective is to make the response fast enough for interactive use. The basic approach is to identify unaffected portions of graphs that represent the interactions among the parts of an assembly. Domain-specific concerns, such as the relative computational expense of identifying paths or loops in the graph, and the expense of propagating values or constraints along a path, determine the right balance to strike between storage of intermediate results versus recomputation. The methods are discussed in the context of two examples: tolerance propagation and kinematic analysis for assemblies. Despite differences in representations and computations for these applications, a number of analogous issues emerge which suggest the utility of a general package for incremental analysis of interacting components.

Nomenclature

n : number of nodes in a graph or a path.
n_a : number of operations between two nodes in a weakly-connected graph.
C_k^n : the binomial coeffecient.
T_{ideal}^i : the ith ideal (nominal) frame transformation in a path.
ΔT^i : the ith deviation transformation in a path.
T_{real}^i : the ith real frame transformation in a path.
$\mathbf{t}_i$: twistspace of joint i in a mechanism.
$\mathbf{w}_i$: wrenches exerted across joint i in a mechanism.
${}^G\mathbf{v}^P$: velocity of point P with respect to body G.
${}^G\omega^A$: angular velocity of body A with respect to body G.
${}^GT^A$: coordinate transformation matrix of body A with respect to body G.

1. Introduction

Although computer-aided tools for engineering analysis and planning abound, their application as an integral part of the design process has been limited by factors inherent in their design. Part of the problem is that such tools have been developed primarily by and for specialists. Consequently, a designer may need to consult a stress analyst to obtain, or interpret the results from, a finite-element solution for a preliminary design. Other limiting factors include the level of detail typically needed for input specifications and the amount of computation required to obtain results. These factors conspire to make engineering analysis tools difficult to use in early design stages where the designer may wish to explore numerous "what if?"

J. S. Gero (ed.), Artificial Intelligence in Design '92, 617–635.

scenarios. Twenty minutes is not unreasonable to wait for a detailed analysis, nor is it a long time to wait for a machining or assembly plan for a part about to be released for production. However, fast response is desired when, for example, evaluating each of several options for the placement of bearing seats on a machined part from the standpoints of stresses, tolerance accumulations and machining complexity.

Recently, a number of researchers have explored symbolic methods as a way of obtaining engineering analysis and planning tools that are better suited for use during early design stages (*e.g.*, Ulrich (1989), Finger and Rinderle (1989), Kannapan and Marshek (1989), Rinderle and Balasubramanium (1990), and Kusiak and Szczerbicki (1990)). Symbolic methods result in tools that work at a higher level of detail than numerical methods, and the results obtained with such tools are often easier to interpret than numerical results. Efficient methods have also been developed for solving the simultaneous sets of equalities and inequalities that result (*e.g.*, Ward (1989) and Serrano (1987)). The advent of powerful general-purpose symbolic processing programs has further encouraged the development of computer-aided tools based on symbolic methods. However, like their numerical counterparts, tools that exploit symbolic methods can be slow. Moreover, the computation time often grows exponentially with the complexity of the system being analyzed.

In this paper, we explore one approach to overcoming the time-consuming nature of engineering analysis tools that employ symbolic methods. As in our previous work on process planning (Kambhampati and Cutkosky, 1990) and fixture planning (Lee, Cutkosky and Kambhampati, 1991), the solution is to exploit incremental methods in which previous results are reused, where possible. In the case of a gradually evolving design, such a system is typically able to reuse most of the recently computed results, modifying them as necessary in response to local design changes and "what if" questions posed by the designer. Although the amount of previous information that can be reused will vary, depending on the nature of the changes wrought upon the design, the average time savings are considerable as compared to recomputing all results anew. In addition, we have found that designers often benefit from seeing the direct consequences of changes in a design made manifest.

We explore incremental approaches to symbolic analysis in the context of two examples from the Next-Cut (Cutkosky and Tenenbaum, 1991) concurrent design system: tolerance propagation and kinematic analysis for assemblies. In both cases, a graph of the assembly is developed and the incremental computation amounts to determining which paths through the nodes and arcs of the graph are affected by design changes. In both examples it is possible to exploit hierarchical representations to broaden the applicability of previous results. Even when the details of a computation are no longer correct, more abstract statements about the graph remain true.

In the following sections we first discuss a few elements of the graph representation that are common to applications such as those presented in this paper. In the following sections we present more detailed descriptions of the tolerance propagation and kinematic analysis, beginning with the representations used in each. In each case we also present an example of incremental computation in response to local design changes. We conclude with a discussion of the issues raised by the examples, including limitations of the adopted methods and proposed extensions or generalizations to cover additional domains.

2. Analysis of cyclic graphs of assemblies

In analyzing an assembly of parts, or a part with a collection of features, it is useful to construct a graph of nodes and arcs, as shown in Fig. 1(a)-(c). The nodes labeled A-G represent assembled bodies or features of parts and the arcs represent relationships such as tolerance constraints or kinematic constraints.

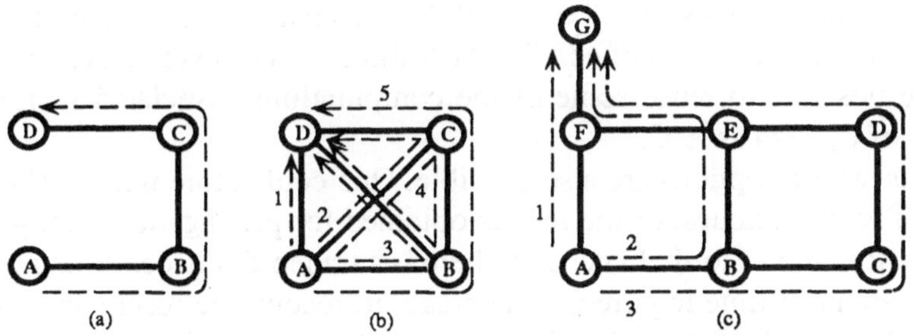

Fig. 1. Paths in a (a) weakly-connected graph (b) strongly-connected graph (c) typical graph

Fig. 1(a) shows a "weakly-connected" graph (Harary, 1969) in which there is only one path for propagating quantities among nodes. In the case of a toleranced part, this would represent a part with no redundant tolerances (perhaps incompletely toleranced); in the case of a mechanism, the graph would represent a serial chain of interacting bodies. For such cases, the number of operations to be performed in propagating quantities from A to D grows linearly with the number of nodes n. It is more common, however, for multiple paths to exist between nodes. In the extreme case, a graph is "strongly-connected" as shown in Fig. 1(b). In this case, the number of possible paths from A to D grows as

$$\sum_{k=0}^{n-2} C_k^{n-2} k!. \tag{1}$$

Since the symbolic propagation of constraints along paths can be time consuming, this exponential growth rate could quickly render the analysis of complex parts too slow for interactive use.

A more typical case is shown in Fig. 1(c). For this case there are evidently three paths from A to G. More generally, if we assume that there are m independent loops for a graph with n nodes then the number of possible paths is between 1 and 2^m. In this case, the number of operations is bounded by n_a in the best case and by

$$\sum_{k=0}^{m}(n_a + k) * C_k^m = (2n_a + m)2^{m-1} \tag{2}$$

in the worst case, where n_a is the number of operations from node A to G in a weakly-connected graph.

A glance at Fig. 1(c) also reveals opportunities for improvements in efficiency. To begin with, paths 1, 2, and 3 share the common subpath F-G and paths 2 and 3 share the common subpaths A-B and E-F-G. Therefore, there is obviously some improvement to be obtained by eliminating redundant computations.

Consider what happens if the design is modified so that the arc connecting nodes D and E is changed. If one is propagating values along paths 1, 2, and 3, then only those nodes "downstream" of D will be affected by the change. In this case, we require the reevaluation of conditions at three of the seven nodes. In addition, it may be possible to reuse some of the computations associated with subpaths upstream of C, such as A-C.

Various storage options are also possible. One could store intermediate results associated with all nodes, or the results obtained for propagations along all independent paths, and so forth. In general, there is a tradeoff between storage demands and the amount of time required, on average, to recompute results in response to a design change. The tradeoff will depend on the domain, including the types of values or constraints being propagated along paths in the graph and the degree to which typical examples in the domain tend toward being strongly-connected or weakly-connected, as in Fig. 1(a) or (b), respectively. In the following sections, we examine such issues in more detail in the context of two applications involving tolerance propagation and kinematic analysis for assemblies of components.

3. Tolerance representation and incremental propagation for assemblies

3.1. TOLERANCE REPRESENTATION AND PROPAGATION

Considerable work has been done in the area of tolerance representation and propagation. Early work developed the important concept of variational geometry (*e.g.*, Requicha (1984), Gossard, Zuffante and Sakurai (1988), and Turner (1987)). However, for propagating tolerances, it is also important to have a representation that is independent of any particular geometric modeling paradigm. This representation should accommodate both linear and angular errors resulting from tolerances and clearances and should address statistical distributions associated with manufacturing processes.

Futhermore, it frequently happens that redundant tolerances are specified among features, resulting in an "over-toleranced" design. While redundant tolerances do not necessarily indicate a design mistake, their occurrence should be detected.

The tolerances among geometric features in a design can be represented as a symbolic constraint graph (Bjorke, 1989; Bernstein and Preiss, 1989; Binford, Frants, Cutkosky and Tsai, 1990) or *Tolerance Network*, TN, defined as follows:

$$TN \equiv G(V,E)$$

where V is the set of features, represented as nodes in a TN; and E is the set of geometric constraints, represented as arcs in the network. The graph can also be extended to represent assemblies of parts with mating features (Lee and Gossard, 1985; Srikanth and Turner, 1990). The symbolic representation of the TN accommodates either interval or statistical tolerance representations and thus provides a uniform representation for design, manufacturing and inspection. Redundant tolerances appear as loops in the TN that are easy to detect by graph traversal. [1]

As an example, consider the simple linkage shown in Fig. 2 (a) which is composed of a base and an arm with an end-effector. A pin joint with clearance connects the base and the arm. From a functional standpoint, we are concerned with the accuracy of the end-effector locus with respect to the base frame. A feature decomposition of the linkage is shown in Fig. 2 (b) where bold lines with arrows show the path for error propagation from the base to the end-effector.

Notice that decomposition down to the feature level is necessary for representing tolerances and mating conditions. Moreover, as this design is over-toleranced, there are two possible paths for propagating errors.

Fig. 3 is a screendump of the tolerance reasoner in Next-Cut (Cutkosky and Tenenbaum, 1991), showing some of the detailed tolerance representations for the linkage in Fig. 2. Window II is a view of the solid model of the linkage. Window I is the TN of the linkage where the two labeled error propagation paths correspond to paths 1 and 2 in Fig. 2 (b). The TN shows that the designer has located the two holes on the arm (Arm-Pin-Hole and Arm-Effector-Hole) with respect to the common datum reference frame of the Arm-Block (constructed by plane-surfaces 12, 13 and 15) using position-tolerances 2 and 3 respectively for manufacturing purposes. However, the *relative* location of the two holes is of more direct importance from a functional standpoint. Therefore, another (redundant) tolerance (Position-tolerance-4) has been established for the location of Arm-Effector-Hole with respect to Arm-Pin-Hole. Consequently, there is a loop formed by paths 1 and 2 from Arm-Effector-Hole to Arm-Pin-Hole (see Fig. 3).

Window III of Fig. 3 shows the corresponding assembly tree and feature decomposition. The relative positions and orientations of features are represented by a frame transformation network based on the TN since each feature has an associated local frame in the product model. Notice that a datum reference frame, such as the two shown in window I of Fig. 3, is represented as a single node in the network though it is composed of multiple datum surfaces.

An ideal, or nominal, geometric relationship among features is represented by a

[1] However, the appearance of loops in a graph does not necessarily indicate over-tolerancing since the tolerances associated with the arcs may apply to orthogonal degrees of freedom.

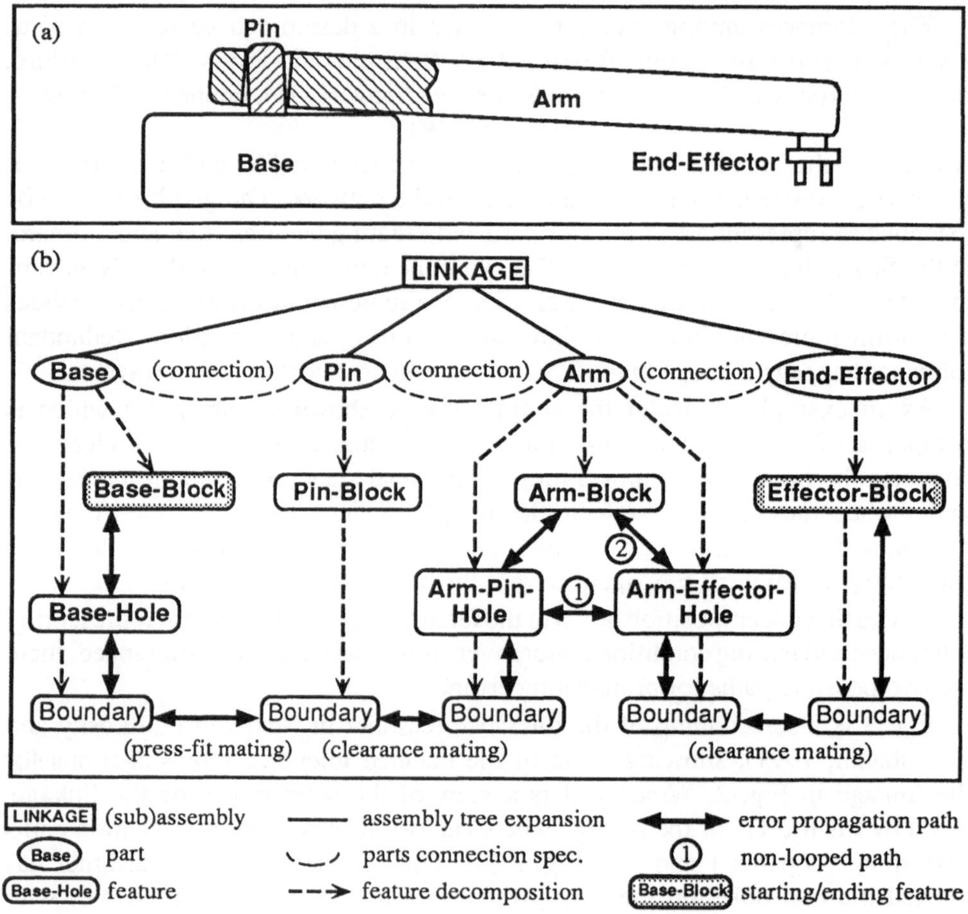

Fig. 2. (a) A linkage with clearance at the pin joint connecting the base and the arm. (b) Feature decomposition and error propagation paths for the linkage.

transformation, T^i_{ideal}. The deviation from the ideal relationship, due to a tolerance or a clearance, is expressed using an error matrix ΔT^i, so that the real transformation of an inter-node arc is given by

$$T^i_{real} = T^i_{ideal} * \Delta T^i \tag{3}$$

Given a non-looped tolerance network, the locus of the ending node is the product of the T^i_{real} in the network:

$$T_{real} = \prod_{i=1}^{n} T^i_{real} = T_{ideal} * \Delta T, \tag{4}$$

where $T_{ideal} = \prod_{i=1}^{n} T^i_{ideal}$ is the ideal transformation from the starting node to

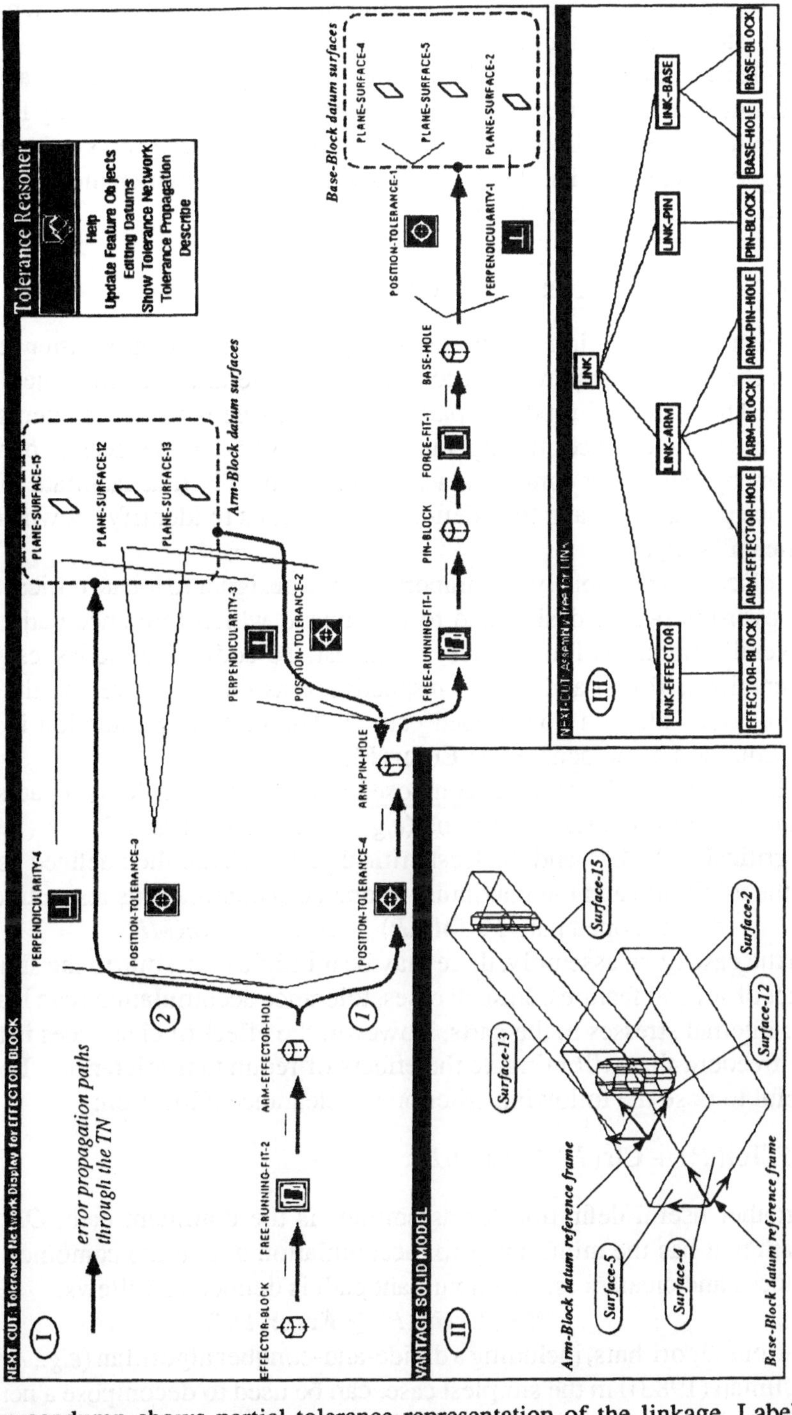

Fig. 3. A screendump shows partial tolerance representation of the linkage. Labels are manually added for explanation.

the ending node based on nominal geometric relationships and ΔT represents the accumulated errors along the TN.

While the computation complexity for tolerance propagation along a non-looped path is linear in the number of arcs, in the worst case the complexity for tolerance propagation through all possible paths in a looped network grows exponentially according to Equation 2. This leads to the development of an incremental propagation method to reduce the complexity.

3.2. Incremental Tolerance Propagation

Incremental computation can be used to provide significantly faster response to design changes as compared to recomputing the tolerance network anew with each change. The increase in speed is particularly valuable when a designer is adjusting tolerances and clearances throughout an assembly to obtain a desired performance. We start by observing some characteristics of the TN that facilitate incremental tolerance propagation and then define two functions to identify "downstream" and "upstream" subpaths.

As discussed previously, it is important for a designer to locate critical tolerances in an over-toleranced design and to understand which tolerance paths dominate the overall tolerances for features of interest. To address these issues, we define two functions, **Tol**(P) and **Clr**(P) respectively, as the tolerance and the clearance accumulations along a non-looped path P. The error accumulation is evaluated using Equation 4 and denoted as **Error**(P).

A looped network can be decomposed into several non-looped paths by using standard graph traversal algorithms (*e.g.*, Mott, Kandel and Baker (1986)). The most critical path, MP, and the least critical path, LP, are then defined as the paths with the minimum and the maximum tolerance accumulations respectively:

$$\forall \text{ non-looped path } P,\ \mathbf{Tol}(LP) \geq \mathbf{Tol}(P) \geq \mathbf{Tol}(MP).$$

In the case of an assembly, there may be multiple paths due to the simultaneous contact of several features. In such cases, tolerance accumulations can lead to binding or internal stresses in the parts. However, the effect of clearances is to provide some freedom that will mitigate the effects of redundant tolerances. Therefore, it is useful to pose the following criterion for clearance allocation:

$$\forall P, \mathbf{Tol}(P) + \mathbf{Clr}(P) \geq \mathbf{Tol}(LP). \tag{5}$$

Another useful definition for assemblies is the dominant path, DP, which is the path that has the minimum error accumulation due to the combined effects of tolerances and clearances. The dominant path is defined as follows:

$$\forall P,\ \mathbf{Error}(P) \geq \mathbf{Error}(DP).$$

Various algorithms, including a divide-and-conquer algorithm (*e.g.*, Aho, Hopcroft and Ullman (1983)) in the simplest case, can be used to decompose a network into non-looped paths and to identify the critical and dominant paths. The computation of errors between any two features can then be obtained by propagating tolerances and clearances along the dominant path among them. If the objective is to allocate clearances in an assembly to avoid binding, then Equation 5 can be used. The basic procedure is summarized below:

Divide-and-conquer algorithm for over-toleranced designs

Identify the starting and ending features for error propagation;
Find all non-looped propagation paths, P_i's;
for each P_i
    calculate **Tol**(P_i) and **Clr**(P_i);
endfor
if (*tolerance propagation*) **then**
    locate *DP* for error propagation;
elseif (*clearance allocation*)
        find *LP*
        for each path P_i
            when **Tol**(P_i) + **Clr**(P_i) < **Tol**(*LP*) relax clearances in P_i;
endif;

A non-looped path *P* is a directed serial chain of tolerances from a starting feature to a destination feature. *DownStream(k, P)* is defined as the subpath that begins at the feature *k* and terminates at the end node of *P*. Similarly, *UpStream(k, P)* is the subpath that starts at the first node of *P* and terminates at the feature *k*.

Clearly, if the feature *k* along *P* is modified, then only *DownStream(k, P)* needs to be reevaluated. This results in at most $O(n)$ complexity for each path with *n* nodes that includes the modified feature. Furthermore, we recall that we are concerned primarily with the tolerance accumulations along the most critical, least critical and dominant paths in an assembly. If the tolerance or clearance associated with the modified feature is relaxed and the feature is on neither *MP* nor *DP*, then it is not necessary to reevaluate the overall tolerances of the end node. Conversely, if the feature is on either *MP* or *DP* and the tolerances become tightened, then it is not necessary to determine whether a new path has become most critical or dominant and it is only necessary to reevaluate error accumulations along *DownStream(k, DP)*. Thus, the incremental computation can reduce the worst-case complexity from exponential to $O(n)$ as it involves only the downstream portions of the relevant paths. The basic incremental procedure is summarized in the Appendix.

Returning to the linkage example in Fig. 3, let us suppose that the designer adjusts the tolerances in the TN, as shown in Window I, to be sure that the endpoint errors are acceptable. In this network there are seven nodes, ten inter-node arcs, and two non-looped paths with six and nine arcs respectively. To evaluate the tolerance accumulation along each path requires $6 + 9 = 15$ computations if no intermediate results are reused. If the designer now tightens Position-tolerance-4, only 5 computations will be required to reevaluate the error locus, since there are 5 tolerances and matings involved in *DownStream(Position-Tolerance-4, P1)*, as opposed to 2 path-finding operations and 15 tolerance propagations for the entire network. A test of the incremental algorithm shows the average computation time for statistical error propagation, using the method of moments, drops by 64%. In the case of a more complex example, the proportional time savings would be greater.

3.3. Discussion

Before leaving this section, a couple of points are worth noting. First, the divide-and-conquer algorithm used to identify the critical paths is only the simplest approach; not the most efficient. Other algorithms, such as the modified Floyd algorithm for the all-pair-shortest-paths problem (Aho et al., 1983) and the A^* heuristic search (Nilsson, 1980), can be used to find the *DP, MP*, and *LP* in a network with a lower average complexity.

In addition, we note that the incremental propagation algorithm listed in the Appendix requires that we store the evaluated result for each intermediate step for possible reuse. For a complex design, this may involve a prohibitively large amount of storage. The trade-off between storage and efficiency is an issue discussed in Section 5.

4. Incremental Kinematic Analysis

4.1. Introduction

A mechanism can be analyzed using a graph in which the nodes represent rigid bodies and the arcs represent kinematic connections, or joint-types. The graph is undirected, connected, and is cyclic in the general case. The use of graphs for kinematic analysis has been addressed in a number of publications including Waldron (1966), Davies (1983), Baker (1980), and Mohamed and Duffy (1985).

The analysis method illustrated in this section emulates the approach an engineer solving kinematics problems on paper would take, identifying critical constraints and saving intermediate results for reuse.

As an example, consider the gripper shown in Fig. 4. The lower half of the figure shows the gripper assembly as represented in Next-Cut (Cutkosky and Tenenbaum, 1991). The boxed nodes in the graph represent the parts that are connected and the lines joining the nodes represent the connections. Only surface-contact or lower-pair kinematic joints are represented in Next-Cut. These joint types are: spherical, planar, cylindric, helical, prismatic, and revolute. In the gripper example, the fingers are attached to the base with revolute joints, and to a common cross-bar with spherical+prismatic joints. The pneumatic cylinder forms a cylindric joint, as the piston can both translate and rotate in the cylinder. (The complete kinematic description of the gripper is given in Fig. 7.)

4.2. Kinematic analysis at different levels of detail

Levels of detail can be exploited to increase the efficiency of the analysis. At the highest level, questions about the number of degrees of freedom can be established

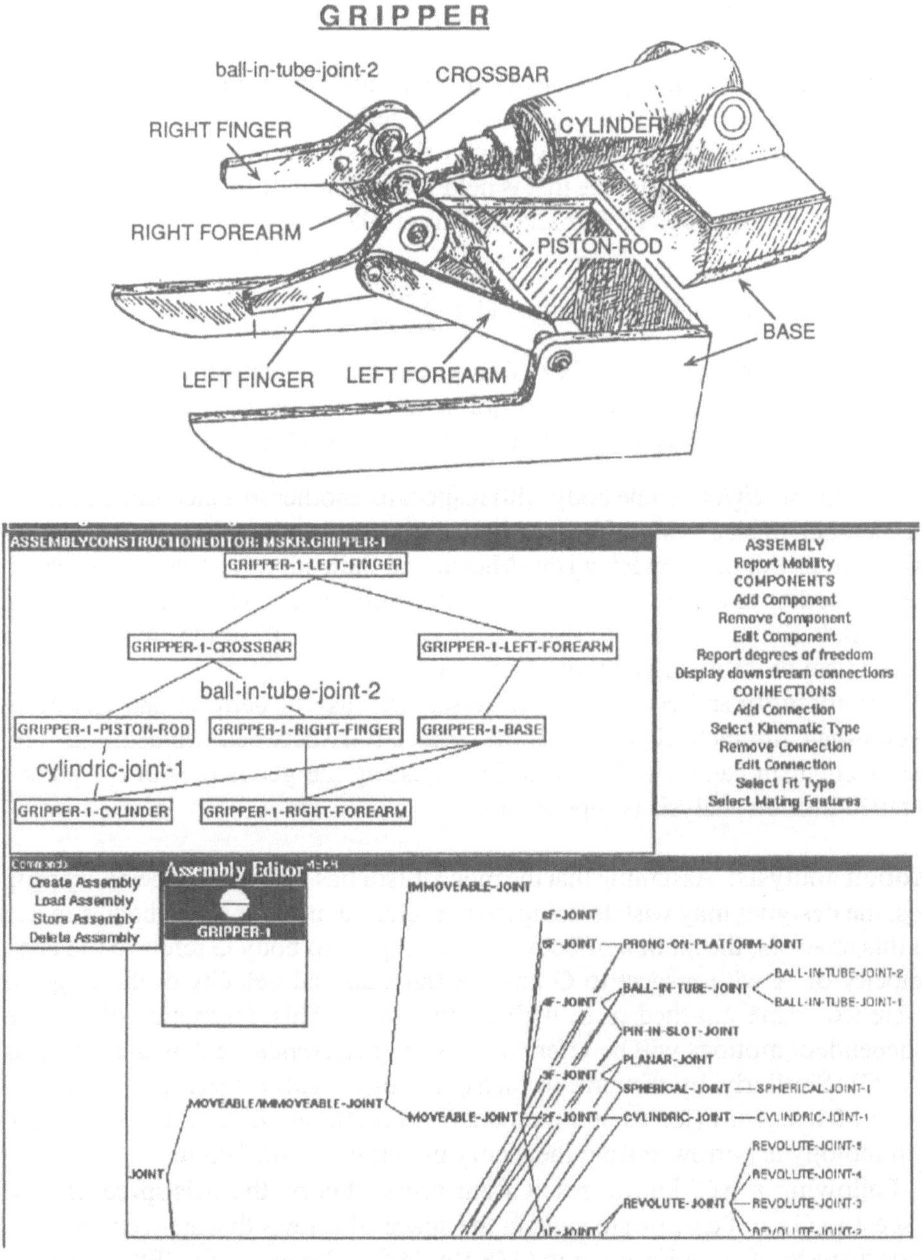

Fig. 4. Representation of assemblies in the Next-Cut system.

just by traversing the graph and keeping track of joint types. With more information, (*e.g.*, link lengths and joint orientations) one can compute the linear and angular velocities of any link. With still more information (*e.g.*, initial conditions) one can do a simulation.

The analyses at the highest level of detail include mobility and connectivity calculations. The *mobility* of a mechanism is defined as the number of independent parameters required to completely specify the positions of all the members of that mechanism with respect to one that is held fixed. The mobility, M, of a mechanism is obtained using Grübler's formula (Phillips, 1984):

$$M \geq 6(n-g-1)+\sum_{i=1}^{g} f_i$$

where n=number of bodies in the mechanism,
g=number of joints in the mechanism, and
f_i=degrees of freedom of the ith joint.

The *connectivity* of one body with respect to another in a mechanism describes the number of degrees of freedom between them, and can be derived from the overall mobility of the mechanism (Mason and Salisbury, Jr., 1985). First consider the two bodies to be fixed and determine the mobility of each subchain connecting them. Sub-chain mobilities greater than 0 are then subtracted from the overall mechanism mobility to yield the connectivity.

The mobility and connectivity analyses are easy to perform and can be used to eliminate impractical designs; if the connectivity between two bodies is wrong, the mechanism cannot be corrected by adjusting the geometries of the parts and detailed motion analysis is superfluous.

Motion analysis: Assuming that the mechanism has the right connectivity properties, the designer may wish to compute the relative motions of the bodies in detail. In this analysis, the *motion* of body A with respect to body G refers to the angular velocity of A with respect to G and the translational velocity of the origin of a reference frame attached to it, with respect to G. This description of its sets of independent motions will be referred to as the *twistspace* of A with respect to G, T(A,G). Similarly, by the term *wrenchspace* of A with respect to G is meant the space of wrenches (forces and moments) exertable upon A by G, W(A,G). This terminology is borrowed from the theory of screws (Ball, 1900).

Following a well-known result from screw theory, the twistspace of A with respect to G can be shown to contain the space of screws that are reciprocal to the wrenchspace of A with respect to G (Roth, 1984; Ohwovoriole, 1980). Therefore, to determine the twistspace of A with respect to G it is necessary to determine the wrenches acting upon it from G and to compute the nullspace of that set. Before considering arbitrary mechanisms, it is useful to examine three special cases: 1. A and G are directly connected by multiple joints in parallel; 2. A and G are connected by a single serial chain; 3. A and G are connected by multiple serial chains, acting purely in parallel.

Case I: In the case where A and G are the only two bodies in an assembly the wrenchspace of A with respect to G is simply the union of the wrenches exertable across the individual physical joints, or contacts, connecting them.

Case II: If A and G are connected by a single serial chain as in Fig. 5, then the wrenches exertable from G to A are the reciprocal screws to the motion screws of A with respect to G. The motion screws are obtained as the union of the motions allowed by each joint in the chain.

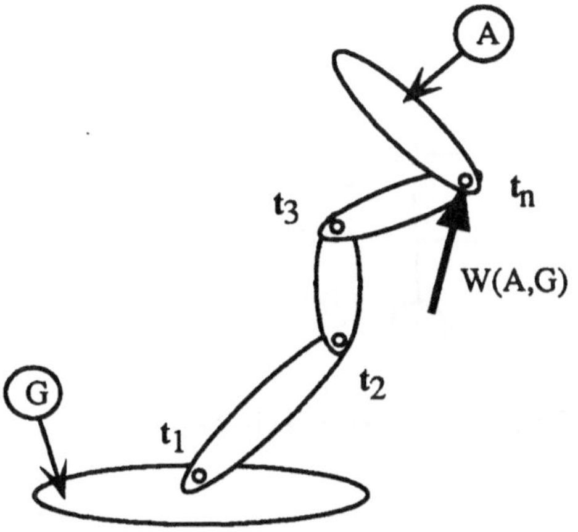

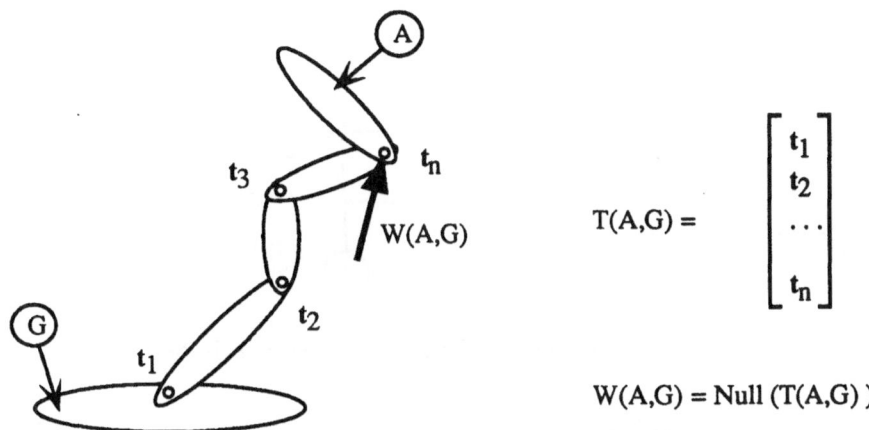

$$T(A,G) = \begin{bmatrix} t_1 \\ t_2 \\ \cdots \\ t_n \end{bmatrix}$$

$$W(A,G) = \text{Null}\,(T(A,G))$$

Fig. 5. A connected to G by a single serial chain.

Case III: If A and G are connected by multiple serial chains acting in parallel then the resultant wrenchspace of A with respect to G is the union of the wrenches (Case I) exerted by each chain (Case II above).

Case IV: In the general case, there may be multiple cross-connected chains, as shown in Fig. 6. A recursive procedure is followed in this case, which is a combination of all the above cases. The recursion principle is as follows: If A and G are connected by a serial chain, and B is the body directly connected to A, then the motion of A is the superposition of the motion made possible by the joint between B and A (*i.e.*, the motion that A would have if B were held stationary and the joint between them were exercised) and the motions that A would have if rigidly connected to B, while B is moved. The resulting motions of A can immediately be obtained only if the motions of B are known. Otherwise, the procedure recurses so that the motions of B are obtained from the bodies connected to it, and so forth until the base link, G, is reached. The algorithm for obtaining the kinematic relationship between two bodies in the general cross-connected case is outlined below. This algorithm does a depth-first traversal of the connections that act upon OF. Implicit in the procedure are cases I-III.

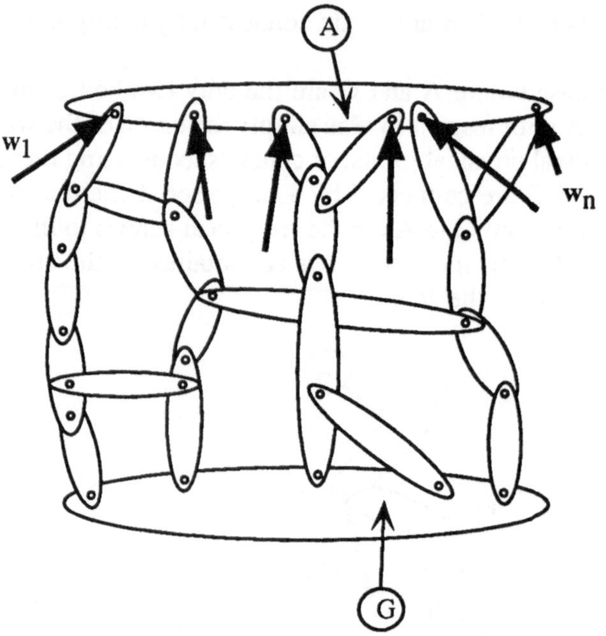

Fig. 6. A connected to G by cross-connected chains.

```
Function Motion(OF,WRT,ASSEMBLY):
 for each physical connection (P.C.) to OF (in ASSEMBLY):
     temporary assembly (T.A.) ← ASSEMBLY with all connections to
                                 OF except P.C., disconnected.
     connecting body (C.B.) ← body connected to OF by P.C.
     if C.B. = WRT then
        Chain-motion (OF,WRT) ← Motion of OF in T.A. due to P.C.
                     (keeping all other connections in T.A. locked)
        goto :Compute-wrenches
     else if no other connection on C.B. besides P.C. then : next P.C.
     else
          connection-motion ← Motion of OF in T.A. due to P.C.(keeping
                              all other connections in T.A. locked)
     endif
     subassembly (S.A.) ← ASSEMBLY with OF removed
     shared-motion ← Motion(C.B.,WRT,S.A.)+OF
     ;; the "+OF" in the line above means 'the motion imparted to OF by the
     ;; motion of C.B., the two regarded as a single rigid body.
     Chain-motion(OF,WRT) ← connection-motion + shared-motion
     :Compute-wrenches
          Chain-wrenches ← Reciprocal-screws(Chain-Motion(OF,WRT))
          W ← Chain-wrenches assembled into a matrix
     [if Reciprocal-screws(W) = null return 'immobile]
endfor
return Reciprocal-screws(W)
end
```

4.3. Efficiency Measures

The algorithm given above can be improved by the application of some domain knowledge. The improvements implemented are as follows:

1. Sufficiency of constraints: If A is shown to be rigidly constrained with respect to G, then connecting them with further chains in parallel will not affect their connectivity. The addition of parallel connections can only decrease the degrees of freedom between two bodies, or leave them unchanged. Applying this principle to the algorithm above, at the end of each loop that iterates over the connections to the body OF, the current state of the body's twistspace is determined and if the body is found to be immobile then the remaining connections are not pursued. This constraint is expressed in the statement in square parentheses. Conversely, if A is fully mobile with respect to G then the addition of more connections to the existing serial chains will not affect it. Adding serially to a chain can only increase a body's freedom or leave it unaffected.

2. Storing intermediate results: During the recursive, depth-first traversal of the system graph, when two chains share identical paths after reaching a certain depth (Fig. 7), the same motion expressions are computed and returned for each path. This redundancy is avoided by storing the expressions in the OF body the first time the common subpath is traversed. Thus, in the case of the gripper example in Fig. 7, the subpaths B-H and F-G-H are shared and the wrenches acting upon B and F can be reused. Storage is also obviously useful when repetitive analyses have to be done, as during the initial kinematic design of a new mechanism.

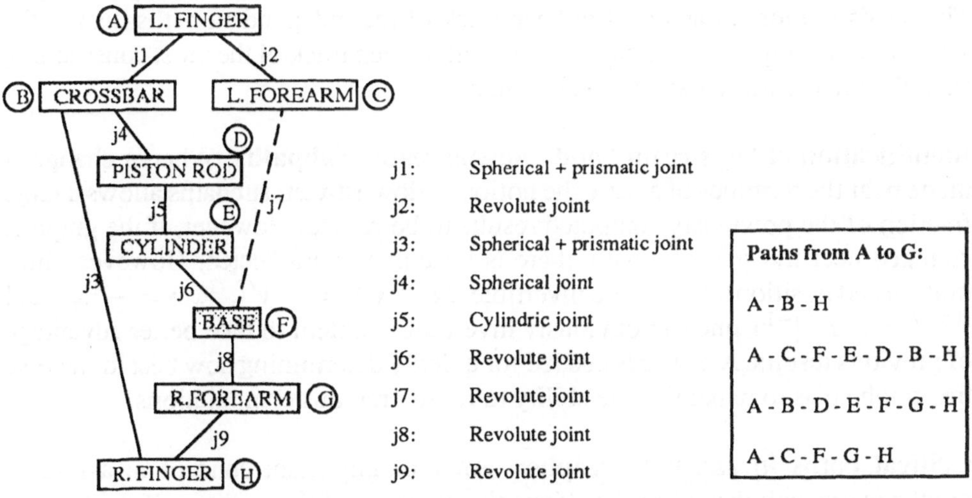

Fig. 7. Paths from A to G in the graph of the gripper assembly.

5. Discussion

In this section, we explore a number of issues that emerge when comparing the approaches adopted for tolerance and kinematic analysis.

Identification of "isolated" subgraphs: When designs become too complex (either in terms of the number of nodes and arcs or in terms of spatial configurations), it is helpful to indicate which portions of the graph are affected by local design changes before proceeding to implement them. If a group of bodies is interconnected by rigid joints, then they can be clumped together as one body for the purpose of kinematic analysis. For example, the shaded nodes in Figure 7 indicate that subset of the gripper that is unaffected by modifications to the left forearm or its associated joints. Similarly, when a tolerance or mating specification is changed, it only affects the subgraph to which the modified specification belongs.

Tradeoff between storage and recomputation: It may prove prohibitive to store detailed information about every node, arc and path in a complex design. Hence the decision of how much and what kind of information to store becomes important.

In the case of kinematics, because of the recursive nature of the algorithm, nullspace computation is performed at each node to derive wrenches and twists. Since this is the most time consuming step, the wrench or twist structures become the prime candidates for storage and reuse.

In the case of tolerances, the most time consuming computation is the propagation of tolerances along a path, if done symbolically. If the propagation is done numerically, then the most time consuming part is the determination of all the independent paths because the number of nodes and arcs for a typical design is large.[2] This leads to more of an effort to keep track of the independent paths than in the kinematics example. In particular, it is useful to keep track of the most constraining (*i.e.*, the most critical and dominant) paths.

Identification of "upstream" and "downstream" subpaths: When a change is made near the terminus of a path, the notion of downstream subpaths allows a large fraction of the previously computed results to be reused. However, if the graph is changed near the start of a path, there is little that is unaffected. However, since motion and position relations are invertible (*i.e.*, ${}^{G}\mathbf{v}^{A} = -{}^{A}\mathbf{v}^{G}$, ${}^{G}\omega^{A} = -{}^{A}\omega^{G}$ and ${}^{A}T^{G} = [{}^{G}T^{A}]^{-1}$), one can obviously invert the problem to take better advantage of previous results. A metric is needed for quickly determining how best to traverse the graph so as to maximize the ability to reuse previous computations.

Critical paths in assembly graphs: Another important concept in assembly graphs is the path that critically affects the outcome of the analysis, if such a path exists. In the case of tolerance analysis, the identification of critical paths played an

[2] unless using statistical tolerances, in which case the propagation is again time consuming.

important role, as discussed in Section 3.2. In the case of kinematics, an analogous concept exists, although it was not used in the implementation: the upper bound on a body's mobility is the number of degrees of freedom of the least mobile chain connecting it to the base.

6. Conclusions

Incremental computation methods for graphs of connected components have been considered in the context of kinematic and tolerance analysis. The methods basically amount to identifying unaffected portions of the graphs so that previous results can be reused. Domain-dependent issues, such as the amount of computation associated with identifying paths and loops, as opposed to computing values at each node, lead to different approaches for traversing the graph and storing intermediate results. Beyond simply storing previous results, improvements in efficiency can be obtained by exploiting analyses at different levels of detail and by identifying critical or "dominant" paths where they exist.

Acknowledgements

Funding for this project has been provided by DARPA under ONR contract N-00014-88-K-0620.

References

Aho, A. V., Hopcroft, J. E. and Ullman, J. D.: 1983, *Data Structures and Algorithms*, Addison-Wesley.

Baker, J. E.: 1980, On relative freedom between links in kinematic chains with cross-jointing, *Mechanism and Machine Theory* **15**, 397–413.

Ball, R. S.: 1900, *The Theory of Screws*, Cambridge University Press.

Bernstein, N. S. and Preiss, K.: 1989, Representation of tolerance information in solid models, *Advances in Design and Automation–1989, Vol I:1989 ASME Design Technical Conference, Computer-Aided and Computational Design*, ASME, pp. 37–48.

Binford, T., Frants, L., Cutkosky, M. R. and Tsai, J.-C.: 1990, Represention and propagation of tolerances for CAD/CAM systems, *Proceedings of the IFIP WG 5.2 Workshop on Geometric Modeling*. Rensselaer Polytech Institute.

Bjorke, O.: 1989, *Computer-Aided Tolerancing*, ASME Press. 2nd ed.

Cutkosky, M. R. and Tenenbaum, J. M.: 1991, Toward a framework for concurrent design, *International Journal of Systems, Automation: Research and Applications* **1**(3), 239–261.

Davies, T.: 1983, Mechanical networks - I, II, III, *Mechanism and Machine Theory* **18**(2), 95–112.

Finger, S. and Rinderle, J.: 1989, A transformational approach to mechanical design using a bond graph grammar, *Proceedings of the First Design Theory and Methodology Conference*.

Gossard, D. C., Zuffante, R. P. and Sakurai, H.: 1988, Representing dimensions, tolerances, and features in MACE systems, *IEEE Computer Graphics and Applications* pp. 51–59.

Harary, F.: 1969, *Graph Theory*, Addison-Wesley.

Kambhampati, S. and Cutkosky, M. R.: 1990, An approach toward incremental and interactive planning for concurrent product and process design, *Proceedings of ASME WAM: Computer Based Approaches to Concurrent Engineering*, pp. 1–8.

Kannapan, S. M. and Marshek, K. M.: 1989, An algebraic and predicate logic approach to representation and reasoning in machine desi gn, *Technical Report 212*, The University of Texas at Austin.

Kusiak, A. and Szczerbicki, E.: 1990, Model-based synthesis in conceptual design.

Lee, K. and Gossard, D. C.: 1985, A hierarchical data structure for representing assemblies: Part 1, *Computer-Aided Design* **17**(1), 15 – 19.

Lee, S. H., Cutkosky, M. R. and Kambhampati, S.: 1991, Incremental and interactive geometric reasoning for fixture and process planning, *in* A. Sharon (ed.), *Proceedings of ASME WAM: Issues in Design/Manufacture Integration*, ASME Winter Annual Meeting, ASME, pp. 7–13.

Mason, M. T. and Salisbury, Jr., J. K.: 1985, *Robot Hands and the Mechanics of Manipulation*, The MIT Press Series in Artificial Intelligence, The MIT Press, Cambridge, Massachusetts; London, England.

Mohamed, M. G. and Duffy, J.: 1985, A direct determination of the instantaneous kinematics of fully parallel robot manipulato rs, *Transactions of the ASME: Journal of Mechanisms, Transmissions and Automation in Design* **107**, 226–229.

Mott, J. L., Kandel, A. and Baker, T. P.: 1986, *Discrete Mathematics for Computer Science and Mathematicians*, Prentice-Hall. 2nd ed.

Nilsson, N. J.: 1980, *Principles of Artificial Intelligence*, SRI international, Tioga Publishing Co.

Ohwovoriole, M.: 1980, *An Extension of Screw Theory and Its Application to the Automation of Industrial Assembly*, PhD thesis, Stanford University.

Phillips, J.: 1984, *Freedom in Machinery*, Vol. 1, Cambridge University Press.

Requicha, A. A.: 1984, Representation of tolerances in solid modelling: Issues and alternateive approaches, *in* M. S. Pickett and J. W. Boyse (eds), *Solid Modeling by Computers*, Plenum Press, pp. 3–22.

Rinderle, J. R. and Balasubramanium, L.: 1990, Automated modelling to support design, Engineering Design Research Center, Carnegie Mellon University.

Roth, B.: 1984, Screws, motors, and wrenches that cannot be bought in a hardware store, *in* M. Brady and R. Paul (eds), *Robotics Research*, The MIT Press, chapter 8, pp. 679–693.

Serrano, D.: 1987, *Constraint Management in Conceptual Design*, PhD thesis, Massachussetts Institute of Technology.

Srikanth, S. and Turner, J. U.: 1990, Toward a unified representation of mechanical assembly, *Engineering with Computers* **6**, 103–112.

Turner, J. U.: 1987, *Tolerances in Computer-Aided Geometric Design*, PhD thesis, Rensselaer Polytechnic Institute.

Ulrich, K.: 1989, *Computation and Pre-Parametric Design*, PhD thesis, Massachusetts Institute of Technology. Tech Report 1043.

Waldron, K. J.: 1966, The constraint analysis of mechanisms, *Journal of Mechanisms* **1**, 101–114.

Ward, A.: 1989, *Quantitative Inference in a Mechanical Compiler*, PhD thesis, Massachussetts Institute of Technology.

Appendix

Algorithm for incremental tolerance propagation

Given the critical and dominant paths of a design, when the design involves only a change of tolerance or mating condition, a, the incremental propagation algorithm follows:

case (a *becomes tighter*):
    ;;; for the Most Critical Path, *MP*
    if ($a \in MP$) **then**
        reevaluate **Tol**(*MP*) = **Tol**(*UpStream*(a, *MP*)) + **Tol**(*DownStream*(a, *MP*));
        ;;;*MP* is the same and we don't reevaluate **Tol**(*UpStream*(a, *MP*)).

```
    else
        for each path P that a ∈ P
            reevaluate Tol(DownStream(a, P));
            Tol(P) = Tol(UpStream(a, P)) + Tol(DownStream(a, P));
            when (Tol(P) < Tol(MP)) replace MP by P;
        endfor;
    endif;
    ;;; for the Dominant Path, DP
    if a ∈ DP then
      reevaluate Error(DP) = Error(UpStream(a, DP)) + Error(DownStream(a, DP));
      ;;; Again, we don't reevaluate Error(UpStream(a, DP))
    else
        for each path P that a ∈ P
            reevaluate Clr(DownStream(a, P));
            Error(P) = Error(UpStream(a, P)) + Error(DownStream(a, P));
            when (Error(P) < Error(DP)) replace DP by P;
        endfor;
    endif;
    break case;
case (a becomes looser):
    ;;; for MP
    if (a ∈ MP) then
      reevaluate Tol(MP) = Tol(UpStream(a, MP)) + Tol(DownStream(a, MP));
      let path P be the path such that Tol(P) = Min(Tol(P_i)), ∀ P_i that a ∉ P_i;
      when (Tol(P) < Tol(MP) replace MP by P;
    endif;
    for each path P that a ∈ P
        reevaluate Tol(P) = Tol(UpStream(a, P)) + Tol(DownStream(a, P));
    endfor;
    ;;; for DP
    if (a ∈ DP) then
      reevaluate Error(DP) = Error(UpStream(a, DP)) + Error(DownStream(a, DP));
      let path P be the path such that Error(P) = Min(Error(P_i)), ∀P_i, a ∉ P_i;
      when (Error(P) < Error(DP) replace DP by P;
    endif;
    for each path P that a ∈ P
        reevaluate Error(P) = Error(UpStream(a, P)) + Error(DownStream(a, P));
    endfor;
    break case;
```

11

ARCHITECTURES FOR DESIGN KNOWLEDGE

STRUCTURING DESIGN KNOWLEDGE ON THE BASIS OF GENERIC COMPONENTS

L. K. ALBERTS, P. M. WOGNUM and N. J. I. MARS
Department of Computer Science
University of Twente
PO Box 217
7500 AE Enschede
The Netherlands

Abstract. A framework is proposed for describing design as a synthesis process, based on *generic components* at different levels of abstraction. These levels range from the system level to the physical level and represent the technical vocabularies used to describe the design. The resulting framework is used for structuring the knowledge needed to execute a design. Furthermore, the possibility of using generic components as basic building blocks for constructing higher-level concepts, in particular *design prototypes*, is discussed. The approach is illustrated with examples from the field of bridge design.

1. Introduction

1.1. PHILOSOPHY

The research described in this paper is part of the Stevin Project, a research programme aimed at supporting the design of technical systems, at the University of Twente. This project was named after the early Dutch physicist Simon Stevin (1548-1620), who was also involved in design, in particular architectural design and civil engineering. Stevin was fascinated with showing that there is a hidden pattern, possibly complex but understandable, behind physical phenomena. In his own words: "Wonder, en is gheen wonder", which means (translated after (Simon, 1969)) "Wonderful, but not incomprehensible". Stevin's motto also describes the philosophy behind the research being conducted within the Stevin Project in a nutshell.

The approach taken in the research presented here adheres to what is generally known as the 'product' school in AI. In other words, our primary interest is not in imitating the process of designing as performed by human designers, but rather in obtaining the same or preferably better results (the products). The *design of technical systems* will be regarded as generating a description of the system that can be physically realised, given a possibly highly abstract functional description and a set of requirements concerning the performance of the system.

J. S. Gero (ed.), Artificial Intelligence in Design '92, 639–656.

1.2. Design as Synthesis

We propose a framework for describing design as a synthesis process, based on generic components, or basic 'building blocks' at different levels of abstraction. These levels range from the system level to the physical level and represent the technical vocabularies used to describe the design. Basically, a generic component specifies the relation(s) between a behaviour, useful for obtaining some required functionality, and physical characteristics of the class of elements that show this behaviour. In other words, it represents a relation between what is required from a designer's point of view and what is physically or technically possible.

Design is generally thought to be inherently linked with creativity. Intuitively this seems to imply that design is more than an extended configuration task. How then can design be described as a synthesis process, based on predefined building blocks? Perhaps this question is answered best with the following quote from the psychologist Kneller:

> "Creative novelty springs largely from the rearrangement of existing knowledge - a rearrangement that is in itself an addition to knowledge. Such rearrangements reveal an unsuspected kinship between 'facts long known but wrongly believed to be strangers to one another'." (from: (Kneller, 1965), p. 4)

This 'rearranging' of known concepts seems to be a task that in many cases may be carried out by a machine. In terms of the framework presented, rearranging existing knowledge entails putting together generic components in such a way, that useful new functions arise.

Although the concept of generic components has originated in work done in the field of VLSI-design (Alberts et al., 1989), we believe that the approach is feasible for any domain of technical system design. To test this hypothesis, cases in different domains are being developed. Some preliminary results in analog circuit design have been described in (Alberts, Wognum and Mars, 1991a). The ideas presented in this paper are illustrated with examples from structural engineering, in particular bridge design.

1.3. Linking up with Other Representational Schemes

Generic components as a knowledge representation scheme have a rather fine granularity. More information encompassing concepts can be defined in terms of generic components. Some possibilities of using such concepts to represent and structure the knowledge of how to synthesize designs from generic components will be discussed. In particular, the possible role of design prototypes (Gero, Maher and Zhang, 1988) and the benefits of composing prototypes from vocabularies of generic components are looked into.

2. A Universal Framework for Design as Synthesis

2.1. Introduction

The design problem as defined in the introduction of this paper is how to obtain a physical description of a technical system given a high-level (abstract) specification. Theoretically, one can envision a continuous decrease in abstraction ranging from one extreme to the other. In other words: the design is gradually taking shape. In practice, however, design will never be of a completely top-down nature. In order to determine what the next refinement action will be, a designer has to have some knowledge of what the possibilities roughly are. Knowledge about which functions can be realised given specific physical properties of the realisation material is propagated upwards. Without such prior knowledge about the feasibility of alternatives, design would result in 'blind search'.

In the framework presented here, this bottom-up knowledge is represented as sets of generic components (that are known to have several possible physical realisations) at different levels of abstraction.

2.2. Technology-Based Layers

The sets of generic components are located at distinct levels of abstraction, ranging from the system level to the physical level. The contents of the layers is based on the technical vocabularies in use, therefore we will speak of *technology-based* layers or levels. Each technology-based layer represents an abstraction of the levels below; only the information 'relevant' for the current level of abstraction is propagated upwards from the levels below.

A typical example of such a hierarchy found in VLSI-design is shown in table I. For a more extensive description of the models that formed the basis for our notion of technology-based layers, see (Alberts, Wognum and Mars, 1991b). It is important to realise that the actual contents of the layers as well as the number of layers will be domain-specific.

2.3. Generic Components

Generic components represent behaviours that are known to be physically realisable. They are generic in the sense that each component stands for a range of alternative realisations. For instance, the conceptual design of a bridge will involve choosing the type of spanning element: arch, beam or (suspension) cable, each of which stands for a whole range of actual physical components. This also implies that the generic components still have to be given their actual shape.

Relevant technical or physical limitations manifest themselves in the values of a specific set of parameters belonging to the generic components. These parameters are used to get a rough impression of the consequences of certain design choices at the current level of abstraction for the final result.

	behaviour	**structure**	**geometry**
architectural level	system specs	CPU, memories	physical partitions
register transfer level	register-transfer operations	ALU, registers	floor-plan
logic level	boolean functions	gates, flip-flops	coarse layout
circuit level	differential equations	transistors, capacitors, resistors	symbolic layout
physical level	diffusion equations	cross-tacs, conductors	boxes, polygons

TABLE I
Adapted from (Walker and Thomas, 1985))

This interplay between physical possibilities and behavioural requirements is formalised in the following definition:

DEFINITION 1. A *generic component* is a triple of the form $< B, S, F >_t$, where index t stands for the technology-based layer the component belongs to, and where: B is the behaviour, S is the structure, and F is the form of the component.

The *behaviour* of a component specifies what a component does. However, these components bring along a number of technical and physical prerequisites that have to be met for the components to behave properly. In other words, the generic components can perform a particular behaviour, but at a certain 'cost'. These prerequisites are specified in the *form*. So, this use of 'form' is not only as the actual shape of the artifact being designed, but also in determining what the effects of physical and technical limitations on the required behaviour are. Finally, the *structure* of a component specifies how it can be linked to other components, i.e., what its connections are, without detailed physical information (which is part of the form!). It describes the relation between the behaviour of a component and its form.

Generic components are divided into basic components, the smallest possible building blocks, and macro-components. *Macro-components* provide for a systematic representation of design knowledge concerning essential basic configurations of generic components (Note that the concept of macro-components is a recursive one). They provide for a more appropriate level of granularity or detail than that of the basic generic components. Macro-components are based on technical principles, not on experience with specific design situations or on heuristics.

In the case of macro-components, it is necessary to distinguish between the *internal* structure of the component and the *external* structure. The first specifies

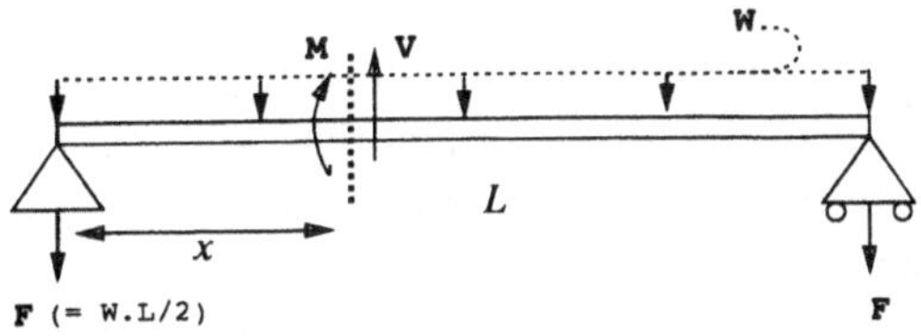

Fig. 1. Structure of generic component 'beam'

how its constituent elements are connected, the second how the macro-component can be connected to other generic components.

As an example of a generic component in bridge design, consider the beam in figure 1. The behaviour of a beam depends on its type of support and the type of load. For a simply supported beam (one support pinned, other rolling) of length L with a uniformly distributed force w (e.g. as a consequence of its own weight), (part of) the behaviour is described in terms of bending and shearing. More specifically:

behaviour :

- shear force $V(x) = w.(L/2 - x)$, where x is the distance from a support, and
- bending moment $M = (L.x - x^2).w/2$.

structure : basically consists of the diagram, as in figure 1, of (external) forces and loads, and (internal) shearing force and bending moment.

form : the information about the shape, L, A (area of cross-section of beam) etc., as well as a number of characteristics like critical load and ultimate strength of the beam. The latter two are good examples of physical limitations imposed by a component.

The critical load and ultimate strength of a structural element are (along with a number of other variables and constants) measures of the quality of that element. They are used to classify the 'material' a design is to be made of. While the strength, for instance, is part of the behaviour, the ultimate strength sets an upper limit to this behaviour as a consequence of the physical peculiarities of a particular element.

2.4. Design as Synthesis Based on Generic Components

The design process can now be described in terms of transformations of the design description within or between the technology-based layers.

The behavioural description of the required system at a certain level is decomposed according to the behaviours of the available generic components. The selected generic components are instantiated and put together in a single configuration. Depending on the structural properties of the components, different configurations may arise. In order to estimate what the restrictions imposed by the generic components imply for the overall system, the resulting configuration has to be parameterised in terms of the form characteristics of the individual components. Macro-components may be helpful in this process, in that they specify combinations of more basic components that implement behaviours that are often used. This simplifies the process of finding a suitable decomposition of the required behaviour and the accompanying configuration of generic components.

Once the physical properties of the resulting construction are deemed satisfactory, the results are translated to the next lower level, where it serves as a new and more specific design (problem) description. Here, macro-components provide for a means for closing the gap in abstraction between generic components of different levels. Macro-components may serve as a first approximation for translating generic components of the current level into configurations of components at the next lower level.

Throughout this process, the performance requirements serve as constraints on the possibilities for configuring generic components into larger structures and for assessing the physical characteristics.

A number of problems cause the synthesis process to be more than just a predetermined configuration problem.

Finding a decomposition into generic components that satisfies the performance requirements at the current level does not guarantee, however, that the final result (at the lowest level) will satisfy as well. Recall that the form at higher levels only provides an estimation of the physical consequences of design decisions. It may appear at lower levels, when the amount of information about the form has increased, that the required performance cannot be met after all. In that case, backtracking to a higher level is required to reconsider, where necessary, the choices made.

Another difficulty is that a chosen configuration of generic components at the current level can often be implemented in several ways by components of the next lower level. This phenomenon is inherently linked with the fact that higher-level components are abstractions of the components at lower levels.

2.5. CASE: BRIDGE DESIGN

A possible inventory of the domain of bridge design according to the framework presented in the previous sections, is given in table II. We have distinguished two technology-based layers.

At the *conceptual* level the design of a bridge is described in terms of basic structural principles and consists of a basic configuration of highly idealized elements. This description serves as a first estimation of the overall load transfer in the bridge and the overall dimensions. A bridge at this level is basically a supported or self-contained span carrying a road across a gap.

conceptual level	(beams) cantilever, fixed, simply supported, continuous suspension cable, stay-cable hinged arch, fixed arch column, abutment truss (joints) pinned, rolling, fixed
construction level	pier, tower, spandrel (wall), ... I-beam, r.c. beam, ..., r.c. slab, ... steel-truss, r.c. truss, trussing materials (ties & struts), girder plates, ... r.c. arch ribs, steel arch ribs, ... reinforcement plate, ...

TABLE II
Technology-based layers in bridge design (r.c. = reinforced concrete)

A description of a bridge at the *construction* level contains all the necessary technical details, such as the actual connections between the different structural elements, the exact properties of the materials used, and additional auxiliary structures such as stiffners. Furthermore, the focus shifts from the properties of the bridge as a whole, to those of the individual construction members that can no longer be assumed to behave ideally. Table II only shows some of the generic components at this level. Although the set of components is finite, it is too large to be covered within the scope of this paper.

We might have added a "manufacturing level" at the lower end of the hierarchy, to describe how for instance an I-beam can be made up of a separate web welded to two flanges, or rolled as one shape. However, for our purposes, illustration of the concepts introduced, two levels will do. Our approach does not require that there is a unique set of technology-based layers.

In general, the behaviour of bridge elements is described in terms of load transfer, resulting in stress (normal and/or shear), and deformation of the elements, known as strain. Besides the overall (shape) dimensions, material specific limitations to stresses and strains are part of the form. Examples are: the yield stress, ultimate and specific strength, and modulus of elasticity. Furthermore, it is useful to distinguish between *internal* and *external* behaviour. The internal behaviour specifies how the load is transferred internally (stress distribution), whereas the external behaviour describes the reaction of the element on connected elements.

For another example of the use of the framework, in the field of analog circuit design, see (Alberts, Wognum and Mars, 1991a).

2.6. Related Research

The concept of defining elementary building blocks from which to generate designs has been suggested by numerous researchers in different domains. Already in 1875, Reuleaux (Reuleaux, 1963) proposed a vocabulary of basic 'kinematic pairs' to describe the behaviour of moving bodies in machines.

In particular the methodic or systematic design approach in mechanical engineering originated in Germany has come up with similar hierarchies as that shown in table I. For instance, in (Rodenacker, 1976), (Koller, 1979) and (Hansen, 1968)) tables with basic or prototypical solutions for elementary mechanical design problems at different levels of abstraction are proposed. Another form of a list of elementary solutions is the 'morphological chart' (e.g. (Cross, 1989)). However, these approaches assume that a human designer will be interpreting these basic elements.

For a computer program to deal with such concepts, a lot of extra information is required. In the field of knowledge-based design systems a number of concepts have been proposed that are related to our notion of generic components.

The research on form-function characteristics of physical components (e.g. (Rinderle et al., 1989)) is particularly interesting, because it may provide us in the end with a technique of generating generic components in an automatic way. The idea of distinguishing between functional and physical aspects of components is also encountered in research on reasoning from first principles, in for instance (Davis, 1983) and (Murthy and Addanki, 1987).

The renewed interest in Reuleaux's approach to kinematics (e.g. (Joskowicz and Addanki, 1988)) suggests that further revenues are to be expected in terms of elementary kinematic components for design. The link with design prototypes (Gero, Maher and Zhang, 1988) will be discussed later.

The most promising results so far, it appears, have been obtained in the field of VLSI-design with the automated (knowledge-based) synthesis of digital systems, based on basic design elements (e.g. (Camposano, 1990)).

Synthesis based on generic components has in common with configuration design that the artifact being designed is assembled from a set of pre-defined components (e.g. (Mittal, Dym and Morjaria, 1986) and (Mittal and Frayman, 1989)).

However, in configuration design the components can only be connected together in certain fixed ways. In our approach there are, in general, infinite many ways to connect generic components within the structural limitations of the individual components. This makes the problem of guiding the generate-and-test process involved in synthesis a lot more complex.

3. Structuring of Design Knowledge

3.1. Introduction

In this section, the use of the framework for the structuring of design knowledge is discussed. The design knowledge is to be centered around the concept of generic components, and classified according to the appropriate level of abstraction: the technology-based layer it is related to. This way, a modular structure for the knowledge-base is obtained, allowing for incremental addition of knowledge without major modifications in the structure imposed on the knowledge already acquired. The design knowledge consists both of generic components and knowledge of how to combine these components to realise the required system specification.

There is an obvious separation between knowledge concerning transformations *within a layer*, and knowledge about transformations *between different layers*. The first will consist primarily of decomposition knowledge. The second is of a more complex nature, since it requires the translation of the current design description in terms of another vocabulary.

Another important division is between *domain* and *strategic* knowledge. The knowledge concerning individual transformations (i.e., design actions) will be called domain knowledge. There is also strategic knowledge, about the preferred sequence of design actions. This knowledge determines which 'path' to follow in a lattice like the one shown in table I. Here, we will focus exclusively on the structuring of domain knowledge.

3.2. Intra-Level Synthesis Knowledge

Part of the design knowledge is contained in the basic generic components, being behaviour-form relations. The macro-components represent decomposition knowledge concerning fundamental combinations of basic components, thus implementing more complex behaviour-form relations.

Extra knowledge is required to determine how a generic component can be useful for a particular *purpose*. To describe this type of knowledge, an additional attribute is introduced.

DEFINITION 2. A *function* of a generic component describes the use or purpose of its behaviour in a particular context.

A function relates the behaviour of a generic component to a particular design context or situation. Thus, it is not part of a generic component! One generic component may have a multitude of functions.

A truss for example has a behaviour which can be described in English (to avoid a profusion of technical details) as: "transferring loads internally as tension and compression stresses, distributed over the web-members, thus resisting bending and buckling." This behaviour may be used for different purposes. Two possible functions of the truss in a bridge design might be: "stiffening a (concrete) slab supporting the road", and "spanning a gap in an arch-like fashion".

In short, the main difference between the behaviour of a generic component and its function in a particular context can be typified as: general versus specific (respectively).

Functions, in general, also assume particular values for the different parameters in the behaviour, structure and form of the generic component. In this example, for instance, the first function assumes that the truss has a straight, beam-like shape and has the same length as the slab. The second function requires that the truss has the shape of an arc and may introduce limitations to the minimal thickness of the truss members.

Another important type of design knowledge belonging to a particular level consists of *technical principles* that are not necessarily related to a particular generic component.

In the case of bridge design, for instance, a well-known structural principle is that in all statically indeterminate structures an increase in the stiffness at any point tends to attract force to that point and to draw it away from other parts. This principle states something about the behaviour of *any* combination of basic elements. In terms of the technology-based layers distinguished in table II, it belongs to the conceptual level.

Another example would be the design heuristic that it is always advisable to aim for (load transfer) paths as simple and direct as possible for the loads to pass to the support. This rule relates to the structure of any configuration of generic components at the conceptual level.

3.3. Inter-Level Synthesis Knowledge

Switching layers implies a change of the technical vocabulary. In general this will involve many-to-many mappings between the generic components from the current vocabulary and the components from another layer. In terms of our framework, the knowledge required for this process is 'located' at the boundary between two particular levels.

Part of the knowledge involved in the synthesis between different levels is concerned with the *relations between* particular groups of *generic components* at the levels involved. In general, a component at the current level will represent an abstraction of a one or more configurations of components at the next lower level. Macro-components may be useful here to describe such configurations.

Another type of knowledge describes the *abstraction* involved in going from one particular level to another. To make a 'translation' it is necessary to know which *assumptions* and *approximations* have been made, or which details have been neglected. This knowledge generalizes over all the behaviours, structures and forms of all generic components at the levels involved.

For instance, in the case of table II we assumed that all the components at the conceptual level behave ideally (in a structural engineering sense). In many situations a uniform distribution of loads is assumed at the construction level, whereas this is seldom the case in actual situations. Here, knowledge is required to determine what the restrictions on a configuration at the construction level are, to make the approximations at the higher level valid. The abstractions and idealisations are only valid within certain ranges.

4. Generic components as the basis for prototype construction

Given the concepts of technology-based layers and generic components, we would like to extend our framework with representational schemes for the types of knowledge discussed in the previous section. Here, the possibilities of one particular scheme are discussed, called design prototype (Gero, Maher and Zhang, 1988), for describing the use of generic components in different situations.

4.1. Design Prototypes in Relation to Generic Components

A *design prototype* is "a conceptual schema for representing a class of a generalised heterogeneous grouping of elements derived from alike design cases that provides the basis for the start and continuation of a design. Design prototypes provide this basis by bringing all the requisite knowledge appropriate to the design situation together in one schema". (Gero, 1990;, p. 30). Important for our discussion is the fact that the proposed schema entails the definition of the function, behaviour and structure of a design prototype, together with relevant design knowledge. There are a number of differences between the function, behaviour and structure of design prototypes and the properties we defined for generic components.

First of all, the definitions are slightly different; e.g. "behaviour" is defined in (Gero, 1990) as: "[description of] how the structure of the artifact achieves its function". In contrast, in generic components the available behaviours form the basis for the definition of a function. The difference could be described as top-down versus bottom-up respectively. Furthermore, the function is part of the prototype and there is no distinction between structure and form.

Function, behaviour and structure in the case of design prototypes only refer to the relevant variables and do not contain relations between these variables; such information is stored separately in the design prototype. This relational knowledge is divided over the behaviour, structure and form in the case of generic components.

The most important difference, however, is the way in which the relevant variables are assigned to either the function, the behaviour or the structure. This last point needs some further explanation, because to a large extend it justifies the definition of a separate concept like generic components next to that of design prototypes.

In the case of generic components, a bottom-up approach is chosen that starts by identifying the relevant design vocabulary. Only after this vocabulary has been determined, are the generic components defined. Whether a variable is deemed part of the behaviour or, instead, of the structure or form, is dictated by the design vocabulary. In contrast, the attributes of a design prototype are defined in a more top-down manner. Basically, its starts with determining required functionalities and working back to technical and possibly highly abstract components that can fulfill these functions.

A design prototype brings together all the requisite knowledge appropriate to a particular [1] *design situation* in one schema. A generic component represents a basic element in a design vocabulary, that reflects the *general* technical principles available, and only contains knowledge related to that particular element. Taking this difference to the extreme, we might classify the knowledge contained in generic components as general, fundamental and theory-based, whereas the prototypes represent specific, situational and experience-based knowledge.

4.2. Integration of Both Concepts in the Framework

Rather than being alternative representation schemas, both concepts are complementary. The generic components can serve as the basic elements of which to construct design prototypes. By constructing prototypes on the basis of generic components, a more systematic way of determining the relevant behavioural and structural variables is obtained. Furthermore, the concept of technology-based layers helps structuring the prototypes at the right level of abstraction, thus facilitating the classification and selection of prototypes. At the same time, design prototypes help structuring and representing situational synthesis knowledge, such as functions, related to the generic components.

In our definition, assigning a function to a generic component restricts its applicability to a particular situation or context. Therefore, the general knowledge contained in a generic component can not be represented by a design prototype schema, because that would require the specification of a function. This justifies the definition of two distinct representation schema's for both concepts.

In the rest of this paper, we will assume a slightly different schema for a prototype from that proposed in (Gero, 1990) or (Tham, Lee and Gero, 1990). To distinguish it from the original, it is called 's-prototype'.

DEFINITION 3. An *s-prototype* assigns one or more functions to a generic component or configuration of components for a particular design situation. The adjustments to the behaviour, structure and form required by the function are represented

[1] though generalised from actual design situations

within the s-prototype as well. Furthermore, it contains the relevant synthesis knowledge, required for that situation.

The main differences are the adherence to our own definitions and the neglect (for the sake of simplicity) of the different types of knowledge distinguished in the original design prototype schema.

Recall that generic components may be either basic or macro-components. Any combination of basic and/or macro-components is allowed in the construction of prototypes. Although macro's describe configurations of more basic generic components, they represent general technical principles. Thus, they differ fundamentally from s-prototypes, which reflect knowledge about specific situations.

4.3. ILLUSTRATIONS

To illustrate the ideas presented in this section, a number of examples of the combined use of both generic components and prototypes are presented. Most functions, behaviours etc. are described in words, or in terms of some of the relevant variables, rather than in the appropriate equations, to avoid a profusion of details. Furthermore, all examples are at the conceptual level in table II.

Suppose the set of generic components consists of the one described in table III. The listed behaviours should be worked out in terms of the relevant stresses and strains. The structures can be described in a similar way as that of the beam in figure 1. Amongst the relevant form variables are: length, width, depth, weight, material_type, and yield stress. These should be related by the appropriate equations. In the case of the arch we need at least an expression for the curvature and radius as well.

Fixed_arch	
behaviour:	(internal) compression, (external) diagonal thrust (on supports)
form:	radius, curvature
Column	
behaviour:	(internal) compression, (external) vertical thrust (on foundation or soil)
Abutment	
behaviour:	(internal) tension, (external) horizontal thrust (on abutting structures)
Road	
behaviour:	(internal) tension and compression (external) uniform distribution of load (over supports)

TABLE III
Generic components for examples

A first example of the use of s-prototypes, given these generic components, is *specialisation of the general case.* A single-span fixed arch is an often recurring part in bridge design. The additional information which is required compared to the generic case (the fixed arch) is assigned to the appropriate 'slots' in an s-prototype schema: (a) the particular purpose of this type of arch is specified in the function; (b) which generic components are used in a prototype, is specified in a slot called "decomposition"; (c) the refinements in the behaviour, structure and form are specified in the slots concerned; and (d) there is a slot for the required situational knowledge. The result might look like the following:

Single-span fixed arch prototype

Function: "carry a load across a gap in a single span"
Decomposition: Arch(B,S,F) **of_type** fixed_arch(B,S,F)
Behaviour: B
Structure: S
Form: F
  where F.radius $\leq R_{max}$ = f(material_type)
   % maximum span, depending on material %
  and F.material_type $\in$ {reinforced concrete, stone, steel, wood}
Situational knowledge: ...

If we want to describe *compound structures*, the situation becomes slightly more complicated. Suppose we want to represent a specific type of bridge, according to the Séjourné concept. In this case a number of generic components are involved, each of which has to be refined in a different way in order to account for the overall features of this type of bridge. The span of this particular type of bridge consists of two single-span arches, side by side, that are supported at the ends by two (vertical) abutments. The road is carried by two rows of six columns at each side, that spring from the arches.

The behaviour of the bridge as a whole has to be deduced from the behaviours of the individual generic components, as well as their particular configuration. The structure describes the way the individual components are to be connected, as well as the distribution of the loads within the bridge. As a consequence of the chosen configuration, there will be a number of restrictions on the forms of the components. These restrictions are specified in the form of the s-prototype, along with the other form information obtained from each component. Typically, (additional) knowledge about the circumstances under which this type of bridge is favourable will be part of the situational knowledge.

Séjourné bridge prototype

Function: "carry road across water or valley in single span"

Decomposition: Arch1(B1,S1,F1) **of_type** fixed_arch
Arch2(B2,..) **of_type** fixed_arch
Road(..) **of_type** road
Column01(..) **of_type** column
..
Column12(..) **of_type** column
Abutment01(..) **of_type** abutment
Abutment02(..) **of_type** abutment

Behaviour: B = f(B1,B2, ...)
Structure: see fig. 2
Form: F = f(F1,....)
where Road.width == 2 * A1.width
and Column01.length == Column06.length ==
Column07.length == Column12.length
andAbutment01.length == Abutment02.length ==
Arch01.height == Arch02.height
..
Situational knowledge: ...

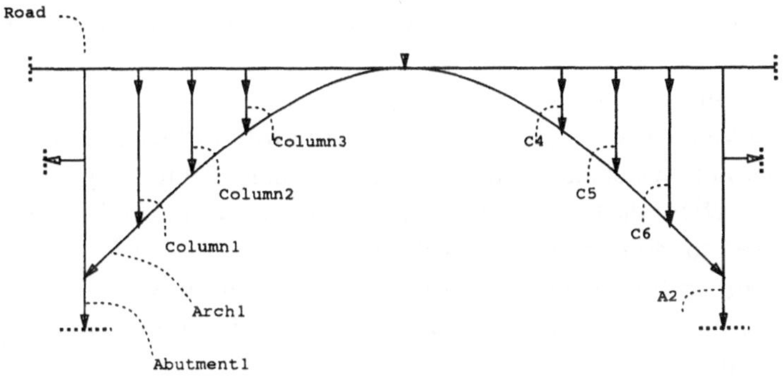

Fig. 2. Structure of a Séjourné bridge-prototype

Note that the resulting s-prototype still describes a generalisation of actual instances of this bridge type. To obtain an actual instance, values have to be assigned to all the variables involved.

4.4. Other work

There have been other proposals for describing the relation between a situation specific function and the underlying general technical or physical principles. One of them, relevant for our framework, is a functional representation scheme presented in (Goel and Chandrasekaran, 1989).

An appealing idea in the work by Goel et al., is the addition of some sort of 'causal explanation' for the way the structural components implement a function, in terms of changes in the system parameters involved. This is labeled the 'causal behaviour' of a component or system and represented as a causal graph.

The causal graph may also contain pointers to relevant domain principles and primitive components. However, the scheme as a whole doesn't seem to provide for a means to describe a function as a special instantiation of a combination of basic generic behaviour-structure(-form) mappings. Furthermore, there is no distinction between a generalized, 'typical' design situation (as represented in a prototype) and a specific design case.

We are looking for a layered representation of design knowledge, ranging from general physical and technical principles to case-specific knowledge. This makes most other existing design representation schemes less suited, since they have no means for explicitly representing these layers.

5. Conclusions

We have presented a framework for describing design as synthesis based on generic components at different levels of abstraction. These levels are characterised by the technology used and range from the system level down to the physical level. Both the number and the contents of these technology-based layers will differ from domain to domain.

The main advantage of the proposed framework for a knowledge-based approach is its universal and systematic nature. Once the basic inferencing and knowledge structures required for the transformations of design descriptions have been defined, they can be used for a large number of domains.

The resulting uniform description of different design processes also allows for comparison, on the basis of which more general design knowledge or principles may be obtained.

Rather than being alternative concepts or representation schemes, generic components and prototypes are complementary. On the one hand, generic components serve as the basis for the systematic construction of design prototypes. On the other hand, design prototypes are used to describe the use of the general knowledge represented in a generic component, in a particular design situation. In other words, generic components bridge the gap between the 'first principle' knowledge of a domain and the situational knowledge described in a design prototype.

6. Further Research

We will continue evaluating design cases in the domains of electrical, mechanical, and structural engineering on the basis of the proposed framework. At the same time, we are working on further implementation of both inferencing and knowledge structures as suggested by the framework. For the translation between different technology-based layers, we are investigating the links with the theory on so-called design grammars (e.g., (Stiny, 1989) and (Coyne, 1990).

In the context of the interrelation between generic components and design prototypes, two items for further research are particularly interesting.

The original concept of design prototypes also permits the construction of higher-level prototypes that refer to other prototypes (Tham, Lee and Gero, 1990). This way, classes of prototypes can be constructed. Another possibility would be to use higher-level prototypes for describing abstract phenomena, for instance: 'stiffening of a bridge'.

It is possible to regard generic components as part of the domain knowledge outside the prototype base, required in 'non-routine' design (Rosenman and Gero, 1989) for creating new prototypes. Although the set of basic generic components may be fixed and limited for a particular domain, there may be numerous useful combinations that can serve as the basis for the automatic generation of new prototypes.

Acknowledgements

This work was supported by a grant from the Netherlands Organization for Scientific Research (NWO). The ideas presented in section 4 of this paper have benefited substantially by discussions with the members of the Design Computing Unit, Department of Architectural and Design Science, the University of Sydney, in particular Mary Lou Maher and Leila Alem.

References

Alberts, L.K., Huijs, C. Mars, N.J.I. and Spaanenburg, L.: 1989, Structuring knowledge in VLSI design based on a universal model, *Proceedings Ninth International Workshop Expert Systems and their applications. Specialized conference Second Generation Expert Systems, Avignon, 29 May–2 June 1989*, EC2, Paris, pp. 131-143.

Alberts, L.K., Wognum, P.M., and Mars, N.J.I.: 1991a, A systematic approach to knowledge-based design, illustrated for discrete circuits design in electronics, *Proc. IJCAI Workshop "AI in Design", August 24th, Sydney*, pp. 59-75.

Alberts, L.K., Wognum, P.M., and Mars, N.J.I.: 1991b, *Design as science instead of art*, Memoranda Informatica 91-10, UT-KBS-91-02, ISSN 0923-1714, University of Twente, Department of Computer Science, Enschede, The Netherlands

Camposano, R.: 1990, From behavior to structure: high-level synthesis, *IEEE Design & Test of Computers*, pp. 8-19.

Coyne, R.: 1990, Logic of design actions, *Knowledge-Based Systems*, 3, pp. 242-257.

Cross, N.: 1989, *Engineering design methods*, John Wiley & Sons, Chichester.

Davis, R.: 1983, Reasoning from first principles in electronic troubleshooting, *Int. J. Man-Machine Studies*, 19, pp. 403-423.

Gero, J.S.: 1990, Design prototypes: a knowledge representation schema for design, *AI Magazine*, pp 26-36.

Gero, J.S., Maher, M.L. and Zhang, W.: 1988, Chunking structural design knowledge as prototypes, *in* J.S. Gero (ed), *A.I. in engineering: Design*, Computational Mechanics Publications,pp. 3-21.

Goel, A. and Chandrasekaran, B.: 1989, Functional representation of designs and redesign problem solving, *Proc. IJCAI'89*, pp. 1388-1394.

Hansen, F.: 1968, *Konstruktionsystematik*, VEB Verlag Technik, Berlin, DDR.

Joskowicz, L. and Addanki, S.: 1988, From kinematics to shape: an approach to innovative design, it AAAI-88, pp. 347-352.

Kneller, G.F.: 1965, *The Art and Science of Creativity*, Holt, Rinehart and Winston Inc.,New York.

Koller, R.: 1979, *Konstruktionsmethode für den Maschinen-, Geräte und Apparatenbau*, Springer-Verla, Berlin

Mittal, S., Dym, C.L., and Morjaria, M.: 1986, PRIDE: an expert system for the design of paper handling systems, *IEEE Computer*, July 1986, pp. 102-114.

Mittal, S. and Frayman, F.: 1989, Towards a generic model of configuration tasks, *Proc. IJCAI'89*, pp. 1395-1401.

Murthy, S.S. and Addanki, S.: 1988, PROMPT: An innovative design tool, *AAAI-87*, pp. 637-642.

Reuleaux, F.: 1963, *The kinematics of machinery: outlines of a theory of machines*, Dover Publications, New York (org. "Theoretische Kinematik: Grundzüge einer Theorie des Machinenwesens", first translation: 1876, Macmillan and Company, London).

Rinderle, J.R. et al.: 1989, Form-function characteristics of electro-mechanical design, *in* S.L. Newsome, W.R. W.R. Spillers and S. Fingers (eds), *Design Theory '88*, Springer Verlag, N.Y., pp. 132-147.

Rodenacker, W.G.: 1976, *Methodisches Konstruieren*, Konstruktionsbücher Band 27, Springer-Verlag, Berlin.

Rosenman, M.A. and Gero, J.S.: 1989, Creativity in design using a prototype approach *Preprints Modeling Creativity and Knowledge-based Creative Design*, Design Computing Unit, University of Sydney, pp. 207-232.

Simon, H.A.: 1969, *The sciences of the artificial*, The MIT Press, Cambridge, MA.

Stiny, G.: 1989, Formal devices for design, *in* S.L. Newsome, W.R. W.R. Spillers and S. Fingers (eds), *Design Theory '88*, Springer Verlag, N.Y., pp. 173-188.

Tham, K.W., Lee, H.S. and Gero, J.S.: 1990, Building envelope design using design prototypes, *AI in Building Design: Progress & Promise*, ASHRAE Symposium, St. Louis, Missouri, 1-33.

Walker, R.A. and Thomas, D.E.: 1985, A model of design representation and synthesis, *Proc. 22nd Design Automation Conference*, ACM/IEEE, pp. 453-458.

EXPLICIT REPRESENTATION OF DESIGN PROCESS KNOWLEDGE

J. TREUR
Artificial Intelligence Group
Department of Mathematics and Computer Science
Vrije Universiteit Amsterdam
De Boelelaan 1081a
1081 HV Amsterdam
The Netherlands

and

P. J. VEERKAMP
Department of Interactive Systems
Centre for Mathematics and Computer Science
PO Box 4079
1009 AB Amsterdam
The Netherlands

Abstract. In this paper a reflective architecture for modelling design tasks is discussed. This architecture makes an explicit distinction between two levels of reasoning. One level represents the reasoning about a design object (using object-level knowledge). The other level explicitly represents the reasoning about why, when and which process steps to undertake (here process or meta-level knowledge is used). The paper gives two examples of specification and representation languages for design tasks based on the two-level reflective architecture as described here: ADDL and DESIRE.

1. Introduction

The emphasis in the development of intelligent CAD systems is traditionally based on the representation of object knowledge. The strategic knowledge about how to direct and control the design process often is either absent or it is represented implicitly. A system with explicitly and separately represented process knowledge is however more modular, and thus easier to develop, debug and modify (e.g., see van Harmelen, 1989, pp. 14). In this paper, we discuss a reflective architecture that distinguishes two separate levels of knowledge representation: a *meta-level* and an *object-level*. The knowledge at the meta-level is used to control the modelling process of a design object. The latter is represented at the object-level. Based on this distinction two specification and representation languages for design tasks have been specified and implemented. One approach involves ADDL (Artifact and Design Description Language), a knowledge representation language for design with features of both logic and object-oriented programming. The other

J. S. Gero (ed.), Artificial Intelligence in Design '92, 677–696.

3. provide a satisfactory representation that comprehensively incorporates and integrates generalised design knowledge to bestow sufficient expressiveness and power which supports the activities mentioned in 1. and 2.

This paper describes a knowledge-based design system that attempts to address these issues in a cohesive and integrated manner. The system utilises design prototypes (Tham et al., 1990; Gero, 1990) to produce feasible designs within the category of routine design, and whose acronym is PROBER (**PRO**totype-**B**as**E**d **R**outine design system).

2. Philosophy of PROBER

PROBER is founded on the following philosophical underpinnings:

1. design is modelled as reasoning among function, behaviour and structure; and
2. design knowledge is schematised, and that appropriate chunks of knowledge are retrieved and applied at various stages as the design description is developed.

Only a summary of the main aspects of the philosophical framework is presented here to provide a sufficient background for discussing the architecture of PROBER. A fuller account of modelling design as reasoning among function, behaviour and structure, and in particular, the role of behaviour in design reasoning, is given in Gero et al. (1991) and Tham (1991). The design prototype as a powerful schema for representing design knowledge is described in Gero (1990) while an elaboration of the processes working in tandem with design prototypes is elaborated in Tham et al. (1990).

2.1. Function, Behaviour and Structure in Design

A designed artifact may be broadly interpreted in terms of function, structure and behaviour. Function reveals the intention of the artifact, structure specifies what the artifact is composed of and how the structure elements are interconnected, while behaviour spells out how the structure of the artifact achieves its function. Design concerns itself with the selection or production of structure to meet functional requirements. In Gero et al. (1991), we have argued that behaviour provides an appropriate platform for reasoning between function and structure in design because of its teleological association with function and its basic nature as manifestations of structure.

2.2. Design as Reasoning among Function, Behaviour and Structure

Designing an artifact may be modelled as the process of traversing the design space (of structure) and choosing structures that produce a design description (or alternatives) which satisfy the requirements. Knowing the behaviours of the structures provides clues to the designer as to the utility of the structures to contribute to the satisfaction of the requirements. Understanding the behaviour of the structure makes the selection of structure less intuitive and more reasoned. Options for satisfaction of a function may be explored and considered by examining the behaviours that contribute towards satisfying the function. By matching the behaviours of the structures with the behaviours that are associated with the functional requirements, a basis for structure selection is established. Design may thus be modelled as a reasoning process where function, behaviour and structure are dominant considerations.

2.3. DESIGN PROTOTYPES AS POWERFUL SCHEMAS FOR REPRESENTING DESIGN KNOWLEDGE AND SUPPORTING DESIGN

The design prototype (Gero, 1990; Tham, 1991) has been developed as a representation schema for collating and integrating pertinent information relating to a design concept, as well as providing the means to operationalise the concept. As such, a design prototype includes pertinent descriptions of function, behaviour and structure as well as embedding knowledge which supports the reasoning pertaining to design synthesis, analysis and refinement.

How does the design prototype support design? Of necessity, design must commence often with incomplete information and before all relevant information is available. Design involves exploration, drawing on already available information as well as seeking pertinent information that facilitates the continuation of design. As a schema, the design prototype permits design concepts to be retrieved based on partial matching of the contents contained in the respective design prototypes. Each retrieved design prototype brings in pertinent information which:

1. enhances elaboration of what the designer already consciously knows of the concept represented by the design prototype;
2. facilitates the exploration of design possibilities within the boundaries of the design prototype; and
3. provides a means to invoke associated concepts which is made possible via the structure elements of a design prototype when they are design prototypes in their own right at a different level of granularity.

The relational, computational and qualitative knowledge within a design prototype supports the synthesis, analysis and refinement of a design description by providing explicit knowledge for reasoning among function, behaviour and structure. In the design prototype paradigm, the evolution of a design description through the interplay of function, behaviour and structure reasoning across the spectrum of retrieved design prototypes, addresses significant issues related to the commencement and continuation of design such as the following:

1. the emergence of additional functions as design progresses;
2. the identification of relevant behaviours at various stages of the design;
3. the generation of structure description, and how structures are interrelated; and
4. the invocation of various knowledge that supports design synthesis, analysis, evaluation and refinement.

Design prototypes are therefore schemas which have sufficient expressiveness and comprehensiveness for representing design knowledge and are powerful enough to support the commencement, continuation, exploration and refinement of a design.

2.4. PROCESSES FOR DESIGNING WITH DESIGN PROTOTYPES

In the design prototypes paradigm, the bulk of the knowledge base is organized as a prototypes base, which is a library of design prototypes. Design is modelled as the process of continually finding appropriate design prototypes, deriving instances from them and synthesising these instances to compose the desired design description. The design description is then evaluated and refined if necessary until the design requirements are satisfied. Processes pertinent to designing using design prototypes are design prototype retrieval, design prototype selection, design prototype instantiation, instance initialisation, instance evaluation, instance refinement, requirements management; and control.

The design process commences by specifying a set of design requirements which are processed to yield retrieval variables on which design prototype retrieval is based. Design prototype selection is then executed as necessary. Selected design prototypes are then instantiated and the specified requirements assigned to the instances. Instance initialisation procures values for the variables, followed by instance evaluation. When the requirements are not satisfied, instance refinement occurs, with refinement actions proposed by the heuristics embedded within the design prototypes. If no further satisfactory refinement is proposed, then requirements management is invoked which either redistributes the requirements to other instances or cause the requirement to be converted to new requirements. New requirements then spawn the next cycle of processing similar to the once just described. Design *synthesis* is modelled by design prototype retrieval and selection, design prototype instantiation (which identifies the relevant variables from the design prototype to be included in the instance, and therefore constructing a parameterised design solution) and instance initialisation (where values are assigned to the variables to provide a specific solution). Design *analysis* occurs during instance evaluation where the actual behaviour of the instance is determined from the values of the relevant structure and behaviour variables). Design *evaluation* corresponds to instance evaluation where the actual performance, or behaviour, of the instance is compared with the requirements. Design *refinement* is modelled by the instance refinement process.

There are other processes which are not fundamental to design prototype-based routine design, but nonetheless, are included for efficiency reasons. Control of the design processes, in the form of planning or scheduling, creates a logical order for the processes, for example, design prototype retrieval, selection and instantiation may be sequenced consecutively; so may instance initialisation, evaluation and refinement; etc. Multiple tasks may be generated during the design process, but sequential processing permits only one task at a time. Record keeping and maintenance processes are essential in design prototype-based routine design which tends to be opportunistic rather than procedural. Where supplementary constructs are utilised, such as instance and requirements networks to facilitate design information storage and retrieval, then the corresponding processes are required, for example, network building and maintenance.

The processes identified are described in more detail and specificity when we discuss the architecture of PROBER.

3. Architecture of PROBER

The architecture of PROBER, Figure 1, reflects the operational organisation and integration of the processes involved in the design paradigm described thus far. The various processes are organised and placed under the jurisdiction of separate modules that assume responsibility for:

1. the control and integration of the respective specific tasks that affect the processes associated with each module; and
2. processing the information that flows through the module which either results in advancing the design state or in generating additional information that helps to advance the state of design.

Protocols for interfacing these modules are designed to co-ordinate information flow among themselves to produce an integrated system.

3.1. DESIGN PROTOTYPE BASE

Central to the paradigm is a repository of design prototypes called the design prototype base. The design prototype base is the equivalent of a knowledge base in an expert system. As the class of design addressed is routine design, feasible design solutions may be derived by instantiating the relevant design prototypes and assigning appropriate values to the variables in the instances. There is no necessity to appeal to external knowledge in the generation of feasible design descriptions. As such, the contents of the design prototypes do not change with design processes.

3.2. PROTOTYPE ENGINE

The prototype engine is responsible for all operations that pertain to design prototypes. These include the indexing of design prototypes to facilitate the retrieval of design prototypes during design, the retrieval of design prototypes in response to design requirements, the selection of appropriate design prototypes from among all the retrieved design prototypes, and the instantiation of design prototypes to generate instances in response to the demands of the design situation. Accordingly, the prototype engine comprises four agents: design prototype indexer, design prototype retriever, design prototype selector and design prototype instantiator.

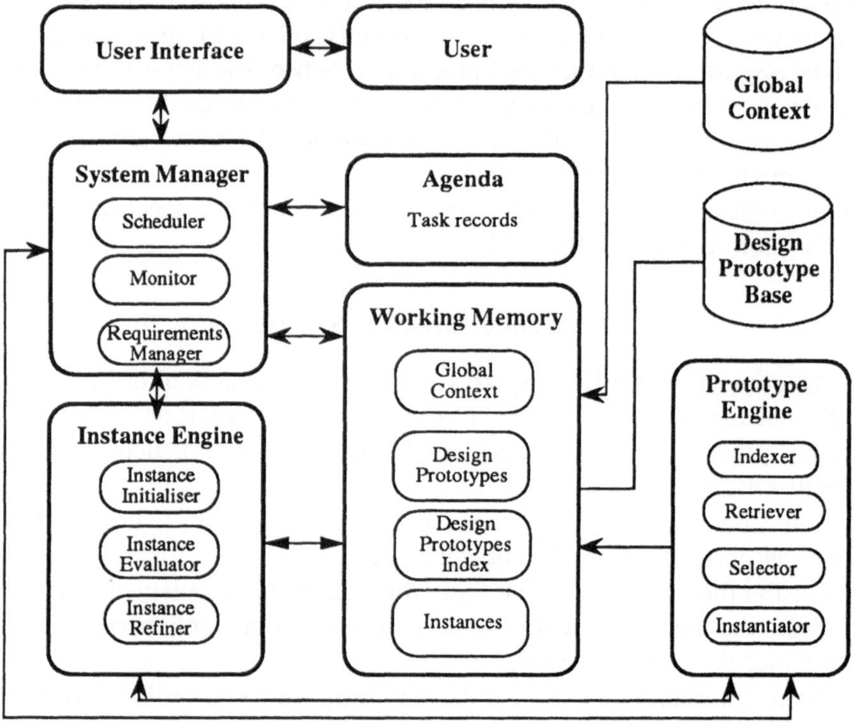

Fig. 1. Architecture of a prototype-based routine design system (arrows indicate data flow)

3.2.1 Design prototype indexer. As a schema, the design prototype may be accessed by its various descriptors. Just as a schema may be invoked by matching part of its contents, a design prototype may be retrieved based on partial matching of its contents. The information contained in a design prototype that provides meaningful indices for retrieval are the labels, functions, behaviours, behaviour variables, structure and structure variables. The prototype base indexer generates an index based on the information described under these index groupings within the design prototypes.

3.2.2. Design prototype retriever. Cognitively, design prototype retrieval matches the process of searching the designer's mind for relevant design prototypes by examining the contents embedded within the ontological boundaries defined by design prototypes. It may be interpreted as the identification of schemata to apply to a situation given a description of a situation.

In PROBER retrieval is based on direct symbolic matching between retrieval criteria and the contents in the design prototype index. Retrieval criteria are derived from requirements which are specified by the user or generated during the design process. The process retrieves all design prototypes that contain matching descriptions specified in the retrieval criteria, producing a list of design prototypes such that:

1. each retrieved design prototype contains descriptors that match at least one retrieval criterion; and
2. together, the retrieved design prototypes match all the retrieval criteria.

Retrieved design prototypes may be classed as perfect matches or partial matches. A perfect match is a design prototype whose contents match all the retrieval criteria. If only a single design prototype is identified, then a unique perfect match is said to occur; where more than one perfect match is found, the matching is non-unique and design prototype selection is required. A partial match is a design prototype that matches at least one, but not all of the retrieval criteria. Partial matches must be combined to meet all the retrieval criteria, an issue discussed in Section 3.2.3.

Retrieved design prototypes provide a basis for the continuance of design in two ways. Firstly, detail information about parts of the design referenced only by name is readily available through the description and knowledge contained within their corresponding design prototypes. Secondly, candidate structures which are potentially capable of meeting design requirements are presented to the designer. These structures are retrieved on the basis of their behaviours or functions, and the designer may examine the contents of their corresponding design prototypes before deciding which ones to select for design development. The retrieved design prototypes provide additional information which may not have been previously made explicit to the design system, and by so doing, supports design exploration.

3.2.3. Design prototype selector. Design prototype retrieval merely identifies the design prototypes that match the requirements, either perfectly or partially. There is no need to select among the retrieved design prototypes when the retrieval is based on label, or when there is only one perfect match, or when, in the absence of a perfect match, there is only one combination of partial matches that matches all the requirements. In such situations, all the retrieved design prototypes are essential in the generation of a feasible design. Design alternatives arise in other situations, and since PROBER does not support the maintenance of multiple design alternatives, the selection of retrieved design prototypes is necessary.

Design prototype selection reduces a set of retrieved design prototypes to either a single design prototype (which is a perfect match) or a set of partially matched design prototypes which together matches all the requirements.

By examining the information in the retrieved design prototypes, the designer can manually pick the combination such that all the requirements are covered. Alternatively, an algorithm for generating these 'perfect' partial match combinations, based on recursively selecting the design prototype that matches the most requirements at each cycle, has been developed and may be applied optionally in PROBER.

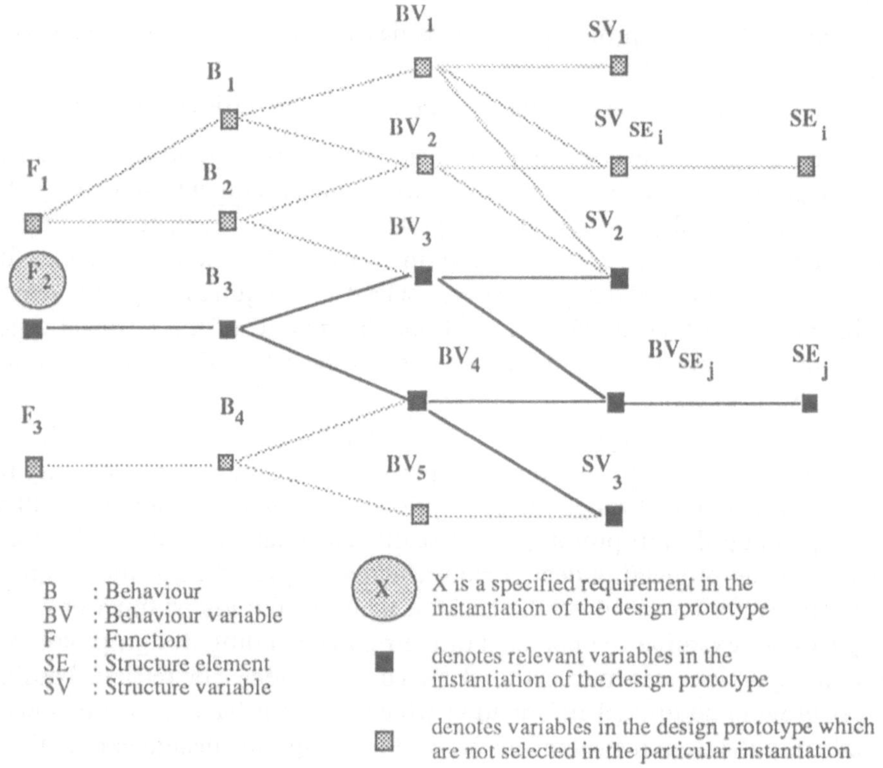

Fig. 2. An instance derived from a design prototype by traversing the dependency network

3.2.4. Design prototype instantiator. Design prototype instantiation consists of generating instances from the selected design prototypes and incorporating essential information that describes and facilitates the production of the instances.

The identification of the relevant variables is accomplished by appealing to the relational knowledge of the design prototype. The relational knowledge may be schematically represented in the form of a dependency network (Tham et al., 1990) in which nodes represent the functions, behaviours, structure elements and their characterising variables, and the links between nodes represent their relations (dependencies). By traversing the dependency network of the design prototype, the pertinent functions, behaviours, structure elements and their variables are extracted. The instance is formed by incorporating this extracted information. Figure 2 shows the dependency network of a design prototype and the relevant variables of an instance generated in response to the requirements of F_2. Values for these variables are decided during instance initialisation.

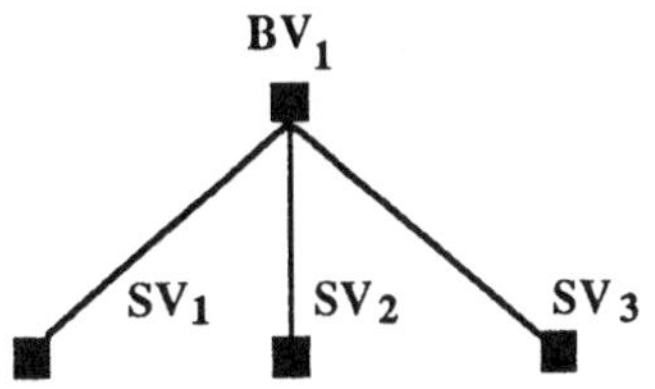

$$BV_1 = \phi\,(SV_1, SV_2, SV_3)$$

↓ reformulation

$$SV_2 = \psi\,(BV_1, SV_1, SV_3)$$

Fig. 3. Reformulation of original formula to compute the value of SV_2

A dependency tree is constructed for each proper tree that may be identified from the traversal of the dependency network in picking out the pertinent variables. The number of dependency trees corresponds to the number of roots that are present in the instantiated portion of the dependency network. Each dependency tree is constructed by extracting the tree associated with each of the roots.

The purpose in constructing the dependency tree is to assist in determining the order in which the variables are to be assigned values. The value of a node on the tree is dependent on the nodes immediately below it, and the strategy is to assign values to the nodes at the bottom of the tree as far as possible, and then applying the computational knowledge to calculate the values of the nodes above. Sometimes, instead of assigning values to all the nodes at the bottom of the tree, values may be assigned to other nodes higher up in the tree. The computational knowledge is then used to calculate the values of the nodes at the lower levels of the tree. This may require the reformulation of the formulae represented in the computational knowledge.

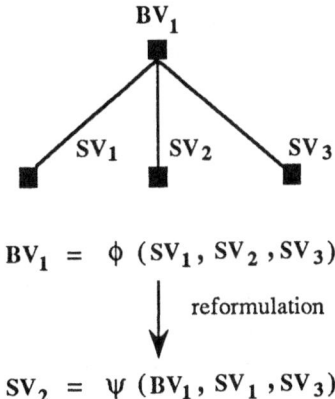

Figure 3 shows an example of the reformulation. Given an original formula which specifies BV_1 as a function of SV_1, SV_2 and SV_3, the assignment of values to BV_1, SV_1 and SV_1 would necessitate the reformulation from $BV_1 = \phi\,(SV_1, SV_2, SV_3)$ to $SV_2 = \psi\,(BV_1, SV_1, SV_3)$ in order that the value of SV_2 may be calculated under the closed world assumption.

The dependency tree also offers a check as to whether a sufficient number of variables have their values assigned before the computational knowledge is applied. For a formula relating n variables, the values for $n - 1$ variables must be known in order to calculate the value of the remaining variable. If this is not the case, the variables whose values are not known are indicated for the designer to assign values for the variables.

Constraints are posted on the instance by the requirements manager (see Section 3.5). When a design prototype is retrieved, the requirements which provide the basis for its retrieval are stored. Upon instantiation of the design prototype, these requirements are posted as constraints on the corresponding variables of the instance.

3.3. Instance Engine

Whereas the prototype engine operates on design prototypes, the instance engine operates on instances, moulding them into a design solution that satisfies the specified requirements. When design prototype instantiation produces an instance, the values of variables are not known. Thus, the instance must be initialised, evaluated and refined whenever necessary to derive a solution through applications of appropriate operations. To facilitate these operations, three agents are incorporated in the instance engine, namely, the instance initialiser, instance evaluator and instance refiner.

3.3.1. Instance initialiser. Instance initialisation is the process of assigning values to the variables of the instance. This may be accomplished by value assignment, value computation, value inheritance or defaults.

The dependency trees provide information pertaining to the preferred order for assigning values to the variables. After values have been assigned, the system then scans the dependency tree to see if the values of other variables can be computed from the current information. Values for variables which can be determined are computed, using formula reformulation where necessary. Those variables for which values cannot be computed due to inadequate information being available are converted into requirements for the next level of design.

3.3.2. Instance evaluator. Evaluation is an essential part of design where the values of the variables are checked against their constraints. (Here, design requirements are modelled as constraints on the corresponding variables. A significant proportion of the variables tend to be behaviour variables which characterise the performance of the artifact). If all the constraints are satisfied, the instance is evaluated to be satisfactory, and its contribution to the design is considered completed. The system can then focus on other instances. If the instance is unsatisfactory, the instance refiner is invoked.

3.3.3. Instance refiner. The instance refiner is invoked for two reasons: firstly to revise the values of the variables in the light of constraints violation, and secondly to improve the performance of the instance. Refinement heuristics to generate plausible suggestions for instance refinement are based on the qualitative knowledge associated with the corresponding design prototype. Such suggestions are only directional in nature, and the designer arrives at a feasible solution incrementally. The basic cycle of the refinement process is the suggestion of values for some of the variables, propagating the effects of the changes in these values to other variables and then evaluating the instance against the constraints or the improvement in the values of the behaviour variables.

3.4. Working Memory

The working memory in this system provides a storage for pertinent context information and design information generated during the design process which comprise the following:

1. an instance base consists of all the instances generated during the design process;
2. a design prototype base constructed by loading in all the design prototypes before the commencement of the design process;
3. a design prototype index; and
4. a global context which stores information which may be relevant to the instances on an across-the-board scale (the exogenous variables specified in a design prototype are mapped onto corresponding variables described in the global context).

3.5. System Manager

The system manager's role is to control problem-solving activities of the system by coordinating the various agents' processes (e.g. design prototype retriever, selector, instantiator, etc.). The system manager consists of the requirements manager, monitor and scheduler. The control strategy corresponds to a blackboard model which is described in Section 4.

3.5.1. Requirements manager. Requirements must be satisfied in producing designs. They are crucial in the prototype-based design system because design processes are generated essentially in response to them and attempts to satisfy them. For example, from these requirements, appropriate design prototypes are retrieved, selected, instantiated and refined. Every requirement should match a variable in the design prototype index. In prototype-based design, there are two different sources of requirements: externally input by the designer and internally generated by the system during the instance refinement. The requirements manager performs the following tasks: requirements screening and requirements assignment.

Requirements screening is used to investigate if the requirements input by the user has a corresponding match in the design prototype index. Only requirements that have corresponding matches are considered, while the unmatched ones are signalled to the designer for modification. This rests on the assumption that all pertinent design knowledge is known a priori in routine design, including the interpretation of requirements. In PROBER the design prototype index provides a complete reference to all variables in all the design prototypes and any requirement should match against one of the indices for routine design. The model of routine design upon which PROBER is based produces solutions through iterative instance refinement. These solutions are feasible, but not guaranteed to be optimal. As such, PROBER excludes the specification of requirements that are cast in terms of optimizing objectives. It permits requirements to be stated in the form of constraints or merely as features that the design solution must possess. Where requirements are formulated as constraints, the requirements manager extracts the variable which is constrained and uses it for matching against the design prototype index. After the design prototype (which is retrieved using this constrained variable) is instantiated, the constraint is posted onto the instance.

After requirements have been screened, they are either assigned to existing instances or used to spawn a design prototype retrieval, selection and instantiation process to derive new instances, and then the requirements are assigned to these instances. In the former case, instances that already have the variable which matches against that in the requirement are indicated and the designer can opt to assign the requirement to any of these instances. If there are no matching variables in the current descriptions of the instances, their corresponding design prototypes are examined to see if the variable exists in the design prototypes. If so, then the designer has the choice of extending the instance by including this variable in the instance, or to direct the system to retrieve other design prototypes in response to the requirement. This would bring in other related variables and cause an update of the dependency trees. In the latter case, the requirement is used to spawn a design prototype retrieval process whereby new design prototypes are introduced into the current design to further the design.

3.5.2. Monitor. The monitor acts as watchdog over the various modules of PROBER and maintains the agenda of design tasks. It receives messages from each agent (for example the design prototype retriever, instance refiner, etc.) whenever they complete or cannot continue their tasks their processes. Each agent's message to the monitor is particular to the nature of the tasks they perform, and relays information about the status of the tasks and descriptions of the current state of the design which is essential for the furtherance of the design. Accordingly the monitor updates the agenda, adding new tasks, deleting completed ones and updating suspended one.

3.5.3. Scheduler. The scheduler's function is to select a task to execute from among a set of executable tasks in the agenda. In the current version of PROBER, scheduling is entirely manual. The list of executable tasks is displayed to the user who selects the task to be performed.

Task scheduling can be developed to be dynamic and opportunistic. This would require the development of scheduling heuristics for computing the priorities of the tasks. Such heuristics may be based on the task's relationships with the currently processing instance, the sequential position of the task in the course of design process and recency. For example, the scheduler may assign a high priority to a task associated with the currently processing instance, schedule a 'design prototype selection' immediately after 'design prototype retrieval' by assigning the next highest priority to the design prototype selection after the design prototype retrieval tasks, or give the highest priority to the most recently generated task. The most highly ranked task is then chosen and the corresponding agent which executes the selected task is invoked.

3.6. AGENDA

The agenda is conceptually similar to the agenda of blackboard systems. In blackboard systems, the agenda contains all currently invoked specialists and complete descriptions of the nodes that triggered them (Hayes-Roth et al, 1988). In PROBER the agenda is a record of all tasks that are generated during the design process. Each task record contains information relevant to the agent which would be invoked to execute the task. At the beginning of the design session, the agenda is empty. As design progresses, the agenda grows, recording every task generated. It maintains, among other information, the status of each task. Design is complete when all tasks have been performed. By recording the cycle number of each task, the agenda provides a record of the development of the design as well.

The agenda of PROBER is divided into three levels of depending on the characteristics of the tasks: system processes, instance focus and instance processes. The system process agenda contains three types of task records corresponding to system level control: design prototype retrieval, design prototype selection and design prototype instantiation. The instance focus agenda task record is used to facilitate the determination of the instance to work on. There are three types of instance process agenda task records which pertain to the initialisation, evaluation and refinement of the instances. Each task record contains information, where relevant, about triggering source, cycle number, record status and variables relevant to the agent that the record is associated with.

3.7. USER INTERFACE

The user interface provides a medium for information exchange between the user and the system. By providing runtime interaction, it serves the following functions:

1. information entry (for example requirements specification and assigning values to variables);
2. display of the contents of the agenda, depicting the status of the tasks, and accepting decisions from the user about the tasks to be performed; and
3. display of design prototypes and instances to enable the user to examine the ones he/she desires.

4. The Basic Control Cycle

As designing with design prototypes involves the different processes discussed earlier, a control mechanism is essential to keep track of and control the design processes. This control mechanism is embedded in the system manager which incorporates the scheduler, monitor and the agenda. It is adopted from the blackboard model (Engelmore and Morgan, 1988) as it supports opportunistic problem-solving. There is no plan associated for the design of an artifact, and the artifact evolves in a manner dependent on both the design prototypes selected and the way they are instantiated. Thus the design is determined dynamically, and is characteristic of the design processes and decisions.

PROBER differs from conventional blackboard systems in the representation of the solution. In conventional blackboard applications, the solution is structured hierarchically, with knowledge sources tied to different levels of the hierarchy. These knowledge sources are triggered whenever the solution corresponding to the level they are associated with changes, such as to satisfy the triggering conditions. Knowledge sources whose execution conditions are met, are subject to selection, whereupon the chosen knowledge source is executed. The solution is changed, triggering other knowledge sources, and the cycle repeats. In PROBER, the solution is represented as independent objects connected as a network which grows dynamically as design proceeds. Instead of knowledge sources, agents perform tasks specified in the prototype and instance engines. Task records are generated by the monitor, and contain information specifying the execution conditions and information which are to be passed to the respective agents. As the agents are invoked, they execute the tasks and produce changes to the design description. They also return information to the monitor which determines the tasks to be posted on the agenda.

The basic control cycle is depicted in Figure 4. It follows the trigger-invoke-schedule-execute-update model common to blackboard systems (Hayes-Roth et al, 1988). The notion of triggering is synonymous to the posting of such agenda task records. Invocation corresponds to checking the current solution state to identify which agenda tasks' executing conditions are satisfied, and subsequently modifying their status to executable. Scheduling selects one of the executable tasks for execution. The execution stage is the most complex of the tasks. The appropriate agent is invoked and the relevant information passed to the agent which then executes the task and updates the solution and the working memory. Information is sent to the monitor for record maintenance and the generation of new agenda task records as necessary.

5. Implementation

This section presents the implementation of PROBER. It outlines the programming paradigm and language adopted, knowledge representation scheme and the state of development of PROBER.

5.1. Programming Paradigm, Language and Environment

The programming paradigm adopted for PROBER is dictated primarily by the operations associated with designing with design prototypes. As the essential notion is one of instantiation and refinement of instances derived from design prototypes, an object-oriented approach is highly suitable. The object-oriented paradigm supports representation of design prototypes as classes, provides the means to operationalise

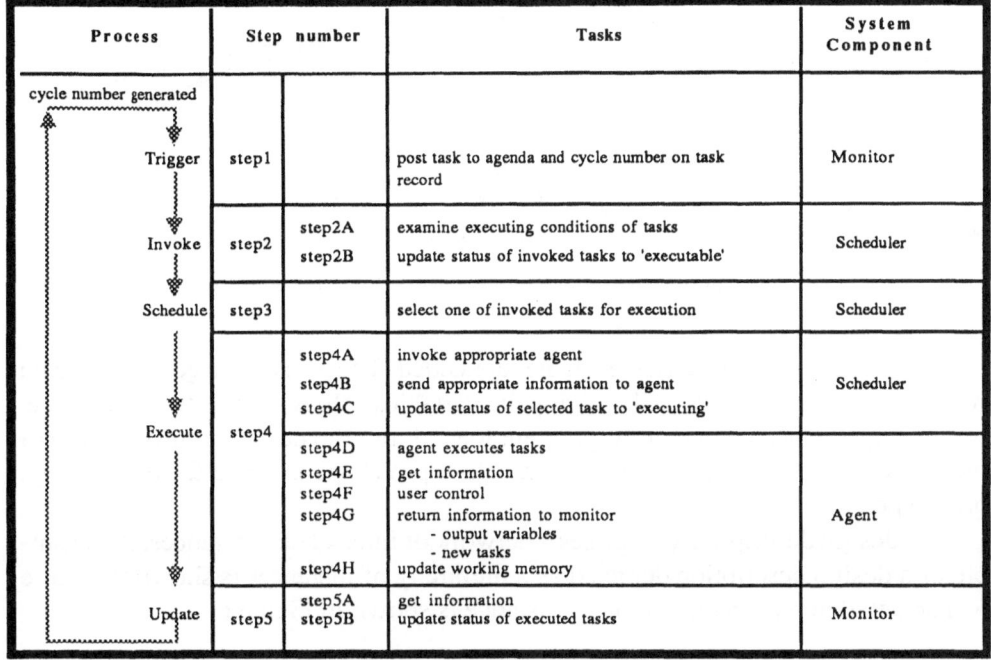

Process	Step number		Tasks	System Component
cycle number generated				
Trigger	step1		post task to agenda and cycle number on task record	Monitor
Invoke	step2	step2A	examine executing conditions of tasks	Scheduler
		step2B	update status of invoked tasks to 'executable'	
Schedule	step3		select one of invoked tasks for execution	Scheduler
Execute	step4	step4A	invoke appropriate agent	Scheduler
		step4B	send appropriate information to agent	
		step4C	update status of selected task to 'executing'	
		step4D	agent executes tasks	Agent
		step4E	get information	
		step4F	user control	
		step4G	return information to monitor - output variables - new tasks	
		step4H	update working memory	
Update	step5	step5A	get information	Monitor
		step5B	update status of executed tasks	

Fig. 4. Basic control cycle of PROBER

the concept by means of methods associated with the classes as well as supports inheritance among classes.

PROBER is developed in a UNIX environment on SUN workstations and is written in C++.

5.2. Representing Knowledge and Design Description

Design knowledge is made available to PROBER in the form of design prototypes, while the instances (of design prototypes) and the related design information produce the design description. Different classes are designed to represent the knowledge and information contained in them.

Each of these classes embeds other classes which provide descriptions of modular segments of the embedding class. This enhances modularity in the development of classes. Some of the embedded classes are general class descriptions that may be shared by other classes, enhancing the re-usability of code. Specified within each class are methods that describe how the information contained in the class may be put to use during the design process. These methods are invoked in response to the design situation, and serve to procure information from other instances and the working memory, generate information to further the design or to post information on other instances and the working memory.

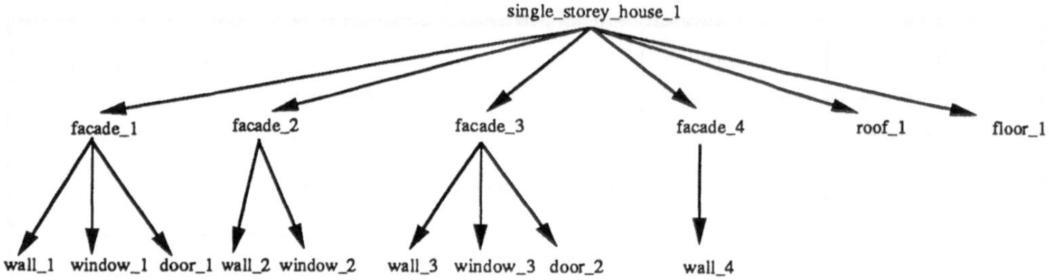

Fig. 5. Hierarchical representation of a design description

These classes are instantiated whenever needed. They serve as a generic schema which has to be made specific by giving the variables values specific to the instance. For example, when design prototypes are loaded into the design prototype base of the working memory, an instance of the class prototype is made for each design prototype.

The design description comprises a network of interrelated instances. An example of a design description depicted as a hierarchy of instances is shown in Figure 5. The detailed representation of an instance is shown in Figure 6.

5.3. Implementing the Agents of PROBER

All agents in all the modules of PROBER are implemented as classes. Describing each agent are the variables of the class, and the operations of each module are defined collectively by the methods associated with each of the class representing the module. Whenever an agent is invoked, an instance is made of its corresponding class, and the salient information for describing the tasks it has to perform is transferred to it from the scheduler (which extracts the information from the respective agenda record). Upon completion of its task, the instance is automatically destroyed by the inbuilt destructor function of the class, freeing memory for use.

In between, the associated methods are called in response to the design tasks required of the instance. These methods usually produce side effects of modifying the information in the working memory, particularly those pertaining to instances. In this way, the design description is elaborated. When the agent has successfully completed its task, its corresponding agenda task record's status is changed to 'completed'. Additionally, another agenda task record may be posted to reflect that other design tasks need be performed.

6. Discussion

This section discusses how PROBER address the issues of design commencement and continuation as well as function, behaviour and structure reasoning in design.

Recent research has alluded to the importance of behaviour as an intermediary to facilitate a deeper reasoning between function and structure (Murthy and Ad-

```
Instance      : single_storey_house_1
Prototype     : single_storey_house
          Structure_Elements:        facade_1, facade_2, facade_3, facade_4, roof_1, floor_1
          Structure_Variables:
                    length                                          value: 12
                    width                                           value: 8
                    height                                          value: 2.4
                    number_of_facades                               value: 4
                    number_of_roofs                                 value: 1
                    number_of_floors                                value: 1
          Behaviour_Variables:
                    interior_sound_level                            value: 51.5
                    winter_heating_requirement                      value: 1157
                    specific_heat_loss_rate                         value: 299
                    winter_heating_degree_hours                     value: 3869
                    winter_balance_point_temperature                value: 10.5
                    specific_conduction_heat_loss_rate              value: 261
                    specific_ventilation_heat_loss_rate             value: 38
                    winter_solar_gain                               value: 2000
                    number_of_air_changes_per_hour                  value: 0.5
                    volume                                          value: 230.4
                    area                                            value: 96
          Exogenous_Variables:
                    exterior_sound_level                            value: 85
                    winter_neutral_temperature                      value: 19.2
                    winter_average_temperature                      value: 5.3
                    winter_temperature_std_dev                      value: 2.7
                    internal_heat_gain                              value: 582
          Constraints:
            Behaviour_Variables:
                    must_be_satisfied:
                    interior_sound_level<55
                    winter_heating_requirement<1200
```

Fig. 6. Description of an instance in PROBER

1987; La Rota et al., 1990; Umeda et al., 1990; Tham et al., 1990; Gero et al., 1991). Given that the goal of design is to generate structure descriptions which satisfy the function requirements, function-structure reasoning is a predominant process. Systems that adopt an explicit function-structure relationship to assist in the choice of structure in response to functional requirements (Freeman and Newell, 1971; Maher and Fenves, 1985; Mackenzie, 1990) subscribe to a direct coupling between function and structure and preclude a deeper reasoning in the consideration of structure vis-a-vis function by forming a superficial bridge between function and structure. The model of routine design upon which PROBER is based provides a framework for formalising and supporting the reasoning between function and structure via behaviour. By embedding descriptions of function, behaviour and structure and explicating the relationships among these variable groups, the design prototype renders itself an appropriate representation schema to be used in conjunction with the model. PROBER contains the mechanism as well as a representation schema which explicitly models design as reasoning among function, behaviour and structure.

As a schema, the design prototype may be invoked by partially matching a portion of its description. Often, design commences and continues with insufficient information. With design prototypes, this information serves as a basis for retrieving design prototypes for consideration. Once invoked, the design prototype brings into the design such other information contained within itself which has not been available to the designer prior to the invocation of the design prototype. PROBER operationalises the essential processes for routine designing with design prototypes. These processes retrieve, select and explore relevant design prototypes, bringing in functions, behaviours and structures throughout the development of the design description. These functions, behaviours and structures may not be explicitly identified prior to the instantiation of the design prototypes, and their emergence which arises only through design exploration is supported. By drawing on design prototypes PROBER possesses sufficient expressive power for design commencement, continuation and exploration.

7. Conclusions

This paper describes PROBER, a system supporting routine design and is based on the following tenets:

1. a theoretical framework whereby design is modelled as reasoning among function, behaviour and structure; and
2. schematisation of design knowledge into comprehensive wholes (design prototypes) which are drawn upon throughout the design process to support design commencement, continuation, exploration and refinement.

Based on these tenets, a process model for routine design that explicitly adopts function, behaviour and structure reasoning as the fundamental mechanism for design development has been developed. This model is translated into an computational model by identifying the processes which operationalise the philosophy embodied within the design prototype paradigm.

As a prototype implementation PROBER provides a plausible computational approach which attempts to address issues such as: how it is possible to commence design when only meagre information is available initially; the emergence of additional functions and behaviours not specifically identified at the onset of design; the exploration of function-structure relationships on a rational, rather than intuitive, basis; and the development of a complex design solution (final structure description) from an initially scanty specification.

Some future directions utilising the concepts embedded within PROBER have been identified. PROBER may be modified to function as a parameterised design generator whereby function, behaviour and structure reasoning is adopted to synthesise an initial parameterised design description which may then serve as an input model to an operations research based paradigm to generate optimal design solutions. The framework for reasoning among function, behaviour and structure may also be utilised with creative design processes, such as mutation and analogy (Gero and Maher, 1991), to yield new design prototypes via the creation and incorporation of new design variables into existing design prototypes.

Acknowledgements

The contribution of Mr. Hyun Soo Lee in designing and implementing parts of PROBER is gratefully acknowledged.

References

Coyne, R. D., Rosenman, M. A., Radford, A. D., Balachandran, M. and Gero, J. S.: 1990, *Knowledge-Based Design Systems*, Addison-Wesley, Reading.

Engelmore, R. S. and Morgan, A. J. (eds): 1988, *Blackboard Systems*, Addison-Wesley, London.

Freeman, P. and Newell, A.: 1971, A model for functional reasoning in design, *IJCAI-71*, pp. 621-640.

Gero, J. S: 1990, Design prototypes: a knowledge representation schema for design, *AI Magazine*, **11**(4), 27-36.

Gero, J. S. and Maher, M. L.: 1991, Mutation and analogy to support creativity in computer-aided design, *in* G. Schmitt (ed.), *Proc. CAAD Futures '91*, ETH, Zurich, pp. 241-249.

Gero, J. S., Tham, K. W. and Lee, H. S.: 1991, Behaviour: a link between function and structure in design, *in* D. Brown, M. Waldron and H. Yoshikawa (eds), *Preprints of the IFIP WG5.2 Working Conference on Intelligent CAD*, Ohio State University, Columbus, pp. 201-230.

Hayes-Roth, B., Hayes-Roth, F., Roscenchein, S. and Cammarata, S.: 1988, Modelling planning as an incremental, opportunistic process, *in* R. S. Engelmore and A. J. Morgan (eds), *Blackboard Systems*, Addison-Wesley, London, pp. 231-244.

La Rota, J. L., Biswas, G. and Basu, P. K.: 1990, A model-based approach to structural design, *in* J. S. Gero (ed.), *Applications of Artificial Intelligence in Engineering V: Design*, CMP/Springer-Verlag, Berlin, pp. 3-22.

Mackenzie, C. A.: 1990, *Function and Structure Relationships and Transformations in Design Processes*, Ph.D Thesis, Department of Architectural and Design Science, University of Sydney, Australia.

Maher, M. L. and Fenves, S. J.: 1985, HI-RISE: an expert system for the preliminary structural design of high rise buildings, *in* J.S. Gero (ed.), *Knowledge-Engineering in Computer-Aided Design*, North-Holland, Amsterdam, pp. 125-164.

Murthy, S. S.and Addanki, S.: 1987, PROMPT: an innovative design tool, *in* J. S. Gero (ed.), *Expert Systems in Computer-Aided Design*, North-Holland, Amsterdam, pp. 323-341.

Tham, K. W.: 1991, *A Model of Routine Design Using Design Prototypes*, PhD Thesis, Department of Architectural and Design Science, University of Sydney, Australia.

Tham, K. W., Lee, H. S. and Gero, J. S.: 1990, Building envelope design using design prototypes, *ASHRAE Transactions*, **96**(2), 508-520.

Umeda, Y., Takeda, H., Tomiyama, T. and Yoshikawa, H.: 1990, Function, behaviour and structure, *in* J. S. Gero (ed.), *Applications of Artificial Intelligence in Engineering V: Design*, CMP/Springer-Verlag, Berlin, pp. 177-194.

EXPLICIT REPRESENTATION OF DESIGN PROCESS KNOWLEDGE

J. TREUR
Artificial Intelligence Group
Department of Mathematics and Computer Science
Vrije Universiteit Amsterdam
De Boelelaan 1081a
1081 HV Amsterdam
The Netherlands

and

P. J. VEERKAMP
Department of Interactive Systems
Centre for Mathematics and Computer Science
PO Box 4079
1009 AB Amsterdam
The Netherlands

Abstract. In this paper a reflective architecture for modelling design tasks is discussed. This architecture makes an explicit distinction between two levels of reasoning. One level represents the reasoning about a design object (using object-level knowledge). The other level explicitly represents the reasoning about why, when and which process steps to undertake (here process or meta-level knowledge is used). The paper gives two examples of specification and representation languages for design tasks based on the two-level reflective architecture as described here: ADDL and DESIRE.

1. Introduction

The emphasis in the development of intelligent CAD systems is traditionally based on the representation of object knowledge. The strategic knowledge about how to direct and control the design process often is either absent or it is represented implicitly. A system with explicitly and separately represented process knowledge is however more modular, and thus easier to develop, debug and modify (e.g., see van Harmelen, 1989, pp. 14). In this paper, we discuss a reflective architecture that distinguishes two separate levels of knowledge representation: a *meta-level* and an *object-level.* The knowledge at the meta-level is used to control the modelling process of a design object. The latter is represented at the object-level. Based on this distinction two specification and representation languages for design tasks have been specified and implemented. One approach involves ADDL (Artifact and Design Description Language), a knowledge representation language for design with features of both logic and object-oriented programming. The other

J. S. Gero (ed.), Artificial Intelligence in Design '92, 677–696.

approach is concerned with DESIRE (a framework for DEsign and Specification of Interacting REasoning modules), a formal specification language for compositional architectures of reasoning systems.

In this paper, we discuss meta-level reasoning in section 2. Section 3 presents an example design problem that is used to explain ADDL and DESIRE in section 4 and section 5 respectively. Finally, section 6 compares the two approaches and concludes the paper.

2. The Use of Meta-Level Reasoning

In this paper we describe an architecture based on a global task analysis of design tasks. One of the main issues is to distinguish process knowledge from object-level knowledge.

2.1. SEPARATING PROCESS KNOWLEDGE FROM OBJECT KNOWLEDGE

In design tasks knowledge on the object being designed is used. We consider this as *object knowledge*: statements referring directly to the object we deal with and not to the problem solving process. The description of the object consists of facts that may be true, false or (as yet) unknown: it is thought of as a picture of the object that should be created.

The object knowledge only specifies the facts and relations that hold or should hold in case of a solution. Separate strategic knowledge is needed and used to generate such a solution: in addition to the object knowledge, *knowledge on the problem solving process* is involved. This knowledge refers to the current state of the process and to (strategic) decisions that could be made to continue the process. It represents information about the state of the design process, but it does not tell anything directly about the object itself. At this *meta-level*, the system can assert (sub)goals that aim at a solution to the design problem or a part of it. The actual satisfaction of a goal takes place at the object-level, although proposals for that can be generated and compared at the meta-level. After satisfaction of a goal, the system evaluates the object-level information state by use of upward reflection and subsequent meta-level reasoning. Based on the evaluation new (sub)goals are asserted, if the design problem has not yet been solved. These new goals are again satisfied at the object-level. The use of reflection enables the designer to evaluate and determine the direction of the design process dynamically. For instance, based on the evaluation of a qualitative model of the design object obtained during conceptual design, the system formulates new goals for fundamental design. The system maintains a high level dialogue with the designer who is ultimately responsible for the direction of the design process.

Separating this knowledge from object knowledge is a basic modelling decision, needed to obtain a clear description (Treur, 1989 and 1991a). The description of the problem solving process deals with the solution steps (to be) taken in the design process and the argumentation for deciding to take a specific solution step. As will be shown in this paper the knowledge involved (to be called *process knowledge* or *strategic knowledge*) can be described in a (declarative) meta-level language.

2.2. A REFLECTIVE ARCHITECTURE

Meta-level architectures (Clancey, 1988; Maes and Nardi, 1988; Treur, 1991a; Weyhrauch, 1980) can be classified into several types of architectures. Fig. 1 illustrates a meta-level architecture. At the object-level reasoning is performed about the application domain and at the meta-level the object-level reasoning is controlled. The main advantage of such architectures is the separation between what the system knows about the world (the object-level) and what the system knows about how this knowledge is applied (the meta-level).

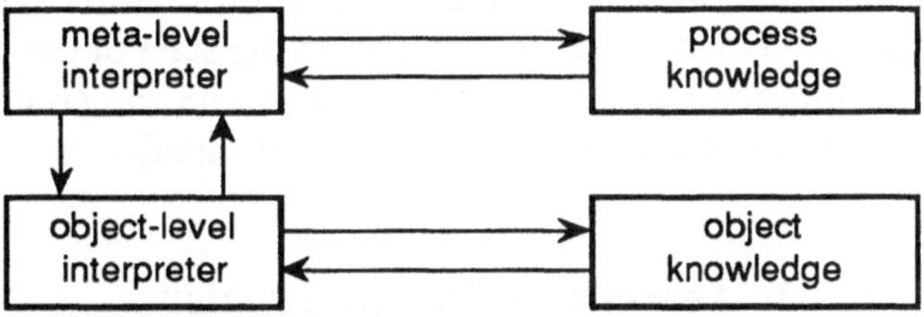

Fig. 1. A meta-level architecture

This paper presents a meta-level architecture where both the object-level and the meta-level knowledge-base are partitioned in *modules*. A module consists of a set of rules that are applicable to an information state. An information state consists of a (partial) description of either a design object or a design process state, depending on the level of reasoning. The application of the rules of an object-level module results in an extended *object information state*. An object information state embodies a (partial) model of the design object. It consists of the entities that describe parts of the design object decomposition and it consists of relationships among these entities. Therefore, the application of an object-level module causes an expansion of the entities and relations in the object information state. Object-level modules represent knowledge about a certain aspect of design. The reasoning involved is a forwardly directed activity. Initially the design object model consists of a minimal description which is gradually extended as the design proceeds.

The application of the rules of a meta-level module cause an extension of a *process information state*. A process information state consists of the following three categories.

A meta-level description of an *object information state*. This upward reflection principle enables meta-modules to evaluate the current design object description: which facts are true, false, or yet undefined.

Design state parameters that provide additional information to object-level literals. An object level literal represents a property of the design object description. A design state parameter adds information on the design process state to this property.

Process parameters that describe the flow of the design process. The names of asserted goals and the names of goals that have been satisfied add up to the process parameters that represent the history of the design process being conducted so far.

Thus, a process information state consists on the one hand of information about the truth values of object-level atoms which is obtained from the object information state by a so called *reflection principle* or reflective transformation (Treur, 1991ab). On the other hand, the process information state consists of other process parameters whose values depend on the state of the design process. These parameters provide additional information about the process state of the design object description, and the design goals being satisfied so far.

3. Analysis of Design Problems

In this section we give a global task analysis of design tasks and introduce an example design task that will be formally specified in more detail in the Sections 4 and 5.

3.1. PROBLEM DECOMPOSITION AND SUBPROBLEM SOLUTION

In design tasks a solution is usually not constructed in one step. A designer chooses an *adequate decomposition* of the problem into components that play a role as subproblems; these subproblems are solved by *generating a solution* to each of them in turn. Thus a solution of the design problem is constructed step by step, by a cyclic process of selecting a next subproblem to be solved and solving it (see Fig. 2). Such an approach is taken by Takeda *et al.* (1990).

A subproblem can be viewed as a subset of the set of all facts for which a truth value should be found. For each subproblem, those logical relations which were prescribed for the entire problem and refer to facts within the given subproblem should be met. However, the solutions of all subproblems only provide a total solution of the problem if the required relations between facts in *different* subproblems are also satisfied. Often problem decomposition is done hierarchically.

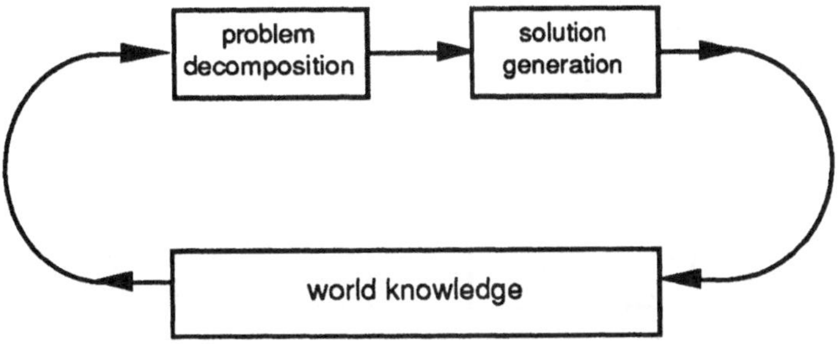

Fig. 2. Cyclic process of generating subproblems and solving them

The construction part of the problem solving process considers one at a time the subproblems generated by the problem decomposition task, and tries to determine a solution for this subproblem. This means that some unknown facts, related to the subproblem will get a truth value (true or false). These truth values should be chosen in such a manner that no requirements are violated. Since these requirements may occur as logical relations between the facts related to the current subproblem and facts related to subproblems solved earlier or to be solved later, not every solution of the subproblem may be extendible to a solution of the entire problem (in combination with solutions of other subproblems). Here the issue of *adequate strategic knowledge* comes in. In principle this process has a defeasible nature: a solution of the current subproblem is proposed, but if it turns out that with such a partial solution no satisfactory total solution can be found, one has to revise this partial solution (belief revision or backtracking). Therefore we consider the partial solutions essentially as *assumptions*.

3.2. A GENERIC TASK MODEL

In principle we distinguish four meta-level subtasks.

1. Decomposing the problem into subproblems
2. Selecting a subproblem to solve
3. Determining possibilities for solving this subproblem
4. Selecting one of these possibilities

For each of these tasks, we create a separate module. The exchange of information between the subtasks is depicted in Fig. 3. Task 1 will need (meta-) information about (the partial description of) the world state in order to determine which subproblems are yet to be solved. It will pass the subproblems to task 2, which will select one and pass it to task 3. Task 3 will reason about the subproblem to be solved. It will also need information about the world state in order to determine candidate solutions ('assumptions') for the given subproblem, and these will be passed to task 4. The assumption selected by task 4 will be 'transformed to a new fact in the world module; this is an exchange of information between a meta-level subtask and an object-level subtask (*downward reflective interaction*). We will explain this subtle but essential point in more detail.

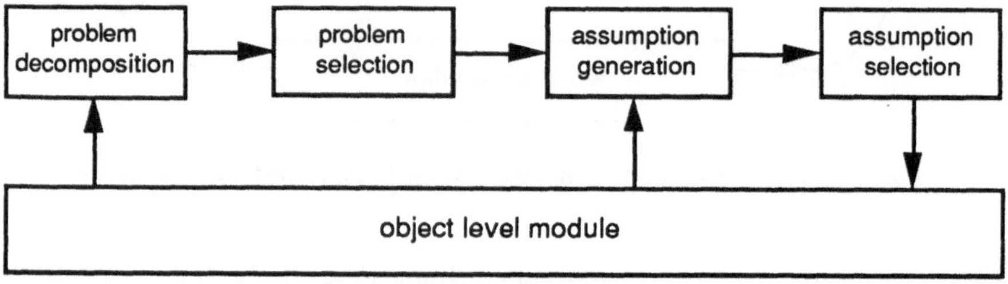

Fig. 3. Global Data Flow

At the meta-level, using strategic (meta-) knowledge it is just derived that some *assumption* about an assignment is preferred over the others. Once this has been derived at the meta-level, the assumption really will be *made* by the system. This means a new fact about the *object* has been created: once the assumption has been made it can be considered as just a simple object-level fact (that was not known before). The (partial) object-level knowledge on the world will thus be extended during the reasoning as additional assumptions are made.

To reason about the current state of the problem solving process, information is needed about this process state. This includes information about which object-level facts already have a truth value true or false, and which object-level facts still have a truth value undefined. To this end there are upward reflective interactions transferring epistemic information on the state of the object-level world to the meta-level.

3.3. META-LEVEL DESIGN CONCEPTS

The following concepts are used at the meta-level. They are used to evaluate and express information about the design process. Below, these are expressed by unary predicate symbols.

`positive(a)` Determines whether a fact `a` has a positive truth value in the object information state.

`negative(a)` evaluates whether a fact `a` has a negative truth value in the object information state.

`unknown(a)` `a` occurs neither as a positive fact nor as a negative fact in the object information state.

`abstract(a)` `a` is a positive fact and the description of `a` as given by the object information state is abstract. It indicates that a qualitative model of the concerned object(s) has been made.

`concrete(a)` `a` is a positive fact and the description of `a` as given by the object information state is concrete. The description of the concerned object(s) consists of a physical structure.

`exact(a)` `a` is a positive fact and the description of `a` as given by the object information state is exact. The description of the concerned object(s) fulfills the given functional specifications.

`goal(a)` `a` is the name of a goal that needs to be solved.

`solved(a)` `a` is the name of a goal that has been solved.

`possible_decomposition(D)` Given a goal, this concept expresses that `D` is a decomposition that may be used to solve this goal (one of the possible decompositions). One may view `D` as a list of components that serve as subgoals.

`selected_decomposition(D)` Expresses that the decomposition `D` has been chosen.

`component_of(X,D)` X is one of the components of `D` (element of the list).

`decomposable(a)` Expresses whether the goal `a` can be decomposed or not.

`goal_to_be_filled(a)` By this concept it is expressed that `a` should be filled in (e.g. by choosing a value for a basic parameter).

`user_positive(a)` Expresses that the user has given a positive answer to the question given by `a`. It gives the designer the opportunity to give his opinion about a possible path towards the solution as proposed by the system. This can be done both for accepting possibilities and for selecting one of the possibilities and both for decomposition and subgoal determination and for assumption determination.

`user_negative(a)` Similar: here the user has given a negative answer.

`possible_assumption(a)` This expresses that `a` is one of the possibilities to fill in a (not decomposable) goal.

`selected_assumption(a)` The possibility `a` has been chosen.

The first three concepts are merely relevant to meta-level reasoning in general rather than only to design problems.

3.4. DESCRIPTION OF A PROBLEM

Presented language constructs in this paper are illustrated by a design problem implemented by a student from a school on mechanical engineering (Algemene Hogeschool Amsterdam). They are part of a small design system for the design of an aquaplaning test. The assignment was done by a tire manufacturer. The system is part of a graduation project. The purpose of the project is the design of a technical device in order to visualize the phenomenon of aquaplaning. Test results must be a picture of the phenomenon of aquaplaning and a picture of the pressure distribution. A tire must be driven with a constant speed over a glass-plate with a variable load from 2650 till 7000 N. A sketch of the device is depicted in Fig. 6. In this paper we focus on the suspension of the designed device.

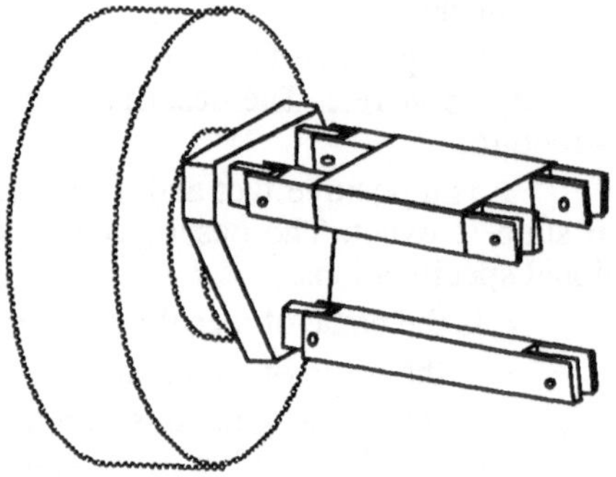

Fig. 4. The suspension of a device for testing the aquaplaning behaviour of tires

During the conceptual stage of design the device consists of the following three functional components: a wheel, a suspension, and a guide. In order to get a load of 7000 N on the tire, there must be a force of 7000 N on the wheel. This is achieved by letting the suspension act as a lever. The physical features, we are dealing with in this example, are forces and moments that have to do with a lever. The load on the wheel requires a certain strength of the suspension. When the suspension is further detailed and the length is is determined, then the moment of the force can be measured, and the tolerances of the suspension can be set. This can be achieved, since the system has knowledge about physical properties.

The suspension must be applicable for tires with different diameters. A parallel construction takes care of the constant orientation of the axis of the wheel. The suspension consists of two parallel beams and a fastening-plate (see Fig. 4). The upper beam is widened to absorb the tractive power caused by the acceleration of the vehicle. The lower beam is adjustable making it easier to change a wheel. Considering the design of the suspension we distinguish three stages (Brown and Chandrasekaran, 1989; Hubka, 1987; Mostow, 1985). The first stage involves the building of an abstract anatomical description of the design object. It consists of a description of the model in terms of function and behaviour. Furthermore, the limitations placed upon the resulting product are given by the designer, e.g. spatial requirements, cost constraints, etc. The next stage deals with the construction of a concrete anatomical description, that is, a description in terms of geometry. Finally, during the last stage the strength of the suspension is verified against the given specifications. For reasons of presentation, this paper only deals with the first stage of the design process.

3.5. ANALYSIS OF THE EXAMPLE DESIGN PROBLEM

Design as a stepwise refinement process (Tomiyama and Yoshikawa, 1987) can easily be regarded as a *goal-driven* activity. The original design problem is a goal that is subdivided in a sequence of subgoals. The subgoals are consecutively satisfied by the designer and/or a design system. However, the whole sequence of goals is *not* known beforehand. The designer dynamically chooses a next goal when the previous goal has been satisfied. This kind of strategic decisions also pop up during the design of the suspension. For instance, the designer has to choose for the upper beam among a single one, a widened one or a double beam depending on the amount of load. Furthermore, the designer has to choose among three possible mechanisms enabling the wheel to slide vertically. This kind of decisions are easily modeled by a reflective system as is shown in Sections 4, 5.

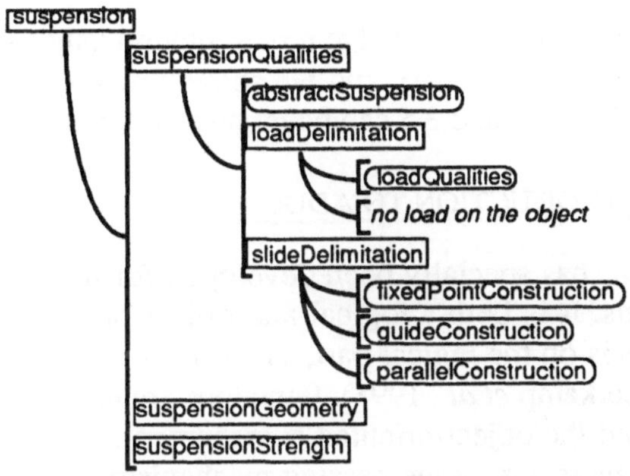

Fig. 5. Goal structure for the example

For reasons of presentation, we only present a working out of the conceptual phase of the design process. A list of goals that are consecutively satisfied describes this process. Goals can be satisfied either by object-level reasoning or by meta-level reasoning. Fig. 5 depicts part of the goal structure of the design of the testing device introduced in section 3. Goal names in the figure that appear in square boxes are solved using meta-level reasoning. Consequently, object-level reasoning takes care of goals appearing in rounded boxes. An outcoming arc of a box indicates that the goal is decomposed in subgoals. A multiple arc indicates a selection among several possible goals. Only one of the involved subgoals needs to be satisfied. An arc without a box refers to a selection where no goal needs to be satisfied. As an example the goal `loadDelimitation` is either composed of the subgoal `loadQualities` or it has no subgoal structure. The figure only depicts a detailed view on the first stage of the design process, also called conceptual design. During this phase, the designer constructs a *qualitative model* of the design object. The goal `suspensionQualities` deals with this phase.

The goal suspensionQualities aims at constructing a qualitative model of the suspension. By the same token, the goals `suspensionGeometry` and `suspensionStrength` stand for fundamental and detailed design respectively. They aim at building a *concrete* and *exact* model of the design object. In this context a model is called concrete when its geometry has been delimitated. Furthermore, a model is called exact when its strength meets the given requirements. A closer inspection of Fig. 5 reveals that only the subtree of goals satisfying the goal `suspensionQualities` has been depicted. The remaining goals have been left out of consideration.

4. Example Specifications Using ADDL

In this section, we explain the knowledge representation language ADDL (Artifact and Design Description Language), and we show how the example design task described in the sections 3.4 and 3.5 can be represented in ADDL.

4.1. A SHORT INTRODUCTION TO ADDL

The language ADDL has specially been developed for implementing *intelligent CAD* (ICAD) systems. It is both a logical and object-oriented language. In this paper, the emphasis is on the logical part. The object-oriented aspects of ADDL are discussed in (Veerkamp *et al.*, 1991). Functions are used to bridge the gap between the logical and the object-oriented environment. A function in an object-level expression triggers a message passing mechanism. Objects are instantiated by using prototype definitions. The manipulation of objects is entirely controlled by the logical (object-level) expressions.

Both the meta-level and the object-level language of the architectures presented in this paper are based on a subset of sorted first order logic. Modules consist of a set of rules that are equivalent to logical implications. The object-level language is based on three-valued logic. The meta-level language only uses the classical truth values and is based on the closed world assumption. If needed, object-level rules should therefore derive explicit negative conclusions while meta-level rules may not. In this paper the emphasis is on the meta-level language. For a thorough discussion on the object-level language, we refer to Veerkamp *et al.*(1991). The meta-level interpreter evaluates the rules of a meta-level module with respect to an process information state. In the sequel a module refers to a meta-level module unless stated otherwise. After application of a module the process information state undergoes a state transition to a next state. It is updated with the information that has been derived from the module.

Modules are called *scenarios* in ADDL and thus there are meta-level and object-level scenarios. Scenarios at both levels consist of a set of rules that have a declarative meaning, i.e., they can be read as true statements about either the world (object-level scenario) or about the process state (meta-level scenario). Actions and computations are left to the object-oriented part. For each design step (either at the process level or at the object level) there is an appropriate scenario that satisfies the concerning goal. Thus there are two distinct interpreters that control the reasoning at the two levels. The concerning interpreter evaluates the rules of a scenario in succession until the outcome solved(<scenarioname>) is met. The flow of control is not only between a meta-level scenario and an object-level scenario as Fig. 2 might suggest. There can also be a mutual flow of control between meta-level scenarios. One meta-level scenario can assert a goal that is solved by another meta-level scenario rather than an object-level scenario. The representation of several aspects of the design process necessitates this mechanism. We distinguish multiple levels at which the designer reasons about the design process (see Fig. 5 and (Veerkamp and ten Hagen, 1991)). In the following section we give five scenarios that represent the knowledge concerned with the design of a suspension. These are four meta-level scenarios and a single object-level scenario. In the sequel a scenario refers to an meta-level scenario unless stated otherwise. A more extensive discussion on ADDL can be found in (Xue *et al.*, 1990)

4.2. ADDL-SPECIFICATION OF THE DESIGN TASK

The scenario `solveSuspension` aims at the satisfaction of the goal named `suspension`. It controls the overall process concerned with the design of a part of the aquaplaning testing device, that is, its suspension. For each presented scenario, a *signature* gives a logical specification of the meta-language that is used in the scenario (the lexicon). The specification is independent of an actual implementation of the meta-language. The signature of a scenario `s` is denoted by Σ(`s`) More details on the signatures and the scenarios are given in (Treur and Veerkamp, 1991). The following knowledge-base belongs to `solveSuspension`.

Rules of solveSuspension:

```
1  IF unknown(isSuspension(ω)) THEN goal(suspensionQualities)
2  IF abstract(isSuspension(X)) THEN goal(suspensionGeometry(X))
3  IF concrete(isSuspension(X)) THEN goal(suspensionStrength(X))
4  IF exact(isSuspension(ω)) THEN solved(suspension)
```

The antecedents of the four rules are all concerned with an object-level literal `isSuspension()`. The first one is true if the literal is neither positively nor negatively present in an object information state. The second one is true if the literal *is* present and if its anatomical description is abstract. The latter means that a qualitative model of the suspension exists in the object information state. The third antecedent and the fourth antecedents are true if a concrete and exact anatomical structure have respectively been constructed. The goals of the rules have been explained in section 3.2. The argument of the goal function (or object-level atom) indicates that the scenario that satisfies the goal is focused on a portion of the object information state. This portion only contains the facts related to a suspension to which the argument `x` has been unified by the antecedent. The consequent of the last rule concludes that the goal `suspension` has been satisfied.

This section deals further with the three scenarios that aim at satisfying the goals in Fig. 7 whose names are enclosed by square boxes. Their names are `suspensionQualities`, `loadDelimitation` and `slideDelimitation` in order of presentation. The purpose of the first is to establish a qualitative model of the design object. The knowledge-base of `solveSuspensionQualities` is.

```
Rules of solveSuspensionQualities:
1 IF unknown(isSuspension(w)) THEN goal(abstractSuspension)
  & goal(loadDelimitation) & goal(slideDelimitation)
2 IF positive(isSuspension(X)) & solved(abstractSuspension)
  THEN abstract(isSuspension(X)) & solved(suspensionQualities)
```

The consequent of the first rule asserts a conjunction of three goals. The first goal is satisfied by an object-level scenario. It creates a qualitative model of a suspension by asserting a set of relationships that describe a suspension in terms of function and behaviour. An extensive discussion on qualitative modelling has been given in (Veerkamp and ten Hagen, 1991) A paper that deals with object-level reasoning about the same example as given in this paper. The second and third goal are satisfied by scenarios that are given further on in this section. The antecedent of the second rule is true if the literal `isSuspension()` is positively present in the object information state and if the goal `abstractSuspension` has been satisfied. It then concludes that the description of the suspension is *abstract* Furthermore, it concludes that goal `suspensionQualities` has been satisfied. The last two conclusions are added to the set of hypotheses that are derived form the scenario's rules. This set is merged with the current process information state when a state transition takes place.

The knowledge-base of the scenario `solveLoadDelimitation` contains two rules of which only one can be applied. It, therefore, inhabits strategic knowledge. There is either a load on the wheel of the vehicle or there is no load.

```
Rules of solveLoadDelimitation:
1 IF userPositive('loadOnWheel')
  THEN goal(loadQualities) & solved(loadDelimitation)
2 IF userNegative('loadOnWheel') THEN solved(loadDelimitation)
```

The antecedent of the first rule queries to the user interface whether or not there is a load of the wheel. The user interface is to be regarded as a kind of object information states that can be queried by the predicate symbol `userPositive`. In this case, if it returns true, the first rule is applicable. If it returns false, the second rule is applicable. Otherwise, the meta-level interpreter waits until an answer has been given. The consequents of both rules satisfy the current goal. Furthermore, the first rule asserts a goal that aims at extending the qualitative model in such a way that it can handle a load on the wheel. It causes the activation of the object-level scenario `solveLoadQualities`. Its knowledge-base consists of the following three rules.

```
Object-rules of solveLoadQualities:
1 IF isWheel(W) & typeOf(F,force)
  THEN loadedObject(W) & isForce(F) & loadOn(F,W)
  & solved(loadDelimitation)
2 IF isWheel(W) & loadOn(F1,W) & typeOf(L,lever)
  & typeOf(F2,force) & typeOf(F3,force)
  THEN isLever(L) & leverLoad(F1,F2,F3,L)
3 IF isWheel(W) & loadOn(F,W) & user('loadOnWheel',M)
  THEN value(magnitude(F),M) & loadQualities
```

Because all important decisions are taken at the meta-level, the object-level performs very little reasoning. The rules simply add new facts to the object information state. The binary built-in predicate symbol `typeOf` instantiates an object given by the first argument of the type given by the second argument. Thus, the unification of the atom `typeOf(F,force)` succeeds if the variable `F` is bound to an object of the type `force`. The first rule further reads as follows: if there is a wheel, then that wheel is a 'loaded' object, there is a force and that force causes the load on the wheel. The atom `isWheel(W)` unifies with a fact in the object information state that has been created during a previous phase of the design process.

The second rule states that the force that causes the load on the wheel is generated by a lever. The generation of the load by the lever generates two more forces whose values can be determined when a concrete anatomical description of the suspension has been made. The last rule asks the user-interface for the value of the load and assigns this value to the attribute `magnitude` of the force on the wheel by the expression: `value(magnitude(F),M)`. Furthermore, it concludes by means of the proposition symbol `loadQualities` that the goal of the scenario has been satisfied. It causes not only a state transition of the object information state, but also a state transition of the process information state because the meta-level interpreter adds the process parameter `solved(loadQualities)` by using a built-in upward reflection.

The last scenario is concerned with the choice for the actual solution of the suspension. There are three possible solutions, i) a construction with a fixed point fastening that provides a very simple but inelegant solution, ii) a construction with a sliding guide, which is a very precise but expensive solution, and iii) a parallel construction that is reasonably precise and still simple. Fig. 6 shows a working out of the last solution. The knowledge-base of `solveSlideDelimitation` is not given due to size limitations. The design process state of the suspension is *abstract* now.

The meta-level interpreter transfers control back to the scenario `solveSus-pension`. Its first rule has now successfully been applied. Its second rule asserts the goal `suspensionGeometry()` aiming at a geometric model of the suspension. Though these are not given in this paper they have been implemented in ADDL.

5. Example specifications using DESIRE

In this section we discuss the framework DESIRE (framework for DEsign and Specification of Interacting REasoning modules) that can be used to design and formally specify complex reasoning tasks and compositional architectures of knowledge-based reasoning systems that perform these tasks. We will show how the example design task introduced in the Section 3 can be specified in DESIRE.

5.1. A SHORT INTRODUCTION TO DESIRE

Within DESIRE, precisely defined notions of a module and of interactions between modules are used. This provides uniformity of specifications of both modules and interactions, which can serve as standardized building blocks and interfaces among them. By combining such building blocks, a compositional architecture is obtained that models the given complex reasoning task. Some of the main characterisistics of DESIRE will be discussed in this section. A more detailed description of DESIRE can be found in (Kowalczyk and Treur, 1990; van Langevelde *et al.*, 1991).

A clear distinction can be made between a description of the *global* architecture and of the *local* level. At the global level the architecture is viewed as a whole: as a number of reasoning modules that are active and have connections and interactions with each others (task model). At the local level specific information and knowledge elements are described.

In DESIRE a reasoning task can be specified in such a manner that not only *static aspects* are covered (data and knowledge involved), but also a complete specification can be given of the *dynamic aspects* (procedural aspects of the behaviour) of the reasoning. Static aspects as well as dynamic aspects can be specified both globally and locally.

A third main characteristic of DESIRE is that an explicit distinction can be made between *object-level* reasoning and *meta-level* reasoning. The connection between dynamic aspects and declarative descriptions of them is modelled by means of reflective interactions (reflection principles; see Treur, 1991ab). The separate strategic knowledge that is needed and used to drive and guide the problem solving process can be explicitly specified in DESIRE as modules that reason at the meta-level.

A fourth main characteristic of DESIRE is that it enables one to specify a generic part of the architecture (*generic task model*) separate from the domain-specific instantiation of it. A more extensive paper on the notion of a generic task model is (Kowalczyk and Treur, 1990).

A fifth main characteristic is that a specification using DESIRE can be expressed in a *formal knowledge level specification language* that, however, is not related to any specific knowledge representation or implementation environment. Given the fact that we use a formal specification language, dedicated editors could be made and actually are available to support this syntax. Our current design environment based on DESIRE contains a formal syntax description, a syntax- and hypertext-based editor, a parser, syntax and semantics checkers.

Given the (knowledge level) specification document, one is free to choose for any environment to build an implementation. Different *implementation environments* are available, and implementation generators that transform a formal specification into a working prototype implementation in one of these environments. One of the implementation environments is based on NEXPERT OBJECT™ and an extension of it written in C. The automatic implementation generator related to this environment is not able to treat variables and functions. A second implementation environment is based on Prolog. Here variables and functions can be treated.

5.2. DESIRE-SPECIFICATION OF A GENERIC TASK MODEL

The generic task model as given in section 3 can be specified in DESIRE as follows. Here for shortness global control flow, data flow and user interactions are left out. For more details, see the report version (Treur and Veerkamp, 1991).

```
reasoning module problem_decomposition
   input signature object_level_facts_signature
      sorts
         FACTS, PROBLEMS;
      relations
         positive, negative, known             : FACTS;
         goal, solved                          : PROBLEMS;
   output signature decompositions-signature
      sorts
         DECOMPOSITIONS;
      relations
         possible_decomposition                : DECOMPOSITIONS;
endmod
reasoning module problem_selection
   input signature problem_selection_input
      sorts
         PROBLEMS, DECOMPOSITIONS;
      relations

reasoning module possible_decomposition,
         user_positive                         : DECOMPOSITIONS;
         solved                                : PROBLEMS;
   output signature goal-signature
      sorts
         PROBLEMS;
      relations
         goal, goal_to_be_filled               : PROBLEMS;
endmod

reasoning module assumption_generation
   input signature selected_problem_signature
      sorts
         PROBLEMS;
      relations
         user_positive, user_negative,
         goal_to_be_filled                     : PROBLEMS;
output signature solution-candidates-signature
      sorts
         SOLUTION_CANDIDATES;
      relations
         possible_assumption                   : SOLUTION_CANDIDATES;
endmod

reasoning module assumption_selection
   input signature assumption_selection_input
      sorts
         SOLUTION_CANDIDATES;
      relations
         user_positive,
         possible_assumption                   : SOLUTION_CANDIDATES;
   output signature selected-solution-signature
      sorts
         PROPOSED_SOLUTIONS;
      relations
         selected_assumption                   : PROPOSED_SOLUTIONS;
endmod
```

The global reasoning pattern will start with module 1 (see Fig. 4). It will cycle through modules 1, 2, 3, 4 until either module 1 fails to determine any subproblems (in that case, the whole problem is solved), or until module 3 appears incapable of determining any solution candidate for a subproblem (in that case, the whole problem is unsolvable by the available strategic knowledge). Using DESIRE this global reasoning pattern can be expressed in the form of a control structure.

5.3. INSTANTIATION OF THE GENERIC TASK MODEL

The DESIRE-specification of the above generic task model can be filled with specific knowledge, resulting in a DESIRE-specification of a knowledge-based design system. For each of the meta-level modules we only give a part of the knowledge that should be added. For an active module the rules are meant to be evaluated exhaustively. For simplicity, we do not use functions here, although they can easily be used in DESIRE. More details can be found in the already mentioned report (Treur and Veerkamp, 1991).

problem decomposition

```
if   positive(X)
then solved(X)

if   unknown(X)
then not solved(X)

if   not solved(suspension)
then possible_decomposition([suspension])

if   goal(suspension)
then possible_decomposition([suspension_qualities,
     suspension_geometry,suspension_strength])
```

(and so on, following Fig. 7)

problem selection

```
if   possible_decomposition(D)
and  user_positive(D)
then selected_decomposition(D)

if   selected_decomposition(D)
and  component_of(X,D)
then one_of_the_subgoals(X)

if   one_of_the_subgoals(suspension)
and  not solved(suspension)
then goal(suspension)

if   one_of_the_subgoals(suspension_geometry)
and  not solved(suspension_geometry)
and  solved(suspension_qualities)
then goal(suspension_geometry)
```

```
if   one_of_the_subgoals(slide_delimitation)
and  not solved(slide_delimitation)
and  solved(load_delimitation)
then goal(slide_delimitation)

if   goal(X)
and  not decomposable(X)
then goal_to_be_filled(X)
```

assumption generation

```
if   goal_to_be_filled(load_qualities)
and  user_positive(load_on_the_object)
then    possible_assumption(load_qualities)
```

assumption selection

```
if   possible_assumption(X)
and  user_positive(X)
then selected_assumption(X)
```

object module

```
if   suspension_qualities
and  suspension_geometry
and  suspension_strength
then suspension

if   load_qualities
then load_delimitation
```

6. Conclusions and Comparisons

In this paper, we discussed an architecture based on a strict separation between object knowledge and control knowledge. When it is applied to a design system, this formalism leads to one level at which the design process is modelled and another level at which the design object is modelled. The formalism has been illustrated by the specification and implementation of an example design system in both ADDL and DESIRE. Both systems have a meta-level architecture although there are slight differences in functionality.

These differences may appear greater than they really are because two different approaches in modelling process knowledge have been used in the sections 4 and 5. The introduction of different modelling approaches serves a dual purpose. On the one hand, it gives some samples of what kind of representations can be used. On the other hand, a single and uniform method for the representation of meta-knowledge has not yet been established. The ADDL specification employs a scenario decomposition method which specifies scenarios in accordance with the (sub) goal structure. Therefore, for each goal there is a distinct scenario. This provides for a *fine-tuned* problem specific architecture.

In the DESIRE specification, the modules are grouped in accordance with the generic role that the applied knowledge plays in the design process. The rules that generate the decomposition are grouped. Also, the rules that select a goal and a decomposition are grouped and so are the rules that generate assumptions. This provides for a *generic* architecture that can be instantiated for more than a single problem. Please note that the used modelling approach is *independent* of the chosen specification language. The modelling approach as being used in ADDL could also have been used in DESIRE and vice versa.

A slight difference is that in the DESIRE specifications more rules are used. Part of this is due to the not essential reason that antecedents of rules in DESIRE do not use disjunctions nor do consequents of rules use conjunctions while they are both used in ADDL. However, some of the rules that are explicitly given in the DESIRE specifications do not occur in the ADDL-specifications. For instance, the explicit treatment of different possibilities for assumptions in the DESIRE-specifications takes six additional rules while this step is omitted in the ADDL-specifications. In the DESIRE specifications this shortcut could be made as well and in ADDL this knowledge could be added: it is a modelling decision not related to the medium that has been used. Such knowledge plays an essential role as soon as possible solutions to a design problem should be compared and an optimal choice should be made. But, if this is not the case the knowledge can be omitted. The same applies to the separation of generating possible decompositions of a goal and selecting one of them. Also this point is a modelling decision with both advantages and disadvantages.

There is an ongoing discussion about which modelling approach is most appropriate for the separation of object and control knowledge. However, it is generally agreed upon that such a separation is necessary for a successful modelling of complex processes such as designing. Therefore, we claim that meta-level architectures such as can be represented in ADDL and DESIRE are indispensable for truly intelligent CAD systems.

Acknowledgements

About the ideas described in this paper fruitful discussions have taken place with the other members of the IIICAD-project; especially with Pieter Geelen and Zsófia Ruttkay discussions have taken place in the context of the example modelling task related to the SISYPHUS-project.

References

Brown, D.C. and Chandrasekaran, B.: 1989, *Design Problem Solving; Knowledge Structures and Control Strategies*, Pitman, London.

Brumsen, H., Pannekeet, J. and Treur, J.: 1990, Modelling dynamic aspects of design processes, *Proc. of the Fourth Eurographics Workshop on Intelligent CAD*, Compiegne.

Clancey, W.J.: 1988, Representing control knowledge as abstract tasks and metarules, *in* Bolc and Coombs (eds), *Expert System Applications.*

Harmelen, F. van: 1989, A classification of meta-level architectures, *in* P. Jackson, H. Reichgelt and F. van Harmelen (eds), *Logic-Based Knowledge Representation*, MIT Press, Cambridge, MA, pp. 13-35.

Hubka, V.: 1987, *Principles of Engineering Design*, Springer-Verlag, Berlin.

Kowalczyk, W. and Treur, J.: 1990, On the use of a formalized generic task model in knowledge acquisition, *in* B.J. Wielinga et al. (eds), *Current Trends in Knowledge Acquisition (Proceedings EKAW-90)*, IOS Press, Amsterdam, pp. 198-221.

Langevelde, I. A. van, Philipsen, A. W. and Treur, J.: 1991, Formal specification of compositional architectures, *Report IR-282*, Department of Mathematics and Computer Science, Free University Amsterdam.

Maes, P. and Nardi, D. (eds) 1988: *Meta-level Architectures and Reflection*, North-Holland, Amsterdam.

Mostow, J.: 1985, Toward better models of the design process, *AI Magazine*, **6**(1), 44-57.

Takeda, H., Veerkamp, P. J., Tomiyama, T. and Yoshikawa, H.: 1990, Modeling design processes, *AI Magazine*, **11**(4), 37-48.

Tan, Y.-H. and Treur, J.: 1991, A bi-modular approach to non-monotonic reasoning, *Proc. of the First World Congress on the Fundamentals of Artificial Intelligence, WOCFAI'91*, Paris, pp. 461-476.

Tomiyama, T. and Yoshikawa, H.: 1987, Extended general design theory, *in* H. Yoshikawa and E. A. Warman (eds), *Proceedings of the IFIP WG 5.2 Working Conference on Design Theory for CAD*, North-Holland, Amsterdam, pp. 95-125.

Treur, J.: 1989, A logical analysis of design tasks for expert systems, *International Journal of Expert Systems*, **2**, 233-253.

Treur, J.: 1991a, On the use of reflection principles in modeling complex reasoning, *International Journal of Intelligent Systems*, **6**, 277-294.

Treur, J.: 1991b, Interaction types and chemistry of generic task models, *Proc. European Knowledge Acquisition Workshop*, EKAW-91, Springer Verlag.

Treur, J. and Veerkamp, P. J.: 1991, Explicit representation of design process knowledge, *Report IR-285*, Department of Mathematics and Computer Science, Free University Amsterdam.

Veerkamp, P. J., Pieters Kwiers, R. S. S. and ten Hagen, P. J. W.: 1991, Design process representation in ADDL, *in* P. J. W. ten Hagen and P. J. Veerkamp (eds), *Intelligent CAD Systems III - Practical Experience and Evaluation*, Springer-Verlag, Berlin, pp. 155-168.

Veerkamp, P. J. and ten Hagen, P. J. W.: 1991, Qualitative reasoning about design objects, *Preprints of the 5th International Conference on the Manufacturing Science and Technology of the Future*, Enschede, The Netherlands.

Veth, B.: 1987, An integrated data description language for coding design knowledge, *in* P. J. W. ten Hagen and T. Tomiyama (eds), *Intelligent CAD Systems I—Theoretical and Methodological Aspects*, Springer-Verlag, Berlin, pp. 295-313.

Weyhrauch, R.W.: 1980, Prolegomena to a theory of mechanized formal reasoning, *Artificial Intelligence*, **13**, pp. 133-170.

Xue, D. , Kiriyama, T., Veerkamp, P. J. and Tomiyama, T.: 1990, Representation and implementation of design knowledge for intelligent CAD—implementational aspects, *Proceedings of the Fourth Eurographics Workshop on Intelligent CAD*, Compiegne.

12

LEARNING IN DESIGN

MACHINE LEARNING TECHNIQUES IN OPTIMAL DESIGN

G. CERBONE
Department of Computer Science
Oregon State University
Corvallis OR 97331
USA

Abstract. Many important application problems can be formalized as constrained non-linear optimization tasks. However, numerical methods for solving such problems are brittle and do not scale well. This paper describes a novel framework for numerical optimization in engineering design that augments the traditional run-time optimization by a three step compilation. First, symbolic learning methods are used to partition the task into sub-problems that can be solved more efficiently. This also produces specialized versions of the optimization tasks that are faster to evaluate than the original. Second, each sub-problem is further simplified by using inductive discovery techniques to reduce the number of independent variables. This introduces a further speedup in the optimization process of the individual functions. Third, a novel ID3-like inductive learning algorithms is used to derive selection rules that associate problem instances to sets of candidate solutions. At run time, the problem solver uses these rules to map the problem instance into a set of efficient numerical gradient-directed optimizations that can be performed in parallel. In the domain of 2-dimensional structural design, this procedure yields a 95% speedup over traditional optimization methods and decreases the dependence of the numerical methods on having a good starting point.

1. Introduction

Many important applications can be formalized as constrained optimization tasks. For example, we are studying the engineering domain of two-dimensional (2-D) structural design. In this task, the goal is to design a structure of minimum weight that bears a set of loads.

Figure 1 shows a solution to a design problem in which there is a single load (L) and two stationary support points (S1 and S2). The solution consists of four members, E1, E2, E3, and E4 that connect the load to the support points. In principle, optimal solutions to problems of this kind can be found by numerical optimization techniques. However, in practice (Vanderplaats 1984) these methods are slow and they can produce different local solutions whose quality (ratio to the global optimum) varies with the choice of starting points. Hence, their applicability to real-world problems is severely restricted.

To overcome these limitations, we propose to augment numerical optimization by first performing a symbolic compilation stage to produce (a) objective functions that are faster to evaluate and that depend less on the choice of the starting point and (b) selection rules that associate problem instances to a set of recommended solutions. These goals are accomplished by successive specializations of the problem class and of the associated objective functions. In the end, this process reduces

J. S. Gero (ed.), Artificial Intelligence in Design '92, 699–717.

the problem to a collection of independent functions that are fast to evaluate, that can be differentiated symbolically, and that represent smaller regions of the overall search space. However, the specialization process can produce a large number of sub-problems. This is overcome by deriving inductively selection rules which associate problems to small sets of specialized independent sub-problems. Each set of candidate solutions is chosen to minimize a cost function which expresses the tradeoff between the quality of the solution that can be obtained from the sub-problem and the time it takes to produce it. The overall solution to the problem, is then obtained by solving in parallel each of the sub-problems in the set and computing the one with the minimum cost.

In addition to speeding up the optimization process, our use of learning methods also relieves the expert from the burden of identifying rules that exactly pinpoint optimal candidate sub-problems. In real engineering tasks it is usually too costly to the engineers to derive such rules. Therefore, this paper also contributes to a further step towards the solution of the knowledge acquisition bottleneck (Feigenbaum 1977) which has somewhat impaired the construction of rule-based expert systems.

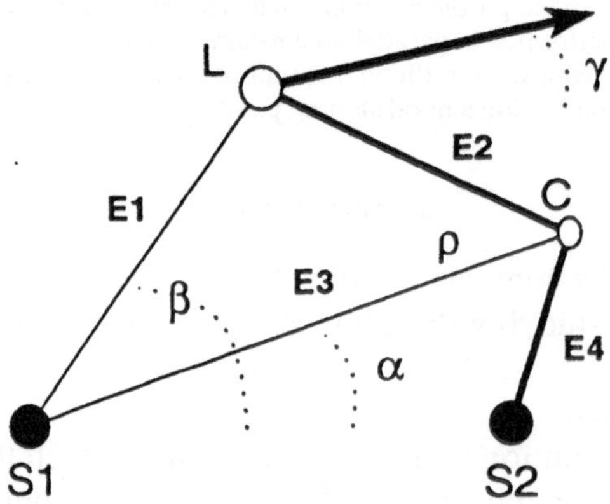

Fig. 1. A solution to a 2-D structural design problem with given topology.

Our optimization schema differs from techniques currently used in the machine learning community. Our approach relies on the specialization of the problem via incorporation of constraints prior to optimization. Braudaway (Braudaway, 1988) designed a system along the same principle. However, to our knowledge, very little work has been done in using learning techniques to speedup numerical optimization tasks. In contrast, the current trend in the machine learning community focuses on methods, such as Explanation Based Learning (EBL) (Ellman 1989), capable of generating rules. In addition, EBL methods have had little success in the task of optimizing numerical procedures. We conjecture that one of the reasons is the dependence of EBL methods on the trace of the problem solver. The trace of a numerical optimizer gives little information on the structure of the problem. Therefore, in mathematical domains, EBL-derived rules are too detailed to produce any appreciable speedup.

The remainder of the paper is organized as follows. Section 2 presents the 2-D structural design task. This is followed in Section 3 by an overview of numerical optimization methods, their limitations, and our solution which is illustrated using a simple example. The machine learning methods are outlined in Section 4. These methods are then applied in Section 5 which illustrates the experiments. These show that, for a certain family of problems, the compilation stage produces a substantial improvement in the performance of the optimization methods. Benefits and limitations of our strategy are summarized in Section 6, which also outlines future work.

2. Task description

Table I describes the 2-dimensional structural design task that we are attacking. Figure 1 shows an example problem in which L is the load and S1 and S2 are two supports. The so-called "topology" is given as a graph structure containing four edges (the members) and four vertices (the load, the two supports, and an intermediate connection point C). The topology does not specify the lengths of the members or the location of C. The topology and the position shown in the figure

TABLE I
The 2-D Design Task.

Given:	A 2-dimensional region R A set of stable points (supports) A set of external loads with application points within R
Find:	The number of members, connectivity, and positions of all intermediate connection points such that the structure has minimum weight and is stable with respect to all external loads.

give the minimum-weight solution. In this solution, 4 members are used and E1 and E3 are in tension (they are being "stretched"), while members E2 and E4 are in compression. Tension members will be referred to as "rods" and indicated by thin lines. Compression members will be referred to as "columns" and indicated by thick lines. The type of members used in the solution is an abstraction that we have used throughout our work. To indicate a configuration of tensile and compressive members that consititutes a solution, we have defined the *stress state*. The stress state is an array of m elements in which each element corresponds to a member. The value of each element in the array is $+1$ if the member is tensile and -1 if the member is compressive.

The weight of a truss can be decreased in at least two ways. First, the engineer can use lighter material. Second, the "shape" can be designed in such a way that, for instance, it uses less material and, hence, it is lighter. In this paper we do not consider the (admittedly) important advances in the science of material but, instead, we focus on the synthesis of shapes that reduce the weight of a truss with a chosen construction material.

The task shown in Table I is actually only one step in the larger problem of designing good structures. In general, structural design proceeds in three steps (Palmer and Sheppard 1970, Vanderplaats 1984). First, the problem solver chooses the topology, which specifies the locations of the loads and supports and the connectivity of the members. Then, the second step is to determine the locations of the connection points (and hence the lengths, locations, internal forces, and cross-sectional areas of the members) so as to minimize the weight of the structure. This is usually accomplished by numerical non-linear optimization techniques. The third and final step in the process optimizes the shapes of the individual members. This can often be accomplished by linear programming.

In addition to focusing only on the first two steps, we have introduced several simplifying assumptions to provide a tractable testbed for developing and testing machine learning methods. Specifically, we assume that structural members are joined by frictionless pins, only statically determinate structures are considered, the cross section of a column is square, columns and rods of any length and cross sectional area are available, and supports have no freedom of movement. A statically determinate structure contains no redundant members, and hence, the geometrical layout completely determines the forces acting in each member.

Given these assumptions, the weight of a candidate solution is usually calculated by a three-step process. The first step is to apply the *method of joints* (Wang and Salmon 1984) to determine the forces operating in each member. Once this is known, the second step is to classify each member as compressive or tensile. This is important, because compressive and tensile members are composed of different materials and have different densities; e.g. concrete columns and high tensile steel rods. The third step is to determine the cross-sectional area of each member. The load that a member can bear is assumed to be linearly proportional to its cross-sectional area. Finally, the weight of each member can be computed as the product of the density of the appropriate material, the length of the member, and the cross-sectional area of the member.

The last two steps can be collapsed into a single parameter k: the ratio of the density per-unit-of-force-borne for compressive members to density per-unit-of-force-borne for tensile members. With this simplification, instead of minimizing the actual weight, we can minimize the following quantity which, with an abuse of notation, we define as

$$Weight = \sum_{\substack{\text{tensile}\\ \text{members}}} \|F_i\|\, l_i + \sum_{\substack{\text{compressive}\\ \text{members}}} k\, \|F_j\|\, l_j.$$

F_i is the force in member i, and l_i is the length of member i. This is the initial objective function for the work described in this paper.

We conclude this section with a brief description of the method of joints, which is one of the methods used to calculate the F_i in statically determinate structures. The method of joints computes these forces by solving a system of linear equations as illustrated, for the problem in Figure 1, in Table II. The matrix of coefficients is called (Wang and Salmon 1984) the *axial* (or *static*) matrix and the vector of givens is defined as the *load vector*. In Figure 1, let $C = (x, y)$, $S1 = (x_1, y_1)$, and $S2 = (x_2, y_2)$, be the cartesian coordinates of the connection point, and the two supports, respectively. In addition, let (x_l, y_l) be the coordinates, and let p and γ be the magnitude and direction of the load L. The internal forces in each member are obtained by first constructing the axial matrix and load vector and then solving the system of equations for the unknown internal forces. Table II shows the symbolic system of equations for the example in Figure 1 with unknown forces F_1, F_2, F_3, and F_4 and with the coordinates of all the points explicitly substituted.

TABLE II

Method of Joints for the example in Figure 1. The product of the *axial matrix* and of the unknown forces F_i equals the *load vector*.

$$\begin{pmatrix} cos(\alpha_1) & cos(\alpha_2) & 0 & 0 \\ sin(\alpha_1) & sin(\alpha_2) & 0 & 0 \\ 0 & cos(\alpha_2+180) & cos(\alpha_3) & cos(\alpha_4) \\ 0 & sin(\alpha_2+180) & sin(\alpha_3) & sin(\alpha_4) \end{pmatrix} \begin{pmatrix} F_1 \\ F_2 \\ F_3 \\ F_4 \end{pmatrix} = \begin{pmatrix} Lcos(\gamma) \\ Lsin(\gamma) \\ 0 \\ 0 \end{pmatrix}$$

where: $cos(\alpha_1) = (x_1 - x_l)/l_1, cos(\alpha_2) = (x - x_l)/l_2$
$cos(\alpha_3) = (x_1 - x)/l_3,\ cos(\alpha_4) = (x_2 - x)/l_4$
$sin(\alpha_1) = (y_1 - y_l)/l_1, sin(\alpha_2) = (y - y_l)/l_2$
$sin(\alpha_3) = (y_1 - y)/l_3,\ sin(\alpha_4) = (y_2 - y)/l_4$

and l_i's are Euclidean distances: $l_1 = \sqrt{(x_1 - x_l)^2 + (y_1 - y_l)^2}$
$l_2 = \sqrt{(x - x_l)^2 + (y - y_l)^2}$
$l_3 = \sqrt{(x - x_1)^2 + (y - y_1)^2}$
$l_4 = \sqrt{(x - x_2)^2 + (y - y_2)^2}$.

Now that we have defined the 2-dimensional design task and formulated it as a non-linear optimization problem, let us turn, in the next section, to a brief review of existing techniques for optimization and to the proposed methods.

3. Knowledge-based Optimization

Classical optimization textbooks (Vanderplaats 1984, Papalambros and Wilde 1988) present a comprehensive survey of optimization methods and of various techniques for conducting the search for an optimal solution. The schema illustrated in Figure 2 is typical of many domain independent non-linear optimization methods. The process is iterative. Starting at some initial point, the objective function is evaluated and the termination criteria are tested. If the test fails, a new point is generated by taking a step, of some chosen length in some chosen direction, away from the current point. Each point defines a set of values for the independent variables in the objective function.

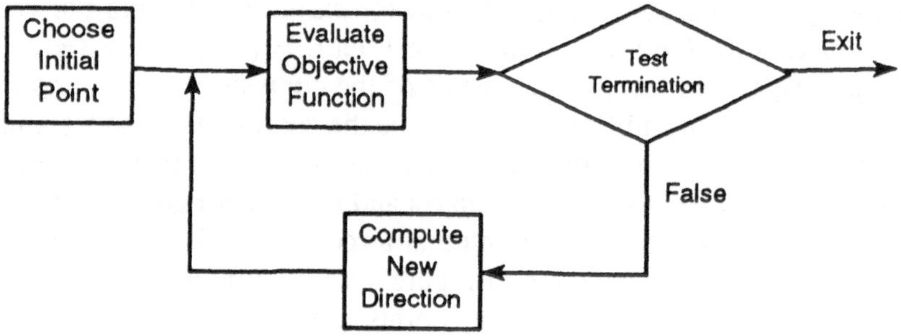

Fig. 2. Traditional optimization schema.

Most optimization algorithms differ primarily in the criteria used to choose the direction along which to optimize. Some optimization methods (e.g., Powell's method (Vanderplaats 1984)) choose the direction and step size using only evaluations of the objective function. Other methods, such as gradient descent and its variations (Papalambros and Wilde 1988), require computation of the partial derivatives of the objective function to choose the new direction of optimization. Still other methods approximate the partial derivatives numerically by evaluating the objective function at many points.

The primary computational expense of numerical optimization methods is the repeated evaluation of the objective function. An advantage of gradient descent methods is that they need to evaluate the objective function less often, because they are able to take larger, and more effective steps. Of course, they incur the additional cost of repeatedly evaluating the partial derivatives of the objective function. Hence, they produce substantial savings only when the reduction in the number of function evaluations offsets the cost of evaluating the derivatives.

In engineering design, the objective function is typically very expensive to evaluate. This slows the numerical optimization process because the speed of numerical optimization is determined by the cost and frequency of evaluating the objective function. For the structural design domain to compute the objective function (volume of each structure) a system of linear equations must be solved. This is typically carried out by algorithms which are cubic in the number of unknowns. This number is usually large in real applications like bridge design. Furthermore, the fact that the constant k is applied only to compressive members makes it impossible to obtain a differentiable closed-form. The signs of the internal forces must be computed before it is possible to determine which members are compressive. This prevents the use of gradient-based optimization methods that

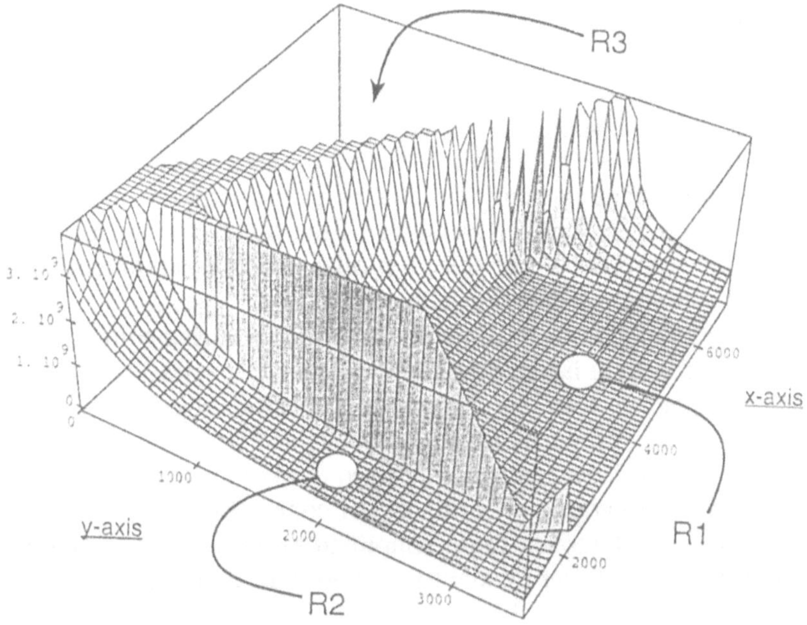

Fig. 3. Volume of the structure in Figure 1.

require fewer evaluations of the objective function – only slower function-based methods are applicable. One measure of the performance of a numerical optimizer is the time it takes to produce a solution. This quantity, however, depends on the choice of the starting point. Therefore, to obtain an accurate measurement, it is necessary to average the values obtained running the optimizer from different starting points.

Moreover, most engineering models are not unimodal. This directly affects the reliability of the solutions because numerical optimizers settle for local minima since they are unable to leap from one region to another to determine the global minimum. As shown in Figure 3, the objective function for the structural design domain is non unimodal. For instance, for the function in Figure 1 gradient methods started with $x = 1500$ and $y = 2000$ reach a local minimum in region R2 while the global minumum is in region R1. A measurement of the reliability can be obtained by taking the ratio (quality) of the local minimum and of the global minimum in controlled experiments in which the absolute minimum can be easily computed. Time and quality induce a tradeoff that can be exploited by defining the function:

$$\texttt{utility(solution)} = \texttt{CPUtime(solution)} * \texttt{CPUcost} + \texttt{quality(solution)}$$

where CPUcost is a positive constant that accounts for the cost of running the optimizer. We have used this definition in the learning stages of our approach to focus the attention of the optimization process on a few candidates that will produce solutions of maximum utility.

As shown in Figure 4, the increased reliability and speed are accomplished by augmenting the traditional run time optimization with a "compilation" stage prior to numerical optimization. The inputs to the compiler are (a) an high level description of the problem, (b) domain knowledge about stress states, and (c) a procedure to generate training examples. Symbolic and inductive techniques are then used to (1) produce simplified versions of the objective function per each stress state, and (2) learn stress state selection rules which map problem instances into sets of candidate stress states of minimum cost.

First, the compiler produces one objective function for each topology and stress state. Each of these functions is a specialized version of the expression of the weight and it is faster to evaluate than the original, less specific, objective function. As an example, the function produced for the topology and stress state in Figure 1 is illustrated in Table III. This expression is a closed form of the weight of a structure as a function of the two cartesian coordinates of connection point C restricted to region R1 in Figure 3. Moreover, these simplified expressions are differentiable and this permits the use of faster gradient-based optimization algorithms.

TABLE III
Partially evaluated objective function for the problem of Figure 1.

$$\begin{aligned} Weight = & \\ & (1.14\ 10^{13}x - 5.66\ 10^{9}x^2 + 8.16\ 10^{5}x^3 + \\ & 3.28\ 10^{13}y - 3.26\ 10^{9}xy + 2.44\ 10^{5}x^2y - \\ & 6.70\ 10^{9}y^2 + 8.16\ 10^{5}xy^2 + 2.44\ 10^{5}y^3 - 4.08\ 10^{16}) / \\ & (1.28\ 10^{1}xy - 2.56\ 10^{4}x + 2.56\ 10^{4}y - 6.40\ y^2 - 2.56\ 10^{7}) \end{aligned}$$

Another obstacle to practical applications of numerical optimization methods is the high dimensionality (number of independent variables) of the problems. Our compilation strategy decreases the dimensionality of optimization problems by searching a set of training examples for relations (regularities) among independent variables. These relations are then used as constraints among variables and are incorporated into the specialized versions of the objective function. This procedure eliminates independent variables with the result of greatly simplifying the optimization process, of enlarging its scope of applicability, and of speeding up run time optimization. For the region R1 in Figure 3, the compiler will determine that if the connection is expressed in polar coordinates ρ and α only the distance ρ from support S1 need be determined (see Figure 1.) This is because, in the analysis of the examples, it will discover that the angle α can be computed as one half of the angle β which is one of the givens of the problem. The final objective function is shown in Table IV which contains only a single variable ρ vs. the two (x and y) in the expression in Table III. This final expression indicates a reduction in dimen-

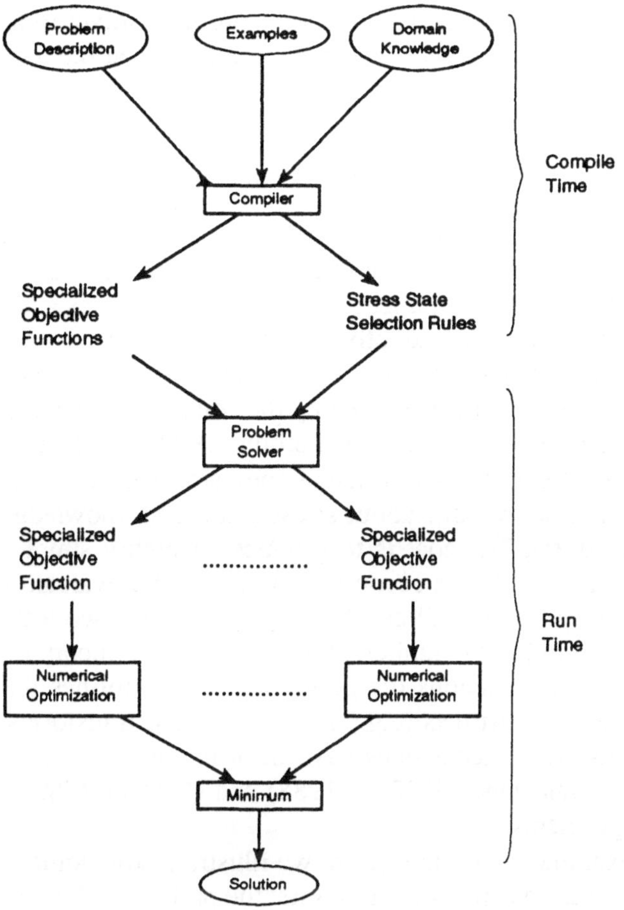

Fig. 4. Proposed numerical optimization framework.

sionality because, at run time, the numerical optimizer will only need to determine the value of ρ to compute the position of the connection point.

Finally, the compiler learns search control knowledge in the form of `IF-THEN-ELSE` rules. This is then used at run time to select stress states that lead quickly to quasi-optimal solutions. The set of stress states is chosen so that the *utility* of the stress states is maximized. The utility is a function that combines the time it takes to produce a solution with its expected quality (ratio to the global minimum.) This function introduces a tradeoff between quality and time that is exploited by the learning algorithm (Cerbone 1992). As an example, for the design problem in Figure 1 whose objective function is shown in Figure 3, the compiler derives search control knowledge that allows the problem solver to focus the attention of the numerical optimizer on regions R1 and R2 when the load is directed toward support S2 and away from support S1.

4. Machine Learning Methods

This section describes in greater detail the symbolic and inductive learning techniques. Inductive learning techniques are used to (a) simplify the optimization process by reducing the number of independent variables and (b) derive the stress state selection rules. The inductive methods rely upon knowledge about the partitioning of the design space and upon a set of training examples that, for many engineering tasks, can be generated by the compiler. A complete discussion of the compilaton stages can be found in (Cerbone 1992, in preparation).

Symbolic Methods. Symbolic techniques are used to incorporate into the objective function knowledge about stress states and knowledge discovered during inductive analysis. The goal is to produce an highly simplified and specialized objective function. This is accomplished by partial evaluation (Futamura 1971), and loop unrolling (Burstall and Darlington 1977) – two techniques widely used in high-end optimizing compilers. Partial evaluation incorporates constant values for variables into functions (or programs) and simplifies them. Loop unrolling unfolds iterative constructs (e.g., `for` loops) and transforms them into sequential programs. These techniques have been implemented using the Mathematica programming language (Wolfram 1988) and (Maeder 1989) which is suitable to numerical problems.

As an example of specialization, we illustrate how domain knowledge is used to specialize the objective function. First the problem solver chooses the topology. This can be simply done by enumerating a few possible configurations. Once the topology is chosen, it can be incorporated into the objective function. This allows us to compute symbolically the axial matrix and the load vector (see Section 2). We then apply symbolic algorithms to solve and simplify the system of equations and to obtain a closed-form expression for the forces. In principle, an infinite number of topologies should be explored; however, Friedland (1971) experimentally demonstrated that only a few of them need be considered to achieve satisfactory solutions.

The second specialization step is to plug in the givens of the problem and par-

TABLE IV
Objective function for the structure in Figure 1 with reduced dimensionality.

$$Weight_{simplified} \approx \left(1.16\ 10^{13}\rho - 5.19\ 10^{9}\rho^2 + 8.19\ 10^{5}\rho^3 - 4.08\ 10^{13}\right)/3.95\rho^2$$

tially evaluate the resulting mixed symbolic/numeric expression. For our examples, the givens of the problems are the loads and supports; however, one may wish to analyze a structure subject to different inputs such as various loading conditions or support locations. In such cases it is possible to leave those values in symbolic form and substitute their numerical values at run time.

The third compilation step is to split the objective function V into cases according to stress state. When the objective function is specialized according to stress state, the result is a collection of special-case objective functions $\{V_1, \ldots, V_n\}$. Because each V_j corresponds to one stress state, it is possible to tell, at compile time, which forces should be multiplied by k. Hence, each V_j is differentiable, and this enables us to employ gradient-based optimization techniques that, typically, are faster than methods based only on evaluating the objective function alone.

Reduction of independent variables. A further speedup and increase in reliability of the numerical optimizers is obtained using inductive methods to decrease the number of independent variables (*dimensionality*) in the numerical optimization problem. The compiler is given a series of examples and uses them to inductively determine which independent variables can be computed as functions of known quantities. For instance, in the design domain, when searching within a region it might turn out to be superfluous to search along all dimensions because there might exist a simple relationship between one of the coordinates and known quantities like the location of loads and supports. These relations are then used as constraints and are incorporated into the objective functions. The result is the reduction of the number of independent variables. This, in turn, produces an even simpler and faster optimization problem. For instance, the function shown in Table III has two independent variables while the corresponding inductively simplified version has only one independent variable and it is shown in Table IV. Hence, the final optimization problem entails a simple linear optimization while the original one has two dimensions.

The variables to be eliminated are determined using an EBL-like approach which employs:

- training examples
- a library of given geometry entities (points, angles, etc.)
- a geometrical domain theory
- known relationships among geometric entities
- *regularities* – a mixture of heuristics and statistical regression techniques.

Each unknown connection point is subject to a compile time heuristic search process that attempts to compute (reformulate) the location as a function of loads and supports.

To see how this works, let us consider again the example problem in Figure 1 which we shall refer to as the "bisector" example. In this example, the connection point `C` is the unknown and the givens are the load `L` and the supports `S1` and `S2`. Moreover, let us assume that a set of training examples has been either provided or derived by the system. The reformulation starts by identifying all geometric objects using the given domain theory. For the bisector example, the system identifies, among others, the following geometric objects:

```
point(S1), point(S2), point(C), point(L),
angle(β, L, S1, S2), angle(α, C, S1, S2),
segment(SG1, S1, S2), ...
```

Predicates such as `point` and `angle` are basic elements of the given geometric domain theory. This means that, given a set of cartesian coordinates, the system is capable of computing each predicate. During the computation of each predicate, the system tags it as *given* or *unknown*. A predicate is *given* if all the entities used to compute it are either givens of the problem (loads or supports) or can be expressed a combination of given predicates. Otherwise, the predicate is tagged as *unknown*. For the bisector example, `point(C)` and all predicates that involve it in their derivation (e.g. `angle(α, C, S1, S2)`) are unknowns, all others are givens.

With this knowledge, the system then tries to relate the unknown geometric entity `point(C)` to as many other entities as possible with the ultimate goal of expressing it only using given geometric entities. This is accomplished by using a blend of EBL and discovery techniques. In the EBL jargon, the geometric knowledge base is the *domain theory*, `point(C)` is the target *concept*, and the operationality criterion is the fact that a concept must be expressed in terms of known geometric objects. To visualize this reformulation step, let us refer to the derivation tree in Figure 5. The rightmost branch indicates that `C` is a connection point and, therefore, it is no longer explored. The leftmost branch, instead, uses a domain rule that reformulates a point in polar coordinates. Intuitively, the domain rule states that a point can be identified by its distance ρ from `S1` and by the angle α between points `C`, `S1`, and `S2`. With this in mind, the system recursively tries to determine `angle(α, C, S1, S2)` and `distance(ρ, C, S1)`. After having exploited all proofs, the system concludes that it is not possible to re-express the `angle` and the `distance` in terms of known entities. If we were to follow EBL strictly, we should conclude that the domain theory is incomplete; that is, it is not powerful enough to bridge the gap between unknowns and givens. This, in turn, implies that the search would terminate concluding that `point(C)` cannot be re-expressed in terms of known geometric objects.

To overcome this problem we have used a discovery approach that fills these knowledge gaps with *eureka* (Burstall and Darlington 1977) steps. Despite the name, however, in our strategy these steps are not arbitrary but inductive For the example in Figure 1, we determine that the angle α between points `C`, `S1`, and `S2` is exactly one-half the angle β between points `L`, `S1`, and `S2`. Once this *regularity* is determined, in contrast with Burstall and Darlington's approach, we test the eureka step against all user provided examples to determine if it is a random occurance or a widespread phenomenon. In the former case, any use of this regularity is abandoned and others (if any) are tried. In the latter case, the regularity is assumed

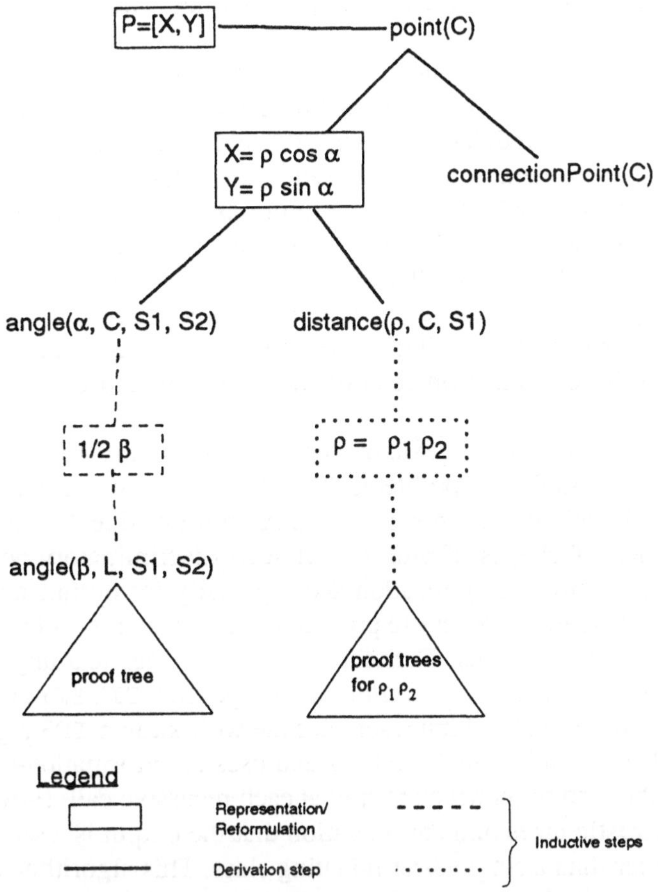

Fig. 5. Decision tree to derive the *concept* `point(C)`.

as a transformation of the unknown geometric entity. This is shown by the node in Figure 5 connected by the dashed lines. The system then subgoals on the geometric entities that were used to recognize the angle β. These are recognized as givens because they were derived from the position of the load and of the supports and the search terminates. The discussion of the branch identified by the dotted is similar to the one above and it is omitted for the sake of brevity.

The actual domain rules used in the geometric theory carry along also information that bridge the gap between the cartesian representation of a point and the polar one. This implies that the x and y coordinates of `C` can be expressed in terms of the angle α and of the distance ρ. In turn, the angle α is substituted by $\frac{\beta}{2}$ which can be computed from the given position of the load and supports. These transformations are considered as constraints and are incorporated into the objective function which is further simplified using the symbolic techniques. The result of the incorporation is shown in Table IV.

Rule derivation. The specialization steps discussed above greatly improve the running time of the optimizers on each objective function but they might introduce a large number of candidate solutions. These, in principle, can be exponential. To overcome this problem, we have devised a new inductive learning method to prune candidates that do not lead to optimal solutions. This method learns search control knowledge in the form of decision trees which can then be quickly transformed into `IF-THEN-ELSE` rules. These design rules associate features of the problem to a few regions in which the global minimum is believed to lie according to the examples given to the learning algorithms. The global solution is then obtained by running the optimizer on each of these regions and by taking the minimum solution.

We have found that most existing learning algorithms are not suitable for learning rules for optimization problems. The main obstacle is the absence of features that allow discrimination among classes. Algorithms like `ID3` implicitely require independence of classes. Features with such discriminatory power are difficult to derive for many real application and especially for optimization tasks. On the other hand, it is relatively easy to provide *shallow* features which can circumscribe a set of possible solutions. Therefore, in devising our learning method we have assumed that all features are *shallow* and proposed `UTILITYID3`, a novel learning algorithms. The algorithm resembles the well-known `ID3` algorithm (Quinlan 1987) in that it builds a decision trees and uses an information-theoretic heuristic to choose the feature on which to split at each recursive call. However, it is new in that the heuristic takes into consideration that the output is a set of recommended actions rather than a single discriminating class. This algorithm is fully described in (Cerbone 1992, in preparation) and (Cerbone 1992).

In addition to the learning algorithm, we have introduced *maximum utility learning set*, a new learning framework. In this framework, a utility is associated to each candidate solution. The problem is to learn a set of actions of maximum utility that covers all given examples. For instance, in the design problem, the utility is a function of the time it takes the numerical optimizer to find a solution. The quality is measured with respect to the globally optimal design. It turns out that this learning problem is $\mathcal{NP}-complete$ (Garey and Johnson 1979). Hence, `UTILITYID3` uses an approximation algorithm to determine a solution.

5. Experiments

To test the efficacy of this approach, we (Cerbone and Dietterich 1991) have solved a series of design problems using an implementation based on Mathematica (Wolfram 1988), and we have measured the impact of the compilation stages on the evaluation of the objective function, on the optimization task, and on the reliability of the optimization method. The measurements presented are averages over five randomly generated designs and, for each design, over 25 randomly generated starting points.

Objective function. The objective function of each design problem was evaluated in four different ways and, for each of them, we averaged the CPU[1] time over the different designs and starting points. The volume was first computed using the traditional, naive, numerical procedure with the method of joints. We then compiled the designs incorporating, in three successive stages, topological information, the givens of the problems, and the stress state. Figure 6 shows the time (per 100 runs) to evaluate the objective function at the various compilation stages. The biggest speedup was obtained with the numerical substitution of values into the symbolic closed form expression obtained and with the specialization to stress states. This suggests that the gain is related to the elimination of arithmetic operations from the original numerical problem.

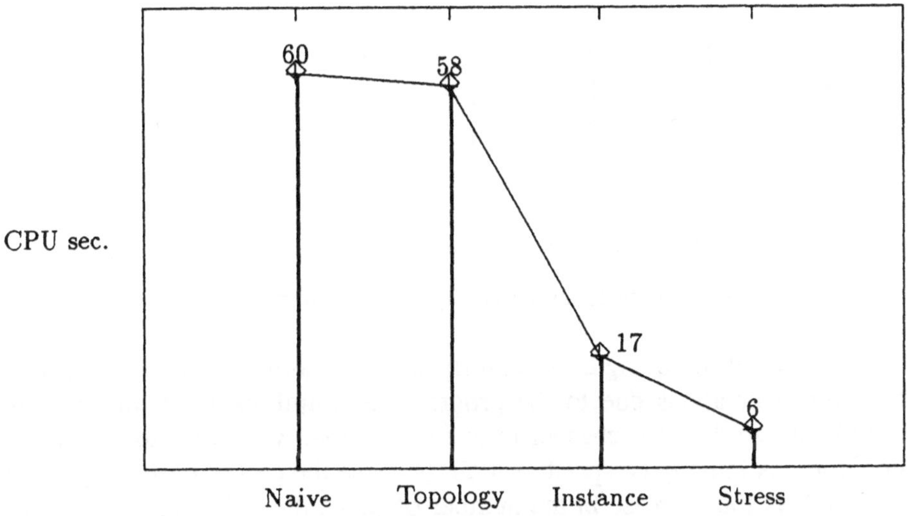

Fig. 6. Influence of the compilation stage on the CPU time per function evaluation.

Optimization. As indicated in Section 1, the running time of the optimizers is influenced by the number of function calls and by the time for each function evaluation. To present the benefits of our approach on the optimization task, we have experimented with two optimization algorithms (a) an optimizer based on Powell's method (Pike 1986) that does not require gradient information and (b) the version of conjugate gradient descent (Press et al. 1988) provided by Mathematica. The graphs in Figures 7 and 8 report, respectively, the number of objective-function calls and the overall CPU time for each optimizer. The values connected by solid

[1] The examples were run on a NeXT Cube with a 68030 board.

lines correspond to cases where the optimizer had no gradient information, while the values connected by dashed lines indicate averages utilizing the conjugate gradient descent method with alternative approximations for the gradient vector.

As expected, the number of evaluations remains constant throughout the compilation stages when the non-gradient is used, while it decreases drastically when we switch to the gradient-based optimization method.

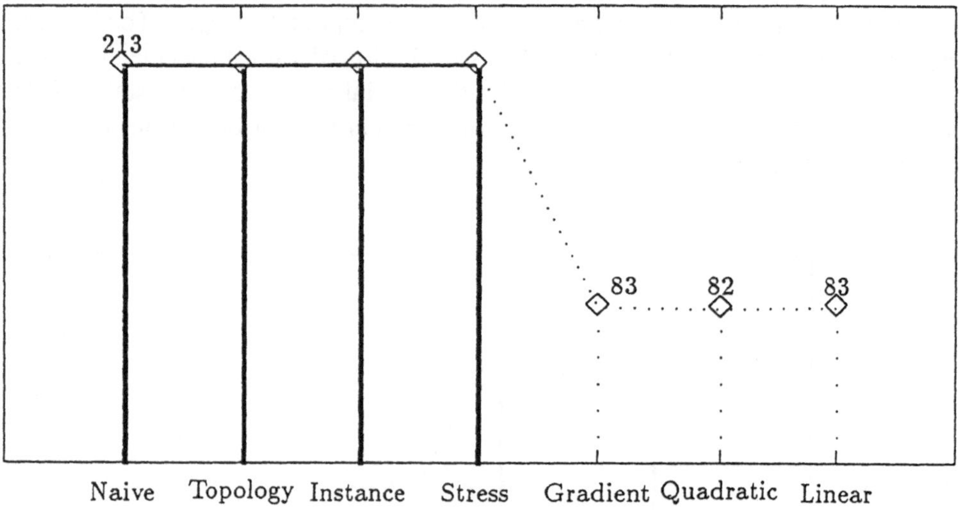

Fig. 7. Influence of the compilation stage on the number of function calls.

The overall CPU time (Figure 8) steadily decreases as well. For the non-gradient method, the decrease is due to the progressive simplification of the objective function itself, so that it is cheaper to evaluate. When we switch to the gradient method, there is initially no speedup at all, because the cost of evaluating the full gradient offsets the decrease in the number of times the objective function must be evaluated. However, additional speedups are obtained by approximating the objective function as a quadratic and as a linear function (by truncating its Taylor series).

We have found experimentally that there is no appreciable difference between the minima reached using the full gradient vector and the minima computed using quadratic approximations of the partial derivatives. However, the precision of the results obtained with the linear approximation is significantly reduced. Depending on the application, this trade of accuracy for speed may be acceptable. If not, the quadratic approximation should be employed.

Another possibility is to employ the linear approximation for the first half of the optimization search, and then switch to the quadratic approximation once the minimum is approached. In other words, the linear approximation can be applied to find a good starting point for performing a more exact search.

Reliability. An optimization method is reliable if it always finds the global min-

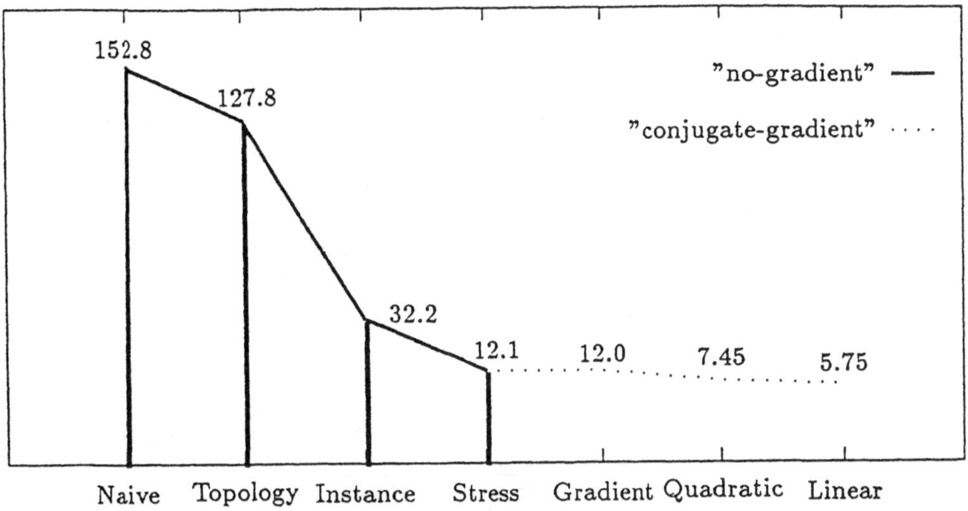

Fig. 8. Influence of the compilation stage on the CPU time.

imum regardless of the starting point of the search. Unfortunately, as shown in Figure 3, the objective function in this task is not unimodal, which means that simple gradient-descent methods will be unreliable unless they are started in the right "basin." It is the user's responsibility to provide such a starting point, and this makes numerical optimization methods difficult to use in practice.

From inspecting graphs like Figure 3, it appears that, over each region corresponding to a single stress state, the objective function is unimodal. We conjecture that this is true for most of 2-D structural design problems. This means that optimization can be started from *any* point within a stress state, and it will always find the same minimum. If this is true, then our "divide-and-conquer" approach of searching each stress state in parallel will be guaranteed to produce the global optimum.

We have tested these hypothesis by performing 20 trials of the following procedure. First, a random starting location was chosen from one of the basins of the objective function that did not contain the global minimum. Next, two optimization methods were applied: the non-gradient method and the conjugate gradient method. Finally, our divide-and-conquer method was applied using, for each of the specialized objective functions V_j, a random starting location that exhibited the corresponding stress state. In all cases, our method found the global minimum while the other two methods converged to some other, local minimum.

6. Concluding Remarks

In this paper we have illustrated how machine learning techniques can be applied to optimal engineering design. This has been accomplished by tackling problems in two different areas:

- speeding up existing numerical methods
- learning a set of candidate optimal solutions.

Table V illustrates the correspondence between these problems and the machine learning techniques used in their solution. Our main contribution is to have shown that ML techniques can be effectively used to overcome some of the drawbacks of numerical optimizers and to increase their efficiency. Another contribution of this paper is to have shown that inductive techniques can complement traditional software engineering approaches in mathematical domains. This greatly reduces the need for knowledge transfer from experts to computer systems. In our approach,

TABLE V
Rows enumerate problems in optimal design. Columns list Machine Learning paradigms. X's indicate the ML paradigm used to solve the problem.

	Symbolic Methods	*Inductive Learning*
Selection Rules		X
Speedup of Numerical Optimizers	X	X

these results required the use of a blend of novel and traditional optimization techniques. First, we have defined a new learning framework which is more appropriate to optimization tasks. This framework involves (a) the requirement that the output of the learning algorithm be a set of alternatives and (b) measures of the cost of obtaining solutions. The learning methods produce sets of minimum cost. Within this framework we have developed algorithms which output `IF-THEN-ELSE` rules that associate problem characteristics (features) to sets of optimal solutions. This is a contribution to basic research in machine learning. Second, we have demonstrated that inductive methods can also be used to simplify numerical problems. In fact we employed a discovery approach to reduce the number of independent variables. Finally, we have used more traditional compiler optimization techniques in a learning framework and merged them with inductive methods. We have shown that the overall result is a drastic speedup of the numerical optimization techniques.

Our approach opens new research directions into the so far unexplored area of applications of machine learning to numerical optimization. It is our hope that, in the medium-to long-term, our techniques will allow the use of specialized numerical optimizers in real-time applications like intelligent CAD systems.

Acknowledgements

The author wishes to thank his advisor, Thomas G.Dietterich, for the discussions that lead to this paper, David G. Ullman and Prasad Tadepalli for comments on related papers, Igor Rivin for information on the internals of Mathematica, and Jerry Keiper for insights into `FindMinimum[]` in Mathematica version 1.2. Ullman is responsible for suggesting the term *stress state*. This research was supported by NASA Ames Research Center under Grant Number NAG 2-630.

References

Braudaway, W.: 1988, Constraint incorporation using constrained reformulation, *Tech.Rep. LCSR-TR-100 Computer Science Dept.*, Rutgers University.

Burstall, R. and Darlington, J.: 1977, A transformation system for developing recursive programs, *Journal of the ACM* **24**(1), 44–67.

Cerbone, G. and Dietterich, T. G.: 1991, Knowledge compilation to speed up numerical optimization, *Proceedings of the Machine Learning Workshop*, pp. 600–604.

Cerbone, G.: 1992, Inductive learning in engineering: A case study, *Proceedings of the Adaptive and Learning Systems Conference of the IEEE Society for Optical Engineers*.

Cerbone, G.: 1992, in preparation, *Machine Learning in Engineering: Techniques to Speed up Numerical Optimization*, PhD thesis, Oregon State University, Corvallis, OR.

Ellman, T.: 1989, Explanation-based learning: A survey of programs and perspectives, *ACM Computing Surveys* **21**(2), 163–222.

Feigenbaum, E.: 1977, The art of artificial intelligence 1: themes and case studies of knowledge engineering, *Tech.Rep. STAN-CS-77-621*, Stanford University, Dept. of Computer Science.

Friedland, L.: 1971, *Geometric Structural Behavior*, PhD thesis, Columbia University at New York, N.Y.

Futamura, Y.: 1971, Partial evaluation of a computation process – an approach to a compiler-compiler, *Systems, Computers, and Controls* **2**(5), 45–50.

Garey, M. J. and Johnson, D. S.: 1979, *Computers and Intractability, A Guide to $\mathcal{NP}$-completeness*, Freeman.

Maeder, R.: 1989, *Programming in Mathematica*, Redwood City, Calif. : Addison-Wesley, Advanced Book Program.

Palmer, A. and Sheppard, D.: 1970, Optimizing the shape of pin-jointed structures, *Proc. of the Institution of Civil Engineers*, pp. 363–376.

Papalambros, P. Y. and Wilde, D. J.: 1988, *Principles of optimal design: modeling and computation.*, Cambridge University Press.

Pike, R. W.: 1986, *Optimization for Engineering Systems*, Van Nostrand.

Press, W. H. et al.: 1988, *Numerical Recipes in C: the art of scientific computing*, Cambridge University Press, Cambridge.

Quinlan, R. J.: 1987, Simplifying decision trees, *International Journal of Man-Machine Studies* **27**, 221–234.

Vanderplaats, G. N.: 1984, *Numerical Optimization Techniques for engineering design with applications*, New York: McGraw Hill.

Wang, C.-K. and Salmon, C. G.: 1984, *Introductory Structural Analysis*, Prentice Hall, New Jersey.

Wolfram, S.: 1988, *Mathematica*, Wolfram Research.

INCREMENTAL LEARNING FOR IMPROVED DECISION SUPPORT IN KNOWLEDGE BASED DESIGN SYSTEMS

K. MILZNER and A. HARBECKE
Lehrstuhl Informatik 1
University of Dortmund
PO Box 500500
D–4600 Dortmund 50
Germany

Abstract. In this paper, LEAR, a concept learning system for decision support in routine design tasks, is presented. LEAR learns incrementally from the instances which are generated during normal design sessions with a host CAD system. To achieve this an existing learning method was adapted to task specific constraints and complemented with several novel approaches to knowledge organization. This includes a flexible distributed concept representation which exploits background knowledge to reduce the complexity of the learning process as well as an efficient method for constructive induction which uses fuzzy sets to determine the most suitable high level attributes.

1. Introduction

In many technical domains knowledge based systems assist engineers in various aspects of design (e.g. Gero, 1988a; Forsyth, 1989; Gero, 1990). The performance of these systems substantially depends on the quality of the incorporated knowledge. To remain useful in the scope of a fast technical evolution this knowledge has to be constantly updated to keep track of new developments. As conventional (manual) knowledge engineering often turned out to be to slow and error-prone considerable effort has been spent in recent years to integrate machine learning capabilities into knowledge based design systems (e.g. Buchanan, 1989).

Also from the cognitive point of view learning is the normal situation in a design process, rather than the exception (Horner and Brown, 1990). Design experts are continually reorganizing their knowledge as part of their normal problem-solving activity. The reorganized knowledge improves subsequent problem-solving by avoiding similar failures in the future. 'Intelligent' systems which take over parts of an engineers work should mimic that behaviour.

In embedding machine learning techniques into real world systems two main subtracks can be identified in literature: a) systems that learn procedural design knowledge or so called *design episodes* consisting of design operators, their sequence and applicability (e.g. *LEAP*, Mitchell et. al., 1985; *REKL*, Simoudis, 1990) and b) systems that learn the evaluation or classification of certain design states or results (e.g. *LIMES*, Herrmann, 1987, 1991). In this paper we will

J. S. Gero (ed.), Artificial Intelligence in Design '92, 719–738.

focus on the latter type, which will be discussed in the context of the LEAR approach (LEArning Result evaluations). LEAR is a concept learning system which was designed for decision support in the result evaluation phase of routine design tasks.

Routine design in contrast to innovative or creative design is that type of design, in which all possible problem solving methods and design structures are known prior to the beginning of a design process. Nevertheless, routine design may still involve complex problem-solving activities (Horner and Brown, 1990; Steier, 1990). The majority of all currently available knowledge based design systems probably is dedicated to routine design in its different instantiations. Common examples are *variant* or *parametric design* in mechanical engineering (e.g. *BRIDGER*, Reich, 1991a, 1991b) and *cell based design* in electrical engineering (e.g. *OASE*, Milzner and Klinke,1990; Milzner, 1991).

LEAR learns incrementally from the design instances which are created during normal design sessions with a host design system. The learning method which is applied is able to deal efficiently with the specific constraints which are imposed on such a system by typical engineering environments. Only a very short learning phase is required to acquire multiple concept descriptions. To achieve this an existing concept learning approach (*STAGGER*, Schlimmer, 1987b, 1987c) was modified and enhanced with several novel methods concerning concept generalization and matching. This includes a novel method for *constructive induction*, which uses fuzzy sets for building the most suited high-level attributes and a *flexible distributed concept representation*, which exploits background knowledge to reduce the complexity of the learning process. Originally, the system was developed for application in the *OASE/SILAS* analog circuit design environment (Milzner and Brockherde,1991; Milzner, 1991). It can, however, easily be adapted to further domains which employ a similar model of the design process.

The paper is organized as follows: Section 2 deals with the task and domain specific considerations which were applied to model the design process and to select appropriate learning mechnisms. Section 3 shortly introduces the STAGGER approach, which formed the starting point of the LEAR development. In Section 4 details of the LEAR system are discussed which includes the method for constructive induction, the use of background knowledge and the handling of irrelevant attributes. Finally in Section 5 some empirical results are presented.

2. Task And Domain Specific Considerations

Although different domain independent models of design processes have been developed (e.g. Chandrasekaran, 1990; Maher, 1990) up to now no unified theory of design could be established. For this reason in this section we will concentrate on the model we used to describe the different stages of a cell-based design processs and the criteria which were applied to select appropriate learning mechanisms.

In literature several approaches to design aids which can help to simplify the modeling of a design task as well as the mapping of task structures to appropriate learning mechanisms are discussed. Gero et. al. propose a representation scheme called *prototypes* which can be used to structure conceptual knowledge and to describe its functional, structural and behavioural properties (e.g. Balachandrian and Gero, 1990). Chandrasekaran introduced the concept of domain independent *generic tasks*, which can be used as building blocks for a multitude of standard tasks (e.g. hierachical classification) in assembling a complete model of a design process as well as basis of explanation based learning techniques (Chandrasekaran, 1989, 1990). Reich suggests a fourstep procedure for building learning systems, which is called M^2LTD (Matching Machine Learning To Design, Reich, 1991). In the remainder of this section we will follow this procedure to explain the task and domain specific considerations LEAR is based upon.

M^2LTD consists of the following four steps: 1) analyse and decompose the design process into smaller tasks (e.g. synthesis, analysis, redesign), 2) identify the representation of design objects, 3) find closely related machine learning paradigms and 4) use domain characteristics to select a suitable program or approach.

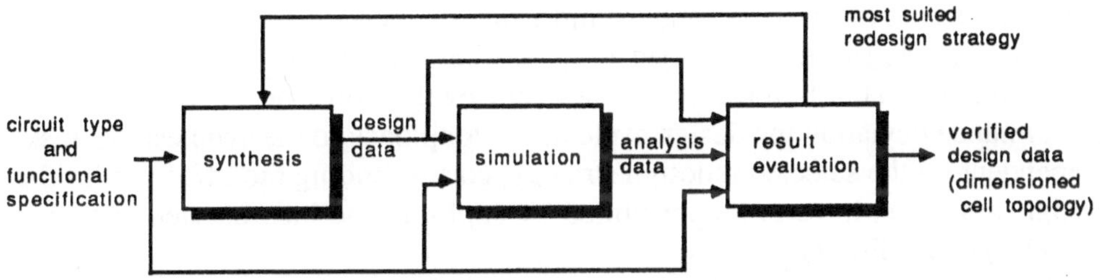

Figure 1: Simplified model of one cycle in a hierarchical design process

2.1. MODELING THE DESIGN PROCESS

Figure 1 shows the basic task structure as it is typically performed by cell-based design systems in the domain of integrated circuit synthesis (e.g. Berkcan et.al., 1988; Sheu et.al., 1990; Milzner, 1991). It is, however, also applicable in other domains (e.g. mechanical engineering: Powell et.al., 1990). This basic task sequence (called *design cycle* in the sequel) consisting of synthesis, simulation and result evaluation is repeated iteratively during the hierarchical decomposition and stepwise refinement of a complex design task.

Input to a design cycle is a specification consisting of a set of attribute-value pairs which describe the basic structure (the *circuit type*) and the functional requirements of the cicuit to be designed. In the synthesis step these requirements are translated by design heuristics and selection rules into a description of an appropriate topology or subcell structure of the circuit. The synthesis step also includes the calculation of all corresponding device sizes and lower level requirements (e.g. transistor dimensions). These design data are feed into a simulator to analyze the performance of the synthesized artefact. The achieved analysis results can also be represented by a simple attribute-value list. Simulation, however, typically is a computationally very complex and thus time consuming and costly task. For this reason and due to the often very large possible solution space in this type of design processes no *generate-and-test strategy* as frequently used in other models of design (e.g. Reich, 1991a) is applicable. Instead of finding *all* possible solutions the goal of the synthesis step is to create the most suitable one. The design data resulting from the synthesis process is described by attribute-value lists which may be partially ordered and include dependencies between attributes.

At the end of each design cycle the quality of the achieved results is checked by matching the functional requirements (the specification) against the analysis data resulting from simulation. If the simulation results satisfy the given specification the design is accepted and the next level of the hierarchical design process is initiated. Otherwise, the failure which caused the unacceptable behaviour has to be classified and the most suited redesign strategy out of several alternatives has to be determined (e.g. *device size optimization* or *topology optimization*). This is the task of the result evaluation step. In the case of a performance deviation the design cycle has to loop back to the synthesis subtask. Dependent on the selected redesign strategy, corresponding modifications to the designed artefact have to be performed by applying the incorporated synthesis knowledge in a different context.

To classify the reason of a failure all available design information (specification, design data, analysis data) have to be taken into account. The correct evaluation of this complex set of information strongly influences the quality of the overall design as well as the design time needed. In manual design, the efficacy of this process largely depends on previous design experience or training. When evaluating design results a human expert typically does not rate the dozens of design parameters one by one but more intuitively by a selective recognition of specific patterns which are composed from single design characteristics. This is due to the large quantities of numerical data and the very complex concept descriptions in technical domains. They often represent a problem size which in full detail is beyond a human's scope. Therefore in an acquisition session it is very difficult for the expert as well as for the knowledge engineer to identify and to formalize this type of implicit knowledge. As a way out knowledge-based design systems often employ simple but not always optimal default redesign strategies instead [e.g. Milzner, 1991].

The learning system LEAR was developed to replace these default procedures by a more problem oriented strategy selection. The qualification of a specific redesign strategy is derived from previously performed designs by generalization. When in learning mode, during a design cycle the processing of the current instance by LEAR is delayed to the end of all necessary redesign cycles. By this means it is possible to explore the effects of the selected redesign strategy and to obtain self-generated critique concerning the correctness of the classification.

2.2. DOMAIN SPECIFIC CONSTRAINTS

From the design model described above domain specific constraints can be derived, which can be used as selection criteria for an appropriate learning method. The most important requirements for a system that learns concept descriptions from examples in a real-world CAD environment are:

- The design instances which are available to the learning system have to be processed incrementally since they are generated during an engineer's normal work. This requires effective mechanisms to correct wrong classifications by modifications to the affected concept descriptions.
- When evaluating the results of a design process typically several alternatives have to be considered. This implies the capability of the learning system to learn a corresponding number of concept descriptions.
- Design instances contain nominal, real-valued and structured attributes. Efficient methods are required to transform real-valued attributes into a symbolic representation.
- The number of attributes in an instance which have to be taken into account is not static due to the dynamic character of a design process.
- A design instance can be very complex (e.g. several hundred attributes). This requires an efficient concept representation and the identification and elemination of irrelevant attributes. Background knowledge like dependencies between attributes should be exploited for a further reduction of the size of concept descriptions and to increase learning efficiency.
- Typically in knowledge-based design systems the knowledge is frequently updated due to the technical evolution. For this reason the learning system must have the ability to deal with a medium term concept drift.
- To be accepted by a CAD-system user the learning process has to run virtually in real time. In learning mode it also should generate the classification of each example without user interaction.

2.3. METHOD SELECTION

Although *learning from examples* is a classic method most of the existing approaches are restricted to only partial aspects of the above requirements. For reasons of limited space only a very coarse review can be given.

Approaches which are based on *version space strategy* as put forward by Mitchell (e.g. Mitchell, 1982), for example, are restricted to only one concept description. A *version space* describes a concept by a subset of the possible instance space which is formed by a maximum specific and a maximum general concept description. This memory efficient concept representation, however, becomes inefficient if the interval partitions of real-valued attributes frequently change, since this requires the adaptation of the complete acquired knowledge.

Learning programs which use *decision trees* for concept representation have been widely studied in the past (e.g. ID3, Quinlan, 1986; GID3, Cheng et.al., 1988). Although manifold aspects of learning like incremental example processing (Utgoff, 1988) and boolean feature discovery (Pagallo and Haussler, 1990) are covered by these approaches, none of them is able to cope with the sum of the above requirements. Especially the handling of frequently changing interval partitions in real-valued attributes results in computational costly operations, as the complete tree has to be reorganized (Utgoff, 1988).

An approach which matches the above requirements in a very high degree is a multi-strategy learning program called STAGGER (Schlimmer, 1987a-c). As the STAGGER approach formed the starting point of development for the presented LEAR system its basic principles will be briefly discussed in the next section.

3. The STAGGER Approach

To facilitate a better understanding of the general approach we now will concentrate on those aspects of STAGGER, which have been adopted directly by the LEAR system. This essentially comprises the example and the concept description languages and the strategy of cooperative learning methods. For a more detailed description of STAGGER please refer to the given literature.

3.1. CONCEPT DESCRIPTION AND MATCHING

In STAGGER all instances which are processed in the learning or matching process are described as a set of attribute-value pairs along with a correct concept classification . The attributes can have nominal or numerical ranges.

Each concept which has to be learned is represented as a set of dually weighted, predictive attributes in a *distributed concept description* (see also Figure 2). These weights are calculated using Bayesian formulae (e.g. Rich, 1983;

Schlimmer, 1987c). The logical sufficiency (*LS*) indicates how much the presence of an attribute in the current instance implies the membership of the instance to the given concept. The logical necessity (*LN*) indicates how much the absence of an attribute implies a nonmembership of the instance to the given concept.

$$LS = \frac{p(\text{matched} \mid \text{example})}{p(\text{matched} \mid \neg\text{example})} \qquad LN = \frac{p(\neg\text{matched} \mid \text{example})}{p(\neg\text{matched} \mid \neg\text{example})} \tag{1}$$

Initially, the concept descriptions consist of all possible attribute-value pairs that exist in the description language of the examples with all weights set to a value of one. As learning proceeds the weights are modified corresponding to the matching of the attributes with the instances by using counts of the different matching situations. Furthermore new, more complex attributes may be formed by constructive induction (see Section 4).

When a new instance E is generated for each possible concept the expectation concerning its membership to a concept C is calculated by multiplying the prior expectation with the *LS* weights of the matched attributes and the *LN* weights of the unmatched attributes.

$$\text{odds}(E \mid C) = \text{prior.odds}(E) \times \prod_{\forall\,\text{matched}} LS \times \prod_{\forall \neg\text{matched}} LN \tag{2}$$

The resulting scores reflect the degree of matching between the instance and the different concept descriptions. The instance is assigned to the concept with the highest score larger than one.

3.2. COOPERATIVE LEARNING

If weight adjustment of attributes would be the only learning method which is available only the special case of *linear separable* concepts could be learned. A linear separable concept is a concept for which the independent determination of of all attribute values is sufficient for a correct discrimination of positive and negative instances of the concept. In a real world domain this is a very unlikely situation. To overcome this severe limitation complementary learning methods are required. For this reason in STAGGER weight adjustment was combined with a method for constructive induction which is based on Boolean learning and a method for partitioning real-valued attributes into concept oriented intervals. This increases the flexibility concerning the concepts which can be learned. Furthermore it enables an incremental learning process that also tolerates noisy data and possible concept drifts (Schlimmer and Granger, 1986).

However, in STAGGER these complementary learning methods still show some deficiencies with respect to the above requirement list. They will be explained in the next section in the context of the alternative approaches which have been developed with LEAR.

4. LEAR

With the learning system LEAR several novel approaches to different aspects of concept learning have been developed. They are oriented at the requirements of a complex technical design environment. The most important improvements of LEAR compared to STAGGER are:

- a novel method of constructive induction, which uses fuzzy sets to form efficiently appropriate high level attributes
- the use of background knowledge to reduce the complexity of the learning process and to increase learning speed
- a concept oriented and more efficient discretization method for real-valued attributes.

These aspects will be discussed in greater detail in the following subsections.

4.1. CONSTRUCTIVE INDUCTION USING FUZZY SETS

Every time LEAR fails in predicting correctly the concept of an instance it is necessary to modify the affected concept descriptions as they are obviously not consistent with the examples. This is achieved by a mechanism of constructive induction which adds newly formed, more suitable complex attributes to the concept descriptions. These newly formed attributes are constructed from the existing ones by Boolean operations (conjunction, disjunction, negation). In each case two concept descriptions have to be adapted in dependency of the failure. The description of the concept to which the instance was assigned has to be specialized as it was to general. On the contrary the target concept has to be generalized as it did not match with the instance. The necessary procedure consists of two basic steps:

1) the selection of suitable attributes

2) the Boolean combination of these attributes.

In STAGGER the selection of the attributes which are suitable for Boolean combination is based on some weak heuristics (Schlimmer, 1987c). This results in an underconstrained set of attributes which in turn means additional time and learning effort. In LEAR a more formal approach is applied. It is based on fuzzy set theory (Zadeh, 1965) and was motivated by the results of an approach to goal-oriented design control which were presented by Felix (1990). The formalism takes into consideration that an attribute with a high *LS*-weight and a low *LN*-weight *supports* a concept when matched, whereas an attribute with low *LS*-weight and high *LN*-weight *distracts* a concept when not matched. Based on that idea we can define a support function s and a distraction function d for a concept C and an attribute e as follows:

$$s_C(e) = \begin{cases} LS * LN, & \text{if } LS \text{ is more important than } LN \\ 0, & \text{else} \end{cases} \tag{3}$$

$$d_C(e) = \begin{cases} LS * LN, & \text{if } LN \text{ is more important than } LS \\ 0, & \text{else} \end{cases} \tag{4}$$

with *LS is more important than LN* $\leftrightarrow$ LS $*$ LN > 1
LN is more important than LS $\leftrightarrow$ LS $*$ LN < 1

These functions can be used to define two fuzzy sets for each concept C :

$$S_C = \{ e \mid e \in E \leftrightarrow s_C(e) > 0 \} \tag{5}$$

$$D_C = \{ e \mid e \in E \leftrightarrow d_C(e) > 0 \} \tag{6}$$

With these definitions an attribute of a concept description is exclusively assigned to one of the above fuzzy sets. However, there may be attributes which belong to neither of the sets. These attributes are considered to be irrelevant for combination and can be characterized by the following equation:

$$e \notin S_C \cup D_C \leftrightarrow LS * LN = 1 \tag{7}$$

The attributes of the concept descriptions are not mutually exclusive with respect to the defined set relations. It is possible, for example, that an attribute supports more than one concept. This means that the concepts compete with each other during the learning process. In terms of the fuzzy sets for two concepts A and B this can be described by

A and B are competitive $\leftrightarrow$ $S_A \cap S_B \neq \emptyset$, $D_A \cap D_B \neq \emptyset$ (8)

A and B are mutually exclusive $\leftrightarrow$ $S_A \cap S_B = \emptyset$, $S_A \cap D_B \neq \emptyset$,

$$D_A \cap S_B \neq \emptyset,$$
$$D_A \cap D_B = \emptyset \tag{9}$$

During the matching process some attributes of a concept description exist in the instance and others do not. This motivates the definition of two more sets. Let X be an instance and $e \in E$ an attribute of the concept description. Then the following sets can be defined for each concept C:

$$M_C = \{ e \mid e \in E \leftrightarrow e \; matches \; X \} \tag{10}$$

$$N_C = \{ e \mid e \in E \leftrightarrow e \neg matches \; X \} \tag{11}$$

With these basic definitions the goal of the learning process, the mutually exclusion of the possible concept descriptions, can be formally described. For all instances which belong to concept A but not to concept B the following relation holds when the learning phase is completed:

$$S_A \supseteq M_A, \; D_A \supseteq N_A, \; S_B \supseteq N_B, \; D_B \supseteq M_B, \tag{12}$$

The task of the constructive induction is to achieve these relations by combining existing attributes of a concept description to more effective complex ones. The selection of the most suitable attributes to be combined is based on the defined functions and sets. To generalize a concept description a new distracting conjunctive attribute and a new supporting disjunctive attribute are added. To specialize a concept description a new distracting disjunctive attribute and a new suppporting conjunctive attribute are added. In either case an appropriate negated attribute is added, too. In terms of the above sets the description of all attributes which are suitable for combination can be formulated as follows:

<u>*generalization of a concept C*</u>

- Add a disjunctive attribute to S_C :
  $$e = x \vee y, \quad x \in M_C \cap S_C, \quad y \in N_C \cap S_C \tag{13}$$
- Add a conjunctive attribute to D_C :
  $$e = x \wedge y, \quad x \in M_C \cap D_C, \quad y \in N_C \cap D_C \tag{14}$$
- Add a negated attribute to S_C :
  $$e = \neg x, \quad x \in N_C \cap D_C \tag{15}$$

<u>specialization of a concept C</u>

- Add a conjunctive attribute to S_C :

  $$e = x \wedge y \,, \quad x \in M_C \cap S_C \,, \quad y \in N_C \cap S_C \tag{16}$$

- Add a disjunctive attribute to D_C :

  $$e = x \vee y \,, \quad x \in M_C \cap D_C \,, \quad y \in N_C \cap D_C \tag{17}$$

- Add a negated attribute to D_C :

  $$e = \neg x \,, \quad x \in M_C \cap S_C \tag{18}$$

The most suitable attributes to be combined are determined from the subsets described by formula (13)-(18) on basis of their supporting or distracting characteristic. In Table I a summary of the selection criteria is given.

Table I : Summary of concept refinement

instance	classific.	function	selection
positive	negative	OR (attrib1, attrib2) AND (attrib1, attrib2) NOT (attrib1)	$u_X(e) \gg 1$ $h_X(e) \ll 1$ $\max(h_X(e))$
negative	positive	AND (attrib1, attrib2) OR (attrib1, attrib2) NOT (attrib1)	$u_X(e) \gg 1$ $h_X(e) \ll 1$ $\min(u_X(e))$

A combination of attributes is not irrevocable. When new attributes are added to a concept description they compete with the attributes from which they were combined. Only those attributes which show a better matching behaviour are retained. If, after having processed some further instances, the combined attributes do not prove to be more effective concerning their membership functions (see formula (3), (4)) they are removed from the concept description by backtracking.

4.2. USE OF BACKGROUND KNOWLEDGE

Most learning programs including STAGGER start their knowledge acquisition process from the scratch. Learning, however, is much more effective if some initial knowledge about the domain (*background knowledge*) is available (e.g. Michalski, 1987). In LEAR two types of background knowledge are taken into account: structured attributes and dependencies between attributes. In knowledge based environments like OASE this knowledge often is directly available from the already existing knowledge bases.

In LEAR, additionally to the language constructs which are available in STAGGER, *structured attributes* can be described. In technical domains structured objects (e.g. different levels of refinement in a design) are a frequently used method of modeling. This background knowledge, which has to be described by an concept tree, is be exploited as a further means to guide the generalization and specialization of concept descriptions.

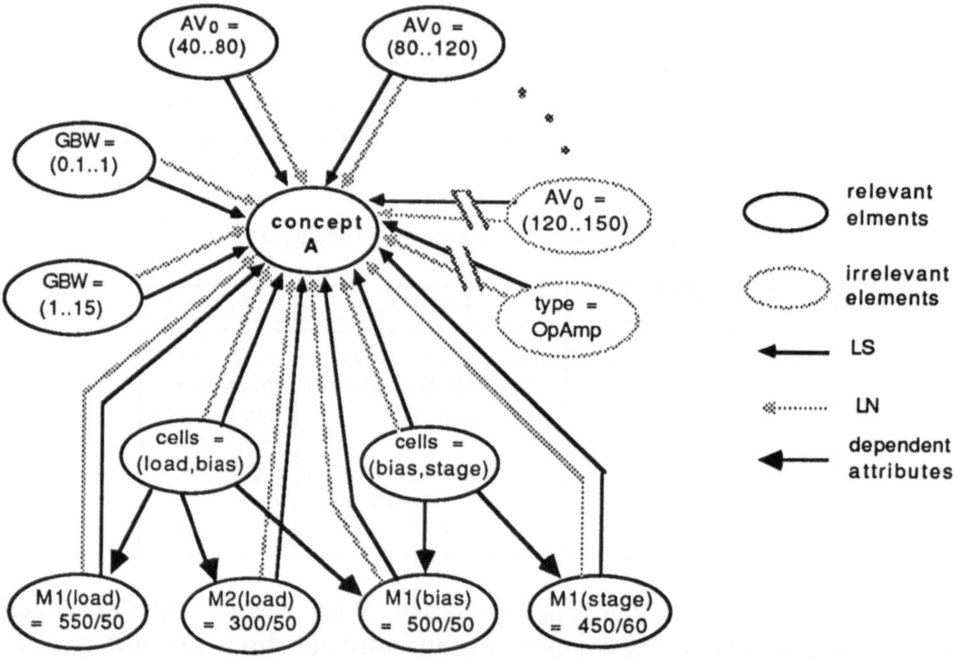

Figure 2: Example of LEAR's distributed concept description

We will now concentrate on the dependencies between the attributes of a concept description. In LEAR this knowledge is represented in the form of a dependency graph. This graph reflects the dependencies between the possible attributes with regard to their current values (see Figure 2 and Figure 3). The attributes on the first level are independent and represent those attributes which are part of every example description. Their values determine which attributes on lower levels of the graph have to be taken into account. This can be illustrated with an example of a cell-based design process. In such a design process for every level of design abstraction a certain set of predefined templates or *cells* is available which can be used as basic building blocks (e.g. (bias, stage) in Figure

3). Dependent on the selections made on higher levels of a design the possible cells on lower levels are constrained to certain subsets (e.g. M1(bias)... M1(stage)). All other (sub)cells known by the system are no longer relevant to the design in question and thus do not have to be considered by the learning system.

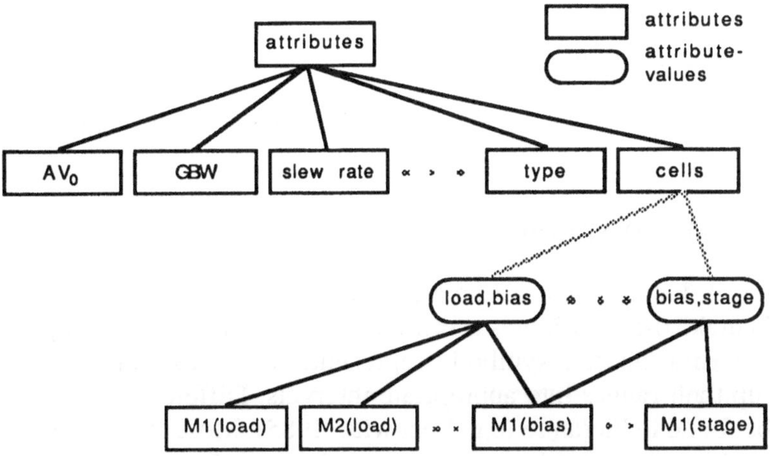

Figure 3: Example of a dependency graph

In LEAR these dependencies are exploited to reduce temporarily the instance space which can be described by the example description language. This is done by reducing the allowed attributes to the set which is available from the current instance. Those attributes which are faded out due to specific value combinations of first level attributes are simply ignored. By this means the size of the concept description can be considerably reduced. This on the one hand increases the matching efficiency and on the other hand prevents negative effects on the concept descriptions as it avoids weight adjustments of attributes which never will occur in a specific type of instances.

During the initiation phase of the concept descriptions the available background knowledge about (static) dependencies between primitive attributes is used to form a corresponding graph of attribute dependencies for the concept descriptions. This graph, however, has to be dynamically adjusted during the learning process due to changes in the concept descriptions caused by constructive induction and the partitioning of real-valued valued attributes.

4.3. IRRELEVANT ATTRIBUTES

If a design did not meet the required specification typically only partial aspects of its realization have to be taken into account as far as the selection of the right redesign strategy is concerned. To reduce the size of the concept descriptions and to avoid unnecessary time and learning effort by focussing temporarily on non-relevant attributes in LEAR such attributes are identified and ignored during the matching and learning process. Only their weights are adjusted (see Figure 2). An attribute is considered to be *irrelevant* if the following weight condition is fulfilled for some e and a sufficient number of instances:

$$|LS - 1| < \varepsilon \;\wedge\; |LN - 1| < \varepsilon \tag{19}$$

As learning proceeds an attribute can become relevant again if condition (19) is not longer fulfilled.

4.4. REAL-VALUED ATTRIBUTES

In a technical domain most of the attributes of an instance will have numeric values. To make these attributes accessible to a learning process they have to be projected onto a discrete symbolic representation. This can be achieved by breaking up their ranges into appropriate intervals. Different approaches to this task can be found in literature (e.g. Lebowitz, 1985). Most of them use statistical measures which do not relate the generated interval partitions to a specific concept. The procedure applied in STAGGER is one exception which does not have this severe deficiency. It determines appropriate subintervals by calculating their "usefulness" concerning the possible concepts. The procedure fails, however, if the concept description contains complex attributes which were combined using an exclusive-or (Schlimmer, 1987a). This results in a considerable restriction of its qualification for technical domains.

In LEAR a novel approach was developed, which is able to break up real-valued attributes into a variable number of concept related subintervals without the described problems. It can be characterized as *looking for heaps of incorrectly assigned numerical values* concerning a specific interval and concept. Each interval in the possible range of a numeric attribute is described by the following set of information:

1) the range of the interval (initially the complete range of the attribute)
2) the number of examples which matched with this interval
3) the concept to which the interval is assigned
4) a list of the values of all negative instances (incorrect assignments) sorted by concepts

The partitioning process is controlled by a number of predefined variables which avoid problems with too small subinterval as described by Schlimmer (1987a). They are assigned with heuristic values which proved to achieve good results in several tests with the LEAR prototype. Examples of these variables are the *maximum number of nonexamples* allowed in an interval (e.g. 10% of all instances which matched with the interval), the *minimum length* of an intervall (e.g. ((upper bound-lower bound)/(5∗number of concepts)), the *minimum number of instances* for a new generated subinterval (e.g. 50% of the instances which matched with this subinterval), etc.

If in the matching process the number of negative instances in an interval exceeds the predefined limits the interval is broken down into suitable subintervals. The partitioning process employs the concept oriented counts of negative instances in the interval description in consideration of the required minimum number of instances in a new subinterval. Compared to the approach presented by Lebowitz (1985) this means that not *all* instances have to be stored to enable the computation of a new interval partition. Only those attribute values, which so far could not be assigned correctly to a specific concept are temporarily recorded. Their number decreases when learning proceeds and new intervals are formed.

If the number of negative instances in an interval is exceeded and this interval cannot be further subdivided with respect to the minimum length requirements it is considered to be irrelevant for the learning process. To adjust the concept descriptions to a new interval partitioning all complex attributes which contain the corresponding primitive numeric attributes are simply duplicated and the interval ranges are projected accordingly onto the newly formed attributes.

5. Experimental Results

Finally, we will present some experimental results which illustrate the performance of the described approach in different applications with varying complexity

5.1. SIMPLE CLASSIFICATION PROBLEMS

For reasons of clarity and to allow a comparison with similar tests which were carried out with the STAGGER system (Schlimmer, 1987b) we first will investigate LEAR´s performance in simple classification tasks. For this purpose a very simple description language for the examples was chosen. It includes, however, characteristical properties of engineering domains like dependencies between attributes as well as real-valued attributes. Table II gives an overview of the attributes and the allowed value ranges of the example description language.

The results of two different tests from the test-phase of LEAR will be presented to show the impact of different concept complexities on the performance of LEAR. The concepts which had be learned are described by :

A) (*triangular* AND *green*) OR (*small* AND *blue*) with *small* = [1...6]
B) (*triangular* AND *red*) XOR (*medium*) with *medium* = [6...12]

Table II: Description language of the examples

color	∈	{red, green, blue}
shape	∈	{triangular, square, sphere}
volume	∈	[0 ... 20], if shape = sphere
area	∈	[0 ... 20], else

To learn these concepts it is necessary to build complex elements from the attributes of the example description by applying different types of Boolean operations including exclusive-or and to break up real-valued attributes into suitable intervals. In each test approximately 50 instances were presented to the learning system. These instances were generated by a random selection from the possible instance space (6 examples / 21 nonexamples in test A and 10 examples / 17 nonexamples in test B). After the processing of each instance the classification error concerning the selected instance set (the *sample*) or the learning success as well as the size of the current concept description were determined. Figure 4 illustrates the results of test A, Figure 5 the results of test B.

As the result of test A shows, LEAR is able to classify over 90% of the instances correctly on the basis of a sample containing only 54 instances. After having processed just 5 instances the novel approach to constructive induction based on fuzzy sets enables LEAR to find a concept description which is suitable for 75% of the sample. The partitioning of numeric attributes, however, starts relatively late after having processed approximately 25 instances. This results from the dependencies between the attributes of the example description which decrease the probability of the appearance of a specific attribute value. The size of the concept description increases corresponding to the formation of new intervals.

The results of test B are not quite as good as those of test A. This is caused by the exclusive-or operator in the concept description. As this is not a standard boolean operator, LEAR had to learn the following concept description:

(((*triangular* AND *red*) AND (NOT *medium*)) OR
((NOT (*triangular* AND *red*)) AND *medium*)) .

This is a comparatively complex concept which is also reflected by the strong increase in the size of the concept description. Additionally the individual attribute-value-pairs of the concept description seem to be contradictory ((*medium*) ↔ (NOT *medium*)). Nevertheless test B proved with a result of nearly 85% correct classifications at the end of the test run that LEAR is able to learn concepts which include numerical attributes in combination with an exclusive-or operator. However, to learn such concept descriptions with a sufficient degree of reliability a higher number of training instances is required.

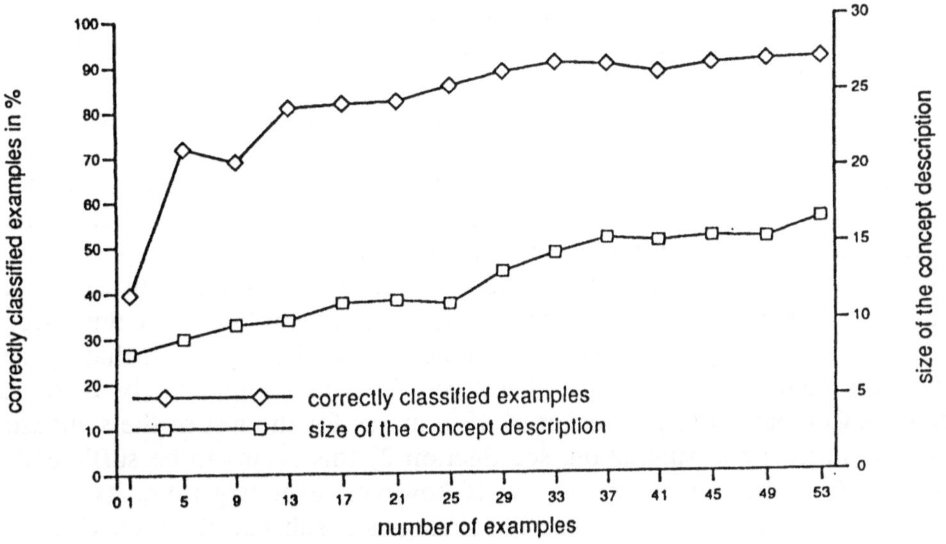

Figure 4: Results of test A

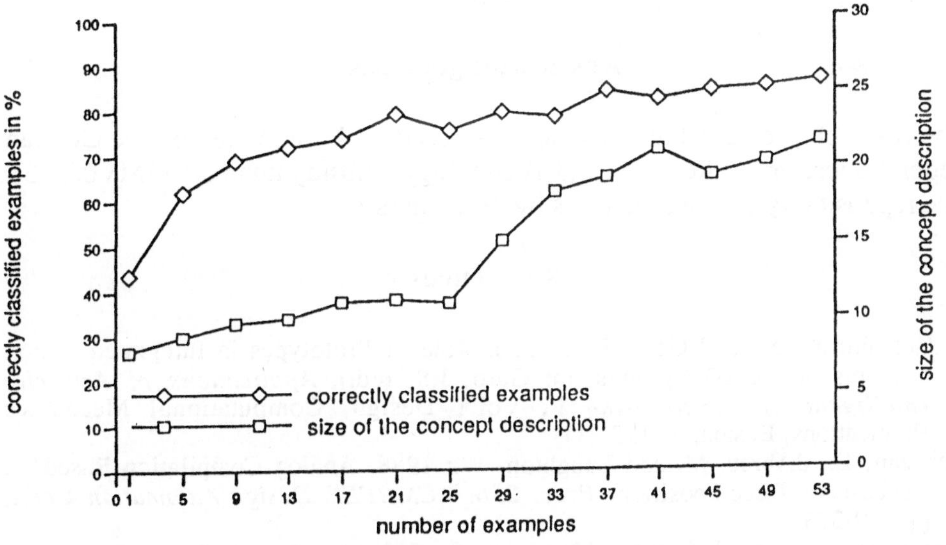

Figure 5: Results of test B

5.2. COMPLEX CAD-ENVIRONMENTS

We will discuss now some results which were achieved with the LEAR system in the OASE and SILAS analog circuit design environment. LEAR was used to classify the results of the synthesis of operational amplifiers. Each of the three concept descriptions *parameter optimization*, *topology optimization* and *out-of-reach* which had to be learned consisted of 287 attributes, whereas each instance was described by approximately 100 attributes. The majority of these attributes (approx. 90%) had real-valued ranges.

As it is very difficult to generate a suffiently large number of training instances with an interactive CAD-environment like OASE and SILAS in a reasonable time, no measures or learning curves can be given. We can, however, present the resulting time and memory requirements. Each of the concepts listed above required about 200 kB of memory to represent the 287 attributes. The time effort required for matching an instance against the concept descriptions including the necessary adaptions in the case of a mismatch was approximately 10 to 15 seconds. Compared to a time effort of 10 minutes for the rest of the synthesis cycle (synthesis and simulation, see Section 2) this seems to be sufficiently efficient from the user´s point of view. If, however, even larger circuits have to be processed or the number of redesign strategies is substantially increased this may result in problems concerning the required time and memory consumption. If this problem occurs, it has to be investigated if a more efficient reimplemetation of LEAR, which currently is implemented using the KEE knowledge engineering environment by IntelliCorp, could relax these runtime problems.

Acknowledgements

The research project which this paper is based on is supported by the German Federal Secretary of Research and Technology (subsidy number: 13MV00343). The reponsibility for the contents is by the authors.

References

Balachandrian, M. and Gero, J.S.: 1990, Role of Prototypes in Integrated Expert Systems and CAD Systems, in: Gero, J.S. (ed.), *Applications of Artificial Intelligence in Engineering V*, Vol.1 Design, Computational Mechanics Publications, Boston, pp.195-211

Berkcan, E., dÁbreu, M. and Laughton, W.: 1988, Analog Compilation Based on Successive Decomposition, *Proc. 25th ACM/IEEE Design Automation Conf.*, pp.369-375

Buchanan, B.G.: 1989, Can Machine Learning Offer Anything to Expert Systems ?, *Machine Learning*, 4, pp.251-254

Chandrasekaran, B.: 1989, Task-Structures, Knowledge Acquisition and Learning, *Machine Learning*, 4, pp.339-345

Chandrasekaran, B.: 1990, Design Problem Solving: A Task Analysis, *AI Magazine*, Vol.11, No.4, pp.59-71

Cheng, J. et. al.: 1988, Improved Decision Trees: A Generalized Version of ID3, *Proc. 5th Int. Workshop on Machine Learning*

Felix;, R.: 1990, Goal-oriented Control of VLSI Design Processes Based on Fuzzy Sets, *Proc. 20th Int. Symposium on Multiple-Valued Logic*, Charlotte, N.C., pp.386-393

Forsyth, Richard (ed.): 1989, *Expert Systems - Principles and Case Studies*, Chapman and Hall

Gero, J.S. (ed.): 1988a, *Artificial Intelligence in Engineering: Design*, ELSEVIER Publishers, New York, Tokyo

Gero, J.S. (ed.): 1988b, *Artificial Intelligence in Engineering: Diagnosis and Learning*, ELSEVIER Publishers, New York, Tokyo

Gero, J.S. (ed.): 1990, *Applications of Artificial Intelligence in Engineering V*, Vol.1 Design, Computational Mechanics Publications, Southampton, Boston

Herrmann, J. and Reusch, B.: 1987, Combining Expert Systems and Machine Learning in CAD Systems for Micro Electronics, *Proc. Int. Workshop on AI Applications to CAD Systems for Electronics*, Oktober 1987

Herrmann, J.: 1991, Learning Analytical Knowledge about VLSI-Design from Observation, *Proc. of the Eighth Int. Workshop on Machine Learning*

Horner, R. and Brown, D.C.: 1990, Knowledge Compilation Using Constraint Inheritance, *in:* Gero, J.S. (ed.): *Applications of Artificial Intelligence in Engineering V*, Vol.1 Design, Computational Mechanics Publications, Boston, pp.161-174

Lebowitz, M.: 1985, Categorizing Numeric Information for Generalization, *Cognitive Science*, No.9, pp.285-308

Maher, M.L.: 1990, Process Models for Design Synthesis, *AI Magazine*, Vol.11, No.4, pp.49-58

Michalski, R.S.: 1987, Learning Strategies and Automated Knowledge Acquisition, *in:* Leonard Bolc (ed.), *Computational Models of Learning*, Springer Verlag, Berlin

Milzner, K. and Klinke, R.: 1990, Synthesis of Analog Circuits using a Blackboard Approach, *Proc. Third International Conf. on Industrial and Engineering Applications of AI and Expert Systems,* Charleston S.C., Vol.I, pp.114-122

Milzner, K.: 1991, An Analog Circuit Design Environment Based on Cooperating Blackboard Systems, *Journal of Applied Intelligence 1*, Kluwer Academic Publishers, Boston, pp.179-194

Milzner, K. and Brockherde, W.: 1991, SILAS: A Knowledge Based Simulation Assistant, *IEEE Journal of Solid-State Circuits*, Vol.26, No.3, pp.310-318

Mitchell, T.M.: 1982, Generalizations as Search, *Artificial Intelligence*, 18(2), 1982, pp.203-226

Mitchell, T.A., Mahadevan, S. and Steinberg, L.I.: 1985, LEAP: A learning apprentice system for VLSI design, *Proc. IJCAI 1985*, Los Angeles, CA., pp.573-580

Pagallo, G. and Haussler, D.: 1990, Boolean Feature Discovery in Empirical Learning, Machine Learning, No.5, pp.71-99

Powell, D.J., Skolnick, M.M. and Tong, S.S.: 1990, EnGENous: A Unified Approach to Design Automation, *in:* Gero, J.S. (ed.): *Applications of Artificial Intelligence in Engineering V*, Vol.1 Design, Computational Mechanics Publications, Boston, pp.137-157

Quinlan, J.R.: 1986, Induction of Decision Trees, *Machine Learning*, No.1, pp.81-106

Reich, Y.: 1991a, Designing integrated learning systems for engineering design, *Proc. of the Eighth Int. Workshop on Machine Learning*, pp.635-639

Reich, Y.: 1991b, Macro and Micro Perspectives of Multistrategy Learning, *Proc. of the First Conf. on Multi-Strategy Learning*, pp.97-112

Rich, E.: *Artificial Intelligence*; McGraw-Hill, 1983

Schlimmer, J.C. and Granger, R.H.: 1986, Incremental Learning from Noisy Data, *Machine Learning 1*, pp. 317-354

Schlimmer, J.C.: 1987a, Learning and Representation Change, *Proc. 6th Int. Conf. on Artificial Intelligence*, pp.511-515

Schlimmer, J.C.: 1987b, Incremental Adjustment of Representations for Learning, *Proc. 4th Int. Workshop on Machine Learning*, pp.79-89

Schlimmer, J.C.: 1987c, *Concept Acquisition through Representational Adjustment*, PhD. thesis, University of California, Irvine

Sheu, B.J., Lee, J.C. and Fung, A.H.: 1990, Flexible architecture approach to knowledge-based analogue IC design, *IEE Proceedings*, Vol.137, Pt.G, No.4, August 1990, pp.266-274

Simoudis, E.: 1990, Learning Redesign Knowledge, *IEEE Trans. on Computer-Aided Design*, Vol.9, No.10, pp.1047-1062

Steier, D.: 1990, Creating a Scientific Community at the Interface Between Engineering Design and AI, *AI Magazine*, Vol.11, No.4, pp.18-22

Utgoff, P.E.: 1988, ID5: An Incremental ID3, *Proc.5th Int. Workshop on Machine Learning*;

Zadeh, L.A.: 1965, Fuzzy Sets and Systems, *Proc. Symposium on System Theory*, Polytechnic Press of the Institute of Brooklyn, New York, pp.29-37

APPLICATION OF A NEURAL NETWORK TO SIMULATE ANALYSIS IN AN OPTIMIZATION PROCESS

J. L. ROGERS* and W. J. LAMARSH II**
***UNISYS*
**NASA Langley Research Center*
Hampton VA 23665
USA

Abstract. Expensive analysis programs are often applied to solve engineering design problems. To obtain an optimal solution typically requires numerous iterations between the analysis program and an optimization program. This often becomes prohibitive due to the amount of computer time required for convergence. Therefore, any new software that could significantly reduce the computer time required to solve a complex design problem would be beneficial. A new experimental software package called NETS/PROSSS has been developed to help meet this need. NETS/PROSSS combines an neural network for simulating the analysis program with an optimization program.

This research project addresses the question: Can a neural network replace a finite element analysis program in the optimization process and will the design converge to a reasonably accurate optimum solution? In this system, the neural network is applied to approximate the results of the finite element analysis program to quickly reach a near-optimal solution. The results obtained for the NETS/PROSSS optimization process could also be used as an initial design in a normal optimization process and converge to an optimum solution with significantly fewer iterations resulting in substantial savings of computer time.

Several questions were addressed while testing the system. These questions include: (1) What is the best way to select training pairs? (2) What is an appropriate number of training pairs to use to train the neural network? (3) Can a user of the system begin the optimization process from different starting points with the neural network and still converge?

1. Introduction

Expensive analysis programs are often applied to solve engineering design problems. To obtain an optimal solution typically requires numerous iterations between the analysis program and an optimization program. This often becomes prohibitive due to the amount of computer time required for convergence. Therefore, any new software that could significantly reduce the computer time required to solve a complex design problem would be beneficial.

This paper describes a new experimental software package called NETS/PROSSS (Rogers and LaMarsh, 1991) which has been developed to help meet this need. PROSSS (Programming System for Structural Synthesis, see Sobieszczanski-Sobieski and Bhat, 1979; Rogers, Sobieszczanski-Sobieski, and Bhat, 1981; and Rogers, 1982) was developed several years ago to provide an

J. S. Gero (ed.), Artificial Intelligence in Design '92, 739–754.

open-ended system for coupling analysis and optimization. Although PROSSS was designed to handle any type of analysis program, most of the work has evolved around structural analysis with the EAL (Engineering Analysis Language, Whetstone, 1980) finite element analysis program. PROSSS includes the CONMIN (Vanderplaats, 1973) optimization program. In NETS/PROSSS, EAL is replaced by NETS (Baffes, 1989), a neural network program developed at NASA Johnson Space Center.

This research project addresses the question: Can a neural network replace a finite element program in the optimization process and will the design converge to a reasonably accurate optimum solution? In this system, the neural network is applied to approximate the results of the finite element analysis program to quickly reach a near-optimal solution. The results obtained for the NETS/PROSSS optimization process could also be used as an initial design in a normal optimization process and converge to an optimum solution with significantly fewer iterations, resulting in substantial savings of computer time.

This paper begins with a description of the cantilevered beam test problem. Next is a description of the NETS/PROSSS system and how it is incorporated into the design optimization process. The next section describes the process that was used to select the training pairs for the neural network. Several questions were raised while testing the system. These questions include: (1) What is the best way to select training pairs? (2) What is an appropriate number of training pairs to use to train the neural network? (3) Can a user of the system begin the optimization process from different stating points with the neural network and still converge? These questions are addressed in separate sections followed by a summary.

2. The Test Problem

The problem to be solved is to optimize the shape of a cantilever beam to minimize the weight (the objective function) while satisfying stress constraints. A 3000 DOF cantilever beam (figure 1) with 1025 joints and 640 3D solid brick elements is used for the finite element model. The model is 40" long, 4" wide, and begins at a height of 8". The problem has five design variables which determine the shape of the beam by modifying the heights of the elements. Each design variable controls the heights of a block of 128 elements. There are 40 cumulative stress (Barthelemy and Riley, 1988) constraints, one for each of 40 stations (16 elements per station) along the beam. A total load of 10,000 pounds is applied in the z-direction distributed at the 25 joints on the end of the beam. The beam is made of steel-like material with a modulus of elasticity of $35.9\text{x}10^6$ psi, a weight per unit volume of .283 lb/in^3, and an allowable stress of $150*10^3$ psi.

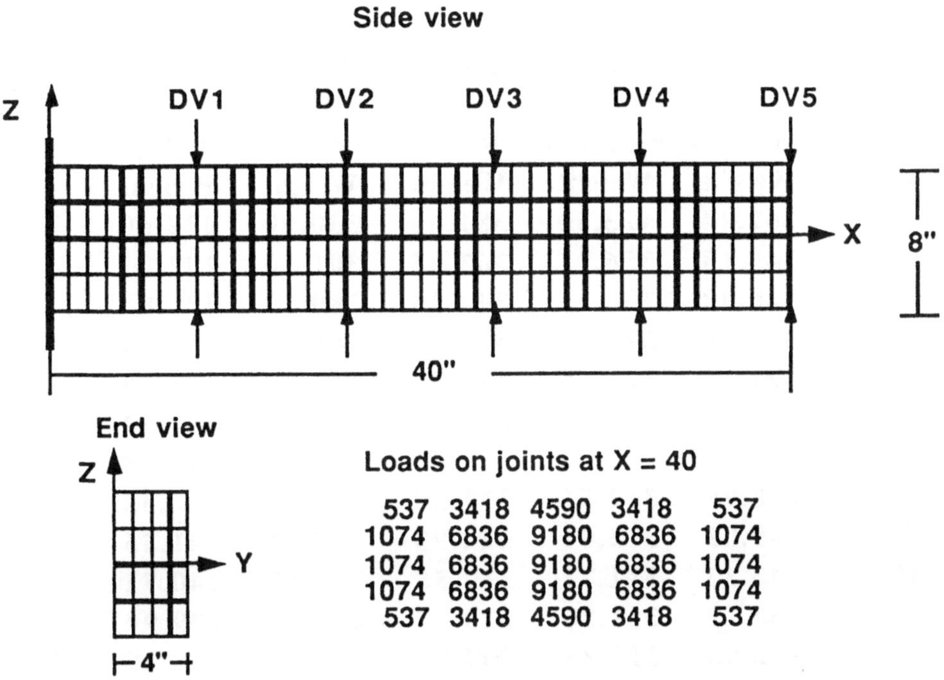

Fig. 1. Test problem and model with loading conditions

The complete optimization process using PROSSS with EAL and CONMIN is traced in table 1. Because the objective function is linear and no analytical gradients are available, PROSSS option 2.2 is chosen. This option uses a piecewise linear analysis with gradients computed by finite difference external to CONMIN. This requires six EAL analyses to be executed in each optimization cycle with each analysis requiring 20 minutes of computing time on a DEC MicroVax. The starting point has all design variables at 8". It takes 18 cycles to converge, with each cycle (six analyses) requiring two hours of computing time. At each cycle, the design variables are limited to a 5% maximum change (5% move limits).

Pairs Range System Cycle	 EAL/ PROSSS Weight	30 Large NETS/ PROSSS Weight	5 Large NETS/ PROSSS Weight	30 Small NETS/ PROSSS Weight	5 Small NETS/ PROSSS Weight
1	362.24	423.50	381.40	330.00	332.60
2	344.13	356.95	363.05	301.20	292.03
3	326.92	338.15	344.20	280.93	272.85
4	310.58	319.55	328.57	261.45	255.28
5	297.23	314.95	325.33	243.08	245.49
6	283.87	314.26	322.43	240.00	244.52
7	271.90	313.30	320.02	239.06	243.18
8	258.99	312.51	317.20	238.42	242.35
9	249.02	312.04	314.93	237.91	241.55
10	243.33	311.71	312.58	237.41	240.71
11	235.13	311.26	310.17	236.90	239.90
12	232.03	310.83	307.04	236.36	239.26
13	224.49	310.45	304.68	235.77	238.69
14	223.69	310.06	301.63	235.32	238.18
15	221.84	309.95	299.83	235.63	237.31
16	219.61	309.70	299.52	235.34	237.66
17	219.61		299.34	235.04	236.75
18	219.61		299.17	234.21	236.47
19				233.97	236.20
20				234.99	235.95
21				235.00	236.65
22				235.00	235.75
23				235.00	236.06
24					236.09
25					236.93
26					236.30
27					236.30

TABLE 1
Optimization results from EAL/PROSSS vs. NETS/PROSSS
(training pairs from interval approach)

The final design is shown in figure 2.

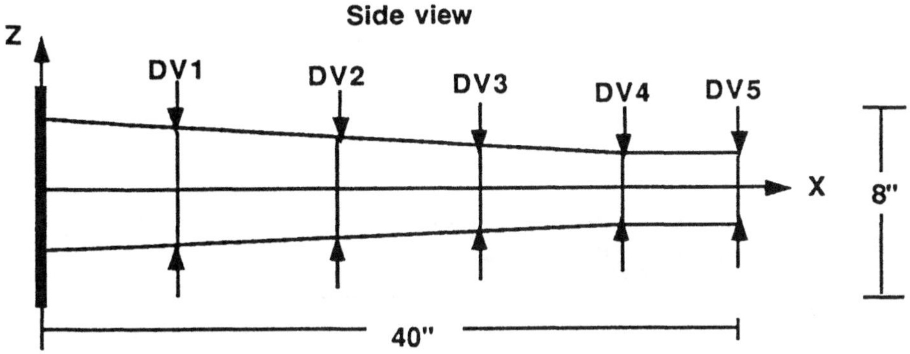

Fig. 2. Beam after optimization

3. NETS/PROSSS

This paper assumes that the reader has a working knowledge of back propagation neural networks (Lippmann, 1987; and Jones and Hoskins, 1987). Only recently have engineers begun to apply neural nets to structural mechanics problems (Rehak, Thewalt, and Doo, 1989; Swift and Batill, 1991; Berke and Hajela, 1991; and VanLuchene and Roufei, 1990). For example, Berke and Hajela use optimum designs of trusses to train a neural network. The input data consist of two lengths and the height for the truss, while the output data consist of the optimized bar areas and the weight. The neural net was then used to determine new optimal truss designs by changing the input data. Van Luchene and Roufei apply a neural network to simulate the finite element analysis of a simply supported rectangular plate. The analysis predicts the location and magnitude of the maximum moment. In their conclusions, the authors suggest that the use of neural networks in solving civil engineering optimization problems is an area that requires additional study. The research presented in this paper addresses that issue by coupling the NETS (Baffes, 1989) back-propagation neural network to simulate finite element analysis to the CONMIN optimization program to form a new system called NETS/PROSSS.

For the test problem described above, the input data for the neural net are the five design variables and the output data from the neural net are the 40 constraints and the objective function. The number of nodes in the hidden layer is 46 (the sum of the input and output nodes). Initially, the are 2116 unknown (5*46+41*46) weights. Training pairs of known input data and known output data are processed through the neural network to estimate the unknown weights. This is called "training the network." Once the network has been trained (ie. the unknown weights have been estimated), new input data (design variables) can be input to the neural network to approximate new output data (constraints and objective function). A flowchart of the entire optimization system is shown in figure 3.

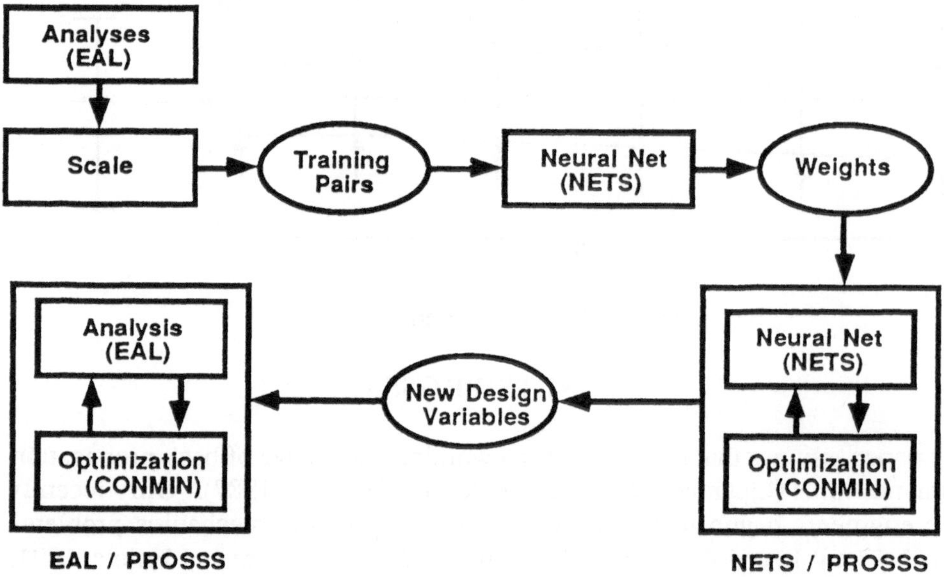

Fig. 3. Flowchart of optimization process with NETS/PROSSS

Different sets of design variables are input into the EAL (Engineering Analysis Language) finite element analysis program to compute the constraint and objective function output data. The known input data (the design variables) and the now known output data (the constraints and objective function) make up the training pairs which are input to the neural network program. The neural network uses the training pairs to create a weight matrix, an input to NETS/PROSSS. The weight matrix essentially indicates the relationship between the input and the output. NETS/PROSSS (flowchart in figure 4) is a system which combines a neural network (NETS) with an optimization program (CONMIN, Vanderplaats, 1973). The neural network in NETS/PROSSS simulates an analysis program by multiplying a vector of input data (the design variables) by the weight matrix to compute the output data (constraints and objective function). This data is then passed into CONMIN for optimization. NETS/PROSSS iterates between NETS and CONMIN until an optimum is reached (the objective function does not change by more than 0.1% in three successive cycles. The new design variables from this approximated design could then be input as a starting design in PROSSS which combines EAL and CONMIN to find a more accurate objective function.

NETS/PROSSS is composed mainly of routines written in FORTRAN. Data is input to the system in routine ICNMN. This data consists of CONMIN input and scaling information. Subroutine OPT2ANL scales the design variables to be between 0.1 and 0.9 for input into the neural network. This routine is problem dependent. Subroutine ANALYNN (a C routine generated by NETS) reads the scaled design variables and the weight matrix generated with the training pairs and calls the neural network (NETS) to simulate the analysis program. The neural network computes scaled values for the constraints and the objective function. Subroutine ANL2OPT unscales this data for input to CONMIN. This routine is also problem dependent. Subroutine EVALSUB computes gradient data using the finite differences. CONMIN performs the optimization in conjunction with a piecewise linear analysis routine called ANALY. If the problem has not converged, then the system loops back to the OPT2ANL routine for another cycle with a new set of design variables.

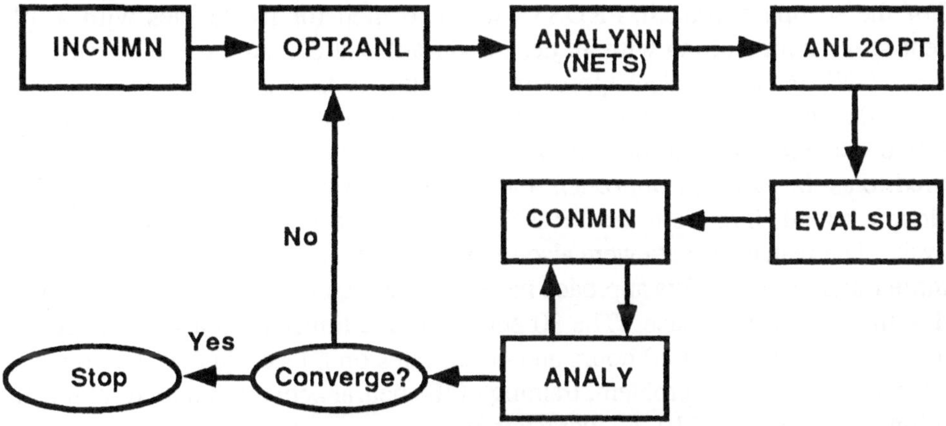

Fig. 4. NETS/PROSSS flowchart
(main program is NNOPT)

4. Process For Selecting Training Pairs

One of the keys to approximating an analysis with a neural network is the generation of the training pairs. Training pairs consist of known inputs and known outputs. Two approaches were tested to generate the training pairs for this project. Different numbers of training pairs were generated for each approach to determine if a neural network trained with a small number of training pairs can produce accurate results. In each of these cases, the number of training pairs is much less than the number of unknowns in the neural network leaving the system underdetermined. This may contradict the conclusion (Carpenter and Barthelemy, 1992) which recommends that the number of training pairs must exceed the number of unknowns. This was done intentionally because of the expense involved in developing training pairs. For this problem, there are 2116 unknowns and it would require over 700 hours of computing time to develop that many training pairs. In each case, the neural network was trained to a .01 RMS (root mean square) error between the scaled known output and scaled computed output. Even though the neural network was trained to a .01 RMS error, individual outputs can be off by as much as 15-20%.

For the first approach, EAL was executed with all five design variables at 4" (the lower limit), 6", 8", 10", and 12". This is referred to as the *interval approach.* This was done in the anticipation of giving the neural network a meaningful range for interpolation. The analysis data obtained when perturbing the designs for finite difference gradient calculation were also saved so the neural network could be trained with as few as five training pairs or as many as 30 for this approach.

For the second approach, PROSSS was executed for five cycles with a 30 percent move limit starting with all design variables at 8". At the end of the five cycles, CONMIN produced an approximate design. A final analysis was executed with this design data as input to generate an additional training pair. One analysis was also executed with all design variables at 4" (the lower bound). This is referred to as the *PROSSS approach* and resulted in seven training pairs, one for the lower bound, five from the PROSSS cycles, and one more from the final analysis. The reanalysis data were also saved from the five PROSSS cycles giving 25 more training pairs. This approach has the advantage of selecting training pairs well within the design space. The 30 percent move limits yield a wide range of values. While it required 10 hours and 40 minutes on a Dec MicroVax to obtain the training pairs for this problem, training of the neural network required, at most, 15 minutes on a Sun SPARC1+ workstation.

5. Effect Of PROSSS Approach vs. Interval Approach To Select Training Pairs

Figure 5 compares the convergence of the original PROSSS optimization to NETS/PROSSS with training pairs generated from the PROSSS approach and the interval approach. For the PROSSS approach, the neural network was trained with 32 training pairs. For the interval approach the neural network was trained with 30 training pairs. The convergence pattern of the neural network trained with the PROSSS approach closely matches that of the original PROSSS run while the convergence of the neural network trained with the interval approach converges to a different design.

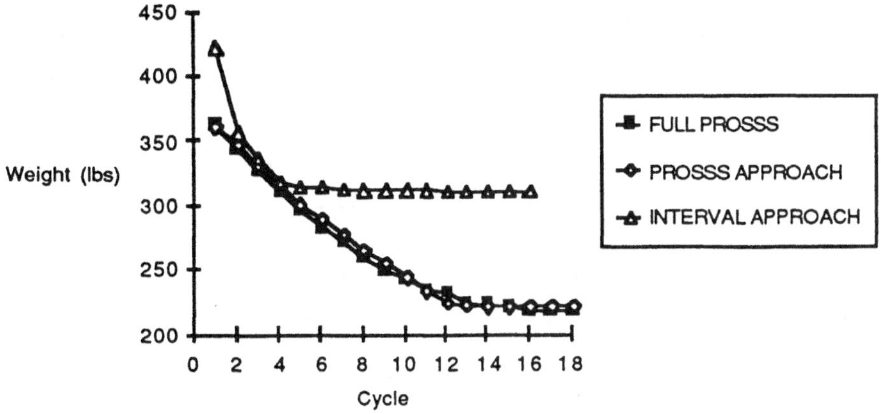

Fig. 5. Comparison of approaches to select training pairs

The results from all the trials with the interval approach can be seen in table 1. Even though the neural network was trained to a .01 RMS error, the resulting approximation was not accurate. This can be easily seen by comparing the weights from cycle 1 in the table. Therefore, whether the neural network (NN) used 5 or 30 training pairs or a large (L) or small (S) range size for the data the convergence results are poor. For the large range, the design variables ranged between 3 and 12, the constraints ranged between -2 and 2, and the objective function ranged between 150 and 950. For the small range, the design variable ranged between 4 and 12, the constraints ranged between -1 and 2, and the objective function ranged between 150 and 550. The design (table 2) is also poor. When the final design was used as a starting point in PROSSS, it produced an error message pertaining to a bad geometry generated by the neural network.

Table 2 compares the final design data from the optimization processes. As can be seen, the optimization results from the neural network trained with 32 training pairs from the PROSSS approach closely resembles the optimization data from the original PROSSS. A final analysis was made to determine how close the approximated objective function was to the objective function computed by the analysis program. The final design variables were used to create the geometry for input to EAL to compute the objective function. The objective function approximated by this neural network (222.1) was very close to that computed by EAL (221.5). The main difference is the time that was taken to compute these numbers. It took 36 hours to obtain the final results using the original PROSSS and only 10-15 seconds to obtain the data using NETS/PROSSS. (Note: This time does not include the ten hours and 40 minutes to collect the data for the training pairs nor the 15 minutes to train the neural network.) On the other hand, the neural network trained with 30 training pairs from the interval approach produced a geometry that was not acceptable.

System	EAL/PROSSS	NETS/PROSSS	NETS/PROSSS
Approach		PROSSS	INTERVAL
Approx. Obj.	219.61	221.50	309.70
Obj. Fun	219.61	222.10	
DV1	6.68	6.58	12.00
DV2	5.53	5.53	4.63
DV3	4.53	4.80	7.84
DV4	4.00	4.00	4.00
DV5	4.00	4.00	4.61
Active	1	1	1
Constraints	8		
	9		
	10	10	
	11	11	
	12	12	
	13	13	
	14	14	
	15	15	
	16	16	
	17	17	
	18	18	
	19	19	
	20	20	
	21		

TABLE 2
Final designs of EAL/PROSSS vs. NETS/PROSSS

6. Effect Of The Number Of Training Pairs

The importance of the effect of the number of training pairs is that the smaller the number of training pairs needed, the less amount of time will be required to obtain them. As discussed in the previous section, the results from NETS/PROSSS, when trained with training pairs from the interval approach, are not very accurate. Therefore, the remainder of the comparisons will concern the results from NETS/PROSSS when trained with training pairs from the PROSSS approach. The time difference between obtaining 7 training pairs (2 hours and 20 minutes) and 32 training pairs (10 hours and 40 minutes) is significant. Figure 6 compares the convergence of the objective function from NETS/PROSSS with a neural network trained with 32 training pairs and a neural network trained with 7 training pairs.

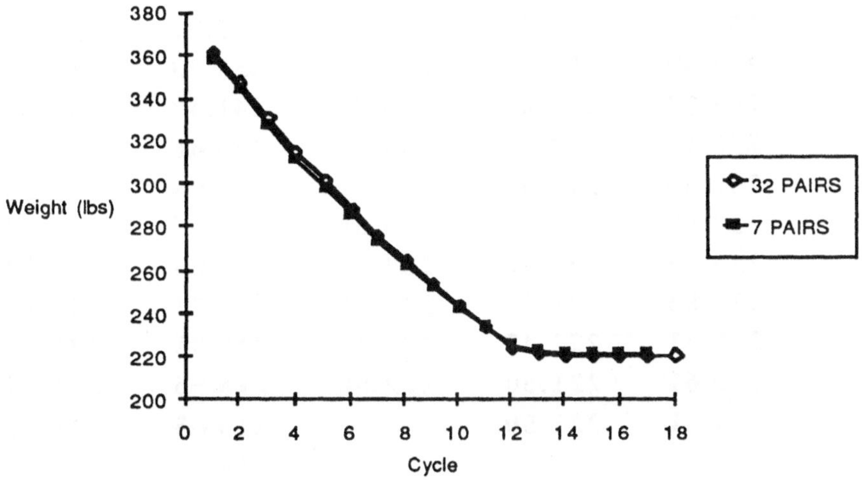

Fig. 6. Comparison of results with 7 and 32 training pairs (PROSSS approach)

A complete comparison of the neural network optimization process from trials with 7 and 32 training pairs and a large (L) and small (S) data range can be seen in table 3.

Pairs Range System	PROSSS	32 Large NETS/ PROSSS	7 Large NETS/ PROSSS	32 Small NETS/ PROSSS	7 Small NETS/ PROSSS
Cycle	Weight	Weight	Weight	Weight	Weight
1	362.24	362.10	359.30	359.10	358.78
2	344.13	347.95	344.55	349.38	350.93
3	326.92	331.75	328.30	337.65	338.16
4	310.58	315.70	312.50	323.80	323.33
5	297.23	301.46	298.47	307.93	307.38
6	283.87	288.82	286.19	293.62	293.64
7	271.90	276.69	274.42	279.86	280.37
8	258.99	265.30	263.31	266.32	267.49
9	249.02	254.39	252.98	253.43	255.44
10	243.33	244.04	243.35	241.47	244.34
11	235.13	234.41	234.29	230.66	234.28
12	232.03	225.32	225.90	220.98	225.30
13	224.49	222.79	223.56	217.96	224.30
14	223.69	221.80	222.65	217.04	222.84
15	221.84	221.52	222.49	215.71	222.12
16	219.61	221.47	222.43	215.95	221.63
17	219.61	221.50	222.50	216.55	221.63
18	219.61	221.50		216.65	221.63
19				216.65	
20				216.65	

TABLE 3
Optimization results from EAL/PROSSS vs. NETS/PROSSS
(trained with PROSSS approach)

7. Effect Of Different Starting Points On Optimization Results

To assess the effect of varying the starting designs without changing the neural net, NETS/PROSSS was started from an infeasible starting point where all the design variables were set to the lower bounds (4"). The final design data is shown in table 4.

Start System	Feasible EAL/ PROSSS	Feasible NETS/ PROSSS	Infeasible NETS/ PROSSS
Approx. Obj.	219.61	221.5	221.3
Obj. Fun.	219.61	222.1	221.82
DV1	6.68	6.58	6.65
DV2	5.53	5.53	5.46
DV3	4.53	4.80	4.81
DV4	4.00	4.00	4.00
DV5	4.00	4.00	4.00
Active Constraints	1	1	1
	8		
	9		
	10	10	10
	11	11	11
	12	12	12
	13	13	13
	14	14	14
	15	15	15
	16	16	16
	17	17	17
	18	18	18
	19	19	19
	20	20	20
	21		

TABLE 4
Final designs of EAL/PROSSS vs. NETS/PROSSS

Figure 7 demonstrates that NETS/PROSSS can converge from either a feasible or an infeasible starting point.

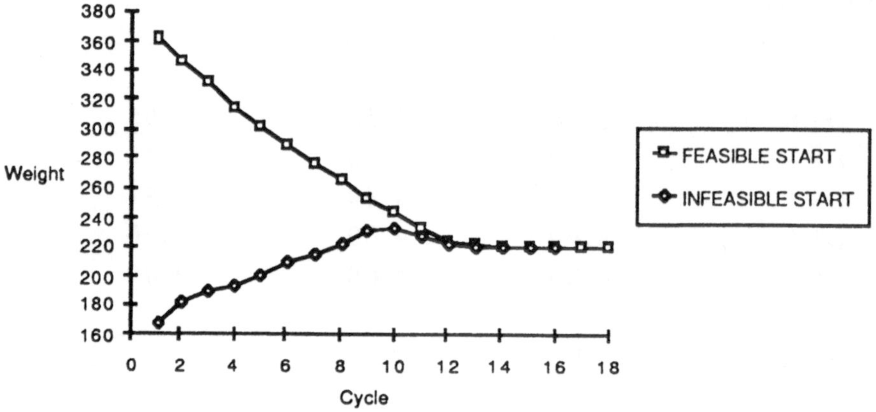

Fig. 7. Comparison of feasible vs. infeasible starting points (PROSSS approach)

8. Summary

NETS/PROSSS is a new experimental software package which has been developed to replace an expensive analysis program with a neural network. This package is to be used in conjunction with some other optimization system which contains a full analysis program. NETS/PROSSS is applied not to obtain the optimal solution, but an approximate one. The results from NETS/PROSSS can then be input to the normal optimization system to obtain the most optimum results.

NETS/PROSSS can produce accurate results depending on the approach to select the training pairs, the number of training pairs, and the range for scaling data for input to the neural network. From the test cases presented in this paper, it appears that the best approach to selecting the training pairs is to take advantage of domain knowledge and identify a region in which the neural network needs to perform well and select training pairs from this region. The interval approach does not take advantage of this knowledge, while the PROSSS approach does by beginning the normal optimization process and collecting input and output data for a few cycles. Using wide move limits (30% for this project) helps spread the training data over a wider range. The number of cycles to use is still a question. Five cycles were used for this project, plus a final analysis and an analysis with design variables set to the lower bounds. This number yielded accurate results.

However, the results were not accurate without the lower bound analysis data. In addition, gradient data can be used if it is available.

Once the neural network has been trained, the user can apply NETS/PROSSS to try different designs and see the effects on the optimization process. This can be done in seconds or minutes as opposed to days or hours resulting in substantial savings of both cost and time.

Although the results from this investigation have been encouraging, there are still questions to be answered which pertain to the generality of these findings. Because this is just one problem, it is not known whether the choices made for this problem are valid for other problems. In addition, the interval approach needs to be re-examined using statistical techniques to incorporate domain knowledge about the starting design point.

References

Baffes, P. T.: 1989, NETS 2.0 User's Guide. LSC-23366, NASA Lyndon B. Johnson Space Center.

Barthelemy, J.-F. M. and Riley, M. F.: 1988, Improved Multilevel Optimization Approach for the Design of Complex Engineering Systems, AIAA Journal, **26**, No. (3), 353-360.

Berke, L. and Hajela, P.: 1991 , Applications of Neural Nets in Structural Optimization, Presented at the NATO/AGARD Advanced Study Institute on "Optimization of Large Structural Systems', Berchtesgaden, Germany.

Carpenter, W. C. and Barthelemy, J.-F. M.: 1992, Comparison of Polynomial Approximations and Artificial Neural nets for response Surfaces in Engineering Optimization. Submitted to the 33rd SDM Conference in Dallas, TX.

Jones, W. P. and Hoskins, J.: 1987, Back-Propagation, *BYTE Magazine*, October, 155- 162.

Lippmann, R. P.: 1987, An Introduction to Computing with Neural Nets, *IEEE ASSP Magazine*, April, 4-22.

Rehak, D. R.; Thewalt, C. R.; and Doo, L. L.: 1989, Neural Network Approaches in Structural Mechanics Computations, Computer Utilization in Structural Engineering, Ed. J. J. Nelson, Jr., ASCE Proceedings from Structural Congress.

Rogers, J. L. Jr.; Sobieszczanski-Sobieski, J.; and Bhat, R. B.: 1981, An Implementation of the Programming Structural Synthesis System (PROSSS). NASA TM 83180.

Rogers, J. L. Jr.: 1982, Combining Analysis with Optimization at Langley Research Center - An Evolutionary Process. " *Proceedings of the Second International ASME Computers in Engineering Conference*, **3** , 83-91, San

Diego, CA.

Rogers, J. L. and Lamarsh, W. J. II: User's Guide for NETS/PROSSS. NASA TM 104166.

Sobieszczanski-Sobieski, J.; and Bhat, R. B.: 1979, Adaptable Structural Synthesis Using Advanced Analysis and Optimization Coupled By a Computer Operating System, *A Collection of Technical Papers on Structures - AIAA / ASME / ASCE / AHS 20th SDM Conference*, 20-71, AIAA Paper No. 79- 0723.

Swift, R. A. and Batill, S. M.: 1991, Application of Neural Networks to Preliminary Design, *A Collection of Technical Papers on Structures - AIAA / ASME / ASCE / AHS 32nd SDM Conference*, AIAA Paper No. 91-1038.

Vanderplaats, G. N.: 1973, CONMIN - A FORTRAN Program for Constrained Function Minimization User's Manual. NASA TM X-62282.

VanLuchene, R. D. and Roufei, S.: 1990, Neural Networks in Structural Engineering, *Microcomputers in Civil Engineering* **5**, 207-215.

Whetstone, W. D.: 1980, EISI - EAL: Engineering Analysis Language, *Proceedings of the Second Conference on Computing in Civil Engineering,* \ ASCE, 276-285.

13

CONCEPTUAL DESIGN

A COMPUTATIONAL MODEL FOR CONCEPTUAL DESIGN BASED ON FUNCTION LOGIC

R. H. STURGES
Department of Mechanical Engineering
Carnegie Mellon University
Pittsburgh PA 15213-3890
USA

Abstract. Function logic and function block diagrams have been successfully employed in preliminary and conceptual design processes for several decades. This paper describes a computational model of this process with extensions of the manual approach. It provides for a systematic identification and definition of form and function variables and identifies a three-level function/allocation/component information structure to represent the state of the design. We outline the inputs, outputs and operations on the form and function variables as a key step prior to the synthesis process. We also illustrate by example how to transfer functional designs across specialist domains.

1. Introduction

Concurrency of product and process design tasks has become recognized as extremely valuable in reducing and predicting life cycle cost. The short-term benefits of such concurrency are reflected in reduced time to market and reduced manufacturing cycle time (Andreason et al., 1988). The longer term benefits accrue from the leverage which well-reasoned design methodologies exert over the product life cycle (Sturges et al,, 1986). The decisions made at the earlier stages of a design have a disproportionately large share of impact than the latter. Thus, it is the preliminary or conceptual stages of the Form-Function synthesis process which we are interested in representing and facilitating. Conceptual design is distinguished from other phases of the design process as illustrated in Figure 1(Westinghouse, 1984). The activities in the preliminary and conceptual phases differ from the latter detailed design work. In mechanical design, conceptual issues are largely functional,with less emphasis on form. Conventional detailed design, on the other hand, requires us to synthesize forms which will not compromise the given functions.

The conceptual phase concerns the problem of coming up with new ideas or new solutions to older problems (Pugh, 1981). Good conceptual design means innovation, and an innovative design comes about when one deliberately tries to create one (Perkins, 1981). For example (Bailey, 1978),

J. S. Gero (ed.), Artificial Intelligence in Design '92, 757–772.

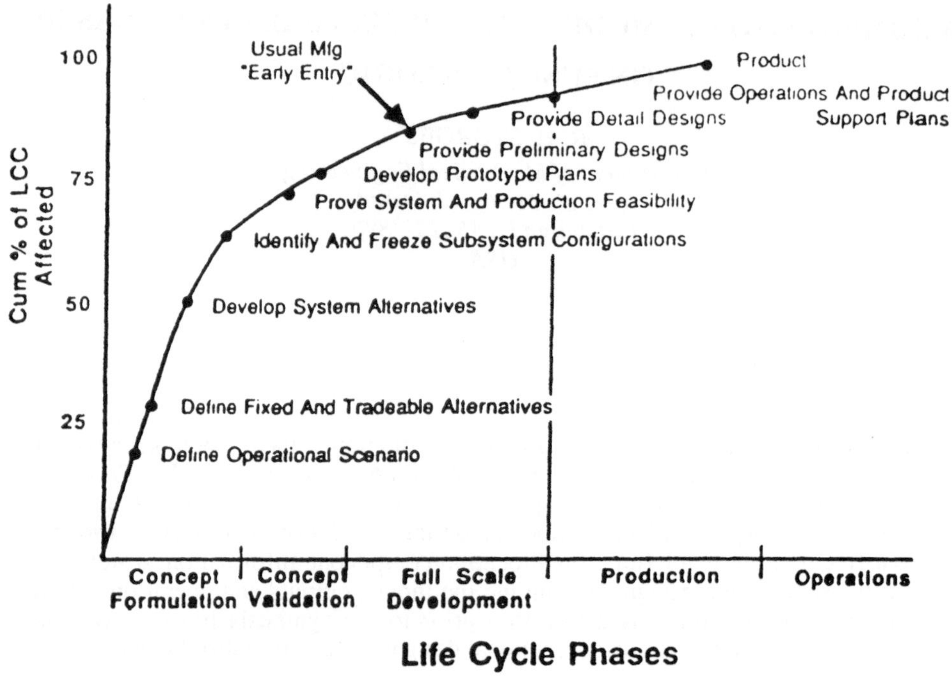

Fig 1. Conceptual design in the life cycle cost of a product

> An engineer carefully studies power losses in a coal-fired plant and is able to increase efficiency by 0.1%. Another engineer studying the same data conceives of the idea of using direct energy conversion to use the waste heat and increases efficiency by 5-10%.

Although the essential result (efficiency) is the same, the functions used to produce it are very different. The former approach represents optimization of a given functional model. The latter represents modification of this functional model. The issue in conceptual design theory is to understand the processes which lead to innovation and to create tools which generate such step changes in function in an orderly and repetitious basis. Processes for prompting innovative design are currently typified by "Buhl's Seven Steps" which are shown in Table 1. Similar approaches are given by Bailey (1978). Processes for managing conceptual design and the process of innovation have long been described in the Value Analysis and Value Engineering literature (e.g., Bytheway, 1964; Ruggles, 1971; Miles, 1982). This method, when applied by a small group of engineers, has been shown to be consistently effective in achieving focused conceptual design goals, such as achieving a given level of product performance, redesigning to reduce costs below a given threshold, etc. A study by the American Ordnance

Association of a sampling of 2000 of its projects revealed improvements in cycle time, reliability, quality and maintainability in excess of 60% (Prendergast,1982). That the process works is not debated. Understanding and translating the process into software tools remains a challenge and an opportunity.

TABLE 1
Buhl's seven steps (Perkins)

Recognition. Recognize that a problem exists and decide to do something about it.
Definition. Define problems in familiar terms and symbols; dissect into sub-problems; determine limitations and restrictions.
Preparation. Compile past experience in the form of data, ideas,opinions, assumptions, etc.
Analysis. Analyze preparatory material in view of defined problems, inter-relations, and evaluation of all information that could bear on the problem.
Synthesis. Develop a solution or solutions from developed information.
Evaluation. Evaluate possible solutions. Verify and check all facets of the solution. Reach a decision.
Presentation. Plan a strategy for convincing others and carry it out.

Tools for carrying out the essence of these conceptual design activities include detailed checklists, data retrieval and management systems, function logic, evaluation methods, and presentation techniques. The single common element in each of the examples above is the expression of a design at the conceptual stage in functional terms (the function block diagram or FBD), and the deliberate manipulation of this functional representation (the function logic process). In this paper we describe a model for conceptual design in which functional representations are extended to include other elements of the manual process and in which the manipulation of the functions is facilitated. We have employed the terminology of the Society of Value Engineers in dealing with functional representations of conceptual design (Bytheway, 1965). Specifically, *function* refers to largely domain independent characteristics or behaviors of elements or groups of elements of a design. The *intent* of a design is expressed by the totality of its functional elements and their structure. The *basic function* refers to the single intended output or use of the product, with *secondary functions* describing necessary but less critical constraints. *Side effects refer* to unintended behaviors which derive from implementation decisions.

2. A Computational Model for Conceptual Design

2.1. OVERALL ARCHITECTURE AND APPROACH

In this section we describe a representation of and a systematic approach to conceptual design based on function logic. The central concept of the approach is to capture design intent through a chain of functional description and reasoning and to highlight the dependencies among the sub-functions of a design. This ability is crucial to life cycle success since the *basis* for a specification changes with time, resources, market, technology,and the evolution of the form of the design itself. This basis and its evolution are rarely present in design representations. Layout and detail drawings tell us "what" but rarely "why." The reverse process of extracting function from form is problematic since critical information must be synthesized to replace that which was discarded along the way. Even a very detailed functional specification gives you only "the answer" while discarding the numerous questions which led to its substance. It is these questions which need to be reexamined when a design changes to meet a new need or requires analysis for improved performance. Also, the interfunctional dependencies of a design ("degree of coupling") are not obvious from its specifications and yet may be highly sensitive to them. Functional dependence is often realized only after a given technology (form of manufacture) or component (form of artifact) is chosen. In short, form-function synthesis processes need not only specifications; they also need a way to systematically produce and manage these early in the evolution of a design.

As mentioned in the introduction, our representation of conceptual design, based on Miles (1982), employs functional descriptions of products or processes according to a set of linguistic and hierarchic rules. Existing artifacts are analyzed by developing their function logic in a bottom-up fashion; new artifacts are synthesized generally in a top-down fashion (Bytheway, 1971). Our extensions to the FBD representation (Sturges et al., 1990a) comprise a three-tiered structure (Figure 2) consisting of: (i) function blocks (compact verb-noun descriptors of what the design does rather than what it *is*) with links to other blocks; (ii) allocations (constraints, performance requirements, specifications and resources (Kantowitz and Sorkin, 1987); and (iii)components (artifacts that satisfy the given function). Design intent is captured and managed during the conceptual design process by each tier of this structure as follows.

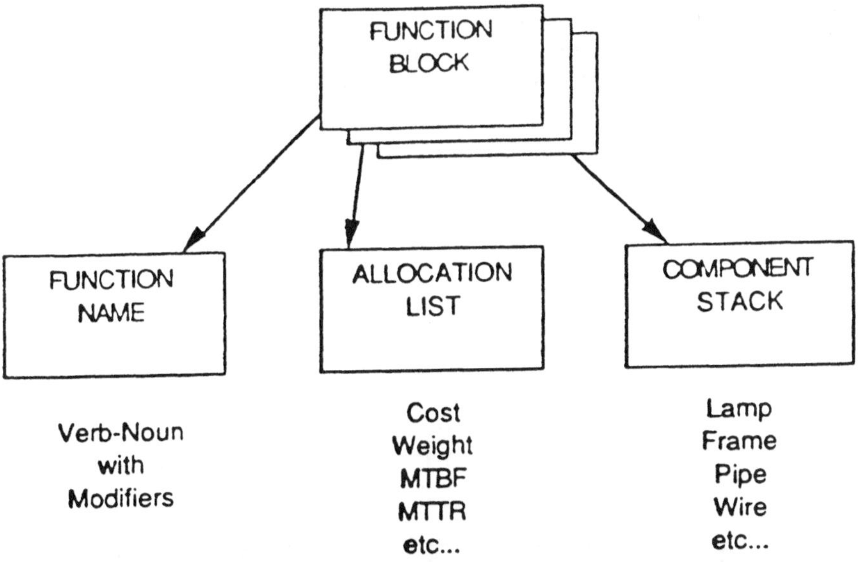

Fig. 2. Three-tier structure for a function logic diagram

2.2. INPUTS

The *problem* in conceptual design is the articulation of the function in sufficient detail to suggest and monitor the development of form through existing knowledge and decision-making methods. Since the process in its manual implementations is reversible, the inputs may be derived from two sources. In forward design, the *inputs* are verbal, syntactic and numeric descriptors which evolve with understanding and negotiation. In reverse design, the inputs comprise the set of such descriptors attendant to each of the given components and subassemblies of an existing design. Since there is little evidence of design proceeding exclusively in the forward mode, descriptors from related but independent designs will always be present (Ullman et al., 1988). In this section we describe a representation of and a systematic approach

2.3. FORM AND FUNCTION VARIABLES

The function variables consist of verb/noun pairs and links between these. The general form of the FBD is shown in Figure 3. The function block (or node) contains the function name (what is done) expressed as a generic noun/verb pair to describe the function of the product or process. The verb must be active; the noun must be measurable. The nodes to the left of a function node represent the reason *why* a function is included: a higher-level function. The nodes to the right are functions describing *how* the function is performed: lower-level functions. Links connect each high-level function with its lower-level function according to this how/why relationship. Links other than how/why have been identified between function blocks, viz.: causal, temporal, informational, alternative and revisional (Sturges and Kilani, 1990). A description of each link is given in Table 2. Strict hierarchy is not required since a more specific function may satisfy more than one less specific function in the diagram.

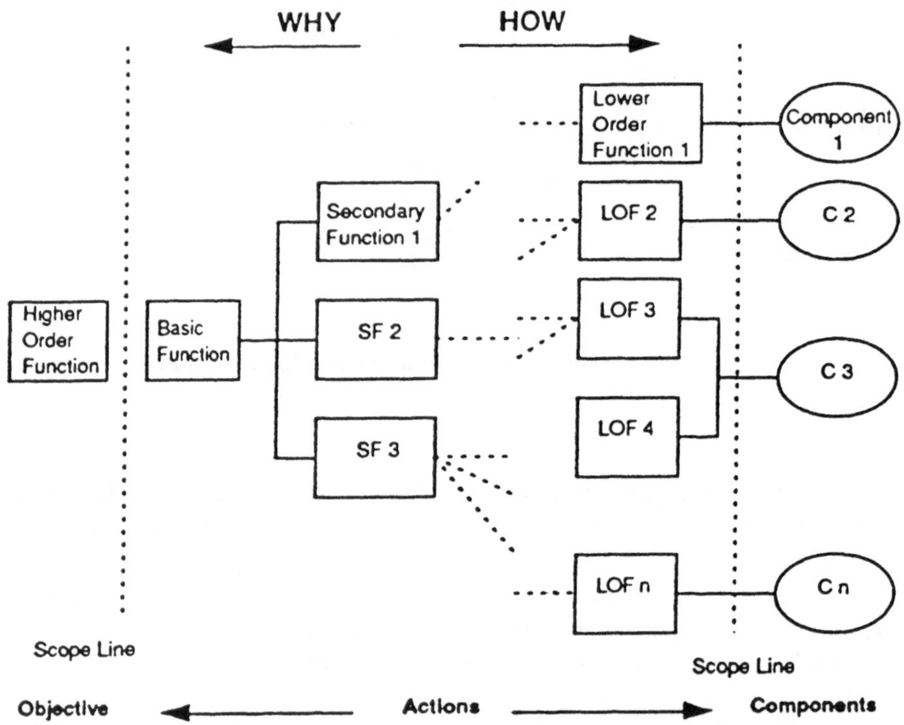

Fig. 3. The general form of a function logic diagram.

The form variables consist of the allocations attached to these verb/noun pairs. The allocation list supplies what needs to be known about the design as it evolves: constraints, performance requirements, specifications and resources. When complete, the allocations specify the behavior of each function and set the constraints for the form of the design. The synthesis of form itself in the component choices and details is not carried out by the conceptual process but, rather, by the parallel process of proposing and testing alternatives. When component decisions are made "early," their characteristics are attached to the related function blocks.Since they provide a compact description of a set of fixed constraints,they are not considered form variables.

TABLE 2
Links used in function block diagrams

And/Or: The "and" links are the conventional links indicated by Bytheway (1971), and are represented as solid lines connecting the blocks. The"or" links suggest a viable alternative solution for decomposing a function block. These links are represented as dotted lines.

Temporal: The "temporal" link connects functions that occur "at the same time,"as suggested by Ruggles (1971). Usually, these function blocks are portrayed vertically and are connected with a dotted/dashed line. An arrow indicates the process flow, or which function must occur before the other. Temporal links are also used to specify a material flow, because the stages in the process occur in time or event sequence.

Causal: The "causal" link is a result of the Miles (1982) side-effect concept and is represented as a solid line with an arrow to indicate which function was created as a side effect. The allocations associated with the side-effect function should be checked with other blocks in the FBD to discover opportunities and/or conflicts. For example, the creation of heat in an engine could be useful in the design of the cabin heating system.

Information: The "information" links indicate which functions are involved in an exchange of some type of information. This is represented as a dashed line with an arrow to indicate the direction of the information flow.

Revision: Tracing the design changes and the thought processes which led to them due to the changes in the design constraints is a valuable, but time consuming effort. We suggest that a record be kept through the use of the "revision"link with a time/date/author stamp before being filed away for future reference. These links are represented as solid, cross-hatched lines. Application of such links can be found in Subramanian (1990).

Chaining: "Chaining" links appear when a function block "decomposes" into only one function rather than several. This suggests that the noun/verb pair was not general enough or that an alternative was considered but discarded as non-viable.

2.4. REPRESENTATION OF FORM AND FUNCTION VARIABLES

First, the *basic function* of the design is established by agreement of the design team. If the basic function cannot be accomplished by a single known component, it is decomposed into several functions which collectively perform the function. These secondary functions may then be translated into components or recursively decomposed. The function decomposition process continues until: (1) the decomposition process is out of scope for the project, (2) there exists a synthesis technique which will complete the decomposition or propose artifacts, or (3) each function can be mapped into a component or structure that will accomplish it directly. Notice that so-called "non-functional"designs are included in this process: purely aesthetic functions such as attract **customer** or **enhance image** are valid and decomposable into artifacts. Such results are for the most part preliminary since practical designs rarely feature a one-to-one correspondence between functions and components.

Second, the allocations for each function block are developed and passed to the neighboring blocks by inheritance rules. Frequently, the allocation list attached to the basic function is only imprecisely known at the beginning of the processes (Paz-Soldan and Rinderle, 1990). As the lower-level functions are satisfied by artifacts or supported functions (which see, below), new categories of information are specified, passed up to the basic function and inherited by certain lower-level functions of the structure. Thus, the allocation list supplies what needs to be known as the design evolves. Its values drive the process of type, number and dimension synthesis through methods outlined in, e.g., Finger and Dixon (1989.)

Third,candidate artifacts and/or supported functions (general purpose models of behavior, coded domain-specific knowledge, formulas and examples) are specified to satisfy the lowest-level functions, which are in turn embodied by the higher-level ones. At this level, the creative process is constrained to issues of domain selection and expression of form through quantitative techniques.

2.5. OPERATIONS PERFORMED ON THE REPRESENTATIONS

2.5.1. *Operations Performed on the Representations*. As mentioned above, the descriptor of a function is a verb/noun pair. The verb must be active,such as "move" or "support." Passive verbs such as "provide" or "allow" are not permitted, since no action is requested. Nouns must be measurable, but cannot specify an artifact. Thus, "load" or "heat" are possible, but "effect" or "bracket" are not. This noun convention avoids domain-specific reasoning and ideation. This apparently simple construction represents a design in a fundamentally abstract form. It requires a deep understanding of the problem at hand and promotes discussion, especially among a group of designers, since it is often difficult to think about a design in this way without practice. The resulting set of descriptors,connected by links, captures the full *intent* of the design in a reasoned hierarchy

Function links represent the relationships between the function descriptors. The primary link between functions in a hierarchical sense is the *how/why* relation as illustrated in Figure 4, the FBD of a mousetrap. If no logical connection can be found, the part is subject to elimination or the function block diagram revised.

Conversely, the process of creating new function blocks through successive decomposition is both guided and encouraged by the discipline of the how/why link. The expanded types (Table 2) represent the design and its reasoning in greater functional detail than the basic how/why logic without adding domain-specific information.An example with the information link is given below.

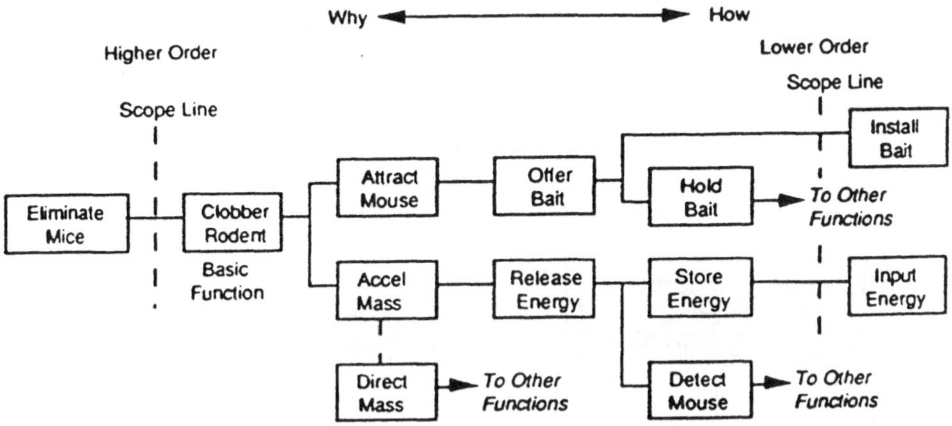

Fig. 4. Preliminary function block diagram of a mousetrap
After Maurer (1984)

The FBD for any given design problem or product does not uniquely specify its physical form. Each verb/noun pair represents a conceptual decision,which is subject to the familiar "brainstorming" and creativity techniques, but is conceptually free from physical constraints. In this way function logic supports and encourages creativity.

2.5.1. *Operations of Form Variables.* During the development of the function logic for a product or process,and ideally before detail design begins, the designer must address the issue of who or what agent will perform each identified function. This process is known as *function allocation.* In addition to resources, the allocation list contains constraints, performance requirements and component specifications. An initial investigation into a more complex control problem, a motor/tachometer speed controller (Figure 5), indicated that there was no means in the function logic representation that could be used to represent the dynamic behavior of a system. This behavior, which includes the gain, natural frequency, damping,etc., is contained therefore in the allocation list.

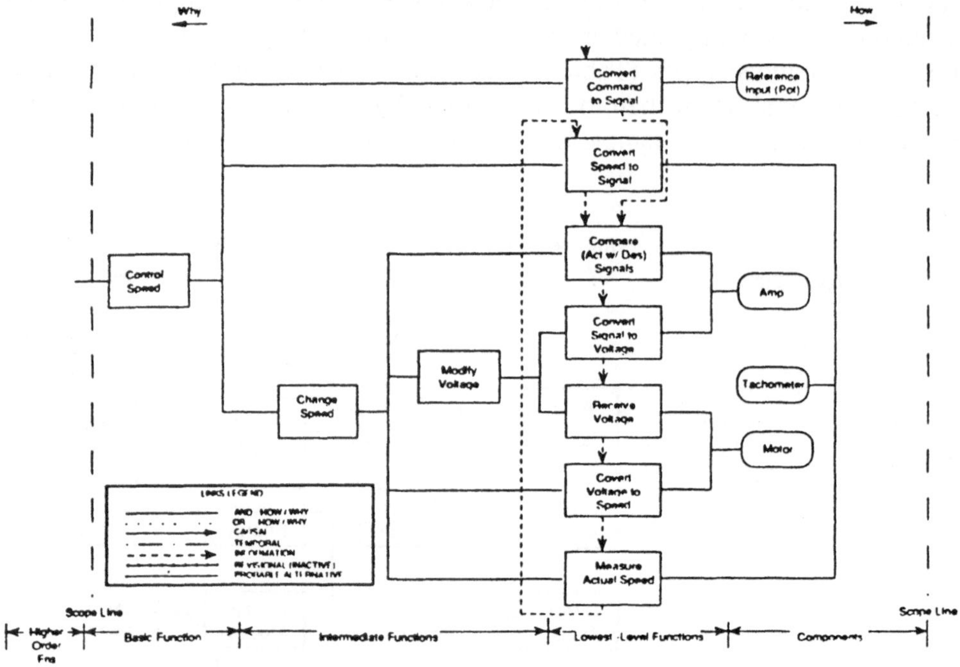

Fig. 5. Amotor/tachometer speed controller FBD

In complex designs, the amount of information to be managed grows rapidly. However, the information that is required for any particular portion of the design will most likely be only a small fraction of the total amount specified. A structure for the allocation list would allow smaller blocks of information to be discarded, or retained and propagated at that level. We have adopted a structure based on the verb classification of signal,energy or material as suggested by Pahl and Beitz (1988). This structure is given in Table 3. This structure also manages the allocation list according to: (1) arithmetic needs; (2) design domain; (3) disinheriting or filtering; or (4) distribution of allocation information.

2.6. CONTROL

In addition to the above rules for FBD analysis and synthesis, one can apply specific stops to detect the existence of linkages in the function structure other than the basic how/why. At the present time, we have discovered a test for the information loops in the functional representation of a design which is independent of implementation domain, but closely related to abstract models of control theory (not to be confused with controlling the form-function process itself).

TABLE 3
Structure of an allocation list
(for position control example)

	Signal	Energy	Material
Cost	x	x	x
Weight	x	x	x
Size	x	x	x
Shape			x
Light		x	
Heat		x	
Humidity		x	
Human Factors	x		
Power		x	
Energy		x	
Voltage			x
Current			x
Speed			x
Load			x
Output:			
Distance			x
Type of Link			x
Maximum Allowable Error		x	
Change in Position	x		
Change in Direction	x		
Type of Controller			x
Design Domain	x		
State Space Equations	x		
Component Specifications:			
Gain of Encoders			x
Motor Characteristics			x
Side Effects			x

While the FBD is a general conceptual design representation, it may not always provide the designer with an intuitive "feel" for system structure or performance. The designer may be more comfortable with a more traditional form, a control loop for example. Since these are two different representations for the same system, it should be possible to create a procedure to transform one to the other. The study of domain crossing in design representation began with the reverse engineering of a model helicopter (Sturges et al., 1990a). This work revealed a linkage which connects functions that share or exchange information. The verb indicates the existence of the information link and its direction, that is, whether the block sends and/or receives the information. The form of this information flow is unspecified, but the associated noun defines the measurable quantity. Typically, the blocks which exhibit an information link are the most specific or elemental, that is, they lie to the far right on the FBD just to the left of their associated components. By creating a classical control block diagram from an FBD,the rules governing the conversion process between the control domain representation and function logic have been determined.

An FBD can also be developed from the control loop. Since there is more information about the design contained in the FBD, certain additional information needs to be extracted from the control loop designer. Thus, more rules are required for this case than for the FBD to control loop conversion process. These rules define how the control loop can be represented on the FBD, and how a control loop can be discovered on a developing FBD. Given a control loop, the governing rules identify which of the corresponding function blocks need to be connected. Simple examples from the control domain, such as a motor/tachometer speed controller of Figure 5, have been used to develop these rules.

For example, a conventional control block diagram was developed from the function block diagram of Figure 5. Figure 6 depicts the transformation process using the set of rules for manipulating the elements of the FBD presented in Table 4. Figure 7 shows the completed control loop. In this case, an FBD has been translated into another design domain: one of signal flows and operators. The FBD is the source of more detailed information than the control block diagram (e.g., constraints and performance requirements). However, the translation back to a usable FBD from the control loop representation is possible with the inverse set of rules presented in Table 5.

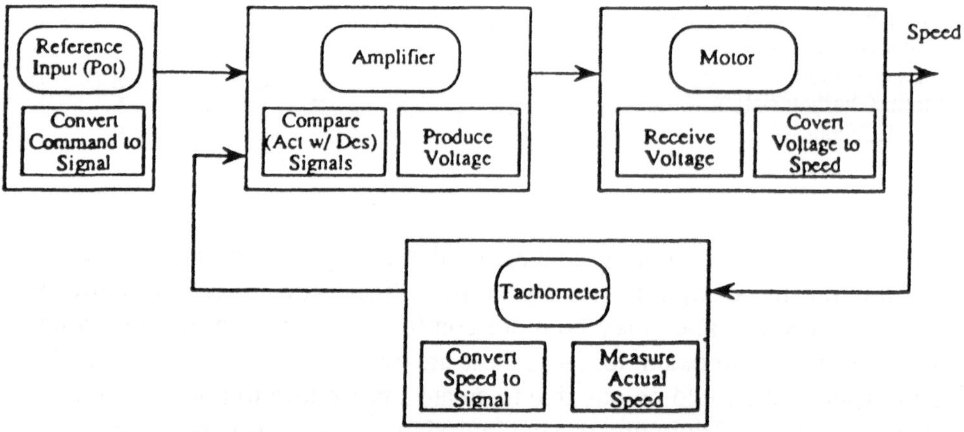

Fig. 6. The transformation of the FBD of Figure 5

TABLE 4

Conversion rules for FBDs to control block diagrams

- Check verbs for input/output properties. (These verbs will have one "input" and one "output" for information; arithmetic functions usually have two "input" ports.)
  - Link matching input and output nouns. If not, check logic or source of information. Avoid synonyms.
  - Indicate the direction of the flow (input to output and output to input) with an arrowhead on the information link.
- Once the information links have been completed, change to control domain:
  - Assign the basic function (leftmost occurrence of a measurable noun) to the output node of the loop.
  - Place the summing junction from the arithmetic function.
  - Attach input and feedback sources ("Receive" or "Determine" and a corresponding noun) to the arithmetic function.
  - Connect control elements to the loop output node by their information flow links. Connect feedback element functions from the desired output to the feedback source located at the summing junction.

TABLE 5

Conversion rules for control block diagrams to FBDs

- Determine the function of each of the control block components by asking "why" they are necessary.
- Draw the control loop on a piece of paper with a Post-it™ to represent the artifact in each block, and placing the lowest-level function over it.
  - The function representing the input source, which might be out of scope,should be considered to exist before the summing junction (e.g., Receive Desired Signal).
  - The information links will adhere to general control loop rules,where each function block contains at least one input and one output node.
  - The output signal from the loop identifies the basic function, at least for the scope of the problem (e.g., Control Velocity). The noun of the output signal corresponds to the noun on the leftmost function.
- Begin the FBD by placing the artifacts to the extreme right and then placing the lowest-level functions to their left.
- The basic function would be placed to the extreme left (inside the scope line).
- Use how/why logic, to develop the intermediate functions, which connect the basic function with the lowest-level functions.
- Check verbs with one "input" and one "output" for information links (arithmetic functions which usually have two "inputs").
- Match the input and output nouns. Place an arrowhead on the information link following the direction of the flow (input to output and output to input).
- Optionally, place information-related functions on a vertical line, top down, in the FBD as follows: Input signal, Feedback signal,Summing junction, Control elements and Feedback elements.
- The corresponding components would be placed to the right of the function (inside scope lines) answering how the function is accomplished.
- Map direction and destination of the information links consistent with the control loop lowest-level functions (not the components).

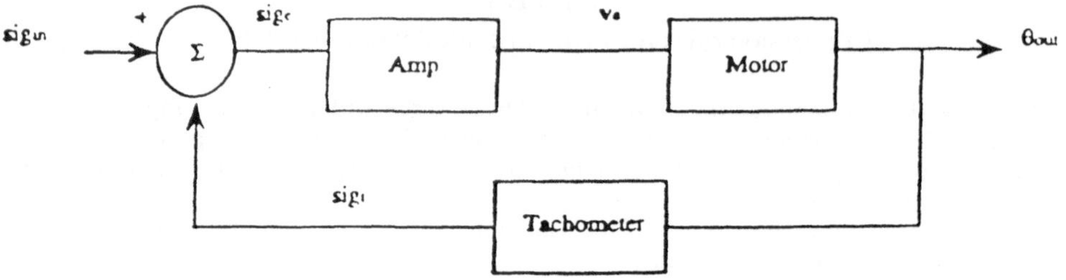

Equivalent Control Block Diagram

Fig. 7. The completed control loop

2.7. OUTPUTS

The principal output is a valid FBD with its allocations satisfied. In some cases the allocations of the lowest level function block will identify a component with definite geometric and material properties. The details of the FBD may include revisions, alternatives, and process-dependent functions. The allocation lists will include behavioral and compliance information which would be used as inputs to optimization and synthesis processes.

3. Conclusions

The essential result of the function logic decomposition process mentioned above is a reasoning structure relating each component to the basic function of the design. The social context in which the FBD was developed depends on the engineers to somehow manage the quantitative data which eventually must be assigned to each component (Subramanian et al, 1989). Also, the how/why linking of functions is insufficient to represent other relationships. Finally, the representation of components as something other than functional and out of scope (i.e., performance and form)artificially divorces abstract functional descriptions from reusable artifacts.

These essential problem needs are addressed by a model of conceptual design for use in a computational environment. As introduced above, the function block is considered as a three-tier structure: (i) the function descriptors and the links which logically connect them, (ii) an allocation list associated with each block, and (iii) the components that jointly satisfy the requirements of the function and the allocation list. In this representation, the function descriptor retains the original noun/verb pair that describes in functional terms what is to be accomplished.

The links which provide the logical connection between the blocks specify the original existence reasoning, but are expanded in type to include information flow requirements, temporal relationships, causal connections, viable and non-viable alternatives and functional revisions. The allocation list attached to a function is a dynamic information structure containing the relevant design specifications, performance requirements, resources and component specifications.

The component level is the functional hook into detailed design analysis and synthesis methods through the supported function structure. Used in reverse, the component level captures functional information about an artifact in a context which includes design intent.

The design intent is captured by the FBD throughout the development of the product since a record is kept of the alternatives that are discarded and the revisions made. The logic behind each design can therefore be understood by both the expert and novice designer at any time. It remains to develop computational systems which can reason within this structure independently of the designer.

References

Andreason, M. M., Kahler, S. and Lund, T. (eds): 1988, *Design for Assembly*, 2nd edn, New York.

Bailey, R. L. (ed.): 1978, *Disciplined Creativity for Engineers*, Ann Arbor Science Publishers, Ann Arbor, MI.

Bytheway, C. W.: 1965, Basic function determination techniques, *Proceedings of the Fifth National Meeting—Society of American Value Engineers*, **11**, April, 21-23.

Bytheway, C. W.: 1971, The creative aspects of FAST diagramming, *Proceedings of the SAVE Conference.*

Finger, S. and Dixon, J. R.: 1989, A review of research in mechanical engineering design. Part I: Descriptive, prescriptive and computer-based models of design processes, *Research in Engineering Design*,**1**, 51-67.

Kantowitz, B. H and Sorkin, R. D.: 1987, Allocation of functions, *in* G.I. Salvendy (ed.), *Handbook of Human Factors*, John Wiley, New York, pp. 355-369.

Miles, L. D. (ed.): 1982, *Techniques of Value Analysis*, 2nd edn, McGraw Hill, New York.

Pahl, G. and Beitz, W.: 1988, *Engineering Design: A Systematic Approach*. Springer-Verlag, New York.

Paz-Soldan, J. P. and Rinderle, J. R.: 1989, The alternate use of abstraction and refinement in conceptual mechanical design, *ASME WAM*, San Francisco, CA, also *EDRC 24-22-90*, Carnegie Mellon University Engineering Design Research Center, September.

Perkins, D. N.: 1981, *The Mind's Best Work,* Harvard University Press, Cambridge, MA.

Pugh, S.: 1981, Concept selection—a method that works, *International Conference on Engineering Design*, ICED, Rome, Italy.

Ruggles, Wayne F.: 1971, FAST—a management planning tool *SAVE Encyclopedia of Value*, **6**, 301.

Sturges, R. H., Dorman, J. G. and Brecker, J. N.: 1986, *Design for Producibility*, Westinghouse Productivity and Quality Center.

Sturges, R. H.,O'Shaughnessy, K. and Kilani, M. I.: 1990a, Representation of aircraft design data for supportability, operability, and producibility evaluations, *EDRC Project Report Number: 14513*,Carnegie Mellon University Engineering Design

Research Center.

Sturges, R. H. and Kilani, M. I.: 1990, A function logic and allocation design environment, *Proceedings for ESD Fourth Annual Expert Systems Conference and Exposition*, Detroit, MI.

Subramanian, E., Podnar, G. and Westerberg, A.: 1989, n-DIM: n-dimensional information modeling—a shared computational environment for design, Carnegie Mellon University, Engineering Design Research Center, September 1989.

Ullman, D. G., Dietterich, T. G. and Stauffer, L. A.: 1988, A model of the mechanical design process based on empirical data, *AI EDAM* 2, 1, 33-52.

Westinghouse Corporate Services Council: 1984, *Report on Life Cycle Costs*, Westinghouse Productivity and Quality Center, Pittsburgh, PA.

USING NETWORK-BASED PROTOTYPES TO SUPPORT CREATIVE DESIGN BY ANALOGY AND MUTATION

F. ZHAO
Department of Civil and Environmental Engineering
Florida International University
Miami Florida 33199 USA

and

M. L. MAHER
Department of Architectural and Design Science
University of Sydney
NSW 2006 Australia

Abstract. In this paper, a model for creative engineering design based on a network-based design prototype representation and a combined analogy and mutation technique is presented. The model distinguishes between a domain-dependent representation of design knowledge and domain independent mutation operators that manipulate the representation. This model extends the representation of design prototypes beyond a parametric expression of function, performance, behavior, and structure to include an explicit representation of the semantics of the relationships between parameters within a prototype. The categorization of design parameters and the explicit representation of the relations of design parameters are the two most important aspects of the network-based prototype representation and can be used to support creative design. The model includes the development of the mutation operator, *COMBINE*, that utilizes both analogical reasoning and mutation to produce creative design ideas. Analogical reasoning is used to find prototypes to be considered and mutation to change a given prototype to include portions of another. The model is illustrated in the domain of structural system design for buildings.

1. Introduction

One important characteristic and an essential element of design is that it is a creative act, as the purpose of design is to improve existing solutions to old problems using new technologies and find solutions to new problems that have not been encountered before. Computers have proven to be powerful tools in assisting designers in processing large amounts of data and managing design information, but the potential for computers to assist designers in creative thinking has not been identified or adequately explored. Computational models of design are needed to extend the use of computers in design.

In this paper, a study of the use of computers to support human designers' creativity is presented. A model for creative engineering design using a network-based design prototype representation and a combined analogy and mutation technique has been developed. The model distinguishes between a representation of domain-dependent design knowledge and domain independent mutation operators that manipulate the representation to produce or suggest new design solutions that are not implied by the existing design knowledge. This model extends the representation of design prototypes beyond a parametric expression of function, performance, behavior, and structure to include an explicit representation of the semantics of the

J. S. Gero (ed.), Artificial Intelligence in Design '92, 773–793.

relationships between parameters within a prototype. The categorization of design parameters and the explicit representation of the relations of design parameters provide a basis for supporting creative design.

The model includes the development of the mutation operator, *COMBINE*, using analogy and mutation to produce creative design ideas. Mutation is guided by a design solution that fails to satisfy a performance requirement. A failed design solution is improved by introducing portions of another design prototype. A relevant prototype is found using graph-based matching criteria derived from the failed design solution. By manipulating the function, behavior, and structure attributes and their relationships, a new graph-based solution description is generated. The model is illustrated in the domain of structural design of buildings.

2. Related Work

There has been a growing interest in developing computational models of creative problem solving and design. Some representative systems include AM (Lenat, 1982; Davis and Lenat, 1982), PROMPT (Murthy and Adanki, 1987), 1stPRINCE (Cagan and Agogino, 1988a; Cagan and Agogino, 1988b; Cagan and Agogino, 1991), EDISON (Dyer, et.al., 1986; Hodges, et.al., 1988), CYCLOPS (Navinchandra, 1991), and CADET (Sycara and Navinchandra, 1991; Navinchandra, et.al., 1991).

AM is a scientific discovery system that (re)discovers elementary mathematical concepts. The main technique used in AM for discovery is syntactic mutation of concepts to result in the semantic change of the concepts. EDISON and PROMPT are innovative design systems for mechanical and physical systems. EDISON relies on the use of analogical reasoning and mutation to create new mechanical devices. Innovative devices are invented by retrieving devices used in a context different from that of the current problem and by mutating a design solution, such as changing the location of the hinges of a door from its side to the top edge. PROMPT's power is expected to come from its knowledge about the physical models of the systems to be designed and the ability of causal reasoning. 1stPRINCE uses optimization techniques and qualitative reasoning to achieve innovation by deriving different properties for a design solution expressed as multi-dimensional integrals in divided intervals. CYCLOPS is a design system that attempts creative design by constraint relaxation and analogical reasoning. Search in a design space is guided using multiple objectives that represent design constraints, exploring new solutions by reformulating the multiple objectives. CYCLOPS also uses past experiences stored as problems and solutions to solve new design problems. CADET is a system that uses analogical reasoning to attempt innovative design of mechanical devices. CADET represents the behaviors of devices using influence graphs that specify how behavior variables are related and how the changes of their values affect each other. Using a graph matching technique, CADET retrieves relevant devices for a given description of the desired behaviors of a device, also expressed as graphs, and, if necessary, combines different devices by combining their influence graphs.

By looking at these systems, it can be seen that three techniques are commonly used: mutation (AM and EDISON), analogy (EDISON, CYCLOPS, and CADET), and reasoning from first principles (PROMPT and 1^{st}PRINCE). Although AM provides some interesting ideas and insights, the generality of mutation in AM is limited due to the special characteristics of its domain. Mutation in EDISON also has the potential problem of being case dependent since no generic definition of mutation is given. EDISON and CYCLOPS both explore the use of contextual information in retrieving relevant design information. EDISON's indexing at different levels of abstraction is also interesting. Reasoning from first principles requires a more structured and precise model of the physical world, which is difficult given the state of art of knowledge representation. It also requires the support of other reasoning techniques such as qualitative reasoning and common sense reasoning. CADET takes advantage of some better understood and expressed aspects of the behavior models of mechanical devices and uses analogical reasoning to manipulate the representation of the behavior models to construct descriptions of the behaviors of new devices.

The above systems have demonstrated to some degree of success that some aspects of creativity can be modeled as well as the usefulness of certain problem solving techniques. They have also improved our understanding of computer creativity and its role in design.

3. Design Knowledge Representation Using Network-Based Prototypes

Design prototypes provide a formalism for organizing design knowledge and are a framework for studying design processes, as proposed by Gero (Gero, 1987; Gero, 1990). Gero states that the generic frame representation is inadequate in representing complex design knowledge and needs to be specialized so that it uniquely reflects characteristics of design knowledge. A prototype is a generalized concept and can be used in various but similar design situations. In the object-oriented sense, a prototype is a class from which an instance can be produced to accommodate the particular design requirements. A prototype organizes the knowledge in the class around the functions, behaviors, and structures of the design elements it represents. Functions identify the intended use and/or goals of the design element, behaviors define the response of the design to its environment, and structure describes its physical existence.

Prototypes can serve as the basis of the commencement and continuation of design by structuring the problem space at any point in the design process as they provide all the available and relevant design alternatives, bringing into bearing the relevant design requirements, constraints, and knowledge to analyze and evaluate the alternatives. For this research, Gero's prototype definition is adopted and forms the basis on which the knowledge representation is implemented.

Previous implementations of design prototypes have been frame-based parametric descriptions of design objects with procedure attachments that encode design process knowledge (Oxman and Gero, 1988; Zhang, 1990; Tham, 1991). Such implementations of prototypes cannot support creative design because as a collection of attributes and design procedures with the relations among the attributes implicitly defined by the procedures, the internal knowledge structure of a prototype is invisible, unaccessible, and unchangeable. This lack of manipulability means that no new design solutions different from those implied by the existing prototypes can be generated. Because creating new design solutions requires reasoning about changes to the solutions generated by instantiating prototypes beyond the expectations of the generalized objects represented by the prototype, the relations among the attributes in a prototype must be explicit, accessible and manipulable.

In order to accommodate creative design, a network-based prototype representation is developed. This extended prototype representation expresses the relations among a prototype's attributes explicitly using a dependency network, in which nodes represent the attributes and the links the relations. Figure 1 shows the graphic form of a partial dependency network of a prototype *rigid-frame* related to the gravity load resistance function and other spatial functions of a rigid frame.

In the graphs, nodes are indicated by symbols in bold face, while the links by arcs with an arrow at one end. The end of the arc that has an arrow is called the *end* of the link, while the other end is called the *start* of the link. The node at the end of a link is called the *dependent node* or simply the *dependent*. The node at the start of the link is called the *factor node* or *factor* for short. *Dependent* and *factor* are relative terms. A dependent is one that depends on its factors, and a factor the one that contributes to its dependent. How a node depends on its factors or how a factor contributes to its dependents is determined by the types of relations that link them together. Links can be binary or hyper links; the former relate a dependent node to one factor node, while the latter relate a dependent node to two or more factor nodes. In the graphical form of a DN, the types of relations are in italic and are either marked directly by the link as in the case of a binary relation, or enclosed in a circle or an ellipse as in the cases of hyper links, or both, also in the cases of hyper links. For instance, node **extend-dimensions = 2D** represents a function of a rigid-frame and has one factor **2D-space** linked together by a relation *because-of*. **2D-space** in turn also has two factors, **height** and **width**, and is linked by the *defined-by* relation.

3.1. NODES IN THE DEPENDENCY NETWORK

The attributes of a prototype are represented by the nodes in the dependency network and are grouped into categories. For the attributes of prototype *rigid-frame* as shown in Figure 1, the categories of the attributes are indicated by the alternately shaded or white areas marked with category names in which they are positioned. Attributes that are enclosed in parentheses are the attributes of the components of the prototype, such as **beams** and **columns**, or the attributes of design objects

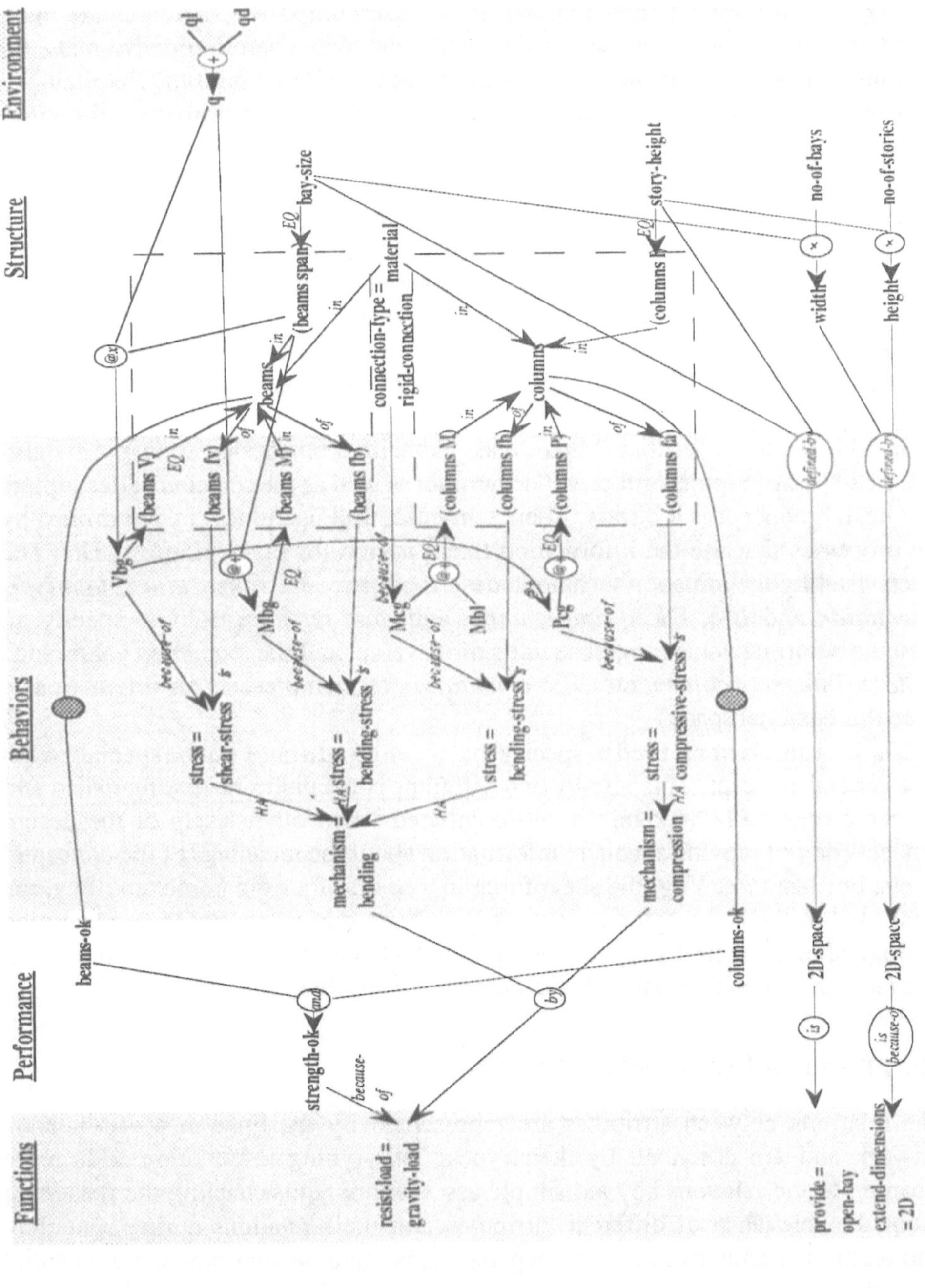

Fig. 1. Partial Dependency Network of Prototype Rigid-Frame under Gravity Load

that are not prototypes, such as **material**. These component attributes are made external to the system prototype that contains them as a part in order to make the information exchange between the system and component prototypes explicit.

The categorization is made based on the authors' interpretation of a rigid frame. Functions are considered to be 'resist-load', 'extend-dimensions', and 'provide'. All attributes describing the stresses, internal forces, and load resisting mechanisms are behavior attributes. External forces (that may also include temperature variations and stresses caused by assembly errors, for instance) belong to the environment category. Attributes describing geometry, topology, components, materials, etc., are structure attributes. The performance category includes attributes that are either exclusively or likely to be used for evaluation purpose. For example, **beams-ok** in Figure 1 is exclusively used for evaluating the design status of the beams.

In addition to the attribute names used as their labels, design concepts are also described by descriptors or facets. The number as well as the contents of descriptors of a design concept is arbitrary. Their semantics and usefulness is determined by the processes that use the information they provide for manipulating a DN. The facets used by the mutation technique described here are: *range*, *units*, *data-type*, *orientation* and *is-a*. Facet *range*, *units*, and *data-type* are used to specify an attribute's normal value range, the units of its value, and the type of its value, such as *float*, *integer*, *boolean*, etc. The *orientation* facet represents an orientation in three dimensional space.

The *is-a* descriptor is used to specify that a design attribute, can be specialization of a general concept. This allows two different concepts to be distinguished and the same ones to be recognized and compared when other facets of the design concepts do not provide adequate information about the meanings of the concepts' labels. For instance, **Vbg**, the shear force in a beam of a rigid frame, and **Pcg**, the axial load in the same column, both of which are due to gravity load, cannot be distinguished by their data types and units, which are the same. They can have an *is-a* facet that specifies that **Vcl** *is-a shear-force* and **Pcg** *is-an axial-force*.

3.2. Links in a Dependency Network

The relations between attributes are represented by the links in a dependency network and are described by their types. This typing infomation adds more meaning to the relations beyond simple associations, thus enabling the reasoning about the relevance of different attributes and their relations during mutation. The relations define two kinds of dependencies: data dependency and existential dependency. A data dependency describes a relationship between the values of two or more attributes. An existential dependency describes a relationship between two attributes, in which the existence of one attribute implies, or affects, the existence of one or more other attributes. A data dependency implies, but does not make explicit, an existential dependency.

3.2.1. Data Dependency Relations

Relations that define a data dependency include *quantitative, qualitative, equivalence, inequality, logical,* and *operational* relations.

1. *Quantitative Relation.* The relations +, -, ×, and / are called quantitative relations and represent a numerical data dependency between one design variable and two or more other variables. These relations show how the value of a dependent attribute is based on those of its factors as defined by the arithmetic operations they represent. For instance, the width of a rigid frame is the product of the bay size and the number of bays as shown in Figure 1.
2. *Qualitative Relations.* The relations @+, @-, @×, and @/ are qualitative relations and are extensions of the quantitative relations. Where quantitative relations describe an exact mathematical expression, qualitative relations describe a complex mathematical relation in a simplified form. The qualitative relationship between two attributes indicates the direction of change in magnitude of one attribute when the related attribute changes in a positive direction. @+ and @× are called 'positive' qualitative relations, while @- and @/ 'negative' qualitative relations. For example, if two attributes are related by a positive qualitative relation, a value increase of one will result in a value increase in the other.
3. *Equivalence Relation.* The equivalence relation is indicated by *EQ* and equates two concepts logically in addition to making the values of the attributes that represent these concepts the same. The direction of the arrow of the link defines the data dependency. In other words, if two nodes are linked by an *EQ* relation, they are logically the same concept, but during design the value of the factor node is given first and the dependent node assumes that value as a consequence. For instance, in Figure 1, **bay-size** and **(beams span)** are linked together by an equivalence link and are therefore equivalent concepts. The implication of this is that any change occurring to **bay-size** will also change **(beams span)** exactly the same way.
4. *Inequality Relations.* The relations >, >=, <, and <= represent a constraint involving two design variables in the form of an inequality expression. A third design variable that is the dependent of such two design variables assumes a truth value that results from the evaluation of the constraint.
5. *Logical Relations. And* and *or* are logical relations representing the dependency of one logical variable on other logical variables. As an example, **strength-ok** is true only if both **columns-ok** and **beams-ok** are true (see Figure 1).
6. *Operational Relations.* An operational link indicates that the value of an attribute represented by the node at the end of the link is computed based on the information provided by its linked factors. This computation is not of a mathematical nature and cannot be expressed by any of the qualitative or quantitative relations. For instance, to determine whether **beams-ok** is true, a procedure performs a check on all the instances of typical beams to see whether they are

individually successfully instantiated. Operational relations are marked by a shaded circle in Figure 1.

3.2.2. *Existential Dependency Relations*

The relations that only describe an existential dependency are: *because-of*, *HA*, *defined-by*, and *is*. Each of these is described below.

1. *Because-of*. This relation represents the causal relation between attributes of a design. For instance, there is a bending moment in the columns of a rigid frame because of the rigid connection between the columns and the beams.
2. *HA*. This relation represents a hueristic association. For instance, load-resisting functions and the mechanisms that provide these functions do not always have a definite correspondence, neither do the types of load-resisting mechanisms and types of stresses. Different types of structures may resist the same type of loads with different mechanisms. Structures that resist loads by the same mechanism may exhibit different stresses. With a mechanical model of a structure, the load-resisting functions, the mechanism and the stresses developed can be analyzed. However, for the purpose of design, their associations are heuristic associations.
3. *Defined-by*. This is a relation that specifies the constitutive elements of a concept in a particular context as in the case of the **2D-space** defined by **story-height** and **bay-size** in Figure 1. The context of this **2D-space** is considered an open bay in a rigid frame. Although the concept's constitutive elements may be redefined for each different context, the concept itself is a generic concept and may have certain restrictions and/or operations that can be applied to its elements.
4. *Is*. Relation *is* relates a generic concept to one that may be considered as its special instance in the given context of a prototype. For instance, **q** is specified as being *gravity-load* and **(columns fb)** *bending-stress*. Although the *is* links appear to point to nodes that represent assertions consisting of both an attribute and a value, it is assumed as default that an *is* link actually points to the value part of the feature. That is that **q** *is* a *gravity-load* and not **resist-load = gravity-load**. This type of information will be useful when, for instance, reasoning about the particular types of loads that are relevant to a certain load resisting function. This information is used in the application of the mutation operator *COMBINE* to formulate requirements on the loading conditions.

3.2.3. *Component-System Relations*

A third group of relations includes the relations that make the link between attributes of a system prototype and its component prototype explicit. Because the instantiation process is encoded procedurally, information of prototype-component interactions is hidden in the procedures and not accessible. DNs provide this information by having *of* and *input* relations to relate a prototype's attributes with its component attributes. Information of such interactions is necessary when the design of a component of a prototype fails and possible remedies at the prototype level are to be found, or when the modifications at the component level can be found in order to improve the performance of the prototype instance.

1. *Of.* The Of relations are represented by links that start from a node representing a component of the prototype and end at a node representing a component attribute, such as the one between attribute **beams** and (**beams fb**) shown in Figure 1. Such relations indicate that the value of the component attribute is a result of the component instantiation, thus *depends* on the component.
2. *In.* The In(put) relations are the other type of relations that relate a component of a prototype with the component's attributes. Similar to *of* relations, they are represented by links pointing to the opposite direction, i.e. from a component attribute to the component. These attributes of the component are called the 'input' of the component because in order to instantiate the component, the value of these attributes must be given and they are determined at the prototype level. For instance, (**beams span**) is an input to **beams** and must be known before beams can be instantiated.

4. A Mutation Operator for Supporting Creative Design

In this study, mutation is guided by a failed design, which specifies an objective for mutation as improving a specified performance criterion. The failed design is an instance of a prototype so that in addition to the design solution, the prototype is available for further reasoning. *COMBINE* is an operator that has an explicit mutation objective and attempts to achieve this objective by combining the description of the structure of a failed design solution with the structure of a different prototype. As a result a new structural form is created and is evaluated to see whether it achieves the mutation objective.

The idea behind the *COMBINE* operator is that certain functions and behaviors of a prototype are related to or result from parts of its structure. By importing parts of the structure of a different prototype, the related functions and behaviors of the different prototype may also manifest in the new solution. If the functions or behaviors that are brought along with the new structural features are not new to the original solution, they can still achieve a mutation objective by strengthening original functions and behaviors, if this is what is desired. To illustrate, consider a rigid frame instance that has an excessive lateral deflection under wind load. In order to improve its performance with respect to its lateral deflection, a braced frame prototype is selected by the *COMBINE* operator to import its structure into the rigid frame instance. The braced frame prototype is chosen because it also has the behavior of deflecting laterally under wind loads and has the function of resisting wind loads. One possible outcome of combining the structure of the rigid

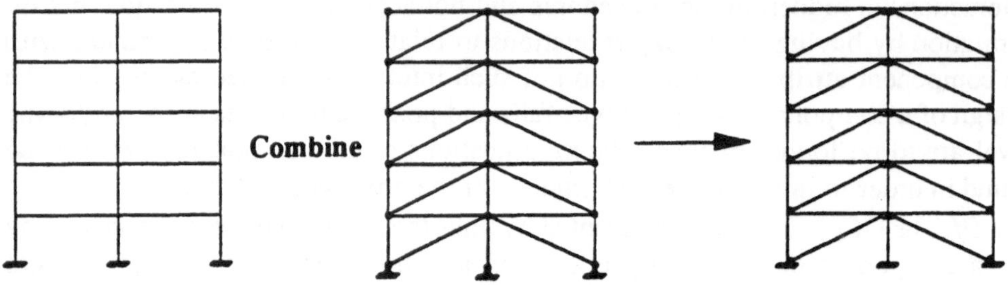

Fig. 2. Combining the Structure of a Rigid Frame Instance with the Structure of a Braced Frame

frame instance with the structure of the braced frame is a rigid frame with each of its panels braced by a diagonal or K bracing as shown in Figure 2.

The lateral deflection of the new solution will be reduced due to the increased stiffness of the frame. This example shows that the *COMBINE* operator can be used to improve the performance of a design solution. The solution is a new one that cannot be generated by merely instantiating the existing prototypes of rigid frames or braced frames.

The *COMBINE* operator is constructed based on two techniques, mutation and analogical reasoning. The analogical reasoning technique used in *COMBINE* is not full analogy, but partial analogy, mainly for the purpose of identifying useful prototypes to solve an existing problem. However, whenever possible, analogy is also employed to produce descriptions of the new solution if sufficient information is available. The mutation technique can be thought as a complementary part of analogy, and it becomes the dominant technique when analogy cannot be carried out effectively. The overall process of applying the *COMBINE* operator consists of the following steps:

1. establish a set of matching criteria for identifying a new prototype;
2. search for matching prototypes using a partial matching technique;
3. identify the key structure variables (KSVs) that determine the basic configuration of the matching prototype for the given solution; and
4. determine the values of KSVs.

4.1. Matching Graph

A set of matching criteria are represented by a graph called the matching graph (MG), in which nodes represent requirements on a matching prototype's attributes and links the relations among these attributes. The reason to relate the relevant attributes together is to make the matching criteria more specific and meaningful. For instance, if a matching prototype is expected to be able to resist gravity load, the load type and the magnitude are also of concern and make the requirement on the

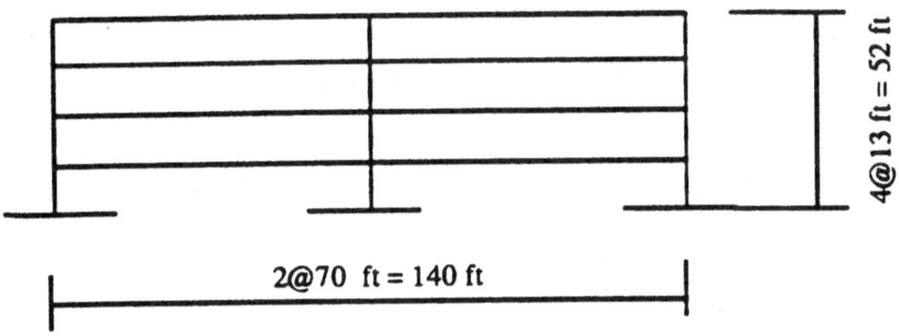

Fig. 3. A Failed Rigid Frame Instance Due to Long Beam Span

load resistance function more specific. On the other hand, if a matching prototype is to be combined with a beam in order to reduce its excessive deflection under the gravity load, the type and the magnitude of the load and the span of the beam are also relevant and important in addition to the deflection attribute since they directly contribute to the deflection. Such information concerning the relevance of different attributes is readily available in the DN of the prototype of which a failed solution is an instance. Therefore, the DN can be used to construct a MG by directly taking out from the DN the subgraphs relevant to the functions, performance, or behaviors of the failed prototype instance as the subgraphs of a MG. The desired values of the attributes represented by the MG nodes are provided by the failed prototype instance.

Consider the following example in which a rigid frame instance is evaluated as unsatisfactory because the design of the beams failed due to the long span (Figure 3). The challenge is then to find a way to strengthen the beams' functions. Given that *COMBINE* can be used to reinforce a prototype instance or change the value of a particular attribute, there are two ways to approach the problem. When a single design constraint involving an attribute that assumes a numerical value is violated, the objective for *COMBINE* can be stated as to increase or decrease the value of that attribute. However, if there are a number of violated design constraints, the objective can be simply stated as to reinforce the prototype instance, where reinforce means to add more structure.

Assume that the objective that has been established is to reinforce the beams in the rigid frame. The MG is constructed by a tree traversal algorithm during which the subgraphs of the prototype *beam* DN that contain functions and their direct supporting structure and environment factors are retrieved and used as the subgraphs of the MG. The traversal starts with the function nodes of the *beam* DN; for each of the functions a MG subgraph is established. Then the subgraphs are expanded by adding to them the nodes that are linked to the functions by relations *is* and *defined-by*. These particular relations are used here because they relate the definitional attributes to the functions, thus making the functions become more specific. The expansion of a subgraph stops when a structure or environment

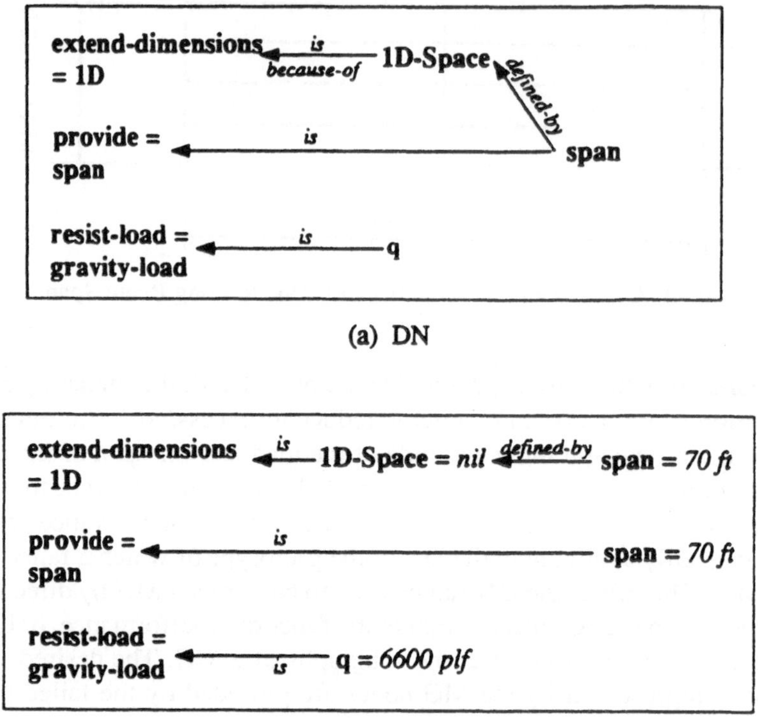

(a) DN

(b) MG

Fig. 4. Partial DN and Matching Graph for a Failed Beam Instance

attribute is reached for all the paths in the subgraph. The paths are not further expanded, since doing so would introduce unnecessary details of the particular solution into the MG. A more detailed description of the algorithm can be found in (Zhao, 1991).

Using this algorithm, a MG is constructed for the beam instances of the given rigid frame. Figure 4 (a) shows the partial DN that contain the functions nodes and used as the basis for constructing the MG, and (b) shows the corresponding MG. The values shown in the MG nodes are obtained from the beam instances or from the rigid frame instance that contain the beams as components.

The above MG specifies that a matching prototype for the beams should be able to resist a linearly distributed load with a magnitude of 6.6 klf, provide the spanning function with a span of 70 feet, and extend one dimension in the space, which is also defined by the span attribute of the beams.

4.2. PARTIAL MATCHING

The second step for applying the *COMBINE* operator is to identify the potentially useful prototypes, called *matches*, that can be later combined with an original solution. Given a MG that contains a set of MG subgraphs, a match is defined as a prototype of which the DN contains at least one subgraph that matches a MG subgraph. In order to match, the nodes in the DN subgraph correspond to nodes in the MG subgraph, and the two subgraphs have the same topology, i.e., the nodes of the DN subgraph are connected together by the same relations that link the nodes that they match in the MG subgraph.

A DN node corresponds to a MG node when certain conditions are satisfied. The conditions require two attributes as represented by a MG node and a DN node being compared are of the same category. In addition, they are required to have the same description, which may be *nil* or include any of the facets *is-a*, *data-type*, *units*, etc. An exception for matching the descriptions is that if the descriptions contain the same *is-a* facet, i.e., the two concepts being compared are the same concept, then the requirement of their having the same units can be removed. The values of two nodes compared are required to be either the same or the value specified by the MG node must fall into the value range defined in the DN node being matched. However, if the MG node value is *any*, the requirement on the value of a matching DN node is removed.

Figure 5 shows the MG constructed for the failed beam instance and a partial DN of a prototype *tied-arch*. Note that the first MG subgraph starting with the requirement that a matching prototype has the function of **extend-dimensions** = **1D** is not matched since the *tied-arch* attribute **extend-dimensions** has a value *2D*. The matched functions of *tied-arch* include **resist-load** = **gravity-load** and **provide** = **span**. The matching algorithm also returns a *matching table* that specifies the correspondence between attributes in the beam MG and the *tied-arch* DN, shown in Table I, where the MG and DN nodes are represented by their names and values (or ranges for DN node **q**).

Although the *tied-arch* DN does not satisfy one of the requirements, i.e. to extend one dimension, it is still a match since here the matching is only partial. However, the more requirements a matching prototype satisfies, the better a match it is. The discrimination of different matching requirements and the evaluation and ranking of matching prototypes is not dealt with here and remains a research topic.

4.3. Identification of Key Structure Variables

An attribute of a prototype is a *key structure attribute* (KSA) if it is of the *structure* category and there exists a path in its DN that connects a function to this attribute without other structure attributes between. For instance, the prototype *beam* has one KSA, i.e., **span**, and *rigid-frame* has four KSAs, including **width**, **height**, **bay-size**, and **story-height** (Figure 6). For *beam*, the key structure variable **span** is a factor shared by two functions **extend-dimensions** = **1D** and **provide** = **span**. The path that connects function **extend-dimensions** = **1D** and **span** has an attribute **1D-space** in between, but **1D-space** is of the *performance* category and therefore does not violate the definition of a key structure attribute. Attribute **q** is not a

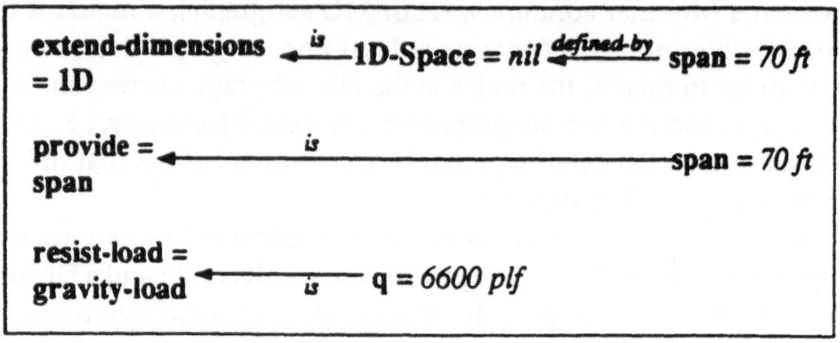

(a) MG for Steel-Beam-1

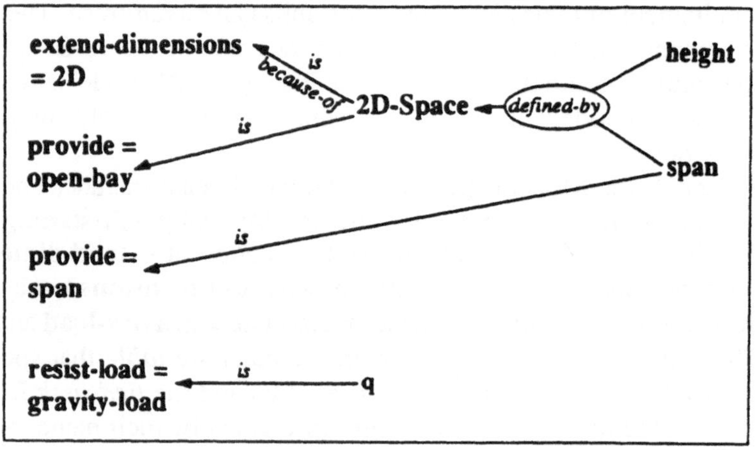

(b) Partial DN of *Tied-Arch*

Fig. 5. The MG for Instance Beam-1 and Partial DN for Prototype Tied-Arch

key structure attribute because it does not belong to the structure category. The functions that the key structure variables are directly related to will also be referred as *structure functions*. A minimal subgraph that consists of a key function and all its connected key structure attributes is called a *key subgraph* and its paths *key paths*. It can be seen that the key subgraphs of a prototype are a subset of the MG if one is constructed.

For a matching prototype, their KSAs are called key structure variables since they are variables in a given *COMBINE* problem and their values are to be determined. The KSVs are of particular interest because their values determine the basic configurations of the instances of a matching prototype that are to be combined with a failed design solution.

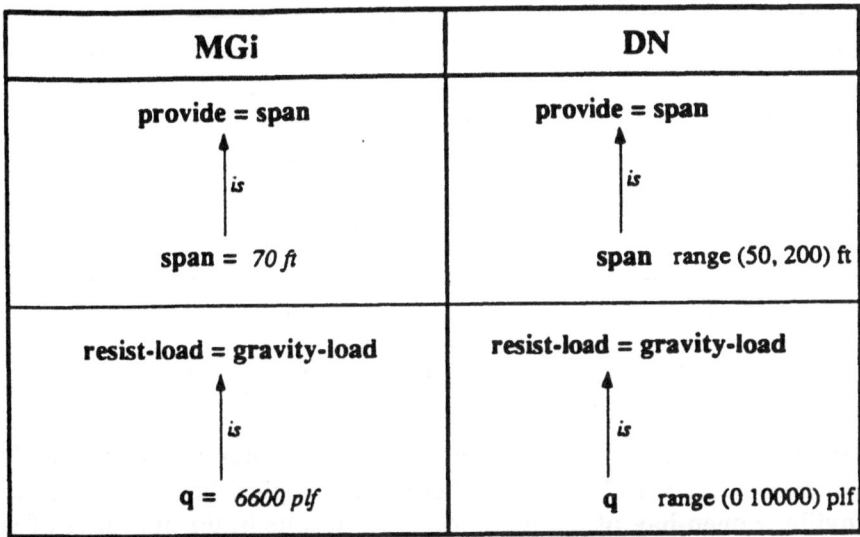

MGi	DN
provide = span ↑ *is* **span** = *70 ft*	**provide = span** ↑ *is* **span** range (50, 200) ft
resist-load = gravity-load ↑ *is* **q** = *6600 plf*	**resist-load = gravity-load** ↑ *is* **q** range (0 10000) plf

TABLE I
The Matching Table Resulted from Matching Beam-1 MG and Tied-Arch DN

4.4. Determination of Values of Key Structure Variables

There are four ways to determine the possible values of a KSV: using the matching result, by matching the key structure graphs of the original prototype and the matching prototype, by mutation, and by graph matching using the contextual information provided by a system prototype and its instance of which the original solution is a part. The four methods are described below.

1. *Use of Matching Results*. As the partial matching establishes the correspondence between the attributes of the original solution and of the match, the values of those KSVs that are matched with the original solution's attributes can be taken as the same as these attributes' values. For *tied-arch* being a match of the beams, the matched KSV is the span of *tied-arch*, **span**, and it is matched with the beam attribute **span**. Therefore, a tied arch when combined with a beam in the original rigid frame, i.e., being integrated into the rigid frame, will have a span of 70 feet.
2. *Matching Key Structure Graphs*. Sometimes, it may happen that the mapping of the attributes of a matching prototype resulting from the partial matching into a solution prototype is inconsistent, i.e. a KSV of the matching prototype is matched with two or more KSAs of the solution prototype. As an example, consider *tied-arch* as the matching prototype for a *rigid-frame* instance, *rigid-frame-2*. Figure 7 shows the two sets of mappings that are possible, one consisting of subgraphs in the shaded areas, and the other boxed. Mapping the factors of function **extend-dimensions = 2D** of the two prototypes results

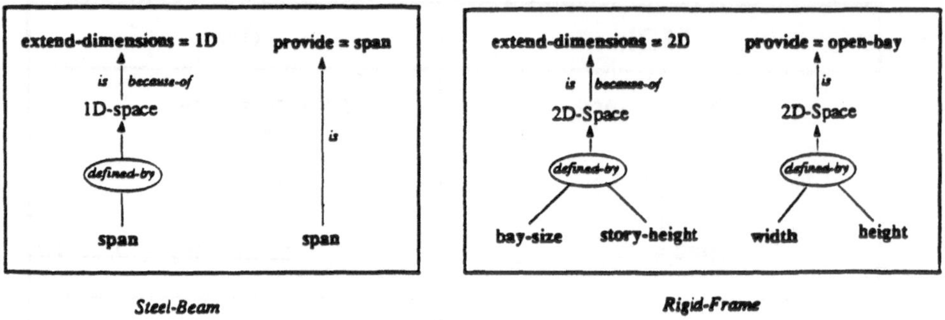

Fig. 6. Key Subgraphs of Beam and Rigid-Frame

in the matching of **span** and **height** of *tied-arch* with **depth** and **height** of *rigid-frame*, respectively. On the other hand, mapping the factors of function **provide = open-bay** of the two prototypes results in the matching of **span** and **height** of *tied-arch* again with different KSVs of *rigid-frame*, **bay-size** and **story-height**, respectively. To resolve the inconsistency, the inconsistent mappings are separated into consistent mappings, each consisting a set of KSAs and KSVs that are uniquely matched and representing a possible configuration of new potential solution. The inconsistency shown in Figure 7 is resolved by generating two sets of mappings, of which the interpretations of the configurations suggested are shown in Figure 8.

3. *Mutation.* If a KSV's value cannot be determined by either of the above two methods, mutation can be used to generate values for it. For *tied-arch* being a match of *beam*, its height cannot be determined by the information provided by either the prototype *beam* or its instances since beams only have one dimension. In this case, mutation can be used to determine the value for the tied arch. A prototype's attribute usually has a default value or knowledge is available to suggest such a value, particularly when the attribute concerns the configuration of a prototype instance. One heuristic that determines the height of a tied arch is that a reasonable height is one fifths of the span. For the above example, the tied arch span has already been determined to be the same as the beam span, i.e., 70 feet. A reasonable height of the tied arch is therefore 14 feet. In order to produce more alternative solutions, this default value is mutated by multiplying it with a factor smaller than 1 and one that is greater than 1. For multipliers 0.5 and 1.5, the new alternative values for the tied arch height are 7 and 21 feet.
4. *Graph Matching Using System Information.* Another method that can be used to determine the value of an unmatched KSV is to use the information provided by the system prototype instance of which the current design solution is a part

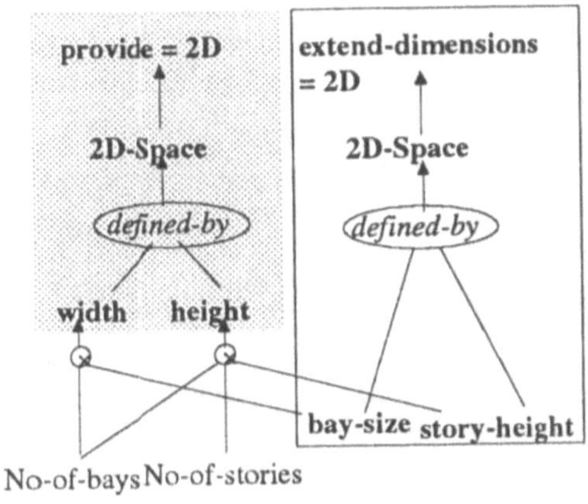

(a) MG of *Rigid-Frame-3*

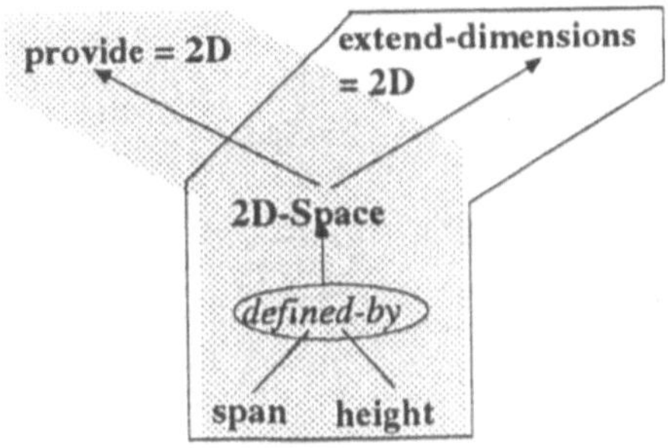

(b) DN of *Tied-Arch*

Fig. 7. Inconsistent Mappings

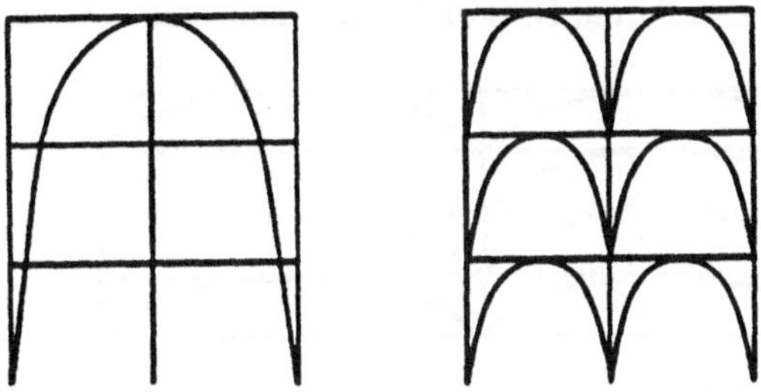

Fig. 8. Dividing Inconsistent Mappings into Consistent Mappings and Their Interpretations

(component). For instance, the beams are part of a rigid frame solution. It can be seen that since **span** of **tied-arch** is matched with **span** of *beam-1*, which is equivalent to **bay-size** in *rigid-frame-1*, the span of *tied-arch* is also matched with **bay-size**. In the same way as **span** is related to **height** in *tied-arch* (indirectly), **bay-size** is related to **story-height** in *rigid-frame*. In addition, **height** and **story-height** are matching concepts as they have the same descriptions. Therefore, it is plausible that **height** of *tied-arch* can be matched with **story-height** of *rigid-frame*. This results in a complete mapping in which the values of all the KSVs are determined, as *tied-arch span = rigid-frame bay-size* and *tied-arch height = rigid-frame story-height*. Figure 10 shows a possible interpretation of this mapping. Even with contextual information available, alternative mappings can still be generated by treating the problem as a context free problem and let *COMBINE* determine a base value for an

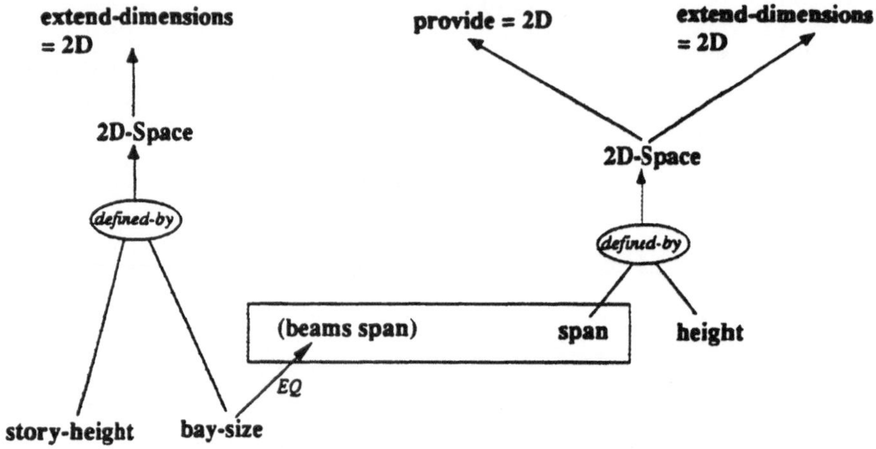

(a) One Key Grapg of *Rigid-Frame* (b) Two Key Graphs of *Tied-Arch*

Fig. 9. Partial DNs of Rigid-Frame and Tied-Arch

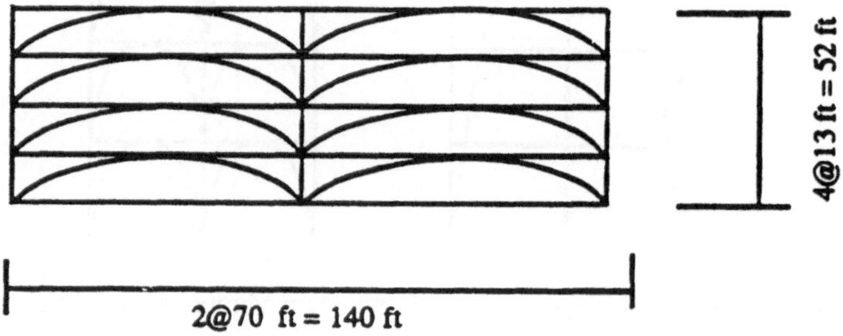

Fig. 10. A Possible Interpretation of Tied-Arch Height = Rigid-Frame Story-Height

unmatched key structure variable and mutate it randomly as discussed before.

5. Conclusions

In this paper, a study on knowledge representation and using mutation and analogical reasoning techniques for creative design has been presented. Computer programs have been developed to implement the network-based prototype representation and the mutation and analogy methods. The dependency networks as well as the mutation operator *COMBINE* are implemented using a frame language, and the methods in Lisp. The overall goal of this research is to identify a model for computer supported creativity using relevant AI techniques, including mutation and analogy. This model is elaborated by developing a representation of design knowledge and a methodology for manipulating this knowledge. The representation of design knowledge is based on an explicit and manipulable representation of dependency knowledge and the methodology is embodied in domain-independent mutation operators.

This model of computational creativity is based mainly on imprecise and empirical knowledge and uses reasoning mechanisms that are of an empirical nature; yet maintains a clear distinction between the domain design knowledge representation and the domain-independent nonroutine design methodology. This distinction makes it possible to extend the model to be used in different engineering domains to solve a variety of design problems. This model has been demonstrated in the domain of structural system design for buildings as being capable of suggesting interesting design solutions that cannot be obtained by conventional design methods.

The potential of prototypes, by Gero's definition, in supporting creative design has also been demonstrated with a dependency network representation. Prototypes are attractive for several reasons:

1. They incorporate design concepts such as *functions*, *performance*, *behaviors*, *structure*, and *performance*, into one uniform representation.
2. In addition to categorizing design concepts, a set of generic relations is identified and used to express the semantics of relationships among the design concepts explicitly. This explicit representation of relations among design concepts overcomes some of the limitations of previous frame-based prototype representations including the arbitrariness of the attributes and lack of manipulability of a prototype's structure, thus providing a basis for prototypes to support nonroutine design.
3. The categorization of prototype attributes is shown to offer advantages in facilitating reasoning in the nonroutine design process. By dealing with certain categories of design information in a particular design task, not only the complexity of reasoning is reduced from the computational viewpoint, but also the relevance of different types of knowledge to the task is made more clear, as are the nature and characteristics of the task.

By developing the knowledge representation, more understanding has been gained of both the design concepts in the structural design domain needed to be represented for the purpose of computer supported creativity, and some important issues involved in using prototypes as the basic representation schema. These issues include, to name a few, the need to substantiate the prototype functions, to represent design constraints and design iterations, to provide interfacing information of the system-subsystem-components, and organization of the prototype hierarchy.

The creative design methodology is implemented as domain independent mutation operators. The identification and implementation of such operators is significant in several ways:

1. The *COMBINE* operator has shown that a design space can be expanded by changing the structure of the design space by combining two prototypes.

2. Given a network-based representation of prototypes, the mutation operator can presently operate on and manipulate a variety of design information including functions, performance, and structure of prototypes. The implication is that the capabilities of the mutation operators can be expanded with more knowledge represented in the prototypes.
3. It has been shown that mutation can be performed in a domain-independent manner by manipulating only the syntactic information provided by the network-based prototype representation. The information used by the mutation operators includes the relations among attributes, the attribute categories, and the attribute specifications. The relations are defined in generic, instead of domain-dependent, terms, and the categories are common to all engineering design domains. The attribute specifications, including the attribute name, value, range, description, factors, dependents and relations, are also an abstract structure of concepts that may be used to represent knowledge of other domains. Their contents, though domain specific, are not referenced in the mutation operators.
4. The methodology has demonstrated the alternate use of mutation and analogy techniques through the *COMBINE* operator so as to maximize the use of the available design information in order to reduce the randomness of mutation operations, but still able to progress even when design information is insufficient by falling back on mutation.

This research has shown with some initial results that mutation techniques can be useful for and have the potential of supporting creative design. Although the potential for producing complete working solutions remains to be seen, they can now model to a certain degree of success the capability of breaking the habitual thinking patterns of human designers and generating alternative interesting suggestions of creative design solutions.

Acknowledgements

The work is supported by the National Science Foundation through grants DDM-8657298 and DMCE 8811877.

References

Cagan, J and Agogino, A. M.: 1988, Innovative design of mechanical structures from first principles, *Artificial Intelligence in Engineering Design, Analysis and Manufacturing*.

Cagan, J and Agogino, A. M.: 1988, Reasoning about mechanical structures from first principles, *Proceedings of the 12th IMACS World Congress on Scientific Computation*, Paris.

Cagan, J and Agogino, A. M.: 1991, Inducing constraint activity in innovative design, *Artificial Intelligence in Engineering Design, Analysis and Manufacturing*, 5(1), 47-61.

Davis, R and Lenat, D. B.: 1982, *Knowledge-based systems in artificial intelligence*, McGraw-Hill, New York.

Dyer, M. G., Flowers, M. and Hodges, J.: 1986, EDISON: An engineering design invention system operating naively, *in* D. Sriram and R. A. Adey (eds), *Proceedings of the First International Conference on Applications of Artificial Intelligence to Engineering Problems*, Springer-Verlag, University of Southampton, UK.

Gero, J.: 1987, Prototypes: a new schema for knowledge-based design, *Working Paper*, Architectural Computing Unit, Department of Architectural Science,University of Sydney, Australia.

Gero, J. S.: 1990, Design prototypes: a knowledge representation schema for design, *AI Magazine*, **11**(4), 26-36.

Hodges, J., Flowers, M., and Dyer, M.: 1988, Knowledge representation for design creativity, *Technical Report*, University of California, Los Angeles, USA.

Lenat, D. B.: 1982, An artificial intelligence approach to discovery in mathematics as heuristic search, *in* Davis, R. and Lenat, D. B. (eds), *Knowledge-Based Systems in AI*, McGraw-Hill, New York.

Murthy, S. S. and Addanki, S.: 1987, PROMPT: An innovative design tool, *Proceedings American Association of Artificial Intelligence National Confenrence*.

Navinchandra, D.: 1991, *Exploration and Innovation in Design: Towards a Computational Model*, Springer-Verlag, New York.

Navinchandra, D., Sycara, K. P., and Narasimhan, S.: 1991, Behavioral synthesis in CADET, a case-based design tool, *Proceedings of the Seventh Conference on Artificial Intelligence Applications*, IEEE, Miami, Florida, USA.

Oxman, R. and Gero, J.S.: 1988, The designers memory: memory-based reasoning in knowledge-based design, *Technical Report*, Department of Architectural Science, University of Sydney, Australia.

Sycara, K. P. and Navinchandra, D.: 1991, Influences: a thematic abstraction for creative use of multiple cases, *Case-Based Reasoning Workshop*, DARPA, Washington DC, pp. 133-144,

Tham, K.W.: 1991, *A Model of Routine Design using Design Prototypes*, PhD thesis, Department of Architecture and Design Science, University of Sydney, Australia.

Zhang, W.: 1990, *Chunking Structural Design Knowledge as Design Prototypes*, PhD thesis, Department of Civil Engineering, Carnegie Mellon University.

Zhao, F.: 1991, *A Knowledge-Based Representation of Mutation for Creative Design*, PhD thesis, Department of Civil Engineering, Carnegie Mellon University.

A DESIGN SUPPORT SYSTEM USING ANALOGY

L. QIAN and J. S. GERO
Design Computing Unit
Department of Architectural and Design Science
University of Sydney, NSW 2006 Australia

Abstract. Analogy is one of the useful tools for creative designs. Most design by analogy systems use within-domain analogies to adapt an old design to a new situation. As design from a between-domain analogy may yield more creative results, it raises issues of how such analogical design can be initiated and what such an approach can do to help the design process. In this paper, we propose a computational model for a design support system that uses between-domain analogy, the system finds analogical mappings through causal links between structure and structure, and between structure and behaviour. The analogical reasoning process is accomplished using the design prototype as the representation.

1. Introduction

Keane (Keane, 1988) defines analogy as a product of certain cognitive processes in which specific coherent aspects (e.g., causally-integrated aspects) of the conceptual structure of one domain are matched with and/or transferred into another domain. In most knowledge-based design systems by analogy such as STRUPLE (Maher and Zhao, 1987), purpose-directed analogy (Kedar-Cabelli, 1988), Argo (Huhns and Acosta, 1987) and behaviour-preserving analogy (Sycara and Navinchandra, 1991) treat analogy as any kind of similarity, from low-level knowledge about structures and structure attributes to higher level knowledge such as explanations, functionalities and causalities. In the first mentioned design systems, source and target designs fall into same design domain - building design, cup design and VLSI digital circuit design. Analogies used in these systems are called within-domain analogies (Vosniadou and Ortony, 1989). The last design system uses a domain-independent approach in which an abstract description of the desired behaviour is an index or transformed index used to retrieve an existing relevant case in a different design domain.

Analogy should not be considered simply as surface similarity, i.e., structure attribute similarity. A designer would not retrieve previous design cases and go through an analogical reasoning process, if all he wants to design is the colour of the door. Analogical design can be more significant if the source and target domains are in different domains. The more remote the source and target domains, the more difficulties there will be for naive designers, and the more explorations there will be, and the greater the likihood of a creative design being produced.

In this paper, we are more interested in between-domain analogy in which the target design domain is different from the source design domain. The important issues raised here are the representation of design knowledge in different domains, the analogical reasoning mechanisms for between-domain analogy and the formal

J. S. Gero (ed.), Artificial Intelligence in Design '92, 795–813.

process of how structure is created from a retrieved design to produce the beginnings of a new design.

Design knowledge is represented here as the conceptual schema design prototype. Design prototypes are chunks of knowledge representing a class of generalised design experiences (Gero, 1990). Design prototypes include shallow, operational knowledge as well as deep causal knowledge. While the shallow-level knowledge can be seen as knowledge that is sufficient for performing a task itself, the deep knowledge captures an underlying causal relationship and behaviour, which can be used to achieve design goals. The design prototype compiles the shallow-level knowledge and the deep knowledge into four groups of variables: function, behaviour, structure and external effect. These four groups interrelate in such a way that the structure of an artifact behaves in some way to achieve a function. We concentrate on function- behaviour-structure paths as units for analogical reasoning in design.

Among many design domains, e.g, mechanical design and architectural design, design functions can be classified by a kind of achievement or a kind of causation. The design prototype provides sufficient and well-organised knowledge to carry out the analogical reasoning for different kinds of designs. A computational model is proposed here which can elaborate analogical mappings between different domains through causal chains from structure to structure and from structure to behaviour, and then to functions. Since the analogical mapping process depends on a type of causation of function achievement, a design support system based on this is extendible to adapt new kinds of design knowledge.

If a new design is created using a within-domain analogy, its feasibility can be evaluated, confirmed and repaired automatically in a design system. However, this is not applicable with between-domain analogies because the new conjecture from the analogical mapping can not provide appropriate new knowledge for evaluation. In this situation, the design system is like a design support tool, it brings a potential design description forward and leaves decision making to the human designer.

In the remainder of this paper the design knowledge representation is described in Section 2; based on this design representation, analogical reasoning mechanisms for between-domain analogies are presented in Section 3 with examples. The paper discusses the implementation of the system in Section 4 and concludes with a discussion of some of the issues this work raises.

2. Design Knowledge Representation

In order to design using analogies, not only shallow-level knowledge is required, but also deep causal knowledge is essential. Design prototypes are a conceptual schema containing necessary and sufficient knowledge to allow this kind of design process to occur.

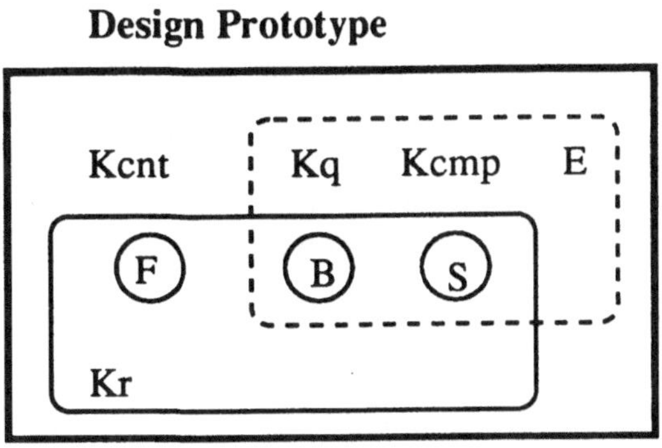

B: behaviour variables
E: exogenous variables
F: function variables
Kcmp: computational knowledge
Kcnt: context knowledge
Kq: qualitative knowledge
Kr: relational knowledge
S: structure variables

Fig. 1. Design prototype schema

2.1. DESIGN PROTOTYPE

A design prototype (Gero, 1990) is a generalised representation which contains four variable groups - function, F, behaviour, B, structure, S and exogenous variables, E, plus related knowledge, Fig. 1. While function describes goals or intentions of the design artifact, structure specifies individual components, their attributes, and relationships between them. The exogenous variables describe the environment affecting the structure of an artifact. They include operations executed by human beings (e.g., push) or natural environment parameters (e.g., temperature) which could affect the structure or behaviour of an artifact. The link between the structure and the function is established by behaviour and it involves deep knowledge such as causal relations.

The related knowledge includes relational knowledge, K_r, which represents the dependencies between function, behaviour and structure, and qualitative knowledge, K_q, to describe the directional effect of changing one variable on other variables. The design prototype also contains computational knowledge, K_{cmp}, that is the quantitative counterpart of qualitative knowledge, and context knowledge, K_{cnt}, that identifies effects from external agents on the design as well as providing values for exogenous variables. Any knowledge about exogenous variables should be compiled into the context knowledge.

2.2. QUALITATIVE CAUSAL KNOWLEDGE

Causality plays an important role in qualitative physics. A systematic set of conventions is used to represent how information propagates through physical artifacts to achieve the overall purposes or functions. This set of conventions is called causality and can be used to analyse, design and diagnose engineered artifacts. Causality is the link between function and structure and this is represented as behaviour in design prototypes.

Qualitative causal knowledge is a representation of the behaviour of a structure.

It expresses directions of causation from one structure to another or from one structure to another behaviour as well as the qualitative measure when a behaviour is caused.

In the design prototype representation, while the dependency knowledge points to which structures produce a certain behaviour, qualitative knowledge describes how structure changes behaviour. The qualitative causal knowledge can be expressed as in equation 1.

$$B\{d_b, q_b\} \leftarrow S\{d_s, q_s, E\} \qquad (1)$$

where
- B: behaviour
- d_b: description of dynamic behaviour (B)
- d_s: description of dynamic structure (S)
- q_b: tendencies of the changes in behaviour (B)
- q_s: tendencies of the changes in structure (S)
- E: external effect
- S: structure
- $\leftarrow$: causal direction

While the tendency is a qualitative measure of some quantities (e.g., increase hinge angle) (Forbus and Gentner, 1990), the description of dynamic behaviour is a notion of mechanism (e.g., motion, coupling) (Dyer et al, 1986).

Qualitative causal knowledge is a higher level representation of K_q, K_{cmp}, K_{cnt}, K_r, and E in order to show the dynamic and static behaviour of a structure. This knowledge sets up the path for analogical reasoning in within- or between-domain designs.

A change in a structure could cause change in another structure, and so on, until a change of a behaviour occurs. This can be expressed as, equation 2:

$$B_n \leftarrow B_{n-1} ... B_1 \leftarrow S_m \leftarrow S_{m-1} \leftarrow ... \leftarrow S_1 \qquad (2)$$

If there are n related behaviour variables and m structure variables in which the mth structure variable has a direct effect on the 1st behaviour variable, the causal propagation is from structure to behaviour. This is called series propagation.

Multiple influences from structure to structure, from structure to behaviour, or from behaviour to behaviour are another kind of causal propagation called parallel propagation. If m structure variables affect one behaviour, this can be expressed by equation 3 for multiple influences from structure to behaviour:

$$B \leftarrow (S_m, S_{m-1}, ..., S_1) \qquad (3)$$

By combining series and parallel causal propagation, structures and behaviours may be represented as influence graphs (Sycara and Navinchandra, 1991).

2.3. Design Classification By Goal Achievement

The concern of design is to determine a structure with its description, capable of producing the functions required. Although design artifacts are very different from one domain to another, the ways in which the structure achieves the design goal can be classified across design domains using the nature of physical operations on the structure. Basically, there are three mechanisms by which a goal is reached.

(i) Structure through its existence.
For example, in architectural design, view can be achieved by the existence of windows in certain locations. The designs of this class can be expressed by transforming equation 1 to equation 4:

$$B\{d_b\} \leftarrow S\{d_s\} \tag{4}$$

(ii) Structure changes through external effects.
The external effect can be an operation by human being. For example, opening a door allows access while closing the door stops access and provides privacy. The physical position of the door is a variable. Depending on the operation applied to the door, a different behaviour occurs and in turn can achieve different functions. The dynamic characteristics can be represented as equation 5 which is as the same as equation 1:

$$B\{d_b, q_b\} \leftarrow S\{d_s, q_s, E\} \tag{5}$$

(iii) Sequence of operations in conjunction with structure.
The third type of design requires more information on how a series operations can attain a function, e.g., software design. Equation 1 can be transformed to equation 6:

$$B\{d_b\} \leftarrow S\{d_s, E\} \tag{6}$$

For different kinds of design, behaviour is represented differently. If a goal is achieved by structure attributes and structure configuration alone, the behaviour of the structure is static in which case there are no behaviour changes over time. If a structure is changeable and it is caused by some operation or other external effect, its behaviour becomes dynamic. This dynamic characteristic can be described with qualitative causal knowledge. The last goal achievement type involves temporal information. The order of operations and contents of the operations are directly used in design.

There could be more kinds of designs which do not fall into the previously mentioned design classes. The aim of classifying design in this way is to help analogical reasoning for between domain designs.

3. Analogical Reasoning

An overview of the analogical reasoning process is shown in Fig. 2. In a design environment, designs can be stored in the form of design prototypes. Assuming that the requirements of a new design given by a user cannot be found by a routine

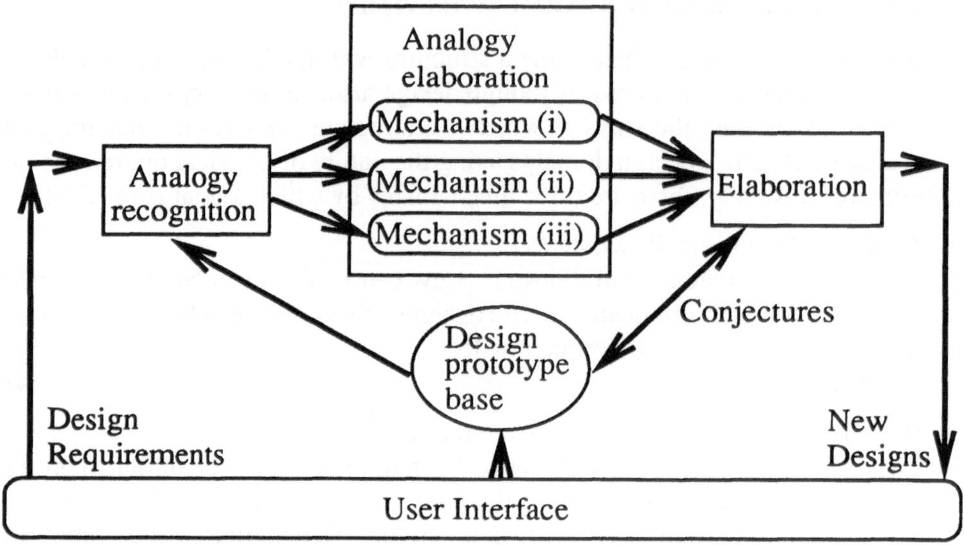

Fig. 2. Analogical reasoning process

design process, then analogical reasoning is used. Note that the arrows shown in the figure indicate control flow directions.

The data flows of the analogical reasoning process are shown in Fig. 3.

We now concentrate on analogical reasoning using the notation of classification by operation as causation mentioned in the previous section. As any goal can be achieved by one of the causation types, a designed artifact can be seen as a composition of causes such as by structure through its existence, structure changes through an operation and/or sequence of operations in conjunction with structures. Corresponding to each kind of design goal achievement there is a mechanism for analogical mapping.

Once a structure conjecture is transferred from the source design to the target design, it must be evaluated. If a new structure is accepted and the overall function of the designed artifact is not violated in the target design domain, the new design is organised into the design prototype base. In the next subsections, more details on each module in the analogical reasoning process are presented.

3.1. Analogical Recognition

When design requirements are raised, they can be decomposed as functions, behaviours, structures and/or operations. A routine design may be initiated. If no solution can be found using routine design processes, design by analogy is one way of producing a solution.

Normally, design requirements specify what is to be designed and what features should be included. For example, design a building to straddle a highway. Here the

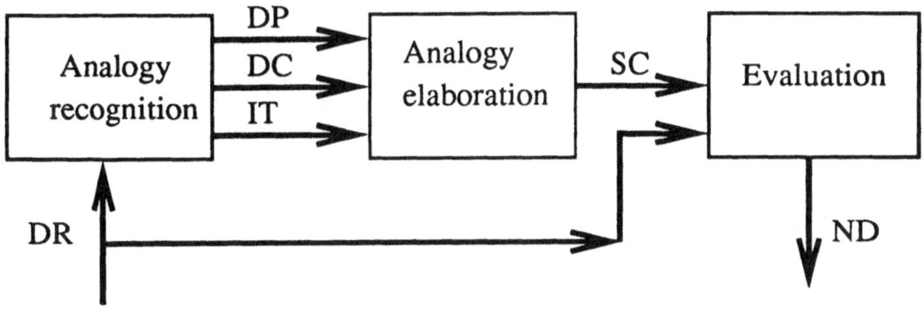

DC:	Design Class represented with causation type	IT:	Index Type
DP:	Design Prototype	ND:	New Design
DR:	Design Requirement	SC:	Structure Conjecture

Fig. 3. Data flows of the analogical reasoning process

building is to be designed and its specified structure is long span. The name of a design artifact often provides the concept of the artifact. This label is an index to retrieve the design prototype as the target design, and in turn to obtain the rest of the properties of the artifact. Design requirements can directly include a potential analogy. For example, *design a door which operates like a curtain.* The statement gives *curtain* as an analogy candidate and restricts the analogy. The analogical recognition process should extract keywords (e.g., operates) and identify what type of causation might be involved.

Sometimes, a designer may just want to design an artifact which is somehow new and cannot be found in the current design solutions using routine design processes. In this case, the analogical recognition process should relax the constraints of the old design and retrieve analogies indexed by any combination of function, behaviour, structure and operation.

As a summary, the analogical recognition process mainly analyses design requirements and extracts key information as an index for analogy retrieval. The index for searching over the design prototype base could be function, behaviour, structure or operation requirements. Once a potential analogy is found, the recognition process identifies the type of causation of the design and guides the analogical mapping process using this type, Fig. 4.

3.2. Qualitative Causal Knowledge Mapping

Different types of designs will have different analogical reasoning mechanisms. In this sense our reasoning engine is not domain-dependent, but type-dependent. In the following sections we will present reasoning mechanisms focusing on the

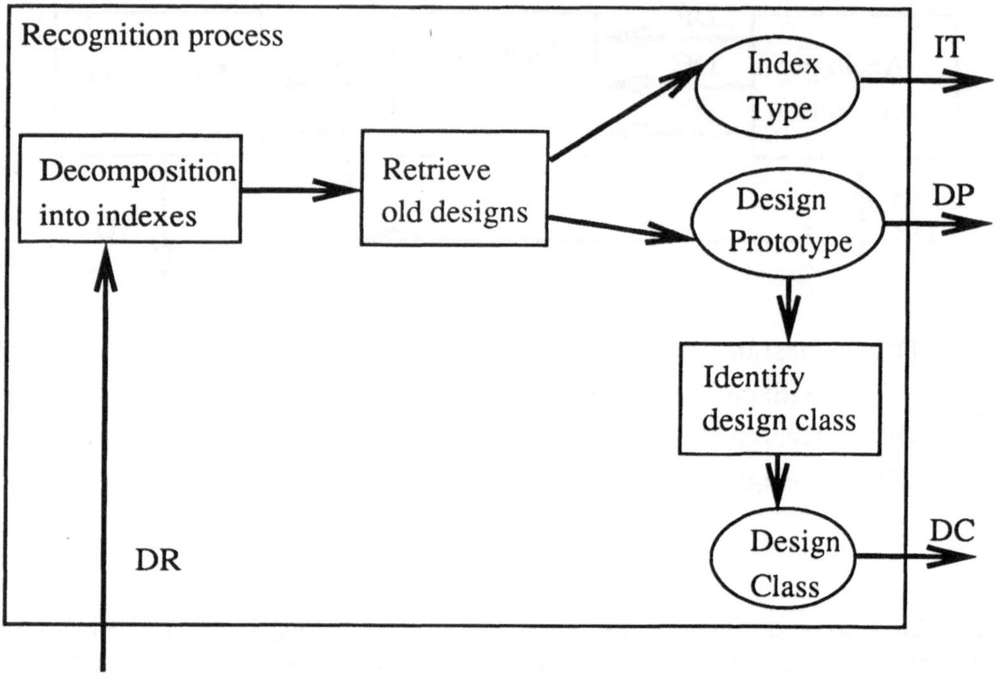

Fig. 4. Recognition process

second type of goal achievement.

In structure mapping theory (Gentner, 1983), structural properties are mapped from a source domain to a target domain. It conveys that a system of relations known to hold in the source also holds in the target by causal links. Our system commences with this theory and develops it for design purposes. The design description consists of individual structure components and configurations and their relations. This can be represented as a graph called a structure graph, Gs, in equation 7, Fig. 5.

$$G_s = \{E, R\} \tag{7}$$

where

E: a set of structure elements or components represented by nodes in the graph; and

R: a set of relations between two components represented by the arcs.

Design by analogy aims to find new components and relations of the components, i.e., it is to find a structure graph for the target design. Note that the structure graph is an undirected graph. Usually, an artifact has many functions each of which is achieved by a set of components with their attributes, structures and configurations through one or more behaviours. The set which includes relevant

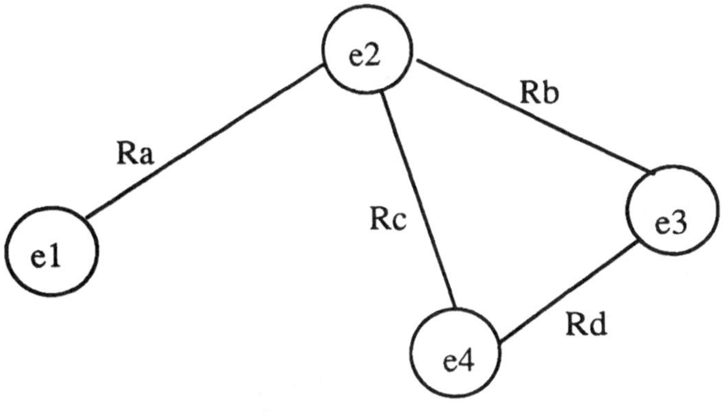

E = { e1, e2, e3, e4 }

R = { Ra, Rb, Rc, Rd }

Fig. 5. An example of a structure graph

function variables, behaviour variables and structure variables is represented as a dependency network (Gero et al, 1991).

3.2.1. *Constructing the behaviour graph*

A set of selected structures can be extracted using the dependency network. This first step reduces working memory in the elaboration of analogical reasoning. However, the structure graph cannot be the working memory for starting the mapping process. We need to add qualitative causal knowledge to the structure graph to convert it into a behaviour graph. As mentioned in equations 2 and 5, description of causation, d, and tendency of change, q, should be added to each component node in the structure graph. The qualitative causal knowledge such as that change on component e_1 will cause a change on component e_2, is expressed as

$$e_2\{d_2, q_2\} - e_1\{d_1, q_1\}$$

and is added to the nodes e_1 and e_2 as square nodes attached to associated circle nodes, Fig. 6. A limited set of indexes is used here. The direction of the causation is given by the directed arc. A behaviour graph is obtained from the structure graph with the addition of qualitative causal knowledge. If a static behaviour is present, that attribute of the component is added in the square node.

The behaviour graph can be abstracted by applying heuristic rules which can be obtained from domain knowledge and common sense knowledge.

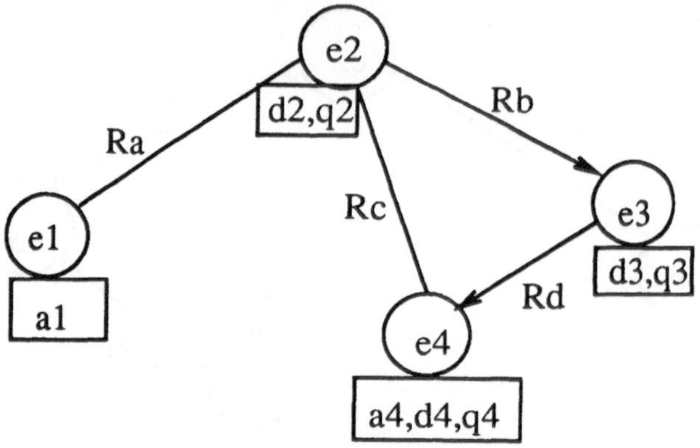

a_l: attribute of element e_l, where $l = 1$ or $l = 4$.
d_j: description of dynamic characteristic, where $2 \leq j \leq 4$
e_i: element i, where $1 \leq i \leq 4$
q_k: tendency of causation, where $2 \leq k \leq 4$

Fig. 6. Behaviour graph

The behaviour graph can be summarised as:

$$G_b = \{(E, A, D, Q), R_d\}$$

where
A: set of attributes
D: set of dynamic descriptions
E: set of elements
G_b: behaviour graph
Q: set of causation tendencies
R_d: set of directed relations between paired nodes.

3.2.2. Mapping

Filtering out elements (circular nodes) and relations (arc labels) from the behaviour graph, only the causal qualitative knowledge is left, Fig. 7. Finding an isomorphism using qualitative causal knowledge between two behaviour graphs is the central idea of mapping structures here. The isomorphism of two graphs is found by matching nodes and corresponding arcs. When qualitative causal knowledge is used for comparison, those qualitative expressions and causal direction are the matching items.

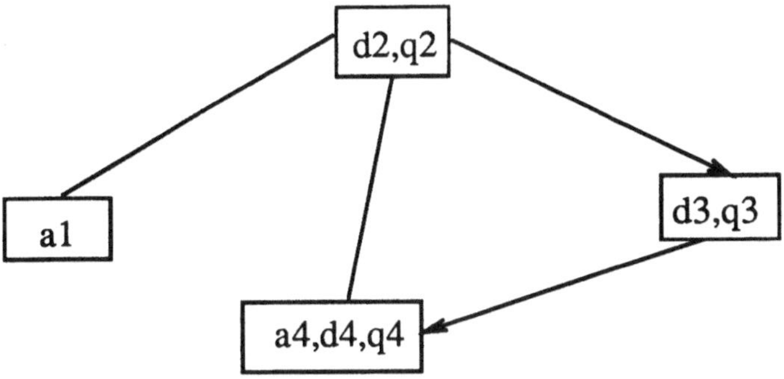

Fig. 7. Qualitative causal knowledge

The two behaviour graphs

$$G_{b_1} = \{(E_1, A_1, D_1, Q_1), R_{d_1}\}$$

and

$$G_{b_2} = \{(E_2, A_2, D_2, Q_2), R_{d_2}\}$$

are isomorphic if (A_1) and (A_2) are identical when (D, Q) are not present, or (D_1, Q_1) and (D_2, Q_2) are identical. In addition, R_{d_1} and R_{d_2} relate the same set of nodes, Fig. 8.

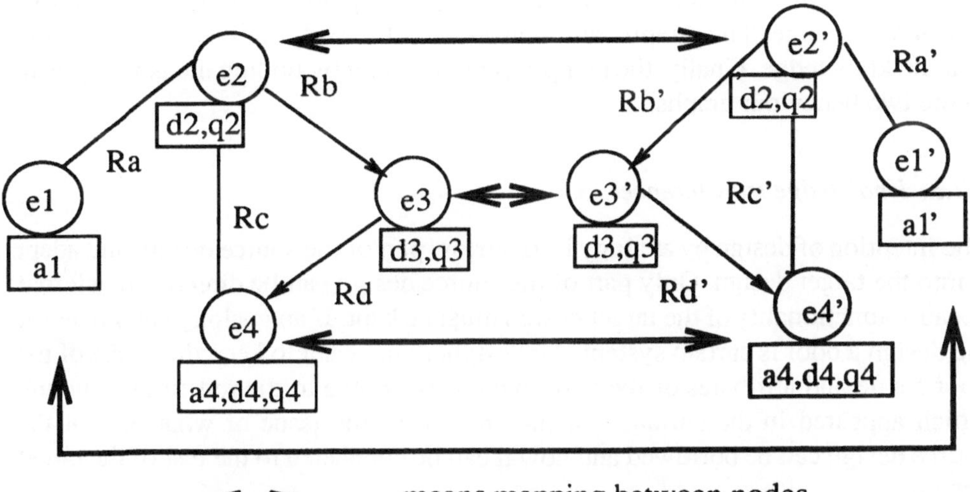

Fig. 8. Isomorphic behaviour graphs

After the qualitative causal knowledge is matched, the element attached to each node in the behaviour graph and the relation labels are retrieved and are mapped to corresponding elements and relations in the other graph.

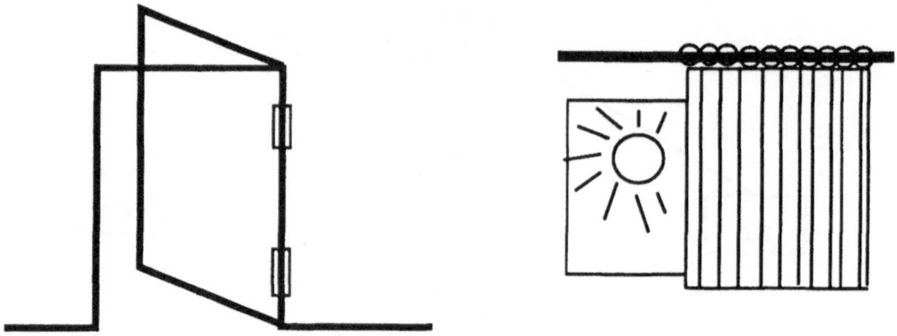

(a) Hinged door (b) Curtain

Fig. 9. Door design as target design and curtain design as source design

If a node in a graph is an abstracted form, the 1-1 correspondence of the isomorphism of two graphs becomes 1-m or n-1. A node could map to a set of nodes.

The example of a door mapping to a curtain system is shown in Fig. 9 and Fig. 10.

In summary, the mapping process first finds useful parts of the structures in terms of structure graphs in both target design and source design included in the structure graph definition. Then within the limited sets of components, the next step is to construct the behaviour graph from the structure graph by applying qualitative causal knowledge. This graph can be abstracted by applying heuristic rules and domain knowledge. Finally, the mapping can be found by finding the isomorphism of the two behaviour graphs.

3.2.3. Knowledge transference

The intention of design by analogy is to borrow part of the source design and adapt it into the target design. Only part of the source design can be directly transferred because functionality of the target design must be kept. If an analogy candidate for the design a door is curtain system, the designer may want to keep the shape of the door frame and attributes of the door, but change hinge to the rod-rings structure which appeared in the curtain system. This raises the issue of what part of the source design can be borrowed and how it can be connected to the rest of the target design.

As mappings on the source and the target designs yield correspondences of the components and the relations, these give the potential of obtaining new designs by substituting part of the target with the source design. Thus, the mapped behaviour

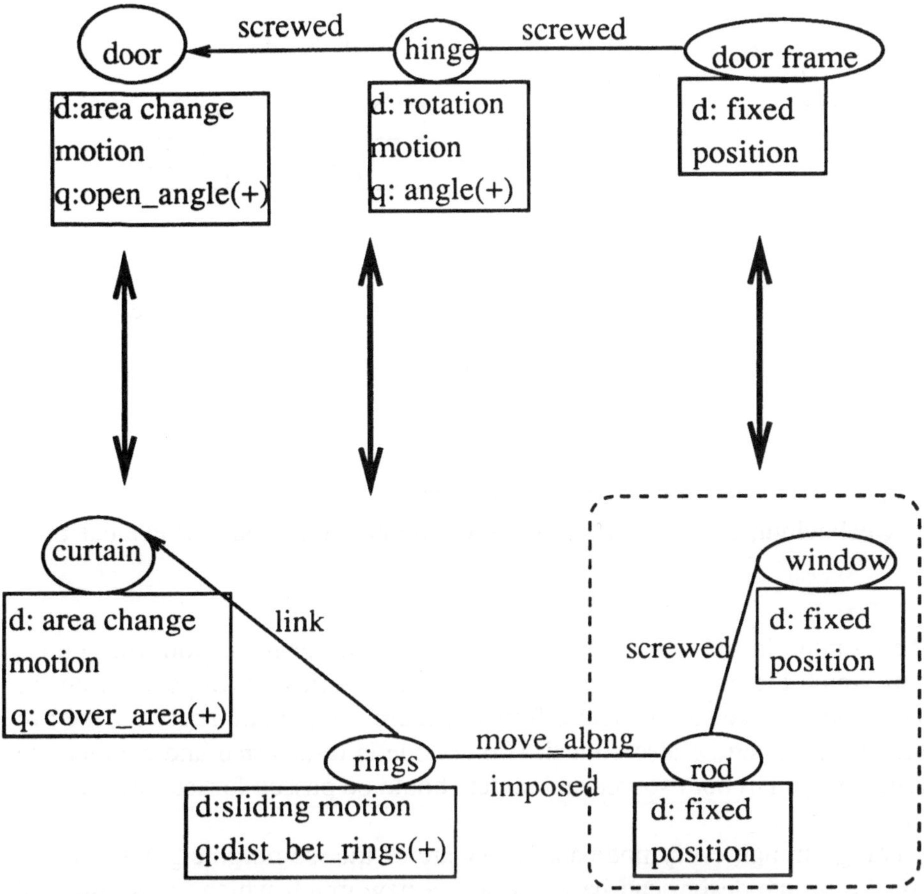

Fig. 10. Mapping between door and curtain

graph in the source design is considered for transference. The explanation of how these structures can achieve goals of the design are also needed during the adaptation after the substitution.

If E_s is a component set on the source design and E_t is a component set for the target design, they can be decomposed as:

$$E_s = E_{s1} \cup E_{s2}$$

and

$$E_t = E_{t1} \cup E_{t2}$$

A new set of components, E_n, after transference from the source design to the target design can be composed as:

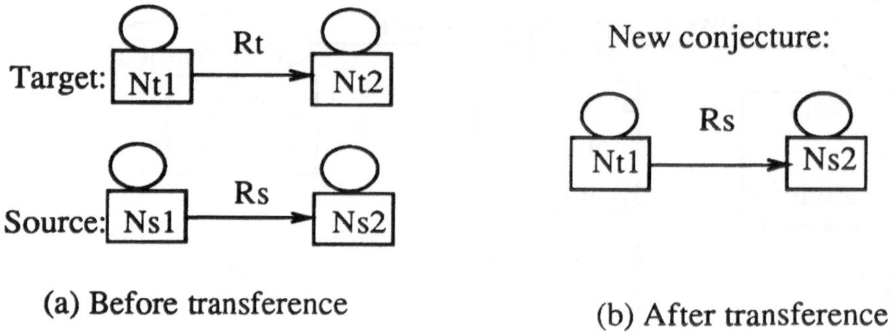

Fig. 11. Relation transference

$$E_n = E_{s1} \cup E_{t2}$$

The new behaviour, G_{b_n}, from the new set of components should be evaluated.

$$G_{b_n} = \{(E_n, A_n, D_n, Q_n), R_{d_n}\}$$

When a node in the target is substituted with another node from the source, relations which are connected to this node must be adjusted. Depending on the adjacent node, if it is also transferred from the source, the relation between the two transferred nodes can be also copied. If one node is transferred and another one is not, the relation in both source and target should be passed for later evaluation, Fig. 11.

By going through every node and every arc of the isomorphic graphs, useful parts of the source design can be found and a new graph which has substituted source parts is constructed. For those abstracted nodes, the final step is to extend each of them down to the lowest level - the element level.

The final behaviour graph gives a conjecture of the new structure for the target design, Fig. 12. The heavy arrow points to the start of the elaboration process according to the index type. The structure conjectures are created which can give to a designer a creative idea about the target design.

In the door design, the conjecture of the new door, after the analogical mapping process and the knowledge transference, is that it is like a sliding door, Fig. 13.

The result of this knowledge transference has only local evaluation which occurs when substitution is considered. The global evaluation of the whole design needs to be executed in the evaluator.

The suggested structures (conjectures) from an analogy should be evaluated with the target domain knowledge. As during the mapping process, a set of structure correspondences and behaviour correspondences are found, not only structure relations should be transferred from source to target but also the explanation of how a structure behaves to achieve a function can be passed. This explanation is described by the dependency network.

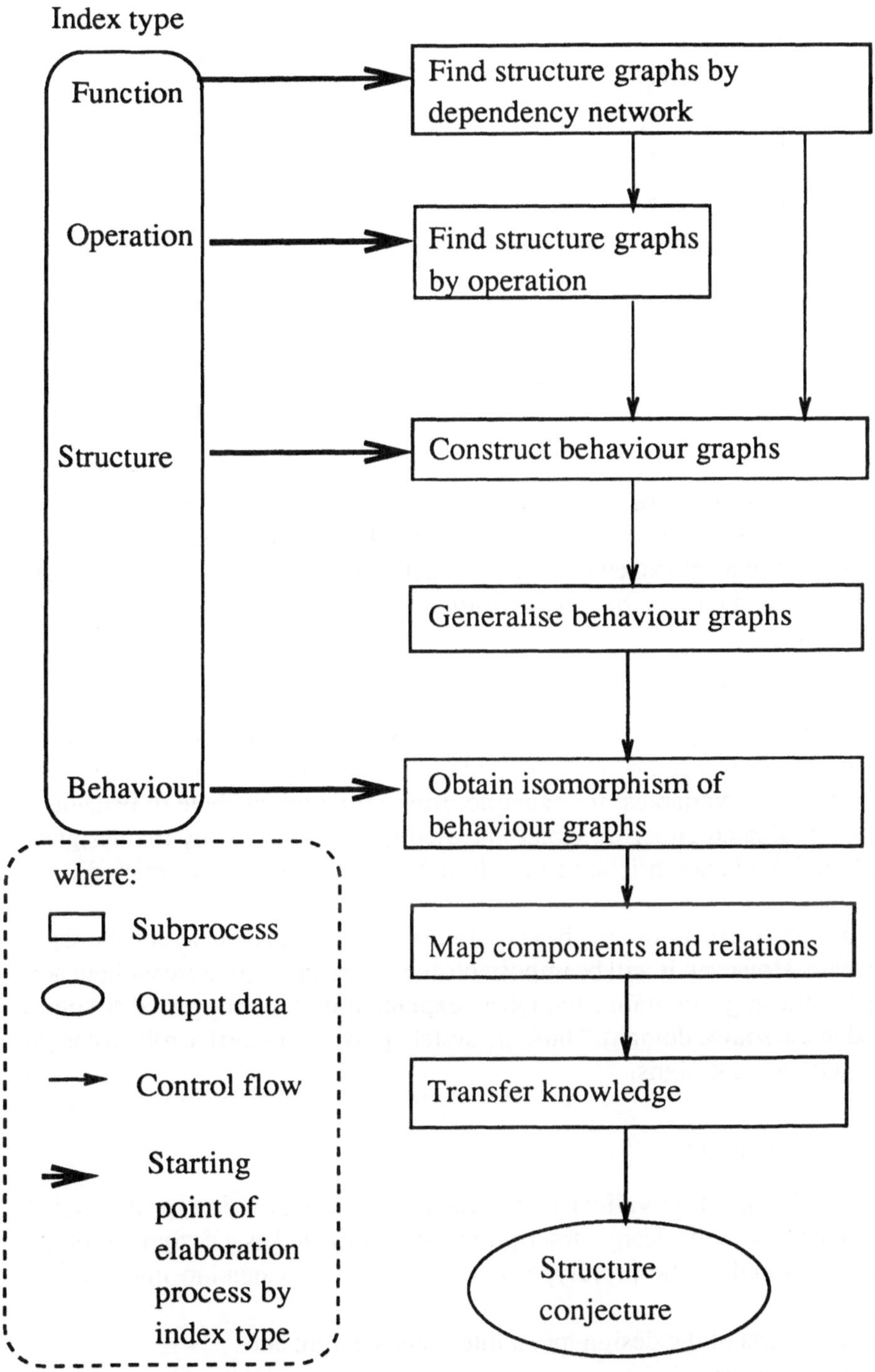

Fig. 12. Second mechanism of analogy elaboration between source and target designs

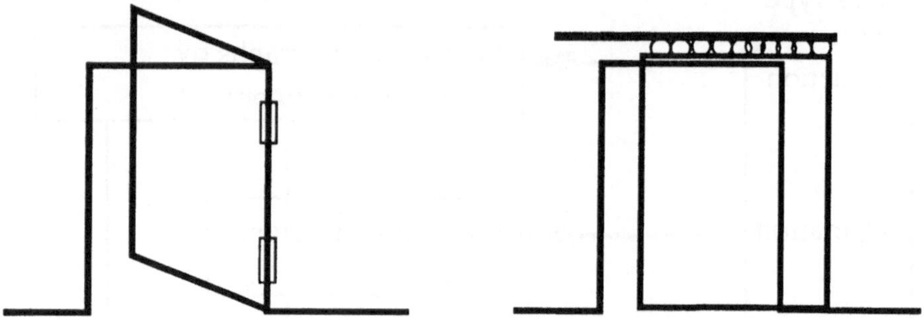

Fig. 13. From hinged door to sliding door

3.3. EVALUATION

Evaluation is the process of comparing the performance of derived behaviour and the required performance of behaviour. While local evaluation confirms each individual design goal, global evaluation judges the interfaces between substructure and substructure and checks the total performance of the design

3.3.1. Local evaluation

Using the transferred dependency network, the system checks every structure variable or its attribute against the target domain knowledge and the requirements. The links between group variables, for example, from structure attribute to behaviour variable, are evaluated either by target knowledge or confirmed by the designer. The transferred explanation is used to help evaluate new structures copied from source domain.

Repair of buggy structures may be automatic if the conjecture is within the same design domain. However, it will be impossible to judge a new structure without any knowledge in the target domain although an explanation of the suggested structure is provided in the source domain. Thus, the system plays a supportive role in design instead of generating designs.

3.3.2. Global evaluation

While the local evaluation validates individual elements of a design, the global evaluation makes sure the design descriptions are completed as a design prototype which can be stored in the prototype base. The global evaluation involves the following tasks:

1. combining parts of the design into a integrated design; and
2. external assistance.

For example, a new design could employ multiple analogies from more than one source domains or have one part borrowed from analogy and another part from mutation. If parts of the design cannot be integrated as a whole design, for example, one key requirement cannot be satisfied and there is no way to repair it automatically, the last step of the evaluation process is to give the current design description to the designer. The interaction between the system and a designer can continue until all requirements are satisfied.

4. Implementation

The computational model proposed in this paper is implemented as a system called DSSUA, **D**esign **S**upport **S**ystem **U**sing **A**nalogy, using Common Lisp on Sun workstations. Design prototypes which come from different domains are stored in the system.

Referring to Fig. 2, Fig. 3 and Fig. 12, the examples demonstrated here are indexed by function and we chose both target and source designs to be the second causation type. Thus, the system used mechanism (*ii*) to perform the analogical mapping and the elaboration process commences from the beginning of the complete algorithm.

The structure and behaviour graphs are represented as matrices. The dimension of the matrix is a list of related components which resulted from dependency network having the index function as the first entry and could be generalised or combined using some rules. For instance, the initial component list for a curtain design is (curtain, rings, rod, window). After the process of generalising the behaviour graph (shown in Fig. 12 and in Fig. 10), the abstracted component list become (curtain, rings, (rod, windows)). While the structure graph is a symmetric matrix with a 1 representing the relationship between one component and another, the behaviour graph considers the causation direction as a 1 from row to column of the matrix, represented as (row, column). Fig. 14 is an example of behaviour graph in the curtain design. There is a 0 in (curtain, rings) and 1 in (rings, curtain) to show the causation directions from changes on the rings to changes on the curtain. If two components, x and y are not related with causal qualitative knowledge, there are two 0's in both (x, y) element and (y, x) element of the matrix, e.g., components (rod window) and rings. Note that (rod window) is a generalised component.

Initially, the component list could be in any order as long as the system keeps a record for mapping the index of the component list to each component name. When the two behaviour graphs are represented as matrices, by using one matrix as a reference it is a simple matter to find the equality of the two matrices or sub-matrices.

The system uses a multiple indexing technique to retrieve a potential source design for the target. As the system groups words of similar meaning together, comparisons of two indexes are based on meanings as well as syntax. It finds the mapping components and relations between two designs which could give a

	(rod window)	curtain	rings
(rod window)	0	0	1
curtain	0	0	0
rings	1	1	0

Fig. 14. Behaviour graph for mapping

designer an inspiration of how the design may be proceed. The user of the system mainly assists the design process in the evaluation stage when there is insufficient domain knowledge.

5. Conclusion

The aim of this paper is to describe a computational model of a design support system that uses analogy. We used the design prototype schema which is considered as a useful, powerful, flexible and feasible design knowledge representation schema to carry on design by within-domain or between-domain analogy. Qualitative causal knowledge embedded in the behaviour of the design prototype plays a major role during the analogical reasoning process for design. By combining static and dynamic behaviours, and the dependencies between behaviours and structures, we developed algorithms to map components or generalised components between two designs.

Designs in different domains include representations of the nature of their goal achievement or the causation types. Analogical mapping commences after it is found that the source and the target designs have the same causation type. Design with the goal achieved by dynamic characteristics of the structure and operation is discussed in this paper. Designs with another kind of goal accomplishment require a related but separate analogical reasoning mechanism.

The result of applying such a system could be a new design description based on requirements or may be some suggestion which provokes the designer who may not have thought to create a design in that particular way. This system is not an automatic design tool; it provides an exploration medium. It is capable of reasoning with between-domain analogies.

More complicated design prototypes are to be tested with more refined mapping algorithms. This includes an algorithm for partial mapping between two behaviour graphs.

References

Dyer, M. G., Flowers, M. and Hodges, J.:1986, Edison: an engineering design invention system operating naively, *in* D. Sriram and R. Adey (eds.), *Applications of Artificial Intelligence in*

Engineering Problems, Springer-Verlag, Berlin, pp.327-361.

Falkenhainer, B., Forbus K. D. and Gentner D.: 1989/90, The structure- mapping engine: algorithm and examples, *Artificial Intelligence* **41**, 1-63.

Forbus D. K. and Gentner D.: 1990, Causal reasoning about quantities, *in* D. S. Weld and J. de Kleer (eds), *Readings in Qualitative Reasoning About Physical Systems*, Morgan Kaufmann, Los Altos, pp.666-677.

Gentner, D.: 1983, Structure-mapping: a theoretical framework for analogy, *Cognitive Science* **7**(2), 55-170.

Gero, J. S.: 1990, Design prototypes: a knowledge representation schema for design, *AI Magazine* **11**(4), 6-36.

Gero, J. S., Tham, K. W. and Lee, H. S.: 1991, Behaviour: a link between function and structure in design, *in* D. C. Brown, H. Yoshikawa and M. Waldron (eds), *IntCAD'91 Preprints*, IFIP, Columbus, Ohio, pp.201- 230.

Huhns, M. N. and Acosta, R. D.: 1988, Argo: an analogical reasoning system for solving design problems, it MCC Technical Report, Number AI/CAD- 092-87, Microelectronics and Computer Technology Corporation, Houston.

Keane, M. T.: 1988, *Analogical Problem Solving*, Ellis Horwood, Chichester.

Kedar-Cabelli, S. T.: 1988, Toward a computational model of purpose-directed analogy, *in* R. S. Michalski, J. G. Carbonell and T. M. Mitchell (eds), *Machine Learning II: An Artificial Intelligence Approach*, Morgan Kaufmann, Los Altos, pp.284-290.

Maher, M. L. and Zhao, F.: 1987, Using experiences to plan the synthesis of new designs, *in* J. S. Gero (ed.), *Expert System in Computer-Aided Design*, North-Holland, Amsterdam, pp.349-369.

Sycara K. and Navinchandra D.: 1991, Index transformation techniques for facilitating creative use of multiple cases, *in* J. S. Gero (ed.), *IJCAI-91 Workshop on AI in Design*, University of Sydney, Sydney, pp.15-20.

Vosniadou, S and Ortony A. (eds): 1989, *Similarity and Analogical Reasoning*, Cambridge University Press, Cambridge.

14

DESIGN PROCESSES

HIERARCHICAL GENERATE-AND-TEST vs CONSTRAINT-DIRECTED SEARCH

A comparison in the context of layout synthesis

U. FLEMMING*, C. A. BAYKAN**, R. F. COYNE*
**Department of Architecture and Engineering Design Research Center*
***The Robotics Institute*
Carnegie Mellon University
Pittsburgh PA 15213
USA

and

M. S. FOX
Department of Industrial Engineering
University of Toronto
Toronto Ontario M5S 1A4
Canada

Abstract. Two systems for layout synthesis, LOOS and WRIGHT, and the approaches underlying them are compared. LOOS uses a form of hierarchical generate-and-test and WRIGHT disjunctive constraint satisfaction, a form of constraint-directed search. LOOS implements a constructive approach that adds objects sequentially, while WRIGHT uses a reductionist approach that satisfies constraints incrementally. The comparisons are based on a series of experiments in which the systems were used to solve identical layout problems and to produce insights at very detailed levels. The conclusions are tentative, as the experiments are still going on at the time of writing.

1. Introduction

The present paper compares two space planning or layout synthesis systems, LOOS and WRIGHT, each of which implements a well-known approach toward solving design problems defined through feasibility constraints. The approach underlying LOOS is a form of *hierarchical generate-and-test*, in which solutions are constructed incrementally and intermediate states evaluated for constraint satisfaction; the overall control is based on these evaluations. WRIGHT implements a form of constraint-directed search called *disjunctive constraint satisfaction*, in which the constraints are incrementally satisfied.

Each approach has been independently implemented in a conceptually clean and clear fashion, and the resulting systems are able to solve *identical layout problems* from various domains. The authors seized the opportunities thus offered and conducted a series of experiments in order to gain concrete insights into the advantages and disadvantages of the underlying approaches that go beyond summary characterizations that dismiss, for example, the first approach as generally inefficient.

J. S. Gero (ed.), Artificial Intelligence in Design '92, 817–838.

The present paper reports some initial results of these experiments. Section 2 characterizes the layout problems solved by the two systems. Sections 3 and 4 briefly describe the systems, and section 5 presents the results of the experiments. Our conclusions are summarized in Section 6.

2. Layout Problems

The layouts considered in this paper are arrangements of rectangles with sides parallel to the axes of an orthogonal system of Cartesian coordinates. The rectangles in a layout can be *loosely packed*; that is, the layout may have holes or an irregular boundary. This class of layouts is interesting across a broad spectrum of applications, domains and disciplines that range from digital and analog electronics design to building and graphic design.

A layout in this class is completely specified if the corner coordinates (or some equivalent set of values) are given for each rectangle in the layout. The problem of finding a feasible set of coordinates gains a considerable degree of complexity from the fact that values for the coordinates cannot be selected independently of each other. For example, the rectangles in a layout often represent physical objects that occupy space and therefore cannot overlap. The overall area available for placing the objects is also often restricted. This creates constraints that may vary with the way in which the objects are placed; an example is shown in Figure 1. These types of constraints have been called in the literature *dependent* (Flemming, 1978) or or *inter-element* (Eastman, 1973) constraints; they indicate that considerations of *structure* or *topology* and discrete decisions about structural or topological variables play a prominent role in layout synthesis, as they do in other design domains dealing with assemblies of discrete parts in 2- or 3-dimensional space.

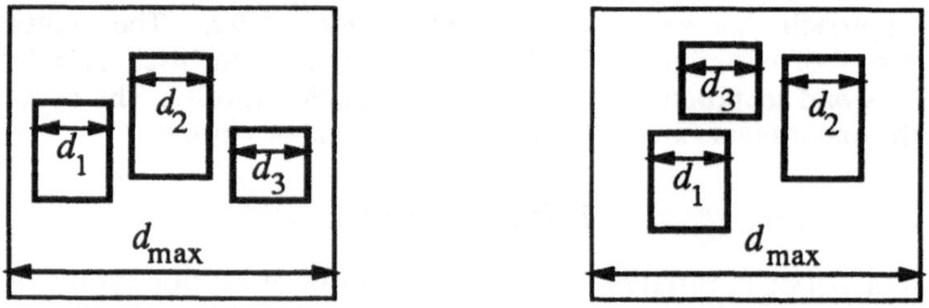

$d_1 + d_2 + d_3 \leq d_{max}$

$d_1 + d_2 \leq d_{max}$
$d_2 + d_3 \leq d_{max}$

Figure 1. Examples of dependent constraints

In addition to the dependent constraints, layouts must satisfy constraints or criteria that are independent of a particular structure. Examples are constraints on the dimensions, area or orientation of an object and required or desired relations between objects (such as adjacency, proximity or physical access).

The general layout problem solved by the two systems can be summarized as follows: Given a set of objects to be allocated and a set of constraints on the shape and placement of these objects, find one or more layouts that satisfy the constraints. The objects to be allocated are called *design units* in the following. Table 1 specifies a very simple layout problem, which will serve as an illustration in succeeding sections. It calls for the design of an efficiency apartment within an area that is accessed from the east and receives natural light from the west [the example is taken from (Flemming, 1979)].

Table 1. Layout Problem 1

Spaces:	Living/sleeping area	Min. dimension	3.60 m
		Min. area	22.00 m^2
	Kitchenette	Min. dimension	1.80 m
		Min. area	4.20 m^2
	Vestibule or hall	Min. dimension	1.20 m
		Max. dimension	6.00 m
	Bathroom	Min. dimension	1.80 m
	Max. extent of overall area from west to east:		7.00 m

Required adjacencies (min. length of shared boundary in brackets)

Living area/vestibule (.90 m)	Living area/western border (3.60 m)
Living area/kitchenette (1.20 m)	Living area/southern border (3.60 m)
Vestibule/eastern border (1.20 m)	Vestibule/bathroom (.70 m)

In solving problems of this kind, both LOOS and WRIGHT perform state-space-search. The principal differences between the two systems stem from contrasting ways in which they represent and handle discrete decisions about structural variables, which result in contrasting ways of setting up and traversing the state space. The following sections briefly describe the two systems. Space limitations prevent us from giving more elaborate descriptions; readers interested in more details are referred to (Flemming, 1988, Flemming et al., 1989) for LOOS and (Baykan, 1991, Baykan & Fox, 1991) for WRIGHT.

3. LOOS

3.1. DESIGN VARIABLES

If given a set of design units, LOOS attempts to find feasible layouts of rectangles in which each rectangle represents one of the design units and no two rectangles overlap. It ultimately tries to determine for each rectangle r values x_r, X_r, y_r, Y_r, which can be interpreted as the coordinates of its corner points or as defining the four lines that lie on the boundary of r (see Figure 2).

The variables handled directly by LOOS are the spatial relations *above, below, to the left* and *to the right*, which are defined as follows: If q and r are two rectangles, then

q is *above* r <=> $y_q \geq Y_r$ $\qquad$ r is *below* q <=> q is *above* r

q is *to the right of* r <=> $x_q \geq X_r$ $\qquad$ r is *to the left of* q <=> q is *to the right of* r

Clearly, q and r *do not overlap* iff at least one of the spatial relations holds between them. But since each of the relations is non-reflexive, non-symmetric and transitive, they cannot be selected independently of each other. We call a

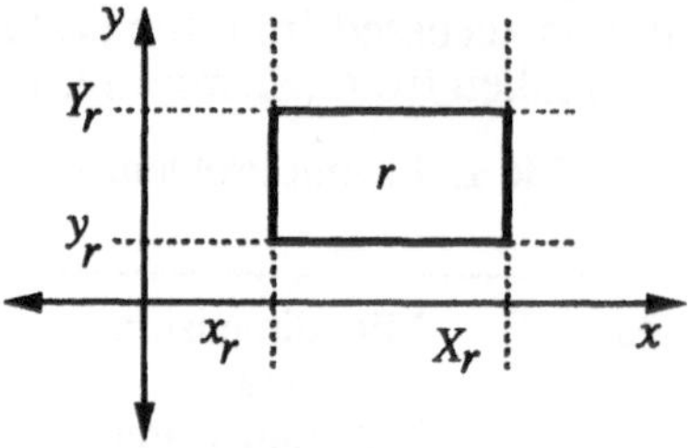

Figure 2. Basic design variables for a rectangle

set of relations that can be simultaneously realized and guarantee non-overlap for each pair of rectangles in a layout a *spatial structure.* The representation of spatial structures used by LOOS is derived from the *wall representation* of rectangular dissections (Flemming, 1978). A rectangular dissection is a layout of rectangles that completely fill the area of a larger rectangle without overlap and holes; an example is shown in Figure 3a. A *wall* in such a configuration is a maximal sequence of connected, collinear line segments separating the rectangles from each other (Figure 3a highlights one such wall). A *wall representation* of a rectangular dissection records all of its walls and the sequence of rectangles bordering each wall from above and below or from the left and right.

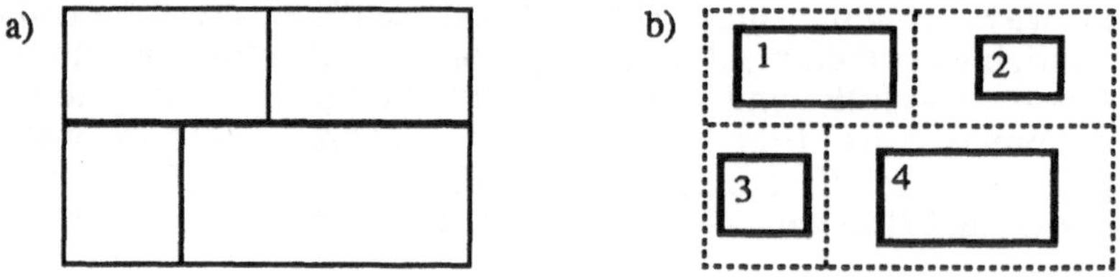

Figure 3. (a) a rectangular dissection and its walls; (b) a loosely packed layout with the same spatial structure

Each wall implies left/right or above/below relations between pairs of rectangles on opposing sides of the wall. Figure 3b demonstrates that the spatial relations implied by the walls of a rectangular dissection can be found also in loosely packed arrangements; for example, rectangles 1 and 2 are above both rectangles 3 and 4. Wall representations, or their equivalents, can consequently also be used to represent the spatial structure of loosely packed layouts if these layouts can be derived from a rectangular dissection by shrinking some rectangles or, conversely, if the layout can be turned into a rectangular dissection by expanding all rectangles until they touch other rectangles on all four sides. Since walls lose their significance in a loosely packed arrangement, we call the gaps separating rectangles from each other *channels*, following the terminology introduced by VLSI designers when they adapted the wall representation to their purposes (Supowit and Slutz, 1984). We indicate channels in a layout by dashed lines as shown in Figure 3b, which also makes the underlying spatial structure immediately recognizable.

The spatial structure of a loosely packed layout cannot be represented directly by a wall representation when the layout contains *non-trivial holes*, which are holes that cannot be eliminated by extending the rectangles in an arrangement until they touch other rectangles (Flemming, 1989); an example is shown in Figure 4a. LOOS circumvents this difficulty by representing non-trivial holes explicitly as rectangles that are marked by a special label to distinguish them from regular rectangles (see Figure 4b). The resulting *marked structures* are able to represent any spatial structure (Flemming, 1989) and form the basis for LOOS, which represents marked structures internally as directed graphs.

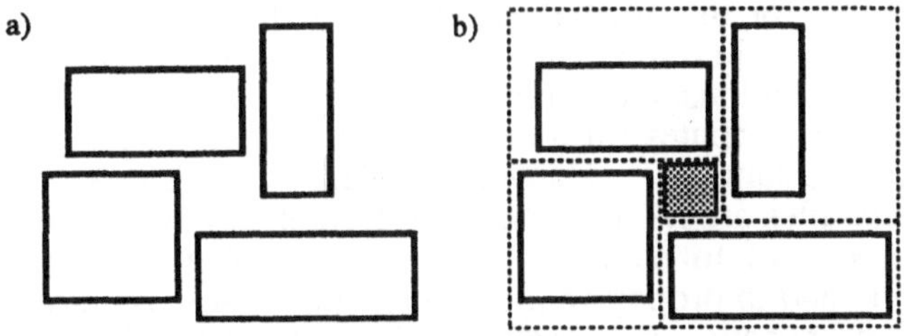

Figure 4. (a) A layout containing a non-trivial hole;
(b) representation of a non-trivial hole by a special rectangle (shown hatched)

In addition to the spatial relations holding between the rectangles it contains, the LOOS representation records explicitly lower and upper bounds for the coordinates of each rectangle; we call the area defined by these values the (dimensional) *range* of the rectangle. Figure 5 depicts the range of one rectangle in a spatial structure with the rectangle drawn in the center of its range. LOOS uses *slacks* to indicate how much a rectangle can move in the *x*- or *y*-direction within its range. The range and slacks are called the *dimensional attributes* of the rectangle. We store these attributes for each rectangle in a marked structure and call the resulting representation a *configuration*. Clearly, a configuration containing some positive slacks represents not a single layout, but a class of layouts because some rectangles can have several, if not infinitely many positions or dimensions, and every combination of these variations defines a different layout.

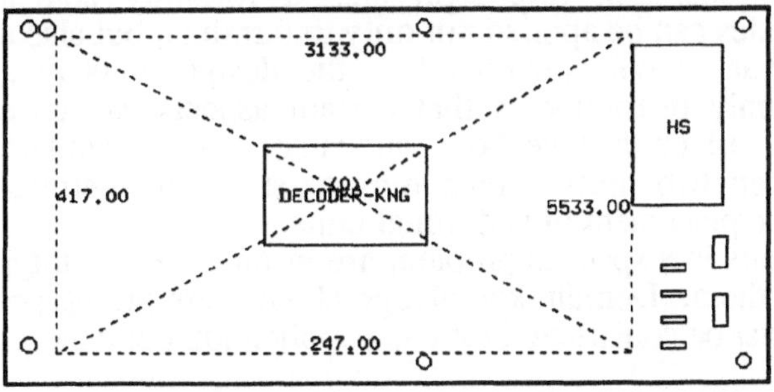

Figure 5. The dimensional range of a rectangle in a configuration

3.2. GENERATION AND PROPAGATION RULES

LOOS constructs and evaluates configurations by using operators or rules that work on configurations. These rules are described in this and the following section.

Generation rules generate marked structures from marked structures by adding one rectangle at a time, starting with a suitable initial structure which may represent nothing more than the enclosing rectangle or a configuration of preplaced objects. Alternative structures are generated if more than one possibility for adding individual rectangles is pursued. This incremental construction allows for intermediate evaluations that can be used for pruning. Figure 6 shows how alternative layouts can be incrementally constructed by successive rule applications, including the insertion of non-trivial holes.

The dimensional attributes for a newly inserted rectangle depend on those of the surrounding rectangles. Conversely, the dimensional bounds of the surrounding rectangles are likely to become tighter through the insertion. LOOS uses *propagation rules* to compute the dimensional attributes for the newly inserted rectangle and to propagate the resulting changes recursively through the structure. The propagation rules can handle rectangles with fixed or variable dimensions. They are also able to take different orientations for a rectangle into account (for example, in terms of its front and back).

3.3. TEST RULES

LOOS is able to evaluate a configuration by application of *test rules*. Each of these rules checks if a favorable or unfavorable condition exists and takes an appropriate action. For example, if a test rule discovers that a particular constraint is violated, it writes an appropriate entry into an *evaluation record* of the configuration; examples of such failing tests are shown in Figure 6.

Test rules can also estimate how well a configuration performs with respect to true criteria (that is, performance aspects that are measured on some sort of scale and differ from constraints which are either satisfied or not). An example is the minimum size of the overall area, for which LOOS is able to estimate lower bounds at any state.

Any aspect that can be evaluated based on the information contained in a configuration can be incorporated into a test rule, and since any configuration produced by a generation rule represents a formally complete layout of rectangles, test rules can be applied not only to terminal, but also to intermediate configurations that do not contain all of the design units in a given layout problem. The only restriction is that certain aspects can be evaluated *with certainty* only if all units have been allocated. An example is a forbidden adjacency between two units, which may exist in an intermediate state, but disappear with the placement of additional units.

The design units in a specific problem are instances of prototypes and inherit constraints from them. Domain knowledge about these prototypes is stored in a hierarchy that must be constructed for each application domain.

3.4. OVERALL ARCHITECTURE AND CONTROL

The original intent behind LOOS was to create a complement to human designers in the form of a system able to *systematically* enumerate solutions to layout problems characterized by diverse and possibly conflicting criteria or constraints. A specific goal was the generation of alternatives with interesting trade-offs in terms of these criteria. The architecture of LOOS reflects this goal and implements a hierarchical generate-and-test (HGT) approach (Stefik et al., 1983). HGT uses intermediate evaluations to guide the search for solutions into promising directions and to avoid the inefficiencies associated with blind generate-and-test. LOOS has strong similarities with and was inspired by DENDRAL, a system able to find chemical structures that are likely to produce a given mass spectrogram (Buchanan et al., 1969).

LOOS comprises five major components:

1. A *preprocessor* that accepts a problem description from the user and performs some initial computations. If some units are preplaced, for example, the preprocessor must construct a starting configuration that describes the spatial relations between these units.
2. A *generator* that accepts any configuration and is able to find all possible ways of adding a new rectangle. It applies the generation and propagation rules under a (rather complicated) control strategy which assures that the spatial relations between already allocated rectangles remain unchanged. This *monotonicity of the spatial relations during generation* makes it possible to evaluate and consequently prune intermediate states of the search space with certainty because constraints that are structure-dependent and not satisfied by an intermediate state cannot be satisfied by a configuration generated from it (exceptions are the constraints mentioned in the previous section that depend on complete configurations).
3. A *tester* that applies the test rules sequentially to evaluate any intermediate or terminal state according to domain-specific constraints or criteria. It is built and works very much like a diagnostic expert system.
4. A *controller* that mediates between generator and tester. After each expansion, it passes the new states to the tester for evaluation; inspects the test results; and terminates the search or selects a new state for expansion and passes it to the generator. In making its decisions, the controller follows a straight-forward branch-and-bound strategy; that is, it selects those and only those states for expansion that have currently the best record. The constraints and criteria are classified as strong, intermediate or weak. The controller counts the number of constraints violated in each class and selects states with the lowest counts, where the counts are ranked lexicographically over the constraint classes.
5. A *postprocessor* that finetunes the solutions thus produced. For example, it may inspect the ranges of rectangles with positive slacks in each terminal state and determine final values for the dimensional coordinates; that is, it selects a specific instance from among the layouts represented by the state. This selection can be the result of some form of optimization, for example, minimization of the total area occupied by the rectangles.

LOOS is able to eliminate certain states *before* they are generated. Even the earliest versions of the generator computed the dimensional range for the new rectangle for each possible expansion and executed only those that could accommodate the new object; that is, all layouts generated could at least be physically realized. Another pregeneration test that we added in the early stages of the system checks whether the insertion under consideration would interrupt a *hard arc*, an arc in the graph representing the spatial structure of the current configuration that has been declared unbreakable; that is, the direct spatial relation represented by that arc between two rectangles cannot be interrupted by putting another rectangle inbetween. Hard arcs can be established by the tester when checking for satisfaction of adjacency constraints and can be effective in preventing the generation of flawed configurations.

More recently, we have added the capability for additional pregeneration tests whose execution remains optional. These tests are restricted at the present time to desired topological properties such as adjacencies including those with the exterior. LOOS is able to determine if these adjacencies are possible *after* a particular application of a generation rule *before* it is applied. We call this mode *constrained generation*. It does not eliminate the need for tests after generation because the pregeneration tests currently performed by LOOS are not comprehensive; in particular, they do not consider constraints that govern the placement of previously allocated rectangles (except for those indicated by hard arcs).

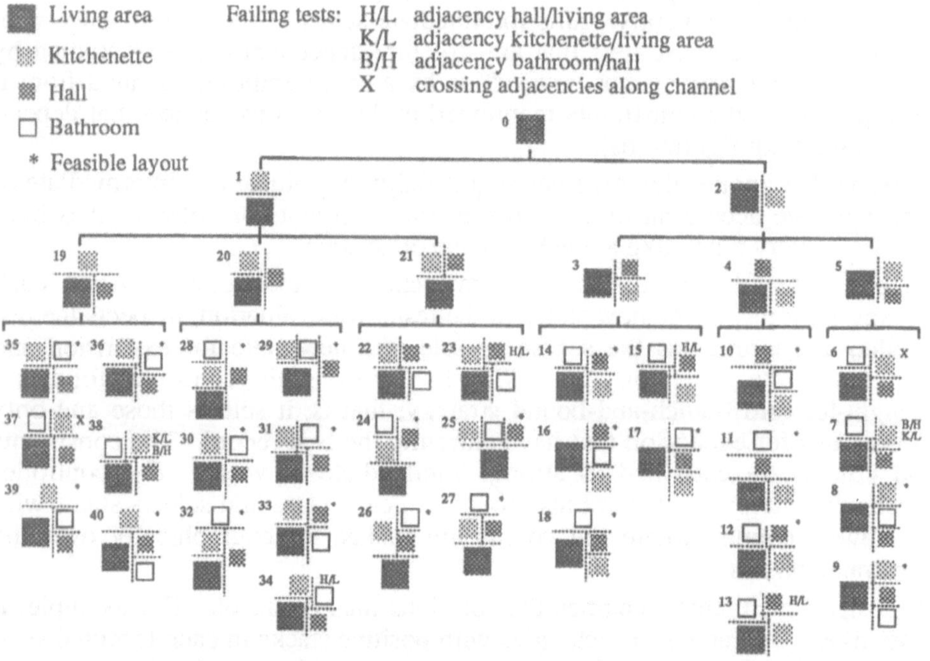

Figure 6. State space for Problem 1 as generated by LOOS

Figure 6 gives a complete trace of how LOOS solves Problem 1 as specified in Table 1 with hard arcs set and pretesting enabled for all required adjacencies. LOOS finds 24 feasible solutions and needs 40 states to find them. The states are numbered in the figure in the order in which they are generated. The figure also lists the constraints violated by a state. Since aside from dimensional constraints, required adjacencies are the only constraints specified for the problem, very few infeasible states are generated: pretesting prevents adjacencies that are satisfied in a state from being interrupted by insertion of a new object (hard arcs) and guarantees that all the adjacencies required for the new object are satisfied after insertion. Exceptions occur when a non-trivial hole is inserted, in which case the current generator does not pretest for hard arcs, but may interrupt adjacencies required for objects already placed (e.g configuration 7). The only other constraint violation occurs when required adjacencies cross each other along a channel and thus cannot be simultaneously satisfied (configurations 6 and 37); the rules that test for required adjacencies always check for this condition, but only *after* generation.

4. WRIGHT

4.1. DESIGN VARIABLES

WRIGHT represents a layout using algebraic equations and inequalities in variables that represent the border lines, dimensions, areas and orientations of the design units. A design unit r is defined by north, south, east and west *lines*. The north and south lines are horizontal, and their values are the coordinates Y_r and y_r; the east and west lines are vertical, and their values are the coordinates X_r and x_r (see Figure 2). A *dimension* is the distance between any two parallel lines or the area of a design unit. The variables in Problem 1 are listed in Table 2 for further reference. The domains of the variables are closed intervals, defined by a minimum and a maximum value.

Table 2. Design units and variables in Problem 1 as defined by WRIGHT

Design unit	*north-ln*	*south-ln*	*west-ln*	*east-ln*	*xdim*	*ydim*	*area*
Apartment	*apN*	*apS*	*apW*	*apE*			
Living room	*lrN*	*lrS*	*lrW*	*lrE*	*lrX*	*lrY*	*lrA*
Vestibule	*vbN*	*vbS*	*vbW*	*vbE*	*vbX*	*vbY*	
Kitchen	*ktN*	*ktS*	*ktW*	*ktE*	*ktX*	*ktY*	*ktA*
Bathroom	*btN*	*btS*	*btW*	*btE*	*btX*	*btY*	

4.2. ATOMIC CONSTRAINTS

The algebraic equations and inequalities that define a layout are called *atomic constraints*. For example, the absolute location of a line is expressed as a binary constraint between the line and a constant. If *apW* is a vertical line, and $100 \leq apW \leq 150$, the location of *apW* defined by this constraint is the grey area in Figure 7. Topology and alignment are expressed by >, ≥ and = relations between two lines. The coordinate system used has the x-axis pointing to the right and the y-axis pointing down.

Figure 8 shows the atomic constraints defining the initial state of Problem 1. These constraints define the relationships between variables belonging to the same design unit. Any change in the bounds of a variable is propagated to the others via the constraints linking them; these relationships are thus maintained in all configurations.

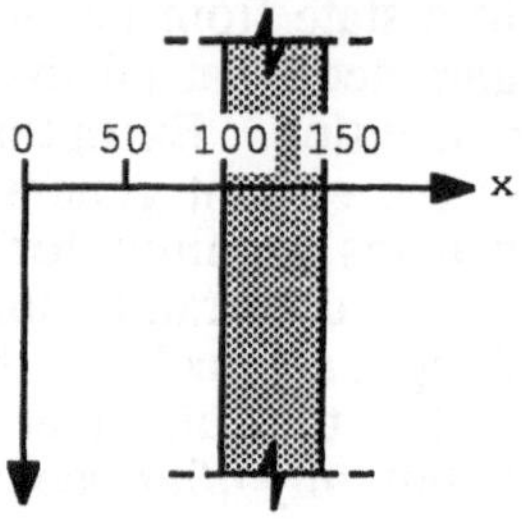

Figure 7. Location of *apW* defined by $apW \in [100, 150]$

$lrN + lrY = lrS$	$lrW + lrX = lrE$
$vbN + vbY = vbS$	$vbW + vbX = vbE$
$ktN + ktY = ktS$	$ktW + ktX = ktE$
$btN + btY = btS$	$btW + btX = btE$
$lrX \times lrY = lrA$	$ktX \times ktY = ktA$

Figure 8. Atomic constraints defining the initial state of Problem 1

$vbW >= ktE$	$ktS > vbN$	$vbE = apE$
$btS = vbN$	$ktS = lrN$	$lrE = vbW$
$lrW = apW$	$lrS = apS$	$btN = apN$
$btE = apE$	$btW = ktE$	$ktN = apN$
$ktW = apW$	$vbS = apS$	

Figure 9. Atomic constraints defining a solution to Problem 1

WRIGHT constructs solutions by asserting atomic constraints. For example, the configuration in state 25 in Figure 12 is defined by adding the constraints given in Figure 9 to the initial state. After adding some constraints, propagation updates variable domains and checks consistency of the constraint set.

A finite set of variables $V = \{v_1, v_2, ..., v_m\}$, each with an associated domain of values, and a set of atomic constraints in these variables $A = \{c_1, c_2, ..., c_n\}$ define a *constraint satisfaction problem* [CSP]. In WRIGHT, every configuration and search state is a CSP. The CSP is consistent if there exist values for all variables that simultaneously satisfy all constraints. During propagation, lower bounds can only increase and upper bounds can only decrease; that is, propagation behaves monotonically. An inconsistency is detected if the upper bound for some variable becomes less than its lower bound. The propagation algorithm used by WRIGHT while running the experiments discussed in this paper is *path-consistency* (Mackworth, 1977).

4.3. DISJUNCTIVE CONSTRAINTS

Design in general and intelligent CAD require a fundamental problem-solving methodology that is able to incorporate arbitrary amounts of knowledge in a principled manner. WRIGHT uses *disjunctive constraint satisfaction* to this end. It provides a formal method for representing expertise uniformly and declaratively in terms of disjunctive and conjunctive combinations of atomic constraints and selects efficient search strategies based on topological and other features of the constraints.

A *disjunctive constraint* is a Boolean combination of atomic constraints. The canonical form of a disjunctive constraint is defined to be its *disjunctive normal*: the top level elements, called *disjuncts*, are connected by an *or* (∨); the second level elements, which are atomic constraints, by an *and* (∧); and there are at most two levels. Thus, a disjunctive constraint C_i has the form

$$C_i = (d_{i1} \vee d_{i2} \vee \ldots \vee d_{ik(i)}),$$

and each disjunct d_j the form

$$d_j = (c_{j1} \wedge c_{j2} \wedge \ldots \wedge c_{jk(j)}).$$

A disjunctive constraint can consist of a single disjunct, and a disjunct can consist of a single atomic constraint.

Each disjunct defines a partial configuration that satisfies a disjunctive constraint in a significantly different way, and the disjuncts, taken together, specify the *structural alternatives considered by* WRIGHT *for satisfying the constraint.* Consider for example two design units that must be adjacent. This can be achieved by placing the first design unit to the north, south, east or west of the second; thus an adjacency requirement can be represented by a disjunctive constraint with four disjuncts. Some requirements of Problem 1 are formulated as disjunctive constraints in Figure 10. *PC-23* formulates the four alternatives of placing the bathroom adjacent to hall. The requirement that the bathroom should be inside the apartment is expressed by *PC-12*, which has a single disjunct.

The requirements which define a problem form a set of disjunctive constraints, $D = \{C_1, C_2, \ldots, C_p\}$, all of which have to be satisfied by a solution. The resulting problem is called a *disjunctive CSP* (DCSP). In WRIGHT, there are no built-in constraints to ensure that design units are non-overlapping or that they are inside the configuration area. All requirements must be explicitly specified. Thus WRIGHT can solve problems containing design units at different *levels of aggregation* and generate tightly and loosely packed configurations by changing the problem requirements.

Figure 11 shows the disjunctive constraints defining Problem 1. Constraints *PC-19—PC-24* express adjacency requirements and are termed *performance constraints*. *PC-9—PC-18* specify that all interior spaces must be inside the apartment and interior spaces should not overlap; they are termed *realizability constraints*. Constraints *OR-1—OR-16* eliminate trivial holes by enforcing that every design unit is adjacent to either another design unit or to the boundary of the envelope on all sides. These latter are called *style* constraints in WRIGHT. Style constraints may also specify which design units can be adjacent to the boundaries of the envelope or occupy corners.

Variables, atomic constraints and disjunctive constraints form a *constraint graph*, which is an and/or network created by the *constraint compiler* at the outset of search. The inputs to the constraint compiler are a taxonomy of

PC-23 bathroom next-to vestibule ≥ 70
$(((vbN + [70,\infty] = btS) \wedge (btN + [70,\infty] = vbS) \wedge (btW = vbE)) \vee$
$((vbN + [70,\infty] = btS) \wedge (btN + [70,\infty] = vbS) \wedge (btE = vbW)) \vee$
$((vbW + [70,\infty] = btE) \wedge (btW + [70,\infty] = vbE) \wedge (btS = vbN)) \vee$
$((vbW + [70,\infty] = btE) \wedge (btW + [70,\infty] = vbE) \wedge (btN = vbS)))$

PC-15 vestibule non-overlap bathroom
$((vbN \geq btS) \vee (btN \geq vbS) \vee ((vbW \geq btE) \wedge (vbS > btN) \wedge (btS > vbN)) \vee$
$((btW \geq vbE) \wedge (vbS > btN) \wedge (btS > vbN)))$

PC-12 bathroom inside apartment
$((btN \geq apN) \wedge (btW \geq apW) \wedge (apE \geq btE) \wedge (apS \geq btS))$

OR-1 bathroom north-adj (livingroom $\vee$ *vestibule* $\vee$ *kitchen* $\vee$ *N)*
$(((btN = lrS) \wedge (btE > lrW) \wedge (lrE > btW)) \vee$
$((btN = vbS) \wedge (btE > vbW) \wedge (vbE > btW)) \vee$
$((btN = ktS) \wedge (btE > ktW) \wedge (ktE > btW)) \vee (btN = apN))$

Figure 10. Expressing spatial relations between design units as disjunctive constraints

prototype design units (also used in LOOS); the templates defining the spatial relations used in constraints; general knowledge about the design domain in the form of desired spatial relations between the prototype design units; and the design unit instances and variables in a problem. For example, given the domain constraint that rooms should not overlap and the design units in Problem 1, the constraint compiler creates the constraints *PC-13—PC-18*, and by using the templates defining the spatial relation non-overlap, it creates their atomic constraints. The design unit taxonomy, the templates defining spatial relations and the domain constraints are represented explicitly and declaratively, are extensible and can be modified by the user through a graphical interface. Users thus can have direct control over the behavior of WRIGHT and apply it to solve layout problems in different domains.

4.4. SOLUTION METHOD AND SEARCH CONTROL

WRIGHT solves the DCSP created by the constraint compiler by sequentially instantiating disjunctive constraints using backtracking search. A disjunctive constraint is instantiated by selecting one of its disjuncts. The atomic constraints in the disjuncts selected in a search path define the configuration. During search, *forward-checking* (Haralick & Elliott, 1980) removes disjuncts that are incompatible with already instantiated disjunctive constraints from further consideration and identifies disjuncts that are satisfied due to transitivity or constraint propagation. The *singleton-disjunct* heuristic instantiates any disjunctive constraint that has only one disjunct left in its domain by immediately asserting the disjunct. The *DCSP* is solved when all disjunctive constraints are instantiated.

Figure 12 shows the search tree that WRIGHT generates as it solves Problem 1. The numbers attached to states indicate the order of generation. The

PC-9 vestibule inside apartment
PC-10 livingroom inside apartment
PC-11 kitchen inside apartment
PC-12 bathroom inside apartment
PC-13 vestibule non-overlap livingroom
PC-14 vestibule non-overlap kitchen
PC-15 vestibule non-overlap bathroom
PC-16 livingroom non-overlap kitchen
PC-17 livingroom non-overlap bathroom
PC-18 kitchen non-overlap bathroom
PC-19 livingroom completely-next-to S
PC-20 livingroom completely-next-to W
PC-21 livingroom next-to vestibule $\geq$ *90*
PC-22 kitchen next-to livingroom $\geq$ *120*
PC-23 bathroom next-to vestibule $\geq$ *70*
PC-24 vestibule next-to E

OR-1 bathroom north-adj (livingroom ∨ *vestibule* ∨ *kitchen* ∨ *N)*
OR-2 bathroom south-adj (livingroom ∨ *vestibule* ∨ *kitchen* ∨ *S)*
OR-3 bathroom east-adj (livingroom ∨ *vestibule* ∨ *kitchen* ∨ *E)*
OR-4 bathroom west-adj (livingroom ∨ *vestibule* ∨ *kitchen* ∨ *W)*
OR-5 kitchen north-adj (vestibule ∨ *livingroom* ∨ *bathroom* ∨ *N)*
OR-6 kitchen south-adj (vestibule ∨ *livingroom* ∨ *bathroom* ∨ *S)*
OR-7 kitchen east-adj (vestibule ∨ *livingroom* ∨ *bathroom* ∨ *E)*
OR-8 kitchen west-adj (vestibule ∨ *livingroom* ∨ *bathroom* ∨ *W)*
OR-9 livingroom north-adj (vestibule ∨ *kitchen* ∨ *bathroom* ∨ *N)*
OR-10 livingroom south-adj (vestibule ∨ *kitchen* ∨ *bathroom* ∨ *S)*
OR-11 livingroom east-adj (vestibule ∨ *kitchen* ∨ *bathroom* ∨ *E)*
OR-12 livingroom west-adj (vestibule ∨ *kitchen* ∨ *bathroom* ∨ *W)*
OR-13 vestibule north-adj (livingroom ∨ *kitchen* ∨ *bathroom* ∨ *N)*
OR-14 vestibule south-adj (livingroom ∨ *kitchen* ∨ *bathroom* ∨ *S)*
OR-15 vestibule east-adj (livingroom ∨ *kitchen* ∨ *bathroom* ∨ *E)*
OR-16 vestibule west-adj (livingroom ∨ *kitchen* ∨ *bathroom* ∨ *W)*

Figure 11. Disjunctive constraints defining Problem 1

constraint identified beneath intermediate states is the disjunctive constraint that is instantiated in that state. States 1,3,5 and 8 have two lists beside them. The top list shows the disjunctive constraints which forward-checking identifies as satisfied. The bottom list shows the disjunctive constraints that are instantiated by the singleton-disjunct heuristic or identified by forward checking after applying the singleton-disjunct heuristic. Solution states have their configuration(s) shown under them. WRIGHT generates 39 states and finds 22 solutions without generating any dead-ends.

The search space expanded by backtracking is the Cartesian product of the domains of all disjunctive constraints and increases exponentially with the number of disjunctive constraints. However, adding constraints reduces the number of solutions and possibly also the number of search states that must be examined because additional constraints increase the probability of inconsistent combinations. A more realistic measure of the effort required to solve a problem is problem "difficulty" according to (Purdom, 1983): *Hard* problems have an exponential number of solutions, and it takes exponential time to solve them by backtracking search; *difficult* problems have an exponentially small number of solutions, but backtracking still takes exponential time; *easy* problems have an exponentially small number of solutions, and there are known procedures for solving them in polynomial time. Problem difficulty is reduced as the ratio of

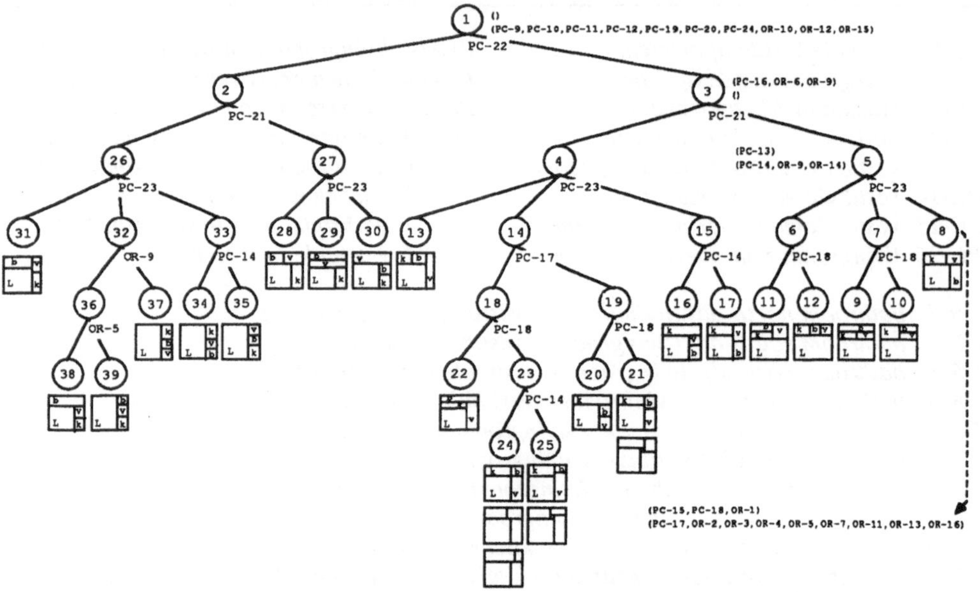

Figure 12. State space for Problem 1 as generated by WRIGHT

constraints to variables increases. Purdom showed that for a subset of difficult problems, backtracking takes polynomial time with dynamic instantiation and exponential time with fixed instantiation. He also conjectured that dynamic instantiation may save exponential time throughout the difficult region even though the resulting times may still be exponential. As they are initially given, layout problems usually do not contain enough constraints to restrict the solutions to an exponentially small set. Baykan (1991) conjectured that by modifying constraints, the designer changes hard problems into difficult ones. These are the problems on which the dynamic control strategy of WRIGHT leads to the greatest reduction in search effort.

WRIGHT selects the disjunctive constraint to instantiate dynamically in each state, using a function of *textures*, which are measures of topological and other features of the constraint graph (Fox et al., 1989). We have defined three texture measures: *Looseness-1* implements a fail-first strategy by selecting a disjunctive constraint that has the fewest active disjuncts; thus the highest probability of failing and terminating the current branch of the search tree using minimum effort. *Looseness-2* calculates the reduction in the domains of interval variables due to satisfying a disjunctive constraint; it is a formulation of fail-first and prune-early strategies that takes into account the sizes of the design units, their current locations and the type of spatial relation in a uniform way. *Interaction* is a measure of the interaction between disjunctive constraints due to shared design units; it favors a constraint that interacts strongly with others, which reduces dead-ends and thrashing (Baykan, 1991).

Textures are applied lexicographically in WRIGHT. Each texture assigns ratings to all future disjunctive constraints and eliminates those with lower values. Dynamic selection using textures reduced search by between 80–90% in difficult problems and by 34–67% in easy problems over random instantiation orders (Baykan, 1991). WRIGHT demonstrates that analysis of problem structure in a domain independent fashion can lead to very good problem solving performance.

5. Comparison

Both LOOS and WRIGHT are portable across workstations running Unix, CommonLISP and X11. LOOS is written in *LISP* and CLOS (Common Lisp Object System). WRIGHT is written in LISP with a few of the critical procedures written in C. In our experiments, both systems were running on single-user DEC 5000/200 machines - LOOS on a machine with 48 megabytes of memory, WRIGHT on a machine with 64 megabytes of memory. But the CPU times given in the following tables should be seen as only broad indicators. Both systems were implemented in the context of research projects with the goal of demonstrating the underlying approach, and computational efficiency received little initial attention beyond attempts to follow established rules of "good programming". WRIGHT has since then undergone determined efforts to improve its computational performance, which caused improvements by several orders of magnitude. LOOS on the other hand keeps expanding in its capabilities without much attention being paid to this aspect; in particular, the abstraction and decomposition capabilities described below may add overhead even when they are not used in solving the problem (as is the case in all examples shown in this paper).

Table 3 gives some basic statistics on the computational efficiency of both LOOS and WRIGHT in solving Problem 1. For LOOS, the results are given for running the system with pretesting enabled and disabled. WRIGHT was run with or without style constraints.

Table 3. Computational performance of LOOS and WRIGHT in solving Problem 1

		No. of states	No. of solutions	CPU time (sec)
LOOS	pretesting enabled	40	24	5.1
	pretesting disabled	51	24	5.9
WRIGHT	with style constraints	39	22	0.2
	without style constraints	41	23	0.1

Given that both systems are based on formalizations that guarantee properties such as completeness of search, it is not surprising that they generate very similar solution sets when solving the same problem. This becomes especially obvious when one realizes that the representation underlying LOOS makes coarser distinctions in terms of spatial relations than WRIGHT; for example, the LOOS solution 30 represents the WRIGHT solutions 21 and 25 (without non-trivial holes) because LOOS suppresses the distinctions made by WRIGHT between these two possibilities; LOOS generates the variants with non-trivial holes, however, as distinct solutions.

The only real difference between the two solution sets is that LOOS generates solution 28, which has no equivalent in the set produced by WRIGHT, and this points to a deeper difference between the two systems. When WRIGHT is run with the no-trivial-hole constraint in Problem 1, it pushes in this particular case every space towards the external boundary and produces layouts that are as densely packed as possible; it also activates for the hall the maximum dimension constraint, which cannot be satisfied for solution 28 under these circumstances. LOOS cannot take maximum dimensions (or maximum areas) into consideration when it evaluates intermediate states because these constraints have consequences only for densely packed arrangements, which LOOS can only generate through its postprocessor, for example, an optimizer that attempts to eliminate trivial holes for a configuration of spatial relations generated by its generator. WRIGHT, on the other hand, can deal with the entire set of units at any level in the search. This difference points to a real distinction between the constructive approach taken by LOOS and the reductionist approach underlying WRIGHT.

However, upper bounds like the ones under consideration here often reflect more general concerns of efficient space planning. LOOS can take these into account during generation; for example, it can estimate the minimum area needed to accommodate placed units for any state and incorporate this as a true criterion into its branch-and-bound strategy because the minimum area can only increase through placement of additional objects; this will be demonstrated in the next example.

As a second problem, we chose a modified version of the problem described in (Flemming, 1978), which calls for the layout of a 3-bedroom apartment in an L-shaped area that borders an open court from the north and east and is accessed from a corridor on its eastern side; the western and southern sides of the available area are blocked from receiving natural light. Table 4 gives a general formulation of the problem that would assure minimum standards of comfort. The court is treated as a unit with variable dimensions and adjacency requirements that assure its position in the south-west corner of the overall area. The general requirement for natural light and ventilation for the major spaces is expressed through adjacencies with the northern border or the court. The adjacencies that force the living room to be adjacent to the entrance side and the master bedroom to the NW corner do not reflect standards of comfort, but general design heuristics that place these larger spaces immediately in the most appropriate zones within the given area.

This general problem formulation is severely underconstrained, and both LOOS and WRIGHT produce an unmanageably large number of feasible solutions. At the time of this writing, we are still experimenting with different ways of handling this problem for both systems. The question is how to introduce general guidelines of good or efficient layout design into the systems.

An obvious way for LOOS is to estimate the minimum overall area for any state as indicated above and take these estimates into account during branch-and-bound. In the current implementation, these estimates come into play after all constraints on the currently placed units have been considered; that is, the controller expands those states that violate not more constraints than any other state and have the lowest area estimates. This left the controller itself completely unchanged. The results produced by this approach are encouraging: LOOS generates 26 feasible "best" solutions in reasonable time (see Table 5), where the solutions are allowed to exceed the minimum area by a preset factor (this allows the generation of solutions that exceed the optimal area by a small

Table 4. Layout problem 2

Context

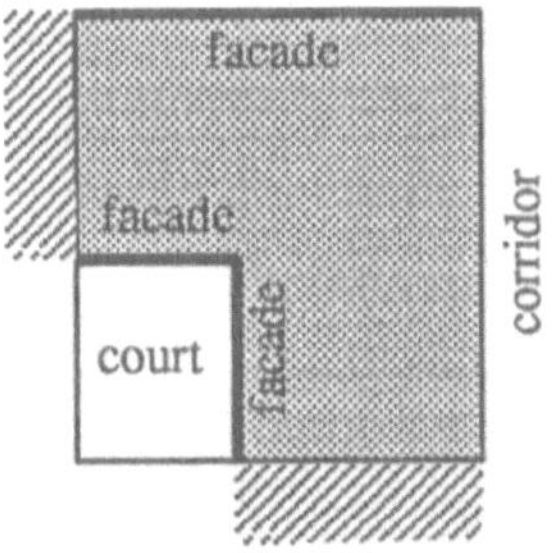

Spaces

Court	Min. dimension	3.60 m		
Living room	Min. dimension	3.60 m	Min. area	22.00 m^2
Master bedroom	Min. dimension	3.30 m	Max. dimension	5.40 m
	Min. area	14.00 m^2		
Bedroom 1	Min. dimension	2.40 m	Max. dimension	4.20 m
	Min. area	7.20 m^2	Max. area	10.00 m^2
Bedroom 2	Min. dimension	2.40 m	Max. dimension	4.20 m
	Min. area	7.20 m^2	Max. area	10.00 m^2
Hall	Min. dimension	1.20 m	Max. dimension	6.00 m
Kitchen	Min. dimension	2.10 m	Max. dimension	5.40 m
	Min. area	7.20 m^2		
Bathroom	Min. dimension	1.80 m	Max. dimension	4.20 m
	Min. area	4.20 m^2		
Max. extent of overall area from north to south:			18.00 m	
Max. extent of overall area from west to east:			12.00 m	

Required adjacencies (min. length of shared boundary in brackets)

Court/southern border (3.60 m)
Court/western border (3.60 m)
Living room/eastern border (3.60m)
Living room/northern border or court (3.60 m)
Living room/hall (.90 m)
Living room/kitchen (.90 m)
Bedroom/hall (.90 m)
Bedroom/northern border or court (1.20 m)
Master bedr./northern border (3.30 m)
Master bedr./western border (3.30 m)
Kitchen/northern border or court (.90 m)

fraction, but may have other advantages).

One approach that can be used in WRIGHT is to declare style constraints that eliminate trivial holes and prevent the hall from being placed towards the outside. In the current problem, this reduces the number of feasible solutions from 3134 to 97 as seen in Table 5 and markedly improves solution quality. But WRIGHT could also adopt the optimizing approach taken by LOOS, which should lead to a reduction in search over the current satisficing formulation, but we have not made this extension. We experimented with the satisficing approach by placing a limit on the maximum area of the apartment. If a reasonable bound is not known at the outset, the system can be run repeatedly with increasing

Table 5. Computational performance of LOOS and WRIGHT in solving Problem 2

		No. of states	No. of solutions	CPU time (sec)
LOOS with area minimization	pretesting enabled	517	26	130.7
	pretesting disabled	831	26	188.8
WRIGHT	underconstrained problem	6930	3134	30.5
	with style constraints	630	97	14.1
	area ≤ 86 m^2	321	7	12.3

values until a good limit has been found. In the present example, an initial limit of 85m^2 generates the solution seen in the top left corner in Figure 13, and a second run with the limit raised to 86m^2 generates the 7 feasible solutions in the same figure.

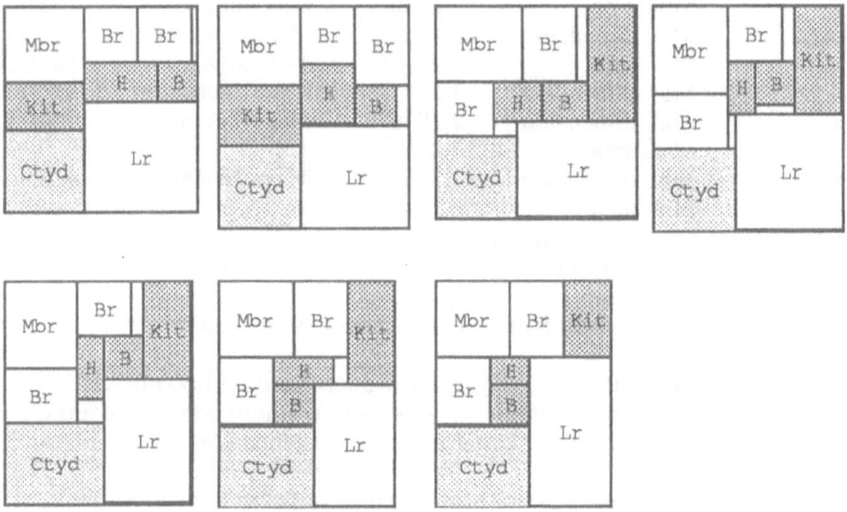

Figure 13. Minimum area solutions generated by WRIGHT for Problem 2

WRIGHT is again significantly faster that LOOS. But it is interesting to note that LOOS's performance improves relatively to WRIGHT, and this indicates a promising direction for further study. At the present time, we can only speculate that the explosion in the number of constraints that have to be explicitly considered by WRIGHT becomes more significant for larger problems. For example, the number of constraints that assure non-overlap in WRIGHT increases with the square of *n*, the number of design units. In LOOS, this constraint is automatically satisfied by the generator, and the number of tests to be performed increases roughly linearly with *n*. Thus, LOOS and WRIGHT may become more similar in computational efficiency as *n* increases, and this would certainly be a counterintuitive result.

WRIGHT is able to compute more accurate minimum values for the overall area due to achieving path-consistency. LOOS on the other hand executes its tests independent of each other and does take into account only a limited set of interactions (like the crossing of adjacencies along a channel mentioned above). As a result, two units may lay claim to the same area without realizing that they cannot occupy it simultaneously, and the first and very rough area estimates implemented for LOOS tend to underestimate the minimum area significantly; this explains why LOOS finds more solutions than WRIGHT. But this is a limitation not so much of the approach itself, but of the current implementation of the tester. Even so, the solutions sets produced by the two systems favor the same overall allocation of objects. This leads us to believe that the current very rough area estimates made by LOOS can be improved without more dramatic changes to the tester and indicates a fruitful direction for further study.

Some important differences between the two systems are not revealed by the two problems described so far. One is that there exist constraints in layout synthesis whose formulation would be difficult in WRIGHT. An example is the requirement that two units be physically accessible from each other, that is, that there be a path between the two units with at least minimal clearance at each point. The difficulty is that an arbitrary number of additional units may be involved in maintaining the path, and this number changes with the layouts themselves. LOOS can handle this constraint without difficulty through a test rule that tries to find this path in a configuration to be evaluated (this is easy because the channel representation indicates all possible paths; LOOS can even find the path with minimal distance along these channels). WRIGHT would have to come up with a disjunct that specifies *all possible paths* between the two units in terms of the various sets of additional units involved. A more promising solution may be to add a test after the DCSP has been solved; that is, WRIGHT may have to include some form of postgeneration test similar to LOOS.

WRIGHT deals with an overconstrained problem by relaxing constraints. It helps greatly if the user has specified which constraints can be relaxed, and in which order to relax the constraints. If these are not specified, the default is to relax style constraints first and performance constraints later. As relaxation possibilities increase, performance gets worse. The generate-and-test approach used by LOOS, on the other hand, does not need such devices because its evaluations can be carried out on feasible or infeasible solutions. A problem occurs, however, when pretesting becomes too strict and prevents the generation of any solution. This is another fruitful area for further study.

In general, the distinction between the generator and tester underlying LOOS enables the system to treat a broad set of constraints and criteria uniformly, while WRIGHT may have to make special provisions for certain classes of constraints. A limitation of the constructivist approach underlying LOOS is that certain aspects have to wait for evaluation until the relevant units have been placed; and the current implementation of the tester does not handle interactions between constraints in a consistent manner. WRIGHT, on the other hand, considers interactions automatically and does not have to delay the satisfaction of constraints for those constraint classes that fit well into its approach.

Another difference between the constructive approach of LOOS and the reductionist approach of WRIGHT stems from differences in degrees of interaction they allow. Every state generated by LOOS represents a formally complete layout that can be understood as such by an observer, who can thus follow the generation process step by step and interrupt at any state. Such an interactive editing capability is currently being developed for LOOS. Furthermore, transitions between modes of generation are easy at any state: a designer may take over and complete the layout through interactive editing, invoke an optimizer to test the potential of an intermediate solution or shift to an iterative improvement strategy. All of this can be done based on the same representation and may use the same rules. For example, any backwards application of a generation rule removes an object and produces again a formally complete layout which can be displayed as such. Removed objects can be reinserted by reapplication of the generation rules, and the tester is a general purpose tool that can evaluate any configuration independently of the way in which it was generated.

In WRIGHT, all design units are placed inside the design envelope at the outset, as it were, and their bounds overlap until their relative locations are determined by the incremental satisfaction of constraints. Thus, displaying an intermediate state poses some problems. On the other hand, an explicit representation of constraints and bounding boxes opens the possibility of constraint propagation in real time, which would change the whole configuration in response to user's actions. Graphical interaction with WRIGHT is one of the research topics being pursued.

Both LOOS and WRIGHT will eventually run into problems as the number of design units increases. Much of the recent work on LOOS has been devoted to this issue and resulted in an expanded version, ABLOOS (Abstraction-based LOOS) (Coyne, 1991, Coyne and Flemming, 1990), which indicates a direction for dealing with this problem that can, in principle, also be used by WRIGHT.

ABLOOS was conceived both as a hierarchical extension of the LOOS approach and as an extensible design framework that is evolving to incorporate a variety of design strategies and methods for producing alternative layouts. It provides designers with an interactive planning capability that allows a layout task to be hierarchically decomposed into subtasks. Each subtask represents a layout problem at a specific level of abstraction (scale or granularity); the subtasks can then be solved and recomposed to achieve an overall solution.

To achieve this, ABLOOS extends the representation underlying LOOS recursively so that a configuration can be decomposed into rectangular components and subcomponents; that is, a rectangle may represent a single design unit or a configuration of rectangles which may in turn represent configurations. This makes it possible to treat a layout as a hierarchy of components and to model decompositions typically found in artifact design; for example, a building is subdivided into floors, a floor into departments, a department into rooms and a room into clusters of furniture or equipment. Such a decomposition defines at the same time a division of the layout problem into tasks and subtasks; that is, it can be used to partition both the *process* and *product* of design. The construct used to decompose uniformly both the layout process and the components to be placed is called a *goal-object* (GOB) in ABLOOS. Each GOB specifies a complete layout task in terms of subgoals represented in turn as GOBs. A problem is decomposed by a designer by specifying a hierarchy of GOBs.

6. Conclusions

Our conclusions are still tentative at the time of this writing because our experiments with the two systems under review have not been completed yet. But we hope that even the short comparisons provided so far indicate how useful this type of close inspection in the context of realistic design problems can be.

Another note of caution has to be added with respect to the generalizations suggested by our findings. Positive aspects surely indicate circumstances in which the overall approach taken by a system (i.e. hierarchical generate-and-test vs. constraint-directed search) is working. But negative aspects have to be interpreted with greater care: they may not indicate shortcomings of the overall approach, but of either the design strategy implemented under the approach (constructive vs. reductionist) or, at an even lower level of abstraction, the particular form in which the design strategy has been implemented (representation, operators, control). We try in the following to keep these levels of distinction in mind when summarizing our findings.

At first sight, our data confirm what is generally known about the two approaches; CDS is computationally more efficient, but less general than HGT in the type of constraints and criteria it can incorporate. More interesting are indications that each approach can overcome some of its limitations by incorporating features of the other approach. CDS can add in principle postgeneration tests to account for constraints that are hard to formulate in the generality required at the outset. HGT can incorporate mechanisms that preclude many infeasible states from being generated.

If pushed hard enough, the two systems (and the approaches they represent) may become roughly equivalent in terms of efficiency when solving equivalent problems. That is, further work may blur the basic differences between the systems. But the actual development of the two systems has been moving in opposite directions. The developers of WRIGHT have been concentrating on identifying heuristic measures of features of the constraint graph that enable efficient search strategies, while the developers of LOOS have been working towards breaking its monolithic problem-solving strategy apart and making the mechanisms used individually available for implementing a broader range of design strategies and modes from a common toolkit. ABLOOS is a first version of a general framework that would allow for this.

These diverging directions may indicate deeper differences between the two approaches than are brought out by a comparison of run-time statistics, however instructive they may be: WRIGHT appears as a precision tool that executes the tasks for which it was designed very well, while LOOS appears as a collection of powerful mechanisms that can be combined to implement various contrasting design strategies, including and especially interactive ones, and allow for easy transitions between strategies.

Acknowledgments

Work on LOOS and ABLOOS has been supported by EDRC, the Engineering Design Research Center at Carnegie Mellon, an NSF-supported Engineering Research Center.

References

Baykan C.A. : 1991, *Formulating spatial layout as a disjunctive constraint satisfaction problem*. Doctoral dissertation, Dept. of Architecture, Carnegie Mellon University, Pittsburgh, PA.

Baykan C.A. and Fox M.S. : 1991, Constraint satisfaction techniques for spatial planning. In P.J.W.ten Hagen, P.J.Veerkamp (Ed.), *Intelligent CAD Systems III Practical Experience and Evaluation*. Berlin: Springer-Verlag.

Baykan C.A. and Fox M.S. : forthcoming, WRIGHT: A constraint-based spatial layout system. In C. Tong and D. Sriram (Eds.), *Artificial Intelligence in Engineering Design*. Academic Press.

Buchanan, B., Sutherland, Georgia and Feigenbaum, E.A. : 1969, HEURISTIC DENDRAL: a program for generating explanatory hypotheses in organic chemistry. In Meltzer, B. and Michie, D. (Ed.), *Machine Intelligence 4*. Edinburgh: Edinburgh University Press.

Coyne, R.F. : 1991, *ABLOOS: An evolving hierarchical design framework*. Doctoral dissertation, Dept. of Architecture, Carnegie Mellon University, Pittsburgh, PA.

Coyne, R.F. and Flemming, U. : 1990, Planning in design synthesis - Abstraction-based LOOS. In J. Gero (Ed.), *Artificial Intelligence in Engineering V. Vol 1: Design (Proceedings of the Fifth International Conference, Boston, MA)*. New York: Springer (Computational Mechanics Publications).

Eastman, Charles M. : 1973, Automated space planning. *Artificial Intelligence*, *4*, 41-64.

Flemming, U. : 1978, Wall representations of rectangular dissections and their use in automated space allocation. *Environment and Planning B*, *5*, 215-232.

Flemming, U. : 1979, Representing an infinite set of solutions through a finite set of principal options. A. Seidel and S. Danford (Eds.), *Proc. 10th Conf. of the Environmental Design Research Association*. Buffalo, NY.

Flemming, U., Coyne, R., Glavin, T. and Rychener, M. : 1988, A generative expert system for the design of building layouts - version 2. In J. Gero (Ed.), *Artificial Intelligence in Engineering: Design (Proceedings of the Third International Conference, Palo Alto, CA)*. New York: Elsevier (Computational Mechanics Publications).

Flemming, U. : 1989, More on the representation and generation of loosely packed arrangements of rectangles . *Environment and Planning B. Planning and Design*, *16*, 327-359.

Flemming, U., Coyne, R. F., Glavin, T., Hung Hsi, Rychener, M. D. : 1989, A generative expert system for the design of building layouts (final report). Report EDRC 48-15-89, Engineering Design Research Center, Carnegie-Mellon University.

Fox M.S., Sadeh N., and Baykan C. : 1989, Constrained heuristic search. *Proceedings of IJCAI-11*. , IJCAI.

Haralick R.M. and Elliott G.L. : 1980, Increasing tree search efficiency for constraint satisfaction problems. *AI*, *14*, 263-313.

Mackworth A.K. : 1977, Consistency in networks of relations. *AI*, *8*, 99-118.

Purdom P.W. : 1983, Search rearrangement backtracking and polynomial average time. *AI*, *21*, 117-133.

Stefik, M. et al. : 1983, Basic concepts for building expert systems. In Hayes-Roth, F. et al. (Eds.), *Building Expert Systems*. Reading, MA: Addison-Wesley.

Supowit, K.J. and Slutz, E.A. : 1984, Placement algorithms for custom VLSI. *Computer Aided Design*, *16*, 46-52.

OPPORTUNISTIC AND GOAL-ORIENTED BEHAVIOUR IN SOFTWARE DESIGN

Combining empirical and theoretical studies in cognitive modelling

S. P. DAVIES
Department of Psychology
University of Nottingham
University Park
Nottingham NG7 2RD UK

and

F. SIMPLICIO-FILHO
Department of Computing
Imperial College of Science, Technology and Medicine
180 Queen's Gate
London SW7 2BZ UK

Abstract: In this paper we discuss two converging strands of research into the nature of design problem solving. Our concerns surround the integration of empirical and theoretically motivated modes of research into the software design activity. We present a model of software design which makes a number of important predictions about the nature of this activity and we discuss independently derived empirical evidence which supports the general characterisation of design that is proposed by this model. We believe that this approach is novel in the sense that there have been few previous attempts to integrate model-based and empirical standpoints within this domain. Our endeavour in this regard suggests that while empirical and model-based accounts may be usefully developed independently, a two pronged approach can provide important insights for both modes of research. Our general argument is that goal-oriented and opportunistic behaviour in design can co-occur and that opportunistic behaviour appears to constitute an appropriate means of dealing with the memory load that is implied by goal-driven models. We contrast this with other work which indicates that opportunistic behaviour is caused by cognitive breakdowns and we suggest that it constitutes the typical mode of cognitive functioning in design and other complex domains. However, such behaviour does not rule out a goal-driven approach and one of our main aims is to show how these different modes of cognitive processing can coexist.

1 Introduction

We begin this paper by outlining a number of important psychological issues relating to descriptions of general modes of cognitive functioning. In this context we discuss evidence for goal-oriented and opportunistic accounts of cognitive control and more specific empirical data stemming from studies in software design and other complex domains. We then describe a model of design behaviour which suggests that opportunistic activities constitute the typical mode of cognitive functioning.We then move on to discuss a study of a software design task which provides broad support for many of the predictions made by the model. Finally, we

J. S. Gero (ed.), Artificial Intelligence in Design '92, 839–860.

look in more detail at the links between the model and the empirical data and we suggest ways in which our dual approach may contribute to a greater understanding of cognitive functioning in the context of design activities and other complex tasks.

2 Evidence for goal-oriented and opportunistic accounts of problem solving and design

One early insight to emerge from cognitive psychology was that problem solving and planning could be considered to be top-down, goal driven and hierarchically structured processes (Miller et al, 1960). From this perspective, problem solving and planning are characterised by the process of successive problem/plan decomposition and refinement. Planning and problem solving are regarded as top-down, focused processes that start from high level goals which are successively refined into achievable actions. More recently an alternative view of the planning/problem solving process has emerged. This view suggests that planning and problem solving are opportunistically mediated, heterachical processes (Hayes-Roth and Hayes-Roth, 1979). Here, in contrast to top-down models, planning is seen as process where interim decisions in the planning space can lead to subsequent decisions at either higher or lower levels of abstraction in the plan hierarchy. At each point during the planning process the planner's current decisions and observations may suggest various opportunities for plan development. For instance, a decision about how to conduct an initially planned activity may highlight constraints on latter activities, causing the planner to refocus attention on that part of the plan. In a similar way, low-level refinements to an abstract plan may suggest the need to replace or modify that plan.

This dichotomy between top-down and opportunistic processing is also evident in a number of empirical studies of software design. Jeffries et al (1981) found that both novice and expert designers decomposed their designs in a top-down fashion - moving between progressive levels of detail until a particular part of the solution could be directly implemented in code. One major difference between novice and expert designers was that novices tended to employ a depth-first search of the solution space - expanding only one part of the solution at progressive levels of detail - while experts adopted a breadth-first approach - synchronously developing many sub-goals at the same level of abstraction before moving to a lower level.

In contrast, a number of more recent, studies have highlighted the opportunistic nature of design tasks. For example, Guindon (1990) has found that software designers often deviate from a top-down, stepwise refinement strategy and tend to mingle high and low-level decisions during a design session. Hence, designers may move from a high level of abstraction - for instance, making decisions about control structure (e.g., central vs distributed) - to lower levels of abstraction, perhaps dealing with implementation issues. Guindon notes that the jumps between these different abstraction levels do not occur in a systematic fashion, as one might expect from hierarchically levelled models, but instead can occur at any point during the evolution of a design.

Similarly, Visser (1990; 1991) has observed many deviations from a hierarchical planning strategy during a longitudinal study of a designer working on a machine tool installation. In a different domain, Ullman, Staffer and Dietterich (1986), studying mechanical engineering design, observed expert designers progressing from systematic to opportunistic behaviours as a design evolved.

One important issue that arises is how we might attempt to explain the range of different results found in previous empirical investigations, and from this account for the dichotomy between top-down and opportunistic strategies. Substantial data supports both views and it would be unreasonable to reject one model in favour of the other. However, these two strategies are often seen to be mutually exclusive. Hayes-Roth and Hayes-Roth (1979) suggest that some tasks are more suited to a top-down approach and others to the adoption of an opportunistic strategy. They claim that "the question is no longer which model is correct, but rather, under what circumstances do planners bring alternative problem solving methods to bear?" (pg 308).

A different view is proposed here. A model of program design is advanced which attempts to integrate existing views by characterising program design tasks as broadly goal oriented with local opportunistic episodes. This view is supported by both a theoretical and an empirical approach. On the one side, an engineering model of cognitive behaviour in design has been developed which suggests that while opportunistic behaviour constitutes the most promising way of succeeding when dealing with unpredictable events, typical of our interaction with the world, it does not totally rule out goal-oriented behaviour. A number of autonomous understanding activities can operate in parallel and this allows the cognitive system to keep to the course of action implied by a plan while undertaking other cognitive activities. The model specifies the mechanisms that underlie this behaviour and suggests to what extent, goal-oriented behaviour can be based on autonomous understanding.

On the other side we discuss an empirical study which suggests that behavioral regularities emerge during the program design process. These regularities appear to have clear top-down, hierarchical and goal-driven characteristics. However, at many points in the evolution of a design, designers can be observed engaging in opportunistically directed activity. This can be contrasted with Hayes-Roth and Hayes-Roth's assumption that choice of strategy will be determined primarily by task characteristics. Rather, both strategies can clearly be seen to be evident in the context of a single task and the existence of opportunistic excursions does not rule out the possibility that the program design task can be broadly described as top-down and hierarchical.

3 A Model of Cognitive Behaviour

Looking at the evidence for both goal-oriented and opportunistic accounts of problem solving and design one might hypothesize that either these strategies are task dependent, that is, one mode is engaged depending on the task being undertaken or they can co-occur, that is, they operate in parallel in some way. In the latter case whether one emerges over the other may depend on observation conditions. In this section we argue for the possibility of the hypothesis of co-occurrence. Moreover, we suggest that this hypothesis is consistent with results from cognitive research. In fact, existing models of and established hypotheses about cognitive functioning point to this co-occurrence. However, this characteristic has been hidden by the lack of a proper notation for discussing it. This co-occurrence may be revealed by the development of a model based on a representation language appropriate for describing distributed systems.

The development of a model able to describe cognitive behaviour faces some representational difficulties. Some of these difficulties have become issues in software and system engineering, in particular in the study of distributed systems. One of the difficulties in this characterization relates to the description of the model's reactive character, through which the system continuously reacts to external and internal stimuli. The reactive behaviour can not be adequately described in terms of simple relationships between inputs and outputs, as commonly used to describe other less complex systems; the allowed combination of inputs and outputs may vary in time and depend on internal states of the system. Such behaviour is better described in terms of the set of allowed sequences of events, actions and conditions restricting them. Another difficulty of this characterization relates to describing concurrence and distribution. In concurrent and distributed systems several processes, which operate concurrently and without hierarchical relationships, should communicate with each other and be controlled. This characteristic can not be adequately described in terms of relationships between states of the system because it is very difficult to get a coherent picture of the system's global states. A system with concurrent components may involve a large number of possible global states which result from the Cartesian product of all sequences of states and events of the system's components.

A visual formalism called hygraph has been proposed to describe complex systems (Harel, 1988). A hygraph-based extension of the standard state-transition diagrams, the statechart, has been proposed to describe the behaviour of reactive systems (Harel & Pnueli, 1987). It provides a means of representing complex relationships between components, such as concurrence and depth, in such a way as to produce a concise diagram. It also allows the representation of a broadcast mechanism for communication between concurrent components. The use of statechart may overcome the representational difficulties above.

The model partially introduced here, uses the statechart representation in order to convey some established hypotheses about cognitive system functioning. It is based on existing models, in particular on the *model human processor* (Card & Newell, 1983) so, it can be described as an attempt to provide a statechart based interpretation of a subset of that model. It also includes an interpretation of some results from Detienne and Soloway (1990) regarding an empirically-derived control architecture for program understanding. A detailed account of this can be found in Simplicio (1992). An advantage of this interpretation is that it provides a basis for discussing these hypotheses in more objective terms. It also permits the elaboration of more complex combinations of behaviour. The aim is to characterize cognitive behaviour in a way that aids understanding and at the same time is amenable to more precise analysis.

The model is based on a description of the underlying cognitive components and their possible interaction. The overall description is presented in figure 1. In the diagram, rounded rectangles represent states; arrows represent transition between states or set of states; dashed lines represent orthogonal, parallel, partition of states. Small doted arrows represent initial states. When a state occurs it may cause further transitions in other components. This is represented by the condition *[in (state)]* related to a transition. It reads: if *state* occurs then the related transition also takes place. The recurring rectangles in the working memory illustrate that a number of activations may be stored simultaneously. Based on this description, sequences of states can be inferred. Sequences of states represent behaviour. Some patterns of behaviour described in the model human processor can be discussed in terms of particular sequences of states. For instance, the memory retrieval behaviour depends on the time elapsed between the storing and retrieval of the information. This period will determine which memory, if any, holds the information. For retrievals done shortly after storing, information may be held in both memories. Given a start state [*w-stored*], which means that there is information in the working memory, a subsequent state [*activated*] in long term memory represents the behaviour in which the information is stored simultaneously in both memories. For retrievals done long after storing, information is retrievable only from long-term memory. This is illustrated by the following sequence: [*stored*, *activated*, *w-clear*]. The state *w-clear* represents the fact that the information is no longer available in the working memory.

Based on the description of the cognitive components and their transitions, we can discuss possible explanations for the mechanisms that underlie cognitive behaviour. In this case, the model can be defined as a plausible explanation of this behaviour. The basic argument developed here, and which emerged from the model, is that two different means to accomplish the understanding process can be distinguished. They are termed autonomous and rational understanding. These notions are certainly not new. For example, the study of Ericson & Simon (1980) advocating the validity of verbal reports for deriving cognitive processes pointed out that only information in "focal attention" can be verbalized. The existence of other "unattended" information is recognized. Also, it is recognized that as experience increases, information processing becomes automatic and less available for verbalization. We relate the distinction between information that can and cannot be verbalized to the distinction between rational and autonomous understanding. In the same vein, distinct forms of understanding have been identified by Schank (1986). Despite this the concepts of autonomous and rational understanding are hardly integrated in cognitive research, we believe, due to observational difficulties. In the development of this model, they emerge as complementary processes, and parts of a more general mechanism of understanding.

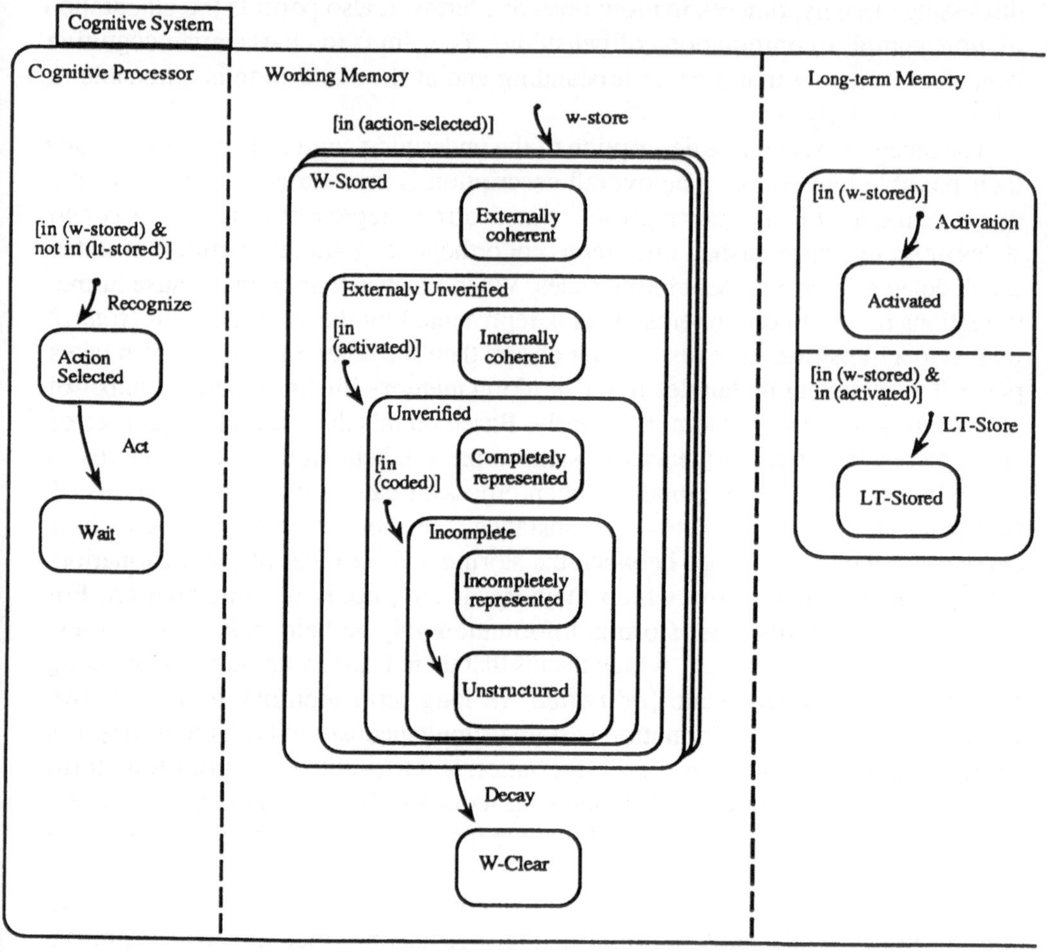

Fig. 1. A State chart representation of cognitive components and their possible interactions.

3.1 DISPERSION AND FOCUSING

Two counteracting patterns of behaviour emerge from figure 1 as the core of cognitive behaviour: dispersion and focusing. On the one hand, during memory retrieval there is no guarantee that activations in long-term memory include the desired information; activations are hypothetical. In figure 1, the relationship between memories is represented in terms of a cyclic reference: the transition activation is initiated by the condition [in (w-stored)]. The transition w-store, on the other hand, is initiated by the condition [in (activated)]. For example, the partial behaviour [w-stored, activated] points to a recursive repetition of itself, leading to cycles that stops when the information stored in the working memory decays and reaches the state w-clear. These cycles produce an effect that we term dispersion tendency. This effect represents the control mechanism involved in the phenomenon of activation spreading described in the model human processor. It suggests that particular information stored in the working memory tends toward dispersion: the previously activated information becomes less accessible and new related information arises continuously. Since there is no guarantee the activation involves the proper information, the cycles increase the chances of an appropriate retrieval. It is important to note that these cycles are autonomous. They will happen independently of the kind of information being activated or its source.

Focusing, on the other hand, is related to the cognitive processor action. Roughly, the sequence [*action-selected*, *wait*] on which the cognitive processor operates, takes place when the state *w-stored* occurs in the working memory. The cyclic reference between the working memory and the cognitive processor, represented by the condition [in (*action-selected*)] in the transition w-store, leads information selected by the cognitive processor to be re-stored in working memory. The model human processor broadly refers to this as a "refresh". A primary effect of this is to focus the system on particular relevant information. A secondary effect is to restrict the dispersion tendency. This restriction happens because dispersion is based on storing new related information in working memory; refreshing old information reduces the chances for dispersion.

3.2 AUTONOMOUS AND RATIONAL UNDERSTANDING

Unlike dispersion, focusing is not an autonomous behaviour. Therefore its action must be driven by mechanisms at a higher abstract level. A second layer of cognitive behaviour may be built based on how one specifies: 1) what the understander is trying to achieve, 2) what cognitive actions may support the attainment of this goal and, 3) what constraints the task faced by the understander imposes on these actions.

First, what the understander is trying to achieve is what we term *understanding target*. The primary understanding target is to build a symbolic representation suitable for storing information in, and retrieving it from long-term memory (Schank, 1986). Input information is said to be understood when a memory structure is found that indicates how and where to store information in long-term memory. The understanding target is achieved when the component *lt-store* in long-term memory is able to store information leading to the state *lt-stored.* For the sake of clarity, we are distinguishing the term *target,* which we relate to the result of understanding as a process, from the term *goal,* which we relate to intentions of the understander. For example, understanding targets may be achieved irrespective of understander goals. In other words, understanding may occur independently of intention.

Second, two different ways to support the attainment of the understanding target may be distinguished: autonomous understanding, which results from immediate match with memory retrievals, and rational understanding which results from the cognitive processor action. During the process of understanding, old structures from long-term memory are called into play. When this retrieval succeeds in matching the information at hand, the understanding target is achieved; the new information "makes sense". However, when a proper memory structure is not immediately found, then understanding requires: finding an applicable old structure, determining to what extent it differs from the current situation and, adapting it to fit the new situation. These actions are carried out by the cognitive processor, to which we relate the term "rational". The cognitive system's ability to achieve the understanding target without involving rational actions is in part due to the dispersion tendency. As a result of dispersion new activations will be produced continuously. If the storing component can find a match between the information at hand and an activation, then it may be stored with the old one. This is represented by the sequence [*w-stored, activated, lt-stored*]. Rational understanding is illustrated by the sequence [*w-stored, action-selected, lt-stored*]. It is relevant to observe that since these two forms of understanding are based on different and concurrent components, one approach does not exclude the other.

Third, in the context of software design, at least two restrictions set out the conditions that ordain what processes are necessary to the attainment of the understanding target: 1) it is restricted to deal with symbolic representations; 2) it involves the notion that information may exhibit different elaboration levels in terms of completeness and consistency.

3.3 ELABORATION LEVELS AND PROMOTIONS

Information being manipulated by the cognitive system may exhibit different levels of completeness and consistency. In ideal terms, at the highest level all relevant information is collected and expressed in a selected representation scheme that is coherent with all other structures available. At the lowest level, information symbolically represented provides simple associations between external stimulus and knowledge eventually activated in memory. In the continuum between these extremes, we distinguish five discrete points that we term elaboration levels. Information may be: unstructured, incompletely-represented, completely-represented, internally-coherent and externally-coherent. High elaboration levels are desirable properties of the information used by the cognitive system. Nevertheless, this is not always required or attainable. In order to provide a proper response to the environment, it may be necessary to elaborate the available information to higher levels. This elaboration we term promotion. Four promotions are conceived. Promotion 1 seeks to lead information from the state *unstructured* to the state *incomplete*. Promotion 2 aims to lead information from *incompletely-represented* state towards *completely-represented*. Upgrading from *completely-represented* state towards *internally-coherent* state is carried out by promotion 3. Finally, from the *internally-coherent* state promotion 4 seeks to lead to the *externally-coherent* state. A detailed description of the promotions is not necessary here. However, it is relevant to observe that understanding may be achieved at any elaboration level and that promotions may involve different patterns of behaviour.

3.4 THE CONTROL MECHANISM OF UNDERSTANDING

The cognitive system components are not equally able to deal with all elaboration levels; some are restricted to handle only lower levels. This is represented in figure 1 by the conditions [in (*state*)] associated with transitions in working memory. The perceptual system activity is restricted to *incomplete* states. This is represented by the condition [in (*coded*)]. Long-term memory retrieval is restricted to *unverified* states by the condition [in (*activated*)]. The cognitive processor activity is related to all states by the condition [in (*action-selected)*]. This indicates that only the cognitive processor is able to deal with all elaboration levels and with promotions 3 and 4. All components may deal with promotion 1, so several patterns of behaviour may lead to this promotion.

Since some promotions may be accomplished by different means, the cognitive system faces a possible choice between them. For example, both sequences [*coded, unstructured, activated, incompletely-represented, lt-stored*] and [*coded, unstructured, action-selected, incompletely-represented, lt-stored*] result in promotion 1. Along these sequences, information is promoted from *unstructured* to *incompletely-represented* states. The difference is that in the first sequence, the promotion is carried out by long-term memory activity. In the second sequence, the promotion is carried out by cognitive processor activity. These sequences are possible because both long-term memory and cognitive processor may deal with those levels of elaboration. However, because the first sequence results from the autonomous behaviour characterized by autonomous understanding it will take place first.

Autonomous understanding comes before rational understanding. However, because they share the same target, there will be no need to pursue the second unless the first fails to accomplish the understanding target. A detailed mechanism to explain how long-term memory informs the cognitive processor that it has failed is not necessary here. The cognitive processor will be aroused to action if long-term memory fails to match information and an activation. This is represented by the condition [not in (lt-stored)] in the cognitive processor in figure 1. Once aroused to action it should recognize what actions are appropriate to deal with the present elaboration level. Understanding may be accomplished by promoting information to higher levels of elaboration. This interpretation is consistent with the control structure proposed by Detienne and Soloway (1990). If the new elaboration level can be still handled by memory activity, then again, autonomous understanding will be attempted first. If autonomous understanding finds its limit of action, in the *unverified* elaboration level, then promotions can be carried out only by rational understanding.

3.5 COGNITIVE BEHAVIOUR IN UNDERSTANDING

To sum up so far, we have examined patterns of behaviour in terms of sequences of states built onto a description of cognitive components. Two layers of behaviour have been discussed. At the basic layer, two counteracting patterns have emerged as relevant: dispersion and focusing. At the second layer, two other patterns, autonomous and rational understanding support the attainment of the understanding target of building a representation able to be stored in and retrieved from long-term memory. The control mechanism of understanding was described by combining these ways of achieving understanding and differences in the nature of the information at hand which may exhibit different elaboration levels for completeness and consistency. This mechanism underlies a third layer in which the cognitive behaviour is discussed.

3.6 GOAL PROCESSING AS UNDERSTANDING-BASED ACTIVITY

Reasoning about goals and ways to achieve them calls for the construction of plans. This notion highlights a functional distinction between control knowledge, characterized in terms of plans, and other uses of knowledge, which can be called problem knowledge. In the context of software design, the distinction between plans and problem knowledge corresponds to distinguishing between 1)reasoning about what to understand, and 2) the actual understanding. At first glance, the importance of consistent plans as a way to succeed in dealing with the world may suggest that the cognitive capability of building and revising plans should be carried out by a special component of the cognitive system, or at least, by a specially adapted mechanism. However, there is no evidence of a component or processor specifically allocated to this function. Goal processing shares the same cognitive resources used for processing other uses of knowledge. The distinction between them is simply functional. Therefore, goal processing is based on the same cognitive mechanisms used for understanding. Plans will have to be understood in the same way that problem knowledge is understood. This view of goal processing amounts to saying that goal-oriented behaviour is based on plan understanding.

Since goal processing and other information processing share the same cognitive resources, they may operate in competition. If goal-oriented behaviour is based on plan understanding then it is supported by autonomous understanding, or when that fails, rational understanding is called into play, so two different situations may be envisaged. First, plans that are frequently used acquire stable representation and effective memory links, so proper activations can increase the chances of autonomous understanding. In addition, usually there is no need to have plans represented in high elaboration levels, so autonomous understanding is enough. Because the process of activation is concurrent with the process of storing in long-term memory, and working memory may hold a number of activations simultaneously, it follows that goal-oriented behaviour based on autonomous understanding may operate in parallel with other processing. In other words, the cognitive system is able to keep the course of actions according to a plan while undertaking other cognitive activities. However, when autonomous understanding fails then goal processing will be based on rational understanding. In this case, the need for goal processing may compete for the use of the cognitive processor with other needs of information processing. This competition is possible because working memory is able to store several activations simultaneously, but the cognitive processor is able to focus on only one at a time.

3.7 SUMMARY

In summary, the above model suggests a view in which cognitive behaviour constitutes an outcome of two complementary means of achieving understanding, that are termed autonomous and rational understanding. Cognitive behaviour can be explained in terms of the mechanisms underlying understanding and a functional distinction between goal processing, in which plans are understood, and processing other uses of knowledge. According to the model, opportunistic strategies does not rule out goal-oriented strategies. Goal processing and other processing can proceed concurrently if they are based on autonomous understanding. This enables the cognitive system, to a certain extent, to keep the course of actions according to plan while undertaking other cognitive activities.

The model presented allows for the analysis of cognitive behaviour in a number of situations, however this is beyond of the scope of this paper. The general argument presented here is that both goal-oriented and opportunistic strategies of problem solving and design can co-occur. The co-occurrence and the mechanisms by which it happens are described in terms of a plausible model of cognitive behaviour. Because the model is a semi-formal representation derived from existing models and established hypotheses about cognitive functioning, we suggest that the conclusion of co-occurrence is also consistent with results from cognitive research. However, due to its theoretical character, this conclusion needs support from empirical results. While we have designed no experiment to support the hypothesis of co-occurrence in terms of the specific predictions of the model, an independent empirical study has shown that this co-occurrence does occur. This study is presented in the next section.

4 An empirical study of program design strategy

The model of software design we have presented above enables one to characterise the design activity as broadly goal-orientated. However, at a lower level of analysis it is predicted that design behaviour will display an opportunistic character. In this section we present an empirical study of the software design activity which provides support for many of the predictions made by the model we have presented here. In this next section we begin by suggesting a number of reasons why previous empirical studies of the software design activity have tended to emphasise an apparently clear dichotomy between goal-driven and opportunistic accounts of the design activity. We then present an empirical study of a software design task in which we have attempted to overcome some of the limitations of previous work. In the final sections of this paper we make an attempt to integrate the findings of this empirical study with the model and we suggest a number of implications of our joint work for general characterisations of the cognitive processes involved in design.

One problem with previous empirical studies of the software design activity is that such studies have failed to observe general statistical regularities in the design process. Moreover, the majority of studies have analysed the verbal protocols generated by a small number of subjects.The validity of drawing implications from these studies for more general characterisations of the design activity may be problematic when one considers the strong impact of individual differences on the adoption of particular planning/problem solving methods (Hayes-Roth, 1979).

A second problem with existing work concerns the identification and definition of levels of abstraction within the solution space. Conflicting views about the nature of the design process may arise because studies have described the same level of abstraction in different ways or different levels of abstraction in the same way.

In the study reported here an attempt has been made to explicitly define a number of different abstraction levels. This is based upon Rist's (1989) analysis of schema creation and focal expansion in program design. Emphasis has been placed upon the derivation of general behavioral regularities from data generated by a reasonably large number of subjects. Data has been collected from a retrospective analysis of code generation rather than from verbal protocols and this means that the imprecision normally involved in classifying salient behavioral aspects of the design activity can be avoided. Collecting data in this way should reduce the inaccuracy stemming from the linearisation effects that are common in subjects' verbalisations about knowledge structures that have a significant temporal and/or spatial dimension (Levelt, 1981).

4.1 THE PROGRAM DESIGN TASK: NONLINEARITIES AND FOCAL EXPANSION

Rist (1989) has proposed a model of program design which traces the evolution of a design through a number of stages. An explicit feature of Rist's model concerns the identification of levels of abstraction in program structure. It is claimed that programs are built from simple knowledge structures that are merged and combined to form more complex structures. At the lowest level of detail , individual fragments of knowledge are combined to form a single line of code. The next stage in the development of a program is to create a programming plan (Rich and Wills, 1990; Soloway and Ehrlich, 1984). Such plans provide 'canned' solutions for common goals such as calculating a running total or reading some data value. Next, these plans need to be merged into the final program structure. Rist is primarily interested in the processes that underlie the plan generation activity and central to his theoretical explanation is the idea of focal expansion.

Focal expansion describes the process of generating a programming plan from a so-called 'focal line'. In terms of Rist's account, each programming plan has an associated focal line that directly encodes the goal of that plan. For instance a 'running total loop plan' will be associated with the focal line 'count:=count+1'. The complete plan will also consist of an initialisation component and some means of reading data values into the plan. The design of a program is seen to progress through various stages beginning with the implementation of a focal line, its extension to form a complete plan and finally to the creation of an entire program through a process of plan merger.

One important issue in the present context relates to the question of whether plans created in a linear order such that the programmer completes one plan before moving onto the next? Or, conversely, are the focal lines of plans instantiated first, providing an abstract skeletal design structure which can later be extended to include other plan elements? A top-down model of design would predict the latter, since focal lines are taken to represent a single level of abstraction within the evolving design structure. Non-focal lines simply extend this plan focus. According to Rist (1989) "The (plan) focus ... marks the start of detailed design in the domain of the program" (p 403).

The study reported here attempts to address the issues outlined above by examining the evolution of high and low-level structures during a design task. Following Rist, it is assumed that focal and non-focal lines represent discrete levels of abstraction within the design hierarchy. One implication of the top-down model of design is that focal lines will be created before non-focal lines. Hence, one level of the design hierarchy will be established before lower-levels in the hierarchy are expanded. In contrast, an opportunistic model would predict that programmers will make many deviations from this hierarchically levelled approach. Hence, there would tend to be a more even distribution of focal and non-focal lines throughout the course of the design task as designers move, at arbitrary points during the task, from one level in the hierarchy to another.

Another way of investigating these issues is to explore more explicitly the nature of the nonlinearities found to exist in code generation.These nonlinearities are key junctures in the coding process where programmers move from one point in a program to insert a new line or to amend an existing structure. Davies (1991) has explored these nonlinearities in the context of plan generation in programming, but an analysis of nonlinearities can also provide more specific evidence for the use of top-down vs opportunistic strategies. Here, interest is directed towards the nonlinearities occurring both within and between hierarchical levels in a design structure. Hence, a number of different categories of nonlinearity are possible; jumps between a focal line and another focal line; jumps between a non-focal line and other non-focal lines (within the program's hierarchical structure). Conversely, jumps may occur between hierarchical structures; from focal line to non-focal line or vice versa. It should be noted that while plan-based accounts of design behaviour make a commitment to a specific form of representation, the model discussed in this paper makes no commitment whatsoever to a particular representational scheme. This is because the model is intended to be generic and we do not wish to be constrained to the adoption of a single representation. However, in the case of this empirical study we have employed a plan-based account to provide a certain measure of formality and to enable us to compare directly the results of this study with previous work.

4.2 PARTICIPANTS, PROCEDURE AND TASKS

Twenty novice and twenty expert subjects took part in this experiment. The expert group consisted of programmers/designers with a number of years industrial experience (4 to 13 years. Mean, 5.6 years) and of teachers of programming and software design, all of whom possessed previous industrial software design experience. The novice group comprised a number of Second year undergraduate students. Members of this group were drawn from the same student cohort and all had been instructed in the basic principles of traditional software design practice and structured programming.Subjects in both groups had experience of the programming language employed for this study - Pascal. All members of the expert group either taught Pascal or used it extensively in their work, while all members of the novice group had attended a first year course in Pascal and had used the language for project work.

Participants were asked to undertake a number of design tasks of varying difficulty. The simplest problem (from Soloway and Ehrlich, 1984) required participants to construct a program that would calculate an average and a running total. More difficult problems were based upon Ratcliffe and Siddiqi's (1985) traffic counting task and a task derived from Rist (1989) which required 2113 weights to be sorted into ascending order. These problems might be considered to be fairly straightforward by professional standards, however the more difficult tasks took between 43 and 78 minutes even for experienced subjects.

Participants were provided with short natural language specification of the problems and were asked to write a program to solve each problem. Participants were told that they could make on-screen notes if desired, but were requested not to use pen and paper. No time limit was imposed on the tasks. Participants were asked to type their programs onto a familiar full-screen editor. All on-screen activities were recorded for further analysis.

The resulting programs were analysed for common plans by three independent raters using a goal hierarchy (Davies, 1991). The plans that were identified varied little between the raters. For each plan, the raters were asked to identify the focal line of that plan based upon the definition provided by Rist (1989). There was high degree of agreement between raters regarding the identification of focal lines. Figure 2 shows a program with plan structures highlighted, illustrating an example of the coding scheme used to classify program statements.

```
[ ].
[I]  Sum:=0;                                  {Running Total Variable Plan}
[I]   Rain:=0;                                {Guard Plan}
[ ].
[ ].
[R] Read (Rain);                              {Guard Plan/Running Total Loop Plan}
[E]  WHILE Rain <> 99999 DO                   {Running Total Loop Plan}
[M]    BEGIN
[E]        IF Rain<0 THEN                     {Guard Plan}
[ ].
[ ].
[M]     ELSE
[M]         BEGIN
[E]            IF Rain=0 THEN                 {Guard Plan}
[F]                Valid := Valid + 1         {Counter Variable Plan}
[M]             ELSE
[M]                BEGIN
[F]                    Valid := Valid + 1     {Counter Variable plan}
[F]                     Rainfall := Rainfall + 1   {Guard Plan}
[M]                 END;
[F]          Sum := Sum + Rain                {Running Total Variable/Loop Plan}
[E]          IF Rain>Max THEN
[E]                  Max := Rain
[M]         END
[E]       Writeln ( 'Please enter next value:' );
[ ].
[ ].
[R]          Read (Rain);                     {Running Total Loop Plan}
[M]     END;
[F] Average := Sum/Valid;                     {Average Plan}
[ ].
```

Fig. 2. Program except illustrating plan structures and statement categorisation. [F] = Focal line; [R] = Read statement; [E] = Extension; [I] = Initialisation and [M] = Miscellaneous.

From the above analysis, and by replying on-screen activity, a retrospective analysis of the temporal distribution of focal and non-focal lines was undertaken. In addition, a number of nonlinearities could be classified. For this purpose a nonlinearity was defined as a jump between one line of code to another. This could be either to edit an existing line or to insert a new line. An analysis was undertaken of the number of nonlinearities occurring between focal lines; between non-focal lines and between focal and non-focal lines and vice versa.

4.3 RESULTS

Figure 3 shows the mean number of focal and non-focal lines (representing different abstraction levels) generated by both novice and expert programmers during the experimental session. The number of lines generated within each block consists of a simple count of the lines produced (focal or non-focal - according to the protocol outlined above) during each 10 minute time period.

The cumulative mean number of focal lines generated by experts and novices over all generation blocks did not differ significantly (t-test) (mean[Novices] = 10.4: mean[Experts] = 9.4). Similarly, no difference was evident between the cumulative mean number of non-focal lines generated by these groups over the same period (t-test (mean[Novices] = 79.1 : mean[Experts] = 72.3).

These data were entered into a pair of two-way ANOVA's - one constructed with the focal line data, and the other with non-focal line data. For the focal line data no significant main effects were evident. However, a significant interaction between generation block and group ($F_{5,228} = 6.73$, $p < 0.001$) was apparent. This interaction appears to reflect the decrement in focal line generation that can be observed in the case of the expert group as the session progressed. The novice group, in contrast, appeared to maintain a fairly constant rate of generation throughout the course of the session. Further support for this finding was provided by instituting multiple pairwise comparisons of means between all adjacent generation blocks using the Newman-Keuls test with a significance level of $p < 0.01$. In the case of the expert group, significant differences were found to exist between blocks t_1 through t_4, with no significant differences between t4 through t6. For the novice group there were no significant differences in an identical range of post hoc comparisons.

Similar findings emerge from the non-focal data. Again, no main effects were apparent. However, there was a significant interaction between generation block and group ($F_{5,228} = 10.32$, $p < 0.001$). Multiple post hoc comparisons of means indicated significant differences between blocks t_2 through t_4 for the expert group, reflecting an increasing rate of non-focal line generation during this period. No significant differences were evident for all other post hoc comparisons.

Figure 4 shows the number of nonlinearities occurring between and within hierarchical level. For this purpose, a between hierarchy jump was classified as a jump between a focal line and a non-focal line or vice versa and a within hierarchy jump as a jump between a focal line and another focal line or between a non-focal line and another non-focal line.

The cumulative mean number of within hierarchy jumps for experts and novices over all generation blocks did not differ significantly (t-test) (mean[Novices] = 12.5: mean[Experts] = 10.2). Similarly, no difference was evident between the cumulative mean number of between hierarchy jumps for these groups over the same period (t-test) (mean[Novices] = 20.9: mean[Experts] = 21.3).

These data were entered into a pair of two-way ANOVA's - one constructed with the within hierarchy data, and the other with between hierarchy data. In the case of the between hierarchy data no main effects were apparent. There was, however, a significant interaction between Group and Generation block ($F_{5,228} = 4.23$, $p < 0.001$), reflecting the decreasing rate of between hierarchy jumps over generation blocks for the expert group. Multiple post hoc comparisons reveal significant differences ($p<0.01$) between means for blocks t_1 through t_4 for the expert group. No other post hoc comparisons proved to be significant.

For the within hierarchy data, there was an interaction between Group and Generation block ($F_{5,228} = 5.89$, $p < 0.001$). Here, multiple post hoc comparisons indicated significant differences between means for blocks t_1 through t_5 for the expert group. No other significant differences were apparent in the case of the novice group in an identical range of comparisons.

4.4 DISCUSSION

These results have a number of implications for the way in which we might attempt to characterise the program design activity. Firstly, the strategy adopted by experienced designers appears to correspond broadly to the adoption of a top-down, hierarchically levelled approach. Hence, the abstract structure of the program, represented by the instantiation of focal lines, is mapped out at an early stage in the evolution of the design. This high level structure provides a framework around which the rest of the program can be built. However, at many points during the evolution of the design, programmers can be seen to engage in opportunistic behaviour - synchronously generating both focal and non-focal structures. Hence, at any particular point during the program design activity, behaviour might legitimately be described as opportunistic. However, this clearly does not rule out the existence of a global top-down design strategy. The findings of this study clearly demonstrate that different forms of strategy may be adopted within the context of a single task, and further, that choice of strategy may not be primarily determined by task characteristics as suggested by existing work (Carroll et al, 1980; Hayes-Roth and Hayes-Roth, 1979).

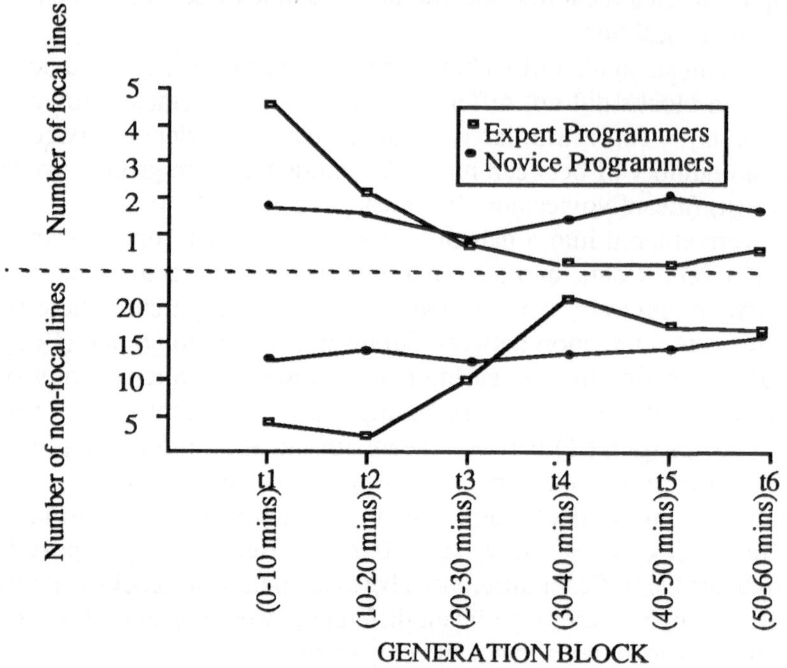

Fig. 3. Mean Number of focal and non-focal lines generated by expert and novice programmers during each generation block

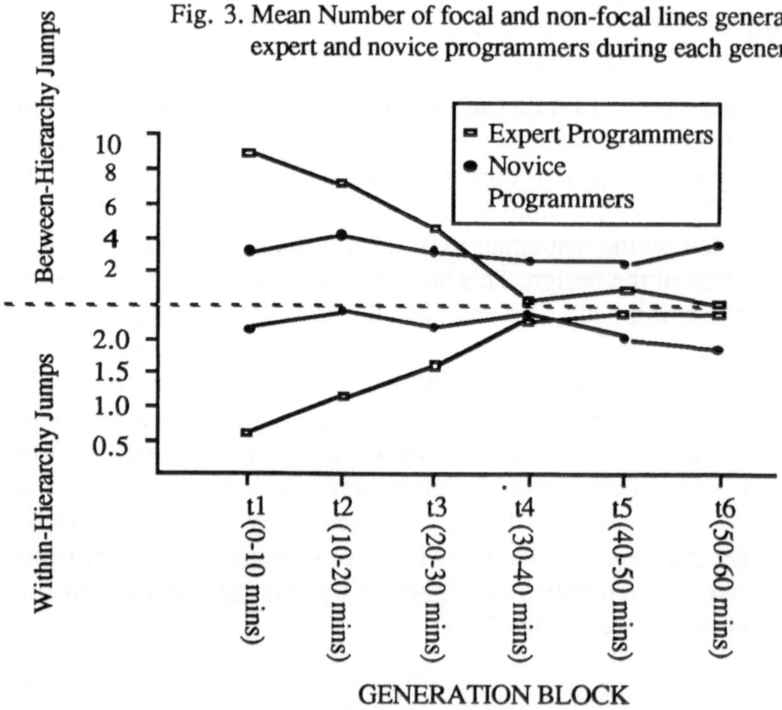

Fig. 4. Mean number of Between and Within-Hierarchy jumps performed by novice and expert programmers during each generation block

The mechanisms which might be thought to underlie the differences that have been found to be associated with different levels of expertise are not clear when one considers previous work. Anderson (1983) suggests that behaviours which might be described as opportunistic deviations from hierarchical problem solving may often arise as a consequence of fairly simple cognitive failures. For instance, subjects may pursue details of a current plan that is inconsistent with higher level goals simply because they have forgotten or have misremembered these goals. Here there is a clear link with Link with Visser's work and with the idea that opportunism can be explained with recourse to plan deviations (Visser, 1990). However, the model of software design presented here suggests that we can explain such behaviour as a natural mode of functioning and that such behaviour does not rule out a characterisation of design as goal driven at higher levels of elaboration.

Thus, as we have suggested, opportunistic behaviour may constitute the more promising way to succeed when dealing with erratic events. Since many world events, including design, are of this nature, opportunistic behaviour is the typical way of reacting, allowing a designer to take advantage of the knowledge that is available at any given time. To behave opportunisticaly corresponds to following the general rule: if something has to be done, it is better to do it at once.

We have suggested that opportunistic and goal orientated behaviours may proceed concurrently if they are both based on autonomous understanding. When autonomous understanding fails to provide the understanding of the plan then goal processing, in terms of rational understanding, is called into play in order to build or revise it. In this case, the need for goal processing may compete with other needs for information processing for the use of the cognitive processor . This competition is possible because working memory is able to store several activations simultaneously, but the cognitive processor is able to focus on only one at a time.

Another issue relating to the link between the model presented here and the empirical study reported above, is the observation that expert designers tend to rely extensively upon the use of external memory sources (VDU screen, notes on paper etc) when developing a design (Davies, 1991). Conversely, novices rely to a much greater extent upon the use of internal memory to develop as much of a solution as possible before transferring it to an external source. Such a strategy may give rise to the opportunistically oriented behaviour evident in the context of the present study. Relying in this way upon internal memory sources means that designers may experience difficulty simultaneously maintaining aspects of an emerging design in memory. In particular, it will prove difficult to map out a global framework at a single level of abstraction - that is, adopt a hierarchically levelled approach. The finding that expert designers rely extensively upon external memory has clear links with the model advanced in this paper since the model predicts that under specified circumstances designers may decide to follow a course of action which involves interrupting their present activity just long enough to take notes in order to guarantee that another aspect of the emerging design may be dealt with later. Moreover, we suggest that taking notes probably requires translating the information into a different representation scheme. Hence, any system intended to support complex design activities should allow for the use of and the translation between a number of different representational schemes.

Another major issue raised by the model presented here is the extent to which behaviour can be described as goal oriented as opposed to being based upon a reactive sequence of behaviours which characterise cognition in terms of the relationship between knowledge in the head and knowledge in the external world. In this regard one major link with the model presented here and existing work in other domains is related to recent work connected with the development of planning models in Artificial Intelligence (Agre and Chapman, 1987; Ambros-Ingerson,1987). In particular, these more recent accounts of the planning process have suggested that actions are often enacted before planning or problem solving sequences are fully articulated. In formulating this alternative account of planning, these models stress the inexorable link between the planning process and the execution of plans.

For example, Ambros-Ingerson introduces the concept of 'knowledge getting acts'. These are invoked during plan execution to obtain information from an external source that may be relevant to the current goal. Models such as this differ significantly from the more classical accounts of planning in that an entire sequence of plans need not be worked out in advance. Rather, the effects of implementing partial plans can be tested against the planner's expectations and information may then be sought from the external world in order to reduce the uncertainty that may be associated with the implementation of particular plans. Once again we can see a number of distinct similarities with the model presented here. However, while these more recent AI planning models have looked at the use of external memory support, we are concerned more with at the other side of this planning process. That is, how and why information might be externalised in the first place. We have suggested that this externalisation process may arise as a direct consequence of the opportunistic strategies that we have both predicted and observed in connection with the software design process.

5 Conclusions

A number of potentially interesting findings emerge from the dual analysis presented here. Firstly, it has been shown that opportunistic episodes may occur at any point during the evolution of a design. However, this does not rule out the existence of an overall top-down strategy. Hence, the clear dichotomy between top-down and opportunistic approaches that is implicated in previous work may be unfounded. It is unlikely that studies involving an analysis of the behaviour of a relatively small number of subjects would make the observation of these regularities possible.

Secondly, it has been shown that the emergence of top-down or opportunistic strategies is not task dependent as suggested by a number of previous studies.

Rather such strategies can co-exist within the context of a single task. However, one form of strategy make take precedence over the other at particular points during the evolution of the design.

The model of software design presented here makes a number of key predictions which are supported by this empirical data, and in particular it suggests that opportunistic and goal-oriented strategies can co-exist. Moreover, it suggests that opportunism may constitute the only economical way of dealing with an erratic world typical of that found in the context of software design. The model proposes a specification of the cognitive mechanisms involved in software design and emphasises the situated nature of design behaviour in terms of the close relationship between the activity of design and the use of external memory sources and information repositories.

Acknowledgements

The authors are listed alphabetically. The second author is supported by Embrapa and the Brazilian National Research Council ' CNPq'

References

Agre, P and Chapman, D.: 1987, Pengi: An implementation of a theory of activity. In Proceedings of the Sixth International Conference on Artificial Intelligence, 268-272. Menlo Park, Calif: American Association for Artificial Intelligence.

Ambros-Ingerson, J. A.: 1987, Relationships between planning and execution. AISB quarterly newsletter, Number 57.

Anderson, J. R.: 1983, The Architecture of Cognition. Harvard University Press, Cambridge, MA.

Card, S. Moran, T. and Newell, A.: 1983, The Psychology of Human-Computer Interaction. Hillsdale, NJ: Lawrence Erlbaum Associates, Inc.

Carroll, J. M., Thomas, J. C., Miller, L. A. and Friedman, H. P.: 1980, Aspects of solution structure in design problem solving. *American Journal of Psychology*, **93**, 2, 269 - 284.

Davies, S. P.; 1991, The role of notation and knowledge representation in the determination of programming strategy: A framework for integrating models of programming behaviour, *Cognitive Science,* **15**, 547- 572.

Detienne, F. and Soloway, E.: 1990, An empirically-derived control structure for the process of program understanding", International Journal of Man-Machine Studies, **33** (3) , 323-342.

Guindon, R.: 1990, Designing the design process: Exploiting opportunistic thoughts. *Human-Computer Interaction*, **5**, 305 - 344.

Harel, D.: 1988, On visual formalisms, *Communications of the ACM,* **31**, 5, 514-530.

Harel, D.: 1987, Statecharts: a visual formalism for complex systems. *Science of Computer Programming*, **8** (3), 231-274.

Harel, D. and Pnueli, A.: 1985, On the development of reactive systems In Logics and Models of concurrent Systems. NATO ASI Series, 13, 477-498, K.R.Apt. (ed), Springer-Verlag.

Hayes-Roth, B.: 1979, Flexibility in executive strategies. N:1170, The Rand Corporation, Santa Monica, California.

Hayes-Roth, B. and Hayes-Roth, F.: 1979, A cognitive model of planning. *Cognitive Science*, **3**, 275-310.

Jeffries, R., Turner, A. A., Polson, P. G. and Atwood, M. E.: 1981, The processes involved in designing software. In J. R. Anderson (Ed.), Cognitive skills and their acquisition. Hillsdale, NJ: Erlbaum, 255 - 283.

Levelt, W. J. M.: 1981, The speaker's linearisation problem. *Philosophical Transactions of the Royal Society* London. 295, 305 - 315.

Lewis, C.: 1990, A research agenda for the nineties in Human-Computer Interaction, *Human-Computer Interaction*, **5**, 125-143.

Miller, G. A., Galanter, E. and Pribram, K. H.: 1960, Plans and the structure of behavior. Holt, Rinehart and Winston, New York.

Newell, A. and Card, S.: 1985, The prospects for psychological science in human-computer interaction. *Human-Computer Interaction*,**1**, 209-242.

Ratcliffe, B. and Siddiqi, J. I. A.,: 1985, An empirical investigation into problem decomposition strategies used in program design. *International Journal of Man-Machine Studies*, **22**, 77 - 90.

Rist, R.: 1989, Schema creation in programming. *Cognitive Science*, **13**, 389 - 414.

Rich, C. and Wills, L. M.: 1990, Recognizing a program's design: A graph-parsing approach. *IEEE Software*, 82 - 89.

Schank, R.: 1986, Explanation Patterns: Understanding mechanically and creatively. Lawrence Erlbaum Associates, Inc.

Simplicio, F.: 1992, Modeling Software Specification Understanding. Research Report 91/28, Imperial College of Science, Technology and Medicine, Department of Computing.

Soloway, E. and Ehrlich, K.: 1984, Empirical studies of programming knowledge. IEEE Transactions on Software Engineering, 10 (5), 595 - 609.

Ullman, D. G., Stauffer, L. A., and Dietterich, T. G.: 1986, Preliminary results of an empirical study on the mechanical design process. Technical report 86-30-9, Oregon State University, Corvallis, Oregon.

Visser, W.: 1990, More or less following a plan during design: Opportunistic deviations in specification. *International Journal of Man-Machine Studies*, **33** (3), 247 - 278.

Visser, W.: 1991, The cognitive psychology viewpoint on design: Examples from empirical studies. In J.Gero (ed.), Artificial Intelligence in Design'91. Butterworth-Heinemann, Oxford.

ON THE RELEVANCE AND TREATMENT OF CATEGORIES IN AI AND DESIGN

S. NEWTON
Department of Architectural and Design Science
The University of Sydney
NSW 2006 Australia

Abstract. The paper establishes a fundamental distinction between two views of categorisation in AI and design. The first view is widely subscribed to in AI and design, and seeks to model categorisation as the model of cognition. The second argues for the primacy of experience in our understanding, where categorisation is considered as an emergent property of cognition. This experiential account leads to the contention that there can be no computable model of cognition. In the light of this second view, the future of artificial intelligence is considered and one possible direction (based on an alternative metaphor to the computational one) is explored.

1. Introduction

"Categorization is not a matter to be taken lightly. There is nothing more basic than categorization to our thought, perception, action, and speech." (Lakoff, 1987: p.5)

Significant confusion has arisen in the context of artificial intelligence (AI) and design over the treatment of categorisation. The source of this confusion can be traced to a rationalist tradition that objectifies the world in an atomistic way, seeking to model categorisation as the model of cognition. The rationalist account is widely subscribed to in AI and design, but does not accord with the way in which categorisation appears to operate for people. An alternative account of cognition (the experiential account) is possible in which categorisation is considered as an emergent property, or quality, and no longer central to cognitive modelling. The distinction between these two views of categorisation (and cognition) clarifies a number of important issues in AI and design.

The aim of this paper is to establish this fundamental distinction, and to do so in order to criticise a cognitive modelling approach based on the computational metaphor. To lend it the broadest brush, the experiential account of cognition (as presented in this paper) argues that the computable model of cognition has been exhausted. The arguments to be considered point towards a future for AI and design grounded in a variety of alternative metaphors for cognition—a return perhaps to fundamentals. An example of one such alternative (the 'communications' metaphor) is considered through a brief description of several research initiatives.

2. The Relevance Of Categories In AI And Design

Linguists and anthropologists commonly assert that the 'reality' we perceive through our various senses is a diffuse continuum (Leach, 1964). The world of a young child contains no distinctive 'things' until such time as the child is taught to discriminate and taught to label those 'things' as separate entities. In this sense,

J. S. Gero (ed.), Artificial Intelligence in Design '92, 861–882.

a *house* is only distinguished from a *teapot* because we have provided separate labels (words) to refer to each.

Colour provides the ideal example of this 'continuum' theory. It is estimated that the human eye physically can discriminate as many as 7.5 million colour variations. Nowhere could you find 7.5 million different labels for these. Rather, the continuum is aggregated into a considerably smaller set of colour terms, and different cultures will compose their own, quite varied sets of terms. Often there are no equivalent colour terms across cultures. In general there are a different number of terms, each relating to a different range of the colour spectrum (Lyons, 1968; Berlin and Kay, 1969). On this basis, the discrete colours we (or our particular culture) recognise are ultimately just a matter of language convention and learned social constructs.

A similar argument can be extended to include everything we recognise and distinguish. Everything is aggregated and categorised in some way. Different cultures inevitably categorise individual instances in different ways. For example, the same physical item of *furniture* may be grouped as a *bench* by one culture, a *bed* by another, and *not a piece of furniture at all* by another. Each culture will carve the world differently, manifest in the particular terms they employ and the usage they make of their language.

The 'continuum' theory is often challenged from a common-sense view-point (Taylor, 1989). Certain categories appear intuitively to occur more regularly and 'naturally', regardless of language. *Trees* are generally distinguished from *elephants*, and a *feather* is generally distinguished from a *waterfall*. Other categories are 'defined' as being distinctive in some way. *Toys* are defined in a different way to *doors*, and a *radio* is not a *cup*. There is a large literature to support both 'natural kinds' and 'nominal kinds' (see Pulman, 1983 for an overview). Both forms of distinction have been challenged in a variety of ways: they tend to be based largely on differences in physical appearance (Rosch *et al*, 1976), they are unclear at the boundaries (Taylor, 1989), and they presume an objectivist metaphysics (Lakoff, 1987). What appears to be an absolute division between two entities is more to do with them being quite removed from each other in a physical sense, in a functional sense, or in the way the objects are used and the activities they permit. The absolute division is considerably less apparent at the boundary between categories: where one 'thing' becomes a similar yet distinctive other 'thing'. And yet it is especially at these edges and boundaries that we focus our attention when we try to make sense of the world (Katz, 1984). The world is understood by breaking it up into packages of entities (categories) that can then be presumed to be discrete. This form of understanding is the 'role' of categorisaton.

But the fact that people do fragment the world, does not then imply that the world 'out there' necessarily is fragmented. Categories merely trace the joints along which we, as a society, elect to carve the world. There is some suggestion that natural categories exist outside the social constructions of language. However, the weight of evidence supports the view that categories are principally a matter of convention (Lakoff, 1987). We create the world as we understand it through categorisation. Categorisation is foundational to human cognition.

"Every time we see something as a *kind* of thing, for example, a tree, we are categorizing. Whenever we reason about *kinds* of things—chairs, nations, illnesses, emotions, any kind of thing at all—we are employing categories. Whenever we intentionally perform any *kind* of action, say something as mundane as writing with a pencil, hammering with a hammer, or ironing clothes, we are using categories... Without the ability to categorize, we could not function at all, either in the physical world or in our social and intellectual lives. An understanding of how we categorize is central to any understanding of how we think and how we function, and therefore central to an understanding of what makes us human." (Lakoff, 1987: p.5)

The "understanding of what makes us human" clearly has application in the context of AI and design, and an understanding of how we categorise will have an equivalent relevance. But it is one thing to recognise categorisation as being central to AI and design, and therefore to require that any proposed model or approach should account for categorisation in some way. It is another thing to recognise categorisation as central to AI and design, and therefore to compose each model or approach around a theory of categorisation. In the former case the treatment of categorisation is a necessary quality of the model. In the latter case categorisation *is* the model.

It is an important distinction to make. Significant confusion has arisen in the context of AI and design over the treatment of categorisation. The source of this confusion can be traced to a rationalist tradition that objectifies the world in an atomistic way, seeking to model categorisation as the model of cognition. The rationalist account is widely subscribed to in AI and design, but does not accord with the way in which categorisation appears to operate for people. The aim of this paper is to emphasise the importance (in terms of the implications and hidden assumptions) of the distinction between categorisation as a quality of cognition and categorisation as a model of cognition.

3. Categorisation As A Model Of Cognition—The Atomistic Treatment

3.1. THE CLASSICAL THEORY OF CATEGORISATION

The so-called classical view of categories is well defined (Smith and Medin, 1981; Lakoff, 1987; Neisser, 1987). There is one fundamental assumption involved, from which a further three basic assumptions follow:

1. Fundamentally, categories are defined in terms of necessary and sufficient features. Where some 'thing' has all of the necessary features it is a member of the category. Without all of the necessary features it is not a member of the category. And hence:
2. All members of a category share the necessary features. The members of a category each have equal status. There are no degrees of membership, and no

members of the category are 'better' members than any others.

3. All features are binary. A feature is either involved in the definition of a category or it is not. An entity either possesses a feature or it does not.
4. Categories have clear boundaries. Once a category is established, entities either are a member of the category or they are not.

This classical theory of categorisation was established during the time of Aristotle, and remained unquestioned until relatively very recently. In fact the pioneering questions were not raised in a recognisable form until the 1970s (most emphatically by Rosch and Lloyd, 1978), although they were possibly anticipated as early as Wittgenstein (1953).

In particular, behavioural experiments during the 1970s indicated that members of a category vary in how good an example (or how typical) they are of their category. A *robin* may be considered very typical in the category *birds*, a *parrot* moderately typical, and a *penguin* atypical. In addition, non-members of a category vary in how good a non-member they are of the category. With respect to *birds*, *cup* is a better non-member than is *butterfly*.

It should be noted that this 'grading' of membership refers only to the behaviour of people in rank-ordering entities as members or otherwise of some category. The concept of graded membership fits well with other observed human behaviours; such as identifying typical entities as members of a category in less time than atypical entities, generating typical entities as members of a category on more occasions than atypical entities, and learning typical members of a category with more ease than atypical members (Barsalou, 1987). Grading also supports the notion introduced previously, of 'reality as a continuum'.

Of course, demonstrating through behavioural experiments that the classical view is flawed in some fundamental way does not within itself provide any alternative structure. There have, however, now been many alternative structures and extensions proposed to the classical theory, in an attempt to explain the various findings of these, and other, behavioural studies.

3.2. Extensions to the Classical Theory of Categorisation

The classical theory of categorisation has not been completely dismissed (Smith and Medin, 1981), but observed human behaviours unquestionably challenge it in a fundamental way. This fundamental challenge calls for a fundamental alternative (Lakoff, 1987). But so entrenched is the classical theory that the necessary alternative is proving to be rather elusive. Instead of fundamental change, we end up with a number of extensions to the classical theory.

3.2.1. Prototypes. Observed human behaviour displays a clear grading of membership, which does not equate with the classical theory of categories. The classical definition of necessary and sufficient features must somehow be eased. The simplest way of doing this is to assert that not all of the defining features of a category are necessary (Rosch, 1978). This allows membership to be graded on the basis of typicality. The more typical, the 'better' the membership of the category. The most typical members are then referred to as prototypes. But this definition of prototypes, as being those members which are somehow 'better', is just a superficial description of the phenomenon. It accounts for observed human behavior but says nothing about the structure of the process, about what actually makes one member more typical than another. Rosch (1978) has observed that "... prototypes themselves do not constitute any particular model of processes, representations, or learning. This point is so often misunderstood that it requires discussion ... to speak of a prototype at all is simply a convenient grammatical fiction; what is really referred to are judgements of degree of prototypicality ... Prototypes do not constitute a theory of representation for categories..."

3.2.2. Exemplars. One way of representing a prototype is through an actual instance of a member. In this sense, categories are not intended to be defined in terms of abstract features, but in terms of actual instances. These defining instances are referred to as exemplars. Membership of the category is then determined by the degree of correspondence between a candidate entity, and the exemplars.

There is considerable evidence from behavioural studies to support the notion of an exemplar (Kahneman and Tversky, 1973; Collins and Loftus, 1975), and this view resolves a number of interesting problems (Smith and Medin, 1981). However, it remains extremely difficult to formulate a structure for exemplars. How, for example, is the degree of correspondence (ie. the similarity) between candidate and exemplar determined?

One method is based on a geometric approach (Shepard, 1964). Members (exemplars) of a category are paired in all possible combinations, including the special case of a member and the category 'type'. (Category type is the name, or label, given to the category containing the members. Thus the members *apple*, *banana* and *grape* may represent the category type, *fruit*.) The task is then to rate the similarity of each pair. These ratings are input to a scaling (clustering) program whose output is a geometric space. Category members are then positioned as points in the space, so that the geometric distance between any two members corresponds as closely as possible to the judged similarity between those members. The exemplars are those members located spatially closest to the category type/name. The geometric distance between a candidate and the category type/name indicates the graded membership of that candidate in that category (hence it is also sometimes referred to as the 'dimensional approach').

Not withstanding the success of this approach in representing perceptual objects (Shepard, 1974), the representation is still descriptive, rather than structural. And the representation works considerably less well (not to say, fails) when dealing in conceptual categories (Tversky, 1977). It would seem inevitable therefore—in the formal, structural sense—that some recall is made to the features of an instance/exemplar.

Features of course are abstractions, and so the original motivation for using exemplars (the intuitive appeal to an actual instance) is lost. The notion of an exemplar may yet provide a useful description of categorisation, but structurally it is encumbered with the same problem as prototypes in general—the problem of having somehow to establish similarity.

3.2.3. Featural. The most structural interpretation of similarity is based on features, and is therefore an abstract representation of any particular member. The features are used to codify the known members, and act as a reference against which a decision about the membership of some candidate entity can be gauged. In its most trivial form, membership is determined on the basis of a candidate having a minimum 'threshold' number of features in common with the category representation.

The most notable departure from the classical approach involved here, is the notion of 'saliency'. The featural approach uses only 'salient' features to represent a category. A salient feature is one that is distinctive of that category. But certain distinctive features occur in only a limited number of cases. For this reason, a probability weight is also included, to indicate those salient features most likely to appear. (The approach is therefore sometimes referred to as the 'probabilistic approach'.) It is the high probability, salient features which best identify members of the category.

Interestingly, the list of features which form the summary representation of a category (the prototype) are not necessarily realisable as an actual instance. There may be no member of the category which matches the prototype on every feature. Thus we have a notion of non-necessary, modal features. A more relaxed view of category membership, but one considerably at odds with the classical approach. Importantly, this relaxed view overcomes many of the problems that embarrassed the classical view.

The major limitation of the approach is that membership is determined on the basis of some critical sum of the weighted features. The best-known application of this process is Tversky's (1977) 'contrast model'. More recently, 'spreading activation' (connectionism) has provided a more dynamic implementation (Collins and Loftus, 1975; Rumelhart and McClelland, 1986). In both cases, membership is determined on the basis of both similarity and dis-similarity. In the contrast model, ratings are summed statically. In spreading activation, weights are applied dynamically to coerce other features into play. Groups of features are formed, and these groups use their combined weights to force incompatible features out of consideration. In this sense, categories 'emerge' over several iterations, and final groupings represent the summary features of an implied category (Coyne *et al*, 1989). Categories are implied by the example entities used as input to the 'training' part of the connectionist approach.

3.2.4. Fuzzy Sets. The featural methods view graded membership in terms of binary features. Features are either present or they are not. The grade of membership for any candidate is determined on the number of features it has in common with the summary representation for the category, and their criticality. In contrast, the fuzzy set approach grades membership in terms of the features themselves (Zadeh, 1982). Features have a graded presence.

Membership functions indicate the set of graded features which comprise the summary representation for the category, and any candidate entities. The theory of fuzzy sets then provides a mechanism that will compare those membership functions, and thereby 'grade' membership of the category. This is an obvious extension to the classical view.

3.3. The Current Agenda for AI-in-Design Research is Extensionalist

Whilst overcoming many of the problems associated with the classical view, the extensions considered above, themselves introduce new and specific problems. There is now an expanding range of subsequent, higher-level extensions. These higher-level extensions build on the prototype, exemplar, featural and fuzzy set approaches. The higher-level extensions address the problems raised by the previous extensions, introducing problems of their own, and so on, adding inexorably to the complexity of the task facing AI and design.

One general example of this 'layering' of complexity extends from the basic notion of a feature. Features come in many forms. As a result, there is a constant difficulty in deciding which features to consider in a category description, and how each feature might be represented. One might argue that each entity can have an infinite range of features (Murphy and Medin, 1985). Any two entities can be arbitrarily similar or dis-similar depending on the features used to describe them (Medin and Wattenmaker, 1987). Some features are binary, some are continuous, some are incommensurable. These, and a number of other problems, have each prompted their own extensions to the notion of a feature.

The most immediate extension is to include within the representation framework relationships between features. Where there is some correlation between features, this appears to influence the efficiency with which entities are categorised. In AI, the solution generally has been to recognise these relationships in the form of higher-level features. Thus 'special' features link the standard features through simple functional dependencies—'is_a', 'has_a', 'equals' (Newton and Logan, 1988). But then dependencies can be variable, so the representation is further extended to accommodate ranges of possible values. The ranges are represented as abstract features which are instantiated as required for each entity—eg. 'slots' and 'slot fillers' (Minsky, 1975).

Unfortunately for the neat and tidy theory builder, even the structured features exhibit significant limitations in categorising certain entities. The relationships tend to be idealised (they very often over-generalise), asymmetrical (the simple functional dependency admits relationships in only one direction), selective (an infinite range of potential relationships are possible), and objectified (there is no access for beliefs or value systems)—to name but a few of the limitations to have emerged.

Of course extensions which tackle these new limitations are also being proposed. Meta-level features such as function and behaviour, in which features relating to function or behavior are somehow separated from other features, and provide a more effective means of categorising otherwise remote entities (Rosenman and Gero, 1989; Rosenman and Sudweeks, 1990). Meta-level features such as multiple views, in which different views of the 'same' category are accommodated in a single representation. Each feature may 'appear' in several views, and each view is composed of different, possibly conflicting representative features (Logan et al, 1990). Meta-level features such as hidden units, where neural networks (from a set of training examples) automatically construct 'micro-features' which make some contextual 'sense' of the input units/features (Cussins, 1990). But with each new extension, new problems emerge.

This entire approach is 'extensionalist'. Take a simple theory such as the classical theory of categorisation, and extend it to accommodate an ever widening range of requirements. Current views of categorisation (as indicated by the status of AI and design representations) may look very different to the classical theory of old, but there are clear and inescapable family traits. And absolutely central in this regard, is the basic element on which all views are grounded—the feature. Here is a basic atom from which the full range of AI representations in design are composed. Indeed, the feature has become so foundational as to be almost beyond further consideration.

But questions are to be raised against the use of features in formalising categorisation. The questions are those being raised against all such atomistic views. Features are intended to be non-degradable structures (although the evidence suggests that they are not). Categories are then defined as a summary set of features. But the practice of defining categories in this way promotes the view that an entity is no more than the sum of its features—a most questionable position (Armstrong et al, 1983). It also suggest that the functions of logic, set union, intersection, and complementation then provide the necessary resources to model the cognitive processes of categorisation.

We have here a compositional, 'atomistic' view of the world. A view which takes the 'given', the axiomatic, the incontrovertible, and constructs the world from the 'bottom-up'. The ultimate goal is a single, unified theory of the universe (Hawking, 1988). It is the 'rationalist tradition', or 'Cartesian legacy', established by Descartes (Coyne and Snodgrass, 1991a). It is the view currently commanding a majority of support. It is pervasive and incredibly well-entrenched. But despite the position of strength, it is very much a view under attack.

The challenge comes principally from within philosophy, social science, anthropology and language theory. But the wave of criticism has already spilled into the domains of AI (Searle, 1980; Dreyfus and Dreyfus, 1985; Winograd and Flores, 1986; Dreyfus, 1990) and design (Schön, 1983; Snodgrass and Coyne, 1990; Coyne and Snodgrass, 1991b). The same challenge motivates this paper. The intention is to present an alternative to the atomistic treatment of categories (the rationalist ontology). This alternative is presented through a description of categories as being language-based and centred on people (the hermeneutic ontology). The hermeneutic ontology is grounded on experience.

4. Categorisation As A Quality Of Cognition—The Experiential View

4.1. The Primacy of Experience in our Understanding

To develop the primacy of experience I base my arguments primarily on the interpretations of Heidegger and hermeneutics provided by Wachterhauser (1986), Winograd and Flores (1986), Suchman (1987), Snodgrass and Coyne (1990), Rorty (1991), and Coyne and Snodgrass (1991a, 1991b).

One of the key consequences of the current challenge to the rationalist tradition hinges on the status of experience in the way we understand and relate to the world—the nature of our 'Being'. In the rationalist tradition theory is of first importance. That is 'theory' in the sense of generalised rules which can be logically validated (or at least tested) in some sense, and also used to predict phenomena. The rules that compose the theory are proposed as an explanatory hypothesis for observed phenomena. Experience, the actual observation of the phenomenon, is furthest removed from theory.

This is a compositional or reductive view of the world. At the lowest level there is sense data—the experience of 'observing' phenomenon. (In fact the rationalist tradition has already recognised the existence of a level beyond sense data. The 'crisis of rationality' which emerged in the 1960s echoed a realisation that the notion of objective sense data is a fallacy. The scientist cannot avoid intervening, and inflicting her or his own background of prejudices and beliefs on the observations being made. In this account the background, or experience, of the observer is an unwanted interference in theorising, and an unwanted 'noise' in our knowledge and understanding.) On the basis of this sense data, logical rules are proposed that explain the information. The rules are intended to represent a testable hypothesis that can be validated more generally, finally achieving the status of theory.

So theory provides the 'essence' of rationalist understanding, and it is in the form of theory that rationalists will aim to manifest their understanding. Consequently much of the research effort in AI is focussed on representation frameworks for 'the rules' and other explicit forms of 'the theory'. "The rationalist orientation not only underlies both pure and applied science but it is also regarded, perhaps because of the prestige and success that modern science enjoys, as the very paradigm of what it means to think and be intelligent." (Winograd and Flores, 1986).

In terms of a hermeneutic ontology this ordering of importance (with theory at the top and experience at the bottom, with experience most removed from understanding) is completely reversed. That is to say, experience is elevated to the pre-eminent and essential form of our understanding. In this case, experience means more than just the 'sense data' of the rationalist tradition. Even sense data is at a level above (more abstract) the 'pure experience'. Pure experience is achieved through being actively involved in some situation. Active involvement here refers to the sense of 'doing without thinking'. To be capable of doing something without being aware of 'thinking', is to truly 'know' what you do, and to truly 'think'. (This is a difficult aspect for the rationalist to accept. We are all of the time seeming to relate our thoughts to the rules of theory. Thinking things through, and reflecting on what we do. Anything beyond this must surely be sub-conscious and 'mystical'. But the rationalist view of thinking as reflection is precisely what is being challenged. Reflection is not our primary mode of thought.)

As we engage in everyday life, everything we know, understand and do is invisible to us. As we sit at a drawing board, holding a pencil and sketching on a piece of paper, the objects and properties of the world do not exist. We merely engage in sketching. We 'know' what it is we do, and we 'think' without intrusion. The objects, the drawing board, the pencil, the room, the world outside the window, the tendons in the sketcher's arm, all melt into an unrecognised background. Heidegger calls this the 'background of readiness-to-hand', and maintains that this is the primary way in which we understand the world—thrown into an action, without reflecting on the action or recognising objects 'out there'.

We become aware of objects (sense data) only as they emerge from the nondescript background—as they are 'brought to mind'. Objects can be brought to mind in a variety of ways. During the act of sketching, for example, an object may fail to function as we anticipate (the pencil lead breaks), an object may interfere with the action (no more room on the paper), the situation may change (a friend enters the office), and so on. This sudden, explicit recognition of objects, as objects, distinct from the background is what we might conventionally call 'conscious thought'. (The notion of conscious thought assumed here however is rather misleading, since the tendency is then to equate readiness-to-hand with the 'mystery' of sub-conscious activity. Readiness-to-hand is part of our Being which may be difficult to describe in conventional, cognivistically loaded terms, but we can decsribe it and we do 'understand' it in terms of other metaphors. The alternative to conscious activity is not necessarily dark and mysterious.) Heidegger calls this sudden recognition an event of 'breaking down' in which objects become 'present-at-hand'. The present-at-hand is already one step removed from the essential thought and understanding of our Being. Heidegger insists that the present-at-hand is therefore meaningless when considered in the absence of an active engagement in a situation. The objects of our thought and understanding only 'exist' in the sense that they emerge from the background of readiness-to-hand (experience).

Once the objects are present-to-hand they become 'sense data'. The data is assigned properties and features that are labelled and categorised. The properties and features are related (through 'rules') to describe phenomena in a way that can then be generalised into theory. On the basis of this account, theory is furthest removed from the way we actually think and understand the world. Theory is also seen to be actively disabling in the way we actually think and understand. Notice, for example, how the speed of sketching, writing, typing or reading slows down as the process becomes present-at-hand. Performance degrades considerably, say, when the professional typist reflects on where the keys on a keyboard are located, or the practised writer endeavours to be aware of how each letter is being formed as it is written. The more abstract and theoretical our engagement within a concernful activity, and the further removed from readiness-to-hand (experience), then the less we understand or know about what it is we do.

What has been here described in relation to 'objects' can equally be applied to other 'entities' such as concepts (where the ready-to-hand would equate with unbridled, non-reflective thought), and motor actions (where the ready-to-hand would equate with spontaneous, reflexive body movements).

Having elevated experience (ie. the readiness-to-hand of concernful activity) to a level of first importance, how might we describe the process of understanding? The process is well described in terms of the hermeneutic circle.

4.2. COGNITION AS INTERPRETATION—THE HERMENEUTIC CIRCLE

Hermeneutics began as the theory of the interpretation of ancient texts. The motivating question concerned how it might be possible for a reader to transcend the intervening cultural and language changes, to understand a text as it would have been understood by the original writer. Fundamental to this question is the possibility of meaning existing within the text, absolute and independent of the reader. If we reject that possibility, because the hermeneutic account of understanding and meaning is bound inexorably to experience and human action, then meaning can only emerge through the act of interpretation. Meaning will be relative to the individual interpreter and a particular situation.

Gadamer (1975) is at pains to establish the counter-dependence between the interpreter and the situation. Interpretation to Gadamer requires (Note: 'requires', not just 'would benefit from') a rich dialectic between the two. The interpreter brings to any situation a set of prejudices and pre-conceptions. They are the culmination of an interpreter's own past history modified by the previous actions of experience. This set of prejudices Gadamer calls the 'effective historical consciousness', or 'horizon'. The prejudices project onto a new situation in the form of expectations and those expectations grant each individual a unique interpretation of the situation.

But the 'situation' has a background also—a background of cultural and social history, manifest in the shared meaning of our language. Without a shared meaning we would be unable to communicate effectively. Since shared meaning will reflect the many individual interpretations being made, the meaning of a language will change over time as it is used. Notions judged by today's standard will be very different to equivalent notions of the past.

Interpretation is seen to be circular, with individual influencing cultural and cultural influencing individual—the 'hermeneutic circle'. There is first the effective historical consciousness (background or horizon) of each individual interpreter. This background is intrinsic to people, and emerges in the form of prejudices or expectations. The expectations of each interpreter are projected onto the current situation, including the current situation as part of the interpreters evolving background. The situation has 'its' own, social and cultural background. Because the interpreter is part of society, the way in which this particular interpreter interprets this particular situation will form part of the situation's evolving background. The two things feed from and amplify one another—the background of the interpreter and the background of the situation. The circularity of interpretation becomes harmonic. Gadamer describes this harmonic as the 'fusing of horizons'.

4.3. Categorisation and Cognitive Models

The description given above is an account of thinking grounded on human experience. It is an account very much in contrast to the atomistic treatment presented in section 3. The contrast is particularly acute in terms of representation. The atomistic account is well suited to representation, including representation as symbols in a computer. But the link to representation goes further. The atomistic account actually requires that categories are stored 'things'. In effect that categories (and everything else we know) are stored representations 'located' somewhere, somehow, in our brains.

The experiential account overturns this basic assumption. The experiential account insists that categories are not generated from stored descriptions, but continuously reconstructed in response to active involvement in the world. There is no such thing as a category in memory. A category is not stored locally within a single neuron; a category is not stored as multiple copies in multiple neurons; a category is not stored in a distributed form over several neurons; a category is not stored at various 'levels' and compiled. The distinction is not between local versus distributed, declarative versus procedural, or compilation versus interpretation. A category is not stored as any 'thing'. Rather, categorisation emerges as a capacity to act. To categorise is to perceive is to understand is to think. The categories we see represented in AI, as scripts, semantic nets, grammars, and neural networks, model only the products of our behaviour. By analogy, they are as far removed from a model of thought as the physical description of a coin is removed from a model of the market economy.

The focus of this paper has been on the treatment and relevance of categories, but its aim is much broader than this. From the two different interpretations of categorisation described in sections 3 and 4, there is a necessary distinction to be made between categorisation as a model of cognition and categorisation as a quality of cognition. This distinction is significantly more fundamental than just the relevance and treatment of categorisation. The distinction holds for AI in general, and the entire rationalist tradition. In the following section the focus on categorisation is suspended in order to engage the broader concerns of AI. The same concerns hold more specifically for AI-in-design.

5. The Future Role Of Artificial Intelligence

5.1. WHAT ARE THE OPTIONS?

AI should neither be surprised nor outraged at the extent to which so much of its endeavours are removed from cognitive modelling. The claim being presented here, that there are no representations stored as such in the brain, has been made for decades by numerous researchers (Gibson, 1966; Piaget, 1970; Bransford *et al*, 1977; Iran-Nejad, 1987; Rosenfield, 1988)—including possibly the most famous, and often misrepresented work of Bartlett (1932). However, perhaps the critics of AI should not be surprised or outraged either. For AI to recognise that there are no representations as such in the brain would be to recognise the intractability of modelling cognition as representations in computers. Where would this leave AI?

There are four possible roles for AI to be considered. In characterising AI research in this simple way there is no intention is to suggest that there are not other, interlaced positions actually taken within the AI-in-design community. Unfortunately, these other positions often appear to be inconsistent.

5.1.1. The psychological role. The focus here is on producing a 'phenomenological' (Schön, 1991) model of cognition. The model would be computable and function in a way directly equivalent to human cognition (however that might be established). This is recognised as a very difficult task for AI, and has been largely abandoned (at least in the immediate term, at least until a more promising model is found, ...). To accept the experiential account of cognition is to accept that cognition occurs in the ready-to-hand, the mechanics of which are critically removed from observation. To observe the mechanics of cognition is to make them explicit within the present-at-hand, and thereby no longer actual cognition. The experiential account insists there can be no phenomenological model of cognition (or more accurately, that we could not 'know' if a model was phenomenological or not).

5.1.2. The cognitive role. The focus here is also on producing a computable (symbolic) model of human cognition. The model would not function in any directly equivalent way to human cognition, but rather model in the sense of an input-output representation. Given the same set of symbolic inputs, the intention would be to produce equivalent outputs in the computer model as would be produced by a person, but with no appeal to equivalent mechanisms. This is a weaker claim than the psychological (phenomenological) approach and has provided the generally more popular role for AI. It can also claim a degree of success. Computers play chess, fly aircraft, and diagnose illnesses in ways that have no direct equivalence to the ways in which people perform the same activities—but computers do 'play', 'fly' and 'diagnose'. Or do they? What actually is being modelled? What, for example, is the chess-playing computer actually 'doing'?

The rationalist would present a 'logical' explanation. If, given exactly the same situation (the same inputs) two systems produce exactly the same behaviour (the same outputs), then there has to be a fundamental correspondence between the two. More specifically, if we consider good chess playing to be intelligent then we must also consider good chess playing computers to be intelligent. The argument works through logical transitivity, if A equals B (A=B) and A implies I (A->I) then B must also imply I (B->I). There can be no further argument. But of course there can be further argument, because one can always challenge a logical argument by questioning the basic assumptions or axioms. In this case, the most questionable assumption concerns the nature of the correspondence between the two systems.

Recall that we are considering an input-output or 'functional' (Schön, 1991) modelling approach, and therefore a correspondence between systems in terms of the input and output. To employ that correspondence as a 'proof' of intelligence is to lodge intelligence unequivocally in the externalities of the system. It is to say that human intelligence is vested in the expressions and behaviour people employ to communicate and demonstrate understanding. That the marks made on a piece of paper, the sounds recorded on a tape, or the moves made in a game of chess *are* the intelligence. This is the 'real' assumption in the rationalist argument presented above. It is a mistaken assumption. (See however the continuing debate arising from Searle's equivalent 'Chinese Room Problem'—Searle, 1980; Boden, 1990.)

But if the input-output approach can produce behavior equivalent to the behaviour of people, what difference does it make if we do or do not call it 'intelligent'? Do the ends not justify the means? Perhaps our concern here is with definitions, and misses the real point. Well, the important difference is that the focus of the input-output approach *is* on the production of a computable model of cognition. That point is inescapable, and if (as it is suggested here) the approach is incapable of intentionally producing such a model, then it is an impoverished role. At the mere mention of cognition, people project all kinds of other human qualities onto the system (friendly, creative, equitable, etc.). Those who subscribe to the cognitive approach are all too willing to allude to these other qualities, and encourage others to make similar connections. It all adds to the appeal.

The experiential account of cognition insists that intelligence is intrinsic to the concernful activity of the entity which creates the product, and not within the product itself. Human intelligence is bound within the background of people engaged in purposeful activities, and we should not expect to be able to make that background explicit. The cognitive approach may produce useful mechanisms in the general sense, but there is no apparent reason why those same mechanisms should in any way model human cognition.

5.1.3. The non-cognitive role. In denying both phenomenological and input-output models of cognition we are denying computable models of cognition in total. This will concern a multitude of researchers in AI, for whom cognitive modelling *is* AI. With no appeal to be doing cognitive modelling, what justification is there left for AI? The justification would have to come from elsewhere, possibly the usefulness of the techniques in practice, possibly the usefulness of the techniques in the design of computer interfaces, possibly the usefulness of the techniques in modelling physical (non-human) systems—a genuinely 'artificial' intelligence.

In this case, the only remaining connection with cognition is the use of natural systems as a rich source of ideas for the development of new computational techniques—scripts, genetic algorithms, neural networks, analogical reasoning, etc. The flow of ideas would not be reciprocated. Cognitive science would not focus on computational models of cognition. The concern, simply stated, is that a correspondence with natural systems should not be employed (implicitly or explicitly) to elevate computational techniques to the realm of cognitive models.

The non-cognitive role is remarkably tame in comparison to popular notions of AI, expert systems, and intelligent knowledge-based systems. It maintains merely that there are various techniques available with the potential to contribute in an unobtrusive way to human activities (such as design), and to enhance human experience. The 'frame' could be seen in terms of a database management system, 'genetic algorithms' could be seen in terms of optimisation techniques, as could 'neural networks', 'conceptual clustering' could be seen as a form of statistical analysis, and so on. This is not a condemnation of the technical work already done in AI or even of the specific techniques currently being developed. But in many ways the non-cognitive descriptions are more representative views of the technologies. The issue is how AI techniques relate to human cognition.

5.1.4. The metaphorical role. This paper takes a narrow, but generally accepted view of cognitive modelling, to refer specifically to human cognition and cognitive models that are computable. This view applies the computational metaphor to human cognition. I am careful to indicate that AI does include broader aspects than just cognitive modelling, and that cognitive modelling can relate to non-computable models, but that the terms used in this paper refer to the narrower interpretation. If the definition of AI is now relaxed to include these broader aspects, one other important role for AI presents itself—the application of other metaphors of cognition. Sternberg (1990) provides an interesting guide to the range of metaphors already applied to intelligence. They include the geographic, biological, epistemological, anthropological, sociological and systems metaphors. Lakoff and Johnson (1980) provide an even more wide ranging source of potential metaphors. For many, paying more attention to the different ways in which cognition can be understood (as opposed to the current focus on techniques) would be no more than a return to the roots of AI. Without this return to fundamentals it is possible, like the field of operations research before it (Ackoff, 1979), that AI will choke itself on techniques.

The 'metaphorical' role is to re-focus AI on alternatives to the computational metaphor of cognition. Computers have a considerably lower status in this genre, but they are not devoid of status. Often the most useful description of a metaphor is a description of the ways in which that metaphor might translate into computer-based systems. For example, a number of current initiatives might usefully fit within a metaphor based on 'communications'. A communications metaphor is presented in the following section of this paper as one example of how an alternative metaphor might operate in practice.

5.2. The Communications Metaphor

There are two important intuitions on what it takes to achieve communications (Habermas, 1991). The first is an intentionalist view, for which the communication process consists in conveying the (intended) meaning or intention of a speaker to an addressee by means of an expression. Expressions are 'conduits' through which communication flows. The second is an intersubjectivist view, for which the communication process consists not in a transfer of ideas, but in two parties reaching a shared understanding about something. Expressions are the 'medium' within which communication is shared.

There are four initiatives to be considered briefly. The first two initiatives are related to the intersubjectivist view of communication, the final two are related to the intentionalist view.

5.2.1. Communication as the sharing of understanding The sharing of understanding is more often related directly to the notion of conversation than to the neutral connotation of communication—particularly the neutral connotation of communication within a computer environment. But the experiential account of understanding puts everything we know within the background of 'readiness-to-hand', and the real sharing of understanding is contingent on a 'fusing of horizons'. The 'cold' neutrality of electronic communication is symptomatic of the extent to which computers intrude upon active communication (causing 'breakdown'). Current trends in computing tend to accentuate the intrusion—personal computers which isolate people physically from each other, with multimedia screens that demand an even greater focus of our attention, and virtual reality machines that cocoon users in special goggles and suits and cuts them from the world (Pask and Curran, 1982). A sharing of understanding demands that computers disappear into the background. (Note: it may be that the 'cold neutrality' is an intrusion to conventional communication, but actively enables communication of a new and different form—witness the rise of email 'dating', electronic news services and electronic communication generally. This may present a further communications metaphor.)

The research agenda for AI is then concerned with the process of drawing computers and computer users out of their exile. This is not a user modelling, user interface, networking, operating system, or miniaturisation problem (although aspects of each will require attention). Rather, the problem is how to put computers everywhere—hundreds of computers per room, from the multitude of computers in light switches, door locks and coffee percolators (activated by 'communications'), through the dozens of 'active' Post-it sized computer screens, tens-of book-sized interactive 'pads', and wall-sized displays. "Hundreds of computers in a room could seem intimidating at first, just as hundreds of volts coursing through wires in the walls once did. But like the wires in walls, these hundreds of computers will come to be invisible to common awareness. People will simply use them unconsciously to accomplish everyday tasks." (Weiser, 1991). Appropriate technologies will no doubt appear, the main hurdle is fitting machines to the human environment (and not vice versa). To engage machines in the communications between people, we must engage them unobtrusively. We need to understand more about human cognition, and better develop the expressly artificial (non-humanlike) intelligence of interacting computers.

5.2.2. Communication as an exchange of expressions The experiential account insists that human or human-like intelligence is intrinsic to the actions of people, and that communication involves a sharing (fusing) of 'horizons'. The objects we recognise and the ideas we express have already, in an important sense, escaped our understanding. So an expression of knowledge is no longer knowledge. Neither can expressions be considered as representations of knowledge, since there is no predictable correspondence between the expression and the knowledge actually exchanged. Isolated from people expressions are meaningless tokens. But in the context of people, with people interpreting the expressions, what were otherwise meaningless tokens become the medium of exchange for knowledge. We might say that expressions provide the currency in a market of knowledge.

The research agenda for AI is then framed around an understanding of how people use forms of expressions in the exchange process. And there are various forms of expression and exchange to be considered (painting, writing, dancing, poetry, etc.). The appeal to computers is not immediate. However, as an unrivalled 'logic machine' the computer may be useful in realising new forms of expression (Bijl, 1989 and 1991). The onus is upon users to establish their own appreciation of the logical mechanisms inherent in the technology—closed world assumptions, consistency, intentionality, etc. Designers of computer systems should aim to be minimally prescriptive in terms of the knowledge being exchanged from the expressions the systems manipulate (Tweed and Bijl, 1989).

5.2.3. Communication as coordinated speech acts Expressions may be meaningless tokens, but they are bound to the purposeful activities in which they arise. In different situations, different intentions 'load' the tokens with particular expectations about how the tokens will be used and how they might be interpretted. These different loadings are not entirely cultural or language based, but rather determine the potential engagements between two people in the course of communication (Habermas, 1979). The loadings reflect a kind of commitment or promise by the speaker, to an addressee, to a particular form of action. In this sense they are referred to as speech acts (Searle, 1969 and 1979). A speech act can count as a promise (commitment to some future action), assertion (commitment to the 'truth' of some utterance), request (attempt to get the listener to do something), declaration (bring about a correspondence between the speech act and 'reality'—eg. pronouncing a couple married), or avowal (expressing a mental 'state', such as apologising or praising).

Under different conditions (managing, coaching, taking decisions, designing, playing, etc.) different patterns of speech acts emerge. The research agenda for AI is then framed around how speech acts are related to one another in each of these different domains. Computer systems could then be charged with coordinating the speech acts: establishing the type of speech act (promise, assertion, etc.), its force (the strength of the promise, assertion, etc.), monitoring completion, automating recurrent communications, displaying networks of conversations, possibly indicating the status of an individuals commitments and requests (Holt et al, 1983; Winograd and Flores, 1986). Importantly, the role of the computer system is entirely passive, unable itself to make a request or commitment. Speech acts have always to be placed in the context of people, more specifically the purposeful activities of people.

5.2.4. Communication as situated actions If speech acts occur within the purposeful activities of people, a further view of communication is at the level of those actions. It is common to view purposeful actions in the sense of 'plans of action'. A plan of action is a rational description of the intent and shared knowledge about typical situations and appropriate actions (Suchman, 1987). Once developed, the plan of action can be 'implemented' with almost mechanical ease. One need only know the current (pre)disposition and intention (goal), to anticipate the course of action.

This view is applied to human-computer interaction as a 'user model', either to anticipate the intentions of a user and provide appropriate responses, or to identify discrepancies between an 'actual' user and a 'desired' user (in the context, say, of a tutorial program). The same view is applied to interpersonal communication as directives for action—decision analysis (Arkes and Hammond, 1986; von Winterfeldt and Edwards, 1986). In both cases the 'plan' is viewed as a representation of the purposeful actions *per se*. Not surprisingly, the planning view of purposeful actions has provided an important (and welcome) grounding for much of the work in AI (see Bobrow and Collins, 1975; Torrance, 1984; Priest, 1991; and in AI-in-design more specifically, Coyne *et al*, 1990; Gero, 1991).

But equivalent objections can be raised against the planning view as are raised against other rationalist interpretations. The objections relate again to the moment-by-moment interactions (thrownness) that give rise to actions—what Suchman (1987) terms the situated nature of action. The agenda for AI research is then to explore the relation of action and situation. The 'complexities' of 'real' action, in a 'real' situation, are then an integral and essential consideration in this respect, and can no longer be treated as extraneous to the problem. "Communication in this sense is not a symbolic process that happens to go on in real-world settings, but a real-world activity in which we make use of language to delineate the collective relevance of our shared environment." (Suchman, 1987: p.180)

6. Summary Comments

A fundamental distinction has been made between the atomistic, rationalist view that seeks to model categorisation as a model of cognition, and the experiential account that insists that categorisation is an emergent property of cognition. To accept the experiential account is to deny the basic assumption that 'things' are stored representations in the brain, and therefore to deny the computational metaphor of cognition. This leaves one of two alternatives for the future role of AI. Either we can treat AI in a non-cognitive way, as no more than a collection of computational techniques, or we can return to basics and reconsider alternative metaphors of cognition. In the latter regard, several research initiatives appear to fit within a communications metaphor, but this does not discount the possibility that more appropriate metaphors may be available.

The imperative is to recognise the distinction being made in this paper, and to question again the prevailing view of AI as focussing on a computational model of cognition. The challenge against the rationalist view is a growing one. The move is a very clear return to focussing again on the purposeful actions of people, and through people to focus on society.

Acknowledgements

The author would like to thank Richard Coyne, and to recognise his work and our conversations as the principal source on which my own interpretations of Heidegger and hermenutics are based. Thanks also to Fay Sudweeks, Mike Rosenman and the paper's referees for their comments on earlier drafts.

References

Ackoff, R.L.: 1979, The future of operational research is past, *Journal of the Operational Research Society*, **30**(2), 93-104.

Arkes, H.R. and Hammond, K.R. (eds): 1986, *Judgement and Decision Making: An Interdisciplinary Reader*, Cambridge University Press, Cambridge.

Armstrong, S.L., Gleitman, L.R. and Gleitman, H.: 1983, On what some concepts might not be, *Cognition*, **13**, 263-308.

Barsalou, L.W.: 1987, The instability of graded structure: implications for the nature of concepts, *in* U. Neisser, (ed), *Concepts and Conceptual Development: Ecological and Intellectual Factors in Categorization*, Cambridge University Press, Cambridge, pp.101-40.

Bartlett, F.C.: 1932, *Remembering—A Study in Experimental and Social Psychology*, Cambridge University Press, Cambridge.

Berlin, B. and Kay, P.: 1969, *Basic Color Terms: Their Universality and Evolution*, University of California Press, Berkley.

Bijl, A.: 1989, *Computer Discipline and Design Practice: Shaping our Future*, Edinburgh University Press, Edinburgh.

Bijl, A.: 1991, Logic in ICAD and use of logic by designers, *IntCAD '91 Preprints*, The Ohio State University, Columbus, Ohio, pp.357-79.

Bobrow, D.G. and Collins, A. (eds): 1975, *Representing and Understanding: Studies in Cognitive Science*, Academic Press, New York.

Boden, M.A.: 1990, Escaping from the chinese room, *in* M.A. Bowden (ed), *The Philosophy of Artificial Intelligence*, Oxford University Press, Oxford, pp.89-104.

Bransford, J.D., McCarrell, N.S., Franks, J.J. and Nitsch, K.E.: 1977, Toward unexplaining memory, *in* R.E. Shaw and J.D. Bransford (eds), *Perceiving, Acting, and Knowing: Toward an Ecological Psychology*, Lawrence Erlbaum, Hillsdale, New Jersey, pp.431-66.

Collins, A.M. and Loftus, E.F.: 1975, A spreading-activation theory of semantic processing, *Psychological Review*, **82**, 407-28.

Coyne, R.D. and Snodgrass, A.: 1991a, What is the philosophical basis of AI in design? *Working Paper*, Faculty of Architecture, University of Sydney.

Coyne, R.D. and Snodgrass, A.: 1991b, Is designing mysterious? challenging the dual knowledge thesis, *Design Studies*, **12**(3), 124-31.

Coyne, R.D. and Yokozawa, M.: 1991, Computer assistance in designing from precedent, *Environment and Planning B*, to appear.

Coyne, R.D., Newton, S. and Sudweeks, F.: 1989, Modelling the emergence of schemas in design reasoning, *Preprints Modeling Creativity and Knowledge-Based Creative Design*, DADS, University of Sydney, pp.173-205.

Coyne, R.D., Rosenman, M.A., Radford, A.D., Balachandran, M. and Gero, J.S.: 1990, *Knowledge-Based Design Systems*, Addison-Wesley, Reading, Massachusetts.

Cussins, A.: 1990, The connectionist construction of concepts, *in* M.A. Bowden (ed), *The Philosophy of Artificial Intelligence*, Oxford University Press, Oxford, pp.368-440.

Dreyfus, H.L.: 1990, *Being-in-the-World: A Commentary on Heidegger's Being and Time, Division I*, MIT Press, Cambridge, Massachusetts.

Dreyfus, H.L. and Dreyfus, S.E.: 1985, *Mind Over Machine*, MacMillan/The Free Press, New York.

Gadamer, H.-G.: 1975, *Truth and Method*, Sheed Ward, London.

Gero, J.S. (ed): 1991, *Artificial Intelligence in Design '91*, Butterworth-Heinemann, Oxford.

Gibson, J.J.: 1966, *The Senses Considered as Perceptual Systems*, Houghton Mifflin, Boston, Massachusetts.

Habermas, J.: 1979, What is universal pragmatics? in J. Habermas, *Communication and the Evolution of Society*, translated by T. McCarthy, Beacon Press, Boston, Massachusetts, pp.1-68.

Habermas, J.: 1991, Comments on John Searle: "Meaning, Communication, and Representation", *in* E. Lepore and R. Van Gulik (eds), *John Searle and His Critics*, Basil Blackwell, Cambridge, Massachusetts, pp.17-29.

Hawking, S.W.: 1988, *A Brief History of Time: From the Big Bang to Black Holes*, Bantam Books, New York.

Holt, A.W., Ramsey, H.R. and Grimes, J.D.: 1983, Coordination system technology as the basis for a programming environment, *Electrical Communication*, **57**(4), 307-14.

Iran-Nejad, A.: 1987, The schema: a long-term memory structure or a transient functional pattern, *in* R.J. Tierney, P.L. Anders and J.N. Mitchell (eds), *Understanding Readers' Understanding: Theory and Practice*, Lawrence Erlbaum, Hillsdale, New Jersey.

Kahneman, D. and Tversky, A.: 1973, On the psychology of prediction, *Psychological Review*, **80**, 237-51.

Katz, M.J.: 1984, *Templets and the Explanation of Complex Patterns*, Cambridge University Press, Cambridge.

Lakoff, G.: 1987, *Women, Fire, and Dangerous Things: What Categories Reveal about the Mind*, University of Chicago Press, Chicago.

Lakoff, G. and Johnson, M.: 1980, *Metaphors We Live By*, University of Chicago Press, Chicago.

Leach, E.: 1964, Anthropological aspects of language: Animal categories and verbal abuse, *in* E.H. Lenneberg (ed), *New Directions in the Study of Language*, MIT Press, Cambridge, Massachusetts, pp.23-6.

Lyons, J.: 1968, *Introduction to Theoretical Linguistics*, Cambridge University Press, Cambridge.

Logan, B.S., Millington, K. and Smithers, T.: 1990, Being economical with the truth: assumption-based context management in the Edinburgh Designer System, *in* J.S. Gero (ed), *Artificial Intelligence in Design '91*, Butterworth-Heinemann, Oxford, pp.423-46.

Medin, D.L. and Wattenmaker, W.D.: 1987, Category cohesiveness, theories, and cognitive archeology, *in* U. Neisser, (ed), *Concepts and Conceptual Development: Ecological and Intellectual Factors in Categorization*, Cambridge University Press, Cambridge, pp.25-62.

Minsky, M.: 1975, A framework for representing knowledge, *in* P.H. Winston, (ed), *The Psychology of Computer Vision*, McGraw Hill, New Yorlk, pp.211-77.

Murphy, G.L. and Medin, D.L.: 1985, The role of theories in conceptual coherence, *Psychological Review*, **92**, 289-316.

Neisser, U.: 1987, Introduction: the ecological and intellectual bases of categorization, *in* U. Neisser, (ed), *Concepts and Conceptual Development: Ecological and Intellectual Factors in Categorization*, Cambridge University Press, Cambridge, pp.1-24.

Newton, S. and Logan, B.S.: 1988, Causation and its effect: the blackguard in CAD's clothing, *Design Studies*, **9**(4), 196-201.

Pask, G. and Curran, S.: 1982, *Microman: Living and Growing with Computers*, Century Publishing, London.

Piaget, J.: 1970, *Genetic Epistemology*, Norton and Company, New York.

Priest, S.: 1991, *Theories of the Mind*, Penguin Books, london.

Pulman S.G.: 1983, *Word Meaning and Belief*, Croom Helm, London.

Rorty, R.: 1991, *Essays on Heidegger and Others: Philosophical Papers, Volume 2*, Cambridge University Press, Cambridge.

Rosch, E.: 1978, Principles of categorization, *in* E. Rosch and B.B. Lloyd (eds), *Cognition and Categorization*, Lawrence Erlbaum, Hillsdale, New Jersey, pp.27-48.

Rosch, E. and Lloyd, B.B. (eds).: 1978, *Cognition and Categorization*, Lawrence Erlbaum, Hillsdale, New Jersey.

Rosch, E., Mervis, C.B., Gray, W.D., Johnson, D.M. and Boyes-Braem, P.: 1976, Basic objects in natural categories, *Cognitive Psychology*, **8**, 382-439.

Rosenfield, I.: 1988, *The Invention of Memory: A New View of the Brain*, Basic Books, New York.

Rosenman, M.A. and Gero, J.S.: 1989, Creativity in design using a prototype approach, *Preprints Modeling Creativity and Knowledge-Based Creative Design*, DADS, University of Sydney, pp.207-32.

Rosenman, M.A. and Sudweeks, F.: 1990, Categorisation and prototype in design, *Working Paper*, Design Computing Unit, University of Sydney.

Rumelhart, D.E. and McClelland, J.L. (eds): 1986, *Parallel Distributed Processing: Explorations in the Microstructure of Cognition, Volume 1: Foundations*, MIT Press, Cambridge, Massachusetts.

Schön, D.: 1983, *The Reflective Practitioner: How Professionals Think in Action*, Basic Books, New York.

Schön, D.: 1991, Designing as a reflective conversation with the materials of a design situation, *Invited Speaker*, AID'91, Edinburgh, June 25-27, 1991.

Searle, J.R.: 1969, *Speech Acts*, Cambridge University Press, Cambridge.

Searle, J.R.: 1979, *Expression and Meaning: Studies in the Theory of Speech Acts*, Cambridge University Press, Cambridge.

Searle, J.R.: 1980, Minds, brains, and programs, *The Behavioural and Brain Sciences*, **3**, 417-57.

Shepard, R.N.: 1964, Attention and the metric structure of the stimulus space, *Journal of Mathematical Psychology*, **1**, 54-87.

Shepard, R.N.: 1974, Representation of structure in similarity data: problems and prospects, *Psychmetrika*, **39**, 373-421.

Smith, E.E. and Medin, D.L.: 1981, *Categories and Concepts*, Harvard University Press, Cambridge, Massachusetts.

Snodgrass, A. and Coyne, R.D.: 1990, Is designing hermeneutical? *Working Paper*, Faculty of Architecture, University of Sydney.

Sternberg, R.J.: 1990, *Metaphors of Mind: Conceptions of the Nature of Intelligence*, Cambridge University Press, Cambridge.

Suchman, L.A.: 1987, *Plans and Situated Actions: The Problem of Human-Machine Communication*, Cambridge University Press, Cambridge.

Taylor, J.R.: 1989, *Linguistic Categorization: Prototypes in Linguistic Theory*, Clarendon Press, Oxford.

Torrance, S. (ed): 1984, *The Mind and the Machine: Philosophical Aspects of Artificial Intelligence*, Ellis Horwood, Chichester.

Tversky, A.: 1977, Features of similarity, *Psychological Review*, **84**, 327-52.

Tweed, C. and Bijl, A.: 1989, MOLE: a reasonable logic for design?, *in* P.J.W. ten Hagen, T. Tomiyama and V. Akman (eds), *Intelligent CAD Systems II: Implementation Issues*, Springer-Verlag, Holland.

von Winterfeldt, D. and Edwards, W.: 1986, *Decision Analysis and Behavioural Research*, Cambridge University Press, Cambridge.

Wachterhauser, B.R. (ed): 1986, *Hermeneutics and Modern Philosophy*, State University of New York Press, Albany.

Weiser, M.: 1991, The computer for the 21st century, *Scientific American*, September 1991, pp.66-75.

Winograd, T. and Flores, F.: 1986, *Understanding Computers and Cognition: A New Foundation for Design*, Addison Wesley, Reading, Massachusetts.

Wittgenstein, L.: 1953, *Philosophical Investigations*, translated by G.E.M. Anscombe, Blackwell, Oxford.

Zadeh, L.A.: 1982, A note on prototype theory and fuzzy sets, *Cognition*, **12**, 291-7.

HEURISTIC DECISION SUPPORT PROBLEMS

Integrating heuristic search and expert systems for the design of continuous-manufacturing products

S. Z. KAMAL
The M. W. Kellogg Co.
601 Jefferson
Houston TX 77210-4557 USA

and

J. W. GARSON* and F. MISTREE**
**Department of Philosophy*
***Department of Mechanical Engineering*
University of Houston
Houston TX 77204-4792 USA

Abstract. In this paper we introduce the heuristic Decision Support Problem (HDSP) as the means for integrating heuristic search and knowledge-based systems to provide decision support in design. Furthermore, we present a unique application of the HDSP to the design of products made by process manufacturing (e.g., gasoline, cosmetics) rather than products made by discrete manufacturing (e.g., machinery). The HDSP provides a format for representing the design problem that is independent of the specific application. The representation of the design problem is not restricted to mathematically exact information. A heuristic search algorithm, called the synthesis engine, is used to provide decision support in the form of solutions. The synthesis engine is able to evaluate feasibility and goodness of competing designs even though design variables and their associated relationships are not mathematically exact. The design of lubricant blends is used as a specific example of continuous-manufacturing products.

1. Our Frame of Reference

The largest applications of AI to design problems consist of various expert systems and knowledge representation schemes. These applications have been developed to help solve specific design problems (Finger and Dixon 1989a; Finger and Dixon 1989b). Two handicaps of adopting this approach are:

- It does not provide a structured/systematic environment for integration across domains and communication across different phases of the design process.
- It requires the development of systems that embody solution techniques specific to the domain of application.

Contrary to the conventional use of expert systems for design we propose a structured environment in which one or more expert systems (in addition to other tools available for evaluation and measuring the goodness of competing designs) may be used to support decisions in design. The basis for this approach is that concurrent product development, i.e., the consideration of issues from different

J. S. Gero (ed.), Artificial Intelligence in Design '92, 883–902.

disciplines and process in the development cycle, requires domain independent model and solution methods. We subscribe to Decision-Based Design (DBD) and assert that decisions are the primary constructs needed to model the design of a product into a generic model of the design process.

Our approach, within the realm of DBD, is called the Decision Support Problem (DSP) Technique. All decisions in the DSP Technique are categorized as Selection, Compromise, or a combination of these. Selection models choice; the emphasis being on the acceptance of certain alternatives through the rejection of others. Selection in design requires a human designer to reduce alternatives to a realistic and manageable number based on different attributes. Compromise models trade-off; the emphasis being on achieving multiple trade-offs through the modification of variables. Compromise in design allows a human designer to pursue a progressive process, requiring generation, evaluation and alteration of different designs based on feasibility and performance considerations. In the DSP technique, decisions relevant to design are represented on a computer by a suite of different Decision Support Problems (DSPs) (Mistree, Smith et al. 1990; Mistree, Hughes et al. 1992).

The heuristic DSP is one of the Decision Support Problems within the DSP Technique and in this paper, we introduce the HDSP model and the synthesis engine. In this paper we focus on the modelling aspects of the HDSP. The synthesis engine is described in greater detail in (Kamal 1990; Kamal, Garson et al. 1991). The role of the HDSP in helping a designer create distinct design concepts is illustrated using as an example, products made by continuous-manufacturing

1.1 OUR FOCUS

In the early stages of original design, decisions must be made without hard information. The HDSP has been developed to model decisions in characterized by soft information. The components of a HDSP are shown schematically in Figure 1. It consists of:

- a *representation model*, the keywords and descriptors, needed to model domain relevant information;
- a *synthesis engine* to process information represented using the keywords and descriptors; and
- an *inductive learning* tool to qualify domain relevant historical information into rules that may be used to evaluate competing designs.

We use the formulation of engine oils (automotive lubricants) as an example of designing continuous-manufacturing products. The design of such products requires that we select appropriate components and determine the *best* mix of these components to maximize the performance characteristics of the blend.

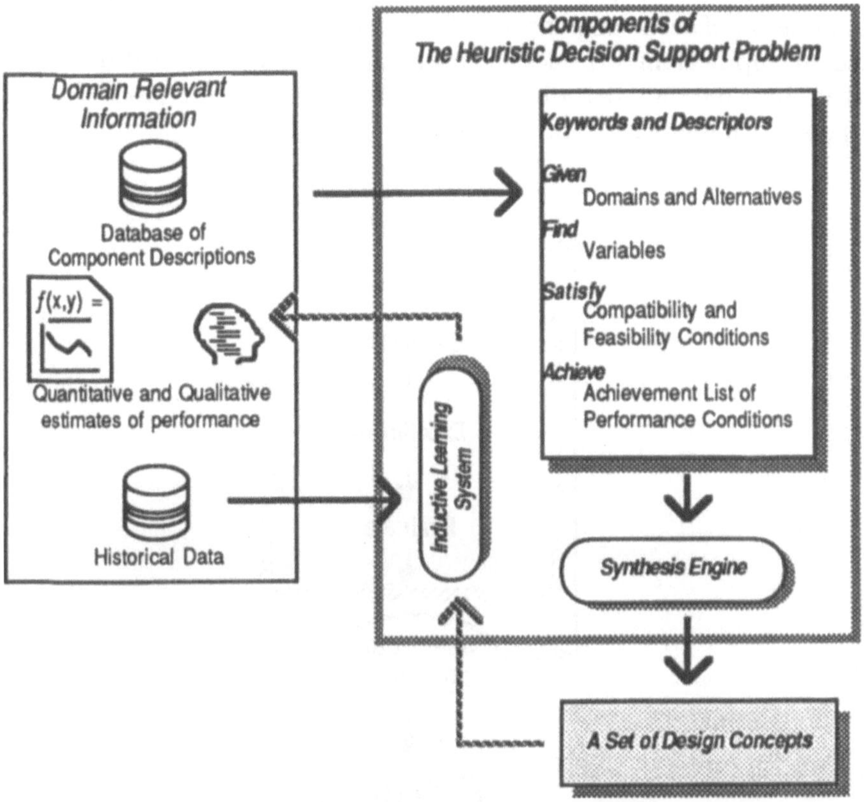

Fig. 1 Components of the heuristic Decision Support Problem

1.2 DESIGNING A CONTINUOUS-MANUFACTURING PRODUCT

Typically, the early stages of design of lubricants evolve around the *ideate* (create) and *test* (evaluate) cycle as shown in Figure 2. This process is rarely structured, is based on intuition and dependent on localized knowledge and expertise. A designer, at this stage of design, tries to identify *design concepts* that may be further developed and tested in detail. The HDSP is applied to the stage of design called *preliminary testing* in Figure 2. It helps generate an *intelligent set* of concepts by utilizing greater amounts of domain relevant information and considering a larger set of design concepts in less time.

A typical automotive engine oil is obtained by blending several additive packages with a lubricant base stock (BS). These additives include a viscosity index improver (VI), a dispersant inhibitor (DI) and detergent (DET). These chemicals are typically packaged as solutions of approximately 50% concentration in a base stock. These components interact with each other and the base stock to give the lubricant blend desired properties. A designer is faced with choosing from several different packages performing a given task and determining the concentration of each component in the blend. These choices are typically made based on empirical mixing rules and compatibility constraints. Blend components used in this study are described below. Readily available properties for each component are summarized in Table 1.

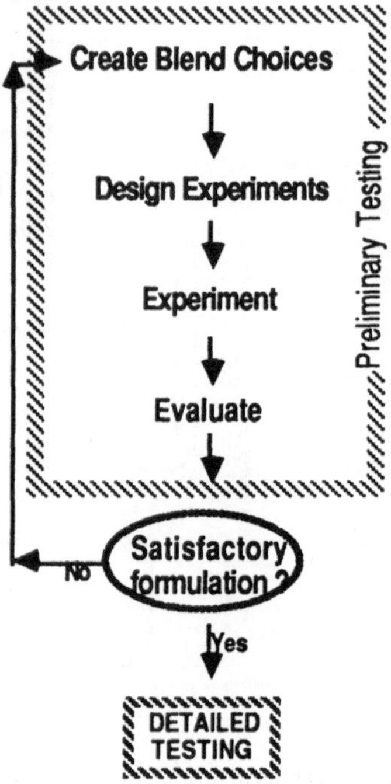

Fig. 2 The create, experiment and evaluate cycle

Base Stock: Three base stocks (BS-1, BS-2, BS-3) are available. The volume fraction of base stock in the blend is restricted to the range 75% - 95%.
Viscosity Index Improver: Three VI packages (VI-1, VI-2, VI-3) are available. These packages contain polymers that stabilize blend viscosity at elevated temperatures. This package also includes a pour point depressant for controlling low temperature properties of the blend. The volume fraction of this additive in the blend is restricted to the range 0% - 10%.
Dispersant Inhibitor: Three DI packages (DI-1, DI-2, DI-3) are available. Dispersants are solubilizing agents that reduce sludge and varnish deposits by keeping contaminants suspended in the oil. This package is a mixture of different chemicals that affect several blend properties. The volume fraction of this additive can have values of *small*, *medium* or *large*. The DI-2 package is not miscible in base stock BS-2 and cannot coexist in a blend.

Detergent: Three alternative detergent packages (DET-1, DET-2, DET-3) are available. Detergents keep engine parts clean from the accumulation of deposits associated with oxidation. This process results in oil thickening leading to difficult cold starts and reduced fuel economy. The volume fraction for each alternative is bounded in the ranges 0% ≤ DET-1 ≤ 0.8%, 0% ≤ DET-2 ≤ 0.5% and 0% ≤ DET-3 ≤ 1.0%.

TABLE 1
Properties of components available for blending

Alternatives	Kinematic Viscosity at 40°C cSt.	Kinematic Viscosity at 100°C cSt.	Pour Point °C	Specific Gravity	Calcium Wt. %	Zinc Wt. %	Sulfur Wt. %	Sulfur Sulfonate Wt. %
BS-1	20.65	4.60	0.0	0.850	NA.	NA.	NA.	NA.
BS-2	32.10	5.50	0.0	0.860	NA.	NA.	NA.	NA.
BS-3	43.33	7.00	0.0	0.880	NA.	NA.	NA.	NA.
VI-1	12600.00	920.00	NA.	0.908	NA.	NA.	NA.	NA.
VI-2	10000.00	1200.00	NA.	0.862	NA.	NA.	NA.	NA.
VI-3	1660.00	465.00	NA.	0.902	NA.	NA.	NA.	NA.
DI-1	1042.00	60.40	-18.0	0.982	2.00	1.30	2.1	NA.
DI-2	936.20	53.00	-18.0	1.021	2.83	1.74	4.1	NA.
DI-3	978.00	104.60	-15.0	1.003	2.40	1.40	3.1	NA.
Det-1	530.00	38.00	-18.0	0.972	2.70	NA.	NA.	NA.
Det-2	480.00	40.00	-1.0	1.220	15.80	NA.	NA.	NA.
Det-3	1575.00	180.00	NA.	1.160	12.00	NA.	NA.	30.0

A feasible blend has to satisfy conditions on kinematic viscosity (at 40°C and 100°C) and the pour point temperature. These values are approximated for the blend using mixing rules based on the volume fraction of each component. In addition, components are blended to minimize sulfur content while maximizing resistance to oxidation. A rough estimate of oxidative resistance can be made based on the quantity of high ph compounds in the blend. These compounds offset formation of reaction products at elevated operating temperatures (greater than 100°C). Effects of oxidation and sulfur content are estimated using heuristics provided as rules.

1.3 COMPLEXITY AND CONTROL HEURISTICS

Providing computer-based decision support for problems in the early stages of design is affected by two types of complexity (Kamal 1990):

- *Complexity in Representation.* Design problems in the early stages of design may not be completely modeled using traditional modeling techniques. The first reason for this is the values-types associated with variables. Design variables may attain values in discrete and nonuniform intervals. The second reason is that the underlying relationships may be highly discontinuous and nonlinear. Relationships may also be expressed as *heuristics*, or *rules*, that cannot be put into exact mathematical expressions.

- *Computational Complexity*. The representational complexity of the problem precludes the use of exact methods and requires that inexact methods be used[1]. Kamal (Kamal 1990) shows that the number of possible solutions for certain types of problems increases exponentially with the size of the problem and the exhaustive search of all possible solutions becomes a NP-Hard problem. Consequently, intelligent search methods must be used to generate and evaluate alternative designs.

Intelligence of the search method depends on its ability to prune and explore the search space utilizing available domain-relevant information. For the HDSP this intelligence is provided through the following:

- The creation of an internal representation scheme for a design space that can be intelligently traversed, defined by standard keywords and descriptors, and based on object-oriented programming concepts.
- Defining a search strategy that gives a designer various levels of control.

In the rest of this paper we discuss the representation model for the HDSP and provide and overview of the search strategy in the context of designing continuous-manufacturing products.in general and a lubricant in particular.

2. The Heuristic DSP

Blend design involves both the selection of components and the determination of their quantities. The HDSP provides a bridge between the designers view of the world and the method used to generate and evaluate solutions. This is done in the form of keywords and descriptors that help model and provide decision support for the following decisions:

- *The Selection Decision:* The components of the system are determined without any consideration being given to the size/amount of each component.
- *The Compromise Decision:* Each of the components in the system is sized or its effective amount in the system is determined.
- *The Concurrent Selection-Compromise Decision:* The components are selected and their size within the system is determined simultaneously.

From a perspective of providing decision support for the design problem the formulation of the HDSP is used

- independent of the mathematical form of the design variables and relationships, and

1 Exact methods involve the creation of an approximate model of the real world to obtain a mathematically exact solution. Inexact methods involve using approximate methods to solve an approximate model.

- the key words and descriptors help customize/control the search methods in a generic manner.

The HDSP consists of various levels of abstraction defined by the keywords and descriptors as shown in Figure 3. The keywords help organize domain relevant information in terms of the descriptors such that it can be used in the solution process.

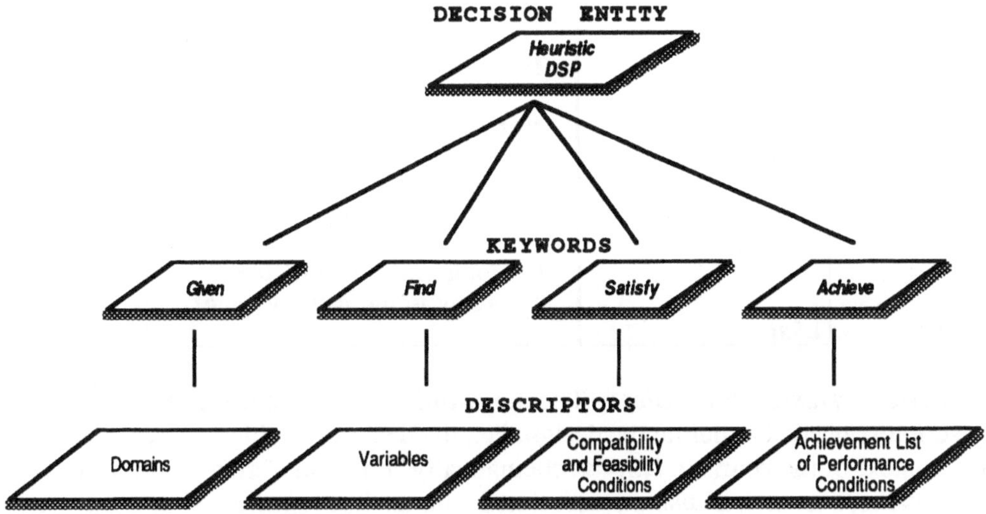

Fig. 3 The heuristic DSP keywords and descriptors

2.1 THE KEYWORDS AND DESCRIPTORS FOR THE HEURISTIC DSP

The keywords and descriptors for the HDSP are summarized in Table 2. The role of the descriptors (domains, variables, compatibility, feasibility and performance conditions, and the achievement list) within the HDSP formulation is defined by the keywords (Given, Find, Satisfy and Achieve).

GIVEN - Reference Data: The keyword *Given* is used to identify domain relevant *reference* data needed for the solution process. The most important descriptor here is called *domains*. It is used to keep track of the domain of values that are attained by various entities. A schematic of the domain data type is shown in Figure 4a. Each domain has a name, type and a list of symbols. Two types of domains, numeric or nonnumeric, may be used. A numeric domain is defined by three quantities, the lower bound, upper bound and the perturbation step size stored in the list of symbols. A nonnumeric domain is defined by a set of symbols, which represent the permissible values that may be attained. Based on this definition of domain a value within the HDSP formulation is defined as shown in Figure 4b. A *value* within the HDSP contains within itself a reference to its value type, i.e., the domain and may represent numeric or nonnumeric information. For instance, the domain of values for base stock are continuous and bounded by 75% to 100%, whereas those for the DI additive are discrete and belong to the set {small, medium, large}.

TABLE 2
The keywords and descriptors for the heuristic DSP

Keywords	*Descriptors*
Given	**Domains**
Find	**Variables**
Satisfy	**Compatibility Conditions** **Feasibility Conditions**
Achieve	**Performance Conditions** according to the **Achievement**
List	

FIND - Design Variables: The keyword `Find` is used to group the descriptors needed to define the design, i.e., the design variables. A *variable* in the context of the heuristic DSP is schematically shown in Figure 4c. A variable has a name, *choice* and *amount* associated with it. Both *choice* and *amount* are instances of value as shown in Figure 4b, i.e., they have an *associated domain* and attain values within this domain. For instance, the variable BS indicates the choice of selected base-stock and its amount in the blend.

SATISFY - Conditions for Compatibility, Feasibility: The restricting relationships within the design space may be modelled using the *compatibility* and *feasibility* conditions grouped under the keyword `Satisfy`. Conversely, these conditions may be used to control the heuristic search process at various levels. The compatibility conditions are used to represent restrictions that make the system infeasible regardless of the amount of each component present in the system. Compatibility conditions help exclude ***a set*** of alternative designs from further consideration. The feasibility conditions depend on the choice of a component and the size/amount in which it is used. These conditions are used to include, or exclude, ***a*** design from further consideration. Feasibility conditions are less restrictive then the compatibility conditions. Subsequently, compatibility conditions may have a greater impact on the solution process than the feasibility conditions. The compatibility and feasibility conditions are analogous to constraints in the traditional single objective mathematical programming (Taha 1987) or *hard-goals* in goal programming (Ignizio 1983). These conditions are treated as Boolean (0/1) indicators of compatibility and feasibility.

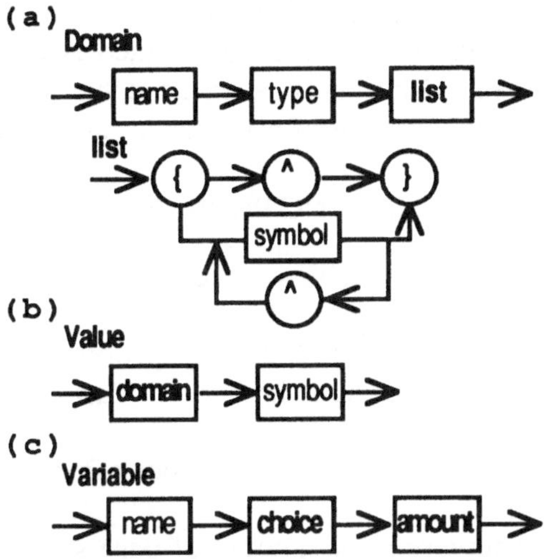

Fig. 4 Domains, values and variables for the synthesis engine

ACHIEVE - The Achievement List of Performance Conditions: The keyword *Achieve* is used to group various measures of performance as a ranked list of performance conditions called the *achievement list.* Performance conditions evaluate some *value* (Figure 4b) and are used to represent the goodness of designs. This value can be numeric or nonnumeric but must belong to a predefined *domain of values* (Figure 4c). The value of the performance condition provides an estimate for comparing competing designs to determine the path of most improvement and/or to stop the search process. The role of the performance conditions as multiple objectives is analogous to *soft-goals* in goal programming (Ignizio 1983).

Traditionally, multiple objectives have been transformed into an aggregate measure of performance. This is done by lumping the objectives into a single utility function. Appropriate weights are used to quantify the importance of each of the objectives. This method is appropriate if all objectives are commensurate with each other, i.e., they are similar in terms of units and range of values. Another approach is to preemptively rank each objective. In this scheme the most important objective is to be achieved first and then the next and so on. The underlying principle in this scheme is that the lower priority objectives should be improved only if the attainment of higher priority objectives are not adversely affected. This scheme is appropriate when objectives are disparate and cannot be compared on the same scale. The concept of preemptive ranking is further explained in the Appendix.

The HDSP is used to provide decision support to a human designer faced with a design problem characterized by information from different sources and different value types. Multiple objectives for the HDSP cannot be combined into a single utility function. Therefore a preemptive or a lexicographically ranked scheme is used and multiple objectives are represented as an ordered ranking of all the performance conditions. This list is called the *achievement list* and is used to keep track of how different measures of performance are achieved.

2.2 THE LUBRICANT BLENDING PROBLEM AS A HEURISTIC DSP

The design variables and relationships associated with the lubricant blending problem follow.

Design Variables: The lubricant blending problem identified earlier is characterized by the following design variables[2]. Various attributes associated with the variables are listed in Tables 1.

1 Design variable to indicate the selected base stock from the available choices - BS-1, BS-2 and BS-3.

2 Design variable to represent the amount of base stock, the value of which varies continuously between 75% to 100% by weight.

3 Design variable to indicate the selected VI additive from the available choices - VI-1, VI-2 and VI-3.

4 Design variable to represent the amount of VI additive, the value of which varies continuously between 0% to 10% by weight.

5 Design variable to indicate the selected DI additive from the available choices - DI-1, DI-2 and DI-3.

6 Design variable to represent the amount of DI additive, the value of which varies discretely and attains the value of *small, medium* or *large.*

7 Design variable to indicate the selected Detergent additive from the available choices - DET-1, DET-2 and DET-3.

8 Design variable to represent the amount of Detergent additive , the values of which varies continuously, based on the choice of additive, i.e., $0\% \leq$ DET-1 $\leq 0.8\%$, $0\% \leq$ DET-2 $\leq 0.5\%$ and $0\% \leq$ DET-3 $\leq 1.0\%$.

In addition to the variables the following relationships must be satisfied:

1 The incompatibility of BS-2 and DI-2.

2 Kinematic viscosity at 40°C must be less than 95.0 cSt.

3 Kinematic viscosity at 100°C must be greater than 10.0 cSt.

4 Total weight of sulfur must be less than 3.5 ppm.

And, the following criteria for performance must be maximized:

1 Kinematic viscosity at high temperatures.

2 Oxidation resistance of the blend.

The lubricant blending problem in terms of the HDSP descriptors and keywords is represented as follows.

---

2 Note that this is a simplistic view of the required design variables, a larger set of variables may have to be used if traditional mathematical programming models are used.

Design Variables

- BS represents the choice and amount of base stock. The amount of this component varies continuously between 75% to 100% by weight.
- VI represents the choice and amount of VI additive. The amount of this additive varies continuously between 0% to 10% by weight.
- DI represents the choice and amount of DI additive. The amount of this additive varies discretely and attains the values of *small*, *medium* or *large*.
- DET represents the choice and amount of detergent additive. The amount of this additive varies continuously based on the choice of additive that is selected as defined in Section 1.2.

Compatibility Conditions

BS_vs_DI - models the incompatibility of BS-2 and DI-2.

Feasibility Conditions

KV40 - kinematic viscosity at 40°C must be less than 95.0 cSt.

KV100 - kinematic viscosity at 100°C must be greater than 10.0 cSt.

Sulfur - total weight of sulfur must be less than 3.5 ppm.

Performance Conditions

KV-HT - kinematic viscosity at high temperatures needs to be maximized.

Oxidation - oxidation resistance of the blend needs to be maximized.

A formal representation of the HDSP follows. Details about this representation format are provided in (Kamal 1990) and a brief explanation is given in the footnotes.

Given *§1*

<***domain:*** (***name:*** perc_BS) (***type:*** 0) (***list:*** {75.0 100.0 5.0})>

<***domain:*** (***name:*** perc_VI) (***type:*** 0) (***list:*** {0.0 10.0 0.1})>

<***domain:*** (***name:*** percent) (***type:*** 0) (***list:*** {0.0 100.0 1.0})>

<***domain:*** (***name:*** norm) (***type:*** 0) (***list:*** {0.0 1.0 0.01})>

<***domain:*** (***name:*** norm-8) (***type:*** 0) (***list:*** {0.0 0.8 0.01})>

<***domain:*** (***name:*** norm-5) (***type:*** 0) (***list:*** {0.0 0.5 0.01})>

<***domain:*** (***name:*** frac_DI) (***type:*** 1) (***list:*** {small moderate large})> *§1.1*

<***domain:*** (***name:*** BS_choices) (***type:*** 1) (***list:*** {BS-1 BS-2 BS-3})>

<***domain:*** (***name:*** VI_choices) (***type:*** 1) (***list:*** {VI-1 VI-2 VI-3})>

<***domain:*** (***name:*** DI_choices) (***type:*** 1) (***list:*** {DI-1 DI-2 DI-3})>

<***domain:*** (***name:*** Det_choices) (***type:*** 1) (***list:*** {Det-1 Det-2 Det-3})>

Find

---

§1 The background information, i.e., the bounds on variables, and the reference data.

§1.1 These are value domains for different variables (and performance conditions). For instance, "perc_BS" defines the domain of values (i.e., the percentage) in which the Base Stocks, listed in Table 1, are added. The minimum value in which a base stock can be added in 75% and the maximum value is 100%.

<**variable:** (**name:** BS) (**choice:** <**domain:** BS_choices> BS-1) (**amount:** <**domain:** perc_BS> 90.0)>
<**variable:** (**name:** VI) (**choice:** <**domain:** VI_choices> VI-2) (**amount:** <**domain:** perc_VI> 0.7)>
<**variable:** (**name:** DI) (**choice:** <**domain:** DI_choices> DI-2) (**amount:** <**domain:** perc_VI> small)>
<**variable:** (**name:** DET) (**choice:** <**domain:** Det_choices> Det-1) (**amount:** <**domain:** norm-8> 0.1)> §2

Satisfy §3

<**comp-cond:** (**name:** BS-Compat) (**type:** 1) (**properties:** <**domain:** NULL> {NULL})> §3.1
<**feas-cond:** (**name:** KV40) (**type:** 0) (**properties:** <**domain:** comp_prop> {KV@40C})>
<**feas-cond:** (**name:** KV100) (**type:** 0) (**properties:** <**domain:** comp_prop> {KV@100C})>
<**feas-cond:** (**name:** environment) (**type:** 1) (**properties:** <**domain:** NULL> {NULL})> §3.2

Achieve §4

{KV-HT OXIDATION} §4.1
where
<**perf-cond:** (**name:** KV-HT (**type:** 0) (**preference:** high) (**properties:** <**domain:** comp_prop> {KV100}) (**value-dom:** <**domain:** kv-value>)>
<**perf-cond:** (**name:** OXIDATION (**type:** 1) (**preference:** high) (**properties:** <**domain:** NULL> NULL) (**value-dom:** <**domain:** symb_1>)> §4.2

---

§2 The name of variables. A variable for each component is declared, e.g., "BS" as the first variable represents the choice (from the domain "BS_choices") and amount (from the domain "perc_BS") of base stock.

§3 The feasibility and compatibility conditions.

§3.1 The compatibility condition about the miscibility of the alternatives BS-2 and DI-2.

§3.2 The feasibility conditions. For instance, "KV40" represents the feasibility of the blend with respect to the Kinematic Viscosity at 40°C.

§4 The achievement list and performance conditions.

§4.1 The achievement list, where the high temperature performance (KV-HT) is ranked higher than oxidation resistance (OXIDATION).

§4.2 The performance conditions. For instance, the second one represents oxidation resistance, it is evaluated using a rule base, i.e., its type=1, a higher value is preferred, and the value type belongs to the domain "norm".

3. Integrating Heuristic Search and Knowledge-Based Systems: The Heuristic DSP Synthesis Engine

The HDSP is solved, using heuristic search, by the synthesis engine. The ability of our synthesis engine to handle design variables and relationships, independent of their mathematical form is a unique and powerful feature. The greatest impact of this feature is the ability to represent relationships as *rules* enabling the use of design criteria as a *knowledge-base*. In the following we first describe how a knowledge base is utilized within the solution process and then provide an overview of the search strategies for the synthesis engine.

3.1 USING KNOWLEDGE-BASED SYSTEMS WITH THE SYNTHESIS ENGINE

The CLIPS[3] (Giarratano 1988) inference engine is embedded within the synthesis engine to support design analysis using rules. This lets a designer specify any of the compatibility, feasibility and performance conditions as rules. A schematic representation of how CLIPS is utilized is shown in Figure 5. The information needed for analysis, i.e., the current values of the design variables and other related information is asserted as facts into the CLIPS working memory[4] before rules are evaluated. The synthesis engine makes CLIPS resolve the rules that match the condition being evaluated by issuing the *run* command. In this phase, embedded functions are used by the rules to access additional information, e.g., reference data, and to change global variable values. The results of the evaluation, following the resolution of a rule-set (when no more rules may be fired), are extracted from the working memory and from external/global variables.

As an example, a simple rule in CLIPS is shown in Figure 6. This rule models the compatibility condition BS_vs_DI for the lubricant blending problem. The facts needed to fire the rules are asserted by the synthesis engine and the rule returns the condition, in this case a 0 indicating incompatibility, through a function called `set_clips_flag`. The clauses before =>[5] must be true for the rule to fire and execute the actions. The rule in Figure 6 fires when the compatibility condition BS_vs_DI needs to be evaluated (indicated by the fact *`(Current-Partition BS_vs_DI)`*), and BS-2 and DI-2 exist together in the blend (indicated by the fact *`(COMP ? ? ? BS-2 $?)`* and *`(COMP ? ? ?`*

---

3 A knowledge-based system development shell.

4 An area in memory where current facts are kept while the CLIPS inference engine evaluates the rules.

5 The symbol => is equivalent to *then* , i.e., the antecedent conditions precede => and the consequent actions follow =>.

DI-2 $?)) as blend components in which case the blend is incompatible. Other examples are given in English text in Figure 7. The first rule establishes that a measure of oxidation resistance is given by the percent weight of Calcium and/or Calcium Sulfonate in the blend. The second rule indicates that in the absence of Calcium and/or Calcium Sulfonate the oxidation resistance is 0.0.

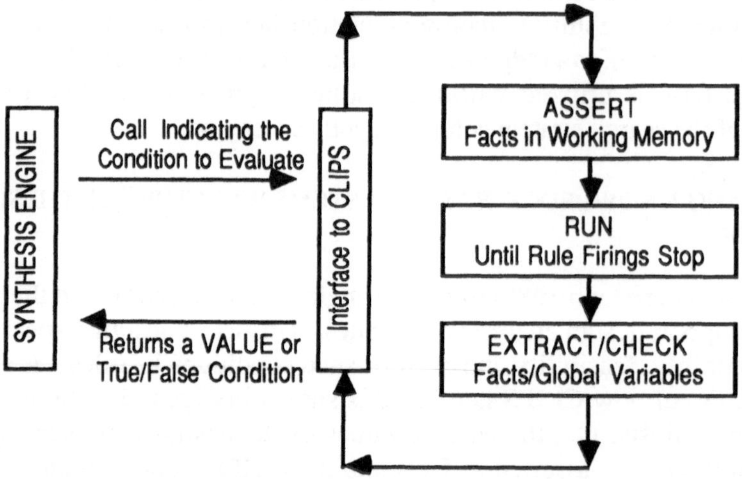

Fig. 5 Utilization of CLIPS from the Synthesis Engine

```
(defrule compat-1 "Check compatibility of BS vs. DI"
       (Current-Partition BS_vs_DI)
        ?f1 <- (COMP ? ? ? BS-2 $?)
        ?f2 <- (COMP ? ? ? DI-2 $?)
=>
        (retract ?f1 ?f2)
        (set_clips_flag 0)
)
```

Fig. 6 A sample rule in CLIPS

a	IF the percentage weight of Sulfur is within limits and Calcium is present or Calcium Sulfonate is present THEN the percentage weight of Calcium and Calcium Sulfonate provides a measure of Oxidation resistance.
b	IF the percentage weight of Sulfur is within limits and there is neither Calcium nor Calcium Sulfonate THEN the measure of Oxidation resistance is 0.0.

Fig. 7 Examples of rules used to approximate the performance condition OXIDATION

3.2 SEARCH STRATEGY FOR THE SYNTHESIS ENGINE

The heuristics used to create the search strategy for the synthesis engine are shown schematically in Figure 8 and described as follows:

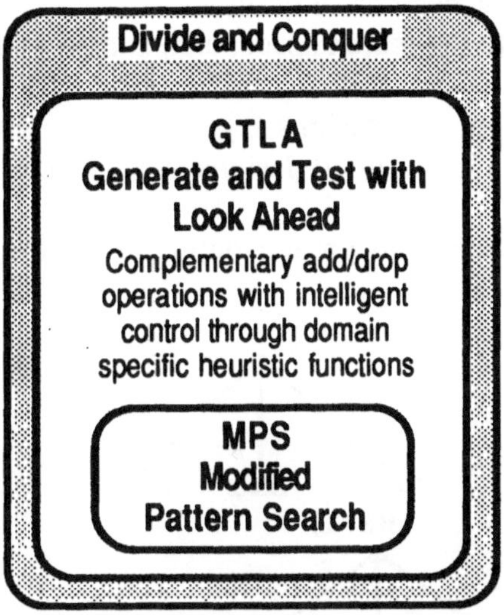

Fig. 8 The heuristics within the synthesis engine

- *The divide and conquer heuristic:* The overall strategy for the heuristic is to generate alternative designs by the systematic partitioning of the problem.
- *The add/drop heuristic:* A Generate and Test with Look Ahead heuristic is developed to generate, evaluate and rate alternative designs. This process is controlled by domain specific heuristics specified as *compatibility*, *feasibility* and *performance* conditions. The role of the compatibility and feasibility conditions is similar to the heuristic functions in the AO^* algorithm (Rich 1983) which help a designer determine the benefit of exploring certain paths and to prune the search space by providing look ahead capabilities. Additionally, a ranked list of designs is maintained as a stack. This stack is updated as better designs emerge.
- *The steepest ascent heuristic:* A Modified Pattern Search (MPS) strategy has been developed to improve the designs with respect to the amount of each component. The MPS is controlled by GTLA and is used for estimating the goodness of alternative designs.

Detailed information about the development and implementation of these heuristics is presented in (Kamal, Garson et al. 1991).

The Generate and Test Heuristic: The GTLA heuristic is used to provide a shell within which alternative designs are generated and improved. In addition to being controllable, it has been shown to be complete and nonredundant (Kamal 1990). A graphical representation of GTLA, based on the lubricant blending example, is provided in Figure 9. Alternative designs generated by GTLA are shown within square parenthesis. The unique features of GTLA are

- recursive depth-first generation of designs,
- the ability to truncate the search space, e.g., subtrees in Figure 9, by looking ahead,
- the ability to ignore specific designs, e.g., nodes in Figure 9, and
- the maintenance of a list of good designs as a ranked list for the purpose of comparison.

These features are described in greater detail in (Kamal 1990) and (Kamal, Garson et al. 1991).

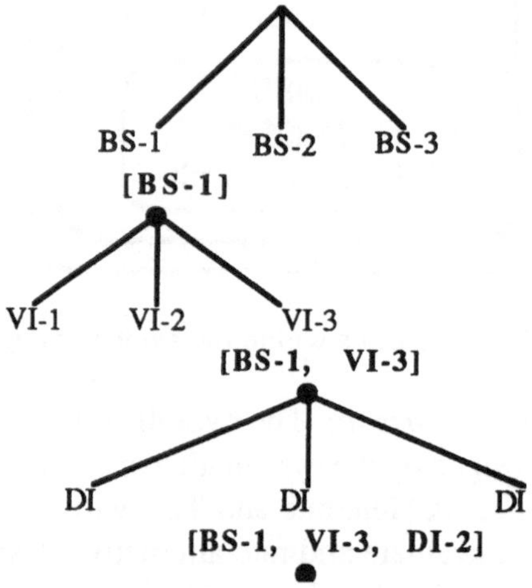

Fig.9 The GTLA heuristic.

3.3 THE MODIFIED PATTERN SEARCH (MPS) HEURISTIC

MPS is used to improve the compatible and feasible designs generated by the GTLA heuristic. MPS is based on the pattern-search method (originally proposed by Hooke and Jeeves (Nicholson 1971)), for function minimization problems. The traditional pattern search heuristic works by defining a direction of improvement, called the pattern, in a multidimensional space. The pattern is found by perturbing each variable and finding the direction of most gain. The search space is explored by making *pattern moves* until the function cannot be improved further. The process of finding a pattern and making pattern moves is repeated until none of the variable can be improved. The speed with which the search space is explored is controlled using an acceleration factor. Features that make the MPS heuristic unique, intelligent, and controllable by a designer are

- using multiobjective criteria for measuring the goodness of designs by lexicographically evaluating the performance conditions listed in the achievement list,
- utilizing the bounds on variables to enhance the pattern movement scheme,

- using a fixed acceleration factor and perturbing the numeric variables according to the designer specified step size and the nonnumeric variables to the next value in the list of permissible values, and
- including a one-dimensional move function to supplement pattern moves for the case where pattern moves cannot be made due to nonuniformities in the search space.

These features are described in greater detail in (Kamal 1990) and (Kamal, Garson et al. 1991).

3.4 AN OVERVIEW OF THE SYNTHESIS ENGINE

An overview of the strategy used within the synthesis engine is schematically presented in Figure 10. The steps in the solution process are shown using a sequential representation for the purpose of simplicity. Designs are generated based on the GTLA heuristic. In Figure 10 points *a, b* and *c* are the points of evaluation where different conditions are evaluated using either C language functions or CLIPS, as shown in Figure 5. The MPS heuristic is used at point *c* to improve selected designs. A list of "good" designs is maintained as a stack which is updated as better designs are generated.

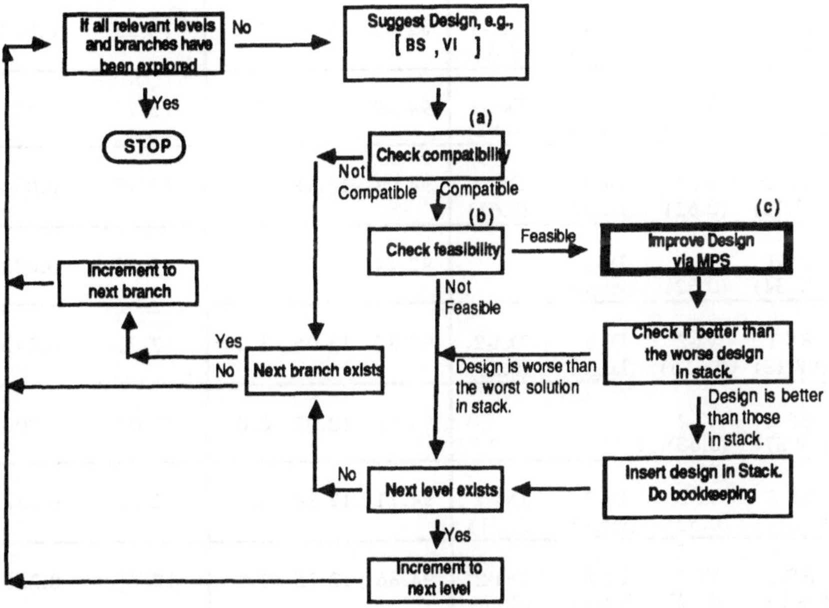

Fig. 10 Conceptual layout of the search strategy for the synthesis engine

4. Designing a Continuous-Manufacturing Product using Heuristic DSP

A sample of 10 designs generated by the synthesis engine for the lubricant blending problems are presented in Table 3. Each row, in Table 3, represents a

TABLE 3
Results for the lubricant blending problem

	VARIABLES Choice (Amount)				FEASIBILITY CONDITIONS			PERFORMANCE CONDITIONS	
No	BS (amount) Parts by Weight	VI (amount) Parts by Weight	DI (amount) Parts by Weight	DET (amount) Parts by Weight	KV40 < 95.0 cSt.	KV100 > 10.0 cSt.	Sulfur < 3.5 ppm.	KV-HT High Value Preferred cSt.	Oxidation High Value Preferred % Wt.
1	BS-3 (97.18)	VI-3 (0.91)	DI-3 (large)	Det-3 (0.91)	81.29	13.71	3.1	13.7	0.562
2	BS-1 (99.26)	VI-2 (0.74)			94.55	13.45	0.0	13.45	0.000
3	BS-3 (97.15)	VI-3 (0.92)	DI-1 (large)	Det-3 (0.92)	82.42	13.37	2.1	13.35	0.560
4	BS-1 (97.67)	VI-2 (0.52)	DI-3 (large)	Det-3 (0.82)	94.49	13.21	3.1	13.2	0.521
5	BS-1 (97.76)	VI-2 (0.62)	DI-3 (large)	Det-1 (0.61)	94.91	13.18	3.1	13.15	0.096
6	BS-1 (98.34)	VI-2 (0.62)	DI-3 (large)		92.44	13.05	3.1	13.05	0.038
7	BS-1 (98.13)	VI-2 (0.615)	DI-3 (large)	Det-2 (0.23)	92.87	13.05	3.1	13.00	0.243
8	BS-2 (99.37)	VI-2 (0.63)			94.82	13.02	0.0	13.00	0.000
9	BS-2 (98.17)	VI-2 (0.52)	DI-3 (large)	Det-2 (0.31)	94.71	12.80	3.1	12.80	0.252
10	BS-1 (98.01)	VI-2 (0.62)	DI-1 (large)	Det-2 (0.36)	94.86	12.75	2.1	12.75	0.255

unique design generated by the synthesis engine identified by the values of the variables, status of feasibility and performance conditions. For instance, the first row represents a design consisting of BS-3 in the amount of 97.18% weight, VI-3 in the amount of 0.91 % weight, DI-3 in the amount of *large*, and DET-3 in the amount of 0.91% weight. The feasibility condition KV40 has a value of 81.29 cSt. The feasibility condition KV100 has a value of 13.71 cSt. The feasibility condition SULFUR has a value of 3.1 ppm. The performance condition KV-HT has value of 13.71 cSt. The performance condition OXIDATION has value of 0.562% wt.

Designs listed in Table 3 are ranked according to the priorities assigned to performance conditions KV-HT and OXIDATION. The first row in Table 3 represents the best design amongst those generated based on the performance conditions. The designs in row 7 and 8 exemplify the preemptive ranking of the performance conditions. These designs have the same value of KV-HT and are ranked on the basis of the OXIDATION performance condition.

5. Closure

Historically, new formulations are made iteratively, largely based on prior experience and previous designs, and experimentally tested for specific measures of performance. A designer, in the early stages of blend design, relies on experimentation due to the unavailability of comprehensive property estimation models, existence of nonuniform variables, and the use of heuristics. Using the synthesis engine a designer is able to model comprehensively the design problem and control the exploration of the design space. We have also shown in (Kamal 1990) that inductive learning can be used as the feedback mechanism to learn qualitative relationships about existing blends and incorporating them into the design model for evaluating the blends being designed by the synthesis engine.

In our opinion, the HDSP augments current practice by providing a structured environment that has the following advantages:

- It is a tool for rapid prototyping in the initial stages of design.
- Helps incorporate different types of information within the decision making process.
- Enables a designer to generate and evaluate simultaneously a large number of design concepts.

The HDSP is not limited to blend design. It can be used in the adaptive design of continuous-manufacturing products. The future? Exploring the efficacy of using the HDSP in other domains and in designing discretely manufactured products.

Acknowledgements

Saiyid Kamal was supported from a grant from the Texas Advanced Technology Program (Grant No. 3652-227). The financial contribution of our corporate sponsor, The BF Goodrich Company, is gratefully acknowledged.

References

Finger, S. and Dixon, J. R.: 1989a, A review of research in mechanical engineering design. Part 1: Descriptive, prescriptive, and computer-based models of design processes, *Research in Engineering Design*, **1**, 51-67.

Finger, S. and Dixon, J. R.: 1989b, A review of research in mechanical engineering design. Part 2: Representations, analysis, and design for the life cycle, *Research in Engineering Design*, **1**, 121-137.

Giarratano, J. C.: 1988, *CLIPS User's Guide*, Version 4.2, Artificial Intelligence Section, Lyndon B. Johnson Space Center, NASA.

Ignizio, J. P.: 1983, Generalized goal programming: an overview, *Computers and Operations Research*, **5**(3), 179-197.

Kamal, S. Z.: 1990, *The Development of Heuristic Decision Support Problems for Adaptive Design*, PhD Dissertation, Department of Mechanical Engineering, University of Houston, Houston, Texas.

Kamal, S. Z., Garson, J., Van Arsdale, W. E. and Mistree, F.: 1991, Development of a synthesis engine for the design of products made by process manufacturing, *Proceedings 1991 IEEE International Conference on Systems, Man and Cybernetics*, University of Virginia, Charlottesville, 1839-1846.

Mistree, F., Hughes, O. F. and Bras, B. A.: 1992, The compromise decision support problem and the adaptive linear programming algorithm, *Structural Optimization: Status and Promise*, AIAA, Washington, DC.

Mistree, F., Smith, W. F., Bras, B. A., Allen, J. K. and Muster, D.: 1990, Decision-based design: a contemporary paradigm for ship design, *Transactions, Society of Naval Architects and Marine Engineers*, Jersey City, New Jersey, 565-597.

Nicholson, T. A. J.: 1971, *Optimization in Industry*, Aldine-Atherton, Inc., Chicago, chap. 6.

Rich, E.: 1983, *Artificial Intelligence*, McGraw-Hill, New York.

Taha, H. A.: 1987, *Operations Research: An Introduction*, Macmillan, New York.

Appendix

A lexicographic minimum (Ignizio 1985) is defined as follows: Given an ordered array, say **a**, of nonnegative elements a_k's, the solution given by $\mathbf{a}^1$ is preferred to $\mathbf{a}^2$ if $a_k^1 < a_k^2$ and all higher-order elements (i.e., $a_1, \ldots, a_{k-1}$) are equal. If no other solution is preferred to **a**, then **a** is the lexicographic minimum. As an example, consider two solutions, $\mathbf{a}^1$ and $\mathbf{a}^2$, where

$\mathbf{a}^1 = (0, 10, 400, 56)$, $\mathbf{a}^2 = (0, 11, 12, 20)$, and $\mathbf{a}^1$ is preferred to $\mathbf{a}^2$.

The value 10 corresponding to $\mathbf{a}_2^1$ is smaller than the value 11 corresponding to $\mathbf{a}_2^2$. Once a preference is established then all higher order elements are assumed to be equivalent.

2+3 MODEL: FRACTAL PROCESSES FOR KNOWLEDGE-BASED ENGINEERING DESIGN

Q. CHEN*
Research Institute of Engineering Mechanics
Dalian University of Technology
Dalian 116023
PR China

Abstract. A novel model, the 2+3 model, for the knowledge—based engineering design using fractal concepts is presented. In this model, complex, large—scale engineering design problem is considered to be a fractal object; the design process solving for this problem is considered to be a fractal process; the fractal process is organized by the basic 2+3 model in statistically self—affine way. The basic 2+3 model includes two basic processes supported by three basic operations. The two basic processes are search for the most promising possible solution subspace and search through the subspace for feasible solution. The three basic operations are pattern operation, symbolic operation, and numerical operation.

1. Introduction

This study has been undertaken to cope with complex, large-scale engineering design problem-solving. The central concern of the research for design methods is to find out the universal laws governing the successful engineering design processes, especially those processes involving *Open Complex Giout System*, in order to provide concepts, tools, and methods for building knowledge-based systems coping with complex, large-scale engineering design problems. A system is open complex giout system (Qian, Yu and Tai, 1990; Qian, 1991; Tai, 1991) if i) there is exchange in mass, energy, or information between the system and its environment; ii) the number of its subsystems is very large; iii) the number of classes of its subsystems is large; and iv) in the hierarchy of the system structure, there are many levels some of which are vague or unknown in advance, and even the number of the levels in the hierarchy is unknown too. Comparing these features with design

* Supported by the National Natural Science Foundation of China.

J. S. Gero (ed.), Artificial Intelligence in Design '92, 903–924.

processes, it is obvious that a complex, large—scale engineering design process corresponds to an open complex giout information system.

Design from pattern, design from logic, and design from both pattern and logic, this is the historical chain of evolution of decision—making methods for engineering design.

The mainstream of design methods in practice has been design from pattern since very early times of human civilization. Patterns may be classified as two groups: natural objects and artifacts. Here artifacts are some existing, typical, successful designs. In essence, design from pattern is the recognition and reorganization of patterns. In this way, the design knowledge from the God on natural objects and those from previous generations of human being on artifacts can be utilized to produce new artifacts. For instance, Nara, the first permanent capital of Japan, was built by using the pattern of Chang'an, the capital of China in the Tang dynasty, 1,200 years ago.

Since 1950s the development of operations research, systems engineering, formal logic, artificial intelligence, and electronic computer leads to great enthusiasm for *systematic design approach*, namely, design from logic. The systematic approach involves three representative research lines: design methodology (Jones, 1963; Alexander, 1963), design optimization, and knowledge-based design.

The validity of systematic design approach is based on such a belief that it is easy to work out a series of logical or arithmetic steps that will arrive at the solution at acceptable cost. However, people has realized that working out such series is very difficult or virtually impossible for complex, large-scale engineering design problems.

2. Features of Engineering Design

An engineering design problem may be stated as follows:

$$\left.\begin{array}{l}\text{Find value for the variable } D\\ \text{such that relationship of the form}\\ R(E,D)\\ \text{is satisfied}\end{array}\right\} \qquad (1)$$

In problem 1, D called *design variable* denotes an artifact, E called *environment parameter* stands for given action of environment, and $R(E,D)$ called *constraint* represents the laws of nature, defined by design variable D, and the requirements in function, safety, cost, and culture D has to satisfy. Obviously problem 1 has a quite universal form, and D, E or $R(E,D)$ may be mathematically indescribable. Of course, problem 1 could be turned into an optimization model in some cases. But before the optimization problem can be stated, we have to find out enough partial solutions for problem 1, which are necessary to constitute the optimization problem. Before we proceed to the author's approach, a brief discussion on the features of engineering design problem seems to be appropriate.

First, engineering design is an ill-structured problem where goal, scope, and permissible operations can not been specified in advance, and become definite gradually as problem-solving process proceeds (Simon, 1973).

Second, engineering design is a "solution-focused" process where human experts generate basic concepts of solution very early , and develop these concepts until a feasible solution is arrived (Lawson, 1979).

Third, engineering design is based on productive logic rather than traditional deductive or inductive logics (March,1976).

Fourth, engineering design involves processing information indescribable in linguistic terms, and with graphic or spatial representations.

Lastly, engineering design is based on two type of knowledge: the discrete knowledge and the organized knowledge (Gero, 1987; Chen and Zhong,1987). The discrete knowledge is used in thinking in terms of logic, whereas the organized knowledge in terms of images.

3. Current Model and Its Limits

The most common model in current knowledge-based systems for engineering design is *production system* performing *heuristic reasoning* directed by *planning*. Planning is a kind of technique for process task decomposition by which the goal of a complex, large-scale problem-solving process is transformed into a hierarchical structure of subgoals of simpler, smaller subprocesses. Heuristic reasoning is a search method using local knowledge on problem solution. These local knowledge termed heuristics can incorporate domain knowledge based on experience in solving similar search problems. Production system is a collection of production rules that is interpreted to produce behavior according to a specified procedure called the production system architecture. Usually, there is a Working Memory of assertion in the architecture, against which the patterns of the rules are matched. A properly specified architecture also has a method for deciding which rule(s) to select in case more than one is matched to the current situation. A continuous stream of behavior is produced by repeating a cycle of pattern matching, conflict resolution, and execution of actions. Here key term is *pattern matching*. Pattern matching is a detailed, piece-by-piece comparison of a template with a configuration of data objects; and is implemented by symbolic string matching in computer language.

The validity of the model is based on such three hypothesis:

i) *physical symbol system* hypothesis: any information processing system can exhibit intelligence, if and only if it can input symbols, output symbols, store symbols, duplicate symbols, build symbolic structures, and shift conditionally;

ii) *complex problem divisibility* hypothesis: any complex problem can be divided into a number of less complex subproblems, then each such subproblem can be divided into a number of simpler subproblems again, and so on;

iii) *rule representation for procedural knowledge* hypothesis: any procedural knowledge including experience-based domain knowledge can be coded into production rules.

Applying the model described above to complex, large-scale engineering design problems such as the preliminary design of cable-stayed bridges runs into the following two inherent difficulties:

i) Uncertainty in problem decomposition. As the complexity of design problems increases, some levels, even the number of levels, in problem hierarchy may become vague or unknown. In this case, complex problem divisibility hypothesis hold no longer and planning technique based on problem decomposition can not work completely.

ii) Combinatorial explosion. As the scale of design problems increases, there are so many combinations of the components at different levels of problem hierarchy that it becomes infeasible to explore them all, even if a definite problem hierarchy can be obtained.

The two inherent difficulties reflect limits of current model, that is, physical symbol system hypothesis, complex problem divisibility hypothesis, and rule representation for procedural knowledge hypothesis can hold only in the case that the problem to solve is simple and small. Unfortunately, most of engineering design problems are not such cases.

Conclusion is clear: we need a new model for complex, large-scale engineering design problem-solving. Some results on this model have been presented. It seems that the most important progress in the research on design model has been the work on prototypes of Gero (1987). In the same year, I presented a three-phase design model in which the concepts and techniques from pattern recognition research are used to solve for engineering design problems (Chen and Zhong, 1987). This work leads to the following 2+3 model.

4. 2+3 Model

4.1. BASIC 2+3 MODEL

I believe that *human design expert's problem-solving process is constituted by a number of basic processes, called basic 2 + 3 model, in some systematic way; basic 2+3 model includes two basic processes supported by three basic operations; the two basic processes are search for the most promising possible solution subspace and search for feasible solution through the subspace; the three basic operations are pattern operation, symbolic operation, and numerical operation.*

The search for the most promising possible solution subspace is to recognize dynamically a subspace from original possible solution space, which is so promising that optimum solution often fall into it, and so small that combinatorial explosion does not take place. In the case that the search for feasible solution through the current subspace can not succeed, it is necessary to search for a new possible solution subspace.

The search for feasible solution through a subspace may be carried out by using current model, that is, production system performing heuristic reasoning directed by planning. Comparing this process with the search through whole of original possible solution space, significant differences are:

i) the scope of possible solution space to search is reduced greatly;

ii) the current possible solution space is a promising one.

The three basic operations implementing the two basic processes are *pattern operation*, *symbolic operation*, and *numerical operation*. Generally speaking, pattern operation is used in the search for the most promising possible solution subspace, symbolic and numerical operations are used in the search through current subspace for feasible solution.

Term *pattern* is used in this paper to stand for a configuration which is constituted by a number of components based on certain relationships. It is the organization property that is the characteristic of a pattern regardless of its degree of abstract. These are the examples of pattern: object, form, structure, image, fashion, artifact, music, and design scheme. Whole of the patterns concerned constitutes a pattern space. In a pattern space, three pattern operations are defined, that is, *pattern recognition*, *pattern addition*, and *pattern association*.

4.1.1. *Pattern Recognition* The engineering design decision — making processes based on pattern operation may be classified into two groups: fast process called pattern recognition and slow process including pattern addition and pattern association.

Carrying out pattern recognition raises the following four issues:

i) *characteristic parameters* describing the key constraints on environment and requirement;

ii) *design patterns* representing the "optimum" value of design variable D under the current circumstances, drawn from existing, successful, typical design schemes or natural objects;

iii) *weight functions* amplifying, or reducing the effect of the individual characteristic parameters;

iv) *algorithm* implementing pattern recognition.

The design configuration generated by pattern recognition is considered as an origin of a most promising subspace of original possible solution space rather than a final solution. Usually pattern recognition could generate a "global optimum" possible solution subspace if enough design patterns have been recorded, key constraints of the problem are seized, and appropriate weight functions are defined. Here, pattern as a concept may be regarded as a kind of generalization of the term pattern used in visual perception, cognitive psychology, or production system.

In fact, the fast design process based on pattern recognition is a type of key technique used by engineering and management experts in various domains in order to cope with complex, large-scale decision-making problems such as engineering design. Depending on degree of abstract of pattern space concerned, pattern recognition may take different form:

i) *Case as pattern*, that is, the previous design schemes giving all the details are selected as pattern space. A typical example was the planning of Nara, the first permanent capital of Japan, as mentioned in the introduction of this paper.

ii) *Structure as pattern*, that is, the structures drawn from the previous design cases or natural objects are selected as pattern space. For instance, the pattern in invention of the first handsaw in the world must be the sawtooth structure of some plant leaves. There are many such instances: bird wing pattern of airplane, fish tail pattern of ship helm, brain pattern of neurocomputer, eggshell pattern of roof, etc.

iii) *Effect as pattern*, that is, the effects from various phenomena are selected as pattern space. For example, the pattern in invention of electric motor is the effect that any conductor in a magnetic field will be subjected to the action of magnetic force. There are many such instances: photoelectricity effect pattern of solar energy cell, nuclear fission effect pattern of nuclear power station, etc.

iv) *Concept as pattern*, that is, the concepts from various cultures are selected as pattern space. For example, the *siheyuan* is the universal layout pattern in Chinese traditional architecture, since it represents the basic concept of Chinese ancient philosophy on the world. Similarly, the steeple pattern in European architecture reflects some religious concepts on the world.

Let us emphasize here:

i) Pattern is used to represent the partial solutions of an engineering design problem instead of the engineering design problem itself.

ii) Pattern is an appropriate representation form for partial solutions, and other organized representation forms such as case and shape may be considered as the instances of pattern.

4. 1. 2. *Pattern Addition* Carrying out pattern recognition demands a similarity measure defined somehow. In general, a pattern with the largest value of the similarity measure is considered to correspond to

current design problem. In some cases, however, pattern recognition is governed by some threshold value associated with the similarity measure: recognition process succeeds only if the largest value of the similarity measure is beyond this threshold value. Hence, recognition process probably does not succeed for a given threshold value . This situation demands creation of new patterns and leads to the slow processes based on pattern operation, namely, pattern addition and pattern association.

Pattern addition creates a new complex pattern by integrating a number of basic patterns. Most of creative engineering designs have been worked out by using this approach: tank as the integration of cannon and armored vehicle; AWACS as the integration of radar, computer, and airplane; etc. The key issue arising in the pattern addition is how to obtain basic patterns to be added. This can be done as follows:

i) Decompose current design problem based on the requirements of function, and obtain a number of new design problems;

ii) Apply fast design process based on pattern recognition to these new design problems.

If the recognition process in stage ii succeeds, a number of patterns corresponding to the new design problems can be obtained and taken as the basic patterns in pattern addition. Otherwise, different decomposition in stage i should be explored. Hence pattern addition often is an iterative, slow procedure. Cable-stayed bridge is a typical case of creating new pattern by using pattern addition. Traditionally, substructure of bridges has been unitary load passing system such as piers of beam bridge, serving as a system against compression, or cables of suspension bridge, serving as a system against tension. As the main span increases, the system against compression demands larger cross section of the superstructure whereas the system against tension more complex anchorage construction. As a result, the long span bridges with unitary load passing system serving as substructure have cost much more than they should. This situation stimulates the creation of new bridge pattern. By decomposing the requirement " passing loads through substructure" into two requirements " passing loads through system against compression" and " passing loads through system against tension" respectively, two new design problems are specified. Applying pattern recognition to the two new problems produces a typical beam bridge pattern plus a typical suspension bridge pattern. Integrating the two basic patterns leads to a new bridge pattern, namely, cable—stayed bridge.

4. 1. 3. *Pattern Association* Pattern association creates a pattern by mapping a member in a pattern space to some member in another pattern space. The key issue arising in pattern association is how to obtain the pattern space serving as target space to be associated. This can be done as follows:

i) Expand current pattern space to produce a larger pattern space;

ii) Wipe out the original pattern subspace from the pattern space newly produced.

Having such target pattern space, pattern recognition can be used to search for the most promising possible solution subspace. Curved cable-stayed bridge is a representative case of creating new pattern by using pattern association. The problem is to build a bridge to cross a deep valley 400 meters wide. Arch bridge, beam bridge, or ordinary straight cable-stayed bridge are not desirable, for they will cost too much due to complex environment. Thus, a new bridge pattern is necessary. Pattern addition, however, does not work in this case. Therefore, pattern association is used to attack this difficult design problem. This process starts with expanding ordinary straight cable-stayed bridge pattern space into arbitrary planar curve cable-stayed bridge pattern space, and then wipes out all straight bridge patterns. As a result, a creative, perfect curved cable-stayed bridge pattern is worked out. This is famous R. A. C highway bridge design of T. Y. Lin.

4. 1. 4. *Symbolic Operation Symbol* is a type of special patterns whose distinction instead of organization is concerned in symbolic operation. The basic task or function of a physical symbol system is to identify the same symbols and to distinguish the different symbols. Symbolic operation is a deductive reasoning process based on the physical symbol system hypothesis by using formal logic. Usually, it is heuristic reasoning of production system that symbolic operation is.

Once pattern operation has found out the most promising possible solution subspace, search through this subspace for feasible solution will begin with taking the typical design configuration standing for the subspace as a goal of production system. Planning technique is used to decompose the goal into a set of subgoals and heuristic rules are evoked to match these subgoals against basic facts. If a subgoal can not obtain enough support from basic facts, the subgoal has to be replaced by a new hypothesis. This often is referred to as backward reasoning, since the reasoning process proceeds from goals to basic facts.

4. 1. 5. *Numerical Operation* Numerical operation is quantitative symbolic operation. After the design variables at higher levels have been determined by pattern operation and symbolic operation, a complex, large-scale engineering design problem often becomes an optimization problem. The key issues arising in numerical operation include mathematical models and algorithms, management of engineering database, and representation and manipulation of graphs. In general, these issues could be coped with by using conventional software techniques.

4. 2. FRACTAL 2+3 MODEL

As mentioned in previous section, human design expert's problem-solving process is constituted by the basic 2+3 model in some systematic way. This systematic way is the central issue to investigate in this section.

In fact, basic 2+3 model is introduced on the assumption that design processes are "integral": the basic processes, searching for the most promising possible solution subspace and searching for feasible solution through the subspace, never embed each other. In other words, a pattern as a whole is either accepted or refused in design process. However, the two basic processes and three basic operations of basic 2+3 model usually embed each other in complex, large-scale engineering design processes, which looks like chaos phenomenon in some nonlinear dynamic systems. How is what we observe on the macroscopic scale related to the "microscopic behavior"? I feel that fractals are essential for the description and understanding of this relation.

Mandelbrot's *fractal geometry* (1982) provides a useful mathematical model for analysis of many natural or physical phenomena.

Mandelbrot offers the following tentative definition of a fractal:

A fractal is by definition a set for which the Hausdorff-Besicovitch dimension strictly exceeds the topological dimension.

In 1986, Mandelbrot has retracted this tentative definition and proposes instead the following:

A fractal is a shape made of parts similar to the whole in some way.

The fractals we discuss may be considered to be sets of points embedded in space. The concept of a distance between points in space is central to the definition of the Hausdorff—Besicovitch dimension and therefore of the fractal dimension D. A simple way to measure the length of curves, the area of surfaces, or the volume of an object is to divide space into small cubes of side δ or small spheres of diameter δ instead. By counting the number of spheres needed to cover the set of points we obtain a measure of the size of the set.

The set S is *statistically self-similar* when S is the union of N distinct subsets each of which is scaled down by r from the original and is identical in all statistical respects to $r(S)$.

The set S is *statistically self-affine* when S is the union of N nonoverlapping subsets $S_1, \ldots, S_n$, each of which is scaled down by a ratio vector $\mathbf{r} = (r_1, \ldots, r_n)$ from the original and is identical in all statistical respects to $\mathbf{r}(S)$.

Comparing fractal shapes with complex, large-scale engineering design problems, I believe that *complex, large-scale engineering design problem is a fractal object; the design process solving for this problem is a fractal process; the fractal process is organized by the basic 2+3 model in statistically self—affine way.*

4. 2. 1. *Engineering Design Problem as Fractal Object* As mentioned previously, design is a "solution-focused" process in which a number of partial solutions gradually evolve and arrive at a complete solution finally. In fact, this is also a construction process of a fractal object as shown in figure 1, where a design example is used to illustrate that a design problem may be a fractal object.

The construction of the design problem starts with very limited constraints available, denoted by set D_1. This starting form is called the *initiator* and may be replaced by a number of sets $T_l(l=1,2,\ldots,i,j,$

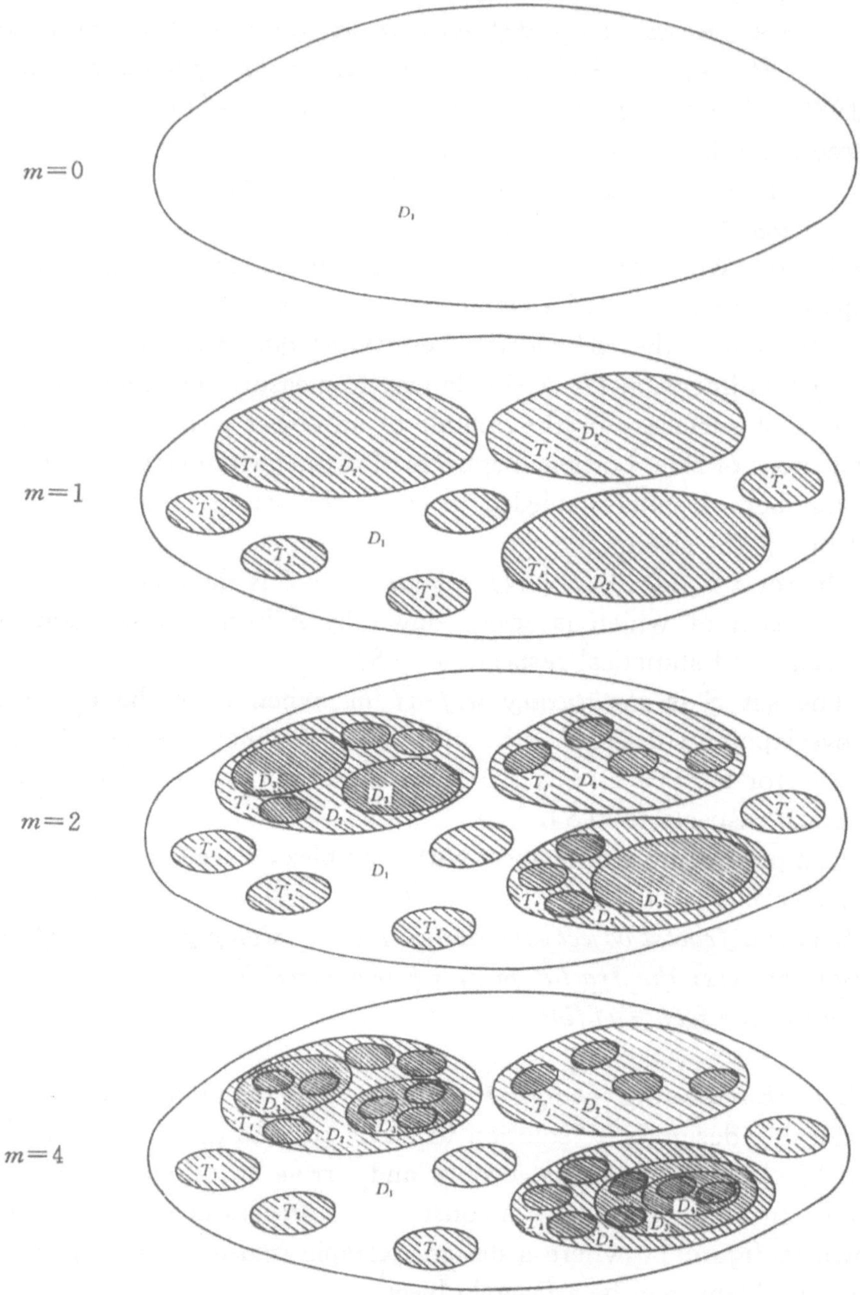

Fig. 1 Construction of a design problem

$k, \ldots n$), which stand for possible partial solutions produced by the pattern operation in the basic $2+3$ model. The initiator is the 0-th generation of the design problem. The construction of the design problem proceeds by replacing the initiator by sets T_l, which are produced by the *generator*, namely, basic $2+3$ model. Thus we obtain the first generation marked $m = 1$ in figure 1, which is a design problem of n subproblems. The next generation is obtained by replacing each set marked D_2 (that is, T_i, T_j, T_k), which correspond to the possible partial solutions not obtaining enough support from the symbolic and numerical operations in basic $2+3$ model, by a scaled-down version of the generator. In other words basic $2+3$ model is used again but the newly produced design problems such as T_i, T_j, and T_k are smaller and/ or simpler than the original marked $m = 0$. Thus in the second generation we have a design problem consisting of more subproblems. By applying a reduced generator, basic $2+3$ model, to all subproblems of a generation of the design problem, which correspond to the possible partial solutions not obtaining enough support from the symbolic and numerical operations, a new generation is obtained.

The fractal description of engineering design problems used here are only loosely based on fractal geometry, since the construction process of a design problem can not proceed on and on. Hence, complex, large-scale engineering design problem should be considered to be a *partially fractal* object.

It is important to realize that one of key issues in the description of a fractal object is its generator, which specifies how a fractal can be constituted from an original form. In complex, large-scale engineering design problem, this generator is basic $2+3$ model.

4.2.2. *Complexity Degree of Engineering Design Problem* The concept "*complexity degree of engineering design problem*" is ambiguous, since one has not been able to give a measure describing inherent complexity of design process quantitatively yet for a long time. Having regarded design problem as a fractal object, *the complexity degree of an engineering design problem may be defined as its Hausdorff-Besicovitch dimension naturally*.

According to Feder (1988), the Hausdorff-Besicovitch dimension D of a set S is the *critical dimension* for which the measure M_d changes from zero to infinity:

$$M_d = \sum \gamma(d)\delta^d = \gamma(d)N(\delta)\delta^d \xrightarrow[\delta\to 0]{} \begin{cases} 0, & d>D; \\ \infty, & d<D. \end{cases} \qquad (2)$$

Here $\gamma(d) = 1$ for lines, squares, and cubes used to cover the set; or $\gamma = \pi/4$ for disks, and $\gamma = \pi/6$ for spheres. The value of M_d, called the d-measure of the set, for $d = D$ is often finite; it is the position of the jump in M_d as a function of d that is important.

The definition (2) of the fractal dimension can be used in practice. Consider again the design problem shown in figure 1. At first, we cover set D_1 standing for the design problem with a set of squares with edge length δ. Each such square contains a piece of information available in design process. Then counting the number of squares needed to cover the design problem gives the number $N(\delta)$. Now we may proceed as implied by equation (2) and calculate $M_d(\delta)$, or we may simply go ahead and find $N(\delta)$ for smaller values of δ. Since it follows from equation (2), that asymptotically in the limit of small δ

$$N(\delta) \sim 1/\delta^D \qquad (3)$$

we may determine the fractal dimension of the design problem by finding the slope of ln $N(\delta)$ plotted as function of ln δ. This dimension D, determined from equation (3) by counting the number of boxes needed to cover the set as a function of the box size, is now called the *box counting dimension*.

4.2.3. *Engineering Design Process as Fractal Process* So far, we have discovered that complex, large-scale engineering design problem is a fractal object. Now we may proceed to give the design process solving for this problem, namely, *fractal* 2 + 3 *model*. A procedure called **F23M** is used to describer the fractal 2+3 model. Procedure **F23M** has three parameters: current design problem *DESIGN*, which is taken as original very limited constraints from environment and requirement in the 0−th generation; previous generation constraint *CONSTRAINT*, which consists of a number of partial solutions of original design problem and is taken as null set in the 0-th generation; and *SOLUTION*, which is composed of all current partial solutions of original design problem and is taken as null set at first. Procedure **F23M** is defined as follows:

Procedure **F23M**(*DESIGN*, *CONSTRAINT*, *SOLUTION*)

1. {*AD*, *ND*} ← **E23M** (*DESIGN*); **E23M** is a procedure implementing the basic 2+3 model. *AD* is the possible partial solution set of *DESIGN* accepted by **E23M** whereas *ND* is the possible partial solution set of *DESIGN* refused by **E23M.**
2. *SOLUTION* ← {*SOLUTION*, *AD*}; Expand current partial solutions of original design problem.
3. **if** *ND* = *NULL* **then return**; Success of **E23M** leads to success of **F23M.**
4. *NEWC* ← {*CONSTRAINT*, *AD*}; Expand current previous generation constraint.
5. **for** $i = 1$ **to** m **do** Solve for new design problems represented by m
6. **begin** components of *ND*.
7. *LOCALC* ← {*DESIGN*, $NC(ND_i)$}; Construct *local constraint* *LOCALC*. $NC(ND_i)$ is a set of new constraints from environment and requirement, introduced by ND_i, the ith possible partial solution of *DESIGN* refused by **E23M.**
8. *NEWD* ← {*LOCALC*, *NEWC*}; Construct new generation of design problems by combining current previous generation constraint and the local constraint.
9. **F23M** (*NEWD*, *NEWC*, *SOLUTION*); Solve for current design problem.
10. **end**

Here relationship *CONSTRAINT* ⊂ *SOLUTION* always holds. To illustrate this, consider again the design problem shown in figure 1. Now assume that problem-solving process is in the first generation; design problems T_j and T_k have been processed. To solve for current design problem T_i, previous generation constraint only includes the partial solutions in the 0-th generation, that is, all of T_l but T_i, T_j, and T_k, independent of the partial solutions accepted in problems T_j and T_k.

5. Implementation of Pattern Recognition

5.1. PATTERN RECOGNITION USING FUZZY THEORY

Assume that the domain discussed, U, consists of all of possible design configurations. A design problem is represented by its N characteristic parameters , that is, key constraints in current problem. Note these key constraints involve not only environment and requirement but also the partial solutions as previous generation constraint in procedure **F23M** presented in section 4. 2. Let the maps of these characteristic parameters on real field be $x_j (j=1,2, \ldots, N)$. M design patterns, denoted by A_i $(i=1,2, \ldots, M)$, and each characteristic parameter of these patterns, denoted by $A_{ij} (i=1,2, \ldots, M; j=1,2, \ldots, N)$ are considered to be fuzzy subsets on U. Take the form of membership function of fuzzy subset A_{ij} as

$$\mu_{ij}(x_j)=e^{-((x_j-a_{ij})/b_{ij})^2} \quad \begin{array}{l} i=1,2,\cdots,M \\ j=1,2,\cdots,N \end{array} \tag{4}$$

Where a_{ij}, $b_{ij}>0$ are determined by statistical approach.

Pattern recognition is carried out based on *maximum subordination principle*. Suppose we have a sample x_j^* $(j=1,2, \ldots, N)$ standing for the current design problem. This sample is considered to be a member of fuzzy subset A_I, that is, the Ith design pattern, if

$$\mu_i(x_1^*, x_2^*, \cdots, x_N^*)=\min_{1\leqslant j\leqslant N} \mu_{ij}(x_j^*) \qquad i=1,2,\cdots,M \tag{5}$$

$$\mu_I(x_1^*, x_2^*, \cdots, x_N^*)=\max_{1\leqslant i\leqslant M} \mu_i(x_1^*, x_2^*, \cdots, x_N^*) \qquad 1\leqslant I\leqslant M \tag{6}$$

This minimizing-maximizing process is rigorous mathematically. In some cases, however, it is not appropriate since too much information concerned with other N-1 characteristic parameters is lost in equation (5). To consider these cases, weight factors may be introduced as functions of characteristic parameters and equation (5) is rewritten as follows:

$$\mu_i(x_1^*, x_2^*, \cdots, x_N^*)=\sum_{j=1}^{N} W_{ij}(x_1^*, x_2^*, \cdots, x_N^*)\mu_{ij}(x_j^*)$$
$$i=1,2,\cdots,M \tag{7}$$

The equations solving for weight functions $W_{ij}(x_1^*, x_2^*, \cdots, x_n^*)$ are as follows:

$$\frac{W_{ij}(x_1^*,x_2^*,\cdots,x_N^*)}{W_{iN}(x_1^*,x_2^*,\cdots,x_N^*)}=\frac{\overline{W}_{ij}\mu_{iN}(x_N^*)}{\overline{W}_{iN}\mu_{ij}(x_j^*)} \qquad \begin{matrix} i=1,2,\cdots,M \\ j=1,2,\cdots,N \end{matrix} \tag{8}$$

$$\sum_{j=1}^{N} W_{ij}(x_1^*,x_2^*,\cdots,x_N^*)=1 \qquad i=1,2,\cdots,M \tag{9}$$

As a result, we have obtained $W_{ij}(x_1^*,x_2^*,\cdots,x_N^*)$ as follows:

$$W_{ij}(x_1^*,x_2^*,\cdots,x_N^*)=\overline{W}_{ij}\frac{\mu_{i1}\mu_{i2}\cdots\mu_{ij-1}\mu_{ij+1}\cdots\mu_{iN}}{\sum_{k=1}^{N}\overline{W}_{ik}\mu_{i1}\mu_{i2}\cdots_{ik-1}\mu_{ik+1}\cdots\mu_{iN}} \tag{10}$$

Where $\mu_{il}=\mu_{il}(x_l{}^*)$.

5. 2. PATTERN RECOGNITION USING NEURAL NETWORK

A neural network consists of a collection of processing elements. Each processing element has many input signals, but only a single output signal. The output signal fans out along many pathways to provide input signals to other processing elements. These pathways connect the processing elements into a network.

There are several important neural network models, such as Hopfield net, Hamming net, Carpenter/Grossberg net, single-layer perceptron, multilayer network, etc. The single-layer Hopfield net is normally used with a binary input and output. It can be used as associative memories to solve problems such as pattern recognition. It is Hopfield net that provides a promising alternative form in order to perform pattern recognition in the 2+3 model. The weights in a Hopfield net are set using the given example patterns in a training stage. Input for a known pattern is introduced to the network. Output of each neuron is fed back to all other neurons via the connection weights. The output pattern is compared with the known pattern, and weights are adjusted if necessary. This process of feeding back neuron output then iterates until the outputs remain nearly unchanged over two iterations.

Modeling the pattern recognition process in the 2+3 model with a

Hopfield net raises the following two issues:

i) Define a design problem in terms of neurons;

ii) Derive connection weights, W_{ij}, of these neurons.

Suppose that there are M design patterns represented by N_c key constraints involving environment, requirement, and partial solutions as previous generation constraint in procedure **F23M**. To use binary network to solve for current design problem, the subset of real field, on which the key constrains are defined, have to be mapped to another set consisting of N_d discrete values. Now we may define a design pattern as a *hyper — character* with the resolution of $N = N_c * N_d$. Thus N neurons are used to constitute a Hopfield net knowing M design patterns, each described by an input vector $X^s = (x_1^s, x_2^s, \cdots, x_N^s)$, $1 \leqslant s \leqslant M$. The network runs as follows:

i) Learn connection weights w_{ij} according to Hebb rule

$$w_{ij} = \begin{cases} \sum_{s=1}^{M} x_i^s x_j^s & i \neq j \\ 0 & i = j \end{cases} \qquad i, j = 1, 2 \cdots, N \tag{11}$$

Here, x_i^s denotes the ith component of the sth pattern, that is, the sth typical design problem.

ii) Input initial sample $X = (x_1, x_2, \cdots, x_N)$ standing for current design problem, and initialize the network

$$u_i(0) = x_i \qquad 1 \leqslant i \leqslant N \tag{12}$$

Here, $u_i(t)$ is the output signal from element i at time t.

iii) Introduce *Lyapunov energy function* and iterate such that the energy of the network is minimized

$$u_i(t+1) = f\left(\sum_{j=1}^{N} w_{ij} u_j(t) + \theta_i \right) \qquad i = 1, 2, \cdots, M \tag{13}$$

Here, $f(y)$ is a transfer function and θ_i is the threshold value or the solution bias of element i.

In the end, the Hopfield network will arrive at an attractor. The attractor is considered to be optimum approximation of current design problem, and the solution of this attractor standing for a typical design problem is regarded to be the solution of current problem.

6. A Case Study

Consider a practical case of complex, large-scale engineering design problem. This is the overall design of Shanghai Baoshan Iron and Steel Complex, the largest iron and steel enterprise in China. The overall design problem involves many design variables such as the sources of ore, the site of the complex, the production capability, the layout of the production lines, the production equipment, the transport way from the sources to the site, the transport mode within complex, etc. Four key constraints are taken account of, that is, the grade of ore, the terrain of territory, the probability of international risk, and the transport cost of ore. These key constraints are the characteristic parameters of current pattern operation. The design pattern space considered comprises all iron and steel complexes in the world.

The typical design configuration generated by the pattern operation (pattern recognition in this stage) is an artifact whose template is Ōita Steelworks of Nippon Steel Corporation of Japan: the sources of ore is Australia; the site of the complex lies in a coastal region of China, which is fairly industrialized and is convenient to construct new ports for transportation of ore; the transport way from the sources to the site is ocean shipping; and other design variables such as the layout of the production lines, the production equipment, and the transport mode within complex are decided by using Ōita Steelworks as reference too. These possible solutions obtained by pattern recognition correspond to the possible partial solutions $T_l(l=1,2,\ldots,i,j,k,\ldots n)$ of the 0-th generation of the design problem as a fractal object, shown in figure 1. In symbolic and numerical operations afterwards, most of T_ls are accepted, others are refused and lead to new design problems like T_i, T_j, and T_k.

" the complex lies in a coastal region of China, which is fairly industrialized and is convenient to construct new ports for transportation of ore" is a partial solution accepted by the basic 2 + 3 model, for example. In symbolic operation, this possible partial solution as a goal to verify is decomposed as four subgoals: i) find a city CITY; ii) near

CITY, constructing new ports for transportation of ore is convenient; iii) CITY has rich technical force to support the further development of the complex; iv) CITY needs a large-scale iron and steel complex. Shanghai, the largest industrial center in China, could satisfy all of these subgoals, and therefore the partial solution on the site of the complex is accepted with CITY=Shanghai.

However, "the port of the complex is constructed by filling sea" is a partial solution refused by the basic 2+3 model, since its subgoals, that rich funds is available and land price is very high, can not be satisfied in symbolic operation. This situation raises the issue of pattern addition or pattern association. Because most of possible partial solutions, especially those at high level, in Ōita Steelworks pattern are accepted by the basic 2+3 model, the similarity measure of current design problem to Ōita Steelworks pattern can be considered to be beyond some threshold value, and therefore Ōita Steelworks pattern as a whole is available. Hence, "the port of the complex" should be considered to be a new design problem, marked by T_k for instance, to solve by the basic 2+3 model in next generation in figure 1 instead of pattern addition or pattern association in current generation.

The design variables of the new design problem marked by T_k include mainly the type of the port and the site of the port. The key constraint of the problem involves the previous generation constraint and the local constraint. The previous generation constraint consists of all of accepted partial solutions of the 0-th generation of design problem. And the local constraint includes the four key constraints in the 0-th generation and some new constraints from environment and requirement, introduced by the refused possible partial solution "the port of the complex". These new constraints include: i) available berth for ocean ship of 10,000 tons displacement; ii) acceptable construction cost of the port; iii) short distance between the large city nearby (Shanghai, in this case) and the port. The design pattern space considered in the scaled-down version of the 2+3 model comprises all ports in the world. Thus we obtain a design problem in the first generation.

The typical design configuration generated by the pattern operation (pattern recognition in this stage) is: the type of the port is the landing stage port; the site of the port is within Yangtze river. This, however, is not an available pattern, since one of two possible partial solutions, " the site of the port is within Yangtze river", can not satisfy the subgoal " the port has available berth for ocean ship of 10, 000 tons displacement" in symbolic operation. This situation leads to pattern addition. Pattern addition begins with decomposing the requirement " the port has available berth for ocean ship of 10,000 tons displacement" denoted by R into two requirements R_1 and R_2. R_1 is " the port has available berth for ocean ship of 5,000 tons displacement" and R_2 is "the port has available berth for ocean ship of 10,000 tons displacement". Adding respectively R_1 and R_2 to design problem " the port of the complex" and deleting R from it leads to two new design problems. Applying pattern recognition to them gives two patterns: pattern 1 is " ordinary port at the harbor nearby"; pattern 2 is "landing stage port within Yangtze river". Pattern addition of the two basic patterns creates a new pattern: the port system of the complex consists of a near port with shallow water berths and a far port with deepwater berths; a cargo ship from Australia discharges first half of the ore it loads at the far port, and then discharges another half at the near port. It is the new pattern that is the solution of the design problem " the port of the complex".

Besides the partial solution "the port of the complex is constructed by filling sea", there are some possible partial solutions refused by the basic 2 + 3 model in the 0-th generation of design problem, such as " production equipment are decided by using Ōita Steelworks as reference". They become new design problems in the first or higher generation and are solved by using the fractal 2+3 model or procedure **F23M**.

7. Conclusions

This paper presents a novel model, the 2+3 model, for the knowledge-based engineering design using fractal concepts. Some significant conclusions on complex, large-scale engineering design problem can be drawn from this study:

1. Scaling fractals may be used in the description of nature of complex, large-scale engineering design problem as an approximation. That is, *design implies fractal problem, fractal process, and fractal logic.*

2. 2+3 model provides an appropriate tool for processing complex, large-scale engineering design problems by using knowledge-based systems.

3. 2 + 3 model can perform creative engineering design by using pattern recognition where structure as pattern, effect as pattern, or concept as pattern are used, or by using pattern addition and pattern association as discussed in section 4. 1.

References

Chen, Q. and Zhong, W. X.: 1987, Frame production, multiple level reasoning, and calling for executable programs, *Proc. of the Workshop on Expert System in Engineering Application*, Shanghai, P. R. China (in Chinese).

Cross, N. (ed.): 1984, *Developments in Design Methodology*, Wiley, New York.

Cross, N.: 1990, The nature and nurture of design ability, *Design Studies*, 11, 127—140.

Feder, J.: 1988, *Fractals*, Plenum Press, New York.

Gero, J. S.: 1987, Prototypes: a new schema for knowledge-based design, *Working Paper*, University of Sydney.

Mandelbrot, B. B.: 1982, *The Fractal Geometry of Nature*, W. H. Freeman, New York.

Qian, X. S., Yu, J. Y. and Tai, R. W.: 1990, A new discipline of science —— the study of open complex giout system and its methodology, *Nature Journal*, 13, 3—10 (in Chinese).

Qian, X. S.: 1991, The open complex giout system, *Pattern Recognition and Artificial Intelligence*, 4, 1—4 (in Chinese).

Simon, H. A.: 1973, The structure of ill-structured problems, *Artificial Intelligence*, 4, 181—200.

Tai, R. W.: 1991, Some remarks on meta-synthetic engineering, *Pattern Recognition and Artificial Intelligence*, 4, 5—10 (in Chinese).

AUTHOR INDEX

AUTHOR ELECTRONIC ADDRESSES

Alberts, L. K., alberts@cs.utwente.nl
Arakawa, M., +81 33 209 9176
Bahler, D., drb@adm.csc.ncsu.edu
Ball, N. R., nrb@eng.cam.ac.uk
Barber, J.,
Bardasz, T., tbardasz@cvzbnet.prime.com
Bauert, F., +44 223 33 2662
Baykan, C. A., cab@cs.cmu.edu
Bhansali, S., bhansali@sumex-aim.stanford.edu
Bhat, R. R., raghu@arc.cmu.edu
Bhatta, S., bhatta@cc.gatech.edu
Birmingham, W. P., wpb@engin.umich.edu
Bowen, J., jabowen@adm.csc.ncsu.edu
Brown, K. N., ken.brown@bristol.ac.uk,
Cagan, J., jcagan@globe.edrc.cmu.edu
Cerbone, G., cerbone@ptolemy.arc.nasa.gov
Cha, J., +86 22 31 8329
Chandrasekaran, B., chandra@cis.ohio-state.edu
Chawla, A.,
csesun!kalyan!chawla@vikram.doe.ernet.in
Chen, Q., +86 411 47 1009
Corne, D. W., dave@aifh.edinburgh.ac.uk
Coyne, R. F., robert.coyne@cad.cs.cmu.edu
Cutkosky, M. R., cutkosky@sunrise.stanford.edu
Davies, S. P., spd@psychology.nott.ac.uk
Demaid, A., +44 908 65 3658
Domeshek, E. A., domeshek@cc.gatech.edu
Faltings, B., faltings@elma.epfl.ch
Finn, D. P., dfinn@vax1.tcd.ie
Flemming, U., ulrich.flemming@cad.cs.cmu.edu
Fox, M. S., msf@phoenix.rose.utoronto.ca
Fromont, B., +3392927190
Gauchel, J., jupp@arc.cmu.edu
Gero, J. S., john@archsci.arch.su.oz.au
Goel, A., goel@pravda.cc.gatech.edu
Gruber, T. R., gruber@sumex-aim.stanford.edu
Guo, W., +86 22 31 8329
Hua, K., hua@lia.di.epfl.ch
Iwasaki, Y., iwasaki@sumex-aim.stanford.edu
Iyengar, G., giridhar@engin.umich.edu
Kamal, S. Z., kamal@mwk.uucp
Klein, M., mklein@atc.boeing.com
Kolodner, J. K., jlk@indigo.cc.gatech.edu
Konkar, R., konkar@sunrise.stanford.edu
Kota, S., kota@ub.cc.umich.edu
Lee, C.-L., cllee@engin.umich.edu
Logan, B. S., brian@aifh.edinburgh.ac.uk
MacCallum, K., ken@cad-centre.strathclyde.ac.uk
MacKellar, B., bonnie@vienna.njit.edu
Maher, M. L., mary@archsci.arch.su.oz.au
Mars, N. J. I., mars@cs.utwente.nl
Milzner, K.,
milzner@jupiter.informatik.uni-dortmund.de
Mistree, F., mececq@uhupvm1.uh.edu
Newton, S., sid@archsci.arch.su.oz.au
Nguyen, G. T., nguyen@imag.fr
Nii, H. P., nii@sumex-aim.stanford.edu
O'Grady, P., ogrady@eos.ncsu.edu
Peckham, J., joan@cs.uri.edu
Qian, L., lena@archsci.arch.su.oz.au
Reddy, G., giri@globe.edrc.cmu.edu
Richards, R. A., buc@jessica.stanford.edu
Rieu, D., +33 76 44 66 75
Rogers, J. L., rogers@lrcn.larc.nasa.gov
Runkel, J. T., jayr@eecs.umich.edu
Sangal, R.,
csesun!kalyan!sangal@vikram.doe.ernet.in
Schmitt, G., schmitt@arch.ethz.ch
Sims Williams, J. H., +44 272 25 1154
Sheppard, S. D., ideas@netserver.stanford.edu
Shi, Z., +86 22 31 8329
Shih, S., shih@arch.ethz.ch
Shimodaira, H., +81 33 508 6085
Smith, I., smith@elgc.epfl.ch
Smithers, T., tim@arti1.vub.ac.be
Sriram, S., sriram@athena.mit.edu
Stroulia, E. eleni@cc.gatech.edu
Sturges, R. H., rs43+@andrew.cmu.edu
Tenenbaum, J. M., jmt@eitech.com
Tham, K. W., bemtkw@nusvm.bitnet
Tommelein, I. D., irist@caen.engin.umich.edu
Treur, J., treur@cs.vu.nl
Tsai, J.-C., tsai@sunrise.stanford.edu
Van Wyk, S., vanwyk@cad.cs.cmu.edu
Veerkamp, P. J., pauljan@cwi.nl
Weber, J. C., weber@eitech.com
Werkman, K. J., keithw@owgvm0.vnet.ibm.com
Wognum, P. M. wognam@cs.utwente.nl
Yamakawa, H., +81 33 209 9176
Young, R. E., young@eos.ncsu.edu
Zhao, F., zhaof@servax.fiu.edu
Zucker, J., j_zucker@vax.acs.open.ac.uk